D0085389

INTERNAL COMBUSTION ENGINE FUNDAMENTALS

McGraw-Hill Series in Mechanical Engineering

Jack P. Holman, *Southern Methodist University*
Consulting Editor

INTERNAL COMBUSTION ENGINE FUNDAMENTALS

John B. Heywood

Professor of Mechanical Engineering
Director, Sloan Automotive Laboratory
Massachusetts Institute of Technology

McGraw-Hill Publishing Company

New York St. Louis San Francisco Auckland Bogotá Caracas
Hamburg Lisbon London Madrid Mexico Milan
Montreal New Delhi Oklahoma City Paris San Juan
São Paulo Singapore Sydney Tokyo Toronto

INTERNAL COMBUSTION ENGINE FUNDAMENTALS

This book was set in Times Roman.
The editors were Anne Duffy and John M. Morriss; the designer
was Joan E. O'Connor; the production supervisor was
Denise L. Puryear. New drawings were done by ANCO.
Project Supervision was done by Santype International Ltd.
R. R. Donnelley & Sons Company was printer and binder.

See acknowledgements on page xxi.

567890 DOCDOC 9321

ISBN 0-07-028637-X

Library of Congress Cataloging-in-Publication Data

Heywood, John B.
   Internal combustion engine fundamentals.

   (McGraw-Hill series in mechanical engineering)
   Bibliography: p.
   Includes index.
   1. Internal combustion engines.   I. Title.   II. Series.
TJ755.H45   1988   621.43   87-15251

ABOUT THE AUTHOR

Dr. John B. Heywood received the Ph.D. degree in mechanical engineering from the Massachusetts Institute of Technology in 1965. Following an additional post-doctoral year of research at MIT, he worked as a research officer at the Central Electricity Generating Board's Research Laboratory in England on magneto-hydrodynamic power generation. In 1968 he joined the faculty at MIT where he is Professor of Mechanical Engineering. At MIT he is Director of the Sloan Automotive Laboratory. He is currently Head of the Fluid and Thermal Science Division of the Mechanical Engineering Department, and the Transportation Energy Program Director in the MIT Energy Laboratory. He is faculty advisor to the MIT Sports Car Club.

Professor Heywood's teaching and research interests lie in the areas of thermodynamics, combustion, energy, power, and propulsion. During the past two decades, his research activities have centered on the operating characteristics and fuels requirements of automotive and aircraft engines. A major emphasis has been on computer models which predict the performance, efficiency, and emissions of spark-ignition, diesel, and gas turbine engines, and in carrying out experiments to develop and validate these models. He is also actively involved in technology assessments and policy studies related to automotive engines, automobile fuel utilization, and the control of air pollution. He consults frequently in the automotive and petroleum industries, and for the U.S. Government.

His extensive research in the field of engines has been supported by the U.S. Army, Department of Energy, Environmental Protection Agency, NASA, National Science Foundation, automobile and diesel engine manufacturers, and petroleum companies. He has presented or published over a hundred papers on

his research in technical conferences and journals. He has co-authored two previous books: *Open-Cycle MHD Power Generation* published by Pergamon Press
in 1969 and *The Automobile and the Regulation of Its Impact on the Environment*
published by University of Oklahoma Press in 1975.

He is a member of the American Society of Mechanical Engineers, an associ-
ate fellow of the American Institute of Aeronautics and Astronautics, a fellow of
the British Institution of Mechanical Engineers, and in 1982 was elected a Fellow
of the U.S. Society of Automotive Engineers for his technical contributions to
automotive engineering. He is a member of the editorial boards of the journals
Progress in Energy and Combustion Science and the *International Journal of
Vehicle Design.*

His research publications on internal combustion engines, power generation,
and gas turbine combustion have won numerous awards. He was awarded the
Ayreton Premium in 1969 by the British Institution of Electrical Engineers. Professor Heywood received a Ralph R. Teetor Award as an outstanding young
engineering educator from the Society of Automotive Engineers in 1971. He has
twice been the recipient of an SAE Arch T. Colwell Merit Award for an outstanding technical publication (1973 and 1981). He received SAE's Horning Memorial
Award for the best paper on engines and fuels in 1984. In 1984 he received the
Sc.D. degree from Cambridge University for his published contributions to
engineering research. He was selected as the 1986 American Society of Mechanical Engineers Freeman Scholar for a major review of "Fluid Motion within the
Cylinder of Internal Combustion Engines."

THIS BOOK IS DEDICATED TO MY FATHER,
Harold Heywood:

I have followed many of the paths he took.

CONTENTS

Appendixes

PREFACE

Internal combustion engines date back to 1876 when Otto first developed the spark-ignition engine and 1892 when Diesel invented the compression-ignition engine. Since that time these engines have continued to develop as our knowledge of engine processes has increased, as new technologies became available, as demand for new types of engine arose, and as environmental constraints on engine use changed. Internal combustion engines, and the industries that develop and manufacture them and support their use, now play a dominant role in the fields of power, propulsion, and energy. The last twenty-five years or so have seen an explosive growth in engine research and development as the issues of air pollution, fuel cost, and market competitiveness have become increasingly important. An enormous technical literature on engines now exists which has yet to be adequately organized and summarized.

This book has been written as a text and a professional reference in response to that need. It contains a broadly based and extensive review of the fundamental principles which govern internal combustion engine design and operation. It attempts to provide a simplifying framework for the vast and complex mass of technical material that now exists on spark-ignition and compression-ignition engines, and at the same time to include sufficient detail to convey the real world dimensions of this pragmatic engineering field. It is the author's conviction that a sound knowledge of the relevant fundamentals in the many disciplines that contribute to this field, as well as an awareness of the extensive practical knowledge base which has been built up over many decades, are essential tools for engine research, development, and design. Of course, no one text can include everything about engines. The emphasis here is on the thermodynamics, combustion physics and chemistry, fluid flow, heat transfer, friction, and lubrication processes relevant to internal combustion engine design, performance, efficiency, emissions, and fuels requirements.

From a fundamental point of view, how the fuel-air mixture within an internal combustion engine cylinder is ignited appropriately organizes the field. From the method of ignition—spark-ignition or compression-ignition—follows each type of engine's important features: fuel requirements, method of mixture preparation, combustion chamber design, details of the combustion process, method of load control, emission formation mechanisms, and performance and efficiency characteristics. While many engine processes (such as intake and exhaust flows, convective heat transfer, and friction) are similar in both types of engines, this distinction is fundamental and lies behind the overall organization of the book.

The book is arranged in four major sections. The first (Chapters 1 to 5) provides an introduction to, and overview of, the major characteristics of spark-ignition and compression-ignition engines, defines the parameters used to describe engine operation, and develops the necessary thermodynamics and combustion theory required for a quantitative analysis of engine behavior. It concludes with an integrated treatment of the various methods of analyzing idealized models of internal combustion engine cycles. The second section (Chapters 6 to 8) focuses on engine flow phenomena. The details of the gas exchange process—intake and exhaust processes in four-stroke and scavenging in two-stroke cycles—and the various methods of supercharging engines—are reviewed. Fuel metering methods for spark-ignition engines and air- and fuel-flow phenomena in intake manifolds are described. The essential features of the various types of fluid motion within the engine cylinder are then developed. These flow processes control the amount of air an engine will induct (and therefore its power), and largely govern the rate at which the fuel-air mixture will burn during combustion.

The third section of the book focuses on engine combustion phenomena. These chapters (9, 10, and 11) are especially important. The combustion process releases the fuel's energy within the engine cylinder for eventual conversion to useful work. What fraction of the fuel's energy is converted depends strongly on how combustion takes place. The spark-ignition and compression-ignition engine combustion processes (Chapters 9 and 10, respectively) therefore influence essentially all aspects of engine behavior. Air pollutants are undesirable byproducts of combustion. Our extensive knowledge of how the major pollutants form during these combustion processes and how such emissions can be controlled is reviewed in Chapter 11.

The last section of the book focuses on engine operating characteristics. First, the fundamentals of engine heat transfer and friction, both of which detract from engine performance, are developed in Chapters 12 and 13. Chapter 14 then focuses on the methods available for predicting important aspects of engine behavior based on realistic models of engine flow and combustion processes. Since the various thermodynamic-based and fluid-mechanic-based models which have been developed over the past fifteen years or so are increasingly used in engine research and development, a knowledge of their basic structure and capabilities is most important. Then, Chapter 15 presents a summary of how the operating characteristics—power, efficiency, and emissions—of spark-ignition and compression-ignition engines depend on the major engine design and operating variables. These final two chapters effectively integrate the analytical under-standing and practical knowledge of individual engine processes together to describe overall spark-ignition and compression-ignition engine behavior. Material on internal combustion engine fuels is distributed appropriately throughout the book. Each chapter is extensively illustrated and referenced, and includes problems for both undergraduate and graduate level courses.

While this book contains much advanced material on engine design and operation intended for the practitioner, each major topic is developed from its beginnings and the more sophisticated chapters have introductory sections to facilitate their use in undergraduate courses. The chapters are extensively cross-referenced and indexed. Thus several arrangements of the material for a course on engines can be followed. For example, an introductory course on internal combustion engines could begin with Chapters 1 and 2, which review the differ-ent types of engines and how their performance is characterized, and continue with the parts of Chapters 3 and 5, which introduce the key combustion concepts necessary to understand the effects of fuel/air ratio, and ideal cycle analysis. Se-lections from the introductory sections of Chapters 6, 9, 10, 11, and 15 could then be used to explain several of the practical and design aspects of spark-ignition and diesel engine intake and exhaust processes, combustion, emissions, and per-formance. A more advanced course would review this introductory material more rapidly, and then move on to those sections of Chapters 4 and 5, which cover fuel-air cycle analysis, a more extensive discussion of engine breathing using addi-tional sections of Chapter 6, and more in-depth treatment of engine combustion and emissions processes based on the appropriate sections of Chapters 9, 10, and 11. Material on engine heat transfer and friction selected from Chapters 12 and 13 could be included next. While Chapter 14 on modeling the thermodynamics and fluid dynamics of real engine processes is primarily intended for the pro-fessional scientist and engineer, material from this chapter along with selections from Chapter 15 could be used to illustrate the performance, efficiency, and emis-sions characteristics of the different types of internal combustion engines. I have also used much of the more sophisticated material in Chapters 6 through 15 for review seminars on individual engine topics and more extensive courses for pro-fessional engineers, an additional important educational and reference opportunity.

Many individuals and organizations have assisted me in various ways as I have worked on this book over the past ten or so years. I am especially indebted to my colleagues in the Sloan Automotive Laboratory at M.I.T., Professors Wai K. Cheng, Ahmed F. Ghoniem, and James C. Keck, and Drs. Jack A. Ekchian, David P. Hoult, Joe M. Rife, and Victor W. Wong, for providing a stimulating environment in which to carry out engine research and for assuming additional burdens as a result of my writing. Many of the Sloan Automotive Laboratory's students have made significant contributions to this text through their research; their names appear in the reference lists. The U.S. Department of Energy provid-ed support during the early stages of the text development and funded the work on engine cycle simulation used extensively in Chapters 14 and 15. I am grateful to Churchill College, Cambridge University, for a year spent as a Richard C. Mellon Visiting Fellow, 1977–78, and the Engineering Department, Cambridge University, for acting as my host while I developed the outline and earlier chapters of the book. The M.I.T. sabbatical leave fund supported my full-time writing for eight months in 1983, and the Mechanical Engineering Department at Imperial College graciously acted as host.

I also want to acknowledge several individuals and organizations who have provided major inputs to this book beyond those cited in the references. Members of General Motors Research Laboratories have interacted extensively with the Sloan Automotive Laboratory over many years and provided valuable advice on engine research developments. Engineers from the Engine Research and Fluid Mechanics Departments at General Motors Research Laboratories reviewed and critiqued the final draft manuscript for me. Charles A. Amann, Head of the Engine Research Department, made especially helpful inputs on engine performance. John J. Brogan of the U.S. Department of Energy provided valuable assistance with the initial organization of this effort. My regular interactions over the years with the Advanced Powertrain Engineering Office and Scientific Research Laboratories of the Ford Motor Company have given me a broad exposure to the practical side of engine design and operation. A long-term relationship with Mobil Research and Development Corporation has provided comparable experiences in the area of engine-fuels interactions. Many organizations and individuals supplied specific material and illustrations for the text. I am especially grateful to those who made available the high-quality photographs and line drawings which I have used and acknowledged.

McGraw-Hill and the author would like to express their thanks to the following reviewers for their useful comments and suggestions: Jay A. Bolt, University of Michigan; Gary L. Borman and William L. Brown, University of Wisconsin at Madison; Dwight Bushnell, Oregon State University; Jerald A. Caton, Texas A & M University; David E. Cole, University of Michigan; Lawrence W. Evers, Michigan Technological University; Samuel S. Lestz, Pennsylvania State University; Willard Pulkrabek, University of Wisconsin; Robert F. Sawyer, University of California at Berkeley; Joseph E. Shepherd, Rensselaer Polytechnic Institute; and Spencer C. Sorenson, The Technical University of Denmark.

Special thanks are due to my secretaries for their faithful and thoughtful assistance with the manuscript over these many years, beyond the "call of duty"; Linda Pope typed an earlier draft of the book, and Karla Stryker was responsible for producing and coordinating subsequent drafts and the final manuscript. My wife Peggy, and sons James, Stephen, and Ben have encouraged me throughout this long and time-consuming project which took many hours away from them. Without their continuing support it would never have been finished; for their patience, and faith that it would ultimately come to fruition, I will always be grateful.

John B. Heywood

ACKNOWLEDGMENTS

The author wishes to acknowledge the following organizations and publishers for permission to reproduce figures and tables from their publications in this text: The American Chemical Society; American Institute of Aeronautics & Astronautics; American Society of Mechanical Engineers; Robert Bosch GmbH, CIMAC, Cambridge University Press; The Combustion Institute; Elsevier Science Publishing Company; G. T. Foulis & Co. Ltd.; General Motors Corporation; Gordon & Breach Science Publishers; The Institution of Mechanical Engineers; The Japan Society of Mechanical Engineers; M.I.T. Press; Macmillan Press Ltd.; McGraw-Hill Book Company; Mir Publishers; Mobil Oil Corporation; Morgan-Grampian Publishers; Pergamon Journals, Inc.; Plenum Press Corporation; The Royal Society of London; Scientific Publications Limited; Society of Automotive Engineers; Society of Automotive Engineers of Japan, Inc.; Society of Tribologists and Lubrications Engineers; Department of Mechanical Engineering, Stanford University.

COMMONLY USED SYMBOLS, SUBSCRIPTS, AND ABBREVIATIONS†

1. SYMBOLS

a	Crank radius
	Sound speed
	Specific availability
a	Acceleration
A	Area
A_C	Valve curtain area
A_{ch}	Cylinder head area
A_e	Exhaust port area
A_E	Effective area of flow restriction
A_i	Inlet port area
A_p	Piston crown area
B	Cylinder bore
	Steady-flow availability
c	Specific heat
c_p	Specific heat at constant pressure
c_s	Soot concentration (mass/volume)
c_v	Specific heat at constant volume
C	Absolute gas velocity

---

† Nomenclature specific to a section or chapter is defined in that section or chapter.

C_s	Swirl coefficient
C_D	Discharge coefficient
	Vehicle drag coefficient
d	Diameter
d_n	Fuel-injection-nozzle orifice diameter
D	Diameter
	Diffusion coefficient
D_d	Droplet diameter
D_{SM}	Sauter mean droplet diameter
D_v	Valve diameter
e	Radiative emissive power
	Specific energy
E_A	Activation energy
f	Coefficient of friction
	Fuel mass fraction
F	Force
g	Gravitational acceleration
	Specific Gibbs free energy
G	Gibbs free energy
h	Clearance height
	Oil film thickness
	Specific enthalpy
h_c	Heat-transfer coefficient
h_p	Port open height
h_s	Sensible specific enthalpy
H	Enthalpy
I	Moment of inertia
J	Flux
k	Thermal conductivity
	Turbulent kinetic energy
k_i^+ , k_i^-	Forward, backward, rate constants for ith reaction
K	Constant
K_c	Equilibrium constant expressed in concentrations
K_p	Equilibrium constant expressed in partial pressures
l	Characteristic length scale
	Connecting rod length
l_T	Characteristic length scale of turbulent flame
L	Piston stroke
L_n	Fuel-injection-nozzle orifice length
L_v	Valve lift
m	Mass
$\dot{m}$	Mass flow rate
m_r	Mass of residual gas
M	Mach number
	Molecular weight

n	Number of moles
	Polytropic exponent
n_R	Number of crank revolutions per power stroke
N	Crankshaft rotational speed
	Soot particle number density
	Turbocharger shaft speed
p	Cylinder pressure
	Pressure
P	Power
$\dot{q}$	Heat-transfer rate per unit area
	Heat-transfer rate per unit mass of fluid
Q	Heat transfer
$\dot{Q}$	Heat-transfer rate
Q_{ch}	Fuel chemical energy release or gross heat release
Q_{HV}	Fuel heating value
Q_n	Net heat release
r	Radius
r_c	Compression ratio
R	Connecting rod length/crank radius
	Gas constant
	Radius
R^+, R^-	One-way reaction rates
R_s	Swirl ratio
s	Crank axis to piston pin distance
	Specific entropy
S	Entropy
	Spray penetration
S_b	Turbulent burning speed
S_L	Laminar flame speed
S_p	Piston speed
t	Time
T	Temperature
	Torque
u	Specific internal energy
	Velocity
u'	Turbulence intensity
u_s	Sensible specific internal energy
u_T	Characteristic turbulent velocity
U	Compressor/turbine impellor tangential velocity
	Fluid velocity
	Internal energy
v	Specific volume
	Velocity
$\mathbf{v}$	Velocity
v_{ps}	Valve pseudo-flow velocity

v_{sq}	Squish velocity
V	Cylinder volume
	Volume
V_c	Clearance volume
V_d	Displaced cylinder volume
w	Relative gas velocity
	Soot surface oxidation rate
W	Work transfer
W_c	Work per cycle
W_p	Pumping work
x, y, z	Spatial coordinates
x	Mass fraction
$\tilde{x}$	Mole fraction
x_b	Burned mass fraction
x_r	Residual mass fraction
y	H/C ratio of fuel
	Volume fraction
Y_α	Concentration of species α per unit mass
Z	Inlet Mach index
α	Angle
	Thermal diffusivity $k/(\rho c)$
β	Angle
γ	Specific heat ratio c_p/c_v
Γ_c	Angular momentum of charge
δ	Boundary-layer thickness
δ_L	Laminar flame thickness
$\Delta \tilde{h}^\circ_{f,i}$	Molal enthalpy of formation of species i
$\Delta \theta_b$	Rapid burning angle
$\Delta \theta_d$	Flame development angle
ε	$4/(4 + y)$: y = H/C ratio of fuel
	Turbulent kinetic energy dissipation rate
η_a	Availability conversion efficiency
η_c	Combustion efficiency
η_C	Compressor isentropic efficiency
η_{ch}	Charging efficiency
η_f	Fuel conversion efficiency
η_m	Mechanical efficiency
η_{sc}	Scavenging efficiency
η_t	Thermal conversion efficiency
η_T	Turbine isentropic efficiency
η_{tr}	Trapping efficiency
η_v	Volumetric efficiency
θ	Crank angle
λ	Relative air/fuel ratio
Λ	Delivery ratio

μ	Dynamic viscosity
μ_i	Chemical potential of species i
v	Kinematic viscosity μ/ρ
v_i	Stoichiometric coefficient of species i
ξ	Flow friction coefficient
ρ	Density
$\rho_{a,0}$, $\rho_{a,i}$	Air density at standard, inlet conditions
σ	Normal stress
	Standard deviation
	Stefan-Boltzmann constant
	Surface tension
τ	Characteristic time
	Induction time
	Shear stress
τ_{id}	Ignition delay time
ϕ	Fuel/air equivalence ratio
Φ	Flow compressibility function [Eq. (C.11)]
	Isentropic compression function [Eq. (4.15b)]
ψ	Molar N/O ratio
	Throttle plate open angle
Ψ	Isentropic compression function [Eq. (4.15a)]
ω	Angular velocity
	Frequency

2. SUBSCRIPTS

a	Air
b	Burned gas
c	Coolant
	Cylinder
C	Compression stroke
	Compressor
cr	Crevice
e	Equilibrium
	Exhaust
E	Expansion stroke
f	Flame
	Friction
	Fuel
g	Gas
i	Indicated
	Intake
	Species i
ig	Gross indicated
in	Net indicated

l	Liquid
L	Laminar
p	Piston
	Port
P	Prechamber
r, θ, z	r, θ, z components
R	Reference value
s	Isentropic
	Stoichiometric
T	Nozzle or orifice throat
	Turbine
	Turbulent
u	Unburned
v	Valve
w	Wall
x, y, z	x, y, z components
0	Reference value
	Stagnation value

3. NOTATION

Δ	Difference
$^{-}$	Average or mean value
$\sim$	Value per mole
[ ]	Concentration, moles/vol
{ }	Mass fraction
$\cdot$	Rate of change with time

4. ABBREVIATIONS

(A/F)	Air/fuel ratio
BC, ABC, BBC	Bottom-center crank position, after BC, before BC
CN	Fuel cetane number
Da	Damköhler number τ_T/τ_L
EGR	Exhaust gas recycle
EI	Emission index
EPC, EPO	Exhaust port closing, opening
EVC, EVO	Exhaust valve closing, opening
(F/A)	Fuel/air ratio
(G/F)	Gas/fuel ratio
IPC, IPO	Inlet port closing, opening
IVC, IVO	Inlet valve closing, opening
mep	Mean effective pressure
Nu	Nusselt number $h_c\, l/k$

ON   Fuel octane number
Re   Reynolds number $\rho u l/\mu$
sfc   Specific fuel consumption
TC, ATC, BTC Top-center crank position, after TC, before TC
We   Weber number $\rho_l u^2 D/\sigma$

CHAPTER

1

ENGINE
TYPES
AND THEIR
OPERATION

1.1 INTRODUCTION AND HISTORICAL PERSPECTIVE

The purpose of internal combustion engines is the production of mechanical power from the chemical energy contained in the fuel. In *internal* combustion engines, as distinct from *external* combustion engines, this energy is released by burning or oxidizing the fuel *inside* the engine. The fuel-air mixture before combustion and the burned products after combustion are the actual working fluids. The work transfers which provide the desired power output occur directly between these working fluids and the mechanical components of the engine. The internal combustion engines which are the subject of this book are spark-ignition engines (sometimes called Otto engines, or gasoline or petrol engines, though other fuels can be used) and compression-ignition or diesel engines.† Because of their simplicity, ruggedness and high power/weight ratio, these two types of engine have found wide application in transportation (land, sea, and air) and power generation. It is the fact that combustion takes place inside the work-

---

† The gas turbine is also, by this definition, an "internal combustion engine." Conventionally, however, the term is used for spark-ignition and compression-ignition engines. The operating principles of gas turbines are fundamentally different, and they are not discussed as separate engines in this book.

producing part of these engines that makes their design and operating characteristics fundamentally different from those of other types of engine.

Practical heat engines have served mankind for over two and a half centuries. For the first 150 years, water, raised to steam, was interposed between the combustion gases produced by burning the fuel and the work-producing piston-in-cylinder expander. It was not until the 1860s that the internal combustion engine became a practical reality.[1, 2] The early engines developed for commercial use burned coal-gas air mixtures at atmospheric pressure—there was no compression before combustion. J. J. E. Lenoir (1822–1900) developed the first marketable engine of this type. Gas and air were drawn into the cylinder during the first half of the piston stroke. The charge was then ignited with a spark, the pressure increased, and the burned gases then delivered power to the piston for the second half of the stroke. The cycle was completed with an exhaust stroke. Some 5000 of these engines were built between 1860 and 1865 in sizes up to six horsepower. Efficiency was at best about 5 percent.

A more successful development—an atmospheric engine introduced in 1867 by Nicolaus A. Otto (1832–1891) and Eugen Langen (1833–1895)—used the pressure rise resulting from combustion of the fuel-air charge early in the outward stroke to accelerate a free piston and rack assembly so its momentum would generate a vacuum in the cylinder. Atmospheric pressure then pushed the piston inward, with the rack engaged through a roller clutch to the output shaft. Production engines, of which about 5000 were built, obtained thermal efficiencies of up to 11 percent. A slide valve controlled intake, ignition by a gas flame, and exhaust.

To overcome this engine's shortcomings of low thermal efficiency and excessive weight, Otto proposed an engine cycle with four piston strokes: an intake stroke, then a compression stroke before ignition, an expansion or power stroke where work was delivered to the crankshaft, and finally an exhaust stroke. He also proposed incorporating a stratified-charge induction system, though this was not achieved in practice. His prototype four-stroke engine first ran in 1876. A comparison between the Otto engine and its atmospheric-type predecessor indicates the reason for its success (see Table 1.1): the enormous reduction in engine weight and volume. This was the breakthrough that effectively founded the internal combustion engine industry. By 1890, almost 50,000 of these engines had been sold in Europe and the United States.

In 1884, an unpublished French patent issued in 1862 to Alphonse Beau de Rochas (1815–1893) was found which described the principles of the four-stroke cycle. This chance discovery cast doubt on the validity of Otto's own patent for this concept, and in Germany it was declared invalid. Beau de Rochas also outlined the conditions under which maximum efficiency in an internal combustion engine could be achieved. These were:

1. The largest possible cylinder volume with the minimum boundary surface
2. The greatest possible working speed

TABLE 1.1
Comparison of Otto four-stroke cycle and Otto-Langen engines[2]

	Otto and Langen	Otto four-stroke
Brake horsepower	2	2
Weight, lb, approx.	4000	1250
Piston displacement, in^3	4900	310
Power strokes per min	28	80
Shaft speed, rev/min	90	160
Mechanical efficiency, %	68	84
Overall efficiency, %	11	14
Expansion ratio	10	2.5

3. The greatest possible expansion ratio
4. The greatest possible pressure at the beginning of expansion

The first two conditions hold heat losses from the charge to a minimum. The third condition recognizes that the greater the expansion of the postcombustion gases, the greater the work extracted. The fourth condition recognizes that higher initial pressures make greater expansion possible, and give higher pressures throughout the process, both resulting in greater work transfer. Although Beau de Rochas' unpublished writings predate Otto's developments, he never reduced these ideas to practice. Thus Otto, in the broader sense, was the inventor of the modern internal combustion engine as we know it today.

Further developments followed fast once the full impact of what Otto had achieved became apparent. By the 1880s several engineers (e.g., Dugald Clerk, 1854–1913, and James Robson, 1833–1913, in England and Karl Benz, 1844–1929, in Germany) had successfully developed two-stroke internal combustion engines where the exhaust and intake processes occur during the end of the power stroke and the beginning of the compression stroke. James Atkinson (1846–1914) in England made an engine with a longer expansion than compression stroke, which had a high efficiency for the times but mechanical weaknesses. It was recognized that efficiency was a direct function of expansion ratio, yet compression ratios were limited to less than four if serious knock problems were to be avoided with the available fuels. Substantial carburetor and ignition system developments were required, and occurred, before high-speed gasoline engines suitable for automobiles became available in the late 1880s. Stationary engine progress also continued. By the late 1890s, large single-cylinder engines of 1.3-m bore fueled by low-energy blast furnace gas produced 600 bhp at 90 rev/min. In Britain, legal restrictions on volatile fuels turned their engine builders toward kerosene. Low compression ratio "oil" engines with heated external fuel vaporizers and electric ignition were developed with efficiencies comparable to those of gas engines (14 to 18 percent). The Hornsby-Ackroyd engine became the most popular oil engine in Britain, and was also built in large numbers in the United States.[2]

In 1892, the German engineer Rudolf Diesel (1858–1913) outlined in his patent a new form of internal combustion engine. His concept of initiating combustion by injecting a liquid fuel into air heated solely by compression permitted a doubling of efficiency over other internal combustion engines. Much greater expansion ratios, without detonation or knock, were now possible. However, even with the efforts of Diesel and the resources of M.A.N. in Ausburg combined, it took five years to develop a practical engine.

Engine developments, perhaps less fundamental but nonetheless important to the steadily widening internal combustion engine markets, have continued ever since.[2–4] One more recent major development has been the rotary internal combustion engine. Although a wide variety of experimental rotary engines have been proposed over the years,[5] the first practical rotary internal combustion engine, the Wankel, was not successfully tested until 1957. That engine, which evolved through many years of research and development, was based on the designs of the German inventor Felix Wankel.[6, 7]

Fuels have also had a major impact on engine development. The earliest engines used for generating mechanical power burned gas. Gasoline, and lighter fractions of crude oil, became available in the late 1800s and various types of carburetors were developed to vaporize the fuel and mix it with air. Before 1905 there were few problems with gasoline; though compression ratios were low (4 or less) to avoid knock, the highly volatile fuel made starting easy and gave good cold weather performance. However, a serious crude oil shortage developed, and to meet the fivefold increase in gasoline demand between 1907 and 1915, the yield from crude had to be raised. Through the work of William Burton (1865–1954) and his associates of Standard Oil of Indiana, a thermal cracking process was developed whereby heavier oils were heated under pressure and decomposed into less complex more volatile compounds. These thermally cracked gasolines satisfied demand, but their higher boiling point range created cold weather starting problems. Fortunately, electrically driven starters, introduced in 1912, came along just in time.

On the farm, kerosene was the logical fuel for internal combustion engines since it was used for heat and light. Many early farm engines had heated carburetors or vaporizers to enable them to operate with such a fuel.

The period following World War I saw a tremendous advance in our understanding of how fuels affect combustion, and especially the problem of knock. The antiknock effect of tetraethyl lead was discovered at General Motors,[4] and it became commercially available as a gasoline additive in the United States in 1923. In the late 1930s, Eugene Houdry found that vaporized oils passed over an activated catalyst at 450 to 480°C were converted to high-quality gasoline in much higher yields than was possible with thermal cracking. These advances, and others, permitted fuels with better and better antiknock properties to be produced in large quantities; thus engine compression ratios steadily increased, improving power and efficiency.

During the past three decades, new factors for change have become important and now significantly affect engine design and operation. These factors are, first, the need to control the automotive contribution to urban air pollution and, second, the need to achieve significant improvements in automotive fuel consumption.

The automotive air-pollution problem became apparent in the 1940s in the Los Angeles basin. In 1952, it was demonstrated by Prof. A. J. Haagen-Smit that the smog problem there resulted from reactions between oxides of nitrogen and hydrocarbon compounds in the presence of sunlight.[8] In due course it became clear that the automobile was a major contributor to hydrocarbon and oxides of nitrogen emissions, as well as the prime cause of high carbon monoxide levels in urban areas. Diesel engines are a significant source of small soot or smoke particles, as well as hydrocarbons and oxides of nitrogen. Table 1.2 outlines the dimensions of the problem. As a result of these developments, emission standards for automobiles were introduced first in California, then nationwide in the United States, starting in the early 1960s. Emission standards in Japan and Europe, and for other engine applications, have followed. Substantial reductions in emissions from spark-ignition and diesel engines have been achieved. Both the use of catalysts in spark-ignition engine exhaust systems for emissions control and concern over the toxicity of lead antiknock additives have resulted in the reappearance of unleaded gasoline as a major part of the automotive fuels market. Also, the maximum lead content in leaded gasoline has been substantially reduced. The emission-control requirements and these fuel developments have produced significant changes in the way internal combustion engines are designed and operated.

Internal combustion engines are also an important source of noise. There are several sources of engine noise: the exhaust system, the intake system, the fan used for cooling, and the engine block surface. The noise may be generated by aerodynamic effects, may be due to forces that result from the combustion process, or may result from mechanical excitation by rotating or reciprocating engine components. Vehicle noise legislation to reduce emissions to the environment was first introduced in the early 1970s.

During the 1970s the price of crude petroleum rose rapidly to several times its cost (in real terms) in 1970, and concern built up regarding the longer-term availability of petroleum. Pressures for substantial improvements in internal combustion engine efficiency (in all its many applications) have become very substantial indeed. Yet emission-control requirements have made improving engine fuel consumption more difficult, and the removal and reduction of lead in gasoline has forced spark-ignition engine compression ratios to be reduced. Much work is being done on the use of alternative fuels to gasoline and diesel. Of the non-petroleum-based fuels, natural gas, and methanol and ethanol (methyl and ethyl alcohols) are receiving the greatest attention, while synthetic gasoline and diesel made from shale oil or coal, and hydrogen could be longer-term possibilities.

It might be thought that after over a century of development, the internal

TABLE 1.2
The automotive urban air-pollution problem

Pollutant	Impact	Mobile source emissions as % of total†	Automobile emissions		Truck emissions††	
			Uncontrolled vehicles, g/km‡	Reduction in new vehicles, % ¶	SI engines, g/km	Diesel, g/km
Oxides of nitrogen (NO and NO$_2$)	Reactant in photochemical smog; NO$_2$ is toxic	40–60	2.5	75	7	12
Carbon monoxide (CO)	Toxic	90	65	95	150	17
Unburned hydrocarbons (HC, many hydrocarbon compounds)	Reactant in photochemical smog	30–50	10	90	17‡‡	3
Particulates (soot and absorbed hydrocarbon compounds)	Reduces visibility; some of HC compounds mutagenic	50	0.5§	40§	n	0.5

† Depends on type of urban area and source mix.

‡ Average values for pre-1968 automobiles which had no emission controls, determined by U.S. test procedure which simulates typical urban and highway driving. Exhaust emissions, except for HC where 55 percent are exhaust emissions, 20 percent are evaporative emissions from fuel tank and carburetor, and 25 percent are crankcase blowby gases.

§ Diesel engine automobiles only. Particulate emissions from spark-ignition engines are negligible.

¶ Compares emissions from new spark-ignition engine automobiles with uncontrolled automobile levels in previous column. Varies from country to country. The United States, Canada, Western Europe, and Japan have standards with different degrees of severity. The United States, Europe, and Japan have different test procedures. Standards are strictest in the United States and Japan.

†† Representative average emission levels for trucks.

‡‡ With 95 percent exhaust emissions and 5 percent evaporative emissions.

n = negligible.

combustion engine has reached its peak and little potential for further improvement remains. Such is not the case. Conventional spark-ignition and diesel engines continue to show substantial improvements in efficiency, power, and degree of emission control. New materials now becoming available offer the possibilities of reduced engine weight, cost, and heat losses, and of different and more efficient internal combustion engine systems. Alternative types of internal combustion engines, such as the stratified-charge (which combines characteristics normally associated with either the spark-ignition or diesel) with its wider fuel tolerance, may become sufficiently attractive to reach large-scale production. The engine development opportunities of the future are substantial. While they present a formidable challenge to automotive engineers, they will be made possible in large part by the enormous expansion of our knowledge of engine processes which the last twenty years has witnessed.

1.2 ENGINE CLASSIFICATIONS

There are many different types of internal combustion engines. They can be classified by:

1. *Application.* Automobile, truck, locomotive, light aircraft, marine, portable power system, power generation

2. *Basic engine design.* Reciprocating engines (in turn subdivided by arrangement of cylinders: e.g., in-line, V, radial, opposed), rotary engines (Wankel and other geometries)

3. *Working cycle.* Four-stroke cycle: naturally aspirated (admitting atmospheric air), supercharged (admitting precompressed fresh mixture), and turbocharged (admitting fresh mixture compressed in a compressor driven by an exhaust turbine), two-stroke cycle: crankcase scavenged, supercharged, and turbocharged

4. *Valve or port design and location.* Overhead (or I-head) valves, underhead (or L-head) valves, rotary valves, cross-scavenged porting (inlet and exhaust ports on opposite sides of cylinder at one end), loop-scavenged porting (inlet and exhaust ports on same side of cylinder at one end), through- or uniflow-scavenged (inlet and exhaust ports or valves at different ends of cylinder)

5. *Fuel.* Gasoline (or petrol), fuel oil (or diesel fuel), natural gas, liquid petroleum gas, alcohols (methanol, ethanol), hydrogen, dual fuel

6. *Method of mixture preparation.* Carburetion, fuel injection into the intake ports or intake manifold, fuel injection into the engine cylinder

7. *Method of ignition.* Spark ignition (in conventional engines where the mixture is uniform and in stratified-charge engines where the mixture is non-uniform), compression ignition (in conventional diesels, as well as ignition in gas engines by pilot injection of fuel oil)

8. *Combustion chamber design.* Open chamber (many designs: e.g., disc, wedge, hemisphere, bowl-in-piston), divided chamber (small and large auxiliary chambers; many designs: e.g., swirl chambers, prechambers)

9. *Method of load control.* Throttling of fuel and air flow together so mixture composition is essentially unchanged, control of fuel flow alone, a combination of these

10. *Method of cooling.* Water cooled, air cooled, uncooled (other than by natural convection and radiation)

    All these distinctions are important and they illustrate the breadth of engine designs available. Because this book approaches the operating and emissions

TABLE 1.3
Classification of reciprocating engines by application

Class	Service	Approximate engine power range, kW	Predominant type		
			D or SI	Cycle	Cooling
Road vehicles	Motorcycles, scooters	0.75–70	SI	2, 4	A
	Small passenger cars	15–75	SI	4	A, W
	Large passenger cars	75–200	SI	4	W
	Light commercial	35–150	SI, D	4	W
	Heavy (long-distance) commercial	120–400	D	4	W
Off-road vehicles	Light vehicles (factory, airport, etc.)	1.5–15	SI	2, 4	A, W
	Agricultural	3–150	SI, D	2, 4	A, W
	Earth moving	40–750	D	2, 4	W
	Military	40–2000	D	2, 4	A, W
Railroad	Rail cars	150–400	D	2, 4	W
	Locomotives	400–3000	D	2, 4	W
Marine	Outboard	0.4–75	SI	2	W
	Inboard motorcrafts	4–750	SI, D	4	W
	Light naval craft	30–2200	D	2, 4	W
	Ships	3500–22,000	D	2, 4	W
	Ships' auxiliaries	75–750	D	4	W
Airborne vehicles	Airplanes	45–2700	SI	4	A
	Helicopters	45–1500	SI	4	A
Home use	Lawn mowers	0.7–3	SI	2, 4	A
	Snow blowers	2–5	SI	2, 4	A
	Light tractors	2–8	SI	4	A
Stationary	Building service	7–400	D	2, 4	W
	Electric power	35–22,000	D	2, 4	W
	Gas pipeline	750–5000	SI	2, 4	W

SI = spark-ignition; D = diesel; A = air cooled; W = water cooled.

Source: Adapted from Taylor.[9]

characteristics of internal combustion engines from a fundamental point of view, the method of ignition has been selected as the primary classifying feature. From the method of ignition—spark-ignition or compression-ignition†—follow the important characteristics of the fuel used, method of mixture preparation, combustion chamber design, method of load control, details of the combustion process, engine emissions, and operating characteristics. Some of the other classifications are used as subcategories within this basic classification. The engine operating cycle—four-stroke or two-stroke—is next in importance; the principles of these two cycles are described in the following section.

Table 1.3 shows the most common applications of internal combustion

---

† In the remainder of the book, these terms will often be abbreviated by SI and CI, respectively.

engines, the predominant type of engine used in each classification listed, and the approximate engine power range in each type of service.

1.3 ENGINE OPERATING CYCLES

Most of this book is about reciprocating engines, where the piston moves back and forth in a cylinder and transmits power through a connecting rod and crank mechanism to the drive shaft as shown in Fig. 1-1. The steady rotation of the crank produces a cyclical piston motion. The piston comes to rest at the top-center (TC) crank position and bottom-center (BC) crank position when the cylinder volume is a minimum or maximum, respectively.† The minimum cylinder volume is called the clearance volume V_c. The volume swept out by the

---

† These crank positions are also referred to as top-dead-center (TDC) and bottom-dead-center (BDC).

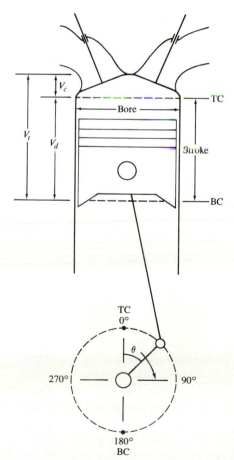

FIGURE 1-1
Basic geometry of the reciprocating internal combustion engine. V_c, V_d and V_t indicate clearance, displaced, and total cylinder volumes.

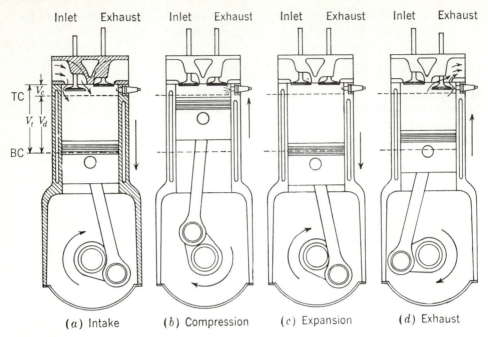

Inlet   Exhaust   Inlet   Exhaust   Inlet   Exhaust   Inlet   Exhaust

(a) Intake   (b) Compression   (c) Expansion   (d) Exhaust

FIGURE 1-2
The four-stroke operating cycle.[10]

piston, the difference between the maximum or total volume V_t and the clearance volume, is called the displaced or swept volume V_d. The ratio of maximum volume to minimum volume is the compression ratio r_c. Typical values of r_c are 8 to 12 for SI engines and 12 to 24 for CI engines.

The majority of reciprocating engines operate on what is known as the *four-stroke cycle*. Each cylinder requires four strokes of its piston—two revolutions of the crankshaft—to complete the sequence of events which produces one power stroke. Both SI and CI engines use this cycle which comprises (see Fig. 1-2):

1. *An intake stroke*, which starts with the piston at TC and ends with the piston at BC, which draws fresh mixture into the cylinder. To increase the mass inducted, the inlet valve opens shortly before the stroke starts and closes after it ends.
2. *A compression stroke*, when both valves are closed and the mixture inside the cylinder is compressed to a small fraction of its initial volume. Toward the end of the compression stroke, combustion is initiated and the cylinder pressure rises more rapidly.
3. *A power stroke*, or expansion stroke, which starts with the piston at TC and ends at BC as the high-temperature, high-pressure, gases push the piston down and force the crank to rotate. About five times as much work is done on the piston during the power stroke as the piston had to do during compression.

As the piston approaches BC the exhaust valve opens to initiate the exhaust process and drop the cylinder pressure to close to the exhaust pressure.

4. *An exhaust stroke,* where the remaining burned gases exit the cylinder: first, because the cylinder pressure may be substantially higher than the exhaust pressure; then as they are swept out by the piston as it moves toward TC. As the piston approaches TC the inlet valve opens. Just after TC the exhaust valve closes and the cycle starts again.

Though often called the Otto cycle after its inventor, Nicolaus Otto, who built the first engine operating on these principles in 1876, the more descriptive four-stroke nomenclature is preferred.

The four-stroke cycle requires, for each engine cylinder, two crankshaft revolutions for each power stroke. To obtain a higher power output from a given engine size, and a simpler valve design, the *two-stroke* cycle was developed. The two-stroke cycle is applicable to both SI and CI engines.

Figure 1-3 shows one of the simplest types of two-stroke engine designs. Ports in the cylinder liner, opened and closed by the piston motion, control the exhaust and inlet flows while the piston is close to BC. The two strokes are:

1. *A compression stroke,* which starts by closing the inlet and exhaust ports, and then compresses the cylinder contents and draws fresh charge into the crankcase. As the piston approaches TC, combustion is initiated.

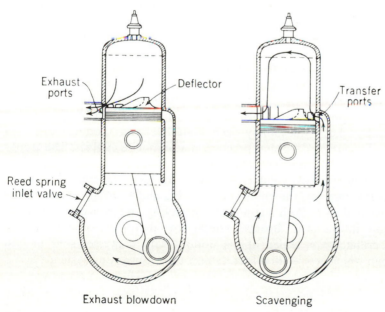

FIGURE 1-3
The two-stroke operating cycle. A crankcase-scavenged engine is shown.[10]

2. *A power or expansion stroke,* similar to that in the four-stroke cycle until the piston approaches BC, when first the exhaust ports and then the intake ports are uncovered (Fig. 1-3). Most of the burnt gases exit the cylinder in an exhaust blowdown process. When the inlet ports are uncovered, the fresh charge which has been compressed in the crankcase flows into the cylinder. The piston and the ports are generally shaped to deflect the incoming charge from flowing directly into the exhaust ports and to achieve effective scavenging of the residual gases.

Each engine cycle with one power stroke is completed in one crankshaft revolution. However, it is difficult to fill completely the displaced volume with fresh charge, and some of the fresh mixture flows directly out of the cylinder during the scavenging process.† The example shown is a *cross-scavenged* design; other approaches use *loop-scavenging* or *uniflow* systems (see Sec. 6.6).

1.4 ENGINE COMPONENTS

Labeled cutaway drawings of a four-stroke SI engine and a two-stroke CI engine are shown in Figs. 1-4 and 1-5, respectively. The spark-ignition engine is a four-cylinder in-line automobile engine. The diesel is a large V eight-cylinder design with a uniflow scavenging process. The function of the major components of these engines and their construction materials will now be reviewed.

The engine cylinders are contained in the engine block. The block has traditionally been made of gray cast iron because of its good wear resistance and low cost. Passages for the cooling water are cast into the block. Heavy-duty and truck engines often use removable cylinder sleeves pressed into the block that can be replaced when worn. These are called *wet liners* or *dry liners* depending on whether the sleeve is in direct contact with the cooling water. Aluminum is being used increasingly in smaller SI engine blocks to reduce engine weight. Iron cylinder liners may be inserted at the casting stage, or later on in the machining and assembly process. The crankcase is often integral with the cylinder block.

The crankshaft has traditionally been a steel forging; nodular cast iron crankshafts are also accepted normal practice in automotive engines. The crankshaft is supported in main bearings. The maximum number of main bearings is one more than the number of cylinders; there may be less. The crank has eccentric portions (crank throws); the connecting rod big-end bearings attach to the crank pin on each throw. Both main and connecting rod bearings use steel-backed precision inserts with bronze, babbit, or aluminum as the bearing materials. The crankcase is sealed at the bottom with a pressed-steel or cast aluminum oil pan which acts as an oil reservoir for the lubricating system.

---

† It is primarily for this reason that two-stroke SI engines are at a disadvantage because the lost fresh charge contains fuel and air.

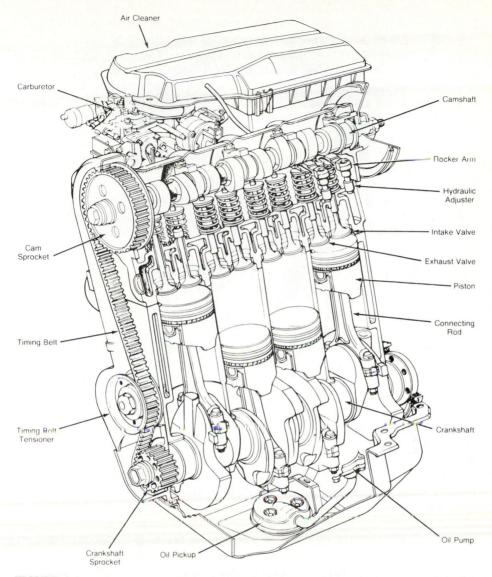

FIGURE 1-4
Cutaway drawing of Chrysler 2.2-liter displacement four-cylinder spark-ignition engine.[11] Bore 87.5 mm, stroke 92 mm, compression ratio 8.9, maximum power 65 kW at 5000 rev/min.

Pistons are made of aluminum in small engines or cast iron in larger slower-speed engines. The piston both seals the cylinder and transmits the combustion-generated gas pressure to the crank pin via the connecting rod. The connecting rod, usually a steel or alloy forging (though sometimes aluminum in small engines), is fastened to the piston by means of a steel piston pin through the rod upper end. The piston pin is usually hollow to reduce its weight.

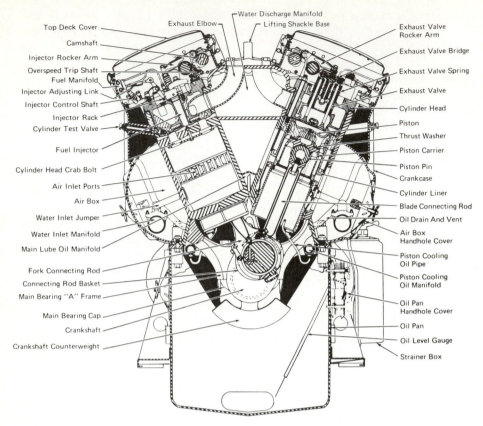

FIGURE 1-5
Cross-section drawing of an Electro-Motive two-stroke cycle diesel engine. This engine uses a uniflow scavenging process with inlet ports in the cylinder liner and four exhaust valves in the cylinder head. Bore 230.2 mm, stroke 254 mm, displaced volume per cylinder 10.57 liters, rated speed 750–900 rev/min. (*Courtesy Electro-Motive Division, General Motors Corporation.*)

The oscillating motion of the connecting rod exerts an oscillating force on the cylinder walls via the piston skirt (the region below the piston rings). The piston skirt is usually shaped to provide appropriate thrust surfaces. The piston is fitted with rings which ride in grooves cut in the piston head to seal against gas leakage and control oil flow. The upper rings are compression rings which are forced outward against the cylinder wall and downward onto the groove face. The lower rings scrape the surplus oil from the cylinder wall and return it to the crankcase. The crankcase must be ventilated to remove gases which blow by the piston rings, to prevent pressure buildup.

The cylinder head (or heads in V engines) seals off the cylinders and is made of cast iron or aluminum. It must be strong and rigid to distribute the gas forces acting on the head as uniformly as possible through the engine block. The cylinder head contains the spark plug (for an SI engine) or fuel injector (for a CI engine), and, in overhead valve engines, parts of the valve mechanism.

The valves shown in Fig. 1-4 are poppet valves, the valve type normally used in four-stroke engines. Valves are made from forged alloy steel; the cooling of the exhaust valve which operates at about 700°C may be enhanced by using a hollow stem partially filled with sodium which through evaporation and condensation carries heat from the hot valve head to the cooler stem. Most modern spark-ignition engines have overhead valve locations (sometimes called valve-in-head or I-head configurations) as shown in Fig. 1-4. This geometry leads to a compact combustion chamber with minimum heat losses and flame travel time, and improves the breathing capacity. Previous geometries such as the L head where valves are to one side of the cylinder are now only used in small engines.

The valve stem moves in a valve guide, which can be an integral part of the cylinder head (or engine block for L-head engines), or may be a separate unit pressed into the head (or block). The valve seats may be cut in the head or block metal (if cast iron) or hard steel inserts may be pressed into the head or block. A valve spring, attached to the valve stem with a spring washer and split keeper, holds the valve closed. A valve rotator turns the valves a few degrees on opening to wipe the valve seat, avoid local hot spots, and prevent deposits building up in the valve guide.

A camshaft made of cast iron or forged steel with one cam per valve is used to open and close the valves. The cam surfaces are hardened to obtain adequate life. In four-stroke cycle engines, camshafts turn at one-half the crankshaft speed. Mechanical or hydraulic lifters or tappets slide in the block and ride on the cam. Depending on valve and camshaft location, additional members are required to transmit the tappet motion to the valve stem; e.g., in in-head valve engines with the camshaft at the side, a push rod and rocker arm are used. A recent trend in automotive engines is to mount the camshaft over the head with the cams acting either directly or through a pivoted follower on the valve. Camshafts are gear, belt, or chain driven from the crankshaft.

An intake manifold (aluminum or cast iron) and an exhaust manifold (generally of cast iron) complete the SI engine assembly. Other engine components specific to spark-ignition engines—carburetor, fuel injectors, ignition systems—are described more fully in the remaining sections in this chapter.

The two-stroke cycle CI engine shown in Fig. 1-5 is of the uniflow scavenged design. The burned gases exhaust through four valves in the cylinder head. These valves are controlled through cam-driven rocker arms. Fresh air is compressed and fed to the air box by a Roots blower. The air inlet ports at the bottom of each cylinder liner are uncovered by the descending piston, and the scavenging air flows upward along the cylinder axis. The fuel injectors are mounted in the cylinder head and are driven by the camshaft through rocker arms. Diesel fuel-injection systems are discussed in more detail in Sec. 1.7.

1.5 SPARK-IGNITION ENGINE OPERATION

In SI engines the air and fuel are usually mixed together in the intake system prior to entry to the engine cylinder, using a carburetor (Fig. 1-6) or fuel-injection system (Fig. 1-7). In automobile applications, the temperature of the air entering

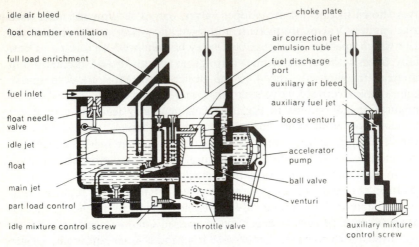

FIGURE 1-6
Cross section of single-barrel downdraft carburetor.[12] (*Courtesy Robert Bosch GmbH and SAE.*)

the intake system is controlled by mixing ambient air with air heated by contact with the exhaust manifold. The ratio of mass flow of air to mass flow of fuel must be held approximately constant at about 15 to ensure reliable combustion. The

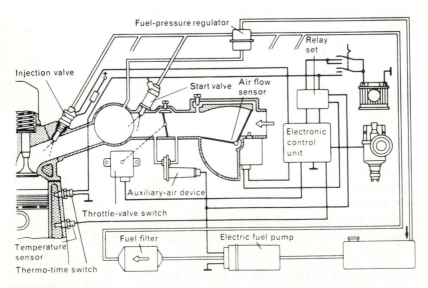

FIGURE 1-7
Schematic drawing of L-Jetronic port electronic fuel-injection system.[12] (*Courtesy Robert Bosch GmbH and SAE.*)

carburetor meters an appropriate fuel flow for the engine air flow in the following manner. The air flow through the venturi (a converging-diverging nozzle) sets up a pressure difference between the venturi inlet and throat which is used to meter an appropriate amount of fuel from the float chamber, through a series of orifices, into the air flow at the venturi throat. Just downstream of the venturi is a throttle valve or plate which controls the combined air and fuel flow, and thus the engine output. The intake flow is throttled to below atmospheric pressure by reducing the flow area when the power required (at any engine speed) is below the maximum which is obtained when the throttle is wide open. The intake manifold is usually heated to promote faster evaporation of the liquid fuel and obtain more uniform fuel distribution between cylinders.

Fuel injection into the intake manifold or inlet port is an increasingly common alternative to a carburetor. With port injection, fuel is injected through individual injectors from a low-pressure fuel supply system into each intake port. There are several different types of systems: mechanical injection using an injection pump driven by the engine; mechanical, driveless, continuous injection; electronically controlled, driveless, injection. Figure 1-7 shows an example of an electronically controlled system. In this system, the air flow rate is measured directly; the injection valves are actuated twice per cam shaft revolution by injection pulses whose duration is determined by the electronic control unit to provide the desired amount of fuel per cylinder per cycle.[12] An alternative approach is to use a single fuel injector located above the throttle plate in the position normally occupied by the carburetor. This approach permits electronic control of the fuel flow at reduced cost.

The sequence of events which take place inside the engine cylinder is illustrated in Fig. 1-8. Several variables are plotted against crank angle through the entire four-stroke cycle. Crank angle is a useful independent variable because engine processes occupy almost constant crank angle intervals over a wide range of engine speeds. The figure shows the valve timing and volume relationship for a typical automotive spark-ignition engine. To maintain high mixture flows at high engine speeds (and hence high power outputs) the inlet valve, which opens before TC, closes substantially after BC. During intake, the inducted fuel and air mix in the cylinder with the *residual* burned gases remaining from the previous cycle. After the intake valve closes, the cylinder contents are compressed to above atmospheric pressure and temperature as the cylinder volume is reduced. Some heat transfer to the piston, cylinder head, and cylinder walls occurs but the effect on unburned gas properties is modest.

Between 10 and 40 crank angle degrees before TC an electrical discharge across the spark plug starts the combustion process. A distributor, a rotating switch driven off the camshaft, interrupts the current from the battery through the primary circuit of the ignition coil. The secondary winding of the ignition coil, connected to the spark plug, produces a high voltage across the plug electrodes as the magnetic field collapses. Traditionally, cam-operated breaker points have been used; in most automotive engines, the switching is now done electronically. A turbulent flame develops from the spark discharge, propagates

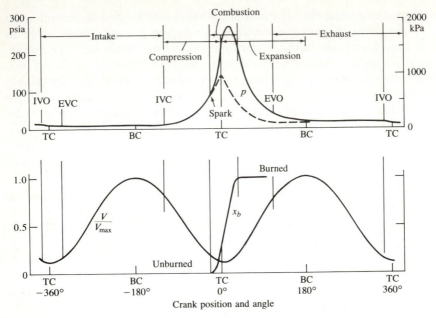

FIGURE 1-8

Sequence of events in four-stroke spark-ignition engine operating cycle. Cylinder pressure p (solid line, firing cycle; dashed line, motored cycle), cylinder volume V/V_{max}, and mass fraction burned x_b are plotted against crank angle.

across the mixture of air, fuel, and residual gas in the cylinder, and extinguishes at the combustion chamber wall. The duration of this burning process varies with engine design and operation, but is typically 40 to 60 crank angle degrees, as shown in Fig. 1-8. As fuel-air mixture burns in the flame, the cylinder pressure in Fig. 1-8 (solid line) rises above the level due to compression alone (dashed line). This latter curve—called the motored cylinder pressure—is the pressure trace obtained from a motored or nonfiring engine.† Note that due to differences in the flow pattern and mixture composition between cylinders, and within each cylinder cycle-by-cycle, the development of each combustion process differs somewhat. As a result, the shape of the pressure versus crank angle curve in each cylinder, and cycle-by-cycle, is not exactly the same.

There is an optimum spark timing which, for a given mass of fuel and air inside the cylinder, gives maximum torque. More advanced (earlier) timing or retarded (later) timing than this optimum gives lower output. Called *maximum*

---

† In practice, the intake and compression processes of a firing engine and a motored engine are not exactly the same due to the presence of burned gases from the previous cycle under firing conditions.

brake-torque (MBT) timing,† this optimum timing is an empirical compromise between starting combustion too early in the compression stroke (when the work transfer is *to* the cylinder gases) and completing combustion too late in the expansion stroke (and so lowering peak expansion stroke pressures).

About two-thirds of the way through the expansion stroke, the exhaust valve starts to open. The cylinder pressure is greater than the exhaust manifold pressure and a *blowdown* process occurs. The burned gases flow through the valve into the exhaust port and manifold until the cylinder pressure and exhaust pressure equilibrate. The duration of this process depends on the pressure level in the cylinder. The piston then *displaces* the burned gases from the cylinder into the manifold during the exhaust stroke. The exhaust valve opens before the end of the expansion stroke to ensure that the blowdown process does not last too far into the exhaust stroke. The actual timing is a compromise which balances reduced work transfer to the piston before BC against reduced work transfer to the cylinder contents after BC.

The exhaust valve remains open until just after TC; the intake opens just before TC. The valves are opened and closed slowly to avoid noise and excessive cam wear. To ensure the valves are fully open when piston velocities are at their highest, the valve open periods often overlap. If the intake flow is throttled to below exhaust manifold pressure, then backflow of burned gases into the intake manifold occurs when the intake valve is first opened.

1.6 EXAMPLES OF SPARK-IGNITION ENGINES

This section presents examples of production spark-ignition engines to illustrate the different types of engines in common use.

Small SI engines are used in many applications: in the home (e.g., lawn mowers, chain saws), in portable power generation, as outboard motorboat engines, and in motorcycles. These are often single-cylinder engines. In the above applications, light weight, small bulk, and low cost in relation to the power generated are the most important characteristics; fuel consumption, engine vibration, and engine durability are less important. A single-cylinder engine gives only one power stroke per revolution (two-stroke cycle) or two revolutions (four-stroke cycle). Hence, the torque pulses are widely spaced, and engine vibration and smoothness are significant problems.

Multicylinder engines are invariably used in automotive practice. As rated power increases, the advantages of smaller cylinders in regard to size, weight, and improved engine balance and smoothness point toward increasing the number of

---

† MBT timing has traditionally been defined as the minimum spark advance for best torque. Since the torque first increases and then decreases as spark timing is advanced, the definition used here is more precise.

cylinders per engine. An upper limit on cylinder size is dictated by dynamic considerations: the inertial forces that are created by accelerating and decelerating the reciprocating masses of the piston and connecting rod would quickly limit the maximum speed of the engine. Thus, the displaced volume is spread out amongst several smaller cylinders. The increased frequency of power strokes with a multicylinder engine produces much smoother torque characteristics. Multicylinder engines can also achieve a much better state of balance than single-cylinder engines. A force must be applied to the piston to accelerate it during the first half of its travel from bottom-center or top-center. The piston then exerts a force as it decelerates during the second part of the stroke. It is desirable to cancel these inertia forces through the choice of number and arrangement of cylinders to achieve a *primary balance*. Note, however, that the motion of the piston is more rapid during the upper half of its stroke than during the lower half (a consequence of the connecting rod and crank mechanism evident from Fig. 1-1; see also Sec. 2.2). The resulting inequality in piston acceleration and deceleration produces corresponding differences in inertia forces generated. Certain combinations of cylinder number and arrangement will balance out these secondary inertia force effects.

Four-cylinder in-line engines are the most common arrangements for automobile engines up to about 2.5-liter displacement. An example of this in-line arrangement was shown in Fig. 1-4. It is compact—an important consideration for small passenger cars. It provides two torque pulses per revolution of the crankshaft and primary inertia forces (though not secondary forces) are balanced. V engines and opposed-piston engines are occasionally used with this number of cylinders.

The V arrangement, with two banks of cylinders set at 90° or a more acute angle to each other, provides a compact block and is used extensively for larger displacement engines. Figure 1-9 shows a V-6 engine, the six cylinders being arranged in two banks of three with a 60° angle between their axis. Six cylinders are usually used in the 2.5- to 4.5-liter displacement range. Six-cylinder engines provide smoother operation with three torque pulses per revolution. The in-line arrangement results in a long engine, however, giving rise to crankshaft torsional vibration and making even distribution of air and fuel to each cylinder more difficult. The V-6 arrangement is much more compact, and the example shown provides primary balance of the reciprocating components. With the V engine, however, a rocking moment is imposed on the crankshaft due to the secondary inertia forces, which results in the engine being less well balanced than the in-line version. The V-8 and V-12 arrangements are also commonly used to provide compact, smooth, low-vibration, larger-displacement, spark-ignition engines.

Turbochargers are used to increase the maximum power that can be obtained from a given displacement engine. The work transfer to the piston per cycle, in each cylinder, which controls the power the engine can deliver, depends on the amount of fuel burned per cylinder per cycle. This depends on the amount of fresh air that is inducted each cycle. Increasing the air density prior to entry into the engine thus increases the maximum power that an engine of given dis-

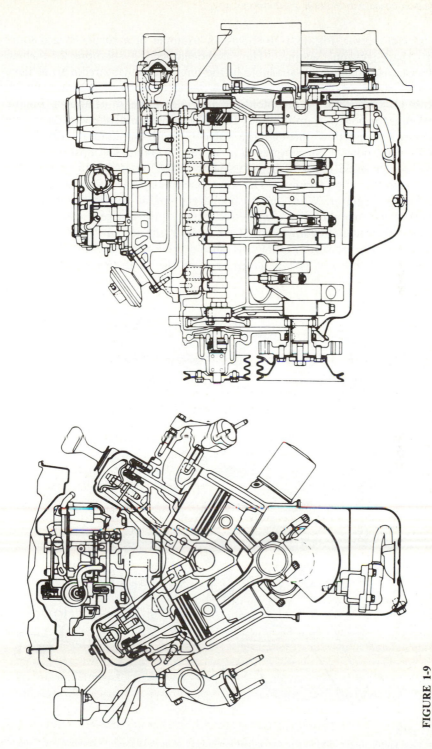

FIGURE 1-9

Cross-section drawings of General Motors 60 degree V-6 spark-ignition engine. [13] Displacement 2.8 liter, bore 89 mm, stroke 76 mm, compression ratio 8.5, maximum power 86 kW at 4800 rev/min.

placement can deliver. Figure 1-10 shows an example of a turbocharged four-cylinder spark-ignition engine. The turbocharger, a compressor-turbine combination, uses the energy available in the engine exhaust stream to achieve compression of the intake flow. The air flow passes through the compressor (2), intercooler (3), carburetor (4), manifold (5), and inlet valve (6) as shown. Engine inlet pressures (or boost) of up to about 100 kPa above atmospheric pressure are typical. The exhaust flow through the valve (7) and manifold (8) drives the turbine (9) which powers the compressor. A wastegate (valve) just upstream of the turbine bypasses some of the exhaust gas flow when necessary to prevent the boost pressure becoming too high. The wastegate linkage (11) is controlled by a boost pressure regulator. While this turbocharged engine configuration has the carburetor downstream of the compressor, some turbocharged spark-ignition engines have the carburetor upstream of the compressor so that it operates at or below atmospheric pressure. Figure 1-11 shows a cutaway drawing of a small automotive turbocharger. The arrangements of the compressor and turbine

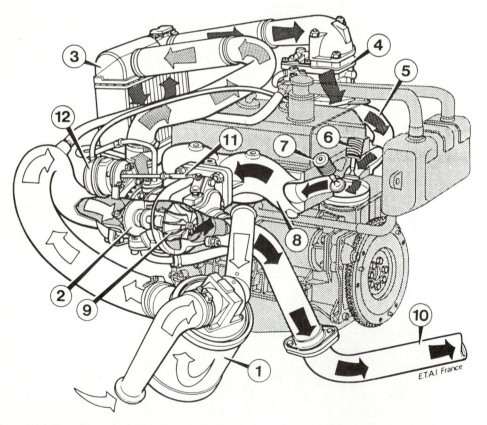

FIGURE 1-10
Turbocharged four-cylinder automotive spark-ignition engine. (*Courtesy Regie Nationale des Usines.*)

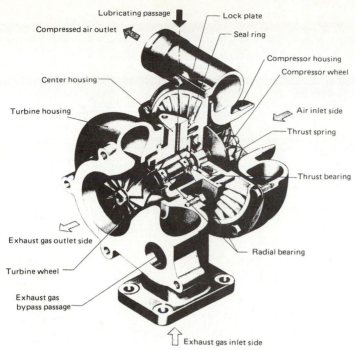

Lubricating passage

Lock plate

Compressed air outlet

Seal ring

Compressor housing

Compressor wheel

Center housing

Turbine housing

Air inlet side

Thrust spring

Thrust bearing

Exhaust gas outlet side

Radial bearing

Turbine wheel

Exhaust gas
bypass passage

Exhaust gas inlet side

FIGURE 1-11
Cutaway view of small automotive engine turbocharger. (*Courtesy Nissan Motor Co., Ltd.*)

rotors connected via the central shaft and of the turbine and compressor flow passages are evident.

Figure 1-12 shows a two-stroke cycle spark-ignition engine. The two-stroke cycle spark-ignition engine is used for small-engine applications where low cost and weight/power ratio are important and when the use factor is low. Examples of such applications are outboard motorboat engines, motorcycles, and chain saws. All such engines are of the carburetor crankcase-compression type which is one of the simplest prime movers available. It has three moving parts per cylinder: the piston, connecting rod, and the crank. The prime advantage of the two-stroke cycle spark-ignition engine relative to the four-stroke cycle engine is its higher power per unit displaced volume due to twice the number of power strokes per crank revolution. This is offset by the lower fresh charge density achieved by the two-stroke cycle gas-exchange process and the loss of fresh mixture which goes straight through the engine during scavenging. Also, oil consumption is higher in two-stroke cycle engines due to the need to add oil to the fuel to lubricate the piston ring and piston surfaces.

The Wankel rotary engine is an alternative to the reciprocating engine geometry of the engines illustrated above. It is used when its compactness and higher engine speed (which result in high power/weight and power/volume ratios), and inherent balance and smoothness, offset its higher heat transfer, and

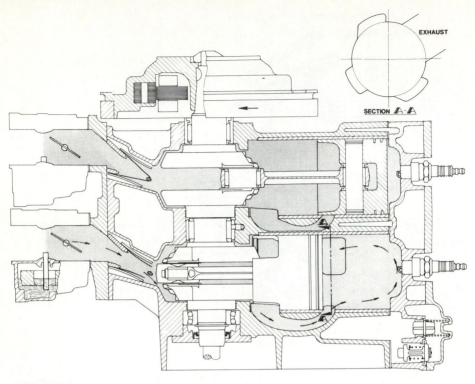

EXHAUST

SECTION *A-A*

FIGURE 1-12
Cutaway drawing of two-cylinder two-stroke cycle loop-scavenged marine spark-ignition engine. Displaced volume 737 cm³, maximum power 41 kW at 5500 rev/min. (*Courtesy Outboard Marine Corporation.*)

its sealing and leakage problems. Figure 1-13 shows the major mechanical parts of a simple single-rotor Wankel engine and illustrates its geometry. There are two rotating parts: the triangular-shaped rotor and the output shaft with its integral eccentric. The rotor revolves directly on the eccentric. The rotor has an internal timing gear which meshes with the fixed timing gear on one side housing to maintain the correct phase relationship between the rotor and eccentric shaft rotations. Thus the rotor rotates and orbits around the shaft axis. Breathing is through ports in the center housing (and sometimes the side housings). The combustion chamber lies between the center housing and rotor surface and is sealed by seals at the apex of the rotor and around the perimeters of the rotor sides. Figure 1-13 also shows how the Wankel rotary geometry operates with the four-stroke cycle. The figure shows the induction, compression, power, and exhaust processes of the four-stroke cycle for the chamber defined by rotor surface AB. The remaining two chambers defined by the other rotor surfaces undergo exactly the same sequence. As the rotor makes one complete rotation, during which the eccentric shaft rotates through three revolutions, each chamber produces one power "stroke." Three power pulses occur, therefore, for each rotor revolution;

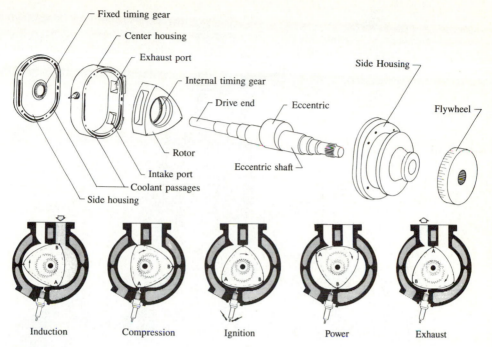

FIGURE 1-13

(a) Major components of the Wankel rotary engine; (b) induction, compression, power, and exhaust processes of the four-stroke cycle for the chamber defined by rotor surface AB. (*From Mobil Technical Bulletin*, Rotary Engines, © *Mobil Oil Corporation, 1971.*)

thus for each eccentric (output) shaft revolution there is one power pulse. Figure 1-14 shows a cutaway drawing of a two-rotor automobile Wankel engine. The two rotors are out of phase to provide a greater number of torque pulses per shaft revolution. Note the combustion chamber cut out in each rotor face, the rotor apex, and side seals. Two spark plugs per firing chamber are often used to obtain a faster combustion process.

1.7 COMPRESSION-IGNITION ENGINE OPERATION

In compression-ignition engines, air alone is inducted into the cylinder. The fuel (in most applications a light fuel oil, though heated residual fuel is used in marine and power-generation applications) is injected directly into the engine cylinder just before the combustion process is required to start. Load control is achieved by varying the amount of fuel injected each cycle; the air flow at a given engine speed is essentially unchanged. There are a great variety of CI engine designs in use in a wide range of applications—automobile, truck, locomotive, marine, power generation. Naturally aspirated engines where atmospheric air is inducted, turbocharged engines where the inlet air is compressed by an exhaust-driven

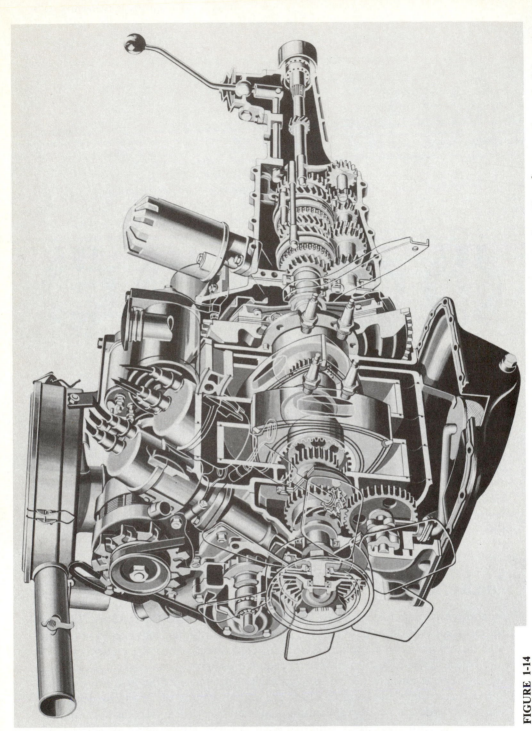

FIGURE 1-14
Cutaway drawing of two-rotor Wankel spark-ignition engine. Displacement of each working chamber 573 cm³, compression ratio 9.4, maximum
... ... 55 kW at 7000 rev/min. (Courtesy Toyo Kogyo Company Ltd.)

turbine-compressor combination, and supercharged engines where the air is compressed by a mechanically driven pump or blower are common. Turbocharging and supercharging increase engine output by increasing the air mass flow per unit displaced volume, thereby allowing an increase in fuel flow. These methods are used, usually in larger engines, to reduce engine size and weight for a given power output. Except in smaller engine sizes, the two-stroke cycle is competitive with the four-stroke cycle, in large part because, with the diesel cycle, only air is lost in the cylinder scavenging process.

The operation of a typical four-stroke naturally aspirated CI engine is illustrated in Fig. 1-15. The compression ratio of diesels is much higher than typical SI engine values, and is in the range 12 to 24, depending on the type of diesel engine and whether the engine is naturally aspirated or turbocharged. The valve timings used are similar to those of SI engines. Air at close-to-atmospheric pressure is inducted during the intake stroke and then compressed to a pressure of about 4 MPa (600 lb/in^2) and temperature of about 800 K (1000°F) during the compression stroke. At about 20° before TC, fuel injection into the engine cylinder commences; a typical rate of injection profile is shown in Fig. 1-15b. The liquid fuel jet atomizes into drops and entrains air. The liquid fuel evaporates; fuel vapor then mixes with air to within combustible proportions. The air temperature and pressure are above the fuel's ignition point. Therefore after a short *delay period*, spontaneous ignition (autoignition) of parts of the nonuniform fuel-air mixture initiates the combustion process, and the cylinder pressure (solid line in Fig. 1-15c) rises above the nonfiring engine level. The flame spreads rapidly through that portion of the injected fuel which has already mixed with sufficient air to burn. As the expansion process proceeds, mixing between fuel, air, and burning gases continues, accompanied by further combustion (see Fig. 1-15d). At full load, the mass of fuel injected is about 5 percent of the mass of air in the cylinder. Increasing levels of black smoke in the exhaust limit the amount of fuel that can be burned efficiently. The exhaust process is similar to that of the four-stroke SI engine. At the end of the exhaust stroke, the cycle starts again.

In the two-stroke CI engine cycle, compression, fuel injection, combustion, and expansion processes are similar to the equivalent four-stroke cycle processes; it is the intake and exhaust pressure which are different. The sequence of events in a loop-scavenged two-stroke engine is illustrated in Fig. 1-16. In loop-scavenged engines both exhaust and inlet ports are at the same end of the cylinder and are uncovered as the piston approaches BC (see Fig. 1-16a). After the exhaust ports open, the cylinder pressure falls rapidly in a blowdown process (Fig. 1-16b). The inlet ports then open, and once the cylinder pressure p falls below the inlet pressure p_i, air flows into the cylinder. The burned gases, displaced by this fresh air, continue to flow out of the exhaust port (along with some of the fresh air). Once the ports close as the piston starts the compression stroke, compression, fuel-injection, fuel-air mixing, combustion and expansion processes proceed as in the four-stroke CI engine cycle.

The diesel fuel-injection system consists of an injection pump, delivery pipes, and fuel injector nozzles. Several different types of injection pumps and

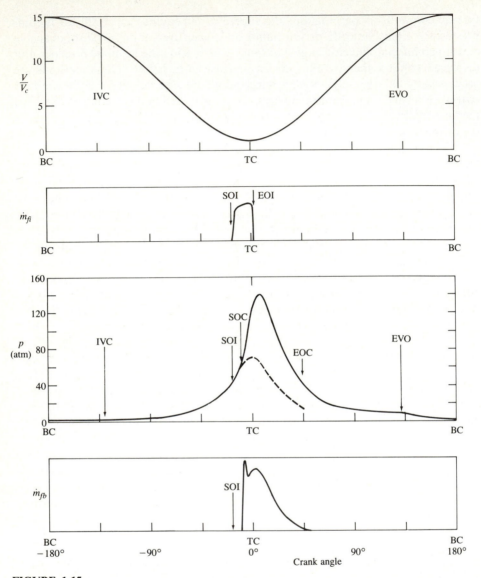

FIGURE 1-15
Sequence of events during compression, combustion, and expansion processes of a naturally aspirated compression-ignition engine operating cycle. Cylinder volume/clearance volume V/V_c, rate of fuel injection $\dot{m}_{fi}$, cylinder pressure p (solid line, firing cycle; dashed line, motored cycle), and rate of fuel burning (or fuel chemical energy release rate) $\dot{m}_{fb}$ are plotted against crank angle.

nozzles are used. In one common fuel pump (an in-line pump design shown in Fig. 1-17) a set of cam-driven plungers (one for each cylinder) operate in closely fitting barrels. Early in the stroke of the plunger, the inlet port is closed and the fuel trapped above the plunger is forced through a check valve into the injection

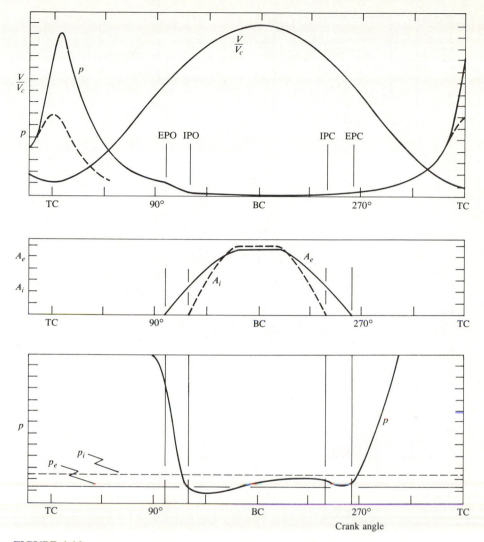

FIGURE 1-16
Sequence of events during expansion, gas exchange, and compression processes in a loop-scavenged two-stroke cycle compression-ignition engine. Cylinder volume/clearance volume V/V_c, cylinder pressure p, exhaust port open area A_e, and intake port open area A_i are plotted against crank angle.

line. The injection nozzle (Fig. 1-18) has one or more holes through which the fuel sprays into the cylinder. A spring-loaded valve closes these holes until the pressure in the injection line, acting on part of the valve surface, overcomes the spring force and opens the valve. Injection starts shortly after the line pressure begins to rise. Thus, the phase of the pump camshaft relative to the engine crankshaft controls the start of injection. Injection is stopped when the inlet port of the pump is uncovered by a helical groove in the pump plunger, because the high

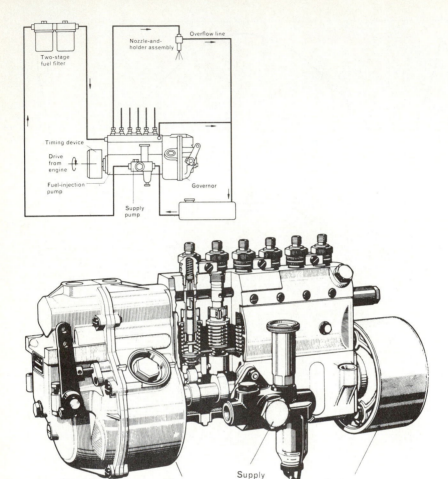

FIGURE 1-17
Diesel fuel system with in-line fuel-injection pump (type PE).[12] (*Courtesy Robert Bosch GmbH.*)

pressure above the plunger is then released (Fig. 1-18). The amount of fuel injected (which controls the load) is determined by the injection pump cam design and the position of the helical groove. Thus for a given cam design, rotating the plunger and its helical groove varies the load.

Distributor-type pumps have only one pump plunger and barrel, which meters and distributes the fuel to all the injection nozzles. A schematic of a distributor-type pump is shown in Fig. 1-19. The unit contains a low-pressure fuel pump (on left), a high-pressure injection pump (on right), an overspeed governor, and an injection timer. High pressure is generated by the plunger which is made to describe a combined rotary and stroke movement by the rotating eccentric disc or cam plate; the rotary motion distributes the fuel to the individual injection nozzles.

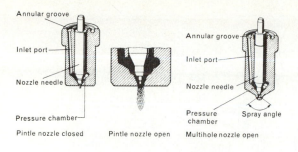

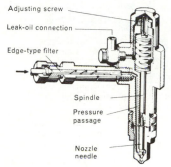

Nozzle-holder assembly with nozzle

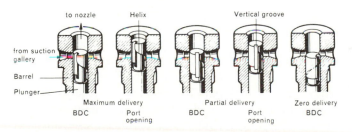

Fuel delivery control (lower helix)

FIGURE 1-18
Details of fuel-injection nozzles, nozzle holder assembly and fuel-delivery control.[12] (*Courtesy Robert Bosch GmbH.*)

Distributor pumps can operate at higher speed and take up less space than in-line pumps. They are normally used on smaller diesel engines. In-line pumps are used in the mid-engine-size range. In the larger diesels, individual single-barrel injection pumps, close mounted to each cylinder with an external drive as shown in Fig. 1-5, are normally used.

1.8 EXAMPLES OF DIESEL ENGINES

A large number of diesel engine configurations and designs are in common use. The very large marine and stationary power-generating diesels are two-stroke

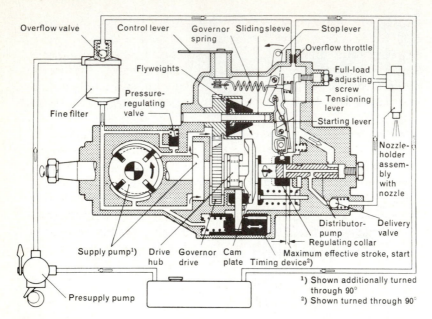

FIGURE 1-19
Diesel fuel system with distributor-type fuel-injection pump with mechanical governor.[12] (*Courtesy Robert Bosch GmbH*.)

cycle engines. Small- and medium-size engines use the four-stroke cycle. Because air capacity is an important constraint on the amount of fuel that can be burned in the diesel engine, and therefore on the engine's power, turbocharging is used extensively. All large engines are turbocharged. The majority of smaller diesels are not turbocharged, though they can be turbocharged and many are. The details of the engine design also vary significantly over the diesel size range. In particular, different combustion chamber geometries and fuel-injection character-istics are required to deal effectively with a major diesel engine design problem—achieving sufficiently rapid fuel-air mixing rates to complete the fuel-burning process in the time available. A wide variety of inlet port geometries, cylinder head and piston shapes, and fuel-injection patterns are used to accomplish this over the diesel size range.

Figure 1-20 shows a diesel engine typical of the medium-duty truck applica-tion. The design shown is a six-cylinder in-line engine. The drawing indicates that diesel engines are generally substantially heavier than spark-ignition engines because stress levels are higher due to the significantly higher pressure levels of the diesel cycle. The engine shown has a displacement of 10 liters, a compression ratio of 16.3, and is usually turbocharged. The engine has pressed-in cylinder liners to achieve better cylinder wear characteristics. This type of diesel is called a *direct-injection* diesel. The fuel is injected into a combustion chamber directly above the piston crown. The combustion chamber shown is a "bowl-in-piston" design, which puts most of the clearance volume into a compact shape. With this size of diesel engine, it is often necessary to use a swirling air flow rotating about the cylinder axis, which is created by suitable design of the inlet port and valve, to achieve adequate fuel-air mixing and fuel burning rates. The fuel injector, shown left-of-center in the drawing, has a multihole nozzle, typically with three to five holes. The fuel jets move out radially from the center of the piston bowl into the (swirling) air flow. The in-line fuel-injection pump is normally used with this type of diesel engine.

Figure 1-21 shows a four-cylinder in-line overhead-valve-cam design automobile diesel engine. The smallest diesels such as this operate at higher engine speed than larger engines; hence the time available for burning the fuel is less and the fuel-injection and combustion system must achieve faster fuel-air mixing rates. This is accomplished by using an *indirect-injection* type of diesel. Fuel is injected into an auxiliary combustion chamber which is separated from the main combustion chamber above the piston by a flow restriction or nozzle. During the latter stages of the compression process, air is forced through this nozzle from the

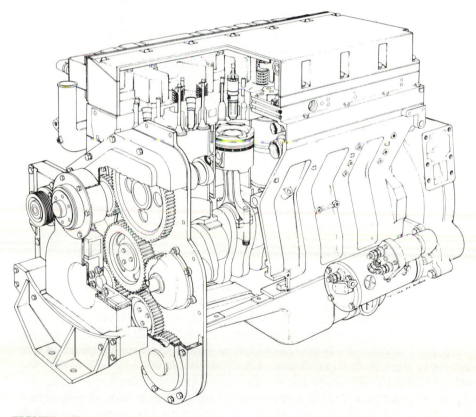

FIGURE 1-20
Direct-injection four-stroke cycle six-cylinder turbocharged Cummins diesel engine. Displaced volume 10 liters, bore 125 mm, stroke 136 mm, compression ratio 16.3, maximum power 168 to 246 kW at rated speed of 2100 rev/min. (*Courtesy Cummins Engine Company, Inc.*)

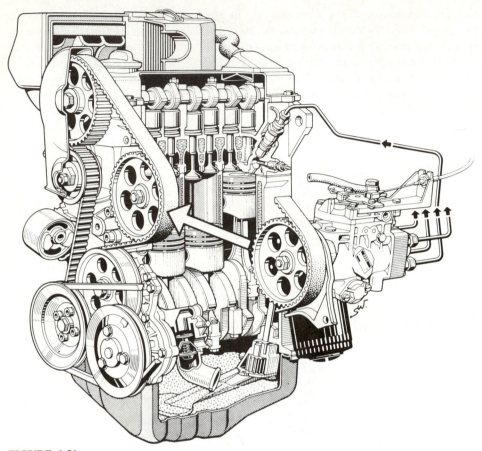

FIGURE 1-21
Four-cylinder naturally aspirated indirect-injection automobile Volkswagen diesel engine.[14] Displaced volume 1.47 liters, bore 76.5 mm, stroke 80 mm, maximum power 37 kW at 5000 rev/min.

cylinder into the prechamber at high velocity. Fuel is injected into this highly turbulent and often rapidly swirling flow in this auxiliary or prechamber, and very high mixing rates are achieved. Combustion starts in the prechamber, and the resulting pressure rise in the prechamber forces burning gases, fuel, and air into the main chamber. Since this outflow is also extremely vigorous, rapid mixing then occurs in the main chamber as the burning jet mixes with the remaining air and combustion is completed. A distributor-type fuel pump, which is normally used in this engine size range, driven off the camshaft at half the crankshaft speed by a toothed belt, is shown on the right of the figure. It supplies high-pressure fuel pulses to the pintle-type injector nozzles in turn. A glow plug is also shown in the auxiliary chamber; this plug is electrically heated prior to and during cold engine start-up to raise the temperature of the air charge and the fuel sufficiently to achieve autoignition. The compression ratio of this engine is 23. Indirect-injection diesel engines require higher compression ratios than direct-injection engines to start adequately when cold.

Diesel engines are turbocharged to achieve higher power/weight ratios. By increasing the density of the inlet air, a given displaced volume can induct more air. Hence more fuel can be injected and burned, and more power delivered, while avoiding excessive black smoke in the exhaust. All the larger diesels are turbocharged; smaller diesels can be and often are. Figure 1-22 shows how a turbocharger connects to a direct-injection diesel.

All the above diesels are water cooled; some production diesels are air cooled. Figure 1-23 shows a V-8 air-cooled direct-injection naturally aspirated

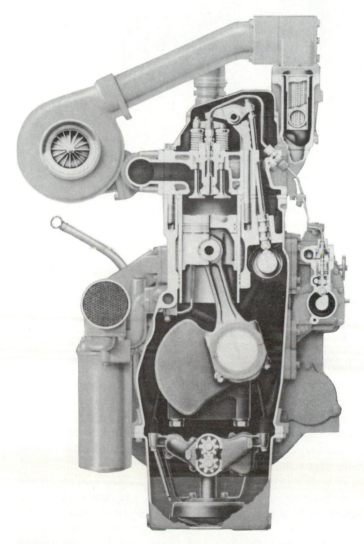

FIGURE 1-22
Turbocharged aftercooled direct-injection four-stroke cycle Caterpillar six-cylinder in-line heavy-duty truck diesel engine. Bore 137.2 mm, stroke 165.1 mm, rated power 200–300 kW and rated speed of 1600–2100 rev/min depending on application. (*Courtesy Caterpillar Tractor Company.*)

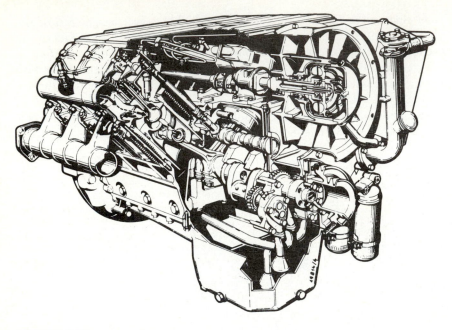

FIGURE 1-23
V-8 air-cooled direct-injection naturally aspirated diesel engine. Displacement 13.4 liter, bore 128 mm, stroke 130 mm, compression ratio 17, maximum rated power 188 kW at rated speed of 2300 rev/min. (*Courtesy Klöcker-Humboldt-Deutz AG.*[15])

diesel. The primary advantage compared to the water-cooled engines is lower engine weight. The fins on the cylinder block and head are necessary to increase the external heat-transfer surface area to achieve the required heat rejection. An air blower, shown on the right of the cutaway drawing, provides forced air convection over the block. The blower is driven off the injection pump shaft, which in turn is driven off the camshaft. The in-line injection pump is placed between the two banks of cylinders. The injection nozzles are located at an angle to the cylinder axis. The combustion chamber and fuel-injection characteristics are similar to those of the engine in Fig. 1-22. The nozzle shown injects four fuel sprays into a reentrant bowl-in-piston combustion chamber.

Diesels are also made in very large engine sizes. These large engines are used for marine propulsion and electrical power generation and operate on the two-stroke cycle in contrast to the small- and medium-size diesels illustrated above. Figure 1-24 shows such a two-stroke cycle marine engine, available with from 4 to 12 cylinders, with a maximum bore of 0.6–0.9 m and stroke of 2–3 m, which operates at speeds of about 100 rev/min. These engines are normally of the crosshead type to reduce side forces on the cylinder. The gas exchange between cycles is controlled by first opening the exhaust valves, and then the piston uncovering inlet ports in the cylinder liner. Expanding exhaust gases leave the cylinder via the exhaust valves and manifold and pass through the turbocharger

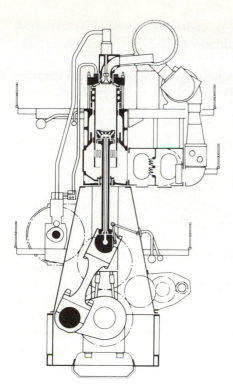

FIGURE 1-24
Large Sulzer two-stroke turbocharged marine diesel engine. Bore 840 mm, stroke 2900 mm, rated power 1.9 MW per cylinder at 78 rev/min, 4 to 12 cylinders. (*Courtesy Sulzer Brothers Ltd.*)

turbine. Compressed air enters via the inlet ports and induces forced scavenging; air is supplied from the turbocharger and cooler. At part load electrically driven blowers cut in to compress the scavenge air. Because these large engines operate at low speed, the motion induced by the centrally injected fuel jets is sufficient to mix the fuel with air and burn it in the time available. A simple open combustion chamber shape can be used, therefore, which achieves efficient combustion even with the low-quality heavy fuels used with these types of engines. The pistons are water cooled in these very large engines. The splash oil piston cooling used in medium- and small-size diesels is not adequate.

1.9 STRATIFIED-CHARGE ENGINES

Since the 1920s, attempts have been made to develop a hybrid internal combustion engine that combines the best features of the spark-ignition engine and the diesel. The goals have been to operate such an engine at close to the optimum compression ratio for efficiency (in the 12 to 15 range) by: (1) injecting the fuel directly into the combustion chamber during the compression process (and thereby avoid the knock or spontaneous ignition problem that limits conventional spark-ignition engines with their premixed charge); (2) igniting the fuel as it mixes with air with a spark plug to provide direct control of the ignition process (and thereby avoid the fuel ignition-quality requirement of the diesel); (3) controlling the engine power level by varying the amount of fuel injected per cycle (with the air flow unthrottled to minimize work done pumping the fresh charge into the cylinder). Such engines are often called *stratified-charge engines* from the need to produce in the mixing process between the fuel jet and the air in the cylinder a "stratified" fuel-air mixture, with an easily ignitable composition at the spark plug at the time of ignition. Because such engines avoid the spark-ignition engine requirement for fuels with a high antiknock quality and the diesel requirement for fuels with high ignition quality, they are usually fuel-tolerant and will operate with a wide range of liquid fuels.

Many different types of stratified-charge engine have been proposed, and some have been partially or fully developed. A few have even been used in practice in automotive applications. The operating principles of those that are truly fuel-tolerant or multifuel engines are illustrated in Fig. 1-25. The combustion chamber is usually a bowl-in-piston design, and a high degree of air swirl is created during intake and enhanced in the piston bowl during compression to achieve rapid fuel-air mixing. Fuel is injected into the cylinder, tangentially into the bowl, during the latter stages of compression. A long-duration spark discharge ignites the developing fuel-air jet as it passes the spark plug. The flame spreads downstream, and envelopes and consumes the fuel-air mixture. Mixing continues, and the final stages of combustion are completed during expansion. Most successful designs of this type of engine have used the four-stroke cycle. This concept is usually called a *direct-injection stratified-charge engine*. The engine can be turbocharged to increase its power density.

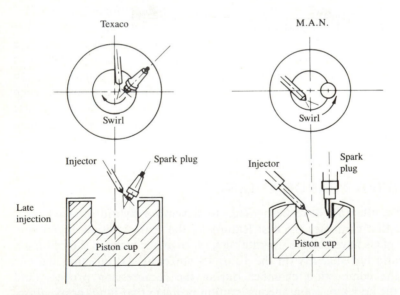

FIGURE 1-25
Two multifuel stratified-charge engines which have been used in commercial practice: the Texaco Controlled Combustion System (TCCS)[16] and the M.A.N.-FM System.[17]

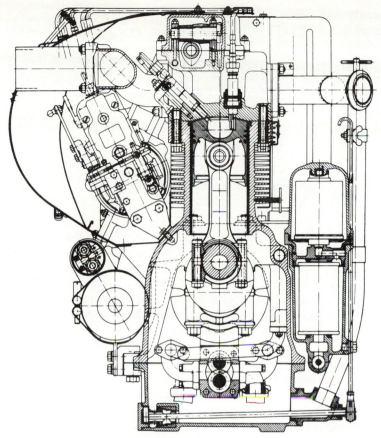

FIGURE 1-26
Sectional drawing of M.A.N. high-speed multifuel four-cylinder direct-injection stratified-charge engine. Bore 94.5 mm, stroke 100 mm, displacement 2.65 liters, compression ratio 16.5, rated power 52 kW at 3800 rev/min.[17]

A commercial multifuel engine is shown in Fig. 1-26. In this particular design, the fuel injector comes diagonally through the cylinder head from the upper left and injects the fuel onto the hot wall of the deep spherical piston bowl. The fuel is carried around the wall of the bowl by the swirling flow, evaporated off the wall, mixed with air, and then ignited by the discharge at the spark plug which enters the chamber vertically on the right. This particular engine is air cooled, so the cylinder block and head are finned to increase surface area.

An alternative stratified-charge engine concept, which has also been mass produced, uses a small *prechamber* fed during intake with an auxiliary fuel system to obtain an easily ignitable mixture around the spark plug. This concept, first proposed by Ricardo in the 1920s and extensively developed in the Soviet Union and Japan, is often called a *jet-ignition* or *torch-ignition* stratified-charge engine. Its operating principles are illustrated in Fig. 1-27 which shows a three-valve

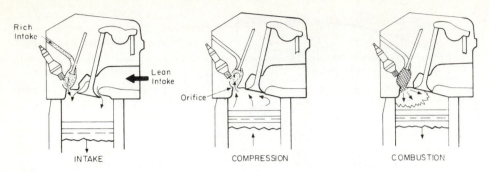

FIGURE 1-27
Schematic of three-valve torch-ignition stratified-charge spark-ignition engine.

carbureted version of the concept.[18] A separate carburetor and intake manifold feeds a fuel-rich mixture (which contains fuel beyond the amount that can be burned with the available air) through a separate small intake valve into the prechamber which contains the spark plug. At the same time, a very lean mixture (which contains excess air beyond that required to burn the fuel completely) is fed to the main combustion chamber through the main carburetor and intake manifold. During intake the rich prechamber flow fully purges the prechamber volume. After intake valve closing, lean mixture from the main chamber is compressed into the prechamber bringing the mixture at the spark plug to an easily ignitable, slightly rich, composition. After combustion starts in the prechamber, rich burning mixture issues as a jet through the orifice into the main chamber, entraining and igniting the lean main chamber charge. Though called a stratified-charge engine, this engine is really a jet-ignition concept whose primary function is to extend the operating limit of conventionally ignited spark-ignition engines to mixtures leaner than could normally be burned.

PROBLEMS

1.1. Describe the major functions of the following reciprocating engine components: piston, connecting rod, crankshaft, cams and camshaft, valves, intake and exhaust manifolds.

1.2. Indicate on an appropriate sketch the different forces that act on the piston, and the direction of these forces, during the engine's expansion stroke with the piston, connecting rod, and crank in the positions shown in Fig. 1-1.

1.3. List five important differences between the design and operating characteristics of spark-ignition and compression-ignition (diesel) engines.

1.4. Indicate the approximate crank angle at which the following events in the four-stroke and two-stroke internal combustion engine cycles occur on a line representing the full cycle (720° for the four-stroke cycle; 360° for the two-stroke cycle): bottom- and top-center crank positions, inlet and exhaust valve or port opening and closing, start of combustion process, end of combustion process, maximum cylinder pressure.

1.5. The two-stroke cycle has twice as many power strokes per crank revolution as the four-stroke cycle. However, two-stroke cycle engine power outputs per unit displaced volume are less than twice the power output of an equivalent four-stroke cycle engine at the same engine speed. Suggest reasons why this potential advantage of the two-stroke cycle is offset in practice.

1.6. Suggest reasons why multicylinder engines prove more attractive than single-cylinder engines once the total engine displaced volume exceeds a few hundred cubic centimeters.

1.7. The Wankel rotary spark-ignition engine, while lighter and more compact than a reciprocating spark-ignition engine of equal maximum power, typically has worse efficiency due to significantly higher gas leakage from the combustion chamber and higher total heat loss from the hot combustion gases to the chamber walls. Based on the design details in Figs. 1-4, 1-13, and 1-14 suggest reasons for these higher losses.

REFERENCES

1. Cummins, Jr., C. L.: *Internal Fire*, Carnot Press, Lake Oswego, Oreg., 1976.
2. Cummins, Jr., C. L.: "Early IC and Automotive Engines," SAE paper 760604 in *A History of the Automotive Internal Combustion Engine*, SP-409, *SAE Trans.*, vol. 85, 1976.
3. Hempson, J. G. G.: "The Automobile Engine 1920–1950," SAE paper 760605 in *A History of the Automotive Internal Combustion Engine*, SP-409, SAE, 1976.
4. Agnew, W. G.: "Fifty Years of Combustion Research at General Motors," *Progress in Energy and Combustion Science*, vol. 4, pp. 115–156, 1978.
5. Wankel, F.: *Rotary Piston Machines*, Iliffe Books, London, 1965.
6. Ansdale, R. F.: *The Wankel RC Engine Design and Performance*, Iliffe Books, London, 1968.
7. Yamamoto, K.: *Rotary Engine*, Toyo Kogyo Co. Ltd., Hiroshima, 1969.
8. Haagen-Smit, A. J.: "Chemistry and Physiology of Los Angeles Smog," *Ind. Eng. Chem.*, vol. 44, p. 1342, 1952.
9. Taylor, C. F.: *The Internal Combustion Engine in Theory and Practice*, vol. 2, table 10-1, MIT Press, Cambridge, Mass., 1968.
10. Rogowski, A. R.: *Elements of Internal Combustion Engines*, McGraw-Hill, 1953.
11. Weertman, W. L., and Dean, J. W.: "Chrysler Corporation's New 2.2 Liter 4 Cylinder Engine," SAE paper 810007, 1981.
12. Bosch: *Automotive Handbook*, 1st English edition, Robert Bosch GmbH, 1976.
13. Martens, D. A.: "The General Motors 2.8 Liter 60° V-6 Engine Designed by Chevrolet," SAE paper 790697, 1979.
14. Hofbauer, P., and Sator, K.: "Advanced Automotive Power Systems—Part 2: A Diesel for a Subcompact Car," SAE paper 770113, *SAE Trans.*, vol. 86, 1977.
15. Garthe, H.: "The Deutz BF8L 513 Aircooled Diesel Engine for Truck and Bus Application," SAE paper 852321, 1985.
16. Alperstein, M., Schafer, G. H., and Villforth, F. J.: "Texaco's Stratified Charge Engine—Multifuel, Efficient, Clean, and Practical," SAE paper 740563, 1974.
17. Urlaub, A. G., and Chmela, F. G.: "High-Speed, Multifuel Engine: L9204 FMV," SAE paper 740122, 1974.
18. Date, T., and Yagi, S.: "Research and Development of the Honda CVCC Engine," SAE paper 740605, 1974.

CHAPTER 2

ENGINE DESIGN AND OPERATING PARAMETERS

2.1 IMPORTANT ENGINE CHARACTERISTICS

In this chapter, some basic geometrical relationships and the parameters commonly used to characterize engine operation are developed. The factors important to an engine user are:

1. The engine's performance over its operating range
2. The engine's fuel consumption within this operating range and the cost of the required fuel
3. The engine's noise and air pollutant emissions within this operating range
4. The initial cost of the engine and its installation
5. The reliability and durability of the engine, its maintenance requirements, and how these affect engine availability and operating costs

These factors control total engine operating costs—usually the primary consideration of the user—and whether the engine in operation can satisfy environmental regulations. This book is concerned primarily with the performance, efficiency, and emissions characteristics of engines; the omission of the other factors listed above does not, in any way, reduce their great importance.

Engine performance is more precisely defined by:

1. The maximum power (or the maximum torque) available at each speed within the useful engine operating range
2. The range of speed and power over which engine operation is satisfactory

The following performance definitions are commonly used:

Maximum rated power. The highest power an engine is allowed to develop for short periods of operation.

Normal rated power. The highest power an engine is allowed to develop in continuous operation.

Rated speed. The crankshaft rotational speed at which rated power is developed.

2.2 GEOMETRICAL PROPERTIES OF RECIPROCATING ENGINES

The following parameters define the basic geometry of a reciprocating engine (see Fig. 2-1):

Compression ratio r_c:

$$r_c = \frac{\text{maximum cylinder volume}}{\text{minimum cylinder volume}} = \frac{V_d + V_c}{V_c} \tag{2.1}$$

where V_d is the displaced or swept volume and V_c is the clearance volume. Ratio of cylinder bore to piston stroke:

$$R_{bs} = \frac{B}{L} \tag{2.2}$$

Ratio of connecting rod length to crank radius:

$$R = \frac{l}{a} \tag{2.3}$$

In addition, the stroke and crank radius are related by

$$L = 2a$$

Typical values of these parameters are: $r_c = 8$ to 12 for SI engines and $r_c = 12$ to 24 for CI engines; $B/L = 0.8$ to 1.2 for small- and medium-size engines, decreasing to about 0.5 for large slow-speed CI engines; $R = 3$ to 4 for small- and medium-size engines, increasing to 5 to 9 for large slow-speed CI engines.

The cylinder volume V at any crank position θ is

$$V = V_c + \frac{\pi B^2}{4}(l + a - s) \tag{2.4}$$

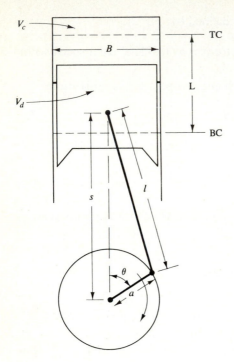

FIGURE 2-1
Geometry of cylinder, piston, connecting rod, and crankshaft where B = bore, L = stroke, l = connecting road length, a = crank radius, θ = crank angle.

where s is the distance between the crank axis and the piston pin axis (Fig. 2-1), and is given by

$$s = a \cos \theta + (l^2 - a^2 \sin^2 \theta)^{1/2} \tag{2.5}$$

The angle θ, defined as shown in Fig. 2-1, is called the *crank angle*. Equation (2.4) with the above definitions can be rearranged:

$$\frac{V}{V_c} = 1 + \tfrac{1}{2} (r_c - 1)[R + 1 - \cos \theta - (R^2 - \sin^2 \theta)^{1/2}] \tag{2.6}$$

The combustion chamber surface area A at any crank position θ is given by

$$A = A_{ch} + A_p + \pi B(l + a - s) \tag{2.7}$$

where A_{ch} is the cylinder head surface area and A_p is the piston crown surface area. For flat-topped pistons, $A_p = \pi B^2/4$. Using Eq. (2.5), Eq. (2-7) can be rearranged:

$$A = A_{ch} + A_p + \frac{\pi B L}{2} [R + 1 - \cos \theta - (R^2 - \sin^2 \theta)^{1/2}] \tag{2.8}$$

An important characteristic speed is the *mean piston speed* $\bar{S}_p$:

$$\bar{S}_p = 2LN \tag{2.9}$$

where N is the rotational speed of the crankshaft. Mean piston speed is often a

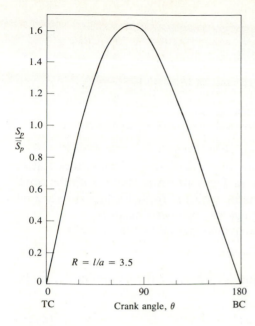

FIGURE 2-2
Instantaneous piston speed/mean piston speed as a function of crank angle for $R = 3.5$.

more appropriate parameter than crank rotational speed for correlating engine behavior as a function of speed. For example, gas-flow velocities in the intake and the cylinder all scale with $\bar{S}_p$. The *instantaneous* piston velocity S_p is obtained from

$$S_p = \frac{ds}{dt} \tag{2.10}$$

The piston velocity is zero at the beginning of the stroke, reaches a maximum near the middle of the stroke, and decreases to zero at the end of the stroke. Differentiation of Eq. (2.5) and substitution gives

$$\frac{S_p}{\bar{S}_p} = \frac{\pi}{2} \sin \theta \left[1 + \frac{\cos \theta}{(R^2 - \sin^2 \theta)^{1/2}} \right] \tag{2.11}$$

Figure 2-2 shows how S_p varies over each stroke for $R = 3.5$.

Resistance to gas flow into the engine or stresses due to the inertia of the moving parts limit the maximum mean piston speed to within the range 8 to 15 m/s (1500 to 3000 ft/min). Automobile engines operate at the higher end of this range; the lower end is typical of large marine diesel engines.

2.3 BRAKE TORQUE AND POWER

Engine torque is normally measured with a dynamometer.[1] The engine is clamped on a test bed and the shaft is connected to the dynamometer rotor. Figure 2-3 illustrates the operating principle of a dynamometer. The rotor is

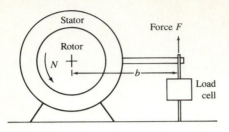

FIGURE 2-3
Schematic of principle of operation of dynamometer.

coupled electromagnetically, hydraulically, or by mechanical friction to a stator, which is supported in low friction bearings. The stator is balanced with the rotor stationary. The torque exerted on the stator with the rotor turning is measured by balancing the stator with weights, springs, or pneumatic means.

Using the notation in Fig. 2-3, if the torque exerted by the engine is T:

$$T = Fb \qquad (2.12)$$

The power P delivered by the engine and absorbed by the dynamometer is the product of torque and angular speed:

$$P = 2\pi N T \qquad (2.13a)$$

where N is the crankshaft rotational speed. In SI units:

$$P(\text{kW}) = 2\pi N(\text{rev/s})T(\text{N}\cdot\text{m}) \times 10^{-3} \qquad (2.13b)$$

or in U.S. units:

$$P(\text{hp}) = \frac{N(\text{rev/min})\ T(\text{lbf}\cdot\text{ft})}{5252} \qquad (2.13c)$$

Note that torque is a measure of an engine's ability to do work; power is the rate at which work is done.

The value of engine power measured as described above is called *brake power* P_b. This power is the usable power delivered by the engine to the load—in this case, a "brake."

2.4 INDICATED WORK PER CYCLE

Pressure data for the gas in the cylinder over the operating cycle of the engine can be used to calculate the work transfer from the gas to the piston. The cylinder pressure and corresponding cylinder volume throughout the engine cycle can be plotted on a p-V diagram as shown in Fig. 2-4. The *indicated work per cycle* $W_{c,i}$† (per cylinder) is obtained by integrating around the curve to obtain the

---

† The term indicated is used because such p-V diagrams used to be generated directly with a device called an engine indicator.

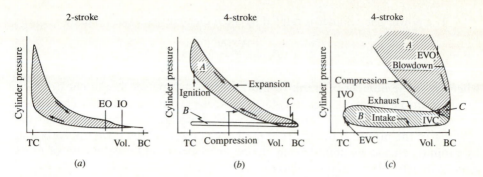

FIGURE 2-4
Examples of p-V diagrams for (a) a two-stroke cycle engine, (b) a four-stroke cycle engine; (c) a four-stroke cycle spark-ignition engine exhaust and intake strokes (pumping loop) at part load.

area enclosed on the diagram:

$$W_{c,i} = \oint p \, dV \tag{2.14}$$

With two-stroke cycles (Fig. 2-4a), the application of Eq. (2.14) is straightforward. With the addition of inlet and exhaust strokes for the four-stroke cycle, some ambiguity is introduced as two definitions of indicated output are in common use. These will be defined as:

Gross indicated work per cycle $W_{c,\text{ig}}$. Work delivered to the piston over the compression and expansion strokes only.

Net indicated work per cycle $W_{c,\text{in}}$. Work delivered to the piston over the entire four-stroke cycle.

In Fig. 2-4b and c, $W_{c,\text{ig}}$ is (area A + area C) and $W_{c,\text{in}}$ is (area A + area C) − (area B + area C), which equals (area A − area B), where each of these areas is regarded as a positive quantity. Area B + area C is the work transfer between the piston and the cylinder gases during the inlet and exhaust strokes and is called the *pumping work* W_p (see Chaps. 5 and 13). The pumping work transfer will be *to* the cylinder gases if the pressure during the intake stroke is less than the pressure during the exhaust stroke. This is the situation with naturally aspirated engines. The pumping work transfer will be *from* the cylinder gases to the piston if the exhaust stroke pressure is lower than the intake pressure, which is normally the case with highly loaded turbocharged engines.†

---

† With some two-stroke engine concepts there is a piston pumping work term associated with compressing the scavenging air in the crankcase.

The power per cylinder is related to the indicated work per cycle by

$$P_i = \frac{W_{c,i} N}{n_R} \qquad (2.15)$$

where n_R is the number of crank revolutions for each power stroke per cylinder. For four-stroke cycles, n_R equals 2; for two-stroke cycles, n_R equals 1. This power is the indicated power; i.e., the rate of work transfer from the gas within the cylinder to the piston. It differs from the brake power by the power absorbed in overcoming engine friction, driving engine accessories, and (in the case of gross indicated power) the pumping power.

In discussing indicated quantities of the four-stroke cycle engine, such as work per cycle or power, the definition used for "indicated" (i.e., gross or net) *should always be explicitly stated.* The gross indicated output, the definition most commonly used, will be chosen where possible in this book for the following reasons. Indicated quantities are used primarily to identify the impact of the compression, combustion, and expansion processes on engine performance, etc. The gross indicated output is, therefore, the most appropriate definition. It represents the sum of the useful work available at the shaft and the work required to overcome all the engine losses. Furthermore, the standard engine test codes[2] define procedures for measuring brake power and friction power (the friction power test provides a close approximation to the total lost power in the engine). The sum of brake power and friction power provides an alternative way of estimating indicated power; the value obtained is a close approximation to the gross indicated power.

The terms brake and indicated are used to describe other parameters such as mean effective pressure, specific fuel consumption, and specific emissions (see the following sections) in a manner similar to that used for work per cycle and power.

2.5 MECHANICAL EFFICIENCY

We have seen that part of the gross indicated work per cycle or power is used to expel exhaust gases and induct fresh charge. An additional portion is used to overcome the friction of the bearings, pistons, and other mechanical components of the engine, and to drive the engine accessories. All of these power requirements are grouped together and called *friction power* P_f.† Thus:

$$P_{ig} = P_b + P_f \qquad (2.16)$$

Friction power is difficult to determine accurately. One common approach for high-speed engines is to drive or motor the engine with a dynamometer (i.e., operate the engine without firing it) and measure the power which has to be

---

† The various components of friction power are examined in detail in Chap. 13.

supplied by the dynamometer to overcome *all* these frictional losses. The engine speed, throttle setting, oil and water temperatures, and ambient conditions are kept the same in the motored test as under firing conditions. The major sources of inaccuracy with this method are that gas pressure forces on the piston and rings are lower in the motored test than when the engine is firing and that the oil temperatures on the cylinder wall are also lower under motoring conditions.

The ratio of the brake (or useful) power delivered by the engine to the indicated power is called the *mechanical efficiency* η_m:

$$\eta_m = \frac{P_b}{P_{ig}} = 1 - \frac{P_f}{P_{ig}} \tag{2.17}$$

Since the friction power includes the power required to pump gas into and out of the engine, mechanical efficiency depends on throttle position as well as engine design and engine speed. Typical values for a modern automotive engine at wide-open or full throttle are 90 percent at speeds below about 30 to 40 rev/s (1800 to 2400 rev/min), decreasing to 75 percent at maximum rated speed. As the engine is throttled, mechanical efficiency decreases, eventually to zero at idle operation.

2.6 ROAD-LOAD POWER

A part-load power level useful as a reference point for testing automobile engines is the power required to drive a vehicle on a level road at a steady speed. Called *road-load power*, this power overcomes the rolling resistance which arises from the friction of the tires and the aerodynamic drag of the vehicle. Rolling resistance and drag coefficients, C_R and C_D, respectively, are determined empirically. An approximate formula for road-load power P_r is

$$P_r = (C_R M_v g + \tfrac{1}{2}\rho_a C_D A_v S_v^2)S_v \tag{2.18a}$$

where C_R = coefficient of rolling resistance $(0.012 < C_R < 0.015)^3$
$\quad\;\; M_v$ = mass of vehicle [for passenger cars: curb mass plus passenger load of 68 kg (150 lbm); in U.S. units W_v = vehicle weight in lbf]
$\quad\;\; g$ = acceleration due to gravity
$\quad\;\; \rho_a$ = ambient air density
$\quad\;\; C_D$ = drag coefficient (for cars: $0.3 < C_D \lesssim 0.5)^3$
$\quad\;\; A_v$ = frontal area of vehicle
$\quad\;\; S_v$ = vehicle speed

With the quantities in the units indicated:

$$P_r(kW) = [2.73 C_R M_v(kg) + 0.0126 C_D A_v(m^2) S_v(km/h)^2] S_v(km/h) \times 10^{-3} \tag{2.18b}$$

or $\quad P_r(hp) = \dfrac{[C_R W_v(lbf) + 0.0025 C_D A_v(ft^2) S_v(mi/h)^2] S_v(mi/h)}{375} \tag{2.18c}$

2.7 MEAN EFFECTIVE PRESSURE

While torque is a valuable measure of a particular engine's ability to do work, it depends on engine size. A more useful relative engine performance measure is obtained by dividing the work per cycle by the cylinder volume displaced per cycle. The parameter so obtained has units of force per unit area and is called the *mean effective pressure* (mep). Since, from Eq. (2.15),

$$\text{Work per cycle} = \frac{Pn_R}{N}$$

where n_R is the number of crank revolutions for each power stroke per cylinder (two for four-stroke cycles; one for two-stroke cycles), then

$$\text{mep} = \frac{Pn_R}{V_d N} \tag{2.19a}$$

For SI and U.S. units, respectively,

$$\text{mep(kPa)} = \frac{P(\text{kW})n_R \times 10^3}{V_d(\text{dm}^3)N(\text{rev/s})} \tag{2.19b}$$

$$\text{mep(lb/in}^2) = \frac{P(\text{hp})n_R \times 396,000}{V_d(\text{in}^3)N(\text{rev/min})} \tag{2.19c}$$

Mean effective pressure can also be expressed in terms of torque by using Eq. (2.13):

$$\text{mep(kPa)} = \frac{6.28 n_R T(\text{N} \cdot \text{m})}{V_d(\text{dm}^3)} \tag{2.20a}$$

or
$$\text{mep(lb/in}^2) = \frac{75.4 n_R T(\text{lbf} \cdot \text{ft})}{V_d(\text{in}^3)} \tag{2.20b}$$

The maximum brake mean effective pressure of good engine designs is well established, and is essentially constant over a wide range of engine sizes. Thus, the actual bmep that a particular engine develops can be compared with this norm, and the effectiveness with which the engine designer has used the engine's displaced volume can be assessed. Also, for design calculations, the engine displacement required to provide a given torque or power, at a specified speed, can be estimated by assuming appropriate values for bmep for that particular application.

Typical values for bmep are as follows. For naturally aspirated spark-ignition engines, maximum values are in the range 850 to 1050 kPa ($\sim$ 125 to 150 lb/in²) at the engine speed where maximum torque is obtained (about 3000 rev/min). At the maximum rated power, bmep values are 10 to 15 percent lower. For turbocharged automotive spark-ignition engines the maximum bmep is in the 1250 to 1700 kPa (180 to 250 lb/in²) range. At the maximum rated power, bmep is in the 900 to 1400 kPa (130 to 200 lb/in²) range. For naturally aspirated four-stroke diesels, the maximum bmep is in the 700 to 900 kPa (100 to 130 lb/in^2) range, with the bmep at the maximum rated power of about 700 kPa (100 lb/in^2). Turbocharged four-stroke diesel maximum bmep values are typically in the range 1000 to 1200 kPa (145 to 175 lb/in^2); for turbocharged aftercooled engines this can rise to 1400 kPa. At maximum rated power, bmep is about 850 to 950 kPa (125 to 140 lb/in^2). Two-stroke cycle diesels have comparable performance to four-stroke cycle engines. Large low-speed two-stroke cycle engines can achieve bmep values of about 1600 kPa.

An example of how the above engine performance parameters can be used to initiate an engine design is given below.

> **Example.** A four-cylinder automotive spark-ignition engine is being designed to provide a maximum brake torque of 150 N·m (110 lbf·ft) in the mid-speed range ($\sim$ 3000 rev/min). Estimate the required engine displacement, bore and stroke, and the maximum brake power the engine will deliver.
>
> Equation (2.20a) relates torque and mep. Assume that 925 kPa is an appropriate value for bmep at the maximum engine torque point. Equation (2.20a) gives
>
> $$V(\text{dm}^3) = \frac{6.28 n_R T_{max}(\text{N·m})}{\text{bmep}_{max}(\text{kPa})} = \frac{6.28 \times 2 \times 150}{925} = 2 \text{ dm}^3$$
>
> For a four-cylinder engine, the displaced volume, bore, and stroke are related by
>
> $$V_d = 4 \times \frac{\pi}{4} B^2 L$$
>
> Assume $B = L$; this gives $B = L = 86$ mm.
>
> The maximum rated engine speed can be estimated from an appropriate value for the maximum mean piston speed, 15 m/s (see Sec. 2.2):
>
> $$\bar{S}_{pmax} = 2 L N_{max} \rightarrow N_{max} = 87 \text{ rev/s (5200 rev/min)}$$
>
> The maximum brake power can be estimated from the typical bmep value at maximum power, 800 kPa (116 lb/in^2), using Eq. (2.19b):
>
> $$P_{bmax}(\text{kW}) = \frac{\text{bmep}(\text{kPa})V(\text{dm}^3)N_{max}(\text{rev/s})}{n_R \times 10^3} = \frac{800 \times 2 \times 87}{2 \times 10^3} = 70 \text{ kW}$$

2.8 SPECIFIC FUEL CONSUMPTION AND EFFICIENCY

In engine tests, the fuel consumption is measured as a flow rate—mass flow per unit time $\dot{m}_f$. A more useful parameter is the *specific fuel consumption* (sfc)—the fuel flow rate per unit power output. It measures how efficiently an engine is using the fuel supplied to produce work:

$$\text{sfc} = \frac{\dot{m}_f}{P} \tag{2.21}$$

With units,

$$\text{sfc(mg/J)} = \frac{\dot{m}_f(\text{g/s})}{P(\text{kW})} \tag{2.22a}$$

or

$$\text{sfc(g/kW} \cdot \text{h)} = \frac{\dot{m}_f(\text{g/h})}{P(\text{kW})} = 608.3 \; \text{sfc(lbm/hp} \cdot \text{h)} \tag{2.22b}$$

or

$$\text{sfc(lbm/hp} \cdot \text{h)} = \frac{\dot{m}_f(\text{lbm/h})}{P(\text{hp})} = 1.644 \times 10^3 \; \text{sfc(g/kW} \cdot \text{h)} \tag{2.22c}$$

Low values of sfc are obviously desirable. For SI engines typical best values of brake specific fuel consumption are about 75 μg/J = 270 g/kW·h = 0.47 lbm/hp·h. For CI engines, best values are lower and in large engines can go below 55 μg/J = 200 g/kW·h = 0.32 lbm/hp·h.

The specific fuel consumption has units. A dimensionless parameter that relates the desired engine output (work per cycle or power) to the necessary input (fuel flow) would have more fundamental value. The ratio of the work produced per cycle to the amount of fuel energy supplied per cycle that can be released in the combustion process is commonly used for this purpose. It is a measure of the engine's efficiency. The fuel energy supplied which can be released by combustion is given by the mass of fuel supplied to the engine per cycle times the heating value of the fuel. The heating value of a fuel, Q_{HV}, defines its energy content. It is determined in a standardized test procedure in which a known mass of fuel is fully burned with air, and the thermal energy released by the combustion process is absorbed by a calorimeter as the combustion products cool down to their original temperature.

This measure of an engine's "efficiency," which will be called the *fuel conversion efficiency* η_f,† is given by

$$\eta_f = \frac{W_c}{m_f Q_{\text{HV}}} = \frac{(P n_R / N)}{(\dot{m}_f n_R / N) Q_{\text{HV}}} = \frac{P}{\dot{m}_f Q_{\text{HV}}} \tag{2.23}$$

where m_f is the mass of fuel inducted per cycle. Substitution for $P/\dot{m}_f$ from Eq. (2.21) gives

$$\eta_f = \frac{1}{\text{sfc } Q_{\text{HV}}} \tag{2.24a}$$

---

† This empirically defined engine efficiency has previously been called thermal efficiency or enthalpy efficiency. The term fuel conversion efficiency is preferred because it describes this quantity more precisely, and distinguishes it clearly from other definitions of engine efficiency which will be developed in Sec. 3.6. Note that there are several different definitions of heating value (see Sec. 3.5). The numerical values do not normally differ by more than a few percent, however. In this text, the lower heating value at constant pressure is used in evaluating the fuel conversion efficiency.

or with units:

$$\eta_f = \frac{1}{\text{sfc(mg/J)}Q_{HV}\text{(MJ/kg)}} \tag{2.24b}$$

$$\eta_f = \frac{3600}{\text{sfc(g/kW·h)}Q_{HV}\text{(MJ/kg)}} \tag{2.24c}$$

$$\eta_f = \frac{2545}{\text{sfc(lbm/hp·h)}Q_{HV}\text{(Btu/lbm)}} \tag{2.24d}$$

Typical heating values for the commercial hydrocarbon fuels used in engines are in the range 42 to 44 MJ/kg (18,000 to 19,000 Btu/lbm). Thus, specific fuel consumption is inversely proportional to fuel conversion efficiency for normal hydrocarbon fuels.

Note that the fuel energy supplied to the engine per cycle is not fully released as thermal energy in the combustion process because the actual combustion process in incomplete. When enough air is present in the cylinder to oxidize the fuel completely, almost all (more than about 96 percent) of this fuel energy supplied is transferred as thermal energy to the working fluid. When insufficient air is present to oxidize the fuel completely, lack of oxygen prevents this fuel energy supplied from being fully released. This topic is discussed in more detail in Secs. 3.5 and 4.9.4.

2.9 AIR/FUEL AND FUEL/AIR RATIOS

In engine testing, both the air mass flow rate $\dot{m}_a$ and the fuel mass flow rate $\dot{m}_f$ are normally measured. The ratio of these flow rates is useful in defining engine operating conditions:

$$\text{Air/fuel ratio } (A/F) = \frac{\dot{m}_a}{\dot{m}_f} \tag{2.25}$$

$$\text{Fuel/air ratio } (F/A) = \frac{\dot{m}_f}{\dot{m}_a} \tag{2.26}$$

The normal operating range for a conventional SI engine using gasoline fuel is $12 \leq A/F \leq 18$ ($0.056 \leq F/A \leq 0.083$); for CI engines with diesel fuel, it is $18 \leq A/F \leq 70$ ($0.014 \leq F/A \leq 0.056$).

2.10 VOLUMETRIC EFFICIENCY

The intake system—the air filter, carburetor, and throttle plate (in a spark-ignition engine), intake manifold, intake port, intake valve—restricts the amount of air which an engine of given displacement can induct. The parameter used to measure the effectiveness of an engine's induction process is the *volumetric efficiency* η_v. Volumetric efficiency is only used with four-stroke cycle engines which have a distinct induction process. It is defined as the volume flow rate of air into the intake system divided by the rate at which volume is displaced by the piston:

$$\eta_v = \frac{2\dot{m}_a}{\rho_{a,i} V_d N}$$

(2.27a)

where $\rho_{a,i}$ is the inlet air density. An alternative equivalent definition for volumetric efficiency is

$$\eta_v = \frac{m_a}{\rho_{a,i} V_d}$$

(2.27b)

where m_a is the mass of air inducted into the cylinder per cycle.

The inlet density may either be taken as atmosphere air density (in which case η_v measures the pumping performance of the entire inlet system) or may be taken as the air density in the inlet manifold (in which case η_v measures the pumping performance of the inlet port and valve only). Typical maximum values of η_v for naturally aspirated engines are in the range 80 to 90 percent. The volumetric efficiency for diesels is somewhat higher than for SI engines. Volumetric efficiency is discussed more fully in Sec. 6.2.

2.11 ENGINE SPECIFIC WEIGHT AND SPECIFIC VOLUME

Engine weight and bulk volume for a given rated power are important in many applications. Two parameters useful for comparing these attributes from one engine to another are:

$$\text{Specific weight} = \frac{\text{engine weight}}{\text{rated power}}$$

(2.28)

$$\text{Specific volume} = \frac{\text{engine volume}}{\text{rated power}}$$

(2.29)

For these parameters to be useful in engine comparisons, a consistent definition of what components and auxiliaries are included in the term "engine" must be adhered to. These parameters indicate the effectiveness with which the engine designer has used the engine materials and packaged the engine components.[4]

2.12 CORRECTION FACTORS FOR POWER AND VOLUMETRIC EFFICIENCY

The pressure, humidity, and temperature of the ambient air inducted into an engine, at a given engine speed, affect the air mass flow rate and the power output. Correction factors are used to adjust measured wide-open-throttle power and volumetric efficiency values to standard atmospheric conditions to provide a more accurate basis for comparisons between engines. Typical standard ambient conditions used are:

Dry air pressure	Water vapour pressure	Temperature
736.6 mmHg	9.65 mmHg	29.4°C
29.00 inHg	0.38 inHg	85°F

The basis for the correction factor is the equation for one-dimensional steady compressible flow through an orifice or flow restriction of effective area A_E (see App. C):

$$\dot{m} = \frac{A_E p_0}{\sqrt{RT_0}} \left\{ \frac{2\gamma}{\gamma - 1} \left[\left(\frac{p}{p_0} \right)^{2/\gamma} - \left(\frac{p}{p_0} \right)^{(\gamma+1)/\gamma} \right] \right\}^{1/2} \tag{2.30}$$

In deriving this equation, it has been assumed that the fluid is an ideal gas with gas constant R and that the ratio of specific heats $(c_p/c_v = \gamma)$ is a constant; p_0 and T_0 are the total pressure and temperature upstream of the restriction and p is the pressure at the throat of the restriction.

If, in the engine, p/p_0 is assumed constant at wide-open throttle, then for a given intake system and engine, the mass flow rate of dry air $\dot{m}_a$ varies as

$$\dot{m}_a \propto \frac{p_0}{\sqrt{T_0}} \tag{2.31}$$

For mixtures containing the proper amount of fuel to use all the air available (and thus provide maximum power), the indicated power at full throttle P_i will be proportional to $\dot{m}_a$, the *dry* air flow rate. Thus if

$$P_{i,s} = C_F P_{i,m} \tag{2.32}$$

where the subscripts s and m denote values at the standard and measured conditions, respectively, the correction factor C_F is given by

$$C_F = \frac{p_{s,d}}{p_m - p_{v,m}} \left(\frac{T_m}{T_s} \right)^{1/2} \tag{2.33}$$

where $p_{s,d}$ = standard dry-air absolute pressure
$\quad\quad p_m$ = measured ambient-air absolute pressure
$\quad\quad p_{v,m}$ = measured ambient–water vapour partial pressure
$\quad\quad T_m$ = measured ambient temperature, K
$\quad\quad T_s$ = standard ambient temperature, K

The rated *brake* power is corrected by using Eq. (2.33) to correct the *indicated* power and making the assumption that friction power is unchanged. Thus

$$P_{b,s} = C_F P_{i,m} - P_{f,m} \tag{2.34}$$

Volumetric efficiency is proportional to $\dot{m}_a/\rho_a$ [see Eq. (2.27)]. Since ρ_a is proportional to p/T, the correction factor for volumetric efficiency, C_F', is

$$C_F' = \frac{\eta_{v,s}}{\eta_{v,m}} = \left(\frac{T_s}{T_m} \right)^{1/2} \tag{2.35}$$

2.13 SPECIFIC EMISSIONS AND EMISSIONS INDEX

Levels of emissions of oxides of nitrogen (nitric oxide, NO, and nitrogen dioxide, NO_2, usually grouped together as NO_x), carbon monoxide (CO), unburned hydrocarbons (HC), and particulates are important engine operating characteristics.

The concentrations of gaseous emissions in the engine exhaust gases are usually measured in parts per million or percent by volume (which corresponds to the mole fraction multiplied by 10^6 or by 10^2, respectively). Normalized indicators of emissions levels are more useful, however, and two of these are in common use. *Specific emissions* are the mass flow rate of pollutant per unit power output:

$$sNO_x = \frac{\dot{m}_{NO_x}}{P} \tag{2.36a}$$

$$sCO = \frac{\dot{m}_{CO}}{P} \tag{2.36b}$$

$$sHC = \frac{\dot{m}_{HC}}{P} \tag{2.36c}$$

$$sPart = \frac{\dot{m}_{part}}{P} \tag{2.36d}$$

Indicated and brake specific emissions can be defined. Units in common use are $\mu g/J$, $g/kW \cdot h$, and $g/hp \cdot h$.

Alternatively, emission rates can be normalized by the fuel flow rate. An *emission index* (EI) is commonly used: e.g.,

$$EI_{NO_x} = \frac{\dot{m}_{NO_x}(g/s)}{\dot{m}_f(kg/s)} \tag{2.37}$$

with similar expressions for CO, HC, and particulates.

2.14 RELATIONSHIPS BETWEEN PERFORMANCE PARAMETERS

The importance of the parameters defined in Secs. 2.8 to 2.10 to engine performance becomes evident when power, torque, and mean effective pressure are expressed in terms of these parameters. From the definitions of engine power [Eq. (2.13)], mean effective pressure [Eq. (2.19)], fuel conversion efficiency [Eq. (2.23)], fuel/air ratio [Eq. (2.26)], and volumetric efficiency [Eq. (2.27)], the following relationships between engine performance parameters can be developed. For power P:

$$P = \frac{\eta_f m_a N Q_{HV}(F/A)}{n_R} \tag{2.38}$$

For four-stroke cycle engines, volumetric efficiency can be introduced:

$$P = \frac{\eta_f \eta_v N V_d Q_{HV} \rho_{a,i}(F/A)}{2} \tag{2.39}$$

For torque T:

$$T = \frac{\eta_f \eta_v V_d Q_{HV} \rho_{a,i}(F/A)}{4\pi} \tag{2.40}$$

For mean effective pressure:

$$\text{mep} = \eta_f \eta_v Q_{HV} \rho_{a,i}(F/A) \tag{2.41}$$

The power per unit piston area, often called the *specific power*, is a measure of the engine designer's success in using the available piston area regardless of cylinder size. From Eq. (2.39), the specific power is

$$\frac{P}{A_p} = \frac{\eta_f \eta_v N L Q_{HV} \rho_{a,i}(F/A)}{2} \tag{2.42}$$

Mean piston speed can be introduced with Eq. (2.9) to give

$$\frac{P}{A_p} = \frac{\eta_f \eta_v \bar{S}_p Q_{HV} \rho_{a,i}(F/A)}{4} \tag{2.43}$$

Specific power is thus proportional to the product of mean effective pressure and mean piston speed.

These relationships illustrate the direct importance to engine performance of:

1. High fuel conversion efficiency
2. High volumetric efficiency
3. Increasing the output of a given displacement engine by increasing the inlet air density
4. Maximum fuel/air ratio that can be usefully burned in the engine
5. High mean piston speed

2.15 ENGINE DESIGN AND PERFORMANCE DATA

Engine ratings usually indicate the highest power at which manufacturers expect their products to give satisfactory economy, reliability, and durability under service conditions. Maximum torque, and the speed at which it is achieved, is usually given also. Since both of these quantities depend on displaced volume, for comparative analyses between engines of different displacements in a given engine category normalized performance parameters are more useful. The following measures, at the operating points indicated, have most significance:[4]

TABLE 2.1
Typical design and operating data for internal combustion engines

	Operating cycle	Compression ratio	Bore, m	Stroke/bore	Rated maximum			Weight/power ratio, kg/kW	Approx. best bsfc, g/kW·h
					Speed, rev/min	bmep, atm	Power per unit volume kW/dm^3		
Spark-ignition engines:									
Small (e.g., motorcycles)	2S,4S	6–11	0.05–0.085	1.2–0.9	4500–7500	4–10	20–60	5.5–2.5	350
Passenger cars	4S	8–10	0.07–0.1	1.1–0.9	4500–6500	7–10	20–50	4–2	270
Trucks	4S	7–9	0.09–0.13	1.2–0.7	3600–5000	6.5–7	25–30	6.5–2.5	300
Large gas engines	2S,4S	8–12	0.22–0.45	1.1–1.4	300–900	6.8–12	3–7	23–35	200
Wankel engines	4S	≈9	0.57 dm^3 per chamber		6000–8000	9.5–10.5	35–45	1.6–0.9	300
Diesel engines:									
Passenger cars	4S	17–23	0.075–0.1	1.2–0.9	4000–5000	5–7.5	18–22	5–2.5	250
Trucks (NA)	4S	16–22	0.1–0.15	1.3–0.8	2100–4000	6–9	15–22	7–4	210
Trucks (TC)	4S	14–20	0.1–0.15	1.3–0.8	2100–4000	12–18	18–26	7–3.5	200
Locomotive, industrial, marine	4S,2S	12–18	0.15–0.4	1.1–1.3	425–1800	7–23	5–20	6–18	190
Large engines, marine and stationary	2S	10–12	0.4–1	1.2–3	110–400	9–17	2–8	12–50	180

1. At maximum or normal rated point:

   Mean piston speed. Measures comparative success in handling loads due to inertia of the parts, resistance to air flow, and/or engine friction.

   Brake mean effective pressure. In naturally aspirated engines bmep is not stress limited. It then reflects the product of volumetric efficiency (ability to induct air), fuel/air ratio (effectiveness of air utilization in combustion), and fuel conversion efficiency. In supercharged engines bmep indicates the degree of success in handling higher gas pressures and thermal loading.

   Power per unit piston area. Measures the effectiveness with which the piston area is used, regardless of cylinder size.

   Specific weight. Indicates relative economy with which materials are used.

   Specific volume. Indicates relative effectiveness with which engine space has been utilized.

2. At all speeds at which the engine will be used with full throttle or with maximum fuel-pump setting:

   Brake mean effective pressure. Measures ability to obtain/provide high air flow and use it effectively over the full range.

3. At all useful regimes of operation and particularly in those regimes where the engine is run for long periods of time:

   Brake specific fuel consumption or *fuel conversion efficiency.*
   Brake specific emissions.

   Typical performance data for spark-ignition and diesel engines over the normal production size range are summarized in Table 2.1.[4] The four-stroke cycle dominates except in the smallest and largest engine sizes. The larger engines are turbocharged or supercharged. The maximum rated engine speed decreases as engine size increases, maintaining the maximum mean piston speed in the range of about 8 to 15 m/s. The maximum brake mean effective pressure for turbo-charged and supercharged engines is higher than for naturally aspirated engines. Because the maximum fuel/air ratio for spark-ignition engines is higher than for diesels, their naturally aspirated maximum bmep levels are higher. As engine size increases, brake specific fuel consumption decreases and fuel conversion efficiency increases, due to reduced importance of heat losses and friction. For the largest diesel engines, brake fuel conversion efficiencies of about 50 percent and indicated fuel conversion efficiencies of over 55 percent can be obtained.

PROBLEMS

2.1. Explain why the brake mean effective pressure of a naturally aspirated diesel engine is lower than that of a naturally aspirated spark-ignition engine. Explain why the bmep is lower at the maximum rated power for a given engine than the bmep at the maximum torque.

2.2. Describe the impact on air flow, maximum torque, and maximum power of changing a spark-ignition engine cylinder head from 2 valves per cylinder to 4 valves (2 inlet and 2 exhaust) per cylinder.

2.3. Calculate the mean piston speed, bmep, and specific power of the spark-ignition engines in Figs. 1-4, 1-9, and 1-12 at their maximum rated power.

2.4. Calculate the mean piston speed, bmep, and specific power of the diesel engines in Figs. 1-20, 1-21, 1-22, 1-23, and 1-24 at their maximum rated power. Briefly explain any significant differences.

2.5. Develop an equation for the power required to drive a vehicle at constant speed up a hill of angle α, in terms of vehicle speed, mass, frontal area, drag coefficient, coefficient of rolling resistance, α, and acceleration due to gravity. Calculate this power when the car mass is 1500 kg, the hill angle is 15 degrees, and the vehicle speed is 50 mi/h.

2.6. The spark-ignition engine in Fig. 1-4 is operating at a mean piston speed of 10 m/s. The measured air flow is 60 g/s. Calculate the volumetric efficiency based on atmospheric conditions.

2.7. The diesel engine of Fig. 1-20 is operating with a mean piston speed of 8 m/s. Calculate the air flow if the volumetric efficiency is 0.92. If (F/A) is 0.05 what is the fuel flow rate, and the mass of fuel injected per cylinder per cycle?

2.8. The brake fuel conversion efficiency of a spark-ignition engine is 0.3, and varies little with fuel type. Calculate the brake specific fuel consumption for isooctane, gasoline, methanol, and hydrogen (relevant data are in App. D).

2.9. You are doing a preliminary design study of a turbocharged four-stroke diesel engine. The maximum rated power is limited by stress considerations to a brake mean effective pressure of 1200 kPa and maximum value of the mean piston speed of 12 m/s.

(*a*) Derive an equation relating the engine inlet pressure (pressure in the inlet manifold at the turbocharger compressor exit) to the fuel/air ratio at this maximum rated power operating point. Other reciprocating engine parameters (e.g., volumetric efficiency, fuel conversion efficiency, bmep, etc.) appear in this equation also.

(*b*) The maximum rated brake power requirement for this engine is 400 kW. Estimate sensible values for number of cylinders, cylinder bore, stroke, and determine the maximum rated speed of this preliminary engine design.

(*c*) If the pressure ratio across the compressor is 2, estimate the overall fuel/air and air/fuel ratios at the maximum rated power. Assume appropriate values for any other parameters you may need.

2.10. In the reciprocating engine, during the power or expansion stroke, the gas pressure force acting on the piston is transmitted to the crankshaft via the connecting rod. List the forces acting on the piston during this part of the operating cycle. Show the direction of the forces acting on the piston on a sketch of the piston, cylinder, connecting rod, crank arrangement. Write out the force balance for the piston (*a*) along the cylinder axis and (*b*) transverse to the cylinder axis in the plane containing the connecting rod. (You are not asked to manipulate or solve these equations.)

2.11. You are designing a four-stroke cycle diesel engine to provide a brake power of 300 kW naturally aspirated at its maximum rated speed. Based on typical values for brake mean effective pressure and maximum mean piston speed, estimate the required engine displacement, and the bore and stroke for sensible cylinder geometry and number of engine cylinders. What is the maximum rated engine speed (rev/min) for your design? What would be the brake torque (N · m) and the fuel flow rate (g/h) at this maximum speed? Assume a maximum mean piston speed of 12 m/s is typical of good engine designs.

2.12. The power per unit piston area P/A_p (often called the specific power) is a measure of the designer's success in using the available piston area regardless of size.

(a) Derive an expression for P/A_p in terms of mean effective pressure and mean piston speed for two-stroke and four-stroke engine cycles.

(b) Compute typical maximum values of P/A_p for a spark-ignition engine (e.g., Fig. 1-4), a turbocharged four-stroke cycle diesel engine (e.g., Fig. 1-22), and a large marine diesel (Fig. 1-24). Table 2-1 may be helpful. State your assumptions clearly.

2.13. Several velocities, time, and length scales are useful in understanding what goes on inside engines. Make estimates of the following quantities for a 1.6-liter displacement four-cylinder spark-ignition engine, operating at wide-open throttle at 2500 rev/min.

(a) The mean piston speed and the maximum piston speed.

(b) The maximum charge velocity in the intake port (the port area is about 20 percent of the piston area).

(c) The time occupied by one engine operating cycle, the intake process, the compression process, the combustion process, the expansion process, and the exhaust process. (*Note*: The word *process* is used here not the word *stroke*.)

(d) The average velocity with which the flame travels across the combustion chamber.

(e) The length of the intake system (the intake port, the manifold runner, etc.) which is filled by one cylinder charge just before the intake valve opens and this charge enters the cylinder (i.e., how far back from the intake valve, in centimeters, one cylinder volume extends in the intake system).

(f) The length of exhaust system filled by one cylinder charge after it exits the cylinder (assume an average exhaust gas temperature of 425°C).

You will have to make several appropriate geometric assumptions. The calculations are straightforward, and only approximate answers are required.

2.14. The values of mean effective pressure at rated speed, maximum mean piston speed, and maximum specific power (engine power/total piston area) are essentially independent of cylinder size for naturally aspirated engines of a given type. If we also assume that engine weight per unit displaced volume is essentially constant, how will the *specific weight* of an engine (engine weight/maximum rated power) at fixed total displaced volume vary with the number of cylinders? Assume the bore and stroke are equal.

REFERENCES

1. Obert, E.F.: *Internal Combustion Engines and Air Pollution*, chap. 2, Intext Educational Publishers, New York, 1973.
2. SAE Standard: "Engine Test Code—Spark Ignition and Diesel," SAE J816b, *SAE Handbook*.
3. Bosch: *Automotive Handbook*, 2nd English edition, Robert Bosch GmbH, Stuttgart, 1986.
4. Taylor, C.F.: *The Internal Combustion Engine in Theory and Practice*, vol. II, MIT Press, Cambridge, Mass., 1968.

CHAPTER
3

THERMOCHEMISTRY
OF FUEL-AIR
MIXTURES

3.1 CHARACTERIZATION OF FLAMES

Combustion of the fuel-air mixture inside the engine cylinder is one of the processes that controls engine power, efficiency, and emissions. Some background in relevant combustion phenomena is therefore a necessary preliminary to understanding engine operation. These combustion phenomena are different for the two main types of engines—spark-ignition and diesel—which are the subject of this book. In spark-ignition engines, the fuel is normally mixed with air in the engine intake system. Following the compression of this fuel-air mixture, an electrical discharge initiates the combustion process; a flame develops from the "kernal" created by the spark discharge and propagates across the cylinder to the combustion chamber walls. At the walls, the flame is "quenched" or extinguished as heat transfer and destruction of active species at the wall become the dominant processes. An undesirable combustion phenomenon—the "spontaneous" ignition of a substantial mass of fuel-air mixture ahead of the flame, before the flame can propagate through this mixture (which is called the end-gas)—can also occur. This autoignition or self-explosion combustion phenomenon is the cause of spark-ignition engine knock which, due to the high pressures generated, can lead to engine damage.

In the diesel engine, the fuel is injected into the cylinder into air already at high pressure and temperature, near the end of the compression stroke. The autoignition, or self-ignition, of portions of the developing mixture of already injected and vaporized fuel with this hot air starts the combustion process, which spreads rapidly. Burning then proceeds as fuel and air mix to the appropriate composition for combustion to take place. Thus, fuel-air mixing plays a controlling role in the diesel combustion process.

Chapters 3 and 4 focus on the thermochemistry of combustion: i.e., the composition and thermodynamic properties of the pre- and postcombustion working fluids in engines and the energy changes associated with the combustion processes that take place inside the engine cylinder. Later chapters (9 and 10) deal with the phenomenological aspects of engine combustion: i.e., the details of the physical and chemical processes by which the fuel-air mixture is converted to burned products. At this point it is useful to review briefly the key combustion phenomena which occur in engines to provide an appropriate background for the material which follows. More detailed information on these combustion phenomena can be found in texts on combustion such as those of Fristrom and Westenberg[1] and Glassman.[2]

The combustion process is a fast exothermic gas-phase reaction (where oxygen is usually one of the reactants). A flame is a combustion reaction which can propagate subsonically through space; motion of the flame relative to the unburned gas is the important feature. Flame structure does not depend on whether the flame moves relative to the observer or remains stationary as the gas moves through it. The existence of flame motion implies that the reaction is confined to a zone which is small in thickness compared to the dimensions of the apparatus—in our case the engine combustion chamber. The reaction zone is usually called the flame front. This flame characteristic of spatial propagation is the result of the strong coupling between chemical reaction, the transport processes of mass diffusion and heat conduction, and fluid flow. The generation of heat and active species accelerate the chemical reaction; the supply of fresh reactants, governed by the convection velocity, limits the reaction. When these processes are in balance, a steady-state flame results.[1]

Flames are usually classified according to the following overall characteristics. The first of these has to do with the composition of the reactants as they enter the reaction zone. If the fuel and oxidizer are essentially uniformly mixed together, the flame is designated as *premixed*. If the reactants are not premixed and must mix together in the same region where reaction takes place, the flame is called a *diffusion* flame because the mixing must be accomplished by a diffusion process. The second means of classification relates to the basic character of the gas flow through the reaction zone: whether it is *laminar* or *turbulent*. In laminar (or streamlined) flow, mixing and transport are done by molecular processes. Laminar flows only occur at low Reynolds number. The Reynolds number (density × velocity × lengthscale/viscosity) is the ratio of inertial to viscous forces. In turbulent flows, mixing and transport are enhanced (usually by a substantial factor) by the macroscopic relative motion of eddies or lumps of fluid which are the characteristic feature of a turbulent (high Reynolds number) flow. A third area of classification is whether the flame is *steady* or *unsteady*. The distinguishing feature here is whether the flame structure and motion change with time. The final characterizing feature is the *initial phase* of the reactants—gas, liquid, or solid.

Flames in engines are unsteady, an obvious consequence of the internal combustion engine's operating cycle. Engine flames are turbulent. Only with substantial augmentation of laminar transport processes by the turbulent convection processes can mixing and burning rates and flame-propagation rates be made fast enough to complete the engine combustion process within the time available.

The conventional spark-ignition flame is thus a premixed unsteady turbulent flame, and the fuel-air mixture through which the flame propagates is in the gaseous state. The diesel engine combustion process is predominantly an unsteady turbulent diffusion flame, and the fuel is initially in the liquid phase. Both these flames are extremely complicated because they involve the coupling of the complex chemical mechanism, by which fuel and oxidizer react to form products, with the turbulent convective transport process. The diesel combustion process is even more complicated than the spark-ignition combustion process, because vaporization of liquid fuel and fuel-air mixing processes are involved too. Chapters 9 and 10 contain a more detailed discussion of the spark-ignition engine and diesel combustion processes, respectively. This chapter reviews the basic thermodynamic and chemical composition aspects of engine combustion.

3.2 IDEAL GAS MODEL

The gas species that make up the working fluids in internal combustion engines (e.g., oxygen, nitrogen, fuel vapor, carbon dioxide, water vapor, etc.) can usually be treated as ideal gases. The relationships between the thermodynamic properties of an ideal gas and of ideal gas mixtures are reviewed in App. B. There can be found the various forms of the ideal gas law:

$$pV = mRT = m \frac{\tilde{R}}{M} T = n\tilde{R}T \tag{3.1}$$

where p is the pressure, V the volume, m the mass of gas, R the gas constant for the gas, T the temperature, $\tilde{R}$ the universal gas constant, M the molecular weight, and n the number of moles. Relations for evaluating the specific internal energy u, enthalpy h, and entropy s, specific heats at constant volume c_v and constant pressure c_p, on a per unit mass basis and on a per mole basis (where the notation $\tilde{u}$, $\tilde{h}$, $\tilde{s}$, $\tilde{c}_v$, and $\tilde{c}_p$ is used) of an ideal gas, are developed. Also given are equations for calculating the thermodynamic properties of mixtures of ideal gases.

3.3 COMPOSITION OF AIR AND FUELS

Normally in engines, fuels are burned with air. Dry air is a mixture of gases that has a representative composition by volume of 20.95 percent oxygen, 78.09 percent nitrogen, 0.93 percent argon, and trace amounts of carbon dioxide, neon, helium, methane, and other gases. Table 3.1 shows the relative proportions of the major constituents of dry air.[3]

TABLE 3.1
Principle constitutents of dry air

Gas	ppm by volume	Molecular weight	Mole fraction	Molar ratio
O_2	209,500	31.998	0.2095	1
N_2	780,900	28.012	0.7905	3.773
Ar	9,300	39.948		
CO_2	300	44.009		
Air	1,000,000	28.962	1.0000	4.773

In combustion, oxygen is the reactive component of air. It is usually sufficiently accurate to regard air as consisting of 21 percent oxygen and 79 percent inert gases taken as nitrogen (often called atmospheric or apparent nitrogen). For each mole of oxygen in air there are

$$\frac{1 - 0.2095}{0.2095} = 3.773$$

moles of atmospheric nitrogen. The molecular weight of air is obtained from Table 3.1 with Eq. (B.17) as 28.962, usually approximated by 29. Because atmospheric nitrogen contains traces of other species, its molecular weight is slightly different from that of pure molecular nitrogen, i.e.,

$$M_{aN_2} = \frac{28.962 - 0.2095 \times 31.998}{1 - 0.2095} = 28.16$$

In the following sections, nitrogen will refer to atmospheric nitrogen and a molecular weight of 28.16 will be used. An air composition of 3.773 moles of nitrogen per mole of oxygen will be assumed.

The density of dry air can be obtained from Eq. (3.1) with $R = 8314.3$ J/kmol · K and $M = 28.962$:

$$\rho(\text{kg/m}^3) = \frac{3.483 \times 10^{-3} p(\text{Pa})}{T(\text{K})} \tag{3.2a}$$

or

$$\rho(\text{lbm/ft}^3) = \frac{2.699 p(\text{lbf/in}^2)}{T(^\circ \text{R})} \tag{3.2b}$$

Thus, the value for the density of dry air at 1 atmosphere (1.0133×10^5 Pa, 14.696 lbf/in^2) and 25°C (77°F) is 1.184 kg/m^3 (0.0739 lbm/ft^3).

Actual air normally contains water vapor, the amount depending on temperature and degree of saturation. Typically the proportion by mass is about 1 percent, though it can rise to about 4 percent under extreme conditions. The *relative humidity* compares the water vapor content of air with that required to saturate. It is defined as:

The ratio of the partial pressure of water vapor actually present to the saturation pressure at the same temperature.

Water vapor content is measured with a wet- and dry-bulb psychrometer. This consists of two thermometers exposed to a stream of moist air. The dry-bulb temperature is the temperature of the air. The bulb of the other thermometer is wetted by a wick in contact with a water reservoir. The wet-bulb temperature is lower than the dry-bulb temperature due to evaporation of water from the wick. It is a good approximation to assume that the wet-bulb temperature is the adiabatic saturation temperature. Water vapor pressure can be obtained from observed wet- and dry-bulb temperatures and a psychrometric chart such as Fig. 3-1.[4] The effect of humidity on the properties of air is given in Fig. 3-2.[5]

The fuels most commonly used in internal combustion engines (gasoline or petrol, and diesel fuels) are blends of many different hydrocarbon compounds obtained by refining petroleum or crude oil. These fuels are predominantly carbon and hydrogen (typically about 86 percent carbon and 14 percent hydrogen by weight) though diesel fuels can contain up to about 1 percent sulfur. Other fuels of interest are alcohols (which contain oxygen), gaseous fuels (natural gas and liquid petroleum gas), and single hydrocarbon compounds (e.g., methane, propane, isooctane) which are often used in engine research. Properties of the more common internal combustion engine fuels are summarized in App. D.

Some knowledge of the different classes of organic compounds and their

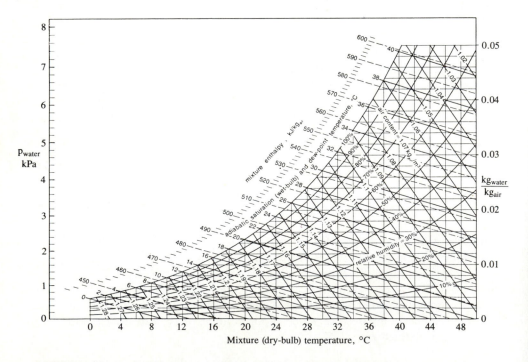

FIGURE 3-1
Psychrometric chart for air-water mixtures at 1 atmosphere. (*From Reynolds.*[4])

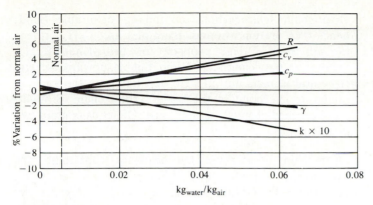

FIGURE 3-2

Effect of humidity on properties of air: R is the gas constant; c_v and c_p are specific heats at constant volume and pressure, respectively; $\gamma = c_p/c_v$; k is the thermal conductivity. (*From Taylor.*[5])

molecular structure is necessary in order to understand combustion mechanisms.[6] The different classes are as follows:

Alkyl Compounds

Paraffins (alk*anes*)

$$C_nH_{2n+2}$$

Single-bonded open-chain saturated hydrocarbon molecules: i.e., no more hydrogen can be added. For the larger molecules straight-chain and branched-chain configurations exist. These are called normal (*n*-) and iso compounds, respectively. Examples: CH_4, methane; C_2H_6, ethane; C_3H_8, propane; C_8H_{18}, *n*-octane and isooctane. There are several "isooctanes," depending on the relative position of the branches. By isooctane is usually meant 2,2,4-trimethylpentane, indicating five carbon atoms in the straight chain (pentane) with three methyl (CH_3) branches located respectively at C-atoms 2, 2, and 4. Radicals deficient in one hydrogen take the name methyl, ethyl, propyl, etc.

Cycloparaffins or *napthenes* (*cyclanes*)

$$C_nH_{2n}$$

Single bond (no double bond) ring hydrocarbons. Unsaturated, since ring can be broken and additional hydrogen added. Examples: C_3H_6, cyclopropane (three C-atom ring); C_4H_8, cyclobutane (four C-atom ring); C_5H_{10}, cyclopentane (five C-atom ring).

Olefins
 (alk*enes*)

H   H
 \   /
  C = C
 /   \
H   H

C_nH_{2n}

Open-chain hydrocarbons containing a double bond; hence they are unsaturated. Examples are: C_2H_4, ethene (or ethylene); C_3H_6, propene (or propylene); C_4H_8, butene (or butylene); .... From butene upwards several structural isomers are possible depending on the location of the double bond in the basic carbon chain. Straight- and branched-chain structures exist. Diolefins contain two double bonds.

Acetylenes
 (alk*ynes*)

$H-C\equiv C-H$

C_nH_{2n-2}

Open-chain unsaturated hydrocarbons containing one carbon-carbon triple bond. First member is acetylene, $H-C\equiv C-H$. Additional members of the alkyne series comprise open-chain molecules, similar to higher alkenes but with each double bond replaced by a triple bond.

Aromatics

        H
        |
        C
      //   \
   HC   CH
    | ||
   HC   CH
      \   //
        C
        |
        H

C_nH_{2n-6}

Building block for aromatic hydrocarbons is the benzene (C_6H_6) ring structure shown. This ring structure is very stable and accommodates additional $-CH_2$ groups in side chains and not by ring expansion. Examples: C_7H_8, toluene; C_8H_{10},   xylene   (several   structural arrangements); .... More complex aromatic hydrocarbons incorporate ethyl, propyl, and heavier alkyl side chains in a variety of structural arrangements.

Alcohols

Monohydric alcohols

        H
        |
  H-C-OH
        |
        H

$C_nH_{2n+1}OH$

In these organic compounds, one hydroxyl ($-OH$) group is substituted for one hydrogen atom. Thus methane becomes methyl alcohol, CH_3OH (also called methanol); ethane becomes ethyl alcohol, C_2H_5OH (ethanol); etc.

3.4   COMBUSTION STOICHIOMETRY

This section develops relations between the composition of the reactants (fuel and air) of a combustible mixture and the composition of the products. Since these relations depend only on the conservation of mass of each chemical element in the reactants, only the relative elemental composition of the fuel and the relative proportions of fuel and air are needed.

If sufficient oxygen is available, a hydrocarbon fuel can be completely oxidized. The carbon in the fuel is then converted to carbon dioxide CO_2 and the hydrogen to water H_2O. For example, consider the overall chemical equation for the complete combustion of one mole of propane C_3H_8:

$$C_3H_8 + aO_2 = bCO_2 + cH_2O \tag{3.3}$$

A carbon balance between the reactants and products gives $b = 3$. A hydrogen balance gives $2c = 8$, or $c = 4$. An oxygen balance gives $2b + c = 10 = 2a$, or $a = 5$. Thus Eq. (3.3) becomes

$$C_3H_8 + 5O_2 = 3CO_2 + 4H_2O \tag{3.4}$$

Note that Eq. (3.4) only relates the elemental composition of the reactant and product species; it does not indicate the process by which combustion proceeds, which is much more complex.

Air contains nitrogen, but when the products are at low temperatures the nitrogen is not significantly affected by the reaction. Consider the complete combustion of a general hydrocarbon fuel of average molecular composition C_aH_b with air. The overall complete combustion equation is

$$C_aH_b + \left(a + \frac{b}{4}\right)(O_2 + 3.773N_2) = aCO_2 + \frac{b}{2}H_2O + 3.773\left(a + \frac{b}{4}\right)N_2 \tag{3.5}$$

Note that only the *ratios* of the numbers in front of the symbol for each chemical species are defined by Eq. (3.5); i.e., only the relative proportions on a molar basis are obtained. Thus the fuel composition could have been written CH_y, where $y = b/a$.

Equation (3.5) defines the *stoichiometric* (or chemically correct or theoretical) proportions of fuel and air; i.e., there is just enough oxygen for conversion of all the fuel into completely oxidized products. The stoichiometric air/fuel or fuel/air ratios (see Sec. 2.9) depend on fuel composition. From Eq. (3.5):

$$\left(\frac{A}{F}\right)_s = \left(\frac{F}{A}\right)_s^{-1} = \frac{(1 + y/4)(32 + 3.773 \times 28.16)}{12.011 + 1.008y}$$

$$= \frac{34.56(4 + y)}{12.011 + 1.008y} \tag{3.6}$$

The molecular weights of oxygen, atmospheric nitrogen, atomic carbon, and atomic hydrogen are, respectively, 32, 28.16, 12.011, and 1.008. $(A/F)_s$ depends only on y; Fig. 3-3 shows the variation in $(A/F)_s$ as y varies from 1 (e.g., benzene) to 4 (methane).

Example 3.1. A hydrocarbon fuel of composition 84.1 percent by mass C and 15.9 percent by mass H has a molecular weight of 114.15. Determine the number of moles of air required for stoichiometric combustion and the number of moles of products produced per mole of fuel. Calculate $(A/F)_s$, $(F/A)_s$, and the molecular weights of the reactants and the products.

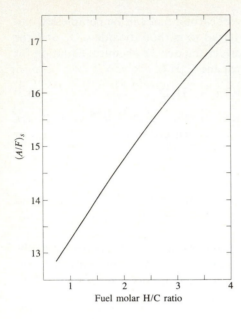

FIGURE 3-3
Stoichiometric air/fuel ratio for air–hydrocarbon fuel mixtures as a function of fuel molar H/C ratio.

Assume a fuel composition $C_a H_b$. The molecular weight relation gives

$$114.15 = 12.011a + 1.008b$$

The gravimetric analysis of the fuel gives

$$\frac{b}{a} = \frac{15.9/1.008}{84.1/12.011} = 2.25$$

Thus

$$a = 8 \qquad b = 18\dagger$$

The fuel is octane C_8H_{18}. Equation (3.5) then becomes

$$\underset{\text{Fuel}}{C_8H_{18}} + \underset{\text{Air}}{12.5(O_2 + 3.773N_2)} = \underset{\text{Products}}{8CO_2 + 9H_2O + 47.16N_2}$$

In moles:

1	$+\ 12.5(1 + 3.773)$	$= 8 + 9 + 47.16$
1	$+\quad 59.66$	$= 64.16$

Relative mass:

$114.15 + 59.66 \times 28.96$	$= 8 \times 44.01 + 9 \times 18.02 + 47.16 \times 28.16$
$114.5\ \ + 1727.8$	$= 1842.3$

---

† Note that for fuels which are mixtures of hydrocarbons, a and b need not be integers.

Per unit mass fuel:

$$1 \quad + \quad 15.14 \quad = 16.14$$

Thus for stoichiometric combustion, 1 mole of fuel requires 59.66 moles of air and produces 64.16 moles of products. The stoichiometric $(A/F)_s$ is 15.14 and $(F/A)_s$ is 0.0661.

The molecular weights of the reactants M_R and products M_P are

$$M_R = \frac{1}{n} \sum n_i M_i = \frac{1}{60.66} (1 \times 114.15 + 59.66 \times 28.96)$$

$$M_P = \frac{1}{n} \sum n_i M_i = \frac{1}{64.16} (8 \times 44.01 + 9 \times 18.02 + 47.16 \times 28.16)$$

or
$$M_R = 30.36 \qquad M_P = 28.71$$

Fuel-air mixtures with more than or less than the stoichiometric air requirement can be burned. With excess air or fuel-lean combustion, the extra air appears in the products in unchanged form. For example, the combustion of isooctane with 25 percent excess air, or 1.25 times the stoichiometric air requirement, gives

$$C_8H_{18} + 1.25 \times 12.5(O_2 + 3.773N_2) = 8CO_2 + 9H_2O + 3.13O_2 + 58.95N_2$$

$$(3.7)$$

With less than the stoichiometric air requirement, i.e., with fuel-rich combustion, there is insufficient oxygen to oxidize fully the fuel C and H to CO_2 and H_2O. The products are a mixture of CO_2 and H_2O with carbon monoxide CO and hydrogen H_2 (as well as N_2). The product composition cannot be determined from an element balance alone and an additional assumption about the chemical composition of the product species must be made (see Secs. 4.2 and 4.9.2).

Because the composition of the combustion products is significantly different for fuel-lean and fuel-rich mixtures, and because the stoichiometric fuel/air ratio depends on fuel composition, the ratio of the actual fuel/air ratio to the stoichiometric ratio (or its inverse) is a more informative parameter for defining mixture composition. The *fuel/air equivalence ratio* ϕ,

$$\phi = \frac{(F/A)_{\text{actual}}}{(F/A)_s} \tag{3.8}$$

will be used throughout this text for this purpose. The inverse of ϕ, the *relative air/fuel ratio* λ,

$$\lambda = \phi^{-1} = \frac{(A/F)_{\text{actual}}}{(A/F)_s} \tag{3.9}$$

is also sometimes used.

For fuel-lean mixtures: $\phi < 1, \lambda > 1$
For stoichiometric mixtures: $\phi = \lambda = 1$
For fuel-rich mixtures: $\phi > 1, \lambda < 1$

When the fuel contains oxygen (e.g., with alcohols), the procedure for determining the overall combustion equation is the same except that fuel oxygen is included in the oxygen balance between reactants and products. For methyl alcohol (methanol), CH_3OH, the stoichiometric combustion equation is

$$CH_3OH + 1.5(O_2 + 3.773N_2) = CO_2 + 2H_2O + 5.66N_2 \qquad (3.10)$$

and $(A/F)_s = 6.47$. For ethyl alcohol (ethanol), C_2H_5OH, the stoichiometric combustion equation is

$$C_2H_5OH + 3(O_2 + 3.773N_2) = 2CO_2 + 3H_2O + 11.32N_2 \qquad (3.11)$$

and $(A/F)_s = 9.00$.

If there are significant amounts of sulfur in the fuel, the appropriate oxidation product for determining the stoichiometric air and fuel proportions is sulfur dioxide, SO_2.

For hydrogen fuel, the stoichiometric equation is

$$H_2 + \tfrac{1}{2}(O_2 + 3.773N_2) = H_2O + 1.887N_2 \qquad (3.12)$$

and the stoichiometric (A/F) ratio is 34.3.

Note that the composition of the products of combustion in Eqs. (3.7) and (3.10) to (3.12) may not occur in practice. At normal combustion temperatures significant dissociation of CO_2 and of H_2O occurs (see Sec. 3.7.1). Whether, at low temperatures, recombination brings the product composition to that indicated by these overall chemical equations depends on the rate of cooling of the product gases. More general relationships for the composition of unburned and burned gas mixtures are developed in Chap. 4.

The stoichiometric (A/F) and (F/A) ratios of common fuels and representative single hydrocarbon and other compounds are given in App. D along with other fuel data.

3.5 THE FIRST LAW OF THERMODYNAMICS AND COMBUSTION†

3.5.1 Energy and Enthalpy Balances

In a combustion process, fuel and oxidizer react to produce products of different composition. The actual path by which this transformation takes place is understood only for simple fuels such as hydrogen and methane. For fuels with more complicated structure, the details are not well defined. Nonetheless, the first law

---

† The approach used here follows that developed by Spalding and Cole.[7]

of thermodynamics can be used to relate the end states of mixtures undergoing a combustion process; its application does not require that the details of the process be known.

The first law of thermodynamics relates changes in internal energy (or enthalpy) to heat and work transfer interactions. In applying the first law to a system whose chemical composition changes, care must be exercised in relating the reference states at which zero internal energy or enthalpy for each species or groups of species are assigned. We are not free, when chemical reactions occur, to choose independently the zero internal energy or enthalpy reference states of chemical substances transformed into each other by reaction.

Consider a system of mass m which changes its composition from reactants to products by chemical reaction as indicated in Fig. 3-4. Applying the first law to the system between its initial and final states gives

$$Q_{R-P} - W_{R-P} = U_P - U_R \tag{3.13}$$

Heat transfer Q_{R-P} and work transfer W_{R-P} due to normal force displacements may occur across the system boundary. The standard thermodynamic sign convention for each energy transfer interaction—positive for heat transfer *to* the system and positive for work transfer *from* the system—is used.

We will consider a series of special processes: first, a *constant volume* process where the initial and final temperatures are the same, T'. Then Eq. (3.13) becomes

$$Q_{R-P} = U'_P - U'_R = (\Delta U)_{V, T'} \tag{3.14}$$

The internal energy of the system has changed by an amount $(\Delta U)_{V, T'}$ which can be measured or calculated. Combustion processes are exothermic [i.e., Q_{R-P} and $(\Delta U)_{V, T'}$ are negative]; therefore the system's internal energy decreases. If Eq. (3.14) is expressed per mole of fuel, then $(\Delta U)_{V, T'}$ is known as the increase in

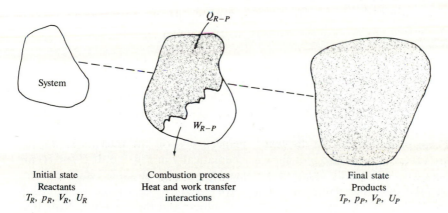

Initial state	Combustion process	Final state
Reactants	Heat and work transfer	Products
T_R, p_R, V_R, U_R	interactions	T_P, p_P, V_P, U_P

FIGURE 3-4
System changing from reactants to products for first law analysis.

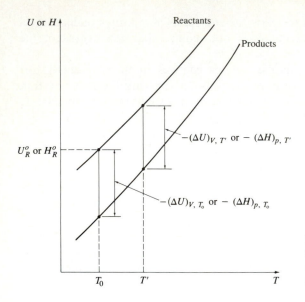

FIGURE 3-5
Schematic plot of internal energy (U)
or enthalpy (H) of reactants and products as a function of temperature.

internal energy at constant volume, and $-(\Delta U)_{V,\,T'}$ is known as the *heat of reaction at constant volume* at temperature T'.

Next, consider a *constant pressure* process where the initial and final temperatures are the same, T'. For a constant pressure process

$$W_{R-P} = \int_{R}^{P} p \, dV = p(V_P - V_R) \tag{3.15}$$

so Eq. (3.13) becomes

$$Q_{R-P} - p(V'_P - V'_R) = U'_P - U'_R$$

or

$$Q_{R-P} = (U'_P + pV'_P) - (U'_R + pV'_R)$$

$$= H'_P - H'_R = (\Delta H)_{p,\,T'} \tag{3.16}$$

The enthalpy of the system has changed by an amount $(\Delta H)_{p,\,T'}$, which can be measured or calculated. Again for combustion reactions, $(\Delta H)_{p,\,T'}$ is a negative quantity. If Eq. (3.16) is written per mole of fuel, $(\Delta H)_{p,\,T'}$ is called the increase in enthalpy at constant pressure and $-(\Delta H)_{p,\,T'}$ is called the *heat of reaction at constant pressure* at T'.

These processes can be displayed, respectively, on the internal energy or enthalpy versus temperature plot shown schematically in Fig. 3-5. If U (or H) for the reactants is arbitrarily assigned a value U^o_R (or H^o_R) at some reference temperature T_0, then the value of $(\Delta U)_{V,\,T_0}$ [or $(\Delta H)_{p,\,T_0}$] fixes the relationship between $U(T)$ or $H(T)$, respectively, for the products and the reactants. Note that the slope of these lines (the specific heat at constant volume or pressure if the diagram is expressed per unit mass or per mole) increases with increasing temperature; also, the magnitude of $(\Delta U)_{V,\,T'}$ [or $(\Delta H)_{p,\,T'}$] decreases with increasing temperature because c_v (or c_p) for the products is greater than for the reactants.

The difference between $(\Delta H)_{p,\,T'}$ and $(\Delta U)_{V,\,T'}$ is

$$(\Delta H)_{p,\,T'} - (\Delta U)_{V,\,T'} = p(V_P - V_R) \tag{3.17}$$

Only if the volumes of the products and reactants in the constant pressure process are the same are $(\Delta H)_{p,\,T'}$ and $(\Delta U)_{V,\,T'}$ equal. If all the reactant and product species are ideal gases, then the ideal gas law Eq. (3.1) gives

$$(\Delta H)_{p,\,T'} - (\Delta U)_{V,\,T'} = \tilde{R}(n'_p - n'_R)T' \tag{3.18}$$

Note that any inert gases do not contribute to $(n'_P - n'_R)$.

With a hydrocarbon fuel, one of the products, H_2O, can be in the gaseous or liquid phase. The internal energy (or enthalpy) of the products in the constant volume (or constant pressure) processes described above in Fig. 3-5 will depend on the relative proportions of the water in the gaseous and liquid phases. The limiting cases of all vapor and all liquid are shown in Fig. 3-6a for a U-T plot. The internal energy differences between the curves is

$$\left|(\Delta U)_{V,\,T',\,H_2O\ liq}\right| - \left|(\Delta U)_{V,\,T',\,H_2O\ vap}\right| = m_{H_2O}\,u'_{fg\ H_2O} \tag{3.19}$$

where m_{H_2O} is the mass of water in the products and $u'_{fg\ H_2O}$ is the internal energy of vaporization of water at the temperature and pressure of the products. Similar

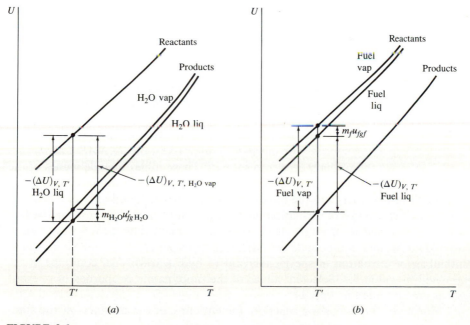

FIGURE 3-6
Schematic plots of internal energy of reactants and products as a function of temperature. (a) Effect of water in products as either vapor or liquid. (b) Effect of fuel in reactants as either vapor or liquid.

curves and relationships apply for enthalpy:

$$|(\Delta H)_{p,\ T',\ H_2O\ liq}| - |(\Delta H)_{p,\ T',\ H_2O\ vap}| = m_{H_2O}\, h'_{fg\ H_2O} \tag{3.20}$$

For some fuels, the reactants may contain the fuel as either liquid or vapor. The U-T (or H-T) line for the reactants with the fuel as liquid or as vapor will be different, as indicated in Fig. 3-6b. The vertical distance between the two reactant curves is $m_f\, u_{fgf}$ (or $m_f\, h_{fgf}$) where the subscript f denotes fuel.

3.5.2 Enthalpies of Formation

For fuels which are single hydrocarbon compounds, or where the precise fuel composition is known, the internal energies or enthalpies of the reactants and the products can be related through the enthalpies of formation of the reactants and products.

The *enthalpy of formation* $\Delta \tilde{h}_f^\circ$ of a chemical compound is the enthalpy increase associated with the reaction of forming one mole of the given compound from its elements, with each substance in its thermodynamic standard state at the given temperature.

The *standard state* is the state at one atmosphere pressure and the temperature under consideration. We will denote the standard state by the superscript $^\circ$.

Since thermodynamic calculations are made as a difference between an initial and a final state, it is necessary to select a *datum state* to which all other thermodynamic states can be referred. While a number of datum states have been used in the literature, the most common datum is 298.15 K (25°C) and 1 atmosphere. We will use this datum throughout this text. Elements at their *reference state* are arbitrarily assigned zero enthalpy at the datum temperature. The reference state of each element is its stable standard state; e.g., for oxygen at 298.15 K, the reference state is O_2 gas.

Enthalpies of formation are tabulated as a function of temperature for all commonly occurring species. For inorganic compounds, the *JANAF Thermochemical Tables* are the primary reference source.[8] These tables include values of the molar specific heat at constant pressure, standard entropy, standard Gibbs free energy (called free energy in the tables), standard enthalpy, enthalpy of formation and Gibbs free energy of formation, and $\log_{10}$ equilibrium constant for the formation of each species from its elements. Some primary references for thermodynamic data on fuel compounds are Maxwell,[9] Rossini et al.,[10] and Stull et al.[11] Enthalpies of formation of species relevant to hydrocarbon fuel combustion are tabulated in Table 3.2. Selected values of thermodynamic properties of relevant species are tabulated in App. D.

For a given combustion reaction, the enthalpy of the products at the standard state relative to the enthalpy datum is then given by

$$H_P^\circ = \sum_{\text{products}} n_i\, \Delta \tilde{h}_{f,i}^\circ \tag{3.21a}$$

TABLE 3.2
Standard enthalpies of formation

Species	State†	$\Delta \tilde{h}_f^{\circ}$, MJ/kmol
O_2	Gas	0
N_2	Gas	0
H_2	Gas	0
C	Gas	0
CO_2	Gas	-393.52
H_2O	Gas	-241.83
H_2O	Liquid	-285.84
CO	Gas	-110.54
CH_4	Gas	-74.87
C_3H_8	Gas	-103.85
CH_3OH	Gas	-201.17
CH_3OH	Liquid	-238.58
C_8H_{18}	Gas	-208.45
C_8H_{18}	Liquid	-249.35

† At 298.15 K (25°C) and 1 atm.

and the enthalpy of the reactants is given by

$$H_R^{\circ} = \sum_{\text{reactants}} n_i \Delta \tilde{h}_{f,i}^{\circ} \tag{3.21b}$$

The enthalpy increase, $(\Delta H)_{p, T_0}$, is then obtained from the difference $(H_P^{\circ} - H_R^{\circ})$. The internal energy increase can be obtained with Eq. (3.17).

> **Example 3.2.** Calculate the enthalpy of the products and reactants, and the enthalpy increase and internal energy increase of the reaction, of a stoichiometric mixture of methane and oxygen at 298.15 K.
>
> The stoichiometric reaction is
>
> $$CH_4 + 2O_2 = CO_2 + 2H_2O$$
>
> Thus, for H_2O gas, from Table 3.2 and Eq. (3.21a, b):
>
> $$H_P^{\circ} = -393.52 + 2(-241.83) = -877.18 \text{ MJ/kmol } CH_4$$
>
> For H_2O liquid:
>
> $$H_P^{\circ} = -393.52 + 2(-285.84) = -965.20 \text{ MJ/kmol } CH_4$$
>
> $$H_R^{\circ} = -74.87 \text{ MJ/kmol } CH_4$$
>
> Hence for H_2O gas:
>
> $$(\Delta H)_P^{\circ} = -877.18 + 74.87 = -802.31 \text{ MJ/kmol } CH_4$$
>
> and for H_2O liquid:
>
> $$(\Delta H)_P^{\circ} = -965.20 + 74.87 = -890.33 \text{ MJ/kmol } CH_4$$

Use Eq. (3.18) to find $(\Delta U)_V^\circ$. With H_2O gas, the number of moles of reactants and products are equal, so

$$(\Delta U)_V^\circ = (\Delta H)_p^\circ = -802.3 \text{ MJ/kmol CH}_4$$

For H_2O liquid:

$$(\Delta U)_V^\circ = -890.33 - 8.3143 \times 10^{-3}(1 - 3)298.15 \text{ MJ/kmol CH}_4$$

$$(\Delta U)_V^\circ = -885.4 \text{ MJ/kmol CH}_4$$

Note that the presence of nitrogen in the mixture or oxygen in excess of the stoichiometric amount would not change any of these calculations.

3.5.3   Heating Values

For fuels where the precise fuel composition is not known, the enthalpy of the reactants cannot be determined from the enthalpies of formation of the reactant species. The *heating value* of the fuel is then measured directly.

The heating value Q_{HV} or calorific value of a fuel is the magnitude of the heat of reaction at constant pressure or at constant volume at a standard temperature [usually 25°C (77°F)] for the complete combustion of unit mass of fuel. Thus

$$Q_{HV_p} = -(\Delta H)_{p, T_0} \tag{3.22a}$$

and

$$Q_{HV_V} = -(\Delta U)_{V, T_0} \tag{3.22b}$$

Complete combustion means that all carbon is converted to CO_2, all hydrogen is converted to H_2O, and any sulfur present is converted to SO_2. The heating value is usually expressed in joules per kilogram or joules per kilomole of fuel (British thermal units per pound-mass or British thermal units per pound-mole). It is therefore unnecessary to specify how much oxidant was mixed with the fuel, though this must exceed the stoichiometric requirement. It is immaterial whether the oxidant is air or oxygen.

For fuels containing hydrogen, we have shown that whether the H_2O in the products is in the liquid or gaseous phase affects the value of the heat of reaction. The term *higher heating value* Q_{HHV} (or gross heating value) is used when the H_2O formed is all condensed to the liquid phase; the term *lower heating value* Q_{LHV} (or net heating value) is used when the H_2O formed is all in the vapor phase. The two heating values at constant pressure are related by

$$Q_{HHV_p} = Q_{LHV_p} + \left(\frac{m_{H_2O}}{m_f}\right)h_{fg \, H_2O} \tag{3.23}$$

where (m_{H_2O}/m_f) is the ratio of mass of H_2O produced to mass of fuel burned. A similar expression with $u_{fg \, H_2O}$ replacing $h_{fg \, H_2O}$ applies for the higher and lower heating value at constant volume.

The heating value at constant pressure is the more commonly used; often the qualification "at constant pressure" is omitted. The difference between the heating values at constant pressure and constant volume is small.

Heating values† of fuels are measured in calorimeters. For gaseous fuels, it is most convenient and accurate to use a continuous-flow atmosphere pressure calorimeter. The entering fuel is saturated with water vapor and mixed with sufficient saturated air for complete combustion at the reference temperature. The mixture is burned in a burner and the combustion products cooled with water-cooled metal tube coils to close to the inlet temperature. The heat transferred to the cooling water is calculated from the measured water flow rate and water temperature rise. The heating value determined by this process is the higher heating value at constant pressure.

For liquid and solid fuels, it is more satisfactory to burn the fuel with oxygen under pressure at constant volume in a bomb calorimeter. A sample of the fuel is placed in the bomb calorimeter, which is a stainless-steel container immersed in cooling water at the standard temperature. Sufficient water is placed in the bomb to ensure that the water produced in the combustion process will condense. Oxygen at 30 atmospheres is admitted to the bomb. A length of firing cotton is suspended into the sample from an electrically heated wire filament to act as a source of ignition. When combustion is complete the temperature rise of the bomb and cooling water is measured. The heating value determined by this process is the higher heating value at constant volume.

The heating values of common fuels are tabulated with other fuel data in App. D. The following example illustrates how the enthalpy of a reactant mixture relative to the enthalpy datum we have defined can be determined from the measured heating value of the fuel.

Example 3.3. Liquid kerosene fuel of the lower heating value (determined in a bomb calorimeter) of 43.2 MJ/kg and average molar H/C ratio of 2 is mixed with the stoichiometric air requirement at 298.15 K. Calculate the enthalpy of the reactant mixture relative to the datum of zero enthalpy for C, O_2, N_2, and H_2 at 298.15 K.

The combustion equation per mole of C can be written

$$CH_2 + \tfrac{3}{2}(O_2 + 3.773N_2) = CO_2 + H_2O + 5.660N_2$$

or $\quad$ 14 kg fuel + $\left.\begin{matrix} 7.160 \text{ kmol} \\ 207.4 \text{ kg} \end{matrix}\right\}$ air → $\left.\begin{matrix} 7.66 \text{ kmol} \\ 221.4 \text{ kg} \end{matrix}\right\}$ products

where $M = 28.962$ for atmospheric air.

The heating value given is at constant volume, $-(\Delta U)_V^\circ$. $(\Delta H)_p^\circ$ is obtained from Eq. (3.18), noting that the fuel is in the liquid phase:

$$(\Delta H)_p^\circ = -43.2 + 8.3143 \times 10^{-3} (7.66 - 7.160) \times \frac{298.15}{14}$$

$$= -43.2 + 0.09 = -43.1 \text{ MJ/kg fuel}$$

---

† Standard methods for measuring heating values are defined by the American Society for Testing Materials.

The enthalpy of the products per kilogram of mixture is found from the enthalpies of formation (with H_2O vapor):

$$h_P = \frac{1(-393.52) + 1(-241.83)}{221.4}$$

$$= 2.87 \text{ MJ/kg}$$

The enthalpy of the reactants per kilogram of mixture is then

$$h_R = h_P - (\Delta h)_P^\circ = 2.87 + \frac{43.1 \times 14}{221.4} = 5.59 \text{ MJ/kg}$$

3.5.4   Adiabatic Combustion Processes

We now use the relationships developed above to examine two other special processes important in engine analysis: constant-volume and constant-pressure adiabatic combustion. For an adiabatic constant-volume process, Eq. (3.13) becomes

$$U_P - U_R = 0 \tag{3.24}$$

when U_P and U_R are evaluated relative to the same datum (e.g., the enthalpies of C, O_2, N_2, and H_2 are zero at 298.15 K, the datum used throughout this text).

Frequently, however, the tables or graphs of internal energy or enthalpy for species and reactant or product mixtures which are available give internal energies or enthalpies relative to the species or mixture value at some reference temperature T_0, i.e., $U(T) - U(T_0)$ or $H(T) - H(T_0)$ are tabulated. Since

$$U_P(T_0) - U_R(T_0) = (\Delta U)_{V, T_0}$$

it follows from Eq. (3.24) that

$$[U_P(T) - U_P(T_0)] - [U_R(T) - U_R(T_0)] = -(\Delta U)_{V, T_0} \tag{3.25}$$

relates the product and reactant states. Figure 3-7 illustrates the adiabatic constant-volume combustion process on a U-T diagram. Given the initial state of the reactants (T_R, V) we can determine the final state of the products (T_P, V).

For an adiabatic constant-pressure combustion process, Eq. (3.13) gives

$$H_P - H_R = 0$$

which combines with Eq. (3.16) to give

$$[H_P(T) - H_P(T_0)] - [H_R(T) - H_R(T_0)] = -(\Delta H)_{p, T_0} \tag{3.26}$$

Figure 3-7 illustrates this process also. Given the initial reactant state (T_R, p) we can determine the final product state (T_P, p).

Note that while in Figs. 3-5, 3-6, and 3-7 we have shown U and H for the reactants and products to be functions of T only, in practice for the products at high temperature U and H will be functions of p and T. The analysis presented here is general; however, to determine the final state of the products in an adia-

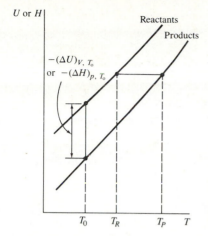

FIGURE 3-7
Adiabatic constant-volume combustion process on U-T diagram or adiabatic constant-pressure combustion process on H-T diagram.

batic combustion process, the constant-volume or constant-pressure constraint must also be used explicitly.

The final temperature of the products in an adiabatic combustion process is called the *adiabatic flame temperature*. Examples of typical adiabatic flame temperatures are shown later in Fig. 3-11.

3.5.5 Combustion Efficiency of an Internal Combustion Engine

In practice, the exhaust gas of an internal combustion engine contains incomplete combustion products (e.g., CO, H_2, unburned hydrocarbons, soot) as well as complete combustion products (CO_2 and H_2O) (see Sec. 4.9). Under lean operating conditions the amounts of incomplete combustion products are small. Under fuel-rich operating conditions these amounts become more substantial since there is insufficient oxygen to complete combustion. Because a fraction of the fuel's chemical energy is not fully released inside the engine during the combustion process, it is useful to define a *combustion efficiency*. The engine can be analyzed as an open system which exchanges heat and work with its surrounding environment (the atmosphere). Reactants (fuel and air) flow into the system; products (exhaust gases) flow out. Consider a mass m which passes through the control volume surrounding the engine shown in Fig. 3-8; the *net chemical energy release* due to combustion within the engine is given by

$$[H_R(T_A) - H_P(T_A)] = m\left(\sum_{i,\,\text{reactants}} n_i \, \Delta \tilde{h}^\circ_{f,i} - \sum_{i,\,\text{products}} n_i \, \Delta \tilde{h}^\circ_{f,i} \right)$$

Enthalpy is the appropriate property since $p_R = p_P = p_{\text{atm}}$. n_i is the number of moles of species i in the reactants or products per unit mass of working fluid and $\Delta \tilde{h}^\circ_{f,i}$ is the standard enthalpy of formation of species i at ambient temperature T_A.

The amount of fuel energy *supplied* to the control volume around the engine which can be released by combustion is $m_f Q_{\text{HV}}$. Hence, the *combustion*

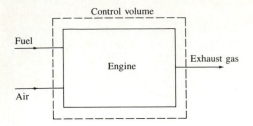

FIGURE 3-8
Control volume surrounding engine.

efficiency—the fraction of the fuel energy supplied which is released in the combustion process—is given by[12]

$$\eta_c = \frac{H_R(T_A) - H_P(T_A)}{m_f Q_{HV}} \tag{3.27}$$

Note that m and m_f could be replaced by the average mass flow rates $\dot{m}$ and $\dot{m}_f$.

Figure 3-9 shows how combustion efficiency varies with the fuel/air equivalence ratio for internal combustion engines. For spark-ignition engines, for lean equivalence ratios, the combustion efficiency is usually in the range 95 to 98 percent. For mixtures richer than stoichiometric, lack of oxygen prevents complete combustion of the fuel carbon and hydrogen, and the combustion efficiency steadily decreases as the mixture becomes richer. Combustion efficiency is little affected by other engine operating and design variables, provided the engine com-

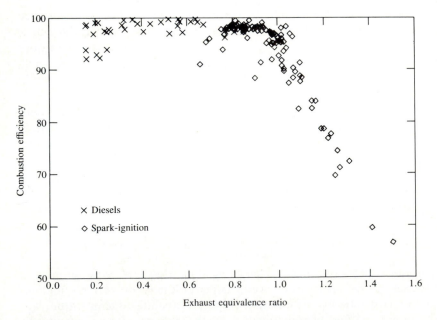

FIGURE 3-9
Variation of engine combustion efficiency with fuel/air equivalence ratio.

bustion process remains stable. For diesel engines, which always operate lean, the combustion efficiency is normally higher—about 98 percent. Details of exhaust gas composition, on which these combustion efficiencies are based, can be found in Sec. 4.9.

3.6 THE SECOND LAW OF THERMODYNAMICS APPLIED TO COMBUSTION

3.6.1 Entropy

In App. B, it is shown how the entropy of a mixture of ideal gases of known composition can be calculated. The discussion earlier relating enthalpies (or internal energies) of reactant and product mixtures applies to entropy also. The standard state entropies of chemical species are tabulated in the JANAF tables relative to zero entropy at 0 K. If the entropies of the elements at a datum temperature are arbitrarily set equal to zero, then the values of the entropy of a reactant mixture of given composition and of the resulting product mixture of given composition are both determined.

3.6.2 Maximum Work from an Internal Combustion Engine and Efficiency

An internal combustion engine can be analyzed as an open system which exchanges heat and work with its surrounding environment (the atmosphere). Reactants (fuel and air) flow into the system; products (exhaust gases) flow out. By applying the second law of thermodynamics to a control volume surrounding the engine, as illustrated in Fig. 3-8, we can derive an expression for the maximum useful work that the engine can deliver.

Consider a mass m of fluid as it passes through the control volume surrounding the engine. The first law gives

$$\Delta Q - \Delta W_U = \Delta H$$

where ΔW_U is the useful work transfer (i.e., non-$p\,dV$ work) to the environment and $\Delta H = H_P - H_R$. Since the heat transfer ΔQ occurs only with the atmosphere at temperature T_A, from the second law

$$\frac{\Delta Q}{T_A} \leq \Delta S$$

These equations combine to give

$$\Delta W_U \leq -(\Delta H - T_A \Delta S) = -\Delta B$$

where B is the steady-flow availability function, $H - T_A S$.[13] Usually $p_R = p_A$ and $T_R = T_A$. The maximum work will be obtained when $p_P = p_A$ and $T_P = T_A$.

TABLE 3.3
Enthalpies and free energies of combustion reactions

Reaction†	$\Delta \tilde{h}^\circ_{298}$, MJ/kmol	$\Delta \tilde{g}^\circ_{298}$, MJ/kmol
$C + O_2 \rightarrow CO_2$	-393.52	-394.40
$H_2 + \frac{1}{2}O_2 \rightarrow H_2O$	-240.91	-232.78
$CH_4 + 2O_2 \rightarrow CO_2 + 2H_2O$	-802.30	-800.76
$CH_4O(l) + \frac{3}{2}O_2 \rightarrow CO_2 + 2H_2O$	-638.59	-685.35
$C_3H_8(g) + 5O_2 \rightarrow 3CO_2 + 4H_2O$	-2044.0	-2074.1
$C_6H_6(l) + \frac{15}{2}O_2 \rightarrow 6CO_2 + 3H_2O$	-3135.2	-3175.1
$C_8H_{18}(l) + \frac{25}{2}O_2 \rightarrow 8CO_2 + 9H_2O$	-5074.6	-5219.9

† H_2O (gas) in products.

Under these conditions,

$$\Delta W_U \leq -[(H - TS)_{P_{T_A, P_A}} - (H - TS)_{R_{T_A, P_A}}] = -(\Delta G)_{T_A, P_A}$$

or $\quad \Delta W_{U\,max} = -(\Delta G)_{T_A, P_A}$ \hfill (3.28)

G is the Gibbs free energy, $H - TS$, and $(\Delta G)_{T_A, P_A}$ is the Gibbs free energy increase in the reaction of the fuel-air mixture to products at atmospheric temperature and pressure. $-(\Delta G)_{T_A, P_A}$ will be a maximum when combustion is complete.

A fundamental measure of the effectiveness of any practical internal combustion engine is the ratio of the actual work delivered compared with this maximum work. This ratio will be called the *availability conversion efficiency* η_a:

$$\eta_a = \frac{\Delta W}{\Delta W_{U\,max}} = -\frac{\Delta W}{(\Delta G)_{T_A, P_A}}$$ \hfill (3.29)

The property *availability* is the maximum useful work transfer that can be obtained from a system atmosphere (or control-volume atmosphere) combination at a given state. This efficiency therefore defines the fraction of the availability of the unburned fuel and air which, passing through the engine and interacting only with the atmosphere, is actually converted to useful work. Availability analysis of engine operation is proving valuable in identifying where the significant irreversibilities or losses in availability occur. This topic is discussed more fully in Sec. 5.7.

$(\Delta G)_{T_A, P_A}$, or $(\Delta g)_{T_A, P_A}$, is not an easy quantity to evaluate for practical fuels; it is the heating value, $-(\Delta h)_{T_A}$, which is usually measured. Values of $(\Delta g)^\circ_{298}$ and $(\Delta h)^\circ_{298}$ for selected fuel combustion reactions are given in Table 3.3. For the pure hydrocarbons they are closely comparable because at 298 K, $\Delta \tilde{s}^\circ \ll \Delta \tilde{h}^\circ/T$. For hydrogen and methanol the differences are larger, however. Because for practical fuels $-(\Delta h)^\circ_{298}$ is measured directly as the heating value of the fuel, it is standard practice to use the following definition of efficiency:

$$\eta_f = \frac{W_c}{m_f Q_{HV}}$$ \hfill (3.30)

which was defined as the *fuel conversion efficiency* in Sec. 2.8. Note that sometimes the higher heating value is used in Eq. (3.30) and sometimes the lower heating value. Whichever value is used should be explicitly stated. The normal practice in internal combustion engine analysis is to use the lower heating value at constant pressure, since the engine overall is a steady flow device and the water in the exhaust is always in vapor form. We will use Q_{LHV_p} in Eq. (3.30) throughout this text. The fuel conversion efficiency is the most commonly used definition of engine efficiency because it uses an easily measured quantity, the heating value, to define the usable fuel energy supplied to the engine. For hydrocarbon fuels, since $\Delta \tilde{h}^\circ \approx \Delta \tilde{g}^\circ$, the fuel conversion efficiency and the availability conversion efficiency are closely comparable in value.

In practice, not all the fuel energy supplied to the engine is released by the combustion process since combustion is incomplete: the combustion efficiency [Eq. (3.27)] is less than unity. It is sometimes useful to separate out the effects of incomplete combustion by defining an efficiency which relates the actual work per cycle to the amount of fuel chemical energy released in the combustion process. We will call this the *thermal conversion efficiency* η_t:

$$\eta_t = \frac{W_c}{H_R(T_A) - H_P(T_A)} = -\frac{W_c}{(\Delta H)_{T_A}} = \frac{W_c}{\eta_c \, m_f \, Q_{\text{HV}}} \tag{3.31}$$

Obviously the fuel conversion, thermal conversion, and combustion efficiencies are related by

$$\eta_f = \eta_c \, \eta_t \tag{3.32}$$

It is important to understand that there is a fundamental difference between availability conversion efficiency as defined by Eq. (3.29) [and the fuel conversion efficiency for internal combustion engines, Eq. (3.30), which closely approximates it] and the efficiency of a thermodynamic heat engine. The second law limit to the availability conversion efficiency is unity. For a thermodynamic heat engine (which experiences heat-transfer interactions with at least two heat reservoirs) the efficiency is limited to a value substantially less than unity by the temperatures of the heat reservoirs available.[13]

3.7 CHEMICALLY REACTING GAS MIXTURES

The working fluids in engines are mixtures of gases. Depending on the problem under consideration and the portion of the engine cycle in which it occurs chemical reactions may: (1) be so slow that they have a negligible effect on mixture composition (the mixture composition is essentially "frozen"); (2) be so rapid that the mixture state changes and the composition remains in chemical equilibrium; (3) be one of the rate-controlling processes that determine how the composition of the mixture changes with time.

3.7.1 Chemical Equilibrium

It is a good approximation for performance estimates in engines to regard the burned gases produced by the combustion of fuel and air as in chemical equilibrium.† By this we mean that the chemical reactions, by which individual species in the burned gases react together, produce and remove each species at equal rates. No net change in species composition results.

For example, if the temperature of a mass of carbon dioxide gas in a vessel is increased sufficiently, some of the CO_2 molecules *dissociate* into CO and O_2 molecules. If the mixture of CO_2, CO, and O_2 is in equilibrium, then CO_2 molecules are dissociating into CO and O_2 at the same rate as CO and O_2 molecules are *recombining* in the proportions required to satisfy the equation

$$CO + \tfrac{1}{2}O_2 = CO_2$$

In combustion products of hydrocarbon fuels, the major species present at low temperatures are N_2, H_2O, CO_2, and O_2 or CO and H_2. At higher temperatures (greater than about 2200 K), these major species dissociate and react to form additional species in significant amounts. For example, the adiabatic combustion of a stoichiometric mixture of a typical hydrocarbon fuel with air produces products with species mole fractions of: $N_2 \sim 0.7$; H_2O, $CO_2 \sim 0.1$; CO, OH, O_2, NO, $H_2 \sim 0.01$; H, O ~ 0.001; and other species in lesser amounts.

The second law of thermodynamics defines the criterion for chemical equilibrium as follows. Consider a system of chemically reacting substances undergoing a constant-pressure, constant-temperature process. In the absence of shear work (and electrical work, gravity, motion, capillarity), the first law gives

$$\delta Q = dH$$

The second law gives

$$\delta Q \leq T \, dS$$

Combining these gives

$$dH - T \, dS \leq 0$$

Since we are considering constant-temperature processes, this equation holds for finite changes:

$$\Delta H - T \, \Delta S = \Delta G \leq 0$$

Thus, reactions can only occur (at constant pressure and temperature) if G ($= H - TS$) for the products is less than G for the reactants. Hence at equilibrium

$$(\Delta G)_{p, \, T} = 0 \tag{3.33}$$

---

† This assumption is not valid late in the expansion stroke and during the exhaust process (see Sec. 4.9). Nor does it take account of pollutant formation processes (see Chap. 11).

Consider a reactive mixture of ideal gases. The reactant species M_a, M_b, etc., and product species M_l, M_m, etc., are related by the general reaction whose stoichiometry is given by

$$v_a M_a + v_b M_b + \cdots = v_l M_l + v_m M_m + \cdots \qquad (3.34a)$$

This can be written as

$$\sum_i v_i M_i = 0 \qquad (3.34b)$$

where the v_i are the stoichiometric coefficients and by convention are positive for the product species and negative for the reactant species.

Let an amount δn_a of M_a react with δn_b of M_b, etc., and produce δn_l of M_l, δn_m of M_m, etc. These amounts are in proportion:

$$\delta n_i = v_i \, \delta n \qquad (3.35)$$

The change in Gibbs free energy of a mixture of ideal gases, at constant pressure and temperature, as the composition changes is given by

$$(\Delta G)_{p,\,T} = \sum_i \tilde{\mu}_i \, \delta n_i \qquad (3.36)$$

where δn_i is the change in number of moles of species i and $\tilde{\mu}$ is the *chemical potential*. The chemical potential, an intensive property, is defined as

$$\tilde{\mu}_i = \left(\frac{\partial G}{\partial n_i} \right)_{p,\,T,\,n_{j(j \neq i)}} \qquad (3.37)$$

It is equal in magnitude to the specific Gibbs free energy at a given temperature and pressure. For an ideal gas, it follows from Eqs. (D.13), (D.15) and (3.37) that

$$\tilde{\mu}_i = \tilde{\mu}_i^\circ(T) + \tilde{R}T \ln \frac{p_i}{p_0} \qquad (3.38)$$

where $\tilde{\mu}_i^\circ$ equals $\tilde{g}_i^\circ$, the standard specific Gibbs free energy of formation. The standard state pressure p_0 is usually one atmosphere.

Substitution in Eq. (3.36) gives, at equilibrium,

$$\sum \left(\tilde{\mu}_i^\circ + \tilde{R}T \ln \frac{p_i}{p_0} \right) \delta n_i = 0$$

or

$$\sum \left(\tilde{\mu}_i^\circ + \tilde{R}T \ln \frac{p_i}{p_0} \right) v_i \, \delta n = 0$$

We can divide by δn and rearrange, to obtain

$$\sum \ln \left(\frac{p_i}{p_0} \right)^{v_i} = -\frac{\left(\sum \tilde{\mu}_i^\circ v_i \right)}{\tilde{R}T} = -\frac{\Delta G^\circ}{\tilde{R}T} = \ln K_p \qquad (3.39)$$

K_p is the equilibrium constant at constant pressure:

$$K_p = \prod_i \left(\frac{p_i}{p_0} \right)^{v_i} \qquad (3.40)$$

It is obtained from the Gibbs free energy of the reaction which can be calculated from the Gibbs free energy of formation of each species in the reaction, as indicated in Eq. (3.39) above. It is a function of temperature only.

In the JANAF tables,[8] to simplify the calculation of equilibrium constants, values of $\log_{10} (K_p)_i$, the equilibrium constants of formation of one mole of each species from their elements in their standard states, are tabulated against temperature. The equilibrium constant for a specific reaction is then obtained via the relation

$$\log_{10} (K_p)_{\text{reaction}} = \sum_i v_i \log_{10} (K_p)_i \qquad (3.41)$$

With the JANAF table values of $(K_p)_i$, the pressures in Eqs. (3.40) and (3.41) must be in atmospheres.

The effect of pressure on the equilibrium composition can be deduced from Eq. (3.40). Substitution of the mole fractions $\tilde{x}_i$ and mixture pressure p gives

$$\prod_i \left(\frac{p_i}{p_0}\right)^{v_i} = \prod_i \left(\tilde{x}_i \frac{p}{p_0}\right)^{v_i} = \left(\frac{p}{p_0}\right)^{\sum_i v_i} \prod_i \tilde{x}_i^{v_i} = K_p$$

If $\sum_i v_i = 0$, changes in pressure have no effect on the composition. If $\sum_i v_i > 0$ (dissociation reactions), then the mole fractions of the dissociation products decrease as pressure increases. If $\sum_i v_i < 0$ (recombination reactions), the converse is true.

An equilibrium constant, K_c, based on concentrations (usually expressed in gram moles per cubic centimeter) is also used:

$$K_c = \prod_i [M_i]^{v_i} \qquad (3.42)$$

Equation (3.40) can be used to relate K_p and K_c:

$$K_p = K_c (\tilde{R} T)^{\sum_i v_i} \qquad (3.43)$$

for $p_0 = 1$ atmosphere. For $\sum_i v_i = 0$, K_p and K_c are equal.

Example 3.4. A stoichiometric mixture of CO and O_2 in a closed vessel, initially at 1 atm and 300 K, is exploded. Calculate the composition of the products of combustion at 2500 K and the gas pressure.

The combustion equation is

$$CO + \tfrac{1}{2}O_2 = CO_2$$

The JANAF tables give $\log_{10} K_p$ (equilibrium constants of formation from the elements in their standard state) at 2500 K of CO_2, CO, and O_2 as 8.280, 6.840, and 0, respectively. Thus, the equilibrium constant for the CO combustion reaction above is, from Eq. (3.41),

$$\log_{10} K_p = 8.280 - 6.840 = 1.440$$

which gives $K_p = 27.5$.

If the degree of dissociation in the products is α (i.e., a fraction α of the CO_2 formed by complete combustion is dissociated), the product composition is

$$CO_2, (1 - \alpha) ; \qquad CO, \alpha; \qquad O_2, \frac{\alpha}{2}$$

For this mixture, the number of moles of reactants n_R is $\frac{3}{2}$; the number of moles of products n_P is $(1 + \alpha/2)$.

The ideal gas law gives

$$p_R V = n_R \tilde{R} T_R \qquad p_P V = n_P \tilde{R} T_P$$

Thus

$$\frac{p_P}{n_P} = \frac{1}{1.5} \times \frac{2500}{300} = 5.555 \text{ atm/mol}$$

The equilibrium relation [Eq. (3.40)] gives

$$\frac{1 - \alpha}{\alpha(\alpha/2)^{1/2}} \left(\frac{n_P}{p_P}\right)^{1/2} = 27.5$$

which can be solved to give $\alpha = 0.074$.

The composition of the products in mole fractions is, therefore,

$$x_{CO_2} = \frac{1 - \alpha}{n_P} = 0.893$$

$$x_{CO} = \frac{\alpha}{n_P} = 0.071$$

$$x_{O_2} = \frac{\alpha/2}{n_P} = 0.037$$

The pressure of the product mixture is

$$p = 5.555 n_p = 5.76 \text{ atm}$$

Example 3.5. In fuel-rich combustion product mixtures, equilibrium between the species CO_2, H_2O, CO, and H_2 is often assumed to determine the burned gas composition. For $\phi = 1.2$, for C_8H_{18}–air combustion products, determine the mole fractions of the product species at 1700 K.

The reaction relating these species (often called the water gas reaction) is

$$CO_2 + H_2 = CO + H_2O$$

From the JANAF tables, $\log_{10} K_p$ of formation for these species at 1700 K are: CO_2, 12.180; H_2, 0; CO, 8.011; $H_2O(g)$, 4.699. The equilibrium constant for the above reaction is, from Eq. (3.41),

$$\log_{10} K_p = 8.011 + 4.699 - 12.180 = 0.530$$

from which $K_p = 3.388$.

The combustion reaction for C_8H_{18}–air with $\phi = 1.2$ can be written

$$C_8H_{18} + \frac{12.5}{1.2} (O_2 + 3.773 N_2) \rightarrow a CO_2 + b H_2O + c CO + d H_2 + 39.30 N_2$$

A carbon balance gives: $\qquad a + c = 8$

A hydrogen balance gives: $\qquad 2b + 2d = 18$

An oxygen balance gives: $\qquad 2a + b + c = 20.83$

The equilibrium relation gives $(bc)/(ad) = 3.388$ (since the equilibrated reaction has the same number of moles as there are reactants or products, the moles of each species can be substituted for the partial pressures).

These four equations can be solved to obtain

$$c^2 - 19.3c + 47.3 = 0$$

which gives $c = 2.89$, $a = 5.12$, $b = 7.72$, and $d = 1.29$. The total number of moles of products is

$$a + b + c + d + 39.3 = 56.3$$

and the mole fractions of the species in the burned gas mixture are

CO_2, 0.0908; $\qquad H_2O$, 0.137; $\qquad CO$, 0.051; $\qquad H_2$, 0.023; $\qquad N_2$, 0.698

Our development of the equilibrium relationship for one reaction has placed no restrictions on the occurrence of simultaneous equilibria. Consider a mixture of N reacting gases in equilibrium. If there are C chemical elements, conservation of elements will provide C equations which relate the concentrations of these N species. Any set of $(N - C)$ chemical reactions, each in equilibrium, which includes each species at least once will then provide the additional equations required to determine the concentration of each species in the mixture. Unfortunately, this complete set of equations is a coupled set of C linear and $(N - C)$ nonlinear equations which is difficult to solve for cases where $(N - C) > 2$. For complex systems such as this, the following approach to equilibrium composition calculations is now more widely used.

Standardized computer methods for the calculation of complex chemical equilibrium compositions have been developed. A generally available and well-documented example is the NASA program of this type.[14] The approach taken is to minimize explicitly the Gibbs free energy of the reacting mixture (at constant temperature and pressure) subject to the constraints of element mass conservation. The basic equations for the NASA program are the following.

If the stoichiometric coefficients a_{ij} are the number of kilomoles of element i per kilomole of species j, b_i^* is the number of kilomoles of element i per kilogram of mixture, and n_j is the number of kilomoles of species j per kilogram of mixture, element mass balance constraints are

$$\sum_{j=1}^{n} a_{ij} n_j - b_i^* = 0 \qquad \text{for } i = 1, 2, \ldots, l \qquad (3.44)$$

The Gibbs free energy per kilogram of mixture is

$$g = \sum_{j=1}^{n} \tilde{\mu}_j n_j \qquad (3.45)$$

For gases, the chemical potential $\tilde{\mu}_j$ is

$$\tilde{\mu}_j = \tilde{\mu}_j^\circ + \tilde{R}T \ln \left(\frac{n_j}{n}\right) + \tilde{R}T \ln p \qquad (3.46)$$

where $\tilde{\mu}_j^\circ$ is the chemical potential in the standard state and p is the mixture pressure in atmospheres. Using the method of lagrangian multipliers, the term

$$G = g + \sum_{i=1}^{l} \lambda_i \sum_{j=1}^{n} (a_{ij} n_j - b_i^*)$$

is defined. The condition for equilibrium then becomes

$$\delta G = \sum_{j=1}^{n} \left(\mu_j + \sum_{i=1}^{l} \lambda_i a_{ij}\right)\delta n_j + \sum_{i=1}^{l} \sum_{j=1}^{n} (a_{ij} n_j - b_i^*)\delta\lambda_i = 0 \qquad (3.47)$$

Treating the variations δn_j and $\delta\lambda_i$ as independent gives

$$\tilde{\mu}_j + \sum_{i=1}^{l} \lambda_i a_{ij} = 0 \qquad \text{for } j = 1, \ldots, n \qquad (3.48)$$

and the original mass balance equation (3.44). Equations (3.44) and (3.48) permit the determination of equilibrium compositions for thermodynamic states specified by a temperature T and pressure p.

In the NASA program, the thermodynamic state may be specified by other pairs of state variables: enthalpy and pressure (useful for constant-pressure combustion processes); temperature and volume; internal energy and volume (useful for constant-volume combustion processes); entropy and pressure, and entropy and volume (useful for isentropic compressions and expansions). The equations required to obtain mixture composition are not all linear in the composition variables and an iteration procedure is generally required to obtain their solution. Once the composition is determined, additional relations, such as those in App. B which define the thermodynamic properties of gas mixtures, must then be used.

For each species, standard state enthalpies $\tilde{h}^\circ$ are obtained by combining standard enthalpies of formation at the datum temperature (298.15 K) $\Delta\tilde{h}_{f298}^\circ$ with sensible enthalpies $(\tilde{h}^\circ - \tilde{h}_{298}^\circ)$, i.e.,

$$\tilde{h}^\circ = \Delta\tilde{h}_{f298}^\circ + (\tilde{h}^\circ - \tilde{h}_{298}^\circ) \qquad (3.49)$$

For the elements in their reference state, $\Delta\tilde{h}_{f298}^\circ$ is zero [the elements important in combustion are C (solid, graphite), $H_2(g)$, $O_2(g)$, $N_2(g)$].

For each species, the thermodynamic quantities specific heat, enthalpy, and entropy as functions of temperature are given in the form:

$$\frac{\tilde{c}_p^\circ}{\tilde{R}} = a_1 + a_2 T + a_3 T^2 + a_4 T^3 + a_5 T^4 \qquad (3.50a)$$

$$\frac{\tilde{h}^\circ}{\tilde{R}T} = a_1 + \frac{a_2}{2} T + \frac{a_3}{3} T^2 + \frac{a_4}{4} T^3 + \frac{a_5}{5} T^4 + \frac{a_6}{T} \qquad (3.50b)$$

$$\frac{\tilde{s}^{\circ}}{\tilde{R}} = a_1 \ln T + a_2 T + \frac{a_3}{2} T^2 + \frac{a_4}{3} T^3 + \frac{a_5}{4} T^4 + a_7 \qquad (3.50c)$$

The coefficients are obtained by least-squares matching with thermodynamic property data from the JANAF tables. Usually two sets of coefficients are included for two adjacent temperature intervals (in the NASA program these are 300 to 1000 K and 1000 to 5000 K) (see Sec. 4.7).

In some equilibrium programs, the species to be included in the mixture must be specified as an input to the calculation. In the NASA program, all allowable species are included in the calculation, though species may be specifically omitted from consideration.

For each reactant composition and pair of thermodynamic state variables, the program calculates and prints out the following:

1. *Thermodynamic mixture properties* (obtained from the equilibrium composition and the appropriate gas mixture rule; see App. B). p, T, ρ, h, s, M, $(\partial \ln V / \partial \ln p)_T$, $(\partial \ln V / \partial \ln T)_p$, c_p, γ_s, and a (sound speed)
2. *Equilibrium composition.* Mole fractions of each species (which are present in significant amounts), $\tilde{x}_i$

Figure 3-10 shows how the equilibrium composition of the products of combustion of isooctane-air mixtures at selected temperatures and 30 atm pressure varies with the equivalence ratio. At low temperatures, the products are N_2, CO_2, H_2O, and O_2 for lean mixtures and N_2, CO_2, H_2O, CO, and H_2 for rich mixtures. As temperature increases, the burned-gas mixture composition becomes much more complex with dissociation products such as OH, O, and H becoming significant.

Figure 3-11 shows adiabatic flame temperatures for typical engine conditions as a function of the equivalence ratio, obtained with the NASA program using the methodology of Sec. 3.5.4. The isooctane-air unburned mixture state was 700 K and 10 atm. Flame temperatures for adiabatic combustion at constant pressure (where p_R and H_R are specified) and at constant volume (where V_R and U_R are specified) are shown. Flame temperatures at constant volume are higher, because the final pressure is higher and dissociation is less. Maximum flame temperatures occur slightly rich of stoichiometric.

3.7.2   Chemical Reaction Rates

Whether a system is in chemical equilibrium depends on whether the time constants of the controlling chemical reactions are short compared with time scales over which the system conditions (temperature and pressure) change. Chemical processes in engines are often not in equilibrium. Important examples of nonequilibrium phenomena are the flame reaction zone where the fuel is oxidized, and the air-pollutant formation mechanisms. Such nonequilibrium processes are controlled by the rates at which the actual chemical reactions which convert

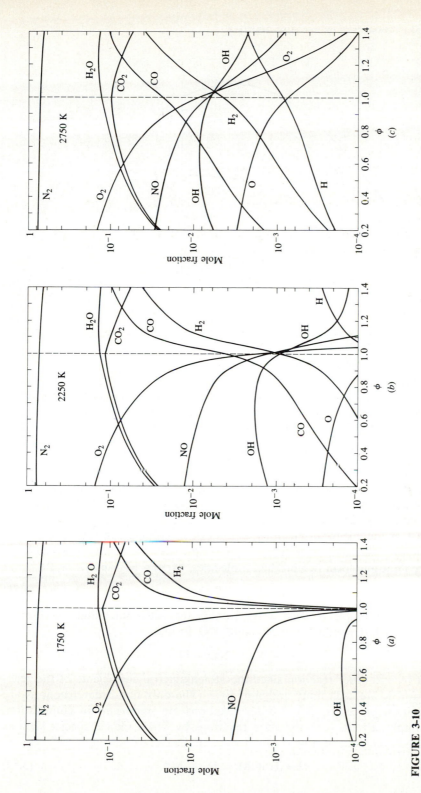

FIGURE 3-10
Mole fractions of equilibrium combustion products of isooctane-air mixtures as a function of fuel/air equivalence ratio at 30 atmospheres and (a) 1750 K; (b) 2250 K; and (c) 2750 K.

93

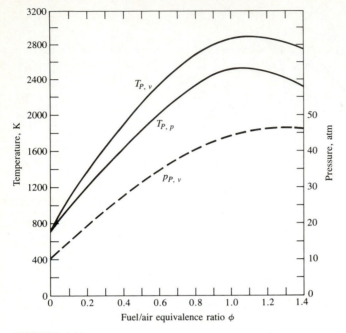

FIGURE 3-11
Equilibrium product temperatures for constant-volume ($T_{P, v}$) and constant-pressure ($T_{P, p}$) adiabatic combustion of isooctane-air mixture initially at 700 K and 10 atm, as a function of fuel/air equivalence ratio. Pressure ($p_{P, v}$) is equilibrium pressure for adiabatic constant-volume combustion.

reactants to products occur. The rates at which chemical reactions proceed depend on the concentration of the reactants, temperature, and whether any catalyst is present. This field is called chemical kinetics and some of its basic relations will now be reviewed.[2]

   Most of the chemical reactions of interest in combustion are binary reactions, where two reactant molecules, M_a and M_b, with the capability of reacting together collide and form two product molecules, M_c and M_d; i.e.,

$$M_a + M_b = M_c + M_d \tag{3.51}$$

An important example of such a reaction is the rate-controlling step in the process by which the pollutant nitric oxide, NO, forms:

$$O + N_2 = NO + N$$

This is a second-order reaction since the stoichiometric coefficients of the reactants v_a and v_b are each unity and sum to 2. (The only first-order reactions are decomposition processes.) Third-order reactions are important in combustion, also. Examples are the recombination reactions by which radical species such as H, O, and OH combine during the final stage of the fuel oxidation process: e.g.,

$$H + H + M = H_2 + M^* \tag{3.52}$$

M is any molecule (such as N_2) which takes part in the collision and carries away the excess energy.

The law of mass action states that the rate at which product species are produced and the rate at which reactant species are removed is proportional to the product of the concentrations of reactant species, with the concentration of each species raised to the power of its stoichiometric coefficient v_i. Thus, for reaction (3.51) above, the reaction rate R^+ in the forward $(+)$ direction, reactants to products, is given by

$$R^+ = -\frac{d[M_a]^+}{dt} = \frac{d[M_c]^+}{dt} = k^+[M_a][M_b]$$ (3.53)

If the reaction can also proceed in the reverse $(-)$ direction, then the backward rate R^- is given by

$$R^- = -\frac{d[M_c]^-}{dt} = \frac{d[M_a]^-}{dt} = k^-[M_c][M_d]$$ (3.54)

k^+ and k^- are the rate constants in the forward and reverse directions for this reaction. The *net rate* of production of products or removal of reactants is, therefore,

$$R^+ - R^- = \frac{d[M_c]^+}{dt} + \frac{d[M_c]^-}{dt} = -\frac{d[M_a]^+}{dt} - \frac{d[M_a]^-}{dt}$$

$$= k^+[M_a][M_b] - k^-[M_c][M_d]$$ (3.55)

These results can be stated more generally as follows. Any reaction can be written as

$$\sum_{i=1}^{n} v_{R_i} M_{R_i} = \sum_{i=1}^{m} v_{P_i} M_{P_i}$$ (3.56)

where v_i is the stoichiometric coefficient of species M_i, subscripts R and P denote reactants and products, respectively, and there are n reactant species and m product species. The forward reaction rate R^+ and the reverse reaction rate R^- are given by

$$R^+ = k^+ \prod_{i=1}^{n} [M_{R_i}]^{v_{R_i}}$$

$$R^- = k^- \prod_{i=1}^{m} [M_{P_i}]^{v_{P_i}}$$ (3.57)

The net rate of removal of reactant species M_{R_i} is

$$-\frac{d[M_{R_i}]}{dt} = v_{R_i}(R^+ - R^-)$$ (3.58a)

and the net rate of production of product species M_{P_i} is

$$\frac{d[M_{P_i}]}{dt} = v_{P_i}(R^+ - R^-) \tag{3.58b}$$

The rate constants, k, usually follow the Arrhenius form:

$$k = A \exp\left(-\frac{E_A}{RT}\right) \tag{3.59}$$

where A is called the frequency or preexponential factor and may be a (moderate) function of temperature; E_A is the activation energy. The Boltzmann factor $\exp(-E_A/RT)$ defines the fraction of all collisions that have an energy greater than E_A—i.e., sufficient energy to make the reaction take place. The functional dependence of k on T and the constants in the Arrhenius form, Eq. (3.59), if that is appropriate, are determined experimentally.

At equilibrium, the forward and reverse reaction rates are equal. Then, from Eq. (3.55), with $R^+ - R^- = 0$:

$$\frac{k^+}{k^-} = \frac{[M_c][M_d]}{[M_a][M_b]} = K_c$$

where K_c is the equilibrium constant based on concentrations defined by Eq. (3.42). It can be related to K_p, the equilibrium constant based on partial pressures, by Eq. (3.43).

The chemical reaction mechanisms of importance in combustion are much more complex than the above illustrations of rate-controlled processes. Such mechanisms usually involve both parallel and sequential interdependent reactions. The methodology reviewed above still holds; however, one must sum algebraically the forward and reverse rates of all the reactions which produce (or remove) a species of interest. In such complex mechanisms it is often useful to assume that (some of) the reactive intermediate species or radicals are in *steady state*. That is, these radicals react so quickly once they are formed that their concentrations do not rise but are maintained in steady state with the species with which they react. The net rate at which their concentration changes with time is set equal to zero.

PROBLEMS

3.1. Isooctane is supplied to a four-cylinder spark-ignition engine at 2 g/s. Calculate the air flow rate for stoichiometric combustion. If the engine is operating at 1500 rev/min, estimate the mass of fuel and air entering each cylinder per cycle. The engine displaced volume is 2.4 liters. What is the volumetric efficiency?

3.2. Calculate the exhaust gas composition of a butane-fueled spark-ignition engine operating with equivalence ratio of 0.9. Assume the fuel is fully burned within the cylinder. Butane is C_4H_{10}.

3.3. The molar composition of dry exhaust gas of a propane-fueled SI engine is given below (water was removed before the measurement). Calculate the equivalence ratio.

$$CO_2 = 10.8\%, \qquad O_2 = 4.5\%, \qquad CO = 0\%, \qquad H_2 = 0\%$$

3.4. Evaluate and compare the lower heating values per unit mass of stoichiometric mixture and per unit volume of stoichiometric mixture (at standard atmospheric conditions) for methane, isooctane, methyl alcohol, and hydrogen. Assume the fuel is fully vaporized.

3.5. The measured engine fuel flow rate is 0.4 g/s, air flow rate is 5.6 g/s, and exhaust gas composition (measured dry) is $CO_2 = 13.0\%$, $CO = 2.8\%$ with O_2 essentially zero. Unburned hydrocarbon emissions can be neglected. Compare the equivalence ratio calculated from the fuel and air flow with the equivalence ratio calculated from exhaust gas composition. The fuel is gasoline with a H/C ratio of 1.87. Assume a H_2 concentration equal to one-third the CO concentration.

3.6. The brake fuel conversion efficiency of an engine is 0.3. The mechanical efficiency is 0.8. The combustion efficiency is 0.94. The heat losses to the coolant and oil are 60 kW. The fuel chemical energy entering the engine per unit time, $\dot{m}_f Q_{HV}$, is 190 kW. What percentage of this energy becomes (a) brake work; (b) friction work; (c) heat losses; (d) exhaust chemical energy; (e) exhaust sensible energy.

3.7. An upper estimate can be made of the amount of NO formed in an engine from considering the equilibrium of the reaction $N_2 + O_2 = 2NO$. Calculate the NO concentration at equilibrium at 2500 K and 30 atm. $\log_{10} K_p = -1.2$ for this reaction at 2500 K. Assume N/O ratio in the combustion products is 15. N_2, O_2, and NO are the only species present.

3.8. Carbon monoxide reacts with air at 1 atm and 1000 K in an exhaust gas reactor. The mole fractions of the exhaust gas-air mixture flowing into the reactor are CO, 3%; O_2, 7%; N_2, 74%; CO_2, 6%; H_2O, 10%.
(a) Calculate the concentration of CO and O_2 in gram moles per cm^3 in the entering mixture.
(b) The rate of reaction is given by

$$d[CO]/dt = -4.3 \times 10^{11} \times [CO][O_2]^{0.25} \exp\left[-E/(RT)\right]$$

[ ] denotes concentration in gram moles per cm^3, $E/R = 20,000$ K. Calculate the initial reaction rate of CO, $d[CO]/dt$: time is in seconds.
(c) The equilibrium constant K_p for the reaction $CO + \frac{1}{2}O_2 = CO_2$ at 1000 K is 10^{10}. Find the equilibrium CO concentration.
(d) Determine the time to reach this equilibrium concentration of CO using the initial reaction rate. (The actual time will be longer but this calculation indicates approximately the time required.)

3.9. The exhaust gases of a hydrogen-fueled engine contain 22.3 percent H_2O, 7.44 percent O_2, and 70.2 percent N_2. At what equivalence ratio is it operating?

3.10. Gas is sampled at 1 atmosphere pressure from the exhaust manifold of an internal combustion engine and analyzed. The mole fractions of species in the exhaust are:

$$H_2O, 0.0468; \qquad CO_2, 0.0585; \qquad O_2, 0.123; \qquad N_2, 0.772$$

Other species such as CO and unburned hydrocarbons can be neglected.
(a) The fuel is a synthetic fuel derived from coal containing only carbon and hydrogen. What is the ratio of hydrogen atoms to carbon atoms in the fuel?

(b) Calculate the fuel/air equivalence ratio at which this engine is operating.

(c) Is the internal combustion engine a conventional spark-ignition or a diesel engine? Explain.

(d) The engine has a displaced volume of 2 liters. Estimate approximately the percentage by which the fuel flow rate would be increased if this engine were operated at its maximum load at this same speed (2000 rev/min). Explain briefly what limits the equivalence ratio at maximum load.

3.11. The following are approximate values of the relative molecular mass (molecular weights): oxygen O_2, 32; nitrogen N_2, 28; hydrogen H_2, 2; carbon C, 12. Determine the stoichiometric fuel/air and air/fuel ratios on a mass basis, and the lower heating value per unit mass of stoichiometric mixture for the following fuels:

Methane (CH_4), isooctane (C_8H_{18}), benzene (C_6H_6), hydrogen (H_2), methyl alcohol (CH_3OH)

Heating values for these fuels are given in App. D.

3.12. Liquid petroleum gas (LPG) is used to fuel spark-ignition engines. A typical sample of the fuel consists of

70 percent by volume propane C_3H_8
5 percent by volume butane C_4H_{10}
25 percent by volume propene C_3H_6

The *higher heating values* of the fuels are: propane, 50.38 MJ/kg; butane, 49.56 MJ/kg; propylene (propene), 48.95 MJ/kg.

(a) Work out the overall combustion reaction for stoichiometric combustion of 1 mole of LPG with air, and the stoichiometric F/A and A/F.

(b) What are the higher and lower heating values for combustion of this fuel with excess air, per unit mass of LPG?

3.13. A spark-ignition engine is operated on isooctane fuel (C_8H_{18}). The exhaust gases are cooled, dried to remove water, and then analyzed for CO_2, CO, H_2, O_2. Using the overall combustion reaction for a range of equivalence ratios from 0.5 to 1.5, calculate the mole fractions of CO_2, CO, H_2, and O_2 in the dry exhaust gas, and plot the results as a function of equivalence ratio. Assume:

(a) that all the fuel is burnt inside the engine (almost true) and that the ratio of moles CO to moles H_2 in the exhaust is 3 : 1, and

(b) that there is no hydrogen in the exhaust for lean mixtures.

For high-power engine operation the air/fuel ratio is 14 : 1. What is the exhaust gas composition, in mole fractions, *before* the water is removed?

REFERENCES

1. Fristrom, R. M., and Westenberg, A. A.: *Flame Structure*, McGraw-Hill, 1965.
2. Glassman, I.: *Combustion*, Academic Press, 1977.
3. Kaye, G. W. C., and Laby, T. H.: *Tables of Physical and Chemical Constants*, Longmans, London, 1973.
4. Reynolds, W. C.: *Thermodynamic Properties in SI*, Department of Mechanical Engineering, Stanford University, 1979.
5. Taylor, C. F.: *The Internal Combustion Engine in Theory and Practice*, vol. 1, MIT Press, Cambridge, Mass., 1960.

6. Goodger, E. M.: *Hydrocarbon Fuels*, Macmillan, London, 1975.
7. Spalding, D. B., and Cole, E. H.: *Engineering Thermodynamics*, 3d ed., Edward Arnold, 1973.
8. *JANAF Thermochemical Tables*, National Bureau of Standards Publication NSRDS-NBS37, 1971.
9. Maxwell, J. B.: *Data Book on Hydrocarbons*, Van Nostrand, New York, 1950.
10. Rossini, F. D., Pitzer, K. S., Arnelt, R. L., Braun, R. M., and Primentel, G. C.: *Selected Values of Physical and Thermodynamic Properties of Hydrocarbons and Related Compounds*, Carnegie Press, Pittsburgh, Pa., 1953.
11. Stull, D. R., Westrum, E. F., and Sinke, G. C.: *The Chemical Thermodynamics of Organic Compounds*, John Wiley, New York, 1969.
12. Matthews, R. D.: "Relationship of Brake Power to Various Energy Efficiencies and Other Engine Parameters: The Efficiency Rule," *Int. J. of Vehicle Design*, vol. 4, no. 5, pp. 491–500, 1983.
13. Keenan, J. H.: *Thermodynamics*, John Wiley, New York, 1941 (MIT Press, Cambridge, Mass., 1970).
14. Svehla, R. A., and McBride, B. J.: "Fortran IV Computer Program for Calculation of Thermodynamic and Transport Properties of Complex Chemical Systems," NASA Technical Note TN D-7056, NASA Lewis Research Center, 1973.

CHAPTER
4

PROPERTIES
OF WORKING
FLUIDS

4.1 INTRODUCTION

The study of engine operation through an analysis of the processes that occur inside the engine has a long and productive history. The earliest attempts at this analysis used the constant-volume and constant-pressure ideal cycles as approximations to real engine processes (see Chap. 5). With the development of high-speed digital computers, the simulation of engine processes has become much more sophisticated and accurate (see Chap. 14). All these engine simulations (from the simplest to the most complex) require models for the composition and properties of the working fluids inside the engine, as well as models for the individual processes—induction, compression, combustion, expansion, and exhaust—that make up the engine operating cycle. This chapter deals with models for the working fluid composition, and thermodynamic and transport properties.

The composition of the working fluid, which changes during the engine operating cycle, is indicated in Table 4.1. The unburned mixture for a spark-ignition engine during intake and compression consists of air, fuel, and previously burned gases. It is, therefore, a mixture of N_2, O_2, CO_2, H_2O, CO, and H_2 for fuel-rich mixtures, and fuel (usually vapor). The composition of the unburned mixture does not change significantly during intake and compression. It is suffi-

TABLE 4.1
Working fluid constituents

Process	Spark-ignition engine	Compression-ignition engine
Intake	Air Fuel† Recycled exhaust‡ Residual gas§	Air Recycled exhaust‡ Residual gas§
Compression	Air Fuel vapor Recycled exhaust Residual gas	Air Recycled exhaust Residual gas
Expansion	Combustion products (mixture of N_2, H_2O, CO_2, CO, H_2, O_2, NO, OH, O, H, ...)	Combustion products (mixture of N_2, H_2O, CO_2, CO, H_2, O_2, NO, OH, O, H, ...)
Exhaust	Combustion products [mainly N_2, CO_2, H_2O, and either O_2 ($\phi < 1$) or CO and H_2 ($\phi > 1$)]	Combustion products (mainly N_2, CO_2, H_2O, and O_2)

† Liquid and vapor in the intake; mainly vapor within the cylinder.
‡ Sometimes used to control NO_x emissions (see Secs. 11.2, 15.3.2, and 15.5.1).
§ Within the cylinder.

ciently accurate to assume the composition is *frozen*. For the compression-ignition engine, the unburned mixture prior to injection contains no fuel; it consists of air and previously burned gas.

The combustion products or burned mixture gases, during the combustion process and much of the expansion process, are close to *thermodynamic equilibrium*. The composition of such mixtures has already been discussed (Sec. 3.7.1). As these combustion products cool, recombination occurs as indicated in Fig. 3-10. Towards the end of the expansion process, the gas composition departs from the equilibrium composition; recombination can no longer occur fast enough to maintain the reacting mixture in equilibrium. During the exhaust process, reactions are sufficiently slow so that for calculating thermodynamic properties the composition can be regarded as *frozen*.

The models used for predicting the thermodynamic properties of unburned and burned mixtures can be grouped into the five categories listed in Table 4.2. The first category is only useful for illustrative purposes since the specific heats of unburned and burned mixtures are significantly different. While the specific heats of the working fluids increase with increasing temperature in the range of interest, a constant-specific-heat model can be matched to the thermodynamic data over a limited temperature range. This approach provides a simple *analytic* model which can be useful when moderate accuracy of prediction will suffice. The appropriateness of frozen and equilibrium assumptions has already been discussed above. Approximations to thermodynamic equilibrium calculations are useful because of

TABLE 4.2
Categories of models for thermodynamic properties

	Unburned mixture	Burned mixture
1.	Single ideal gas throughout operating cycle with c_v (and hence c_p) constant	
2.	Ideal gas; $c_{v,u}$ constant	Ideal gas; $c_{v,b}$ constant
3.	Frozen mixture of ideal gases; $c_{v,i}(T)$	Frozen mixture of ideal gases; $c_{v,i}(T)$
4.	Frozen mixture of ideal gases; $c_{v,i}(T)$	Approximations fitted to equilibrium thermodynamic properties
5.	Frozen mixture of ideal gases; $c_{v,i}(T)$	Mixture of reacting ideal gases in thermodynamic equilibrium

Note: Subscripts *i*, *u*, and *b* denote species *i* in the gas mixture, the unburned mixture, and burned mixture properties, respectively.

the savings in computational time, relative to full equilibrium calculations, which can result from their use.

Values of thermodynamic properties of unburned and burned mixtures relevant to engine calculations are available from charts, tables, and algebraic relationships developed to match tabulated data. A selection of this material is included in this chapter and App. D. The references indicate additional sources.

4.2 UNBURNED MIXTURE COMPOSITION

The mass of charge trapped in the cylinder (m_c) is the inducted mass per cycle (m_i), plus the residual mass (m_r) left over from the previous cycle. The residual fraction (x_r) is

$$x_r = \frac{m_r}{m_c} \tag{4.1}$$

Typical residual fractions in spark-ignition engines range from 20 percent at light load to 7 percent at full load. In diesels, the residual fraction is smaller (a few percent) due to the higher compression ratio, and in naturally aspirated engines is approximately constant since the intake is unthrottled. If the inducted mixture is fuel and air (or air only), then the *burned gas fraction* (x_b) in the unburned mixture during compression equals the residual fraction.

In some engines, a fraction of the engine exhaust gases is recycled to the intake to dilute the fresh mixture for control of NO_x emissions (see Sec. 11.2). If the percent of exhaust gas recycled (%EGR) is defined as the percent of the total intake mixture which is recycled exhaust,†

$$EGR(\%) = \left(\frac{m_{EGR}}{m_i}\right) \times 100 \qquad (4.2)$$

where m_{EGR} is the mass of exhaust gas recycled, then the burned gas fraction in the fresh mixture is

$$x_b = \frac{m_{EGR} + m_r}{m_c} = \left(\frac{EGR}{100}\right)(1 - x_r) + x_r \qquad (4.3)$$

Up to about 30 percent of the exhaust can be recycled; the burned gas fraction during compression can, therefore, approach 30 to 40 percent.

The composition of the burned gas fraction in the unburned mixture can be calculated as follows. The combustion equation for a hydrocarbon fuel of average molar H/C ratio y [e.g., Eq. (3.5)] can be written per mole O_2 as

$$\varepsilon\phi C + 2(1 - \varepsilon)\phi H_2 + O_2 + \psi N_2 \rightarrow n_{CO_2}CO_2 + n_{H_2O}H_2O$$

$$+ n_{CO}CO + n_{H_2}H_2 + n_{O_2}O_2 + n_{N_2}N_2 \quad (4.4)$$

where ψ = the molar N/O ratio (3.773 for air)

$$\varepsilon = \frac{4}{4 + y}$$

y = the molar H/C ratio of the fuel
ϕ = fuel/air equivalence ratio
n_i = moles of species i per mole O_2 reactant

The n_i are determined using the following assumptions:

1. For lean and stoichiometric mixtures ($\phi \leq 1$) CO and H_2 can be neglected.
2. For rich and stoichiometric mixtures ($\phi \geq 1$) O_2 can be neglected.
3. For rich mixtures, either (a) the water gas reaction

$$CO_2 + H_2 = CO + H_2O$$

---

† An alternative definition of percent EGR is also used based on the ratio of EGR to fresh mixture (fuel and air):

$$EGR^*(\%) = \left(\frac{m_{EGR}}{m_a + m_f}\right) \times 100$$

The two definitions are related by

$$\frac{EGR^*}{100} = \frac{EGR}{100 - EGR} \qquad \text{and} \qquad \frac{EGR}{100} = \frac{EGR^*}{100 + EGR^*}$$

can be assumed to be in equilibrium with the equilibrium constant $K(T)$:

$$K(T) = \frac{n_{H_2O} n_{CO}}{n_{CO_2} n_{H_2}}$$

where $K(T)$ can be determined from a curve fit to JANAF table data:[8]

$$\ln K(T) = 2.743 - \frac{1.761 \times 10^3}{T} - \frac{1.611 \times 10^6}{T^2} + \frac{0.2803 \times 10^9}{T^3} \qquad (4.5)$$

where T is in K, or (b) K can be assumed constant over the normal engine operating range. A value of 3.5 is often assumed (see Sec. 4.9), which corresponds to evaluating the equilibrium constant at 1740 K.

The n_i obtained from an element balance and the above assumptions are shown in Table 4.3. The value of c is obtained by solving the quadratic:

$$(K - 1)c^2 - c\{K[2(\phi - 1) + \varepsilon\phi] + 2(1 - \varepsilon\phi)\} + 2K\varepsilon\phi(\phi - 1) = 0 \qquad (4.6)$$

The mole fractions are given by

$$\tilde{x}_i = \frac{n_i}{n_b}$$

where $n_b = \sum_i n_i$ is given in the bottom line of Table 4.3.

While Eq. (4.4) is for a fuel containing C and H only, it can readily be modified for alcohols or alcohol-hydrocarbon blends. For a fuel of molar composition $CH_y O_z$, the reactant mixture

$$CH_y O_z + \frac{1}{\phi}\left(1 + \frac{y}{4} - \frac{z}{2}\right)(O_2 + \psi N_2)$$

can be rearranged per mole of O_2 reactant as

$$\zeta\phi\varepsilon C + 2\zeta\phi(1 - \varepsilon)H_2 + O_2 + \left(1 - \frac{\varepsilon z}{2}\right)\zeta\psi N_2 \qquad (4.7a)$$

TABLE 4.3
Burned gas composition under 1700 K

	n_i, moles/mole O_2 reactant	
Species	$\phi \leq 1$	$\phi > 1$†
CO_2	$\varepsilon\phi$	$\varepsilon\phi - c$
H_2O	$2(1 - \varepsilon)\phi$	$2(1 - \varepsilon\phi) + c$
CO	0	c
H_2	0	$2(\phi - 1) - c$
O_2	$1 - \phi$	0
N_2	ψ	ψ
Sum: n_b	$(1 - \varepsilon)\phi + 1 + \psi$	$(2 - \varepsilon)\phi + \psi$

† c defined by Eq. (4.6).

where

$$\zeta = \frac{2}{2 - \varepsilon z(1 - \phi)} \tag{4.7b}$$

If we write

$$\phi^* = \zeta\phi \quad \text{and} \quad \psi^* = \left(1 - \frac{\varepsilon z}{2}\right)\zeta\psi \tag{4.7c}$$

the reactant expression (4.7a) becomes

$$\phi^*\varepsilon C + 2\phi^*(1 - \varepsilon)H_2 + O_2 + \psi^* N_2$$

which is identical in form to the reactant expression for a hydrocarbon fuel (4.4). Thus Table 4.3 can still be used to give the composition of the burned gas residual fraction in the unburned mixture, except that ϕ^* replaces ϕ and ψ^* replaces ψ in the expressions for n_i.

Now consider the unburned mixture. The number of moles of fuel per mole O_2 in the mixture depends on the molecular weight of the fuel, M_f. If the average molecular formula of the fuel is $(CH_y)_\alpha$ then

$$M_f = \alpha(12 + y)$$

The fresh fuel-air mixture (not yet diluted with EGR or residual),

$$\varepsilon\phi C + 2(1 - \varepsilon)\phi H_2 + O_2 + \psi N_2$$

then becomes

$$\frac{4}{M_f}(1 + 2\varepsilon)\phi(CH_y)_\alpha + O_2 + \psi N_2$$

The unburned mixture (fuel, air, and a burned gas fraction), per mole O_2 in the mixture, can be written:

$$(1 - x_b)\left[\frac{4}{M_f}(1 + 2\varepsilon)\phi(CH_y)_\alpha + O_2 + \psi N_2\right]$$

$$+ x_b(n_{CO_2} + n_{H_2O} + n_{CO} + n_{H_2} + n_{O_2} + n_{N_2})$$

The number of moles of each species in the unburned mixture, per mole O_2, is summarized in Table 4.4. The mole fractions of each species are obtained by dividing by the total number of moles of unburned mixture n_u,

$$n_u = (1 - x_b)\left[\frac{4(1 + 2\varepsilon)\phi}{M_f} + 1 + \psi\right] + x_b n_b \tag{4.8}$$

where n_b is given in Table 4.3.

The molecular weights of the (low-temperature) burned and unburned

TABLE 4.4
Unburned mixture composition

Species	$\phi \leq 1$	$\phi > 1$
	n_i, moles/mole O_2 reactant	
Fuel	$4(1 - x_b)(1 + 2\varepsilon)\phi/M_f$	
O_2	$1 - x_b\phi$	$1 - x_b$
N_2	ψ	ψ
CO_2	$x_b\varepsilon\phi$	$x_b(\varepsilon\phi - c)$
H_2O	$2x_b(1 - \varepsilon)\phi$	$x_b[2(1 - \varepsilon\phi) + c]$
CO	0	$x_b c$
H_2	0	$x_b[2(\phi - 1) - c]$
Sum†	n_u	n_u

† Given by Eq. (4.8).

mixture can now be determined. The mass of mixture (burned or unburned) per mole O_2 in the mixture, m_{RP}, is given by

$$m_{RP} = 32 + 4\phi(1 + 2\varepsilon) + 28.16\psi \tag{4.9}$$

The molecular weight of the burned mixture, M_b, is therefore

$$M_b = \frac{m_{RP}}{n_b} \tag{4.10}$$

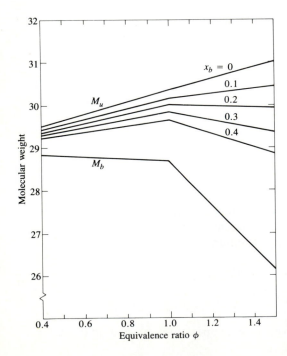

FIGURE 4-1
Molecular weight of unburned and low-temperature burned isooctane-air mixtures as a function of fuel/air equivalence ratio and burned gas fraction.

TABLE 4.5
Factors for relating properties on molar and mass basis

Quantity, per mole O_2 in the mixture	General equation‡		Equation for C_8H_{18}-air mixtures
Moles of burned mixture n_b	$n_b = (1 - \varepsilon)\phi + 1 + \psi,$ $n_b = (2 - \varepsilon)\phi + \psi,$	$\phi \leq 1$ $\phi > 1$	$n_b = 0.36\phi + 4.773$ $n_b = 1.36\phi + 3.773$
Moles of unburned mixture n_u	$(1 - x_b)\left[\dfrac{4(1 + 2\varepsilon)\phi}{M_f} + 1 + \psi\right] + x_b n_b$	$\begin{cases}\phi \leq 1\\ \phi > 1\end{cases}$	$n_u = 0.08\phi + 4.773$ $+0.28x_b\,\phi$ $n_u = 0.08\phi + 4.773$ $+x_b(1.28\phi - 1)$
Mass of mixture† (burned or unburned)	$m_{RP} = 32 + 4\phi(1 + 2\varepsilon) + 28.16\psi$		$138.2 + 9.12\phi$
Mass of air†	$32 + 28.16\psi$		138.2

† Units: kg/kmol or lbm/lb·mol.

‡ For hydrocarbon fuels, ψ for air $= 3.773$; for fuels containing oxygen, ϕ^* and ψ^* given by Eq. (4.7c) are substituted for ϕ and ψ, respectively.

The molecular weight of the unburned mixture, M_u, is

$$M_u = \frac{m_{RP}}{n_u} \tag{4.11}$$

Figure 4-1 gives M_u and M_b for a range of ϕ and x_b for air, isooctane, burned gas mixtures.

Frequently, thermodynamic properties of unburned and burned mixtures are expressed per unit mass of *air* in the *original* mixture (for burned mixture this is the mixture before combustion). To obtain properties in these units, we need the mass of original air, per mole O_2 in the mixture, which is

$$(32 + 28.16\ \psi)$$

with units of kilograms per kilomole or pound-mass per pound-mole.

Table 4.5 summarizes the factors needed to relate properties expressed on a molar and a mass basis.

4.3 GAS PROPERTY RELATIONSHIPS

The individual species in the unburned and burned gas mixtures can with sufficient accuracy be modeled as ideal gases. Ideal gas relationships are reviewed in App. B. The most important relationships for property determination for engine calculations are summarized below.

Since internal energy and enthalpy are functions of temperature only, the specific heats at constant volume and constant pressure are given by

$$c_v = \left(\frac{\partial u}{\partial T}\right)_v = \frac{du}{dT} = c_v(T) \tag{4.12a}$$

$$c_p = \left(\frac{\partial h}{\partial T}\right)_p = \frac{dh}{dT} = c_p(T) \tag{4.12b}$$

and

$$u - u_0 = \int_{T_0}^{T} c_v \, dT \tag{4.13a}$$

$$h - h_0 = \int_{T_0}^{T} c_p \, dT \tag{4.13b}$$

The entropy $s(T, v)$ or $s(T, p)$ is given by

$$s - s_0 = \int_{T_0}^{T} c_v \frac{dT}{T} + R \ln \frac{v}{v_0} \tag{4.14a}$$

$$s - s_0 = \int_{T_0}^{T} c_p \frac{dT}{T} - R \ln \frac{p}{p_0} \tag{4.14b}$$

The integrals in Eqs. (4.14a, b) are functions of temperature only, and are useful in evaluating entropy changes and in following isentropic processes. If we define

$$\Psi(T) = \int_{T_0}^{T} c_v(T) \frac{dT}{T} \tag{4.15a}$$

and

$$\Phi(T) = \int_{T_0}^{T} c_p(T) \frac{dT}{T} \tag{4.15b}$$

then

$$s - s_0 = \Psi + R \ln \left(\frac{v}{v_0}\right) \tag{4.16a}$$

$$s - s_0 = \Phi - R \ln \left(\frac{p}{p_0}\right) \tag{4.16b}$$

Thus, for example, the entropy change between states (T_1, p_1) and (T_2, p_2) is

$$s_2 - s_1 = \Phi_2 - \Phi_1 - R \ln \left(\frac{p_2}{p_1}\right) \tag{4.17}$$

For an isentropic process,

$$\ln \left(\frac{p_2}{p_1}\right) = \frac{\Phi_2 - \Phi_1}{R} \tag{4.18}$$

In these equations, the units of u and h can be on a per unit mass or molar basis [i.e., joules per kilogram (British thermal units per pound-mass) or joules per kilomole (British thermal units per pound-mole)]; similarly, s, c_v, c_p, R, Ψ, and Φ can be in joules per kilogram-kelvin (British thermal units per pound-mass-degree Rankine) or joules per kilomole-kelvin (British thermal units per pound-mole-degree Rankine).

For gas mixtures, once the composition is known, mixture properties are determined either on a mass or molar basis from

$$u = \sum x_i u_i \qquad \tilde{u} = \sum \tilde{x}_i \tilde{u}_i \qquad (4.19a)$$

$$h = \sum x_i h_i \qquad \tilde{h} = \sum \tilde{x}_i \tilde{h}_i \qquad (4.19b)$$

$$s = \sum x_i s_i \qquad \tilde{s} = \sum \tilde{x}_i \tilde{s}_i \qquad (4.19c)$$

and

$$c_v = \sum x_i c_{v,i} \qquad \tilde{c}_v = \sum \tilde{x}_i \tilde{c}_{v,i} \qquad (4.20a)$$

$$c_p = \sum x_i c_{p,i} \qquad \tilde{c}_p = \sum \tilde{x}_i \tilde{c}_{p,i} \qquad (4.20b)$$

4.4 A SIMPLE ANALYTIC IDEAL GAS MODEL

While the first category of model listed in Table 4.2 is too inaccurate for other than illustrative purposes, the second category—constant but different specific heats for the unburned and burned gas mixtures—can with careful choice of specific heat values be made much more precise. The advantages of a simple analytic model may be important for certain problems.

Figure 4-2 shows an internal energy versus temperature plot for a stoichiometric mixture. It is a quantitative version of Fig. 3-5. The unburned mixture line is for a burned gas fraction of 0.1. The fuel is isooctane. Data to construct such graphs can be obtained from charts or tables or computer programs (see Secs. 4.5 to 4.7). The units for u are kilojoules per kilogram of air in the original mixture (the units of the charts in Sec. 4.5). The datum is zero enthalpy for O_2, N_2, H_2, and C (solid) at 298 K. Note that the specific heats of the unburned and burned mixtures (the slopes of the lines in Fig. 4-2) are a function of temperature; at high temperatures, the internal energy of the burned mixture is a function of temperature and pressure.

However, the temperature range of interest for the unburned mixture in an SI engine is 400 to 900 K (700 to 1600°R); for the burned gas mixture, the extreme end states are approximately 2800 K, 35 atm (5000°R, 500 lb/in² abs) and 1200 K, 2 atm (2200°R, 30 lb/in² abs). Linear approximations to the unburned and burned mixture curves which minimize the error in u over the temperature (and pressure) ranges of interest are shown as dashed lines. The error in T for a given u is less than 50 K.

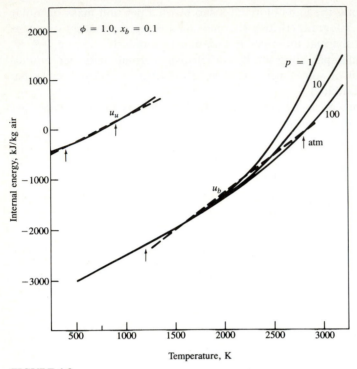

FIGURE 4-2
Internal energy versus temperature plot for stoichiometric unburned and burned gas mixtures: iso-octane fuel; unburned residual fraction 0.1.

The basis for this ideal gas model is

$$u_u = c_{v,u} T_u + h_{f,u} \qquad h_u = c_{p,u} T_u + h_{f,u} \qquad (4.21a, b)$$

$$u_b = c_{v,b} T_b + h_{f,b} \qquad h_b = c_{p,b} T_b + h_{f,b} \qquad (4.22a, b)$$

where $h_{f,u}$ and $h_{f,b}$ are the enthalpies of formation of unburned and burned gas mixture, respectively, at 0 K.

Then, for a constant-volume adiabatic combustion process,

$$u_u = u_b$$

or $$c_{v,u} T_u + h_{f,u} = c_{v,b} T_b + h_{f,b}$$

If we solve for T_b and use the relations $(R_b/R_u) = (M_u/M_b)$ and $c_v/R = 1/(\gamma - 1)$, we obtain

$$T_b = (\gamma_b - 1)\left(\frac{M_b}{M_u}\right)\left(\frac{T_u}{\gamma_u - 1} + \frac{\Delta h_f}{R_u}\right) \qquad (4.23)$$

where $\Delta h_f = h_{f,u} - h_{f,b}$.

For a constant-pressure adiabatic combustion process,

$$h_u = h_b$$

and it can similarly be shown that

$$T_b = \frac{\gamma_b - 1}{\gamma_b} \left(\frac{M_b}{M_u} \right) \left(\frac{\gamma_u}{\gamma_u - 1} T_u + \frac{\Delta h_f}{R_u} \right) \tag{4.24}$$

To use the model, suitable values of γ_u, γ_b, M_u, (M_b/M_u), and $\Delta h_f/R_u$ must be determined. Values for M_u and M_b can be obtained from Eqs. (4.10) and (4.11).† Values of γ_u, γ_b, and $\Delta h_f/R_u$ can be obtained from graphs such as Fig. 4-2 (see Example 4.1 below). Values for γ_u, γ_b, and $\Delta h_f/R_u$ are available in the literature (e.g., Refs. 1 and 2) for a range of ϕ and x_b. However, values used for computations should always be checked over the temperature range of interest, to ensure that the particular linear fit to $u(T)$ used is appropriate.

Example 4.1. Determine the values of γ_u, γ_b, and $\Delta h_f/R_u$ which correspond to the straight-line fits for $u_u(T)$ and $u_b(T)$ in Fig. 4-2.
Equations for the straight lines in Fig. 4-2 are

$$u_u \text{ (kJ/kg air)} = 0.96 T(\text{K}) - 700$$

and

$$u_b \text{ (kJ/kg air)} = 1.5 T(\text{K}) - 4250$$

From Table 4.5, for isooctane fuel with $\phi = 1.0$ and $x_b = 0.1$, the number of moles of unburned mixture per mole O_2 in the mixture is

$$n_u = 0.08 \times 1 + 4.773 + 0.28 \times 0.1 \times 1 = 4.881$$

The mass of air per mole O_2 in the mixture is 138.2. Thus, the number of moles of unburned mixture per unit mass of air in the original mixture is

$$\frac{4.881}{138.2} = 0.0353$$

The molar specific heat of the unburned mixture $\tilde{c}_{v,u}$ is therefore

$$\tilde{c}_{v,u} = \frac{0.96}{0.0353} = 27.2 \text{ kJ/kmol} \cdot \text{K}$$

Since $\tilde{R} = 8.314$ kJ/kmol·K,

$$\gamma_u = \frac{27.2 + 8.314}{27.2} = 1.31$$

The number of moles of burned mixture per mole O_2 is (from Table 4.5)

$$n_b = 0.36 \times 1 + 4.773 = 5.133$$

---

† The error in ignoring the effect of dissociation on M_b is small.

The number of moles of burned mixture per unit mass of air in the original mixture is

$$\frac{5.133}{138.2} = 0.0371$$

The molar specific heat $\tilde{c}_{v,b}$ is therefore

$$\tilde{c}_{v,b} = \frac{1.5}{0.0371} = 40.4 \text{ kJ/kmol} \cdot \text{K}$$

and γ_b is

$$\gamma_b = \frac{40.4 + 8.314}{40.4} = 1.21$$

To find $\Delta h_f / R_u$, R_u is given by

$$R_u = 8.314 \times 0.0353 = 0.293 \text{ kJ/kg air} \cdot \text{K}$$

and so

$$\frac{\Delta h_f}{R_u} = \frac{(-700) - (-4250)}{0.293} = 1.2 \times 10^4 \text{ K}$$

4.5 THERMODYNAMIC CHARTS

One method of presenting thermodynamic properties of unburned and burned gas mixtures for internal combustion engine calculations is on charts. Two sets of charts are in common use: those developed by Hottel et al.[3] and those developed by Newhall and Starkman.[4,5] Both these sets of charts use U.S. units. We have developed a new set of charts in SI units, following the approach of Newhall and Starkman. Charts are no longer used extensively for engine cycle calculations; computer models for the thermodynamic properties of working fluids have replaced the charts. Nonetheless, charts are useful for illustrative purposes, and afford an easy and accurate method where a limited number of calculations are required. The charts presented below are for isooctane fuel, and the following equivalence ratios: $\phi = 0.4, 0.6, 0.8, 1.0, 1.2$.

4.5.1 Unburned Mixture Charts

The thermodynamic properties of each unburned fuel-air mixture are represented by two charts. The first chart is designed to relate the mixture temperature, pressure, and volume at the beginning and at the end of the compression process; the second gives the mixture internal energy and enthalpy as functions of temperature.

The following assumptions are made:

1. The compression process is reversible and adiabatic.

TABLE 4.6
Unburned mixture composition for charts

Equivalence ratio ϕ	(F/A)	Kilograms of mixture per kilogram of air	Moles of mixture per mole of O_2	Kilomole of mixture per kilogram of air	$n_u \tilde{R}$,† J/kg air $\cdot$ K
0.4	0.0264	1.0264	$4.805 + 0.112x_b$	$0.0348 + 0.00081x_b$	289
0.6	0.0396	1.0396	$4.821 + 0.168x_b$	$0.0349 + 0.00122x_b$	290
0.8	0.0528	1.0528	$4.837 + 0.224x_b$	$0.0350 + 0.00162x_b$	291
1.0	0.0661	1.0661	$4.853 + 0.280x_b$	$0.0351 + 0.00203x_b$	292
1.2	0.0792	1.0792	$4.869 + 0.536x_b$	$0.0352 + 0.00388x_b$	292

† For $x_b = 0$. Error in neglecting x_b is usually small.

2. The fuel is in the vapor phase.
3. The mixture composition is homogeneous and frozen (no reactions between the fuel and air).
4. Each species in the mixture can be modeled as an ideal gas.
5. The burned gas fraction is zero.†

It proves convenient to assign zero internal energy or enthalpy to the unburned mixture at 298.15 K. Internal energy and enthalpies relative to this datum are called *sensible internal energy u_s or sensible enthalpy h_s*. By sensible we mean changes in u or h which result from changes in temperature alone, and we exclude changes due to chemical reaction or phase change.

Table 4.6 provides the basic composition data for the unburned mixture charts. Equations (4.13a, b) provide the basis for obtaining the $u_{s,u}(T)$ and $h_{s,u}(T)$ curves shown in Fig. 4-3.

Equations (4.15) and (4.16) provide the basis for following a reversible adiabatic (i.e., isentropic) compression process. Between end states 1 and 2, we obtain, per kilogram of air in the mixture,

$$\Psi(T_2) = \Psi(T_1) - n_u \tilde{R} \ln \left(\frac{v_2}{v_1}\right) \tag{4.25a}$$

$$\Phi(T_2) = \Phi(T_1) + n_u \tilde{R} \ln \left(\frac{p_2}{p_1}\right) \tag{4.25b}$$

where n_u is the number of moles of unburned mixture per kilogram of air. Values

† This assumption introduces negligible error into calculations of the compression process for mixtures with normal burned gas fractions, since the major constituent of the residual is N_2. The burned gas fraction must, however, be included when the unburned mixture properties are related to burned mixture properties in a combustion process.

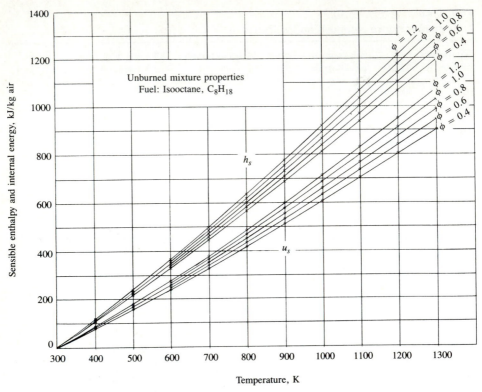

FIGURE 4-3
Sensible enthalpy and internal energy of unburned isooctane-air mixtures as function of temperature.
Units: kJ/kg air in mixture.

of n_u and $n_u \tilde{R}$ are given in Table 4.6. $\Psi(T)$ and $\Phi(T)$ are given in Fig. 4-4. Note that v, p, and T are related by

$$p(\text{Pa})v(\text{m}^3/\text{kg air}) = n_u \tilde{R}(\text{J/kg air} \cdot \text{K})T(\text{K}) \tag{4.26}$$

Example 4.2. The compression process in an internal combustion engine can be modeled approximately as adiabatic and reversible (i.e., as an isentropic process). A spark-ignition engine with a compression ratio of 8 operates with a stochiometric fuel vapor–air mixture which is at 350 K and 1 atm at the start of the compression stroke. Find the temperature, pressure, and volume per unit mass of air at the end of the compression stroke. Calculate the compression stroke work.

Given $T_1 = 350$ K at the start of compression, find T_2 at the end of compression using the isentropic compression chart, Fig. 4-4, and Eq. (4.25a). For $T_1 = 350$ K, $\Psi_1 = 150$ J/kg air $\cdot$ K. From Eq. (4.25a),

$$\Psi_2(T_2) = \Psi_1(T_1) - n_u \tilde{R} \ln\left(\frac{v_2}{v_1}\right) = 150 - 292 \ln\left(\frac{1}{8}\right) = 757 \text{ J/kg air} \cdot \text{K}$$

Figure 4-4 then gives

$$T_2 = 682 \text{ K}$$

The ideal gas law [Eq. (4.26)] gives

$$v_1 = \frac{292 \times 350}{1 \times 1.013 \times 10^5} = 1.0 \text{ m}^3/\text{kg air}$$

and

$$p_2 = p_1\left(\frac{T_2}{T_1}\right)\left(\frac{v_1}{v_2}\right) = \frac{682}{350} \times 8 = 15.5 \text{ atm}$$

$$v_2 = \frac{1.0}{8} = 0.125 \text{ m}^3/\text{kg air}$$

Note that p_2 can also be obtained from Fig. 4-4 and Eq. (4.25*b*):

$$\ln\left(\frac{p_2}{p_1}\right) = \frac{\Phi_2 - \Phi_1}{n_u \tilde{R}} = \frac{980 - 180}{292} = 2.74$$

$$p_2 = 15.5 \text{ atm} = 1.57 \text{ MPa}$$

The compression stroke work, assuming the process is adiabatic and using the data in Fig. 4-3, is

$$-W_{1-2} = u_s(T_2) - u_s(T_1) = 350 - 40 = 310 \text{ kJ/kg air}$$

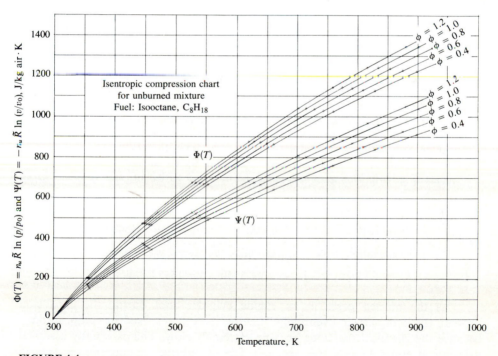

FIGURE 4-4
Isentropic compression functions, Φ and Ψ, as function of temperature for unburned isooctane-air mixtures. Units: J/kg air · K.

4.5.2 Burned Mixture Charts

The primary burned mixture charts are for the products of combustion at high temperatures, i.e., for the working fluid during the expansion process. The following assumptions are made:

1. Each species in the mixture can be modeled as an ideal gas.
2. The mixture is in thermodynamic equilibrium at temperatures above 1700 K; the mixture composition is frozen below 1700 K.
3. *Datum.* At the datum state of 298.15 K (25°C or 77°F) and 1 atm the chemical elements in their naturally occurring form (N_2, O_2, H_2 as diatomic gases and C as solid graphite) are assigned zero enthalpy and entropy.

The charts were prepared with the NASA equilibrium program described in Sec. 3.7.[9,10] The C/H/O/N ratio of the mixture is specified for each chart. The extensive properties (internal energy, enthalpy, entropy, and specific volume) are all expressed per unit mass of air in the original mixture; i.e., they correspond to the combustion of 1 kg of air with the appropriate mass of fuel. The mass basis for the unburned and burned mixture charts are the same.

Figures 4-5 to 4-9 are property charts for the high-temperature burned gas; each is a plot of internal energy versus entropy for a particular fuel and equivalence ratio. Lines of constant temperature, pressure, and specific volume are drawn on each chart. An illustration of the use of these charts follows.

Example 4.3. The expansion process in an internal combustion engine, following completion of combustion, can be modeled approximately as an adiabatic and reversible process (i.e., isentropic). Under full-load operation, the pressure in the cylinder of a spark-ignition engine at top-center immediately following combustion is 7100 kPa. Find the gas state at the end of the expansion stroke and the expansion stroke work. The compression ratio is 8, the mixture is stoichiometric, and the volume per unit mass of air at the start of expansion is 0.125 m³/kg air.

Locate $p_1 = 7100$ kPa and $v_1 = 0.125$ m³/kg air on the $\phi = 1.0$ burned gas chart (Fig. 4-8). This gives $T_1 = 2825$ K, $u_1 = -5$ kJ/kg air, and $s_1 = 9.33$ kJ/kg air · K. The gas expands at constant entropy to $v_2 = 8 \times v_1 = 1$ m³/kg air. Following a constant entropy process from state 1 on Fig. 4-8 gives

$$T_2 = 1840 \text{ K}, \qquad p_2 = 570 \text{ kPa}, \qquad \text{and} \qquad u_2 = -1540 \text{ kJ/kg air}$$

The expansion stroke work, assuming the process is adiabatic, is

$$W_{1-2} = -(u_2 - u_1) = 1540 - 5 = 1535 \text{ kJ/kg air}$$

As the burned gases in an engine cylinder cool during the expansion process, the composition eventually "freezes"—becomes fixed in composition—because the chemical reactions become extremely slow. This is usually assumed to occur at about 1700 K (see Sec. 4.9). The equilibrium assumption is then no longer valid. For lean and stoichiometric mixtures this distinction is not important because the mole fractions of dissociated species below this temperature are

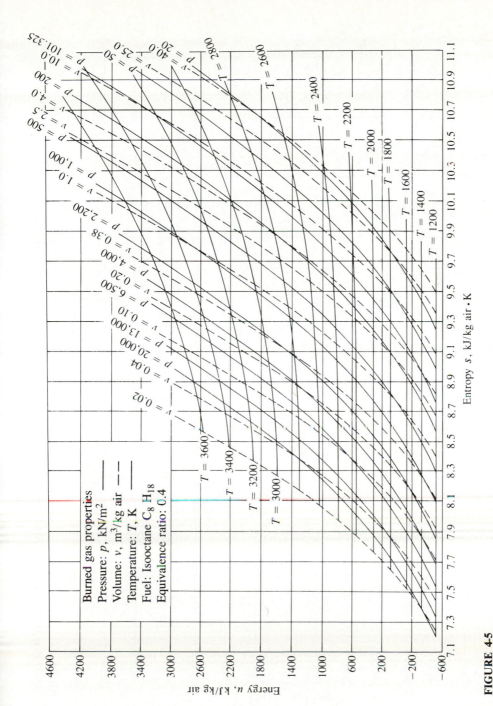

FIGURE 4-5
Internal energy versus entropy chart for equilibrium burned gas mixture, isooctane fuel; equivalence ratio 0.4.

117

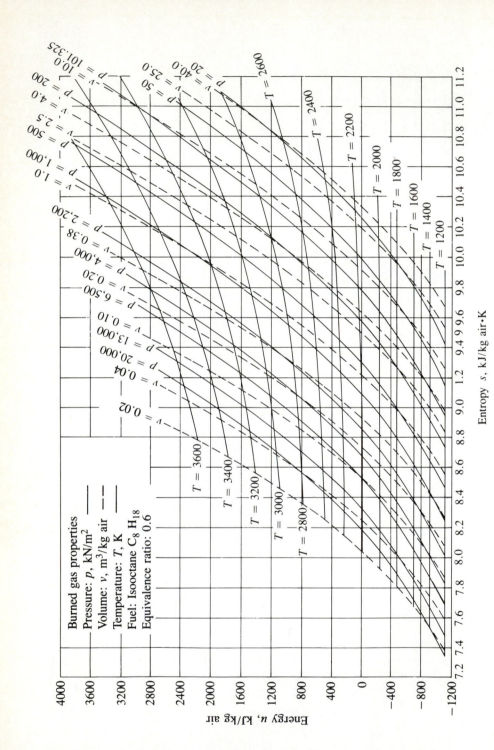

FIGURE 4-6

Internal energy versus entropy chart for equilibrium burned gas mixture, isooctane fuel; equivalence ratio 0.6.

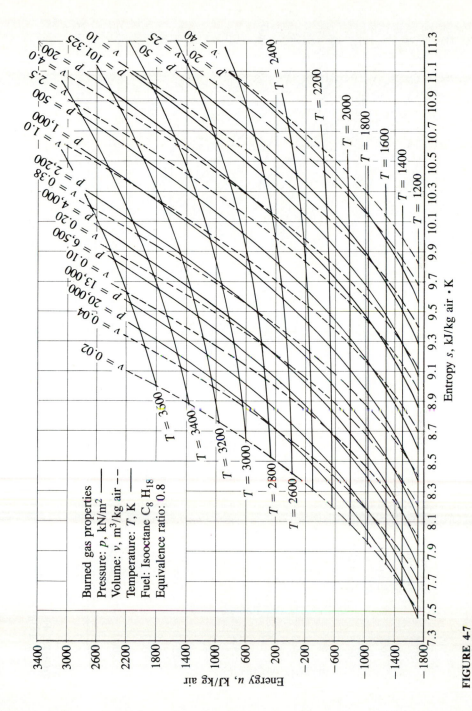

FIGURE 4-7
Internal energy versus entropy chart for equilibrium burned gas mixture, isooctane fuel; equivalence ratio 0.8.

119

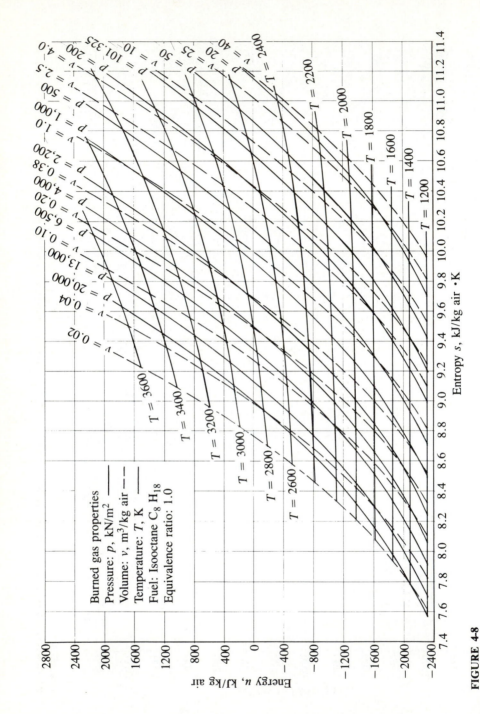

FIGURE 4-8

Internal energy versus entropy chart for equilibrium burned gas mixture, isooctane fuel; equivalence ratio 1.0.

120

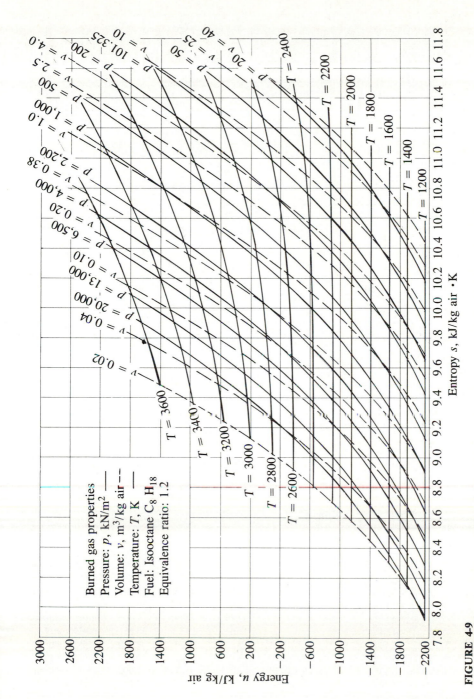

FIGURE 4-9
Internal energy versus entropy chart for equilibrium burned gas mixture, isooctane fuel; equivalence ratio 1.2.

121

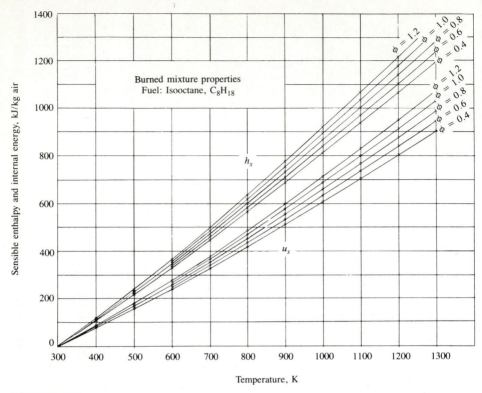

FIGURE 4-10
Sensible enthalpy and internal energy of low-temperature burned gases as function of temperature, isooctane fuel. Units: kJ/kg air in original mixture.

TABLE 4.7
Frozen burned gas composition: C_8H_{18}-air combustion

ϕ	CO_2	H_2O	CO	H_2	O_2	N_2	Sum	Units
0.4	0.0521	0.0586	—	—	0.122	0.767	1.000	mole fractions
	1.85	2.08	—	—	4.34	27.3	35.6	mol/kg air‡
0.6	0.0770	0.0866	—	—	0.0802	0.756	1.000	mole fractions
	2.78	3.13	—	—	2.89	27.3	36.1	mol/kg air‡
0.8	0.101	0.113	—	—	0.0395	0.746	1.000	mole fractions
	3.70	4.14	—	—	1.45	27.3	36.6	mol/kg air‡
1.0	0.125	0.140	—	—	—	0.735	1.000	mole fractions
	4.64	5.2	—	—	—	27.3	37.1	mol/kg air‡
1.2†	0.0905	0.138	0.0516	0.0224	—	0.698	1.000	mole fractions
	3.54	5.38	2.02	0.876	—	27.3	39.1	mol/kg air‡

† $K(T)$ in Eq. (4.6) evaluated at 1740 K; $K = 3.5$.
‡ Note mol/kg air; multiply by 10^{-3} for kmol/kg air.

small. For rich mixtures, a frozen composition must be selected and used because the mole fractions of CO_2, CO, H_2O, and H_2 would continue to change if equilibrium is assumed as the temperature decreases. Internal energy and enthalpy, per kilogram of air in the original mixture, of the frozen burned mixture are plotted against temperature in Fig. 4-10. The assumed frozen burned mixture compositions are listed in Table 4.7. These are sensible internal energies and enthalpies, given relative to their values at 298.15 K.

4.5.3 Relation between Unburned and Burned Mixture Charts

We now address the questions: Given unburned mixture at T_1, p_1, v_1, what is the state of the burned mixture following (1) constant-volume adiabatic combustion or (2) constant-pressure adiabatic combustion?

The datum for internal energy and enthalpy for the unburned mixture in Fig. 4-3 is different from the datum for internal energy and enthalpy for the burned mixture. For the unburned mixture, zero internal energy and enthalpy for the *mixture* at 298.15 K was assumed. For the burned mixture, zero enthalpy for the gaseous species O_2, N_2, and H_2, and C (solid graphite) at 298.15 K was assumed. These data can be related through the enthalpies of formation, from O_2, N_2, H_2, and C, of each species in the unburned mixture.

If $\Delta \tilde{h}^\circ_{f,i}$ is the enthalpy of formation of species i at 298.15 K, per kilomole, and $\Delta h^\circ_{f,u}$ is the enthalpy of formation of the unburned mixture at 298.15 K, per kilogram of air in the original mixture, then

$$\Delta h^\circ_{f,u} = \sum_i n_i \Delta \tilde{h}^\circ_{f,i} \qquad (4.27)$$

where n_i is the number of kilomoles of species i per kilogram of air. The unburned mixture enthalpy h_u, with the same datum as the burned mixture enthalpy, is therefore given by the sum of the sensible enthalpy $h_{s,u}$ and $\Delta h^\circ_{f,u}$:

$$h_u = h_{s,u} + \Delta h^\circ_{f,u} \qquad (4.28)$$

Similarly, the internal energy u_u is given by

$$u_u = u_{s,u} + \Delta u^\circ_{f,u} \qquad (4.29)$$

$\Delta u^\circ_{f,u}$ can be obtained from

$$\Delta u^\circ_{f,u} = \sum_i n_i \Delta \tilde{u}^\circ_{f,i} \qquad (4.30)$$

Alternately, Eq. (3.18) can be used to obtain $\Delta u^\circ_{f,u}$ from $\Delta h^\circ_{f,u}$:

$$\Delta u^\circ_{f,u} = \Delta h^\circ_{f,u} - (n_P - n_R)\tilde{R}T \qquad (4.31)$$

Enthalpies and internal energies of formation of the relevant burned gas species and individual fuel compounds are given in Table 4.8 and App. D. Values of n_i are obtained from Tables 4.4 and 4.7. Following the procedure used in Example 4.4 below, expressions for $\Delta h^\circ_{f,u}$ and $\Delta u^\circ_{f,u}$ in kilojoules per kilogram of

TABLE 4.8
Standard enthalpies and internal energies of formation†

	$\Delta \tilde{h}^{\circ}_{f,i}$, MJ/kmol	$\Delta \tilde{u}^{\circ}_{f,i}$, MJ/kmol
CO_2	-393.5	-393.5
H_2O (gas)	-241.8	-240.6
CO	-110.5	-111.7
C_8H_{18} (gas)	-224.1	-204.3

† At 298.15 K. $\Delta \tilde{h}^{\circ}_{f,i}$ for O_2, N_2, and H_2 are zero by definition.
Sources: JANAF tables,[8] Rossini *et al.*[16]

air can be obtained. For the charts of Figs. 4-3 and 4-5 to 4-9, these expressions are:

$\phi = 0.4$:

$$\Delta h^{\circ}_{f,u} = -51.9 - 1181x_b \qquad \Delta u^{\circ}_{f,u} = -47.3 - 1183x_b$$

$\phi = 0.6$:

$$\Delta h^{\circ}_{f,u} = -77.8 - 1771x_b \qquad \Delta u^{\circ}_{f,u} = -70.9 - 1774x_b$$

$\phi = 0.8$:

$$\Delta h^{\circ}_{f,u} = -103.8 - 2361x_b \qquad \Delta u^{\circ}_{f,u} = -94.6 - 2365x_b \qquad (4.32)$$

$\phi = 1.0$:

$$\Delta h^{\circ}_{f,u} = -129.7 - 2951x_b \qquad \Delta u^{\circ}_{f,u} = -118.2 - 2956x_b$$

$\phi = 1.2$:

$$\Delta h^{\circ}_{f,u} = -155.6 - 2759x_b \qquad \Delta u^{\circ}_{f,u} = -141.9 - 2769x_b$$

Example 4.4. Calculate $\Delta h^{\circ}_{f,u}$, the enthalpy of formation of the unburned mixture, and $\Delta u^{\circ}_{f,u}$, the internal energy of formation of the unburned mixture, for a C_8H_{18}-air mixture with $\phi = 1.0$ and burned gas fraction x_b.

Table 4.4 gives the moles of each species in the unburned mixture, per mole O_2 with $\phi = 1.0$, as

$$C_8H_{18}, 0.08(1 - x_b) \qquad CO_2, 0.64x_b$$

$$O_2, 1 - x_b \qquad H_2O, 0.72x_b$$

$$N_2, 3.773 \qquad CO \text{ and } H_2, 0$$

Table 4.5 gives the mass of air per mole O_2 as 138.2 kg/kmol. Thus the number of kilomoles of each species per kilogram of air is

$$C_8H_{18}, 5.787 \times 10^{-4} (1 - x_b) \qquad CO_2, 4.629 \times 10^{-3}x_b$$

$$O_2, 7.233 \times 10^{-3} (1 - x_b) \qquad H_2O, 5.208 \times 10^{-3}x_b$$

$$N_2, 2.729 \times 10^{-2} \qquad CO \text{ and } H_2, 0$$

With $\Delta \tilde{h}_{f,i}^{\circ}$ from Table 4.8, Eq. (4.27) gives

$$\Delta h_{f,u}^{\circ} = 5.787 \times 10^{-4} \times (-224.1 \times 10^{6})(1 - x_b)$$
$$+ x_b[4.629 \times 10^{-3} \times (-393.5 \times 10^{6}) + 5.208 \times 10^{-3} \times (-241.8 \times 10^{6})]$$

$$\Delta h_{f,u}^{\circ} = (-129.7 - 2951x_b) \times 10^{3} \qquad \text{J/kg air}$$

With $\Delta \tilde{u}_{f,i}^{\circ}$ from Table 4.8, Eq. (4.30) gives

$$\Delta u_{f,u}^{\circ} = 5.787 \times 10^{-4} \times (-204.3 \times 10^{6})(1 - x_b)$$
$$+ x_b[4.629 \times 10^{-3} \times (-393.5 \times 10^{6}) + 5.208 \times 10^{-3} \times (-240.6 \times 10^{6})]$$

$$\Delta u_{f,u}^{\circ} = (-118.2 - 2956x_b) \times 10^{3} \qquad \text{J/kg air}$$

Alternatively, we can determine $\Delta u_{f,u}^{\circ}$ from $\Delta h_{f,u}^{\circ}$ using Eq. (4.31). For this calculation, the "product" gas is the unburned mixture and the "reactant" gas is the mixture of elements from which the unburned mixture is formed. The number of gaseous moles in the unburned mixture n_P, per mole O_2 in the original mixture, is (from Table 4.5 for $\phi \leq 1$)

$$n_P = (1 - x_b)\left[4(1 + 2\varepsilon)\frac{\phi}{M_f} + 1 + \psi\right] + x_b[(1 - \varepsilon)\phi + 1 + \psi]$$

The elemental reactant mixture from which the unburned mixture is formed is, from Eq. (4.4),

$$\varepsilon\phi C + 2(1 - \varepsilon)\phi H_2 + O_2 + \psi N_2$$

Thus, n_R, the moles of *gaseous* elements, is

$$n_R = 2(1 - \varepsilon)\phi + 1 + \psi$$

For air, $\psi = 3.773$; for C_8H_{18} fuel, $\varepsilon = 0.64$ and $M_f = 114$. For $\phi = 1$,

$$n_P - n_R = -0.64 + 0.28x_b \qquad \text{moles/mole } O_2$$

and

$$(n_P - n_R)\tilde{R}T = (-0.64 + 0.28x_b) \times 8.3143 \times 10^{3} \times \frac{298.15}{138.2}$$

or

$$(n_P - n_R)\tilde{R}T = (-11.5 + 5.0x_b) \times 10^{3} \qquad \text{J/kg air}$$

Since

$$\Delta u_{f,u}^{\circ} = \Delta h_{f,u}^{\circ} - (n_P - n_R)\tilde{R}T$$
$$\Delta u_{f,u}^{\circ} = (-129.7 - 2951x_b) \times 10^{3} - (-11.5 + 5.0x_b) \times 10^{3}$$
$$\Delta u_{f,u}^{\circ} = (-118.2 - 2956x_b) \times 10^{3} \qquad \text{J/kg air}$$

The combustion process links the unburned and burned mixture properties as follows:

For an adiabatic constant-volume combustion process,

$$u_b = u_u = u_{s,u} + \Delta u_{f,u}^{\circ} \qquad (4.33)$$

and

$$v_b = v_u$$

Thus, given $u_{s,u}$ and v_u, the state of the burned mixture can be determined from the appropriate burned mixture chart.

For an adiabatic constant-pressure combustion process,

$$h_b = h_u = h_{s,u} + \Delta h^\circ_{f,u} \tag{4.34}$$

Since

$$u_b = h_b - pv_b$$

given $h_{s,u}$ and p, u_b and v_b must be found by trial and error along the specified constant-pressure line on the appropriate burned mixture chart.

Example 4.5. Calculate the temperature and pressure after constant-volume adiabatic combustion and constant-pressure adiabatic combustion of the unburned mixture (with $\phi = 1.0$ and $x_b = 0.08$) at the state corresponding to the end of the compression process examined in Example 4.2.

The state of the unburned mixture at the end of the compression process in Example 4.2 was

$$T_u = 682 \text{ K}, \qquad u_{s,u} = 350 \text{ kJ/kg air}, \qquad p_u = 1.57 \text{ MPa}, \qquad v_u = 0.125 \text{ m}^3/\text{kg air}$$

For an adiabatic *constant-volume* combustion process [Eq. (4.33)],

$$u_b = u_u = u_{s,u} + \Delta u^\circ_{f,u}$$

For $\phi = 1.0$, $\Delta u^\circ_{f,u}$ is given by Eq. (4.32) as

$$\Delta u^\circ_{f,u} = -118.2 - 2956x_b = -118.2 - 236.5 = -355 \text{ kJ/kg air}$$

Hence

$$u_b = 350 - 355 = -5 \text{ kJ/kg air}$$

Also

$$v_b = v_u = 0.125 \text{ m}^3/\text{kg air}$$

Locating (u_b, v_b) on the burned gas chart (Fig. 4-8) gives

$$T_b = 2825 \text{ K}, \qquad p_b = 7100 \text{ kPa}$$

For a *constant-pressure* combustion process [Eq. (4.34)],

$$h_b = h_u = h_{s,u} + \Delta h^\circ_{f,u}$$

For $\phi = 1.0$, $\Delta h^\circ_{f,u}$ is given by Eq. (4.32) as

$$\Delta h^\circ_{f,u} = -129.7 - 2951x_b = -129.7 - 236 = -366 \text{ kJ/kg air}$$

At $T_u = 682 \text{ K}$, $h_{s,u} = 465 \text{ kJ/kg air}$, so

$$h_b = 465 - 366 = 99 \text{ kJ/kg air}$$

Since $p_b = p_u = 1.57 \text{ MPa}$, the internal energy u_b is given by

$$u_b = h_b - p_b v_b = 99 - 1.57 \times 10^3 v_b \qquad \text{kJ/kg air}$$

A trial-and-error solution for v_b and u_b along the $p = 1570$ kPa line on Fig. 4-8 gives

$$u_b = -655 \text{ kJ/kg air}, \qquad T_b = 2440 \text{ K}, \qquad v_b = 0.485 \text{ m}^3/\text{kg air}$$

(Use the ideal gas law to estimate p, T, or v more accurately.)

4.6 TABLES OF PROPERTIES AND COMPOSITION

Tables of thermodynamic properties of air are useful for analysis of motored engine operation, diesels and compressors. Keenan, Chao, and Kaye's *Gas Tables*[6] are the standard reference for the thermodynamic properties of air at low pressures (i.e., at pressures substantially below the critical pressure when the ideal gas law is accurate). These gas tables are in U.S. and SI units. A set of tables for air in SI units has been prepared by Reynolds[7] following the format of the Keenan *et al.* tables. A condensed table of thermodynamic properties of air, derived from Reynolds, is given in App. D. It contains:

h = enthalpy, kJ/kg
u = internal energy, kJ/kg
$$\Psi = \int_0^T \left(\frac{c_v}{T}\right) dT, \text{ kJ/kg} \cdot \text{K}$$
$$\Phi = \int_0^T \left(\frac{c_p}{T}\right) dT, \text{ kJ/kg} \cdot \text{K}$$
p_r = relative pressure
v_r = relative volume
c_p = specific heat at constant pressure, kJ/kg $\cdot$ K
c_v = specific heat at constant volume, kJ/kg $\cdot$ K
γ = ratio of specific heats

all as a function of T(K).

Φ is the standard state entropy at temperature T and 1 atm pressure, relative to the entropy at 0 K and 1 atm pressure. The entropy at pressures other than 1 atm is obtained using Eq. (4.14b).

The relative pressure p_r is defined by

$$\ln p_r = \frac{\Phi}{R} \tag{4.35}$$

and is a function of T only. Along a given isentropic, it follows from Eq. (4.18) that the ratio of actual pressures p_2 and p_1 corresponding to temperatures T_2 and T_1 is equal to the ratio of relative pressures, i.e.,

$$\left(\frac{p_2}{p_1}\right)_{s=\text{const}} = \left(\frac{p_{r_2}}{p_{r_1}}\right) \tag{4.36}$$

This affords a means of determining T_2, for an isentropic process, given T_1 and p_2/p_1 (see Example 4.6).

The relative volume v_r is defined by

$$v_r = \frac{RT}{p_r} \tag{4.37}$$

The units are selected so that v_r is in cubic meters per kilogram when T is in kelvins and p_r is in pascals. Along a given isentropic, the ratio of actual volumes V_2 and V_1 (for a fixed mass) at temperatures T_2 and T_1, from Eq. (4.37), is equal to the ratio of relative volumes

$$\left(\frac{V_2}{V_1}\right)_{s=\text{const}} = \left(\frac{v_{r_2}}{v_{r_1}}\right) \tag{4.38}$$

This affords a means of determining T_2 for an isentropic process, given T_1 and V_2/V_1 (see Example 4.6).

Tables giving the composition and thermodynamic properties of combustion products have been compiled. They are useful sources of property and species concentrations data in burned gas mixtures for a range of equivalence ratios, temperatures, and pressures. Summary information on four generally available sets of tables is given in Table 4.9. The most extensive set of tables of combustion product composition and thermodynamic properties is the AGARD set, *Properties of Air and Combustion Products with Kerosene and Hydrogen Fuels*, by Banes et al.[12] Note, however, that their enthalpy datum differs from the usual datum (enthalpy for O_2, N_2, H_2, and C is zero at 298.15 K). The elements in their reference state at 298.15 K were assigned arbitrary positive values for enthalpy to avoid negative enthalpies for the equilibrium burned gas mixture.

Example 4.6. In a diesel engine, the air conditions at the start of compression are $p_1 = 1$ atm and $T_1 = 325$ K. At the end of compression $p_2 = 60$ atm. Find the temperature T_2 and the compression ratio V_1/V_2.

Air tables (see App. D), at $T_1 = 325$ K, give

$$p_{r_1} = 97.13 \quad \text{and} \quad v_{r_1} = 960.6$$

Use Eq. (4.36),

$$\frac{p_{r_2}}{p_{r_1}} = \frac{p_2}{p_1} = 60$$

to give

$$p_{r_2} = 5828$$

Tables then give

$$T_2 = 992 \text{ K} \quad \text{and} \quad v_{r_2} = 48.92$$

The compression ratio is given by

$$\frac{V_1}{V_2} = \frac{v_{r_1}}{v_{r_2}} = \frac{960.6}{48.92} = 19.6$$

TABLE 4.9
Tables of properties of air and combustion products

Source	Properties P, composition C	Mixture	Units	ϕ range	T range	p range	Enthalpy datum
Keenan, Chao, and Kaye[6]	P	Air	U.S., SI		100–3600 K	Low	$h = 0$ at 0 K
		$(CH_2)_n$-air		0.25, 0.5, 1.0	100–2000 K	Low	$\tilde{h}_b = 0$ at 0 K
Reynolds[7]	P	Air	SI		200–1500 K	Low	
General Electric[11]	P and C	Air-$(CH_2)_n$	U.S.	0.25–4	600–5000°R	0.01–30 atm	h of C, H_2, N_2, O_2 zero at 0°R
AGARD[12]	P and C	Air	SI		100–6000 K	1–800 atm	Arbitrary, to keep $h_b > 0$
		$(CH_2)_n$-air		0.2–2			
		H_2-air		0.2–2			

129

4.7 COMPUTER ROUTINES FOR PROPERTY AND COMPOSITION CALCULATIONS

When large numbers of computations are being made or high accuracy is required, engine process calculations are carried out on a computer. Relationships which model the composition and/or thermodynamic properties of unburned and burned gas mixtures have been developed for computer use. These vary considerably in range of application and accuracy.

The most complete models are based on polynomial curve fits to the thermodynamic data for each species in the mixture and the assumptions that (1) the unburned mixture is frozen in composition and (2) the burned mixture is in equilibrium. The approach used as the basis for representing JANAF table thermodynamic data[8] in the NASA equilibrium program[9,10] (see Sec. 3.7) will be summarized here because it is consistent with the approach used throughout to calculate unburned and burned mixture properties.

For each species i in its standard state at temperature $T(K)$, the specific heat $\tilde{c}_{p,i}$ is approximated by

$$\frac{\tilde{c}_{p,i}}{\tilde{R}} = a_{i1} + a_{i2}\,T + a_{i3}\,T^2 + a_{i4}\,T^3 + a_{i5}\,T^4 \tag{4.39}$$

The standard state enthalpy of species i is then given by

$$\frac{\tilde{h}_i}{\tilde{R}T} = a_{i1} + \frac{a_{i2}}{2}\,T + \frac{a_{i3}}{3}\,T^2 + \frac{a_{i4}}{4}\,T^3 + \frac{a_{i5}}{5}\,T^4 + \frac{a_{i6}}{T} \tag{4.40}$$

The standard state entropy of species i at temperature $T(K)$ and pressure 1 atm, from Eq. (4.14), is then

$$\frac{\tilde{s}_i}{\tilde{R}} = a_{i1}\ln T + a_{i2}\,T + \frac{a_{i3}}{2}\,T^2 + \frac{a_{i4}}{3}\,T^3 + \frac{a_{i5}}{4}\,T^4 + a_{i7} \tag{4.41}$$

Values of the coefficients a_{ij} for CO_2, H_2O, CO, H_2, O_2, N_2, OH, NO, O, and H from the NASA program are given in Table 4.10. Two temperature ranges are given. The 300 to 1000 K range is appropriate for unburned mixture property calculations. The 1000 to 5000 K range is appropriate for burned mixture property calculations. Figure 4-11 gives values of c_p/R for the major species, CO_2, H_2O, O_2, N_2, H_2, and CO, as a function of temperature.

4.7.1 Unburned Mixtures

Polynomial functions for various fuels (in the vapor phase) have been fitted to the functional form:[13–15]

$$\tilde{c}_{p,f} = A_{f1} + A_{f2}\,t + A_{f3}\,t^2 + A_{f4}\,t^3 + \frac{A_{f5}}{t^2} \tag{4.42}$$

TABLE 4.10
Coefficients for species thermodynamic properties

Species	T range, K	a_{i1}	a_{i2}	a_{i3}	a_{i4}	a_{i5}	a_{i6}	a_{i7}
CO_2	1000–5000	0.44608(+1)	0.30982(−2)	−0.12393(−5)	0.22741(−9)	−0.15526(−13)	−0.48961(+5)	−0.98636(0)
	300–1000	0.24008(+1)	0.87351(−2)	−0.66071(−5)	0.20022(−8)	0.63274(−15)	−0.48378(+5)	0.96951(+1)
H_2O	1000–5000	0.27168(+1)	0.29451(−2)	−0.80224(−6)	0.10227(−9)	−0.48472(−14)	−0.29906(+5)	0.66306(+1)
	300–1000	0.40701(+1)	−0.11084(−2)	0.41521(−5)	−0.29637(−8)	0.80702(−12)	−0.30280(+5)	−0.32270(0)
CO	1000–5000	0.29841(+1)	0.14891(−2)	−0.57900(−6)	0.10365(−9)	−0.69354(−14)	−0.14245(+5)	0.63479(+1)
	300–1000	0.37101(+1)	−0.16191(−2)	0.36924(−5)	−0.20320(−8)	0.23953(−12)	−0.14356(+5)	0.29555(+1)
H_2	1000–5000	0.31002(+1)	0.51119(−3)	0.52644(−7)	−0.34910(−10)	0.36945(−14)	−0.87738(+3)	−0.19629(+1)
	300–1000	0.30574(+1)	0.26765(−2)	−0.58099(−5)	0.55210(−8)	−0.18123(−11)	−0.98890(+3)	−0.22997(+1)
O_2	1000–5000	0.36220(+1)	0.73618(−3)	−0.19652(−6)	0.36202(−10)	−0.28946(−14)	−0.12020(+4)	0.36151(+1)
	300–1000	0.36256(+1)	−0.18782(−2)	0.70555(−5)	−0.67635(−8)	0.21556(−11)	−0.10475(+4)	0.43053(+1)
N_2	1000–5000	0.28963(+1)	0.15155(−2)	−0.57235(−6)	0.99807(−10)	−0.65224(−14)	−0.90586(+3)	0.61615(+1)
	300–1000	0.36748(+1)	−0.12082(−2)	0.23240(−5)	−0.63218(−9)	−0.22577(−12)	−0.10612(+4)	0.23580(+1)
OH	1000–5000	0.29106(+1)	0.95932(−3)	−0.19442(−6)	0.13757(−10)	0.14225(−15)	0.39354(+4)	0.54423(+1)
NO	1000–5000	0.31890(+1)	0.13382(−2)	−0.52899(−6)	0.95919(−10)	−0.64848(−14)	0.98283(+4)	0.67458(+1)
O	1000–5000	0.25421(+1)	−0.27551(−4)	−0.31028(−8)	0.45511(−11)	−0.43681(−15)	0.29231(+5)	0.49203(+1)
H	1000–5000	0.25(+1)	0.0	0.0	0.0	0.0	0.25472(+5)	−0.46012(0)

Source: NASA Equilibrium Code.[9]

131

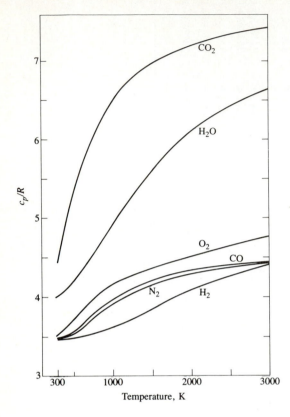

FIGURE 4-11
Specific heat at constant pressure, c_p/R, as function of temperature for species CO_2, H_2O, O_2, N_2, H_2, and CO. (*From JANAF tables.*[8])

$$\tilde{h}_f = A_{f1} t + A_{f2} \frac{t^2}{2} + A_{f3} \frac{t^3}{3} + A_{f4} \frac{t^4}{4} - \frac{A_{f5}}{t} + A_{f6} + A_{f8} \qquad (4.43)$$

where $t = T(K)/1000$. A_{f6} is the constant for the datum of zero enthalpy for C, H_2, O_2, and N_2 at 298.15 K. For a 0 K datum, A_{f8} is added to A_{f6}. For pure hydrocarbon compounds, the coefficients A_{fi} were found by fitting Eqs. (4.42) and (4.43) to data from Rossini *et al.*[16] Values for relevant pure fuels are given in Table 4.11. The units for $\tilde{c}_{p,f}$ are cal/gmol · K, and for $\tilde{h}_f$ are kcal/gmol.

Multicomponent fuel coefficients were determined as follows.[14] Chemical analysis of the fuel was performed to obtain the H/C ratio, average molecular weight, heating value, and the weight percent of aromatics, olefins, and total paraffins (including cycloparaffins). The fuel was then modeled as composed of a representative aromatic, olefin, and paraffin hydrocarbon. From atomic conservation of hydrogen and carbon and the chemical analysis results, component molar fractions and average carbon numbers can be determined. Table 4.11 gives values for the coefficients A_{f1} to A_{f8} for typical petroleum-based fuels. The units of the coefficients give $\tilde{c}_{p,f}$ and $\tilde{h}_f$ in cal/gmol · K and kcal/gmol, respectively, with $t = T(K)/1000$.

TABLE 4.11

Coefficients for polynomials [Eqs. (4.42) and (4.43)] for fuel enthalpy and specific heat

Fuel	Formula	Molecular weight	$(A/F)_s$	$(F/A)_s$	A_{f1}	A_{f2}	A_{f3}	A_{f4}	A_{f5}	A_{f6}	A_{f8}
Methane	CH_4	16.04	17.23	0.0580	−0.29149	26.327	−10.610	1.5656	0.16573	−18.331	4.3000
Propane	C_3H_8	44.10	15.67	0.0638	−1.4867	74.339	−39.065	8.0543	0.01219	−27.313	8.852
Hexane	C_6H_{14}	86.18	15.24	0.0656	−20.777	210.48	−164.125	52.832	0.56635	−39.836	15.611
Isooctane	C_8H_{18}	114.2	15.14	0.0661	−0.55313	181.62	−97.787	20.402	−0.03095	−60.751	20.232
Methanol	CH_3OH	32.04	6.47	0.1546	−2.7059	44.168	−27.501	7.2193	0.20299	−48.288	5.3375
Ethanol	C_2H_5OH	46.07	9.00	0.1111	6.990	39.741	−11.926	0	0	−60.214	7.6135
Gasoline	$C_{8.26}H_{15.5}$	114.8	14.64	0.0683	−24.078	256.63	−201.68	64.750	0.5808	−27.562	17.792
	$C_{7.76}H_{13.1}$	106.4	14.37	0.0696	−22.501	227.99	−177.26	56.048	0.4845	−17.578	15.235
Diesel	$C_{10.8}H_{18.7}$	148.6	14.4	0.0694	−9.1063	246.97	−143.74	32.329	0.0518	−50.128	23.514

Units of A_{fi} such that $\bar{h}_f$ is in kcal/gmol and $\bar{c}_{p,f}$ is in cal/gmol·K with $t = T(K)/1000$.

A_{f6} gives enthalpy datum at 298.15 K; $(A_{f6} + A_{f8})$ gives enthalpy datum at 0 K.

133

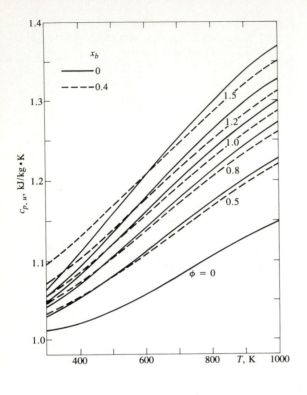

FIGURE 4-12
Specific heat at constant pressure of unburned gasoline, air, burned gas mixtures as function of temperature, equivalence ratio, and burned gas fraction. Units: kJ/kg mixture·K.

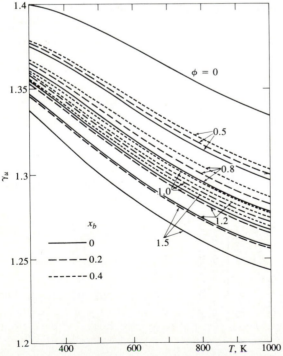

FIGURE 4-13
Ratio of specific heats, $\gamma_u = c_{p,u}/c_{v,u}$, of unburned gasoline, air, burned gas mixtures as function of temperature, equivalence ratio, and burned gas fraction.

The thermodynamic properties of the unburned mixture can now be obtained. With the moles of each species per mole O_2, n_i, determined from Table 4.4, and the mass of mixture per mole O_2, m_{RP}, determined from Table 4.5, the unburned mixture properties are given by

$$c_{p,u} = \frac{1}{m_{RP}} \sum_i n_i \tilde{c}_{p,i} \tag{4.44a}$$

$$h_u = \frac{1}{m_{RP}} \sum_i n_i \tilde{h}_i \tag{4.44b}$$

$$s_u = \frac{1}{m_{RP}} \left\{ \sum_i n_i \left[\tilde{s}_i^\circ - \tilde{R} \ln \left(\frac{n_i}{n_u} \right) \right] - n_u \tilde{R} \ln p \right\} \tag{4.44c}$$

where p is in atmospheres.

Figures 4-12 and 4-13, obtained with the above relations, show how $c_{p,u}$ and $\gamma_u (= c_{p,u}/c_{v,u})$ vary with temperature, equivalence ratio, and burned gas fraction, for a gasoline-air mixture.

4.7.2 Burned Mixtures

The most accurate approach for burned mixture property and composition calculations is to use a thermodynamic equilibrium program at temperatures above about 1700 K and a frozen composition below 1700 K. The properties of each species at high and low temperatures are given by polynomial functions such as Eqs. (4.39) to (4.41) and their coefficients in Table 4.10. The NASA equilibrium program (see Sec. 3.7) is readily available for this purpose and is well documented.[9, 10] The following are examples of its output.

Figure 3-10 showed species concentration data for burned gases as a function of equivalence ratio at 1750, 2250, and 2750 K, at 30 atm. Figure 4-14 shows the burned gas molecular weight M_b, and Figs. 4-15 and 4-16 give $c_{p,b}$ and γ_b as functions of equivalence ratio at 1750, 2250, and 2750 K, at 30 atm. Figures 4-17 and 4-18 show $c_{p,b}$ and γ_b as a function of temperature and pressure for selected equivalence ratios for mixtures lean and rich of stoichiometric.[17] For rich mixtures ($\phi > 1$), for $T > 2000$ K, $c_{p,b}$ and γ_b are equilibrium values. For 1200 K $\leq T \leq 2000$ K, "frozen" composition data are shown where the gas composition is in equilibrium at the given T and p but is frozen as c_p and c_v are computed. Below about 1500 K, fixed composition data are shown corresponding to a value of 3.5 for the water-gas equilibrium constant which adequately describes exhaust gases (see Sec. 4.9).

Because the computational time involved in repeated use of a full equilibrium program can be substantial, simpler equilibrium programs and approximate fits to the equilibrium thermodynamic data have been developed. The approach usually used is to estimate the composition and/or properties of undissociated combustion products and then to use iterative procedures or corrections to account for the effects of dissociation.

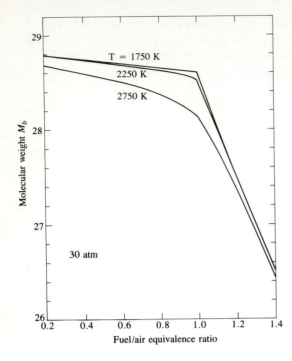

FIGURE 4-14
Molecular weight of equilibrium burned gases as a function of equivalence ratio at $T = 1750$, 2250, and 2750 K, and 30 atm. Fuel: isooctane.

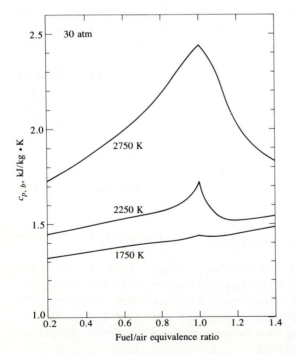

FIGURE 4-15
Specific heat at constant pressure of equilibrium burned gases as a function of equivalence ratio at $T = 1750$, 2250, and 2750 K, and 30 atm. Fuel: isooctane. Units: kJ/kg mixture·K.

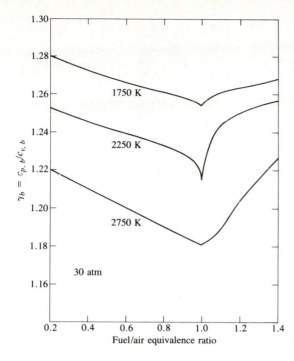

FIGURE 4-16
Ratio of specific heats, $\gamma_b = c_{p,b}/c_{v,b}$, for equilibrium burned gases as a function of equivalence ratio at $T = 1750$, 2250, and 2750 K, and 30 atm. Fuel: iso-octane.

A computer program for calculating properties of equilibrium combustion products, designed specifically for use in internal combustion engine applications, has been developed by Olikara and Borman and is readily available.[18] The fuel composition ($C_nH_mO_lN_k$), fuel/air equivalence ratio, and product pressure and temperature are specified. The species included in the product mixture are: CO_2, H_2O, CO, H_2, O_2, N_2, Ar, NO, OH, O, H, and N. The element balance equations and equilibrium constants for seven nonredundant reactions provide the set of 11 equations required for solution of these species concentrations (see Sec. 3.7). The equilibrium constants are curve fitted from data in the JANAF tables.[8] The initial estimate of mole fractions to start the iteration procedure is the non-dissociated composition. Once the mixture composition is determined, the thermodynamic properties and their derivatives with respect to temperature, pressure, and equivalence ratio are computed. This limited set of species has been found to be sufficiently accurate for engine burned gas calculations, and is much more rapid than the extensive NASA equilibrium program.[9, 10]

Several techniques for estimating the *thermodynamic properties* of high-temperature burned gases for engine applications have been developed. One commonly used approach is that developed by Krieger and Borman.[19] The internal energy and gas constant of undissociated combustion products were first described by polynomials in gas temperature. The second step was to limit the range of T and p to values found in internal combustion engines. Then the deviations between the equilibrium thermodynamic property data published by Newhall and Starkman[4, 5] and the calculated nondissociated values were fitted

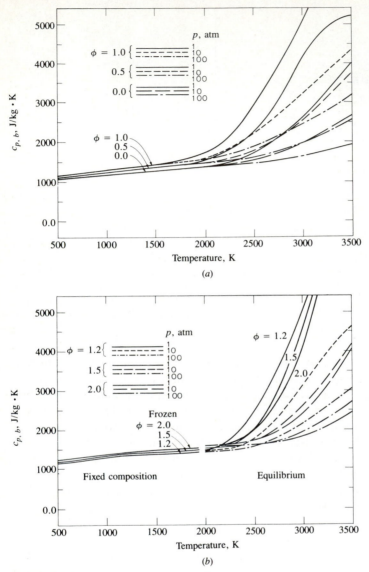

FIGURE 4-17
Specific heat at constant pressure for equilibrium, frozen, and fixed composition burned gases as a function of temperature and pressure: (a) equivalence ratio $\phi \leq 1.0$; (b) equivalence ratio $\phi > 1$. Units: J/kg mixture·K. Fuel: C_nH_{2n}.

by an exponential function of T, p, and ϕ. For $\phi \leq 1$, a single set of equations resulted. For $\phi \geq 1$, sets of equations were developed, each set applying to a specific value of equivalence ratio (see Ref. 19). In general, the fit for internal energy is within $2\frac{1}{2}$ percent over the pressure and temperature range of interest, and the error over most of the range is less than 1 percent. For many applica-

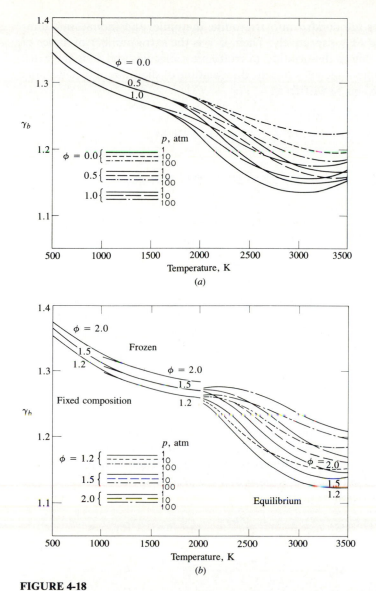

FIGURE 4-18
Ratio of specific heats, $\gamma_b = c_{p,b}/c_{v,b}$, for equilibrium, frozen, and fixed composition burned gases as a function of temperature and pressure: (a) equivalence ratio $\phi \leq 1.0$; (b) equivalence ratio $\phi > 1$. Fuel: C_nH_{2n}.

tions, the undissociated equations for thermodynamic properties are sufficiently accurate.

An alternative approach for property calculations, applicable to a wide range of hydrocarbon and alcohol fuels, is used extensively in the author's laboratory.[20] With this method, the products of combustion of hydrocarbon (or alcohol)-air mixtures are divided into triatomic, diatomic, and monatomic molecules, M_3, M_2, and M_1, respectively. Then, if Y is the extra number of moles of diatomic molecules due to dissociation of triatomic molecules and U is the extra number of monatomic molecules due to dissociation of diatomic molecules, the combustion reaction can be written as

$$\varepsilon\phi C + 2(1 - \varepsilon)\phi H_2 + O_2 + \psi N_2 \rightarrow$$

$$[(2 - \varepsilon)\phi - 2Y]M_3 + [1 - \phi + 3Y - U + \psi]M_2 + 2UM_1$$

$$\text{for } \phi \leq 1 \qquad (4.45)$$

$$[(2 - \varepsilon\phi)\phi - 2Y]M_3 + [2(\phi - 1) + 3Y - U + \psi]M_2 + 2UM_1$$

$$\text{for } \phi > 1$$

The method is based upon a fitting of data obtained from sets of detailed chemical equilibrium calculations to this functional form. Two general dissociation reactions:

$$2M_3 = 3M_2 \qquad \text{and} \qquad M_2 = 2M_1$$

are then used with fitted equilibrium constants $K_1(T)$ and $K_2(T)$ to calculate the relative species concentrations. This approach has been developed to give equations for enthalpy which sum the translational, rotational, and vibrational contributions to the specific heat, and the enthalpy of formation:

$$m_{RP} h = \frac{R}{2}(8N_3 + 7N_2 + 5N_1)T + R(3N_3 + N_2)\frac{T_v}{\exp(T_v/T) - 1} + m_{RP} h_f$$

$$(4.46)$$

where N_3, N_2, and N_1 are the number of moles of triatomic, diatomic, and monatomic molecules respectively per mole O_2 reactant, T_v is a fitted vibrational temperature, m_{RP} is the mass of products per mole O_2 reactant [Eq. (4.9)], and h_f is the average specific enthalpy of formation of the products.

The molecular weight is given by

$$M_b = \frac{m_{RP}}{1 + (1 - \varepsilon)\phi + \psi + Y + U} \qquad \text{for } \phi \leq 1$$

$$(4.47)$$

or

$$M_b = \frac{m_{RP}}{(2 - \varepsilon)\phi + \psi + Y + U} \qquad \text{for } \phi > 1$$

U and Y are found using an approximate solution to the equations obtained by applying the fitted equilibrium constants to the dissociation reactions; h_f is obtained by fitting a correction to the undisssociated products enthalpy of formation. Equations are presented for the partial derivatives of enthalpy h and density ρ with respect to T, p, and ϕ.[20] These relationships have been tested for fuels with H/C ratios of 4 to 0.707, equivalence ratios 0.4 to 1.4, pressures 1 to 30 atm, and temperatures 1000 to 3000 K. The error for burned mixture temperatures relevant to engine calculations is always less than ± 10 K. The errors in density are less than ± 0.2 percent.

4.8   TRANSPORT PROPERTIES

The processes by which mass, momentum, and energy are transferred from one point in a system to another are called rate processes. In internal combustion engines, examples of such processes are evaporation of liquid fuel, fuel-air mixing, friction at a gas/solid interface, and heat transfer between gas and the walls of the engine combustion chamber. In engines, most of these processes are turbulent and are therefore strongly influenced by the properties of the fluid flow. However, turbulent rate processes are usually characterized by correlations between dimensionless numbers (e.g., Reynolds, Prandtl, Nusselt numbers, etc.), which contain the fluid's transport properties of viscosity, thermal conductivity, and diffusion coefficient as well as the flow properties.

The simplest approach for computing the transport properties is based on the application of kinetic theory to a gas composed of hard-sphere molecules. By analyzing the momentum flux in a plane Couette flow,† it can be shown (Chapman and Cowling, Ref. 21, p. 218) that the viscosity μ of a monatomic hard-sphere gas [where $\mu = \tau/(du/dx)$, τ being the shear stress and (du/dx) the velocity gradient] is given by

$$\mu = \frac{[5/(16\sqrt{\pi})](m\tilde{k}T)^{1/2}}{d^2} \tag{4.48}$$

where m is the mass of the gas molecule, d is the molecular diameter, and $\tilde{k}$ is Boltzmann's constant, 1.381×10^{-23} J/K.

For such a gas, the viscosity varies as $T^{1/2}$, but will not vary with gas pressure or density. Measurements of viscosity show it does only vary with temperature, but generally not proportionally to $T^{1/2}$. The measured temperature dependence can only be explained with more sophisticated models for the intermolecular potential energy than that of a hard sphere. Effectively, at higher temperatures, the higher average kinetic energy of a pair of colliding molecules requires that they approach closer to each other and experience a greater repulsive force to be deflected in the collision. As a result, the molecules appear to be smaller spheres as the temperature increases.

An expression for the thermal conductivity k of a monatomic hard-sphere gas [$k = \dot{q}/(dT/dx)$, where $\dot{q}$ is the heat flux per unit area and dT/dx is the temperature gradient] can be derived from an analysis of the thermal equivalent of plane Couette flow (Ref. 21, p. 235):

$$k = \frac{[75/(64\sqrt{\pi})](\tilde{k}^3 T/m)^{1/2}}{d^2} \tag{4.49}$$

---

† In Couette flow, the fluid is contained between two infinite plane parallel surfaces, one at rest and one moving with constant velocity. In the absence of pressure gradients, the fluid velocity varies linearly across the distance between the surfaces.

which has the same temperature dependence as μ. Equations (4.48) and (4.49) can be combined to give

$$k = \tfrac{5}{2}\mu c_v$$

since, for a monatomic gas, the specific heat at constant volume is $3\tilde{k}/(2m)$. This simple equality is in good agreement with measurements of μ and k for monatomic gases.

The above model does not take into account the vibrational and rotational energy exchange in collisions between polyatomic molecules which contribute to energy transport in gases of interest in engines. Experimental measurements of k and μ show that k is less than $\tfrac{5}{2}\mu c_v$ for such polyatomic gases, where c_v is the sum of the translational specific heat and the specific heat due to internal degrees of freedom. It was suggested by Eucken that transport of vibrational and rotational energy was slower than that of translational energy. He proposed an empirical expression

$$k = \frac{9\gamma - 5}{4}\,\mu c_v$$

or

$$\mathrm{Pr} = \frac{\mu c_p}{k} = \frac{4\gamma}{9\gamma - 5}$$

(4.50)

where Pr is the Prandtl number, which is in good agreement with experimental data.

A similar analysis of a binary diffusion process, where one gas diffuses through another, leads to an expression for the binary diffusion coefficient D_{ij}. D_{ij} is a transport property of the gas mixture composed of species i and j, defined by Fick's law of molecular diffusion which relates the fluxes of species i and j, Γ_{xi} and Γ_{xj}, in the x direction to the concentration gradients, dn_i/dx and dn_j/dx (n is the molecular number density):

$$\Gamma_{xi} = -D_{ij}\left(\frac{dn_i}{dx}\right) \qquad \Gamma_{xj} = -D_{ij}\left(\frac{dn_j}{dx}\right)$$

The binary diffusion coefficient for a mixture of hard-sphere molecules is (Ref. 21, p. 245)

$$D_{ij} = \frac{3}{16nd^2}\left(\frac{2\tilde{k}T}{\pi m_{ij}}\right)^{1/2}$$

(4.51)

where m_{ij} is the reduced mass $m_i m_j/(m_i + m_j)$.

A more rigorous treatment of gas transport properties, based on more realistic intermolecular potential energy models, can be found in Hirschfelder et al.,[22] who also present methods for computing the transport properties of mixtures of gases. The NASA computer program "Thermodynamic and Transport Properties of Complex Chemical Systems"[10] computes the viscosity, thermal conductivity, and Prandtl number in addition to the thermodynamic calculations described in Secs. 3.7 and 4.7 for high-temperature equilibrium and frozen gas composition mixtures. The procedures used in the NASA program to compute these transport properties are based on the techniques described in Hirschfelder *et al.*[22] The NASA program has been used to compute the transport properties of hydrocarbon-air combustion products.[17] These quantities are functions of temperature T, equivalence ratio ϕ, and (except for viscosity) pressure p. Approximate correlations were then fitted to the calculated data of viscosity and Prandtl number. The principal advantage of these correlations is computational speed. For Prandtl number ($\mu c_p/k$), it was found convenient to use γ, the specific heat ratio (c_p/c_v), as an independent variable. Values of γ and c_p then permit determination of the thermal conductivity.

The viscosity of hydrocarbon-air combustion products over the temperature range 500 up to 4000 K, for pressures from 1 up to 100 atm, for $\phi = 0$ up to $\phi = 4$ is shown in Fig. 4-19. The viscosity as a function of temperature of hydrocarbon-air combustion products differs little from that of air. Therefore, a power law based on air viscosity data was used to fit the data:

$$\mu_{\text{air}}(\text{kg/m} \cdot \text{s}) = 3.3 \times 10^{-7} \times T^{0.7} \tag{4.52}$$

where T is in kelvins. The viscosity of combustion products is almost indepen-

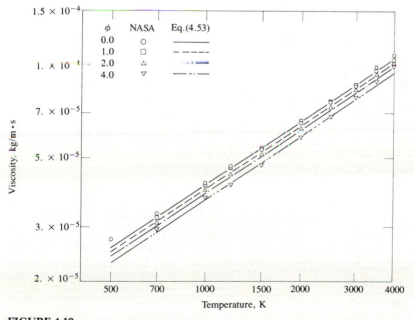

FIGURE 4-19
Viscosity, kg/m · s, of combustion products as a function of temperature and equivalence ratio. Equations shown are (4.52) and (4.53).

dent of pressure. This correlation was corrected to include the effect of the equivalence ratio ϕ on the viscosity of hydrocarbon-air combustion products:

$$\mu_{prod} = \frac{\mu_{air}}{1 + 0.027\phi} \tag{4.53}$$

Figure 4-19 shows that the viscosity predicted using Eqs. (4.52) and (4.53) is very close to the viscosity values calculated with the NASA program. There is less than 4 percent error.

The Prandtl number of hydrocarbon-air combustion products has also been correlated over the above ranges of temperatures, pressures, and equivalence ratios. Since the expression for Prandtl number of a monatomic hard-sphere molecule gas is a function of γ, a second-order polynomial of γ was used to curve-fit the calculated Prandtl number data. A good fit to the data for lean combustion product mixtures was the following:

$$Pr = 0.05 + 4.2(\gamma - 1) - 6.7(\gamma - 1)^2 \qquad \phi \leq 1 \tag{4.54}$$

The values of Pr predicted with Eq. (4.54) are within 5 percent of the equilibrium Pr values calculated with the NASA program. For rich mixtures the following equation is a good fit to the equilibrium values of Pr using equilibrium values of γ, for temperatures greater than 2000 K:

$$Pr = \frac{0.05 + 4.2(\gamma - 1) - 6.7(\gamma - 1)^2}{1 + 0.015 \times 10^{-6}(\phi T)^2} \qquad 1 < \phi \leq 4 \tag{4.55}$$

The predicted values of Pr in this case are also close to the calculated values of Pr, with less than 10 percent error. Equation (4.55) is also a reasonable fit to the frozen values† of Pr for rich mixtures, using frozen values of γ, for the temperature range 1200 to 2000 K. As there are no data for Pr of rich mixtures at low temperatures, we suggest that where a fixed composition for the mixture is appropriate (e.g., during the exhaust process in an internal combustion engine), Eq. (4.55) can also be used with fixed composition values of γ.

The Prandtl number can be obtained from the above relations if γ is known. The thermal conductivity can be obtained from the Prandtl number if values of μ and c_p are known. Values of γ_b and $c_{p,b}$ as functions of temperature, pressure, and equivalence ratio are given in Figs. 4-15 to 4-18.

Since the fundamental relations for viscosity and thermal conductivity are complicated, various approximate methods have been proposed for evaluating these transport properties for gas mixtures. A good approximation for the vis-

---

† In the NASA program, "frozen" means the gas composition is in equilibrium at the given T and p, but is frozen as c_p, c_v, and k are computed.

cosity of a multicomponent gas mixture is

$$\mu_{\text{mixt}} = \sum_{i=1}^{v} \frac{\tilde{x}_i^2}{\tilde{x}_i^2/\mu_i + 1.385 \sum_{j=1, j \neq i}^{v} \tilde{x}_i \tilde{x}_j (\tilde{R}T/pM_i D_{ij})} \qquad (4.56)$$

where $\tilde{x}_i$ and M_i are the mole fraction and molecular weight of the ith species, μ_i is the viscosity of the ith species, v is the number of species in the mixture, and D_{ij} is the binary diffusion coefficient for species i and j.[22]

4.9 EXHAUST GAS COMPOSITION

While the formulas for the products of combustion used in Sec. 3.4 are useful for determining unburned mixture stoichiometry, they do not correspond closely to the actual burned gas composition. At high temperatures (e.g., during combustion and the early part of the expansion stroke) the burned gas composition corresponds closely to the equilibrium composition at the local temperature, pressure, and equivalence ratio. During the expansion process, recombination reactions simplify the burned gas composition. However, late in the expansion stroke and during exhaust blowdown, the recombination reactions are unable to maintain the gases in chemical equilibrium and, in the exhaust process, the composition becomes frozen. In addition, not all the fuel which enters the engine is fully burned inside the cylinder; the combustion inefficiency even when excess air is present is a few percent (see Fig. 3-9). Also, the contents of each cylinder are not necessarily uniform in composition, and the amounts of fuel and air fed to each cylinder of a multicylinder engine are not exactly the same. For all these reasons, the composition of the engine exhaust gases cannot easily be calculated.

It is now routine to measure the composition of engine exhaust gases. This is done to determine engine emissions (e.g., CO, NO_x, unburned hydrocarbons, and particulates). It is also done to determine the relative proportions of fuel and air which enter the engine so that its operating equivalence ratio can be computed. In this section, typical engine exhaust gas composition will be reviewed, and techniques for calculating the equivalence ratio from exhaust gas composition will be given.

4.9.1 Species Concentration Data

Standard instrumentation for measuring the concentrations of the major exhaust gas species has been developed.[23] Normally a small fraction of the engine exhaust gas stream is drawn off into a sample line. Part of this sample is fed directly to the instrument used for unburned hydrocarbon analysis, a flame ionization detector (FID). The hydrocarbons present in the exhaust gas sample are burned in a small hydrogen-air flame, producing ions in an amount proportional to the number of carbon atoms burned. The FID is effectively a carbon atom counter. It is calibrated with sample gases containing known amounts of hydrocarbons. Unburned hydrocarbon concentrations are normally expressed as a mole fraction or volume fraction in parts per million (ppm) as C_1. Sometimes results are expressed as ppm propane (C_3H_8) or ppm hexane (C_6H_{14}); to convert these to ppm C_1 multiply by 3 or 6, respectively. Older measurements of unburned hydrocarbons were often made with a nondispersive infrared (NDIR) analyzer, where the infrared absorption by the hydrocarbons in a sample cell was used to determine their concentration.[23] Values of HC concentrations in engine exhaust gases measured by an FID are about two times the equivalent values measured by an NDIR analyzer (on the same carbon number basis, e.g., C_1). NDIR-obtained concentrations are usually multiplied by 2 to obtain an estimate of actual HC concentrations. Substantial concentrations of oxygen in the exhaust gas affect the FID measurements. Analysis of *unburned* fuel-air mixtures should be done with special care.[23] To prevent condensation of hydrocarbons in the sample line (especially important in diesel exhaust gas), the sample line is often heated.

NDIR analyzers are used for CO_2 and CO concentration measurements. Infrared absorption in a sample cell containing exhaust gas is compared to absorption in a reference cell. The detector contains the gas being measured in two compartments separated by a diaphragm. Radiation not absorbed in the sample cell is absorbed by the gas in the detector on one side of the diaphragm. Radiation not absorbed in the reference cell is absorbed by the gas in the other half of the detector. Different amounts of absorption in the two halves of the detector result in a pressure difference being built up which is measured in terms of diaphragm distention. NDIR detectors are calibrated with sample gases of known composition. Since water vapor IR absorption overlaps CO_2 and CO absorption bands, the exhaust gas sample is dried with an ice bath and chemical dryer before it enters the NDIR instrument.

Oxygen concentrations are usually measured with paramagnetic analyzers. Oxides of nitrogen, either the amount of nitric oxide (NO) or total oxides of nitrogen (NO + NO_2, NO_x), are measured with a chemiluminescent analyzer. The NO in the exhaust gas sample stream is reacted with ozone in a flow reactor. The reaction produces electronically excited NO_2 molecules which emit radiation as they decay to the ground state. The amount of radiation is measured with a photomultiplier and is proportional to the amount of NO. The instrument can also convert any NO_2 in the sample stream to NO by decomposition in a heated stainless steel tube so that the total NO_x (NO + NO_2) concentration can be determined.[23] Gas chromatography can be used to determine all the inorganic species (N_2, CO_2, O_2, CO, H_2) or can be used to measure the individual hydrocarbon compounds in the total unburned hydrocarbon mixture. Particulate emissions are measured by filtering the particles from the exhaust gas stream onto a previously weighed filter, drying the filter plus particulate, and reweighing.

SPARK-IGNITION ENGINE DATA. Dry exhaust gas composition data, as a function of the fuel/air equivalence ratio, for several different multi- and single-cylinder automotive spark-ignition engines over a range of engine speeds and loads are shown in Fig. 4-20. The fuel compositions (gasolines and isooctane) had H/C ratios ranging from 2.0 to 2.25. Exhaust gas composition is substantially

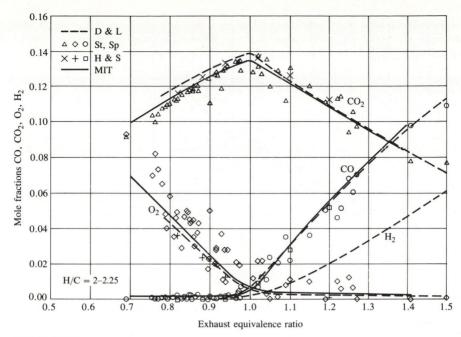

FIGURE 4-20
Spark-ignition engine exhaust gas composition data in mole fractions as a function of fuel/air equivalence ratio. Fuels: gasoline and isooctane, H/C 2 to 2.25. (*From D'Alleva and Lovell,*[24] *Stivender,*[25] *Harrington and Shishu,*[26] *Spindt,*[27] *and data from the author's laboratory at MIT.*)

different on the lean and the rich side of the stoichiometric air/fuel or fuel/air ratios; thus, the fuel/air equivalence ratio ϕ (or its inverse, the relative air/fuel ratio λ) is the appropriate correlating parameter. On the lean side of stoichiometric, as ϕ decreases, CO_2 concentrations fall, oxygen concentrations increase, and CO levels are low but not zero (~ 0.2 percent). On the rich side of stoichiometric, CO and H_2 concentrations rise steadily as ϕ increases and CO_2 concentrations fall. O_2 levels are low (~ 0.2 to 0.3 percent) but are not zero. At stoichiometric operation, there is typically half a percent O_2 and three-quarters of a percent CO.

Fuel composition has only a modest effect on the magnitude of the species concentrations shown. Measurements with a wide range of liquid fuels show that CO concentrations depend only on the equivalence ratio or relative fuel/air ratio (see Fig. 11-20).[26] A comparison of exhaust CO concentrations with gasoline, propane (C_3H_8), and natural gas (predominantly methane, CH_4) show that only with the high H/C ratio of methane, and then only for CO $\geq$ 4 percent, is fuel composition significant.[28] The values of CO_2 concentration at a given ϕ are slightly affected by the fuel H/C ratio. For example, for stoichiometric mixtures with 0.5 percent O_2 and 0.75 percent CO, as the H/C ratio decreases CO_2 concentrations increase from 13.7 percent for isooctane (H/C = 2.25), to 14.2 to 14.5 percent for typical gasolines (H/C in range 2–1.8), to 16 for toluene (H/C = 1.14).[29]

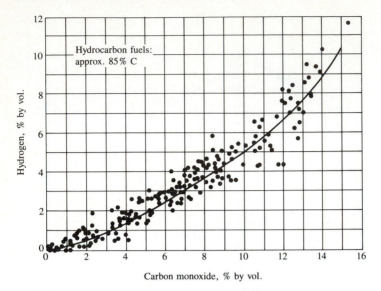

FIGURE 4-21
Hydrogen concentration in spark-ignition engine exhaust as a function of carbon monoxide concentration. Units: percent by volume.[30]

Unburned hydrocarbon exhaust concentrations vary substantially with engine design and operating conditions. Spark-ignition engine exhaust levels in a modern low-emission engine are typically of the order of 2000 ppm C_1 with liquid hydrocarbon fuels, and about half that level with natural gas and propane fuels.

Hydrogen concentrations in engine exhaust are not routinely measured. However, when the mixture is oxygen-deficient—fuel rich—hydrogen is present with CO as an incomplete combustion product. Figure 4-21 summarizes much of the available data on H_2 concentrations plotted as a function of CO.[30]

DIESEL EXHAUST DATA. Since diesels normally operate significantly lean of stoichiometric ($\phi \leq 0.8$) and the diesel combustion process is essentially complete (combustion inefficiency is ≤ 2 percent), their exhaust gas composition is straightforward. Figure 4-22 shows that O_2 and CO_2 concentrations vary linearly with the fuel/air equivalence ratio over the normal operating range. Diesel emissions of CO and unburned HC are low.

4.9.2 Equivalence Ratio Determination from Exhaust Gas Constituents

Exhaust gas composition depends on the relative proportions of fuel and air fed to the engine, fuel composition, and completeness of combustion. These relationships can be used to determine the operating fuel/air equivalence ratio of an engine from a knowledge of its exhaust gas composition. A general formula for

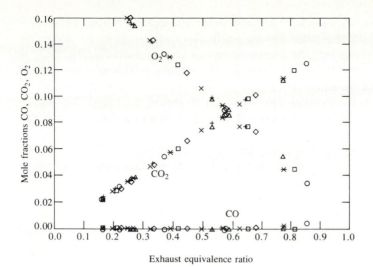

FIGURE 4-22
Exhaust gas composition from several diesel engines in mole fractions on a dry basis as a function of fuel/air equivalence ratio.[31]

the composition of fuel can be represented as $C_nH_mO_r$. For conventional petroleum-based fuels, oxygen will be absent; for fuels containing alcohols, oxygen will be present. The overall combustion reaction can be written as

$$\text{Fuel} + \text{oxidizer} \rightarrow \text{products}$$

The fuel is $C_nH_mO_r$; the oxidizer is air $(O_2 + 3.773N_2)$. The products are CO_2, H_2O, CO, H_2, O_2, NO_x, N_2, unburned hydrocarbons (unburned fuel and products of partial fuel reaction), and soot particles (which are mainly solid carbon). The amount of solid carbon present is usually sufficiently small (≤ 0.5 percent of the fuel mass) for it to be omitted from the analysis. The overall combustion reaction can be written explicitly as

$$C_nH_mO_r + \frac{n_{O_2}}{\phi}(O_2 + 3.773N_2) = n_P(\tilde{x}_{C_aH_b}C_aH_b + \tilde{x}_{CO}CO + \tilde{x}_{CO_2}CO_2$$

$$+ \tilde{x}_{O_2}O_2 + \tilde{x}_{N_2}N_2 + \tilde{x}_{NO}NO$$

$$+ \tilde{x}_{NO_2}NO_2 + \tilde{x}_{H_2O}H_2O + \tilde{x}_{H_2}H_2) \qquad (4.57)$$

where ϕ is the measured equivalence ratio $[(F/A)_{\text{actual}}/(F/A)_{\text{stoichiometric}}]$, n_{O_2} is the number of O_2 molecules required for complete combustion $(n + m/4 - r/2)$, n_P is the total number of moles of exhaust products, and $\tilde{x}_i$ is the mole fraction of the ith component.

There are several methods for using Eq. (4.57) to determine ϕ, the equivalence ratio, depending on the amount of information available. Normally CO_2, CO, O_2, NO_x concentrations as mole fractions and unburned hydrocarbon (as mole fraction or ppm C_1, i.e., $\tilde{x}_{CH_{b/a}}$) are measured. The concentration of the inorganic gases are usually measured *dry* (i.e., with H_2O removed) or *partially dry*. Unburned hydrocarbons may be measured *wet* or *dry* or *partially dry*. NO_x is mainly nitric oxide (NO); its concentration is usually sufficiently low (<0.5 percent) for its effect on equivalence ratio determination to be negligibly small. Thus, in Eq. (4.57) there are seven unknowns which are: ϕ, $\tilde{x}_{H_2}$, $\tilde{x}_{H_2O}$, $\tilde{x}_{N_2}$, n_P, a, b. (There will be additional unknowns if the measurements listed above are incomplete.)

To solve for these unknowns we need seven additional equations. We can obtain five equations using an atomic balance for each element and the definition of mole fraction, as follows:

Carbon balance:

$$n = n_P(a\tilde{x}_{C_aH_b} + \tilde{x}_{CO} + \tilde{x}_{CO_2}) \tag{4.58}$$

Hydrogen balance:

$$m = n_P(b\tilde{x}_{C_aH_b} + 2\tilde{x}_{H_2O} + 2\tilde{x}_{H_2}) \tag{4.59}$$

Oxygen balance:

$$r + \frac{2n_{O_2}}{\phi} = n_P(\tilde{x}_{CO} + 2\tilde{x}_{CO_2} + \tilde{x}_{NO} + 2\tilde{x}_{O_2} + \tilde{x}_{H_2O}) \tag{4.60}$$

Nitrogen balance:

$$\frac{7.546 n_{O_2}}{\phi} = n_P(2\tilde{x}_{N_2} + \tilde{x}_{NO}) \tag{4.61}$$

Mole fractions add up to 1:

$$\tilde{x}_{C_aH_b} + \tilde{x}_{CO} + \tilde{x}_{H_2} + \tilde{x}_{H_2O} + \tilde{x}_{N_2} + \tilde{x}_{NO} + \tilde{x}_{CO_2} + \tilde{x}_{O_2} = 1 \tag{4.62}$$

An additional assumption is made, based on available exhaust gas composition data, that CO_2, CO, H_2O, and H_2 concentrations are related by

$$\frac{\tilde{x}_{CO}\,\tilde{x}_{H_2O}}{\tilde{x}_{CO_2}\,\tilde{x}_{H_2}} = K \tag{4.63}$$

where K is a constant.† Values of 3.8[24, 25] and 3.5[27] are commonly used for K. The difference between these values has little effect on the computed magnitude of ϕ. To complete the analysis, various assumptions are made concerning the composition and relative importance of the unburned hydrocarbons. The most common approaches are summarized below.

---

† Equation (4.63) is often described as assuming a specific value for the water-gas reaction equilibrium constant. In fact K is an empirical constant determined from exhaust gas composition data.

OXYGEN BALANCE AIR/FUEL AND EQUIVALENCE RATIOS. For fuels comprised of carbon and hydrogen only, when all species are measured with the *same background moisture* (wet, dry, or partially dry), the following expression based on the ratio of measured and computed oxygen-containing species to measured carbon-containing species gives the air/fuel ratio. It has been assumed that the unburned hydrocarbons have the same C/H ratio as the fuel:[32]

$$\left(\frac{A}{F}\right) = 4.773 \left(\frac{M_{air}}{M_f}\right) \frac{(CO_2) + (CO)/2 + (H_2O)/2 + (NO)/2 + (NO_2) + (O_2)}{(HC) + (CO) + (CO_2)}$$

(4.64)

where ( ) are molar concentrations (all with the same background moisture) in percent, $M_{air} = 28.96$, $M_f = 12.01 + 1.008y$ where y is the H/C ratio of the fuel, (HC) is molar percent unburned hydrocarbons as C_1, and

$$(H_2O) = 0.5y \frac{(CO_2) + (CO)}{(CO)/[K(CO_2)] + 1}$$

(4.65)

Since nitrogen oxides collectively comprise less than 0.5 percent of the exhaust mixture, their concentrations can be omitted with negligible error.

The fuel/air equivalence ratio ϕ is obtained from the ratio of the stoichiometric air/fuel ratio [Eq. (3.6)] and Eq. (4.64) above.

CARBON BALANCE AIR/FUEL AND EQUIVALENCE RATIOS. When oxygen analysis is not available, for fuels comprised of carbon and hydrogen only, a carbon balance air/fuel ratio may be employed:[25, 32]

$$\frac{A}{F} = \frac{M_{air}}{M_f} \left[\frac{100 + (HC) - (CO)/2 + 3(H_2O)/2 - (H_2O)_a}{(HC) + (CO) + (CO_2)} - \frac{y}{2} \right]$$

(4.66)

The symbols are as defined above. (H_2O) is the molar percent water in the combustion products defined by Eq. (4.65) and $(H_2O)_a$ is the molar percent water vapor at the analyzers.

This carbon balance (A/F) is sensitive to moisture concentration at the analyzers. The use of ice bath exhaust sample chillers generally reduces the $(H_2O)_a$ term to less than 1 percent and little accuracy is then lost by neglecting it. For completely "wet" analysis (uncondensed), $(H_2O)_a = (H_2O)$, and Eq. (4.66) is accurate. For partially dry exhaust gas analysis, knowledge of the dew point of the mixture will provide the $(H_2O)_a$ term by reference to steam tables.

The fuel/air equivalence ratio ϕ is obtained from the ratio of the stoichiometric air/fuel ratio [Eq. (3.6)] and Eq. (4.66) above.

EQUIVALENCE RATIO BASED ON WET HC AND DRY INORGANIC GAS ANALYSIS. Engine exhaust gas composition is often determined by analyzing a fully dried sample stream for CO_2, CO, O_2, and NO_x, and a fully wet (uncondensed) stream with an FID for unburned hydrocarbons. Equations (4.64) and (4.66) are not applicable under these circumstances. The following equations define the exhaust gas composition and equivalence ratio under these conditions. The notation $\tilde{x}_i$ denotes the wet mole fraction of species i and $\tilde{x}_i^*$ denotes the dry mole fraction of species i. Equations (4.57) to (4.62), with Eq. (4.63) to relate CO_2, CO, H_2O, and H_2 concentrations and the assumption that $b/a = m/n$, were used to derive these results. The equations apply for a fuel of composition $C_mH_nO_r$.

The fuel/air equivalence ratio is given by

$$\phi = \frac{2n_{O_2}}{n_P\,\tilde{x}_{H_2O} + n_P(1 - \tilde{x}_{H_2O})(\tilde{x}_{CO}^* + 2\tilde{x}_{CO_2}^* + 2\tilde{x}_{O_2}^* + \tilde{x}_{NO}^* + 2\tilde{x}_{NO_2}^*) - r} \qquad (4.67)$$

where the wet and dry mole fractions are related by

$$\tilde{x}_i = (1 - \tilde{x}_{H_2O})\tilde{x}_i^*$$

and

$$n_P = \frac{n}{\tilde{x}_{CH_{b/a}} + (1 - \tilde{x}_{H_2O})(\tilde{x}_{CO}^* + \tilde{x}_{CO_2}^*)}$$

$$\tilde{x}_{H_2O} = \frac{m}{2n}\,\frac{\tilde{x}_{CO}^* + \tilde{x}_{CO_2}^*}{[1 + \tilde{x}_{CO}^*/(K\tilde{x}_{CO_2}^*) + (m/2n)(\tilde{x}_{CO}^* + \tilde{x}_{CO_2}^*)]} \qquad (4.68)$$

$$\tilde{x}_{H_2} = \frac{\tilde{x}_{H_2O}\,\tilde{x}_{CO}^*}{K\tilde{x}_{CO_2}^*}$$

$$\tilde{x}_{N_2} = \frac{3.773n_{O_2}}{\phi n_P} - (1 - \tilde{x}_{H_2O})\frac{(\tilde{x}_{NO}^* + \tilde{x}_{NO_2}^*)}{2}$$

Note that $\tilde{x}_{CH_{b/a}}$ is the measured (wet) HC concentration as a mole fraction C_1 (ppm $C_1 \times 10^{-6}$): $\tilde{x}_{CH_{b/a}} = a\tilde{x}_{C_aH_b}$. Figure 4-23 shows wet exhaust gas concentrations, based on the MIT measured dry concentrations of CO_2, CO, O_2 shown in Fig. 4-20, and wet HC concentration, as well as Eqs. (4.68).

For lean mixtures, varying the value of K between 1.5 and 5.5 had a negligible effect on the value of ϕ computed from Eq. (4.67). For stoichiometric mixtures, varying K from 2.5 to 4.5 varied the computed ϕ by 2 to 3 percent. For $\phi \approx 1.2$, varying K from 2.5 to 4.5 varied the computed ϕ by 3 to 4 percent. The error in ϕ involved in omitting NO_x is 0.2 percent for an NO_x level of 1000 ppm, increasing to 1 percent for an NO_x level of 5000 ppm. The sensitivity of the computed ϕ to errors in the measurements of CO_2, CO, and O_2 is modest within the normal range of ϕ used. A 2 percent error in CO_2 or CO or O_2 at $\phi \approx 1$ gives about a 0.1 percent error in computed ϕ. For leaner and richer mixtures, the error in ϕ increases for errors in measured CO_2 concentration, and CO ($\phi > 1$) and O_2 ($\phi < 1$) concentrations, but is still significantly less than the measurement error in fuel and air flow.

4.9.3   Effects of Fuel/Air Ratio Nonuniformity

Neither the masses of air inducted into the different cylinders of a multicylinder engine per cycle nor the masses of fuel which enter the different cylinders per

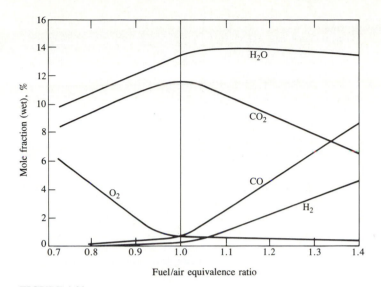

FIGURE 4-23
Wet exhaust gas species concentrations as a function of fuel/air equivalence ratio, based on the dry exhaust gas composition data in Fig. 4-20 and Eqs. (4.68).

cycle are exactly equal. In addition, mixing of fuel and air within each cylinder is not necessarily completely uniform. Thus the exhaust gas composition may correspond to a distribution in the fuel/air ratio in the unburned mixture about the mean value. For example, if the mean fuel/air ratio is stoichiometric, extra oxygen will be contributed by any cylinders operating lean of the average and extra carbon monoxide by any cylinders rich of the average, so that the exhaust gas will have higher levels of O_2 and CO (and a lower level of CO_2) relative to an engine operating with identical fuel/air ratios in each cylinder.

Eltinge has developed a method for defining this nonuniformity in the fuel/air ratio distribution for spark-ignition engines which operate close to stoichiometric.[29] A function $f(x)$ for the fuel/air ratio distribution ($x = F/A$) was assumed, with standard deviation S_x. For each value of x (i.e., F/A), complete utilization of the available oxygen was assumed (i.e., no exhaust HC) and the CO_2, H_2O, CO, H_2 concentrations were related by Eq. (4.63) (with $K = 3.5$). The concentrations of each species for each (F/A) were weighted by the distribution function $f(x)$ and summed to produce the average exhaust concentration. (A correction to allow for the presence of unburned HC in the exhaust was also developed.) Figure 4-24 shows one set of results, for a fuel H/C ratio of 1.8 (typical of gasoline), for a normal distribution in the fuel/air ratio. For each mean (F/A) and maldistribution parameter S_x (the standard deviation of the F/A distribution) the corresponding dry concentrations of CO_2, CO, and O_2 are shown. This type of information can be used to define the fuel-air mixture nonuniformities in spark-ignition engines operating relatively close to stoichiometric. For diesel engines the variations of major exhaust gas species concentrations with fuel/air equiva-

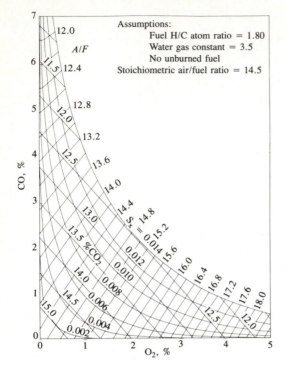

FIGURE 4-24
Computed relationship between dry exhaust gas composition (CO_2, CO, and O_2), air/fuel ratio, and maldistribution parameter S_x. Fuel: $(CH_{1.8})_n$. The correction to be added to the burned gas (F/A) which allows for the measured unburned hydrocarbon concentration is $4.7 \times 10^{-7} \times$ HC (ppm C_1) (*From Eltinge.*[29])

lence ratio are linear, so the effects of any nonuniformities are not apparent in this manner (see Fig. 4-22).

4.9.4   Combustion Inefficiency

Internal combustion engine exhaust contains combustible species: CO, H_2, unburned hydrocarbons, and particulates. When their concentrations are known, the combustion efficiency η_c given by Eq. (3.27) can be calculated. The chemical energy carried out of the engine in these combustibles represents the combustion inefficiency, $1 - \eta_c$:

$$1 - \eta_c = \frac{\sum_i x_i Q_{HV_i}}{[\dot{m}_f/(\dot{m}_a + \dot{m}_f)]Q_{HV_f}} \tag{4.69}$$

where the x_i are the mass fractions of CO, H_2, HC, and particulates, respectively, the Q_{HV_i} are the lower heating values of these species, and the subscripts f and a denote fuel and air. The heating values for CO (10.1 MJ/kg) and H_2(120 MJ/kg) are well defined. The composition of the unburned HC is not usually known. However, the heating values of hydrocarbons are closely comparable, so the fuel heating value (typically 42 to 44 MJ/kg) is used. The particulates (only present in diesels) are soot with some adsorbed hydrocarbons; usually the mass fraction is low enough for their contribution to be small, and a heating value for solid carbon (32.8 MJ/kg) can be used. Combustion efficiency data as a function of equivalence ratio have already been presented in Fig. 3-9.

PROBLEMS

4.1. (a) Calculate the low-temperature burned gas composition resulting from the combustion of 7 g/s air with 0.48 g/s ethane (C_2H_6). Assume K = 3.5 in Eq. (4.6).

(b) Calculate the low-temperature burned gas composition for the combustion of 7 g/s air with 0.48 g/s ethanol (C_2H_6O). Assume K = 3.5 in Eq. (4.6). (Note the large difference in burned gas composition due to this difference in fuel.)

4.2. To evaluate the accuracy of the simple analytic ideal gas model, use the results of Example 4.1 and Eqs. (4.23) (constant-volume adiabatic combustion) and (4.24) (constant-pressure adiabatic combustion) to calculate T_b for a stoichiometric isooctane-air mixture. Compare this result with that obtained using Figs. 4-3 and 4-8. Assume the following unburned mixture conditions: $T = 700$ K, $v = 0.125$ m^3/kg air, $p = 15$ atm, and $x_b = 0.1$.

4.3. (a) In Fig. 4-1, why does M_u decrease as x_b increases?

(b) In Fig. 4-14, why does M_b decrease as T increases?

(c) In Fig. 4-14, why does M_b decrease as ϕ increases?

4.4. Show how, for an ideal gas with fixed composition, the molecular weight M is related to the specific heats c_p and c_v, and $\tilde{R}$. Use Figs. 4-15 and 4-16 to calculate M_b for $T_b = 1750$, 2250, and 2750 K, and $\phi = 1.0$. Compare these results with values of M_b obtained from Fig. 4-14 and explain any differences.

4.5. EGR, exhaust gas recirculation, is often used to reduce NO_x by acting as a diluent in the intake mixture.

(a) For low-temperature isooctane-air combustion products at $\phi = 1.0$, determine the percentage of the burned gases' average specific heat at constant pressure which comes from each component in the burned gas mixture.

(b) Compare the specific heat at constant pressure of isooctane-air combustion products at $\phi = 1.0$ to that of air, both at 1750 K. This difference is one reason why EGR instead of leaner fuel-air mixture is used to control NO_x emissions.

4.6. Explain qualitatively the causes of the trends in the curves in Fig. 4-23 as ϕ is increased from 1.0 and decreased from 1.0.

4.7. Compare the O_2, CO_2, and CO data in Fig. 4-22 with the predictions of the elemental balance in Table 4.3. Explain any differences. Assume diesel fuel is a hydrocarbon with H/C ratio of 2.

4.8. The following exhaust data were obtained from a four-stroke cycle spark-ignition engine. CO, CO_2, and NO_x molar concentrations are all measured fully dry; HC is measured fully wet as ppm C_1. Determine the fuel/air equivalence ratio ϕ for the following three sets of data. Make the following assumptions: (1) the constant K in Eq. (4.63) equals 3.5; (2) the fuel composition is $C_8H_{15.12}$; (3) the unburned hydrocarbon H/C ratio is the same as that of the fuel; and (4) NO_x is entirely NO.

(a) CO_2 14.0%, CO 0.64%, O_2 0.7%, NO_x 3600 ppm, HC 3200 ppm.

(b) CO_2 13.8%, CO 3.05%, O_2 0%, NO_x 1600 ppm, HC 3450 ppm.

(c) CO_2 12.5%, CO 0.16%, O_2 4.0%, NO_x 4600 ppm, HC 2100 ppm.

4.9. The exhaust from a spark-ignition engine has the following composition (in mole fractions):

$$CO_2 = 0.12; \ H_2O = 0.14; \ CO = 0.01; \ H_2 = 0.005;$$

$$N_2 = 0.7247; \ C_8H_{18} = 0.0003$$

The fuel is isooctane, C_8H_{18}; as shown above, a small fraction of the fuel escapes from the cylinder unburned. The lower heating value of isooctane is 44.4 MJ/kg, of carbon monoxide is 10.1 MJ/kg, and of hydrogen is 120 MJ/kg. The atomic weights of the elements are: $C = 12$, $O = 16$, $H = 1$, $N = 14$.

(a) Calculate the combustion *inefficiency* in the engine; i.e., the percentage of the entering fuel enthalpy which is not fully released in the combustion process and leaves the engine in the exhaust gases (for this problem, the exhaust can be assumed to be at room temperature).

(b) What fraction of this inefficiency is due to the unburned fuel emissions?

4.10. A 2-liter displacement four-cylinder engine, operating at 2000 rev/min and 30 percent of its maximum power at that speed, has the following exhaust composition (in percent by volume or mole percent):

CO_2, 11%; H_2O, 11.5%; CO, 0.5%; H_2, 0%; O_2, 2%; unburned hydrocarbons (expressed as CH_2, i.e., with a molecular weight of 14), 0.5%; N_2, 74.5%

The fuel is $(CH_2)_n$ with a heating value of 44 MJ/kg. The atomic weights of the elements are $C = 12$, $H = 1$, $O = 16$, $N = 14$. The heating values of CO and HC are 10 and 44 MJ/kg, respectively.

(a) Is the engine a diesel or spark-ignition engine? Is there enough oxygen in the exhaust to burn the fuel completely? Briefly explain.

(b) Calculate the fraction of the input fuel energy ($m_f Q_{HV}$) which exits the engine unburned as (1) CO and (2) unburned HC.

(c) An inventor claims a combustion efficiency of 100 percent can be achieved. What percentage improvement in engine specific fuel consumption would result?

4.11. A gasoline engine operates steadily on a mixture of isooctane and air. The air and fuel enter the engine at 25°C. The fuel consumption is 3.0 g/s. The output of the engine is 50 kW. The temperature of the combustion products in the exhaust manifold is 660 K. At this temperature, an analysis of the combustion products yields the following values (on a dry volumetric basis):

CO_2, 11.4%; O_2, 1.6%; CO, 2.9%; N_2, 84.1%

(a) Find the composition in moles (number of moles per mole of isooctane) of the reactants and the reaction products.

(b) Determine the heat-transfer rate from the working fluid as the working fluid passes through the engine.

Constants for the calculations:

	Enthalpy of formation, kJ/kmol	Sensible enthalpy at 660 K, kJ/kmol
C_8H_{18}	−259280	—
CO_2	−393522	15823
CO	−110529	10789
H_2O	−241827	12710
O_2	—	11200
N_2	—	10749

4.12. A direct-injection four-stroke cycle diesel engine is used to provide power for pumping water. The engine operates at its maximum rated power at 2000 rev/min,

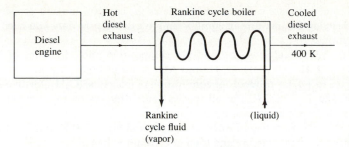

FIGURE P4-12

with an equivalence ratio of 0.8 and an air flow of 0.5 kg/s. The gross indicated fuel conversion efficiency is 45 percent, and the heat losses from the working fluid to the engine coolant and elsewhere within the engine are 280 kW. Diesel fuel has a heating value of 42 MJ/kg and stoichiometric fuel/air ratio of 0.067. Fuel and air enter the engine at ambient conditions. The mechanical efficiency of the diesel engine is 85 percent.

(a) Calculate the rated brake power of the engine, the average sensible enthalpy of the exhaust gases as they leave the engine, and the average exhaust gas temperature.

(b) Since the exhaust gas temperature is significantly above ambient, the advantages of using the diesel exhaust gas stream to heat the boiler of a Rankine cycle (see sketch) and generate additional power are to be explored. If the exhaust gases *leave* the Rankine-cycle system boiler at 400 K and 30 percent of the heat transferred from the exhaust gas stream in the boiler is converted to power at the Rankine-cycle power plant drive shaft, calculate the additional power obtained and the brake fuel conversion efficiency of the combined cycle system (diesel plus Rankine cycle).

4.13. A diesel engine has a compression ratio of 22:1. The conditions in the cylinder at the start of compression are $p = 101.3$ kPa and $T = 325$ K. Calculate the pressure and temperature at the end of compression, assuming the compression process is isentropic:

(a) Assume the cylinder contains an ideal gas with $\gamma = 1.4$ and $R = 287$ J/kg·K.

(b) Assume the cylinder contains air which may be regarded as a semiperfect gas (use the gas tables).

(c) Compare the work of compression in (a) and (b) above.

In practice, heat losses reduce the final compression temperature by 100 K. For a diesel engine operating at an equivalence ratio of 0.75 (full load):

(d) Calculate the ratio of heat loss during compression to the fuel energy added per cycle.

4.14. While the geometric compression ratio of an engine is V_{max}/V_{min}, the actual compression process starts somewhere between bottom-center and when the inlet valve closes, and it is conditions at time of spark (for an SI engine) or fuel injection (for a CI engine) that determine ignition. At low engine speed, compression starts about the time when the inlet valve closes. With this assumption, for the diesel engine of Prob. 4.13, calculate the air pressure and temperature at the start of injection. The inlet valve closes at 30° after BC; injection commences 15° before TC. Use the gas tables. Compare your answers with those of Prob. 4.13(b).

4.15. Use an equilibrium computer code (which calculates the composition and properties of chemically reacting gas mixtures in equilibrium) to calculate the data you need for the following graphs:

(a) Values of c_p, γ, molecular weight, and gas composition (mole fractions of N_2, CO_2, H_2O, CO, H_2, O_2, OH, O, H, and NO) as a function of the equivalence ratio ($\phi = 0.2$ to 1.4) for products of combustion of isooctane (C_8H_{18}) with air at $p = 40$ atm and $T = 2500$ K. Put all species concentrations on the same graph. Use a log scale for the composition axis.

(b) Unburned mixture consisting of isooctane vapor and air at 700 K and 20 atm is burned first at constant pressure and then at constant volume.

(1) Calculate the enthalpy and internal energy of isooctane vapor at 700 K in cal/gmol; also calculate the volume per unit mass of mixture (cm^3/g) for $\phi = 0.2, 0.4, 0.6, 0.8, 1.0, 1.2, 1.4$.

(2) Use these data and the equilibrium program to calculate the temperature attained after combustion at constant pressure, and temperature and pressure attained after combustion at constant volume. Plot these temperatures and pressures against the equivalence ratio ϕ.

Thermodynamic data for isooctane vapor

T, K	$\tilde{c}_p$, cal/mol·K	$\tilde{h} - \tilde{h}_{298}$, kcal/mol	$\Delta\tilde{h}_f^\circ$, kcal/mol
298	45.14	0.00	−53.57
700	85.66	27.02	−62.79

4.16. A heavy wall bomb with a volume of 1000 cm^3 contains a mixture of isooctane with the stoichiometric air requirement at $p = 101.3$ kPa and $T = 25°C$. The mixture is then ignited with a spark. Find the pressure and temperature of the equilibrium combustion products just after combustion is complete (i.e., before heat losses to the wall are significant). Assume the burned gases are uniform.

4.17. A gas engine, running on a gaseous mixture of butane, C_4H_{10}, and air has the following conditions in the cylinder prior to constant-volume adiabatic combustion: pressure, 6.48×10^5 N/m^2; temperature, 600 K. The charge composition is air plus 50 percent of the stoichiometric quantity of butane fuel. Calculate the pressure and temperature at the end of combustion using the data given below.

For air	T, K	$\tilde{u}$, J/gmol
	298	6,161
	600	12,596

For butane	T, K	$\tilde{u}$, J/gmol
	298	−77
	600	38,424

Internal energy of combustion of butane at 298 K is $\Delta \tilde{u} = -2.659$ MJ/gmol. Extract from gas tables for products of combustion for 50 percent stoichiometric fuel:

T, K	$\tilde{u}$, J/gmol
298	6,293
2075	54,227
2080	54,380

REFERENCES

1. Komiyama, K., and Heywood, J. B.: "Predicting NO_x Emissions and the Effects of Exhaust Gas Recirculation in Spark-Ignition Engines," SAE paper 730475, *SAE Trans.*, vol. 82, 1973.
2. Danieli, G., Ferguson, C., Heywood, J., and Keck, J. "Predicting the Emissions and Performance Characteristics of a Wankel Engine," SAE paper 740186, *SAE Trans.*, vol. 83, 1974.
3. Hottel, H. C., Williams, G. C., and Satterfield, C. N.: *Thermodynamic Charts for Combustion Processes*, John Wiley, 1949. See also charts in C. F. Taylor, *The Internal Combustion Engine in Theory and Practice*, vol. 1, MIT Press, 1960.
4. Newhall, H. K., and Starkman, E. S.: "Thermodynamic Properties of Octane and Air for Engine Performance Calculations," in *Digital Calculations of Engine Cycles, Progress in Technology*, vol. TP-7, pp. 38–48, SAE, 1964.
5. Starkman, E. S., and Newhall, H. K.: "Thermodynamic Properties of Methane and Air, and Propane and Air for Engine Performance Calculations," SAE paper 670466, *SAE Trans.*, vol. 76, 1967.
6. Keenan, J. H., Chao, J., and Kaye, J.: *Gas Tables*, 2d ed., John Wiley, 1983.
7. Reynolds, W. C.: *Thermodynamic Properties in SI; Graphs, Tables, and Computational Equations for Forty Substances*, Department of Mechanical Engineering, Stanford University, 1979.
8. *JANAF Thermochemical Tables*, 2d ed., NSRDS-NB537, U.S. National Bureau of Standards, June 1971.
9. Gordon, S., and McBride, B. J.: "Computer Program for the Calculation of Complex Chemical Equilibrium Composition, Rocket Performance, Incident and Reflected Shocks, and Chapman-Jouguet Detonations," NASA publication SP-273, 1971 (NTIS number N71-37775).
10. Svehla, R. A., and McBride, B. J.: "Fortran IV Computer Program for Calculation of Thermodynamic and Transport Properties of Complex Chemical Systems," NASA technical note TND-7056, 1973 (NTIS number N73-15954).
11. Fremont, H. A., *et al.*: *Properties of Combustion Gases*, General Electric Company, Cincinnati, Ohio, 1955.
12. Banes, B., McIntyre, R. W., and Sims, J. A.: *Properties of Air and Combustion Products with Kerosene and Hydrogen Fuels*, vols. I–XIII, Propulsion and Energetics Panel, Advisory Group for Aerospace Research and Development (AGARD), NATO, published by Bristol Siddeley Engines Ltd., Filton, Bristol, England, 1967.
13. Hires, S. D., Ekchian, A., Heywood, J. B., Tabaczynski, R. J., and Wall, J. C.: "Performance and NO_x Emissions Modelling of a Jet Ignition Prechamber Stratified Charge Engine," SAE paper 760161, *SAE Trans.*, vol. 85, 1976.
14. LoRusso, J. A.: "Combustion and Emissions Characteristics of Methanol, Methanol-Water, and Gasoline-Methanol Blends in a Spark Ignition Engine," S. M. Thesis, Department of Mechanical Engineering, MIT, May 1976.
15. By, A., Kempinski, B., and Rife, J. M.: "Knock in Spark-Ignition Engines," SAE paper 810147, 1981.

16. Rossini, F. D., Pitzer, K. S., Arnett, R. L., Braun, R. M., and Primentel, G. C.: *Selected Values of Physical and Thermodynamic Properties of Hydrocarbons and Related Compounds*, Carnegie Press, Pittsburgh, Pa., 1953.
17. Mansouri, S. H., and Heywood, J. B.: "Correlation for the Viscosity and Prandtl Number of Hydrocarbon-Air Combustion Products," *Combust. Sci. Technology*, vol. 23, pp. 251–256, 1980.
18. Olikara, C., and Borman, G. L.: "A Computer Program for Calculating Properties of Equilibrium Combustion Products with Some Applications to I.C. Engines," SAE paper 750468, 1975.
19. Krieger, R. B., and Borman, G. L.: "The Computation of Apparent Heat Release for Internal Combustion Engines," in *Proc. Diesel Gas Power*, ASME paper 66-WA/DGP-4, 1966.
20. Martin, M. K., and Heywood, J. B.: "Approximate Relationships for the Thermodynamic Properties of Hydrocarbon-Air Combustion Products," *Combust. Sci. Technology*, vol. 15, pp. 1–10, 1977.
21. Chapman, S., and Cowling, T. G.: *The Mathematical Theory of Non-Uniform Gases*, Cambridge University Press, Cambridge, 1955.
22. Hirschfelder, J. O., Curtiss, C. F., and Bird, R. B.: *Molecular Theory of Gases and Liquids*, John Wiley, 1954.
23. Patterson, D. J., and Henein, N. A.: *Emissions from Combustion Engines and Their Control*, Ann Arbor Science Publishers, 1972.
24. D'Alleva, B. A., and Lovell, W. G.: "Relation of Exhaust Gas Composition to Air-Fuel Ratio," *SAE J.*, vol. 38, no. 3, pp. 90–96, March 1936.
25. Stivender, D. L.: "Development of a Fuel-Based Mass Emission Measurement Procedure," SAE paper 710604, 1971.
26. Harrington, J. A., and Shishu, R. C.: "A Single-Cylinder Engine Study of the Effects of Fuel Type, Fuel Stoichiometry and Hydrogen-to-Carbon Ratio on CO, NO, and HC Exhaust Emissions," SAE paper 730476, 1973.
27. Spindt, R. S., "Air-Fuel Ratios from Exhaust Gas Analysis," SAE paper 650507, *SAE Trans.*, vol. 74, 1965.
28. Fleming, R. D., and Eccleston, D. B.: "The Effect of Fuel Composition, Equivalence Ratio, and Mixture Temperature on Exhaust Emissions," SAE paper 710012, 1971.
29. Eltinge, L.: "Fuel-Air Ratio and Distribution from Exhaust Gas Composition," SAE paper 680114, *SAE Trans.*, vol. 77, 1968.
30. Leonard, L. S.: "Fuel Distribution by Exhaust Gas Analysis," SAE paper 379A, 1961.
31. Bishop, R. P.: "Combustion Efficiency in Internal Combustion Engines," B.S. Thesis, Department of Mechanical Engineering, MIT, February 1985.
32. The Engine Test Code Subcommittee of the General Technical Committee, *General Motors Automotive Engine Test Code for Four Cycle Spark Ignition Engines*, 6th ed., 1975.

CHAPTER
5

IDEAL
MODELS OF
ENGINE
CYCLES

5.1 INTRODUCTION

The operating cycle of an internal combustion engine can be broken down into a sequence of separate processes: intake, compression, combustion, expansion, and exhaust. With models for each of these processes, a simulation of a complete engine cycle can be built up which can be analysed to provide information on engine performance. Models of individual engine processes at various levels of approximation have been developed. In this chapter we consider the simplest set of models which provide useful insights into the performance and efficiency of engines. The cycles analysed are commonly called the constant-volume, constant-pressure, and limited-pressure cycles; each title describes the approximation made for the engine combustion process.† The description of more accurate simulations of engine processes is deferred until Chap. 14.

For each engine cycle, a choice of models for working fluid thermodynamic properties must be made. These models have been reviewed in Chap. 4. Ideal

---

† These cycles are also referred to by the titles Otto cycle, Diesel cycle, and dual cycle, respectively, for historical reasons. The more descriptive titles used above are preferred because they avoid the often-made assumption that the SI or Otto engine is best approximated by the constant-volume cycle, and the CI or diesel engine is best approximated by the constant-pressure cycle. These assumptions are not necessarily correct.

engine cycle models combined with a simple ideal gas model (specific heats constant throughout the engine cycle—model 1 in Table 4.2) provide analytic results and are useful for illustrative purposes; we will call these cycles *ideal gas standard cycles*. Ideal engine cycles combined with more realistic models of working fluid properties (a frozen mixture of ideal gases for the unburned mixture and an equilibrium mixture for the burned mixture—model 5 in Table 4.2) are called *fuel-air cycles* and provide more quantitative information on engine operation.

An internal combustion engine is not a heat engine in the thermodynamic definition of the term. It is not a closed system. The working fluid does not execute a thermodynamic cycle. The temperature changes which occur around minimum and maximum cylinder volumes are not primarily a result of heat-transfer interactions. An engine can best be analyzed as an open system which exchanges heat and work with its surrounding environment (the atmosphere). Reactants (fuel and air) flow into the system; products (exhaust gases) flow out. (An overall second law analysis of the engine from this point of view has already been presented in Sec. 3.6.) Thus, the cycles discussed in this chapter are not thermodynamic cycles. Rather, each is a consecutive sequence of processes through which we can follow the state of the working fluid as the engine executes a complete operating cycle.

5.2 IDEAL MODELS OF ENGINE PROCESSES

The sequence of processes which make up a typical SI and CI engine operating cycle has been described in Sec. 1.3. To illustrate these processes, cylinder pressure (p) and volume (V) data from a throttled four-stroke cycle SI engine are plotted as a p-V diagram in Fig. 5-1. The cycle can be divided into compression,

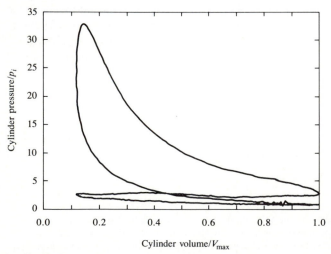

FIGURE 5-1
Pressure-volume diagram of firing spark-ignition engine. $r_c = 8.4$, 3500 rev/min, $p_i = 0.4$ atm, $p_e = 1$ atm, $\text{imep}_n = 2.9$ atm.

TABLE 5.1
Ideal models of engine processes

Process	Assumptions
Compression (1-2)	1. Adiabatic and reversible (hence isentropic)
Combustion (2-3)	1. Adiabatic
	2. Combustion occurs at
	(*a*) Constant volume
	(*b*) Constant pressure
	(*c*) Part at constant volume and part at constant pressure (called limited pressure)
	3. Combustion is complete ($\eta_c = 1$)
Expansion (3-4)	1. Adiabatic and reversible (hence isentropic)
Exhaust (4-5-6)	1. Adiabatic
and	2. Valve events occur at top- and bottom-center
intake (6-7-1)	3. No change in cylinder volume as pressure differences across open valves drop to zero
	4. Inlet and exhaust pressures constant
	5. Velocity effects negligible

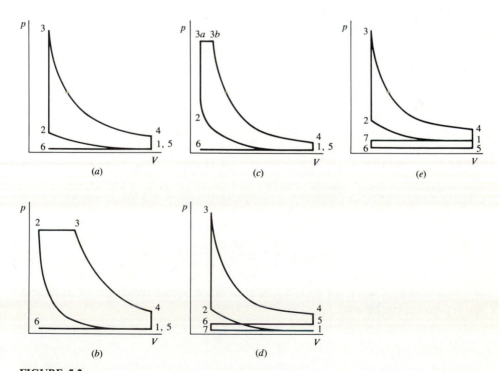

FIGURE 5-2
Pressure-volume diagrams of ideal cycles. Unthrottled operation: (*a*) constant-volume combustion; (*b*) constant-pressure combustion; (*c*) limited-pressure combustion. (*d*) Throttled constant-volume cycle; (*e*) supercharged constant-volume cycle.

combustion, expansion, exhaust, and intake processes. Sets of assumptions which simplify each of these processes to a form convenient for analysis are given in Table 5.1.

Pressure-volume diagrams for the constant-volume, constant-pressure, and limited-pressure cycles for unthrottled engine operation are illustrated in Fig. 5-2a to c. Throttled engine operation ($p_i < p_e$) and supercharged engine operation ($p_i > p_e$) are shown in Fig. 5-2d and e. In each cycle 1-2 is the compression process, 2-3 is the combustion process, 3-4 (or 2-4 in the constant-pressure cycle) is the expansion process, 4-5-6 is the exhaust process, and 6-7-1 is the intake process.

The most critical assumption in determining how useful these ideal cycles are as indicators of engine performance is the form assumed for the combustion process. The real engine combustion process occupies a finite crank angle period (between about 20 and 70 crank angle degrees), and the spark or fuel-injection timing may be retarded from its optimum advance to closer to TC. The constant-volume cycle is the limiting case of infinitely fast combustion at TC; the constant-pressure cycle would correspond to slow and late combustion; the limited-pressure cycle lies in between.

5.3 THERMODYNAMIC RELATIONS FOR ENGINE PROCESSES

The overall engine operating parameters of greatest interest which can be determined from a thermodynamic analysis of the engine operating cycle are:

The indicated fuel conversion efficiency $\eta_{f,i}$:

$$\eta_{f,i} = \frac{W_{c,i}}{m_f Q_{LHV}} \tag{5.1}$$

(which, since the combustion efficiency is unity, is equal to the indicated thermal conversion efficiency $\eta_{t,i}$; see Sec. 3.6.2)

The indicated mean effective pressure (imep):

$$\text{imep} = \frac{W_{c,i}}{V_d} = \frac{m_f Q_{LHV} \eta_{f,i}}{V_d} \tag{5.2}$$

$W_{c,i}$, the indicated work per cycle, is the sum of the compression stroke work and the expansion stroke work:

$$W_{c,i} = W_C + W_E \tag{5.3}$$

Using the notation of Fig. 5-2 to define the endpoints of each engine process, the following relationships are obtained by applying the first and second laws of thermodynamics to the cylinder contents:

Compression stroke:

$$\frac{v_1}{v_2} = r_c \tag{5.4}$$

Since the process is adiabatic and reversible

$$s_2 = s_1 \tag{5.5}$$

The compression work is

$$W_C = U_1 - U_2 = m(u_1 - u_2) \tag{5.6}$$

Combustion process:
For the constant-volume cycle,

$$v_3 = v_2 \qquad u_3 - u_2 = 0 \tag{5.7a, b}$$

For the constant-pressure cycle,

$$p_3 = p_2 \qquad h_3 - h_2 = 0 \tag{5.7c, d}$$

For the limited-pressure cycle,

$$v_{3a} = v_2 \qquad\qquad p_{3b} = p_{3a} \tag{5.7e, f}$$

and $\qquad\qquad u_{3a} - u_2 = 0 \qquad h_{3b} - h_{3a} = 0 \tag{5.7g, h}$

Expansion stroke:
For the constant-volume cycle,

$$\frac{v_4}{v_3} = r_c \qquad s_4 = s_3 \tag{5.8a, b}$$

and the expansion work is

$$W_E = U_3 - U_4 = m(u_3 - u_4) \tag{5.9}$$

For the constant-pressure cycle,

$$p_3 = p_2 \qquad \frac{v_4}{v_2} = r_c \qquad s_4 = s_3 \tag{5.10a, b, c}$$

and the expansion work is

$$\begin{aligned} W_E &= U_3 - U_4 + p_2(V_3 - V_2) \\ &= m[(u_3 - u_4) + p_2(v_3 - v_2)] \\ &= m[(h_3 - h_4) + p_4 v_4 - p_2 v_2] \end{aligned} \tag{5.11}$$

For the limited-pressure cycle,

$$v_4/v_{3a} = r_c \qquad p_{3b} = p_{3a} \qquad s_4 = s_{3b} \qquad \text{(5.12a, b, c)}$$

and the expansion work is

$$
\begin{aligned}
W_E &= U_{3b} - U_4 + p_3(V_{3b} - V_{3a}) \\
&= m[(u_{3b} - u_4) + p_3(v_{3b} - v_{3a})] \\
&= m[(h_{3b} - h_4) + p_4 v_4 - p_3 v_{3a}] \qquad \text{(5.13)}
\end{aligned}
$$

The indicated fuel conversion efficiency is found by substitution into Eqs. (5.3) and (5.1):

For the constant-volume cycle:

$$\eta_{f,i} = \frac{m[(u_3 - u_4) - (u_2 - u_1)]}{m_f Q_{LHV}} \qquad \text{(5.14)}$$

For the constant-pressure cycle:

$$\eta_{f,i} = \frac{m[(h_3 - h_4) - (u_2 - u_1) + p_4 v_4 - p_2 v_2]}{m_f Q_{LHV}} \qquad \text{(5.15)}$$

For the limited-pressure cycle:

$$\eta_{f,i} = \frac{m[(h_{3b} - h_4) - (u_2 - u_1) + p_4 v_4 - p_3 v_{3a}]}{m_f Q_{LHV}} \qquad \text{(5.16)}$$

The state of the mixture at point 1 in the cycle depends on the intake mixture properties and the residual gas properties at the end of the exhaust stroke.

When the exhaust valve opens at point 4, the cylinder pressure is above the exhaust manifold pressure and a *blowdown* process occurs. In the ideal exhaust process model, this blowdown occurs with the piston stationary at BC. During this blowdown process, the gas which remains inside the cylinder expands isentropically. The gases escaping from the cylinder undergo an unrestrained expansion or throttling process which is irreversible. It is assumed that the kinetic energy acquired by each gas element as it is accelerated through the exhaust valve is dissipated in a turbulent mixing process in the exhaust port into internal energy and flow work. Since it is also assumed that no heat transfer occurs, the enthalpy of each element of gas after it leaves the cylinder remains constant.

These processes are illustrated on an *h-s* diagram in Fig. 5-3. The gas remaining in the cylinder expands isentropically along the line 4-5. The first element of gas which leaves the cylinder at point 4 enters the exhaust manifold at state *a* on the pressure = p_e line. An element that leaves the cylinder at an intermediate state *b* on the expansion line 4-5 would enter the exhaust manifold at state *c*. At the end of the blowdown process the gas in the cylinder and the last

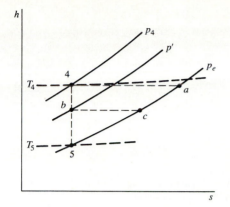

FIGURE 5-3
Enthalpy-entropy diagram of gas state during exhaust process. See text for explanation.

gas to leave have the same state—5. There is, therefore, a gradient in temperature within the exhausted gas. The temperature of the first element exhausted, T_a, is slightly less than T_4; the temperature of the last element exhausted is T_5.

A displacement of gas out of the cylinder follows the blowdown process as the piston moves from BC to TC. If heat-transfer and kinetic energy dissipation effects are neglected, no change in thermodynamic state of the gas occurs. In this *displacement* process, the mass of gas within the cylinder at the end of the blowdown process is further decreased by the ratio V_5/V_6.

The mass of residual gas m_r in the cylinder at point 6 in the cycle is obtained by first determining the state of the gas (T_5, v_5) at the end of the blowdown process following an isentropic expansion from p_4 to p_e and then by reducing the cylinder volume to the clearance volume V_6. The residual mass fraction is thus given by

$$\frac{m_r}{m} = x_r = \frac{v_4/v_5}{r_c} = \frac{v_2}{v_5} \tag{5.17}$$

The average state of the exhausted gas can be determined by considering the open system defined by the piston face, cylinder walls, and cylinder head, shown in Fig. 5-4. Applying the first law of thermodynamics for an open system gives

$$U_6 - U_4 = p_e(V_4 - V_6) - H_e \tag{5.18a}$$

where H_e is the enthalpy of the mass of gas exhausted from the cylinder. The average specific exhaust enthalpy is, therefore,

$$\bar{h}_e = \frac{m_4 u_4 - m_6 u_6 + p_e V_d}{m_4 - m_6} \tag{5.18b}$$

which, with $p = p_e$, defines the average exhausted-gas state.

The mixture temperature at the end of the intake stroke and at the start of the compression stroke (point 1 in Fig. 5-2) can now be determined, again using

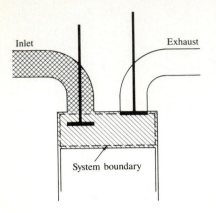

FIGURE 5-4
Definition of system boundary for thermodynamic analysis of ideal cycle processes.

the open system in Fig. 5-4. Application of the first law between points 6 and 1 gives

$$U_1 - U_6 = -p_i(V_1 - V_6) + (m_1 - m_6)h_i \tag{5.19a}$$

or
$$m_1 u_1 - m_6 u_6 = -p_i(V_1 - V_6) + (m_1 - m_6)h_i \tag{5.19b}$$

·or
$$m_1 h_1 = m_6 h_6 + (m_1 - m_6)h_i + V_2(p_i - p_e) \tag{5.19c}$$

where h_i is the specific enthalpy of the inlet mixture and $p_1 = p_i$.

Note that when $p_i < p_e$, part of the residual gas in the cylinder at the end of the exhaust stroke will flow into the intake system when the intake valve opens. This flow will cease when the cylinder pressure equals p_i. However, provided no heat transfer occurs, this backflow will not affect Eqs. (5.19) above, since the flow of residual through the intake valve is a constant enthalpy process.

In many engines, the closing of the exhaust valve and the opening of the intake valve overlap. Flow of exhausted gases from the exhaust system through the cylinder into the intake system can then occur. Equations (5.18) and (5.19) would have to be modified to account for valve overlap.

In the four-stroke engine cycle, work is done on the piston during the intake and the exhaust processes. The work done by the cylinder gases on the piston during exhaust is

$$W_e = p_e(V_2 - V_1) \tag{5.20}$$

The work done by the cylinder gases on the piston during intake is

$$W_i = p_i(V_1 - V_2) \tag{5.21}$$

The net work to the piston over the exhaust and intake strokes, the *pumping work*, is

$$W_p = (p_i - p_e)(V_1 - V_2) \tag{5.22}$$

which, for the cylinder gas system, is negative for $p_i < p_e$ and positive for $p_i > p_e$.

The pumping mean effective pressure (pmep) is usually defined as a positive quantity. Thus:

For $p_i < p_e$: $$\text{pmep} = p_e - p_i \qquad (5.23a)$$

For $p_i > p_e$: $$\text{pmep} = p_i - p_e \qquad (5.23b)$$

The net and gross indicated mean effective pressures are related by

$$\text{imep}_n = \text{imep}_g - (p_e - p_i) \qquad (5.24)$$

The net indicated fuel conversion efficiency is related to the gross indicated fuel conversion efficiency by

$$\eta_{f,\text{in}} = \eta_{f,\text{ig}}\left(1 - \frac{p_e - p_i}{\text{imep}_g}\right) \qquad (5.25)$$

5.4 CYCLE ANALYSIS WITH IDEAL GAS WORKING FLUID WITH c_v AND c_p CONSTANT

If the working fluid in these ideal cycles is assumed to be an ideal gas, with c_v and c_p constant throughout the engine operating cycle, the equations developed in the previous section which describe engine performance and efficiency can be further simplified. We will use the notation of Fig. 5-2.

5.4.1 Constant-Volume Cycle

The compression work (Eq. 5.6) becomes

$$W_C = mc_v(T_1 - T_2) \qquad (5.26)$$

The expansion work (Eq. 5.9) becomes

$$W_E = mc_v(T_3 - T_4) \qquad (5.27)$$

The denominator in Eq. (5.14), $m_f Q_{\text{LHV}}$, can be related to the temperature rise during combustion. For the working fluid model under consideration, the $U(T)$ lines for the reactants and products on a U-T diagram such as Fig. 3-5 are parallel and have equal slopes, of magnitude c_v. Hence, for a constant-volume adiabatic combustion process

$$mc_v(T_3 - T_2) = m_f Q_{\text{LHV}} \qquad (5.28)\dagger$$

---

† Note that if insufficient air is available for complete combustion of the fuel, Eq. (5.28) must be modified. The right-hand side of the equation should then be $\eta_c m_f Q_{\text{LHV}}$, where η_c is the combustion efficiency given by Eq. (3.27).

Note that the heating values at constant volume and constant pressure are the same for this working fluid. For convenience we will define

$$Q* = \frac{m_f Q_{LHV}}{m} \tag{5.29}$$

$Q*$ is the specific internal energy (and enthalpy) decrease, during isothermal combustion, per unit mass of working fluid.

The relation for indicated fuel conversion efficiency (Eq. 5.14) becomes

$$\eta_{f,i} = \frac{(T_3 - T_4) - (T_2 - T_1)}{T_3 - T_2} = 1 - \frac{T_4 - T_1}{T_3 - T_2} \tag{5.30}$$

Since 1-2 and 3-4 are isentropic processes between the same volumes, V_1 and V_2,

$$\frac{T_2}{T_1} = \left(\frac{V_1}{V_2}\right)^{\gamma-1} = r_c^{\gamma-1} = \left(\frac{V_4}{V_3}\right)^{\gamma-1} = \frac{T_3}{T_4}$$

where $\gamma = c_p/c_v$. Hence:

$$\frac{T_4}{T_1} = \frac{T_3}{T_2}$$

and Eq. (5.30) can be rearranged as

$$\eta_{f,i} = 1 - \frac{1}{r_c^{\gamma-1}} \tag{5.31}$$

Values of $\eta_{f,i}$ for different values of γ are shown in Fig. 5-5. The indicated fuel conversion efficiency increases with increasing compression ratio and decreases as γ decreases.

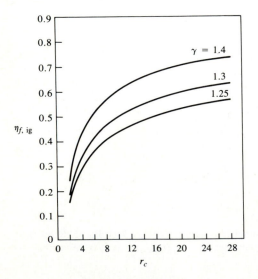

FIGURE 5-5
Ideal gas constant-volume cycle fuel conversion efficiency as a function of compression ratio; $\gamma = c_p/c_v$.

The indicated mean effective pressure, using Eqs. (5.2) and (5.31), becomes

$$\frac{\text{imep}}{p_1} = \left(\frac{Q^*}{c_v T_1}\right)\left(\frac{1}{\gamma - 1}\right)\left(\frac{r_c}{r_c - 1}\right)\left(1 - \frac{1}{r_c^{\gamma-1}}\right) \qquad (5.32)$$

The dimensionless numbers r_c, γ, and $Q^*/(c_v T_1)$ are sufficient to describe the characteristics of the constant-volume ideal gas standard cycle, relative to its initial conditions p_1, T_1.

It is useful to compare the imep—a measure of the effectiveness with which the displaced volume of the engine is used to produce work—and the maximum pressure in the cycle, p_3. The ratio p_3/p_1 can be determined from the ideal gas law applied at points 2 and 3, and the relation

$$\frac{T_3}{T_2} = 1 + \frac{Q^*}{c_v T_1 r_c^{\gamma-1}} \qquad (5.33)$$

obtained from Eq. (5.28). Equations (5.32) and (5.33) then give

$$\frac{\text{imep}}{p_3} = \frac{1}{(\gamma - 1)r_c^\gamma}\left(\frac{r_c}{r_c - 1}\right)\frac{1 - 1/r_c^{\gamma-1}}{c_v T_1/Q^* + 1/r_c^{\gamma-1}} \qquad (5.34)$$

A high value of imep/p_3 is desirable. Engine weight will increase with increasing p_3 to withstand the increasing stresses in components.

The indicated fuel conversion efficiency and the ratios imep/p_1 and imep/p_3 for this ideal cycle model do not depend on whether the cycle is throttled or supercharged. However, the relationships between the working fluid properties at points 1 and 6 do depend on the degree of throttling or supercharging. For throttled engine operation, the residual gas mass fraction x_r can be determined as follows. From Eq. (5.17), since state 5 corresponds to an isentropic expansion from state 4 to $p = p_e$, x_r is given by

$$x_r = \frac{(p_e/p_4)^{1/\gamma}}{r_c} = \frac{(p_e/p_i)^{1/\gamma}(p_1/p_4)^{1/\gamma}}{r_c}$$

Since

$$\frac{p_1}{p_4} = \frac{p_1}{p_2}\frac{p_2}{p_3}\frac{p_3}{p_4} = \frac{1}{r_c^\gamma}\frac{T_2}{T_3}r_c^\gamma = \left(1 + \frac{Q^*}{c_v T_1 r_c^{\gamma-1}}\right)^{-1}$$

it follows that

$$x_r = \frac{1}{r_c}\frac{(p_e/p_i)^{1/\gamma}}{[1 + Q^*/(c_v T_1 r_c^{\gamma-1})]^{1/\gamma}} \qquad (5.35)$$

The residual mass fraction increases as p_i decreases below p_e, decreases as r_c increases, and decreases as $Q^*/(c_v T_1)$ increases.

Through a similar analysis, the temperature of the residual gas T_6 can be determined:

$$\frac{T_6}{T_1} = \left(\frac{p_e}{p_i}\right)^{(\gamma-1)/\gamma}\left(1 + \frac{Q^*}{c_v T_1 r_c^{\gamma-1}}\right)^{1/\gamma} \qquad (5.36)$$

The mixture temperature at point 1 in the cycle can be related to the inlet mixture temperature, T_i, with Eq. (5.19). For a working fluid with c_v and c_p constant, this equation becomes

$$c_p T_1 = x_r c_p T_6 + (1 - x_r)c_p T_i - \frac{RT_1}{r_c}\left(\frac{p_e}{p_i} - 1\right) \qquad (5.37)$$

Use of Eqs. (5.36) and (5.37) leads to the relation

$$\frac{T_1}{T_i} = \frac{1 - x_r}{1 - 1/(\gamma r_c)[p_e/p_i + (\gamma - 1)]} \qquad (5.38)$$

Extensive results for the constant-volume cycle with $\gamma = 1.4$ can be found in Taylor.[1]

5.4.2 Limited- and Constant-Pressure Cycles

The constant-pressure cycle is a limited-pressure cycle with $p_3 = p_2$. For the limited-pressure cycle, the compression work remains

$$W_C = mc_v(T_1 - T_2) \qquad (5.39)$$

The expansion work, from Eq. (5.13), becomes

$$W_E = m[c_v(T_{3b} - T_4) + p_3(v_{3b} - v_{3a})] \qquad (5.40)$$

For the combustion process, Eqs. (5.7g, h) give

$$m_{f2-3a} Q_{LHV} = mc_v(T_{3a} - T_2) \qquad (5.41a)$$

$$m_{f3a-3b} Q_{LHV} = mc_p(T_{3b} - T_{3a}) \qquad (5.41b)$$

or $\qquad\qquad m_f Q_{LHV} = m[c_v(T_{3a} - T_2) + c_p(T_{3b} - T_{3a})] \qquad (5.41c)$

for a working fluid with c_v and c_p constant throughout the cycle.

Combining Eqs. (5.1), (5.3), and (5.39) to (5.41) and simplifying gives

$$\eta_{f,i} = 1 - \frac{T_4 - T_1}{(T_{3a} - T_2) + \gamma(T_{3b} - T_{3a})}$$

Use of the isentropic relationships for the working fluid along 1-2 and 3b-4, with the substitutions

$$\alpha = \frac{p_3}{p_2} \qquad \beta = \frac{V_{3b}}{V_{3a}} \qquad (5.42a, b)$$

leads to the result

$$\eta_{f,i} = 1 - \frac{1}{r_c^{\gamma-1}}\left[\frac{\alpha\beta^\gamma - 1}{\alpha\gamma(\beta - 1) + \alpha - 1}\right] \qquad (5.43)$$

For $\beta = 1$ this result becomes the constant-volume cycle efficiency (Eq. 5.31). For $\alpha = 1$, this result gives the constant-pressure cycle efficiency as a special case.

The mean effective pressure is related to p_1 and p_3 via

$$\frac{\text{imep}}{p_1} = \frac{Q^*}{c_v T_1 (\gamma - 1)} \left(\frac{r_c}{r_c - 1}\right) \eta_{f,i} \tag{5.44}$$

$$\frac{\text{imep}}{p_3} = \frac{1}{\alpha r_c^\gamma} \left(\frac{Q^*}{c_v T_1}\right) \left(\frac{1}{\gamma - 1}\right) \left(\frac{r_c}{r_c - 1}\right) \eta_{f,i} \tag{5.45}$$

5.4.3 Cycle Comparison

The above expressions are most useful if values for γ and $Q^*/(c_v T_1)$ are chosen to match real working fluid properties. Figure 5-5 has already shown the sensitivity of $\eta_{f,i}$ for the constant-volume cycle to the value of γ chosen. In Sec. 4.4, average values of γ_u and γ_b were determined which match real working fluid properties over the compression and expansion strokes, respectively. Values for a stoichiometric mixture appropriate to an SI engine are $\gamma_u \approx 1.3$, $\gamma_b \approx 1.2$. However, analysis of pressure-volume data for real engine cycles indicates that pV^n, where $n \approx 1.3$, is a good fit to the expansion stroke p-V data.[2] Heat transfer from the burned gases increases the exponent above the value corresponding to γ_b. A value of $\gamma = 1.3$ for the entire cycle is thus a reasonable compromise.

Q^*, defined by Eq. (5.29), is the enthalpy decrease during isothermal combustion per unit mass of working fluid. Hence

$$Q^* = \left(\frac{m_f}{m_a}\right) Q_{\text{LHV}} \left(\frac{m_a}{m}\right) \tag{5.46}$$

A simple approximation for (m_a/m) is $(r_c - 1)/r_c$; i.e., fresh air fills the displaced volume and the residual gas fills the clearance volume at the same density. Then, for isooctane fuel, for a stoichiometric mixture, Q^* is given by 2.92×10^6 $(r_c - 1)/r_c$ J/kg air. For $\gamma = 1.3$ and an average molecular weight $M = 29.3$, $c_v = 946$ J/kg·K. For $T_1 = 333$ K, $Q^*/(c_v T_1)$ becomes $9.3 (r_c - 1)/r_c$. For this value of $Q^*/(c_v T_1)$ all cycles would be burning a stoichiometric mixture with an appropriate residual gas fraction.

Pressure-volume diagrams for the three ideal cycles for the same compression ratio and unburned mixture composition are shown in Fig. 5-6. For each cycle, $\gamma = 1.3$, $r_c = 12$, $Q^*/(c_v T_1) = 9.3(r_c - 1)/r_c = 8.525$. Overall performance characteristics for each of these cycles are summarized in Table 5.2. The constant-volume cycle has the highest efficiency, the constant-pressure cycle the lowest efficiency. This can be seen from Eq. (5.43) where the term in square brackets is equal to unity for the constant-volume cycle and greater than unity for the limited- and constant-pressure cycles. The imep values are proportional to $\eta_{f,i}$ since the mass of fuel burned per cycle is the same in all three cases.

As the peak pressure p_3 is decreased, the ratio of imep to p_3 increases. This ratio is important because imep is a measure of the useful pressure on the piston, and the maximum pressure chiefly affects the strength required of the engine structure.

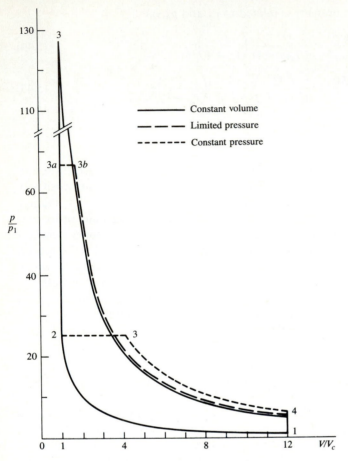

FIGURE 5-6
Pressure-volume diagrams for constant-volume, limited-pressure, and constant-pressure ideal gas standard cycles. $r_c = 12$, $\gamma = 1.3$, $Q^*/(c_v T_1) = 9.3(r_c - 1)/r_c = 8.525$, $p_{3a}/p_1 = 67$.

TABLE 5.2
Comparison of ideal cycle results

	$\eta_{f,i}$	$\dfrac{\text{imep}}{p_1}$	$\dfrac{\text{imep}}{p_3}$	$\dfrac{p_{max}}{p_1}$
Constant volume	0.525	16.3	0.128	128
Limited pressure	0.500	15.5	0.231	67
Constant pressure	0.380	11.8	0.466	25.3

$\gamma = 1.3$; $r_c = 12$; $Q^*/(c_v T_1) = 8.525$.

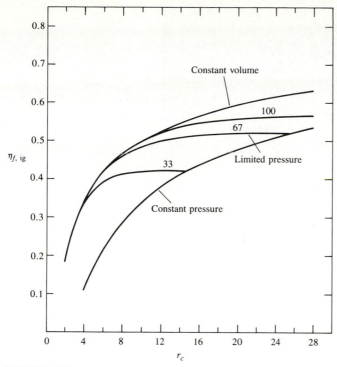

FIGURE 5-7
Fuel conversion efficiency as a function of compression ratio, for constant-volume, constant-pressure, and limited-pressure ideal gas cycles. $\gamma = 1.3$, $Q^*/(c_v T_1) = 9.3(r_c - 1)/r_c$. For limited-pressure cycle, $p_3/p_1 = 33, 67, 100$.

A more extensive comparison of the three cycles is given in Figs. 5-7 and 5-8, over a range of compression ratios. For all cases $\gamma = 1.3$ and $Q^*/(c_v T_1) = 9.3(r_c - 1)/r_c$. At any given r_c, the constant-volume cycle has the highest efficiency and lowest imep/p_3. For a given maximum pressure p_3, the constant-pressure cycle has the highest efficiency (and the highest compression ratio). For the limited-pressure cycle, at constant p_3/p_1, there is little improvement in efficiency and imep above a compression ratio of about 8 to 10 as r_c is increased.

Example 5.1 shows how ideal cycle equations relate residual and intake conditions with the gas state at point 1 in the cycle. An iterative procedure is required if intake conditions are specified.

Example 5.1. For $\gamma = 1.3$, compression ratio $r_c = 6$, and a stoichiometric mixture with intake temperature 300 K, find the residual gas fraction, residual gas temperature, and mixture temperature at point 1 in the constant-volume cycle for $p_e/p_i = 1$ (unthrottled operation) and 2 (throttled operation).

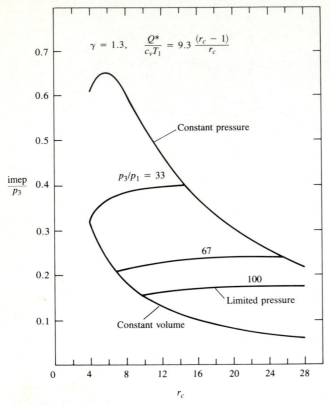

FIGURE 5-8
Indicated mean effective pressure (imep) divided by maximum cycle pressure (p_3) as a function of compression ratio for constant-volume, constant-pressure, and limited-pressure cycles. Details same as Fig. 5-7.

For a stoichiometric mixture, for isooctane,

$$Q^* = \frac{m_f}{m} Q_{LHV} = \left(\frac{m_f}{m_i}\right)\left(\frac{m_i}{m}\right) Q_{LHV} = \frac{44.38}{16.14}(1 - x_r) = 2.75(1 - x_r) \qquad \text{MJ/kg}$$

For $\gamma = 1.3$, $c_v = 946$ J/kg·K and

$$\frac{Q^*}{c_v T_1} = \frac{2.75 \times 10^6}{946 T_1}(1 - x_r) = \frac{2910}{T_1}(1 - x_r) \qquad (a)$$

Equations (5.35), (5.36), and (5.38), for $r_c = 6$ and $\gamma = 1.3$, become

$$x_r = \frac{1}{6} \frac{(p_e/p_i)^{0.769}}{[1 + Q^*/(c_v T_1 \times 6^{0.3})]^{0.769}} \qquad (b)$$

$$\frac{T_r}{T_1} = \left(\frac{p_e}{p_i}\right)^{0.23}\left(1 + \frac{Q^*}{c_v T_1 \times 6^{0.3}}\right)^{0.769} \qquad (c)$$

$$\frac{T_1}{300} = \frac{1 - x_r}{1 - [1/(1.3 \times 6)](p_e/p_i + 0.3)} \tag{d}$$

A trial-and-error solution of Eqs. (a) to (d) is required. It is easiest to estimate x_r, solve for T_1 from (d), evaluate $Q^*/(c_v T_1)$ from (a), and check the value of x_r assumed with that given by (b).

For $(p_e/p_i) = 1$ (unthrottled operation) the following solution is obtained:

$$x_r = 0.044, \qquad T_1 = 344 \text{ K}, \qquad \frac{Q^*}{c_v T_1} = 8.1, \qquad T_r = 1316 \text{ K}$$

For $(p_e/p_i) = 2$ the following solution is obtained:

$$x_r = 0.082, \qquad T_1 = 391 \text{ K}, \qquad \frac{Q^*}{c_v T_1} = 6.8, \qquad T_r = 1580 \text{ K}$$

5.5 FUEL-AIR CYCLE ANALYSIS

A more accurate representation of the properties of the working fluid inside the engine cylinder is to treat the unburned mixture as frozen in composition and the burned gas mixture as in equilibrium. Values for thermodynamic properties for these working fluid models can be obtained with the charts for unburned and burned gas mixtures described in Sec. 4.5, or the computer codes summarized in Sec. 4.7. When these working fluid models are combined with the ideal engine process models in Table 5.1, the resulting cycles are called fuel-air cycles.[1] The sequence of processes and assumptions are (with the notation of Fig. 5-2):

1-2 Reversible adiabatic compression of a mixture of air, fuel vapor, and residual gas without change in chemical composition.

2-3 Complete combustion (at constant volume or limited pressure or constant pressure), without heat loss, to burned gases in chemical equilibrium.

3-4 Reversible adiabatic expansion of the burned gases which remain in chemical equilibrium.

4-5-6 Ideal adiabatic exhaust blowdown and displacement processes with the burned gases fixed in chemical composition.

6-7-1 Ideal intake process with adiabatic mixing between residual gas and fresh mixture, both of which are fixed in chemical composition.

The basic equations for each of these processes have already been presented in Sec. 5.3. The use of the charts for a complete engine cycle calculation will now be illustrated.

5.5.1 SI Engine Cycle Simulation

The mixture conditions at point 1 must be known or must be estimated. The following approximate relationships can be used for this purpose:[3]

$$x_r = \left\{1 + \frac{T_r}{T_i}\left[r_c\left(\frac{p_i}{p_e}\right) - \left(\frac{p_i}{p_e}\right)^{(\gamma-1)/\gamma}\right]\right\}^{-1} \tag{5.47}$$

$$T_1 = T_r r_c \, x_r\left(\frac{p_i}{p_e}\right) \tag{5.48}$$

where $T_r = 1400$ K and $(\gamma - 1)/\gamma = 0.24$ are appropriate average values to use for initial estimates.

Given the equivalence ratio ϕ and initial conditions T_1 (K), $p_1 = p_i$ (Pa), and v_1 (m^3/kg air), the state at point 2 at the end of compression through a volume ratio $v_1/v_2 = r_c$ is obtained from Eq. (4.25a) and the isentropic compression chart (Fig. 4-4). The compression work W_C (J/kg air) is found from Eq. (5.6) with the internal energy change determined from the unburned mixture chart (Fig. 4-3).

The use of charts to relate the state of the burned mixture to the state of the unburned mixture prior to combustion, for adiabatic constant-volume and constant-pressure combustion, has already been illustrated in Sec. 4.5.3.

For the *constant-volume cycle*,

$$u_3 = u_{s2} + \Delta u_{f,u}^\circ \qquad \text{J/kg air} \tag{5.49}$$

where u_{s2} is the sensible internal energy of the unburned mixture at T_2 from Fig. 4-3 and $\Delta u_{f,u}^\circ$ is the internal energy of formation of the unburned mixture [given by Eq. (4.32)]. Since $v_3 = v_2$, the burned gas state at point 3 can be located on the appropriate burned gas chart (Figs. 4-5 to 4-9).

For the *constant-pressure cycle*,

$$h_3 = h_{s2} + \Delta h_{f,u}^\circ \qquad \text{J/kg air} \tag{5.50}$$

Since $p_3 = p_2$, the burned gas state at point 3 can be located (by iteration) on the high-temperature burned gas charts, as illustrated by Example 4.5.

For the *limited-pressure cycle*, application of the first law to the mixture between states 2 and 3b gives

$$h_{3b} = u_{3b} + p_3 v_{3b} = u_2 + p_3 v_2 = u_{s2} + \Delta u_{f,u}^\circ + p_3 v_2 \qquad \text{J/kg air} \tag{5.51}$$

Since p_3 for a limited-pressure cycle is given, point 3b can be located on the appropriate burned gas chart.

The expansion process 3-4 follows an isentropic line from v_3 to v_4 ($v_4 = v_1$) on the burned mixture charts. Equation (5.9) [or (5.11) or (5.13)] now gives the expansion work W_E. The state of the residual gas at points 5 and 6 in the cycle is obtained by continuing this isentropic expansion from state 4 to $p = p_e$. The residual gas temperature can be read from the equilibrium burned gas chart; the residual gas fraction is obtained from Eq. (5.17). If values of T_r and x_r were assumed at the start of the cycle calculation to determine T_1, the assumed values can be checked against the calculated values and an additional cycle computation carried out with the new calculated values if required. The convergence is rapid.

The indicated fuel conversion efficiency is obtained from Eq. (5.1). The indicated mean effective pressure is obtained from Eq. (5.2). The volumetric efficiency (see Sec. 2.10) for a four-stroke cycle engine is given by

$$\eta_v = \frac{r_c(1 - x_r)}{v_1 \rho_{a,i}(r_c - 1)} \tag{5.52}$$

where $\rho_{a,i}$ is the inlet air density (in kilograms per cubic meter) and v_1 is the chart mixture specific volume (in cubic meters per kilogram of air in the original mixture).

Example 5.2. Calculate the performance characteristics of the constant-volume fuel-air cycle defined by the initial conditions of Examples 4.2, 4.3, and 4.5. The compression ratio is 8; the fuel is isooctane and the mixture is stoichiometric; the pressure and temperature inside the cylinder at the start of compression are 1 atm and 350 K, respectively. Use the notation of Fig. 5-2a to define the states at the beginning and end of each process.

Example 4.2 analyzed the compression process:

$$T_1 = 350 \text{ K}, \qquad p_1 = 101.3 \text{ kPa}, \qquad v_1 = 1 \text{ m}^3/\text{kg air}, \qquad u_{s1} = 40 \text{ kJ/kg air}$$

$$T_2 = 682 \text{ K}, \qquad p_2 = 1.57 \text{ MPa}, \qquad v_2 = 0.125 \text{ m}^3/\text{kg air}, \qquad u_{s2} = 350 \text{ kJ/kg air}$$

$$W_{1\text{-}2} = W_C = -310 \text{ kJ/kg air}$$

Example 4.5 analyzed the constant-volume adiabatic combustion process (it was assumed that the residual gas fraction was 0.08):

$$u_{b3} = u_{u2} = u_{s,u2} + \Delta u_{f,u}^\circ = -5 \text{ kJ/kg air}, \qquad s_3 = 9.33 \text{ kJ/kg air} \cdot \text{K}$$

$$v_3 = v_2 = 0.125 \text{ m}^3/\text{kg air}, \qquad T_3 = 2825 \text{ K}, \qquad p_3 = 7100 \text{ kPa}$$

Example 4.3 analyzed the expansion process, from these conditions after combustion at TC, to the volume v_4 at BC of 1 m^3/kg air:

$$T_4 = 1840 \text{ K}, \qquad p_4 = 570 \text{ kPa}, \qquad u_4 = -1540 \text{ kJ/kg air}$$

$$W_{3\text{-}4} = W_E = 1535 \text{ kJ/kg air}$$

To check the assumed residual gas fraction, the constant entropy expansion process on the chart in Fig. 4-8 is continued from state 4 to the exhaust pressure p_5 of 1 atm = 101.3 kPa. This gives $v_5 = 4.0$ m^3/kg air and $T_5 = 1320$ K. The residual fraction from Eq. (5.17) is

$$x_r = \frac{v_2}{v_5} = \frac{0.125}{4.0} = 0.031$$

which is significantly different from the assumed value of 0.08. The combustion and expansion calculations are now repeated with the new residual fraction of 0.031 (the compression process will not be changed significantly and the initial temperature is assumed fixed):

$$u_{b3} = 350 - 118.2 - 2956 \times 0.031 = 140 \text{ kJ/kg air}$$

With $v_3 = 0.125 \text{ m}^3/\text{kg air}$, Fig. 4-8 gives

$$p_3 = 7270 \text{ kPa}, \qquad T_3 = 2890 \text{ K}$$

Expand at constant entropy to $v_4 = 1 \text{ m}^3/\text{kg air}$:

$$p_4 = 595 \text{ kPa}, \qquad T_4 = 1920 \text{ K}, \qquad u_4 = -1457 \text{ kJ/kg air}$$

$$W_{3-4} = W_E = 1597 \text{ kJ/kg air}$$

Continue expansion at constant entropy to the exhaust pressure, $p_5 = 1$ atm:

$$v_5 = 4 \text{ m}^3/\text{kg air}, \qquad T_5 = 1360 \text{ K}$$

Equation (5.17) now gives the residual fraction

$$x_r = \frac{v_2}{v_5} = \frac{0.125}{4} = 0.031$$

which agrees with the value assumed for the second iteration.
    The fuel conversion efficiency can now be calculated:

$$\eta_{f,i} = \frac{W_E + W_C}{m_f Q_{\text{LHV}}}$$

where
$$m_f = \text{kg fuel/kg air at state } 1 = \left(\frac{F}{A}\right)(1 - x_r)$$

Thus

$$\eta_{f,i} = \frac{1597 - 310}{0.0661 \times (1 - 0.031) \times 44.4 \times 10^3} = 0.45$$

The indicated mean effective pressure is

$$\text{imep} = \frac{W_E + W_C}{V_d} = \frac{1597 - 310}{1 - 0.125} = 1470 \text{ kPa}$$

or
$$\frac{\text{imep}}{p_1} = 14.6$$

5.5.2 CI Engine Cycle Simulation

With a diesel engine fuel-air cycle calculation, additional factors must be taken into account. The mixture during compression is air plus a small amount of residual gas. At point 2 liquid fuel is injected into the hot compressed air at temperature T_2; as the fuel vaporizes and heats up, the air is cooled. For a constant-volume mixing process which is adiabatic overall, the mixture internal energy is unchanged, i.e.:

$$m_f[u_{fg} + c_{v,f}(T_{2'} - T_0)] + m_a c_{v,a}(T_{2'} - T_2) = 0 \qquad (5.53)$$

where m_f is the mass of fuel injected, u_{f_g} is the latent heat of vaporization of the fuel, $c_{v,f}$ is the specific heat at constant volume of the fuel vapor, $T_{2'}$ is the mixture temperature (assumed uniform) after vaporization and mixing is complete, m_a is the mass of air used, and $c_{v,a}$ is the specific heat at constant volume of air. Substitution of typical values for fuel and air properties gives $(T_2 - T_{2'}) \approx$ 70 K at full load. Localized cooling in a real engine will be greater.

The limited-pressure cycle is a better approximation to the diesel engine than the constant-pressure or constant-volume cycles.

Note that because nonuniformities in the fuel/air ratio exist during and after combustion in the CI engine, the burned gas charts which assume uniform composition will not be as accurate an approximation to working fluid properties as they are for SI engines.

5.5.3 Results of Cycle Calculations

Extensive results of constant-volume fuel-air cycle calculations are available.[1, 3, 4] Efficiency is little affected by variables other than the compression ratio r_c and equivalence ratio ϕ. Figures 5-9 and 5-10 show the effect of variations in these two parameters on indicated fuel conversion efficiency and mean effective pressure. From the available results, the following conclusions can be drawn:

1. The effect of increasing the compression ratio on efficiency at a constant equivalence ratio is similar to that demonstrated by the constant γ constant-volume cycle analysis (provided the appropriate value of γ is used; see Fig. 5-19).

2. As the equivalence ratio is decreased below unity (i.e., the fuel-air mixture is made progressively leaner than stoichiometric), the efficiency increases. This occurs because the burned gas temperatures after combustion decrease, decreasing the burned gas specific heats and thereby increasing the effective value of γ over the expansion stroke. The efficiency increases because, for a given volume-expansion ratio, the burned gases expand through a larger temperature ratio prior to exhaust; therefore, per unit mass of fuel, the expansion stroke work is increased.

3. As the equivalence ratio increases above unity (i.e., the mixture is made progressively richer than stoichiometric), the efficiency decreases because lack of sufficient air for complete oxidation of the fuel more than offsets the effect of decreasing burned gas temperatures which decrease the mixture's specific heats.

4. The mean effective pressure, from Eq. (5.2), is proportional to the product $\phi \eta_{f,i}$. This exhibits a maximum between $\phi \approx 1.0$ and $\phi \approx 1.1$, i.e., slightly rich of stoichiometric. For ϕ less than the value corresponding to this maximum, the decreasing fuel mass per unit displaced volume more than offsets the increasing fuel conversion efficiency. For ϕ greater than this value, the decreasing fuel conversion efficiency (due to decreasing combustion efficiency) more than offsets the increasing fuel mass.

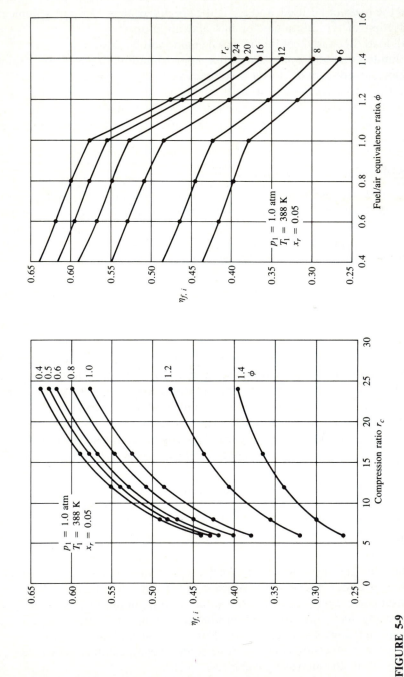

182

FIGURE 5-9
Fuel-air cycle results for indicated fuel conversion efficiency as a function of compression ratio and equivalence ratio. Fuel: octene; $p_1 = 1$ atm, $T_1 = 388$ K, $x_r = 0.05$. (*From Edson and Taylor.*[4])

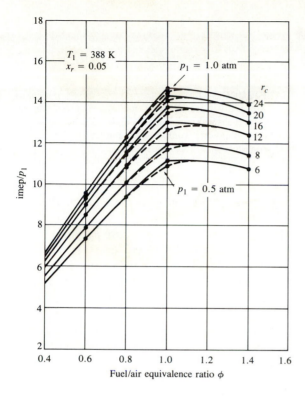

$T_1 = 388$ K
$x_r = 0.05$

$p_1 = 1.0$ atm

r_c
24
20
16
12
8
6

$p_1 = 0.5$ atm

imep/p_1

Fuel/air equivalence ratio ϕ

FIGURE 5-10
Fuel-air cycle results for indicated mean effective pressure as a function of equivalence ratio and compression ratio. Fuel: octene; $p_1 = 1$ atm, $T_1 = 388$ K, $x_r = 0.05$. (*From Edson and Taylor.*[4])

5. Variations in initial pressure, inlet temperature, residual gas fraction, and atmospheric moisture fraction have only a modest effect on the fuel conversion efficiency. The effects of variations in these variables on imep are more substantial, however, because imep depends directly on the initial charge density.

6. Comparison of results from limited-pressure and constant-volume fuel-air cycles[1] shows that placing a realistic limit on the maximum pressure reduces the advantages of increased compression ratio on both efficiency and imep.

5.6 OVEREXPANDED ENGINE CYCLES

The gas pressure within the cylinder of a conventional four-stroke engine at exhaust valve opening is greater than the exhaust pressure. The available energy of the cylinder gases at this point in the cycle is then dissipated in the exhaust blowdown process. Additional expansion within the engine cylinder would increase the indicated work per cycle, as shown in Fig. 5-11, where expansion continues beyond point 4' (the conventional ideal cycle exhaust valve opening point) at $V_{4'} = r_c V_c$ to point 4 at $V_4 = r_e V_c$. The exhaust stroke in this overexpanded cycle is 4-5-6. The intake stroke is 6-1. The area 14'451 has been added

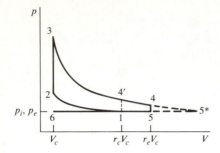

FIGURE 5-11
Pressure-volume diagram for overexpanded engine cycle (1234561) and Atkinson cycle (1235*61). r_c and r_e are volumetric compression and expansion ratios, respectively.

to the conventional cycle p-V diagram area, for the same fuel input, thereby increasing the engine's efficiency.

Complete expansion within the cylinder to exhaust pressure p_e (point 5*) is called the *Atkinson cycle*. Unthrottled operation is shown in Fig. 5-11; throttled operating cycles can also be generated. Many crank and valve mechanisms have been proposed to achieve this additional expansion. For example, it can be achieved in a conventional four-stroke cycle engine by suitable choice of exhaust valve opening and intake valve closing positions relative to BC. If the crank angle between exhaust valve opening and BC on the expansion stroke is less than the crank angle between BC and intake valve closing on the compression stroke, then the actual volumetric expansion ratio is greater than the actual volumetric compression ratio (these *actual* ratios are both less than the *nominal* compression ratio with normal valve timing).

The effect of overexpansion on efficiency can be estimated from an analysis of the ideal cycle shown in Fig. 5-11. An ideal gas working fluid with specific heats constant throughout the cycle will be assumed. The indicated work per cycle for the overexpanded cycle is

$$W_{c,i} = m[(u_3 - u_4) - (u_2 - u_1) - p_1(v_4 - v_1)] \tag{5.54}$$

The isentropic relations for 1-2 and 3-4 are

$$\frac{T_2}{T_1} = r_c^{\gamma-1} \qquad \frac{T_3}{T_4} = r_e^{\gamma-1}$$

With Eq. (5.33) to relate T_3 and T_2, the following expression for indicated fuel conversion efficiency can be derived from Eqs. (5.1), (5.29), and (5.54):

$$\eta_{f,i} = 1 - \frac{1}{(rr_c)^{\gamma-1}}\left\{1 + \frac{c_v T_1}{Q^*}\, r_c^{\gamma-1}[1 - \gamma r^{\gamma-1} + (\gamma - 1)r^\gamma]\right\} \tag{5.55}$$

where

$$r = \frac{r_e}{r_c}$$

Note that the efficiency given by Eq. (5.55) is a function of load (via Q^*), and is a maximum at maximum load. This contrasts with the ideal constant-volume cycle efficiency [Eq. (5.31)], which is independent of load. The ratio r_e/r_c for complete expansion is given by

$$r^\gamma = 1 + \frac{Q^*}{c_v T_1 r_c^{\gamma-1}} \tag{5.56}$$

The effect of overexpansion on fuel conversion efficiency is shown in Fig. 5-12 for $r_c = 4$, 8, and 16 with $\gamma = 1.3$. The ratio of overexpanded cycle efficiency to the standard cycle efficiency is plotted against r. The Atkinson cycle (complete expansion) values are indicated by the transition from a continuous line to a dashed line. Significant increases in efficiency can be achieved, especially at low compression ratios.

One major disadvantage of this cycle is that imep and power density decrease significantly because only part of the total displaced volume is filled with fresh charge. From Eqs. (5.2), (5.29), and the relations $V_d = V_1(r_e - 1)/r_c$ and

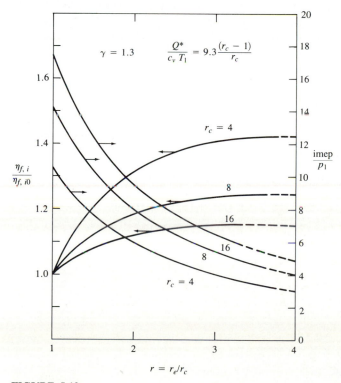

FIGURE 5-12
Indicated fuel conversion efficiency and mean effective pressure for overexpanded engine cycle as a function of r_e/r_c. Efficiencies given relative to $r_e = r_c$ value, $\eta_{f,io}$. $\gamma = 1.3$, $Q^*/(c_v T_1) = 9.3(r_c - 1)/r_c$. Solid to dashed line transition marks the complete expansion point (Atkinson cycle).

$p_1 V_1 = mRT_1$ it follows that imep for the overexpanded cycle is given by

$$\frac{\text{imep}}{p_1} = \left(\frac{Q^*}{c_v T_1}\right)\left(\frac{1}{\gamma - 1}\right)\left(\frac{r_c}{r_e - 1}\right)\eta_{f,i} \tag{5.57}$$

Values of imep/p_1 are plotted in Fig. 5-12 as a function of $r(=r_e/r_c)$. The substantial decrease from the standard constant-volume cycle values at $r = 1$ is clear.

5.7 AVAILABILITY ANALYSIS OF ENGINE PROCESSES

5.7.1 Availability Relationships

Of interest in engine performance analysis is the amount of *useful work* that can be extracted from the gases within the cylinder at each point in the operating cycle. The problem is that of determining the maximum possible work output (or minimum work input) when a system (the charge within the cylinder) is taken from one specified state to another in the presence of a specified environment (the atmosphere). The first and second laws of thermodynamics together define this maximum or minimum work, which is best expressed in terms of the property of such a system-environment combination called *availability*[5] or sometimes *exergy*.[6, 7]

Consider the system-atmosphere combination shown in Fig. 5-13. In the absence of mass flow across the system boundary, as the system changes from state 1 to state 2, the first and second laws give

$$W_{1\text{-}2} = -(U_2 - U_1) + Q_{1\text{-}2}$$

$$Q_{1\text{-}2} \le T_0(S_2 - S_1)$$

Combining these two equations gives the *total work* transfer:

$$W_{t,\,1\text{-}2} \le -[(U_2 - U_1) - T_0(S_2 - S_1)] \tag{5.58}$$

The work done by the system against the atmosphere is not available for pro-

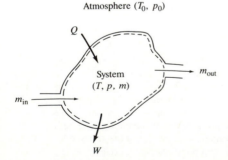

FIGURE 5-13
System-atmosphere configuration for availability analysis.

ductive use. It must, therefore, be subtracted from the total work to obtain the *useful work transfer*:

$$W_{U,1\text{-}2} \leq -[(U_2 - U_1) + p_0(V_2 - V_1) - T_0(S_2 - S_1)] \tag{5.59}$$

The *maximum* useful work will be obtained when the final state of the system is in thermal and mechanical equilibrium with the atmosphere.† The *availability* of this system which is in communication with the atmosphere

$$A = U + p_0 V - T_0 S \tag{5.60}$$

is thus the property of the system-atmosphere combination which defines its capacity for useful work. The useful work such a system-atmosphere combination can provide, as the system changes from state 1 to state 2, is less than or equal to the change in availability:

$$W_{U,\,1\text{-}2} \leq -(A_2 - A_1) \tag{5.61}$$

When mass flow across the system boundary occurs, the availability associated with this mass flow is

$$B = H - T_0 S \tag{5.62}$$

B is usually called the *steady-flow availability function.*

With these relations, an availability balance for the gas working-fluid system around the engine cycle can be carried out. For any process between specified end states which this system undergoes (interacting only with the atmosphere), the change in availability ΔA is given by

$$\Delta A = A_{\text{in}} - A_{\text{out}} - A_{\text{destroyed}} \tag{5.63}$$

The availability transfers in and out occur as a result of work transfers, heat transfers, and mass transfers across the system boundary. The availability transfer associated with a work transfer is equal to the work transfer. The availability transfer dA_Q associated with a heat transfer δQ occurring when the system temperature is T is given by

$$dA_Q = \delta Q \left(1 - \frac{T_0}{T} \right) \tag{5.64}$$

since both an energy and entropy transfer occurs across the system boundary. The availability transfer associated with a mass transfer is given by Eq. (5.62).

---

† The issue of chemical equilibrium with the atmosphere must also be considered. Attainment of chemical equilibrium with the environment requires the capacity to extract work from the partial pressure differences between the various species in the working fluid and the partial pressures of those same species in the environment. This would require such devices as ideal semipermeable membranes and efficient low input pressure, high pressure ratio, expansion devices (which are not generally available for mobile power plant systems). Inclusion of these additional steps to achieve full equilibrium beyond equality of temperature and pressure is inappropriate.[8]

Availability is destroyed by the irreversibilities that occur in any real process. The availability destroyed is given by

$$A_{\text{destroyed}} = T_0 \, \Delta S_{\text{irrev}} \tag{5.65}$$

where ΔS_{irrev} is the entropy increase associated with the irreversibilities occurring within the system boundary.[7, 8]

5.7.2 Entropy Changes in Ideal Cycles

The ideal models of engine processes examined earlier in this chapter provide useful illustrative examples for availability analysis. First, however, we will consider the variation in the entropy of the cylinder gases as they proceed through these ideal operating cycles.

For an adiabatic reversible compression process, the entropy is constant. For the combustion process in each of the ideal gas standard cycles, the entropy increase can be calculated from the relations of Eq. (4.14) (with constant specific heats):

$$s - s_0 = c_v \ln \left(\frac{T}{T_0} \right) + R \ln \left(\frac{v}{v_0} \right) = c_p \ln \left(\frac{T}{T_0} \right) - R \ln \left(\frac{p}{p_0} \right)$$

For the constant-volume cycle:

$$S_3 - S_2 = m(s_3 - s_2) = mc_v \ln \left(\frac{T_3}{T_2} \right) \tag{5.66a}$$

For the constant-pressure cycle:

$$S_3 - S_2 = m(s_3 - s_2) = mc_p \ln \left(\frac{T_3}{T_2} \right) \tag{5.66b}$$

For the limited-pressure cycle:

$$S_{3b} - S_2 = c_v \ln \left(\frac{T_{3a}}{T_2} \right) + c_p \ln \left(\frac{T_{3b}}{T_{3a}} \right) = c_v \ln \alpha + c_p \ln \beta \tag{5.66c}$$

with α and β defined by Eq. (5.42).

Since the expansion process, after combustion is complete, is adiabatic and reversible, there is no further change in entropy, 3 to 4 (or 3b to 4). Figure 5-14 shows the entropy changes that occur during each process of these three ideal engine operating cycles, calculated from the above equations, on a T-s diagram. The three cycles shown correspond to those of the p-V diagrams of Fig. 5-6 with $r_c = 12$, $\gamma = 1.3$, and $Q^*/(c_v T_1) = 8.525$. Since the combustion process was assumed to be adiabatic, the increase in entropy during combustion clearly demonstrates the irreversible nature of this process.

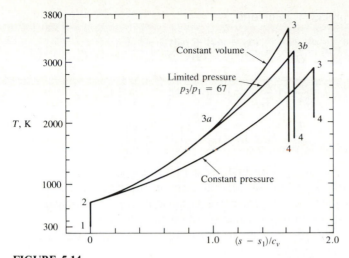

FIGURE 5-14
Temperature-entropy diagram for ideal gas constant-volume, constant-pressure, and limited-pressure cycles. Assumptions same as in Fig. 5-6.

5.7.3   Availability Analysis of Ideal Cycles

An availability analysis for each process in the ideal cycle illustrates the magnitude of the availability transfers and where the losses in availability occur.[9] In general, for the system of Fig. 5-4 in communication with an atmosphere at p_0, T_0 as indicated in Fig. 5-13, the change in availability between states i and j during the portion of the cycle when the valves are closed is given by

$$A_j - A_i = m(a_j - a_i) = m[(u_j - u_i) + p_0(v_j - v_i) - T_0(s_j - s_i)] \qquad (5.67)$$

The appropriate normalizing quantity for these changes in availability is the thermomechanical availability of the fuel supplied to the engine cylinder each cycle, $m_f(-\Delta g_{298})$† (see Sec. 3.6.2). However, it is more convenient to use $m_f(-\Delta h_{298})‡ = m_f Q_{LHV}$ as the normalizing quantity since it can be related to the temperature rise during combustion via Eq. (5.28). As shown in Table 3.3, these two quantities differ by only a few percent for common hydrocarbon fuels. Equation (5.67), with Eq. (5.29), then becomes

$$\frac{A_j - A_i}{m_f Q_{LHV}} = \frac{m(a_j - a_i)}{m_f Q_{LHV}} = \frac{a_j - a_i}{Q^*} \qquad (5.68)$$

---

† Δg_{298} is the Gibbs free energy change for the combustion reaction, per unit mass of fuel.
‡ Δh_{298} is the enthalpy change for the combustion reaction, again per unit mass of fuel.

The compression process is isentropic, so:

$$\frac{A_2 - A_1}{m_f Q_{LHV}} = \frac{a_2 - a_1}{Q^*} = \frac{(u_2 - u_1) + p_0(v_2 - v_1)}{Q^*}$$

$$= \frac{c_v T_1}{Q^*}\left[\left(\frac{T_2}{T_1} - 1\right) - (\gamma - 1)\left(1 - \frac{V_2}{V_1}\right)\right]$$

$$= \frac{c_v T_1}{Q^*}\left[(r_c^\gamma - 1) - (\gamma - 1)\left(1 - \frac{1}{r_c}\right)\right] \tag{5.69}$$

where we have assumed $p_0 = p_1$. The first term in the square brackets is the compression stroke work transfer. The second term is the work done by the atmosphere on the system, which is subtracted because it does not increase the *useful work* which the system-atmosphere combination can perform.

During combustion, for the constant-volume cycle, the volume and internal energy remain unchanged (Eqs. 5.7a, b). Thus

$$\frac{A_3 - A_2}{m_f Q_{LHV}} = \frac{a_3 - a_2}{Q^*} = -\frac{T_0(s_3 - s_2)}{Q^*}$$

$$= -\frac{c_v T_0}{Q^*}\ln\left(\frac{T_3}{T_2}\right) = -\frac{c_v T_0}{Q^*}\ln\left(1 + \frac{Q^*}{c_v T_1 r_c^{\gamma-1}}\right) \tag{5.70}$$

This loss in availability results from the increase in entropy associated with the irreversibilities of the combustion process. This lost or destroyed availability, as a fraction of the initial availability of the fuel-air mixture, decreases as the compression ratio increases (since T_2 increases as the compression ratio increases, T_3/T_2 decreases for fixed heat addition) and increases as Q^* decreases [e.g., when the mixture is made leaner; see Eq. (5.46)]. The changes in availability during combustion for the constant-pressure and limited-pressure cycles are more complex because there is a transfer of availability out of the system equal to the expansion work transfer which occurs.

For the constant-volume cycle expansion stroke:

$$\frac{A_4 - A_3}{m_f Q_{LHV}} = \frac{a_4 - a_3}{Q^*} = \frac{(u_4 - u_3) + p_0(v_4 - v_3)}{Q^*}$$

$$= \frac{c_v T_3}{Q^*}\left[\left(\frac{T_4}{T_3} - 1\right) + (\gamma - 1)\left(\frac{p_0}{p_3}\right)\left(\frac{V_4}{V_3} - 1\right)\right]$$

$$= -\left[\left(1 + \frac{c_v T_1 r_c^{\gamma-1}}{Q^*}\right)\left(1 - \frac{1}{r_c^{\gamma-1}}\right) - (\gamma - 1)\frac{c_v T_1}{Q^*}\left(\frac{r_c - 1}{r_c}\right)\right] \tag{5.71}$$

The availability of the exhaust gas at state 4 relative to its availability at (T_1, p_1) is given by

$$\frac{A_4 - A_1}{m_f Q_{LHV}} = \frac{c_v T_1}{Q^*}\left[\left(\frac{T_4}{T_1} - 1\right) - \frac{T_0}{T_1}\ln\left(\frac{T_4}{T_1}\right)\right] \tag{5.72}$$

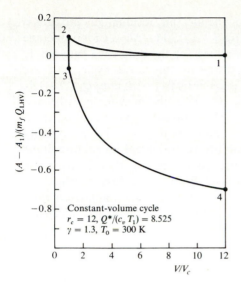

FIGURE 5-15
Availability of cylinder charge relative to availability at state 1 for constant-volume ideal gas cycle as a function of cylinder volume. Availability made dimensionless by $m_f Q_{LHV}$. Assumptions as in Fig. 5-6.

The availability of the gases inside the cylinder relative to their availability at (T_1, p_1) over the compression and expansion strokes of the constant-volume operating cycle example used in Figs. 5-6 and 5-14 is shown in Fig. 5-15. Equations (5.69) and (5.71), with T_2 and T_4 replaced by temperatures intermediate between T_1 and T_2 and T_3 and T_4, respectively, were used to compute the variations during compression and expansion. Table 5.3 summarizes the changes in availability during each process and the availability of the cylinder gases, at the beginning and end of each process, relative to the datum for the atmosphere

TABLE 5.3
Availability changes in constant-volume cycle

Process or state	$\dfrac{A_j - A_i}{m_f Q_{LHV}}$	$\dfrac{A_i}{m_f Q_{LHV}}$
1		1.0294
1-2	0.0976	
2		1.1270
2-3	−0.1710	
3		0.9560
3-4	−0.6237	
4		0.3323
Fuel conversion efficiency $\eta_{f,i}$	0.526	
Availability conversion efficiency $\eta_{a,i}$	0.511	

$r_c = 12$, $\gamma = 1.3$, $Q^*/(c_v T_1) = 8.525$, $T_0 = 300$ K, $T_1 = 333$ K.

(1 atm, 300 K). The availability at state 1 of the fuel, air, residual-gas mixture is $(1.0286 + 0.0008)m_f Q_{LHV}$. 1.0286 is the ratio $(-\Delta g_{298}^\circ)/(-\Delta h_{298}^\circ)$ for isooctane (see Table 3.3). The second number, 0.0008, allows for the difference between T_1 and T_0. Because both work-transfer processes in this ideal cycle case are reversible, the fuel conversion efficiency $\eta_{f,i}$ is given by $(A_3 - A_4)/(m_f Q_{LHV}) - (A_2 - A_1)/(m_f Q_{LHV})$. It is, of course, equal to the value obtained for $r_c = 12$ and $\gamma = 1.3$ from the formula for efficiency (Eq. 5.31), obtained previously. The availability conversion efficiency is $\eta_{f,i}/1.0286$. Note that it is the availability destroyed during combustion, plus the inability of this ideal constant-volume cycle to use the availability remaining in the gas at state 4, that decrease the availability conversion efficiency below unity. Both these loss mechanisms decrease in magnitude, relative to the fuel availability, as the compression ratio increases. This is the fundamental reason why engine indicated efficiency increases with an increasing compression ratio.

5.7.4   Effect of Equivalence Ratio

The fuel-air cycle with its more accurate models for working fluid properties can be used to examine the effect of variations in the fuel/air equivalence ratio on the availability conversion efficiency. Figure 5-16 shows the temperature attained and the entropy rise that occurs in constant-volume combustion of a fuel-air mixture of different equivalence ratios, following isentropic compression from ambient temperature and pressure through different volumetric compression ratios.[8] The entropy increase is the result of irreversibilities in the combustion process and mixing of complete combustion products with excess air. The significance of these combustion-related losses—the destruction of availability that occurs in this process—is shown in Fig. 5-17 where the availability after constant-volume combustion divided by the availability of the initial fuel-air mixture is shown as a function of equivalence ratio for compression ratios of 12 and 36.[8] The loss of

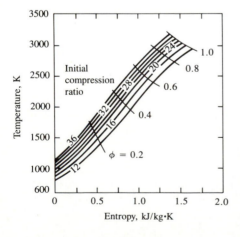

FIGURE 5-16

Temperature and entropy of combustion products after constant-volume combustion following isentropic compression from ambient conditions through specified compression ratio as a function of compression ratio and equivalence ratio. (*From Flynn et al.[8]*)

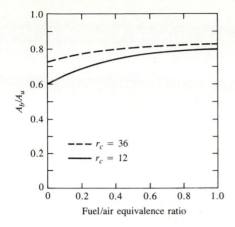

Fuel/air equivalence ratio

FIGURE 5-17
Availability of combustion products after constant-volume combustion relative to availability before combustion following isentropic compression from ambient through specified compression ratio as a function of equivalence ratio. (*From Flynn et al.*[8])

availability increases as the equivalence ratio decreases.† The combustion loss is a stronger function of the rise in temperature and pressure which occurs than of the change in the specific heat ratio that occurs.

Why then does engine efficiency increase with a decreasing equivalence ratio as shown in Fig. 5-9? The reason is that the expansion stroke work transfer, as a fraction of the fuel availability, increases as the equivalence ratio decreases; hence, the availability lost in the exhaust process, again expressed as a fraction of the fuel availability, decreases. The increase in the expansion stroke work as the equivalence ratio decreases more than offsets the increase in the availability lost during combustion; so the availability conversion efficiency (or the fuel conversion efficiency which closely approximates it) increases.

5.8 COMPARISON WITH REAL ENGINE CYCLES

To put these ideal models of engine processes in perspective, this chapter will conclude with a brief discussion of the additional effects which are important in real engine processes.

A comparison of a real engine p-V diagram over the compression and expansion strokes with an equivalent fuel-air cycle analysis is shown in Fig. 5-18.[4] The real engine and the fuel-air cycle have the same geometric compression ratio, fuel chemical composition and equivalence ratio, residual fraction and mixture density before compression. Midway through the compression stroke,

---

† This is consistent with the ideal gas standard cycle result (Eq. 5.70). As ϕ decreases, so does $Q^*/(c_v T_1)$. The factor which multiplies the natural logarithm (which increases) has a greater impact than the logarithmic term (which decreases).

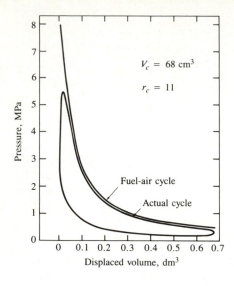

$V_c = 68 \text{ cm}^3$

$r_c = 11$

Fuel-air cycle

Actual cycle

FIGURE 5-18
Pressure-volume diagram for actual spark-ignition engine compared with that for equivalent fuel-air cycle. $r_c = 11$. (*From Edson and Taylor.*[4])

the pressure in the fuel-air cycle has been made equal to the real cycle pressure.†
The compression stroke pressures for the two cycles essentially coincide. Modest differences in pressure during intake and the early part of the compression process result from the pressure drop across the intake valve during the intake process and the closing of the intake valve 40 to 60° after BC in the real engine. The expansion stroke pressures for the engine fall below the fuel-air cycle pressures for the following reasons: (1) heat transfer from the burned gases to the walls; (2) finite time required to burn the charge; (3) exhaust blowdown loss due to opening the exhaust valve before BC; (4) gas flow into crevice regions and leakage past the piston rings; (5) incomplete combustion of the charge.

These differences, in decreasing order of importance, are described below. Together, they contribute to the enclosed area on the p-V diagram for a properly adjusted engine with optimum timing being about 80 percent of the enclosed area of an equivalent fuel-air cycle p-V diagram. The indicated fuel conversion or availability conversion efficiency of the actual engine is therefore about 0.8 times the efficiency calculated for the fuel-air cycle.[1] Use is often made of this ratio to estimate the performance of actual engines from fuel-air cycle results.

1. *Heat transfer.* Heat transfer from the unburned mixture to the cylinder walls has a negligible effect on the p-V line for the compression process. Heat transfer from the burned gases is much more important (see Chap. 12). Due to heat transfer during combustion, the pressure at the end of combustion in the real

---

† Note that in the fuel-air cycle with idealized valve timing, the compression process starts immediately after BC. In most engines, the charge compression starts later, close to the time that the inlet valve closes some 40 to 60° after BC. This matching process is approximate.

cycle will be lower. During expansion, heat transfer will cause the gas pressure in the real cycle to fall below an isentropic expansion line as the volume increases. A decrease in efficiency results from this heat loss.

2. *Finite combustion time.* In an SI engine with spark-timing adjusted for optimum efficiency, combustion typically starts 10 to 40 crank angle degrees before TC, is half complete at about 10° after TC, and is essentially complete 30 to 40° after TC. Peak pressure occurs at about 15° after TC (see Fig. 1-8). In a diesel engine, the burning process starts shortly before TC. The pressure rises rapidly to a peak some 5 to 10° after TC since the initial rate of burning is fast. However, the final stages of burning are much slower, and combustion continues until 40 to 50° after TC (see Fig. 1-15). Thus, the peak pressure in the engine is substantially below the fuel-air cycle peak pressure value, because combustion continues until well after TC, when the cylinder volume is much greater than the clearance volume. After peak pressure, expansion stroke pressures in the engine are higher than fuel-air cycle values in the absence of other loss mechanisms, because less work has been extracted from the cylinder gases. A comparison of the constant-volume and limited-pressure cycles in Fig. 5-6 demonstrates this point.

   For spark or fuel-injection timing which is retarded from the optimum for maximum efficiency, the peak pressure in the real cycle will be lower, and expansion stroke pressures after the peak pressure will be higher than in the optimum timing cycle.

3. *Exhaust blowdown loss.* In the real engine operating cycle, the exhaust valve is opened some 60° before BC to reduce the pressure during the first part of the exhaust stroke in four-stroke engines and to allow time for scavenging in two-stroke engines. The gas pressure at the end of the expansion stroke is therefore reduced below the isentropic line. A decrease in expansion-stroke work transfer results.

4. *Crevice effects and leakage.* As the cylinder pressure increases, gas flows into crevices such as the regions between the piston, piston rings, and cylinder wall. These crevice regions can comprise a few percent of the clearance volume. This flow reduces the mass in the volume above the piston crown, and this flow is cooled by heat transfer to the crevice walls. In premixed charge engines, some of this gas is unburned and some of it will not burn. Though much of this gas returns to the cylinder later in the expansion, a fraction, from behind and between the piston rings, flows into the crankcase. However, leakage in a well-designed and maintained engine is small (usually less than one percent of the charge). All these effects reduce the cylinder pressure during the latter stages of compression, during combustion, and during expansion below the value that would result if crevice and leakage effects were absent.

5. *Incomplete combustion.* Combustion of the cylinder charge is incomplete; the exhaust gases contain combustible species. For example, in spark-ignition engines the hydrocarbon emissions from a warmed-up engine (which come largely from the crevice regions) are 2 to 3 percent of the fuel mass under

normal operating conditions; carbon monoxide and hydrogen in the exhaust contain an additional 1 to 2 percent or more of the fuel energy, even with excess air present (see Sec. 4.9). Hence, the chemical energy of the fuel which is released in the actual engine is about 5 percent less than the chemical energy of the fuel inducted (the combustion efficiency, see Sec. 3.5.5, is about 95 percent). The fuel-air cycle pressures after combustion will be higher because complete combustion is assumed. In diesel engines, the combustion inefficiency is usually less, about 1 to 2 percent, so this effect is smaller.

SUMMARY. The effect of all these loss mechanisms on engine efficiency is best defined by an availability balance for the real engine cycle. A limited number of such calculations have been published (e.g., Refs. 8, 10, and 11). Table 5.4 shows the magnitude of the loss in availability (as a fraction of the initial availability) that occurs due to real cycle effects in a typical naturally aspirated diesel engine.[10] The combustion and exhaust losses are present in the ideal cycle models also (they are smaller, however[9]). The loss in availability due to heat losses, flow or aerodynamic losses, and mechanical friction are real engine effects.

Figure 5-19 shows standard and fuel-air cycle efficiencies as a function of the compression ratio compared with engine indicated efficiency data. The top three sets of engine data are for the best efficiency air/fuel ratio. Differences in the data are in part due to different fuels [(12) isooctane; (13) gasoline; (14) propane] which affect efficiency slightly through their different composition and heating values (see Table D.4). They also result from different combustion chamber shapes which affect the combustion rate and heat transfer. The trends in the data with increasing compression ratio and the $\phi = 0.8$ fuel-air cycle curve (which corresponds approximately to the actual air/fuel ratios used) are similar. The factor of 0.8 relating real engine and fuel-air cycle efficiencies holds roughly. At compression ratios above about 14, however, the data show that the indicated efficiency of actual engines is essentially constant. Increasing crevice and heat

TABLE 5.4
Availability losses in naturally aspirated diesel

Loss mechanism	Loss, fraction of fuel availability
Combustion	0.225
Exhaust	0.144
Heat transfer	0.135
Aerodynamic	0.047
Mechanical friction	0.048
Total losses	0.599
Availability conversion efficiency (brake)	0.401

Source: Traupel.[10]

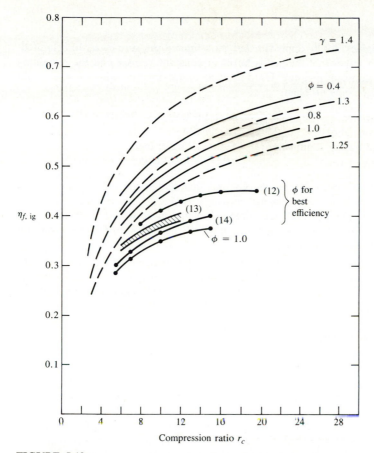

FIGURE 5-19
Indicated fuel conversion efficiency as a function of compression ratio for ideal gas constant-volume cycle (dashed lines, $\gamma = 1.25$, 1.3, 1.4) and fuel-air cycle (solid lines, $\phi = 0.4$, 0.8, 1.0). Also shown are available engine data for equivalence ratios given: best efficiency ϕ;[12-14] $\phi = 1$.[14]

losses offset the calculated ideal cycle efficiency increase as the compression ratio is raised above this value. The standard ideal gas cycle analysis results, with an appropriate choice for the value of γ (1.25 to 1.3), correspond closely to the fuel-air cycle analysis results.

The ideal cycle provides a convenient but crude approximation to the real engine operating cycle. It is useful for illustrating the thermodynamic aspects of engine operation. It can also provide approximate estimates of trends as major engine parameters change. The weakest link in these ideal cycles is the modeling of the combustion processes in SI and CI engines. None of the models examined in this chapter are sufficiently close to reality to provide accurate predictions of engine performance. More sophisticated models of the spark-ignition and diesel engine operating cycles have been developed and are the subject of Chap. 14.

PROBLEMS

5.1. Many diesel engines can be approximated by a limited-pressure cycle. In a limited-pressure cycle, a fraction of the fuel is burnt at constant volume and the remaining fuel is burnt at constant pressure. Use this cycle approximation with $\gamma = c_p/c_v = 1.3$ to analyze the following problem:

Inlet conditions:	$p_1 = 1.0$ bar, $T_1 = 289$ K
Compression ratio:	15 : 1
Heat added during combustion:	43,000 kJ/kg of fuel
Overall fuel/air ratio:	0.045 kg fuel/kg air

(a) Half of the fuel is burnt at constant volume, then half at constant pressure. Draw a p-V diagram and compute the fuel conversion efficiency of the cycle.

(b) Compare the efficiency and peak pressure of the cycle with the efficiency and peak pressure that would be obtained if all of the fuel were burnt at constant pressure or at constant volume.

5.2. It is desired to increase the output of a spark-ignition engine by either (1) raising the compression ratio from 8 to 10 or (2) increasing the inlet pressure from 1.0 atm to 1.5 atm. Using the constant-volume cycle as a model for engine operation, which procedure will give:

(a) The highest pressure of the cycle?

(b) The highest efficiency?

(c) The highest mep?

Assume $\gamma = 1.3$ and $(m_f Q_{HV})/(mc_v T_1) = 9.3(r_c - 1)/r_c$.

5.3. When a diesel engine, originally designed to be naturally aspirated, is turbocharged the fuel/air equivalence ratio ϕ at full load must be reduced to maintain the maximum cylinder pressure essentially constant. If the naturally aspirated engine was designed for $\phi = 0.75$ at full load, estimate the maximum permissible value of ϕ for the turbocharged engine at full load if the air pressure at the engine inlet is 1.6 atm. Assume that the engine can be modeled with the limited-pressure cycle, with half the injected fuel burned at constant volume and half at constant pressure. The compression ratio is 16. The fuel heating value is 42.5 MJ/kg fuel. Assume $\gamma = c_p/c_v = 1.35$, that the air temperature at the start of compression is 325 K, and $(F/A)_{stoich} = 0.0666$.

5.4. A spark-ignition engine is throttled when operating at part load (the inlet pressure is reduced) while the fuel/air ratio is held essentially constant. Part-load operation of the engine is modeled by the cycle shown in Fig. 5-2d; the inlet air is at pressure p_1, the exhaust pressure is atmospheric p_a, and the ambient temperature is T_a. Derive an expression for the *decrease* in net indicated fuel conversion efficiency due to throttling from the ideal constant-volume cycle efficiency and show that it is proportional to $(p_a/p_1 - 1)$. Assume mass fuel $\ll$ mass air.

5.5. (a) Use the ideal gas cycle with constant-volume combustion to describe the operation of an SI engine with a compression ratio of 9. Find the pressure and temperature at points 2, 3, 4, and 5 on Fig. 5-2a. Assume a pressure of 100 kPa and a temperature of 320 K at point 1. Assume $m_f/m = 0.06$, $c_v = 946$ J/kg·K, $\gamma = 1.3$. Q_{LHV} for gasoline is 44 MJ/kg.

(b) Find the indicated fuel conversion efficiency and imep for this engine under these operating conditions.

5.6. Use a limited-pressure cycle analysis to obtain a plot of indicated fuel conversion efficiency versus p_3/p_1 for a compression ratio of 15 with light diesel oil as fuel. Assume $m_f/m = 0.04$, $T_1 = 45°C$. Use $\gamma = 1.3$ and $c_v = 946$ J/kg·K.

5.7. Explain why constant-volume combustion gives a higher indicated fuel conversion efficiency than constant-pressure combustion for the same compression ratio.

5.8. Two engines are running at a bmep of 250 kPa. One is an SI engine with the throttle partially closed to maintain the correct load. The second engine is a naturally aspirated CI engine which requires no throttle. Mechanical friction mep for both engines is 100 kPa. If the intake manifold pressures for the SI and CI engines are 25 kPa and 100 kPa respectively, and both exhaust manifold pressures are 105 kPa, use an ideal cycle model to estimate and compare the *gross* imep of the two engines. You may neglect the pressure drop across the valves during the intake and exhaust processes.

5.9. (a) Plot net imep versus p_i for 20 kPa $< p_i <$ 100 kPa for a constant-volume cycle using the following conditions: $m_f/m = 0.06$, $T_1 = 40°C$, $c_v = 946$ J/kg·K, $\gamma = 1.3$, $r_c = 9.5$, $Q_{LHV} = 44$ MJ/kg fuel. Assume $p_e = 100$ kPa.
(b) What additional information is necessary to draw a similar plot for the engine's indicated torque, and indicated power?

5.10. (a) Draw a diagram similar to those in Fig. 5-2 for a supercharged cycle with constant-pressure combustion.
(b) Use the ideal gas cycle with constant-pressure combustion to model an engine with a compression ratio of 14 through such a supercharged cycle. Find the pressure and temperature at points corresponding to 2, 3, 4, and 5 in Fig. 5-2. Assume a pressure of 200 kPa and temperature of 325 K at point 1, and a pressure of 100 kPa at points 5 and 6. $m_f/m = 0.03$ and the fuel is a light diesel oil.
(c) Calculate the gross and net indicated fuel conversion efficiency and imep for this engine under these operating conditions.

5.11. Use the appropriate tables and charts to carry out a constant-pressure fuel-air cycle calculation for the supercharged engine described in Prob. 5.10. Assume the same initial conditions at point 1, with $\phi = 0.4$ and a residual gas fraction of 0.025. A single cycle calculation is sufficient.
(a) Determine the pressure and temperature at points 2, 3, 4, and 5. Calculate the compression stroke, expansion stroke, and pumping work per cycle per kg air.
(b) Find the gross and net indicated fuel conversion efficiency and imep.
(c) Compare the calculated residual gas fraction with the assumed value of 0.025.

5.12. One method proposed for reducing the pumping work in throttled spark-ignition engines is *early intake valve closing* (EIVC). The ideal cycle p-V diagram shown illustrates the concept. The EIVC cycle is 1-2-3-4-5-6-7-8-1 (the conventional throttled cycle is 1-2-3-4-5-6-7*-1). With EIVC, the inlet manifold is held at a pressure p_i (which is higher than the normal engine intake pressure, p_i^*), and the inlet valve is closed *during* the inlet stroke at 8. The trapped fresh charge and residual is then *expanded* to the normal cycle (lower) intake pressure, p_i^*. You can assume that both cycles have the same mass of gas in the cylinder, temperature, and pressure at state 1 of the cycle.
(a) On a sketch of the intake and exhaust process p-V diagram, shade in the area that corresponds to the difference between the pumping work of the EIVC cycle and that of the normal cycle.
(b) What value of p_i and V_{EIVC} will give the maximum reduction in pumping work for the EIVC cycle.

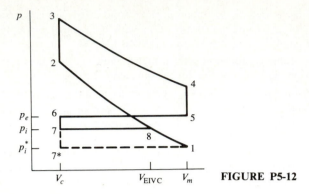

FIGURE P5-12

(c) Derive an expression for this maximum difference in pumping work between the normal cycle and the EIVC cycle in terms of p_e, p_i^*, V_c, and V_m. You can make the appropriate ideal cycle assumptions.

5.13. Calculate the following parameters for a constant-volume fuel-air cycle (Fig. 5-2a):
  (a) The pressures and temperatures at states 1, 2, 3, 4, 5, and 6
  (b) The indicated fuel conversion efficiency
  (c) The imep
  (d) The residual fraction
  (e) The volumetric efficiency
    Inlet pressure $= 1$ atm, exhaust pressure $= 1$ atm, inlet temperature $= 300$ K, compression ratio $= 8 : 1$, equivalence ratio $= \phi = 1$.
  Calculate the above parameters (points a–e) using the SI units charts. Use 44.4 MJ/kg for heating value of the fuel. *Hint:* Start the calculations using the residual mass fraction 0.03 and the residual gas temperature 1370 K.

5.14. The cycle 1-2-3-4-5-6-1 is a conventional constant-volume fuel-air cycle with a compression ratio of 8. The fuel is isooctane, C_8H_{18}, with a lower heating value of 44.4 MJ/kg. The gas state at 1 is $T_1 = 300$ K, $p_1 = 1$ atmosphere with an equivalence ratio of 1.0 and zero residual fraction. The specific volume at state 1 is 0.9 m³/kg air in the mixture. The temperature at the end of compression at state 2 is 600 K.
  (a) Find the indicated fuel conversion efficiency and mean effective pressure of this fuel-air cycle model of a spark-ignition engine.
  (b) The efficiency of the cycle can be increased by increasing the expansion ratio r_e while maintaining the same compression ratio r_c (cycle 1-2-3-4A-5A-6-1). (This

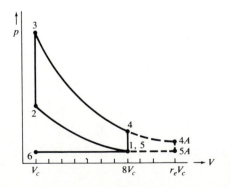

FIGURE P5-14

can be done with valve timing.) If the expansion ratio r_e is 12, while the compression ratio and other details of the cycle remain the same as in (a), what is the indicated efficiency and mean effective pressure (based on the new, larger, displaced volume) of this new engine cycle?

5.15. In spark-ignition engines, exhaust gas is recycled to the intake at part load to reduce the peak burned gas temperatures and lower emissions of nitrogen oxides.

(a) Calculate the reduction in burned gas temperature that occurs when, due to exhaust gas recycle, the burned gas fraction in the unburned gas mixture (x_b) inside the cylinder is increased from 10 percent (the normal residual fraction) to 30 percent. Assume combustion occurs at top-center, at constant volume, and is adiabatic. Conditions at the end of compression for both cases are: $T = 700$ K, $p = 1000$ kPa, $v = 0.2$ m³/kg air in the original mixture; the equivalence ratio is 1.0. The fuel can be modeled as isooctane.

(b) The compression ratio is 8. The compression stroke work is 300 kJ/kg air in the original mixture. Find the indicated work per cycle for the compression and expansion strokes, per kilogram of air in the original mixture, for these two cases.

(c) Briefly explain how you would increase the work per cycle with 30 percent burned gas fraction in the unburned mixture to the value obtained with 10 percent burned gas fraction, with fixed engine geometry. (A qualitative answer, only, is required here.)

5.16. The following cycle has been proposed for improving the operation of a four-stroke cycle engine. Its aim is to expand the postcombustion cylinder gases to a lower pressure and temperature by extending the expansion stroke, and hence extract more work per cycle.

The cycle consists of: (1) an intake stroke; (2) a compression stroke, where the inlet valve remains open (and the cylinder pressure is constant) for the first portion of the stroke; (3) a combustion process, which occurs rapidly close to top-center; (4) an expansion stroke, where the exhaust valve remains closed until the end of the stroke; (5) an exhaust stroke, where the cylinder pressure blows down to the exhaust pressure rapidly and most of the remaining combustion products are expelled as the piston moves from the BC to the TC position. Thus, for this engine concept, the compression ratio r_c (ratio of cylinder volume at inlet valve closing to clearance volume) is less than the expansion ratio r_e (ratio of cylinder volume at exhaust valve opening to clearance volume).

(a) Sketch a p-V diagram for the cylinder gases for this cycle operating unthrottled.

(b) Using the charts in SI units developed for fuel-air cycle calculations, carry out an analysis of an appropriate ideal model for this cycle where the compression ratio r_c is 8 and the expansion ratio r_e is (1) 8; (2) 16. Assume the following:

Pressure in the cylinder at inlet valve close 1 atm
Mixture temperature at inlet valve close 300 K
Mixture equivalence ratio $= 1.0$
Fuel: isooctane C_8H_{18}
Lower heating value $= 44.4$ MJ/kg
Residual gas mass fraction at inlet valve close 0.05
Stoichiometric fuel/air ratio $= 0.066$

Calculate the indicated work per cycle per kg of air in the original mixture (the standard chart units) and the indicated mean effective pressure for these two expansion ratios. Base the mean effective pressure on the volume displaced by the piston during the *expansion* stroke. Tabulate your answers. (*Note:* You are given the initial conditions for the cycle calculation; changing the value of r_e requires only modest changes in the cycle calculation.)

(c) Comment briefly on the effect of increasing the ratio r_e/r_c above 1.0 with this concept on engine efficiency and specific power (power per unit engine weight). Additional calculations are not required.

5.17. In a direct-injection stratified-charge (DISC) engine fuel is injected into the engine cylinder just before top-center (like a diesel); a spark discharge is then used to initiate the combustion process. A four-stroke cycle version of this engine has a displaced volume of 2.5 liters and a compression ratio of 12. At high load, the inlet pressure is boosted by a compressor to above atmospheric pressure. The compressor is geared directly to the engine drive shaft. The exhaust pressure is 1 atm. This DISC engine is to replace an equal displacement conventional naturally aspirated spark-ignition (SI) engine, which has a compression ratio of 8.

(a) Draw qualitative sketches of the appropriate constant-volume ideal cycle pressure-volume diagrams for the complete operating cycles for these two engines at maximum load.

(b) Use available fuel-air results to estimate how much the DISC engine inlet pressure must be boosted above atmospheric pressure by the compressor to provide the same maximum gross indicated power as the naturally aspirated SI engine. The SI engine operates with an equivalence ratio of 1.2; the DISC engine is limited by smoke emissions to a maximum equivalence ratio of 0.7.

(c) Under these conditions, will the brake powers of these engines be the same, given that the mechanical rubbing friction is the same? Briefly explain.

(d) At part load, the SI engine operates at an equivalence ratio of 1.0 and inlet pressure of 0.5 atm. At part load the DISC engine has negligible boost and operates with an inlet pressure of 1.0 atm. Use fuel-air cycle results to determine the equivalence ratio at which the DISC engine must be operated to provide the same net indicated mean effective pressure as the SI engine. What is the ratio of DISC engine net indicated fuel conversion efficiency to SI engine efficiency at these conditions?

5.18. The earliest successful reciprocating internal combustion engine was an engine developed by Lenoir in the 1860s. The operating cycle of this engine consisted of two strokes (i.e., one crankshaft revolution). During the first half of the first stroke, as the piston moves away from its top-center position, fuel-air mixture is drawn into the cylinder through the inlet valve. When half the total cylinder volume is filled with fresh mixture, the inlet valve is closed. The mixture is then ignited and burns rapidly. During the second half of the first stroke, power is delivered from the high-pressure burned gases to the piston. With the piston in its bottom-center position, the exhaust valve is opened. The second stroke, the exhaust stroke, completes the cycle as the piston returns to top-center.

(a) Sketch a cylinder pressure versus cylinder volume diagram for this engine.

(b) Using the charts in SI units developed for fuel-air cycle calculations, carry out a cycle analysis and determine the indicated fuel conversion efficiency and mean effective pressure for the Lenoir engine. Assume the following:

Inlet pressure = 1 atm
Inlet mixture temperature = 300 K
Mixture equivalence ratio = 1.0

Fuel: isooctane C_8H_{18}
Lower heating value $= 44.4$ MJ/kg
Clearance volume negligible

(c) Compare these values with typical values for the constant-volume fuel-air cycle. Explain (with thermodynamic arguments) why the two cycles have such different indicated mean effective pressures and efficiencies.

(d) Explain briefly why the real Lenoir engine would have a lower efficiency than the value you calculated in (b) (the actual *brake* fuel conversion efficiency of the engine was about 5 percent).

5.19. Estimate from fuel-air cycle results the indicated fuel conversion efficiency, the indicated mean effective pressure, and the maximum indicated power (in kilowatts) at wide-open throttle of these two four-stroke cycle spark-ignition engines:

A six-cylinder engine with a 9.2-cm bore, 9-cm stroke, compression ratio of 7, operated at an equivalence ratio of 0.8
A six-cylinder engine with an 8.3-cm bore, 8-cm stroke, compression ratio of 10, operated at an equivalence ratio of 1.1

Assume that actual indicated engine efficiency is 0.8 times the appropriate fuel-air cycle efficiency. The inlet manifold pressure is close to 1 atmosphere. The maximum permitted value of the mean piston speed is 15 m/s. Briefly summarize the reasons why:

(a) The efficiency of these two engines is approximately the same despite their different compression ratios.

(b) The maximum power of the smaller displacement engine is approximately the same as that of the larger displacement engine.

5.20. The constant-volume combustion fuel-air cycle model can be used to estimate the effect of changes in internal combustion engine design and operating variables on engine efficiency. The following table gives the major differences between a diesel and a spark-ignition engine both operating at half maximum power.

	Diesel engine	Spark-ignition engine
Compression ratio	16 : 1	9 : 1
Fuel/air equivalence ratio	0.4	1.0
Inlet manifold pressure	1 atm	0.5 atm

(a) Use the graphs of fuel-air cycle results (Figs. 5-9 and 5-10) to estimate the ratio of the diesel engine *brake* fuel conversion efficiency to the spark-ignition engine *brake* fuel conversion efficiency.

(b) Estimate what percentage of the higher diesel brake fuel conversion efficiency comes from:
   (1) The higher diesel compression ratio
   (2) The leaner diesel equivalence ratio
   (3) The lack of intake throttling in the diesel compared with the spark-ignition engine

The values of fuel conversion efficiency and mean effective pressure given in the graphs are *gross* indicated values (i.e., values obtained from $\int p\,dV$ over the compression and expansion strokes only).

You may assume, if necessary, that for the real engines, the gross indicated efficiency and gross indicated mean effective pressure are 0.8 times the fuel-air cycle values. Also, the mechanical rubbing friction for each engine is 30 percent of the *net* indicated power or mep.

REFERENCES

1. Taylor, C. F.: *The Internal Combustion Engine in Theory and Practice*, vol. 1: *Thermodynamics, Fluid Flow, Performance*, 2d ed., chaps. 2 and 4, 1966.
2. Lancaster, D. R., Krieger, R. B., and Lienesch, J. H.: "Measurement and Analysis of Engine Pressure Data," SAE paper 750026, *SAE Trans.*, vol. 84, 1975.
3. Edson, M. H.: "The Influence of Compression Ratio and Dissociation on Ideal Otto Cycle Engine Thermal Efficiency," *Digital Calculations of Engine Cycles*, SAE Prog. in Technology, vol. 7, pp. 49–64, 1964.
4. Edson, M. H., and Taylor, C. F.: "The Limits of Engine Performance—Comparison of Actual and Theoretical Cycles," *Digital Calculations of Engine Cycles*, SAE Prog. in Technology, vol. 7, pp. 65–81, 1964.
5. Keenan, J. H.: *Thermodynamics*, John Wiley, New York, 1941; MIT Press, Cambridge, Mass., 1970.
6. Haywood, R. W.: "A Critical Review of the Theorems of Thermodynamic Availability, with Concise Formulations; Part 1. Availability," *J. Mech. Engng Sci.*, vol. 16, no. 3, pp. 160–173, 1974.
7. Haywood, R. W.: "A Critical Review of the Theorems of Thermodynamic Availability, with Concise Formulations; Part 2. Irreversibility," *J. Mech. Engng Sci.*, vol. 16, no. 4, pp. 258–267, 1974.
8. Flynn, R. F., Hoag, K. L., Kamel, M. M., and Primus, R. J.: "A New Perspective on Diesel Engine Evaluation Based on Second Law Analysis," SAE paper 840032, *SAE Trans.*, vol. 93, 1984.
9. Clarke, J. M.: "The Thermodynamic Cycle Requirements for Very High Rational Efficiencies," paper C53/76, Institution of Mechanical Engineers, *J. Mech. Engng Sci.*, 1974.
10. Traupel, W.: "Reciprocating Engine and Turbine in Internal Combustion Engineering," in *Proc. CIMAC Int. Congr. on Combustion Engines*, Zurich, pp. 39–54, 1957.
11. Clarke, J. M.: "Letter: Heavy Duty Diesel Fuel Economy," *Mech. Engng*, pp. 105–106, March 1983.
12. Caris, D. F., and Nelson, E. E.: "A New Look at High Compression Engines," *SAE Trans.*, vol. 67, pp. 112–124, 1959.
13. Kerley, R. V., and Thurston, K. W.: "The Indicated Performance of Otto-Cycle Engines," *SAE Trans.*, vol. 70, pp. 5–37, 1962.
14. Bolt, J. A., and Holkeboer, D. H.: "Lean Fuel-Air Mixtures for High-Compression Spark-Ignition Engines," *SAE Trans.*, vol. 70, p. 195, 1962.

CHAPTER
6

GAS EXCHANGE PROCESSES

This chapter deals with the fundamentals of the gas exchange processes—intake and exhaust in four-stroke cycle engines and scavenging in two-stroke cycle engines. The purpose of the exhaust and inlet processes or of the scavenging process is to remove the burned gases at the end of the power stroke and admit the fresh charge for the next cycle. Equation (2.38) shows that the indicated power of an internal combustion engine at a given speed is proportional to the mass flow rate of air. Thus, inducting the maximum air mass at wide-open throttle or full load and retaining that mass within the cylinder is the primary goal of the gas exchange processes. Engine gas exchange processes are characterized by overall parameters such as volumetric efficiency (for four-stroke cycles), and scavenging efficiency and trapping efficiency (for two-stroke cycles). These overall parameters depend on the design of engine subsystems such as manifolds, valves, and ports, as well as engine operating conditions. Thus, the flow through individual components in the engine intake and exhaust system has been extensively studied also. Supercharging and turbocharging are used to increase air flow through engines, and hence power density. Obviously, whether the engine is naturally aspirated or supercharged (or turbocharged) significantly affects the gas exchange processes. The above topics are the subject of this chapter.

For spark-ignition engines, the fresh charge is fuel, air, and (if used for emission control) recycled exhaust, so mixture preparation is also an important goal of the intake process. Mixture preparation includes both achieving the appropriate mixture composition and achieving equal distribution of air, fuel, and recycled exhaust amongst the different cylinders. In diesels, only air (or air plus recycled exhaust) is inducted. Mixture preparation and manifold flow phenomena are discussed in Chap. 7. A third goal of the gas exchange processes is to set up the flow field within the engine cylinders that will give a fast-enough combustion process for satisfactory engine operation. In-cylinder flows are the subject of Chap. 8.

6.1 INLET AND EXHAUST PROCESSES IN THE FOUR-STROKE CYCLE

In a spark-ignition engine, the intake system typically consists of an air filter, a carburetor and throttle or fuel injector and throttle or throttle with individual fuel injectors in each intake port, and intake manifold. During the induction process, pressure losses occur as the mixture passes through or by each of these components. There is an additional pressure drop across the intake port and valve. The exhaust system typically consists of an exhaust manifold, exhaust pipe, often a catalytic converter for emission control, and a muffler or silencer. Figure 6-1 illustrates the intake and exhaust gas flow processes in a conventional spark-ignition engine. These flows are pulsating. However, many aspects of these flows can be analysed on a quasi-steady basis, and the pressures indicated in the intake system in Fig. 6-1a represent time-averaged values for a multicylinder engine.

The drop in pressure along the intake system depends on engine speed, the flow resistance of the elements in the system, the cross-sectional area through which the fresh charge moves, and the charge density. Figure 6-1d shows the inlet and exhaust valve lifts versus crank angle. The usual practice is to extend the valve open phases beyond the intake and exhaust strokes to improve emptying and charging of the cylinders and make the best use of the inertia of the gases in the intake and exhaust systems. The exhaust process usually begins 40 to 60° before BC. Until about BC the burned cylinder gases are discharged due to the pressure difference between the cylinder and the exhaust system. After BC, the cylinder is scavenged by the piston as it moves toward TC. The terms *blowdown* and *displacement* are used to denote these two phases of the exhaust process. Typically, the exhaust valve closes 15 to 30° after TC and the inlet valve opens 10 to 20° before TC. Both valves are open during an *overlap period*, and when $p_i/p_e < 1$, backflow of exhausted gas into the cylinder and of cylinder gases into the intake will usually occur. The advantage of valve overlap occurs at high engine speeds when the longer valve-open periods improve volumetric efficiency. As the piston moves past TC and the cylinder pressure falls below the intake pressure, gas flows from the intake into the cylinder. The intake valve remains open until 50 to 70° after BC so that fresh charge may continue to flow into the cylinder after BC.

In a diesel engine intake system, the carburetor or EFI system and the throttle plate are absent. Diesel engines are more frequently turbocharged. A

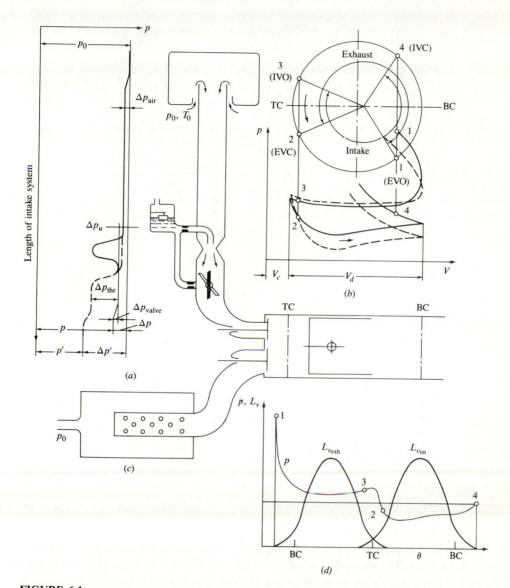

FIGURE 6-1
Intake and exhaust processes for four-stroke cycle spark-ignition engine: (a) intake system and average pressures within it; (b) valve timing and pressure-volume diagrams; (c) exhaust system; (d) cylinder pressure p and valve lift L_v versus crank angle θ. Solid lines are for wide-open throttle, dashed lines for part throttle; p_0, T_0, atmospheric conditions; Δp_{air} = pressure losses in air cleaner; Δp_u = intake losses upstream of throttle; Δp_{thr} = losses across throttle; Δp_{valve} = losses across the intake valve.[1]

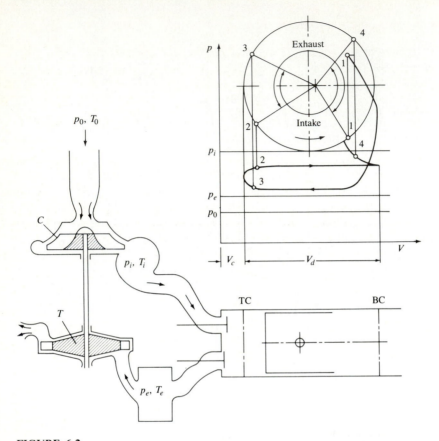

FIGURE 6-2
Intake and exhaust process for turbocharged four-stroke cycle engine. The turbocharger compressor C raises air pressure and temperature from ambient p_0, T_0 to p_i, T_i. Cylinder pressure during intake is less than p_i. During exhaust, the cylinder gases flow through the exhaust manifold to the turbocharger turbine T. Manifold pressure p_e may vary during the exhaust process and lies between cylinder pressure and ambient.[1]

similar set of diagrams illustrating the intake and exhaust processes for a turbocharged four-stroke diesel is shown in Fig. 6-2. When the exhaust valve opens, the burned cylinder gases are fed to a turbine which drives a compressor which compresses the air prior to entry to the cylinder.

Due to the time-varying valve open area and cylinder volume, gas inertia effects, and wave propagation in the intake and exhaust systems, the pressures in the intake, the cylinder, and the exhaust during these gas exchange processes vary in a complicated way. Analytical calculation of these processes is difficult (see Secs. 7.6.2 and 14.3 for a review of available methods). In practice, these processes are often treated empirically using overall parameters such as volumetric efficiency to define intake and exhaust system performance.

6.2 VOLUMETRIC EFFICIENCY

Volumetric efficiency is used as an overall measure of the effectiveness of a four-stroke cycle engine and its intake and exhaust systems as an air pumping device. It is defined [see Sec. 2.10, Eq. (2.27)] as

$$\eta_v = \frac{2\dot{m}_a}{\rho_{a,0} V_d N} \tag{6.1}$$

The air density $\rho_{a,0}$ can be evaluated at atmospheric conditions; η_v is then the overall volumetric efficiency. Or it can be evaluated at inlet manifold conditions; η_v then measures the pumping performance of the cylinder, inlet port, and valve alone. This discussion will cover unthrottled (wide-open throttle) engine operation only. It is the air flow under these conditions that constrains maximum engine power. Lesser air flows in SI engines are obtained by restricting the intake system flow area with the throttle valve.

Volumetric efficiency is affected by the following fuel, engine design, and engine operating variables:

1. Fuel type, fuel/air ratio, fraction of fuel vaporized in the intake system, and fuel heat of vaporization
2. Mixture temperature as influenced by heat transfer
3. Ratio of exhaust to inlet manifold pressures
4. Compression ratio
5. Engine speed
6. Intake and exhaust manifold and port design
7. Intake and exhaust valve geometry, size, lift, and timings

The effects of several of the above groups of variables are essentially quasi steady in nature; i.e., their impact is either independent of speed or can be described adequately in terms of mean engine speed. However, many of these variables have effects that depend on the unsteady flow and pressure wave phenomena that accompany the time-varying nature of the gas exchange processes.

6.2.1 Quasi-Static Effects

VOLUMETRIC EFFICIENCY OF AN IDEAL CYCLE. For the ideal cycles of Fig. 5-2d and e, an expression for volumetric efficiency can be derived which is a function of the following variables: intake mixture pressure p_i, temperature T_i, and fuel/air ratio (F/A); compression ratio r_c; exhaust pressure p_e; and molecular weight M and γ for the cycle working fluid. The overall volumetric efficiency is

$$\eta_v = \frac{m_a}{\rho_{a,0} V_d} = \frac{m(1 - x_r)}{\rho_{a,0}[1 + (F/A)]} \frac{r_c}{(r_c - 1)V_1}$$

where m is the mass in the cylinder at point 1 in the cycle. Since

$$p_i V_1 = m \frac{\tilde{R}}{M} T_1 \quad \text{and} \quad p_{a,0} = \rho_{a,0} \frac{\tilde{R}}{M_a} T_{a,0}$$

and Eq. (5.38) relates T_1 to T_i, the above expression for η_v can be written

$$\eta_v = \left(\frac{M}{M_a}\right)\left(\frac{p_i}{p_{a,0}}\right)\left(\frac{T_{a,0}}{T_i}\right) \frac{1}{[1 + (F/A)]} \left\{\frac{r_c}{r_c - 1} - \frac{1}{\gamma(r_c - 1)}\left[\left(\frac{p_e}{p_i}\right) + (\gamma - 1)\right]\right\} \quad (6.2)$$

For $(p_e/p_i) = 1$, the term in $\{ \; \}$ is unity.

EFFECT OF FUEL COMPOSITION, PHASE, AND FUEL/AIR RATIO. In a spark-ignition engine, the presence of gaseous fuel (and water vapor) in the intake system reduces the air partial pressure below the mixture pressure. For mixtures of air, water vapor, and gaseous or evaporated fuel we can write the intake manifold pressure as the sum of each component's partial pressure:

$$p_i = p_{a,i} + p_{f,i} + p_{w,i}$$

which with the ideal gas law gives

$$\frac{p_{a,i}}{p_i} = \left[1 + \left(\frac{\dot{m}_f}{\dot{m}_a}\right)\left(\frac{M_a}{M_f}\right) + \left(\frac{\dot{m}_w}{\dot{m}_a}\right)\left(\frac{M_a}{M_w}\right)\right]^{-1} \quad (6.3)$$

The water vapor correction is usually small (≤ 0.03). This ratio, $p_{a,i}/p_i$, for several common fuels as a function of $(\dot{m}_f/\dot{m}_a)$ is shown in Fig. 6-3. Note that $(\dot{m}_f/\dot{m}_a)$ only equals the engine operating fuel/air ratio if the fuel is fully vaporized.

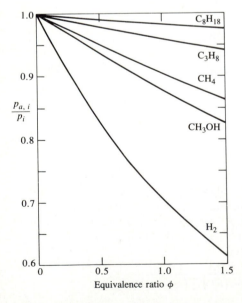

FIGURE 6-3
Effect of fuel (vapor) on inlet air partial pressure. Ratio of air inlet pressure $p_{a,i}$ to mixture inlet pressure p_i versus fuel/air equivalence ratio ϕ for iso-octane vapor, propane, methane, methanol vapor, and hydrogen.

For conventional liquid fuels such as gasoline the effect of fuel vapor, and therefore fuel/air ratio, is small. For gaseous fuels and for methanol vapor, the volumetric efficiency is significantly reduced by the fuel vapor in the intake mixture.

FRACTION FUEL VAPORIZED, HEAT OF VAPORIZATION, AND HEAT TRANSFER. For a constant-pressure flow with liquid fuel evaporation and with heat transfer, the steady-flow energy equation is

$$[\dot{m}_a h_a + (1 - x_e)\dot{m}_f h_{f,L} + x_e \dot{m}_f h_{f,V}]_A = \dot{Q} + (\dot{m}_a h_a + \dot{m}_f h_{f,L})_B \qquad (6.4)$$

where x_e is the mass fraction evaporated and the subscripts denote: a, air properties; f, fuel properties; L, liquid; V, vapor; B before evaporation; A after evaporation. Approximating the change in enthalpy per unit mass of each component of the mixture by $c_p \Delta T$, and with $h_{f,V} - h_{f,L} = h_{f,LV}$ (the enthalpy of vaporization), Eq. (6-4) becomes

$$T_A - T_B = \frac{(\dot{Q}/\dot{m}_a) - x_e(F/A)h_{f,LV}}{c_{p,a} + (F/A)c_{f,L}} \qquad (6.5)$$

Since $c_{f,L} \approx 2c_{p,a}$ the last term in the denominator can often be neglected.

If no heat transfer to the inlet mixture occurs, the mixture temperature decreases as liquid fuel is vaporized. For complete evaporation of isooctane, with $\phi = 1.0$, $T_A - T_B = -19°C$. For methanol under the same conditions, $T_A - T_B$ would be $-128°C$. In practice heating occurs; also, the fuel is not necessarily fully evaporated prior to entry to the cylinder. Experimental data show that the decrease in air temperature that accompanies liquid fuel evaporation more than offsets the reduction in air partial pressure due to the increased amount of fuel vapor: for the same heating rate, volumetric efficiency with fuel vaporization is higher by a few percent.[2]

The ideal cycle equation for volumetric efficiency [Eq. (6.2)] shows that the effect of gas temperature variations, measured at entry to the cylinder, is through the factor $(T_{a,0}/T_i)$. Engine test data indicate that a square root dependence of volumetric efficiency on temperature ratio is closer to real engine behavior. The square root dependence is a standard assumption in engine test data reduction (see Sec. 2.12).

EFFECT OF INLET AND EXHAUST PRESSURE RATIO AND COMPRESSION RATIO. As the pressure ratio (p_e/p_i) and the compression ratio are varied, the fraction of the cylinder volume occupied by the residual gas *at the intake pressure* varies. As this volume increases so volumetric efficiency decreases. These effects on ideal-cycle volumetric efficiency are given by the { } term in Eq. (6.2). For $\gamma = 1.3$ these effects are shown in Fig. 6-4.

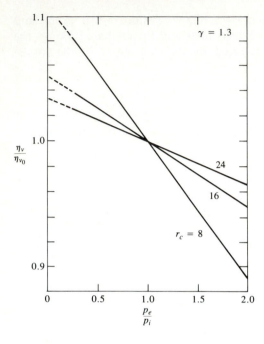

FIGURE 6-4
Effect of exhaust to inlet pressure ratio on ideal-cycle volumetric efficiency.

6.2.2 Combined Quasi-Static and Dynamic Effects

When gas flows unsteadily through a system of pipes, chambers, ports, and valves, both friction, pressure, and inertial forces are present. The relative importance of these forces depends on gas velocity and the size and shape of these passages and their junctions. Both quasi-steady and dynamic effects are usually significant. While the effects of changes in engine speed, and intake and exhaust manifold, port and valve design are interrelated, several separate phenomena which affect volumetric efficiency can be identified.

FRICTIONAL LOSSES. During the intake stroke, due to friction in each part of the intake system, the pressure in the cylinder p_c is less than the atmospheric pressure p_{atm} by an amount dependent on the square of the speed. This total pressure drop is the sum of the pressure loss in each component of the intake system: air filter, carburetor and throttle, manifold, inlet port, and inlet valve. Each loss is a few percent, with the port and valve contributing the largest drop. As a result, the pressure in the cylinder during the period in the intake process when the piston is moving at close to its maximum speed can be 10 to 20 percent lower than atmospheric. For each component in the intake (and the exhaust) system, Bernoulli's equation gives

$$\Delta p_j = \xi_j \rho v_j^2$$

where ξ_j is the resistance coefficient for that component which depends on its geometric details and v_j is the local velocity. Assuming the flow is quasi-steady, v_j is related to the mean piston speed $\bar{S}_p$ by

$$v_j A_j = \bar{S}_p A_p$$

where A_j and A_p are the component minimum flow area and the piston area, respectively. Hence, the total quasi-steady pressure loss due to friction is

$$p_{\text{atm}} - p_c = \sum \Delta p_j = \sum \xi_j \rho v_j^2 = \rho \bar{S}_p^2 \sum \xi_j \left(\frac{A_p}{A_j}\right)^2 \qquad (6.6)$$

Equation (6.6) indicates the importance of large component flow areas for reducing frictional losses, and the dependence of these losses on engine speed. Figure 6-5 shows an example of the pressure losses due to friction across the air cleaner, carburetor, throttle, and manifold plenum of a standard four-cylinder

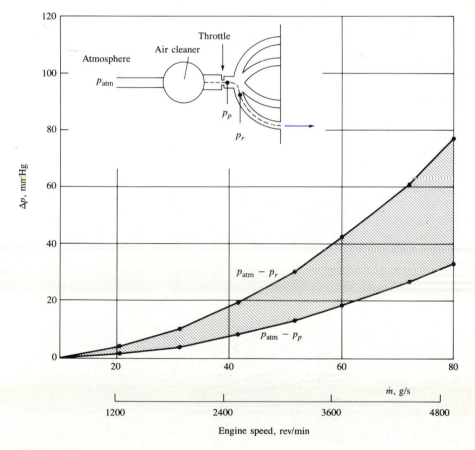

FIGURE 6-5
Pressure losses in the intake system of a four-stroke cycle spark-ignition engine determined under steady flow conditions.[3] Stroke = 89 mm. Bore = 84 mm.

automobile engine intake system. These steady flow tests, conducted over the full engine speed range,[3] show that the pressure loss depends on speed squared.

Equivalent flow-dependent pressure losses in the exhaust system result in the exhaust port and manifold having average pressure levels that are higher than atmospheric. Figure 6-6 shows the time-averaged exhaust manifold gauge pressure as a function of inlet manifold vacuum (which varies inversely to load) and

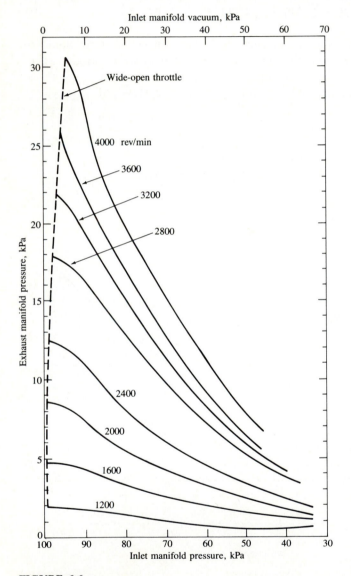

FIGURE 6-6
Exhaust manifold pressure as a function of load (defined by inlet manifold vacuum) and speed, four-stroke cycle four-cylinder spark-ignition engine.[4]

speed for a four-cylinder automobile spark-ignition engine.[4] At high speeds and loads the exhaust manifold operates at pressures substantially above atmospheric.

RAM EFFECT. The pressure in the inlet manifold varies during each cylinder's intake process due to the piston velocity variation, valve open area variation, and the unsteady gas-flow effects that result from these geometric variations. The mass of air inducted into the cylinder, and hence the volumetric efficiency, is almost entirely determined by the pressure level in the inlet port during the short period before the inlet valve is closed.[5] At higher engine speeds, the inertia of the gas in the intake system as the intake valve is closing increases the pressure in the port and continues the charging process as the piston slows down around BC and starts the compression stroke. This effect becomes progressively greater as engine speed is increased. The inlet valve is closed some 40 to 60° after BC, in part to take advantage of this ram phenomenon.

REVERSE FLOW INTO THE INTAKE. Because the inlet valve closes after the start of the compression stroke, a reverse flow of fresh charge from the cylinder back into the intake can occur as the cylinder pressure rises due to piston motion toward TC. This reverse flow is largest at the lowest engine speeds. It is an inevitable consequence of the inlet valve closing time chosen to take advantage of the ram effect at high speeds.

TUNING. The pulsating flow from each cylinder's exhaust process sets up pressure waves in the exhaust system. These pressure waves propagate at the local sound speed relative to the moving exhaust gas. The pressure waves interact with the pipe junctions and ends in the exhaust manifold and pipe. These interactions cause pressure waves to be reflected back toward the engine cylinder. In multicylinder engines, the pressure waves set up by each cylinder, transmitted through the exhaust and reflected from the end, can interact with each other. These pressure waves may aid or inhibit the gas exchange processes. When they aid the process by reducing the pressure in the exhaust port toward the end of the exhaust process, the exhaust system is said to be *tuned.*[6]

The time-varying inlet flow to the cylinder causes expansion waves to be propagated back into the inlet manifold. These expansion waves can be reflected at the open end of the manifold (at the plenum) causing positive pressure waves to be propagated toward the cylinder. If the timing of these waves is appropriately arranged, the positive pressure wave will cause the pressure at the inlet valve at the end of the intake process to be raised above the nominal inlet pressure. This will increase the inducted air mass. Such an intake system is described as *tuned.*[6]

Methods which predict the unsteady flows in the intake and exhaust systems of internal combustion engines with good accuracy have been developed. These methods are complicated, however, so more detailed discussion is deferred to Chap. 14.

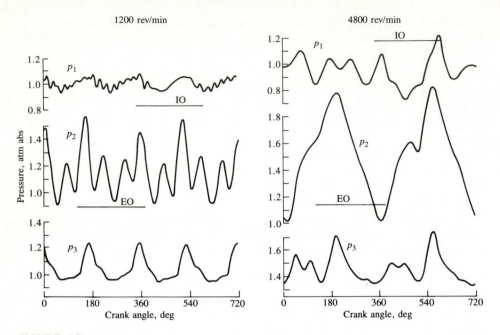

FIGURE 6-7
Instantaneous pressures in the intake and exhaust manifolds of a four-stroke cycle four-cylinder spark-ignition engine, at wide-open throttle. Locations: p_1, intake manifold runner 150 mm from cylinder 1; p_2, exhaust manifold runner 200 mm from cylinder 1; p_3, exhaust manifold runner 700 mm from cylinder 1. IO and EO, intake and exhaust valve open periods for that cylinder, respectively.[3] Stroke = 89 mm. Bore = 84 mm.

Examples of the pressure variations in the inlet and exhaust systems of a four-cylinder automobile spark-ignition engine at wide-open throttle are shown in Fig. 6-7. The complexity of the phenomena that occur is apparent. The amplitude of the pressure fluctuations increases substantially with increasing engine speed. The primary frequency in both the intake and exhaust corresponds to the frequency of the individual cylinder intake and exhaust processes. Higher harmonics that result from pressure waves in both the intake and exhaust are clearly important also.

6.2.3 Variation with Speed, and Valve Area, Lift, and Timing

Flow effects on volumetric efficiency depend on the velocity of the fresh mixture in the intake manifold, port, and valve. Local velocities for quasi-steady flow are equal to the volume flow rate divided by the local cross-sectional area. Since the intake system and valve dimensions scale approximately with the cylinder bore, mixture velocities in the intake system will scale with piston speed. Hence, volumetric efficiencies as a function of speed, for different engines, should be compared at the same mean piston speed.[7] Figure 6-8 shows typical curves of

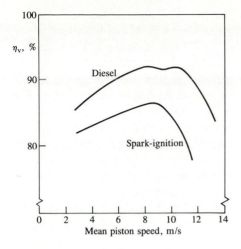

FIGURE 6-8
Volumetric efficiency versus mean piston speed
for a four-cylinder automobile indirect-injection
diesel[8] and a six-cylinder spark-ignition engine.[9]

volumetric efficiency versus mean piston speed for a four-cylinder automobile
indirect-injection diesel engine[8] and a six-cylinder spark-ignition engine.[9] The
volumetric efficiencies of spark-ignition engines are usually lower than diesel
values due to flow losses in the carburetor and throttle, intake manifold heating,
the presence of fuel vapor, and a higher residual gas fraction. The diesel curve
with its double peak shows the effect of intake system tuning.

The shape of these volumetric efficiency versus mean piston speed curves
can be explained with the aid of Fig. 6-9. This shows, in schematic form, how the

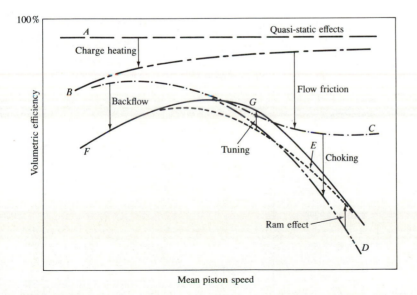

FIGURE 6-9
Effect on volumetric efficiency of different phenomena which affect the air flow rate as a function of
speed. Solid line is final η_v versus speed curve.

effect on volumetric efficiency of each of the different phenomena described in this section varies with speed. Non-speed-dependent effects (such as fuel vapor pressure) drop η_v below 100 percent (curve A). Charge heating in the manifold and cylinder drops curve A to curve B. It has a greater effect at lower engine speeds due to longer gas residence times. Frictional flow losses increase as the square of engine speed, and drop curve B to curve C. At higher engine speeds, the flow into the engine during at least part of the intake process becomes choked (see Sec. 6.3.2). Once this occurs, further increases in speed do not increase the flow rate significantly so volumetric efficiency decreases sharply (curve C to D). The induction ram effect, at higher engine speeds, raises curve D to curve E. Late inlet valve closing, which allows advantage to be taken of increased charging at higher speeds, results in a decrease in η_v at low engine speeds due to backflow (curves C and D to F). Finally, intake and/or exhaust tuning can increase the volumetric efficiency (often by a substantial amount) over part of the engine speed range, curve F to G.

   An example of the effect on volumetric efficiency of tuning the intake manifold runner is shown in Fig. 6-10. In an unsteady flow calculation of the gas exchange processes of a four-cylinder spark-ignition engine, the length of the intake manifold runners was increased successively by factors of 2. The 34-cm length produces a desirable "tuned" volumetric efficiency curve with increased low-speed air flow and flat mid-speed characteristics. While the longest runner further increases low-speed air flow, the loss in η_v at high speed would be unacceptable.[10] Further discussion of intake system tuning can be found in Sec. 7.6.2.

   Figure 6-11 shows data from a four-cylinder spark-ignition engine[3] which illustrates the effect of varying valve timing and valve lift on the volumetric efficiency versus speed curve. Earlier-than-normal inlet valve closing reduces backflow losses at low speed and increases η_v. The penalty is reduced air flow at high speed. Later-than-normal inlet valve closing is only advantageous at very high

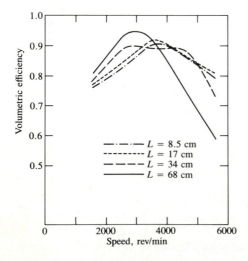

FIGURE 6-10
Effect of intake runner length on volumetric efficiency versus speed for 2.3-dm³ four-cylinder spark-ignition engine.[10]

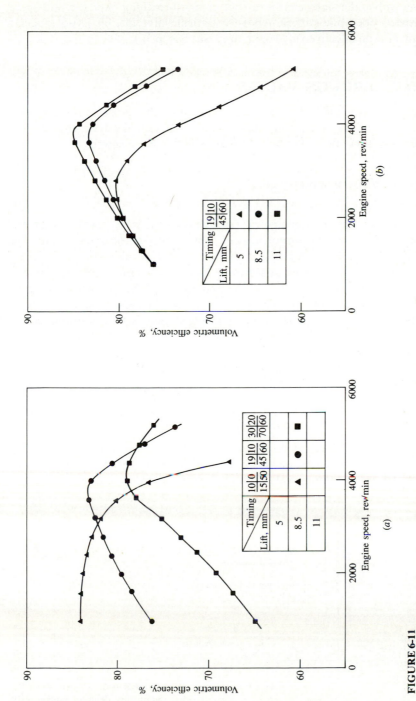

FIGURE 6-11
Effect of (a) valve timing and (b) valve lift on volumetric efficiency versus speed curves. Four-cylinder 1.6-dm³ displacement spark-ignition engine at wide-open throttle, firing conditions, $(A/F) = 13$, MBT ignition timing. Timing numbers are: inlet valve opens (before TC) top left, closes (after BC) bottom left; exhaust valve opens (before BC) bottom right, closes (after TC) top right.³ Stroke = 89 mm.

speeds. Low valve lifts significantly restrict engine breathing over the mid-speed and high-speed operating ranges. Above a critical valve lift, lift is no longer a major constraint on effective valve open area (see Sec. 6.3).

6.3 FLOW THROUGH VALVES

The valve, or valve and port together, is usually the most important flow restriction in the intake and the exhaust system of four-stroke cycle engines. The characteristics of flows through poppet valves will now be reviewed.

6.3.1 Poppet Valve Geometry and Timing

Figure 6-12 shows the main geometric parameters of a poppet valve head and seat. Figure 6-13 shows the proportions of typical inlet and exhaust valves and ports, relative to the valve inner seat diameter D. The inlet port is generally circular, or nearly so, and the cross-sectional area is no larger than is required to achieve the desired power output. For the exhaust port, the importance of good valve seat and guide cooling, with the shortest length of exposed valve stem, leads to a different design. Although a circular cross section is still desirable, a rectangular or oval shape is often essential around the guide boss area. Typical valve head sizes for different shaped combustion chambers in terms of cylinder bore B are given in Table 6.1.[11] Each of these chamber shapes (see Secs. 10.2 and 15.4 for a discussion of spark-ignition and diesel combustion chamber design) imposes different constraints on valve size. Larger valve sizes (or four valves compared with two) allow higher maximum air flows for a given cylinder displacement.

Typical valve timing, valve-lift profiles, and valve open areas for a four-stroke cycle spark-ignition engine are shown in Fig. 6-14. There is no universally accepted criterion for defining valve timing points. Some are based upon a spe-

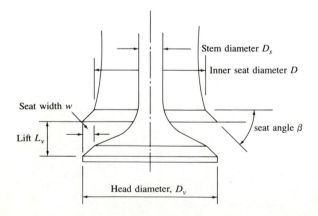

Stem diameter D_s

Inner seat diameter D

Seat width w

Lift L_v

seat angle β

Head diameter, D_v

FIGURE 6-12
Parameters defining poppet valve geometry.

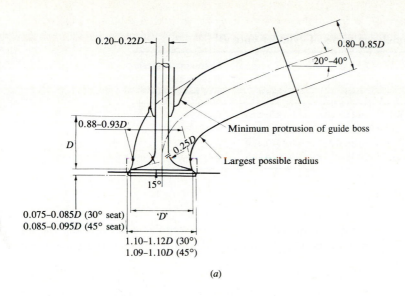

(a)

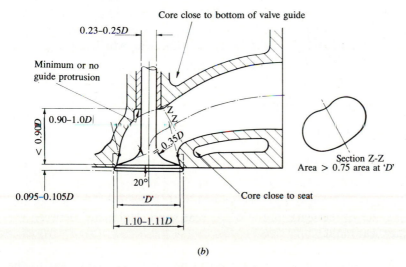

(b)

FIGURE 6-13
Shape, proportions, and critical design areas of typical inlet (top) and exhaust (bottom) valves and ports.[11]

cific lift criterion. For example, SAE defines valve timing events based on reference valve-lift points:[13]

1. **Hydraulic lifters.** Opening and closing positions are the 0.15-mm (0.006-in) valve-lift points.
2. **Mechanical lifters.** Valve opening and closing positions are the points of 0.15-mm (0.006-in) lift plus the specified lash.

TABLE 6.1
Valve head diameter in terms of cylinder bore B^{11}

Combustion chamber shape†	Inlet	Exhaust	Approximate mean piston speed, max power, m/s
Wedge or bathtub	0.43–0.46B	0.35–0.37B	15
Bowl-in-piston	0.42–0.44B	0.34–0.37B	14
Hemispherical	0.48–0.5B	0.41–0.43B	18
Four-valve pent-roof	0.35–0.37B	0.28–0.32B	20

† See Fig. 15-15.

Alternatively, valve events can be defined based on angular criteria along the lift curve.[12] What is important is when significant gas flow through the valve-open area either starts or ceases.

The instantaneous valve flow area depends on valve lift and the geometric details of the valve head, seat, and stem. There are three separate stages to the flow area development as valve lift increases,[14] as shown in Fig. 6-14b. For low valve lifts, the minimum flow area corresponds to a frustrum of a right circular cone where the conical face between the valve and the seat, which is perpendicular to the seat, defines the flow area. For this stage:

$$\frac{w}{\sin \beta \cos \beta} > L_v > 0$$

and the minimum area is

$$A_m = \pi L_v \cos \beta \left(D_v - 2w + \frac{L_v}{2} \sin 2\beta \right) \tag{6.7}$$

where β is the valve seat angle, L_v is the valve lift, D_v is the valve head diameter (the outer diameter of the seat), and w is the seat width (difference between the inner and outer seat radii).

For the second stage, the minimum area is still the slant surface of a frustrum of a right circular cone, but this surface is no longer perpendicular to the valve seat. The base angle of the cone increases from $(90 - \beta)°$ toward that of a cylinder, $90°$. For this stage:

$$\left[\left(\frac{D_p^2 - D_s^2}{4D_m} \right) - w^2 \right]^{1/2} + w \tan \beta \geqslant L_v > \frac{w}{\sin \beta \cos \beta}$$

and

$$A_m = \pi D_m [(L_v - w \tan \beta)^2 + w^2]^{1/2} \tag{6.8}$$

where D_p is the port diameter, D_s is the valve stem diameter, and D_m is the mean seat diameter $(D_v - w)$.

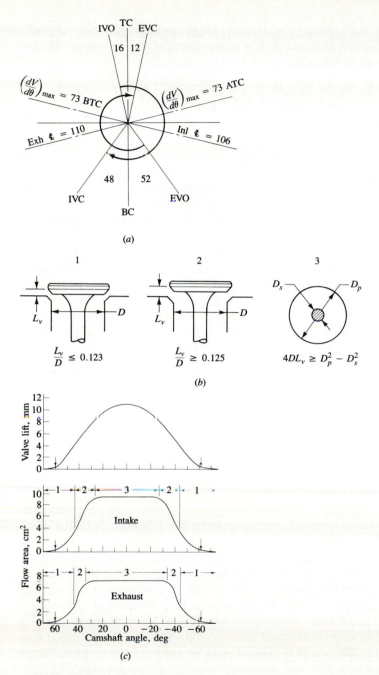

FIGURE 6-14
(a) Typical valve timing diagram for high-speed 2.2-dm³ four-cylinder spark-ignition engine. (b) Schematic showing three stages of valve lift. (c) Valve-lift curve and corresponding minimum intake and exhaust valve open areas as a function of camshaft angle. Inlet and exhaust valve diameters are 3.6 and 3.1 cm, respectively.[12]

Finally, when the valve lift is sufficiently large, the minimum flow area is no longer between the valve head and seat; it is the port flow area minus the sectional area of the valve stem. Thus, for

$$L_v > \left[\left(\frac{D_p^2 - D_s^2}{4D_m} \right)^2 - w^2 \right]^{1/2} + w \tan \beta$$

then

$$A_m = \frac{\pi}{4} (D_p^2 - D_s^2) \tag{6.9}$$

Intake and exhaust valve open areas corresponding to a typical valve-lift profile are plotted versus camshaft angle in Fig. 6-14c. These three different flow regimes are indicated. The maximum valve lift is normally about 12 percent of the cylinder bore.

Inlet valve opening (IVO) typically occurs 10 to 25° BTC. Engine performance is relatively insensitive to this timing point. It should occur sufficiently before TC so that cylinder pressure does not dip early in the intake stroke. Inlet valve closing (IVC) usually falls in the range 40 to 60° after BC, to provide more time for cylinder filling under conditions where cylinder pressure is below the intake manifold pressure at BC. IVC is one of the principal factors that determines high-speed volumetric efficiency; it also affects low-speed volumetric efficiency due to backflow into the intake (see Sec. 6.2.3). Exhaust valve opening (EVO) occurs 50 to 60° before BC, well before the end of the expansion stroke, so that blowdown can assist in expelling the exhaust gases. The goal here is to reduce cylinder pressure to close to the exhaust manifold pressure as soon as possible after BC over the full engine speed range. Note that the timing of EVO affects the cycle efficiency since it determines the effective expansion ratio. Exhaust valve closing (EVC) ends the exhaust process and determines the duration of the valve overlap period. EVC typically falls in the range 8 to 20° after TC. At idle and light load, in spark-ignition engines (which are throttled), it therefore regulates the quantity of exhaust gases that flow back into the combustion chamber through the exhaust valve under the influence of intake manifold vacuum. At high engine speeds and loads, it regulates how much of the cylinder burned gases are exhausted. EVC timing should occur sufficiently far after TC so that the cylinder pressure does not rise near the end of the exhaust stroke. Late EVC favors high power at the expense of low-speed torque and idle combustion quality. Note from the timing diagram (Fig. 6-14a) that the points of maximum valve lift and maximum piston velocity (Fig. 2-2) do not coincide.

The effect of valve geometry and timing on air flow can be illustrated conceptually by dividing the rate of change of cylinder volume by the instantaneous minimum valve flow area to obtain a *pseudo flow velocity* for each valve:[12]

$$v_{\text{ps}} = \frac{1}{A_m} \frac{dV}{d\theta} = \frac{\pi B^2}{4A_m} \frac{ds}{d\theta} \tag{6.10}$$

where V is the cylinder volume [Eq. (2.4)], B is the cylinder bore, s is the distance

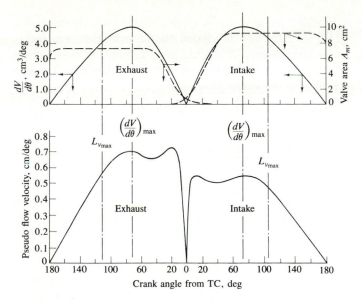

FIGURE 6-15
Rate of change of cylinder volume $dV/d\theta$, valve minimum flow area A_m, and pseudo flow velocity as function of crank angle for exhaust and inlet valves of Fig. 6-14.[12]

between the wrist pin and crank axis [see Fig. 2-1 and Eq. (2.5)] and A_m is the valve area given by Eqs. (6.7), (6.8), or (6.9). Instantaneous pseudo flow velocity profiles for the exhaust and intake strokes of a four-stroke four-cylinder engine are shown in Fig. 6-15. Note the appearance of two peaks in the pseudo flow velocity for both the exhaust and intake strokes. The broad peaks occurring at maximum piston velocity reflect the fact that valve flow area is constant at this point. The peaks close to TC result from the exhaust valve closing and intake valve opening profiles. The peak at the end of the exhaust stroke is important since it indicates a high pressure drop across the valve at this point, which will result in higher trapped residual mass. The magnitude of this exhaust stroke pseudo velocity peak depends strongly on the timing of exhaust valve closing. The pseudo velocity peak at the start of the intake stroke is much less important. That the pseudo velocities early in the exhaust stroke and late in the intake stroke are low indicates that phenomena other than quasi-steady flow govern the flow rate. These are the periods when exhaust blowdown and ram and tuning effects in the intake are most important.

6.3.2 Flow Rate and Discharge Coefficients

The mass flow rate through a poppet valve is usually described by the equation for compressible flow through a flow restriction [Eqs. (C.8) or (C.9) in App. C]. This equation is derived from a one-dimensional isentropic flow analysis, and real gas flow effects are included by means of an experimentally determined discharge coefficient C_D. The air flow rate is related to the upstream stagnation pressure p_0 and stagnation temperature T_0, static pressure just downstream of the flow restriction (assumed equal to the pressure at the restriction, p_T), and a reference area A_R characteristic of the valve design:

$$\dot{m} = \frac{C_D A_R p_0}{(RT_0)^{1/2}} \left(\frac{p_T}{p_0}\right)^{1/\gamma} \left\{\frac{2\gamma}{\gamma-1}\left[1-\left(\frac{p_T}{p_0}\right)^{(\gamma-1)/\gamma}\right]\right\}^{1/2} \tag{6.11}$$

When the flow is choked, i.e., $p_T/p_0 \leq [2/(\gamma+1)]^{\gamma/(\gamma-1)}$, the appropriate equation is

$$\dot{m} = \frac{C_D A_R p_0}{(RT_0)^{1/2}} \gamma^{1/2}\left(\frac{2}{\gamma+1}\right)^{(\gamma+1)/2(\gamma-1)} \tag{6.12}$$

For flow into the cylinder through an intake valve, p_0 is the intake system pressure p_i and p_T is the cylinder pressure. For flow out of the cylinder through an exhaust valve, p_0 is the cylinder pressure and p_T is the exhaust system pressure.

The value of C_D and the choice of reference area are linked together: their product, $C_D A_R$, is the effective flow area of the valve assembly A_E. Several different reference areas have been used. These include the valve head area $\pi D_v^2/4$,[7] the port area at the valve seat $\pi D_p^2/4$,[15] the geometric minimum flow area [Eqs. (6.7), (6.8), and (6.9)], and the curtain area $\pi D_v L_v$,[16] where L_v is the valve lift. The choice is arbitrary, though some of these choices allow easier interpretation than others. As has been shown above, the geometric minimum flow area is a complex function of valve and valve seat dimensions. The most convenient reference area in practice is the so-called valve curtain area:

$$A_C = \pi D_v L_v \tag{6.13}$$

since it varies linearly with valve lift and is simple to determine.

INLET VALVES. Figure 6-16 shows the results of steady flow tests on a typical inlet valve configuration with a sharp-cornered valve seat.[16] The discharge coefficient based on valve curtain area is a discontinuous function of the valve-lift/diameter ratio. The three segments shown correspond to different flow regimes as indicated. At very low lifts, the flow remains attached to the valve head and seat, giving high values for the discharge coefficient. At intermediate lifts, the flow separates from the valve head at the inner edge of the valve seat as shown. An abrupt decrease in discharge coefficient occurs at this point. The discharge coefficient then increases with increasing lift since the size of the separated region remains approximately constant while the minimum flow area is increasing. At high lifts, the flow separates from the inner edge of the valve seat as well.[14, 17] Typical maximum values of L_v/D_v are 0.25.

An important question is whether these steady flow data are representative of the dynamic flow behavior of the valve in an operating engine. There is some evidence that the "change points" between different flow regimes shown in Fig. 6-16 occur at slightly different valve lifts under dynamic operation than under

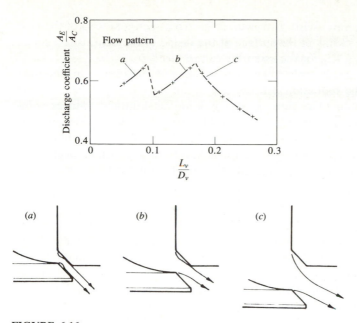

FIGURE 6-16
Discharge coefficient of typical inlet poppet valve (effective flow area/valve curtain area) as a function of valve lift. Different segments correspond to flow regimes indicated.[16]

steady flow operation. Also, as has been discussed in Sec. 6.2.2, the pressure upstream of the valve varies significantly during the intake process. However, it has been shown that over the normal engine speed range, steady flow discharge-coefficient results can be used to predict dynamic performance with reasonable precision.[14, 18]

In addition to valve lift, the performance of the inlet valve assembly is influenced by the following factors: valve seat width, valve seat angle, rounding of the seat corners, port design, cylinder head shape. In many engine designs the port and valve assembly are used to generate a rotational motion (swirl) inside the engine cylinder during the induction process, or the cylinder head can be shaped to restrict the flow through one side of the valve open area to generate swirl. Swirl production is discussed later, in Section 8.3. Swirl generation significantly reduces the valve (and port) flow coefficient. Changes in seat width affect the L_v/D_v at which the shifts in flow regimes illustrated in Fig. 6-16 occur. C_D increases as seat width decreases. The seat angle β affects the discharge coefficient in the low-lift regime in Fig. 6-16. Rounding the upstream corner of the valve seat reduces the tendency of the flow to break away, thus increasing C_D at higher lifts. At low valve lifts, when the flow remains attached, increasing the Reynolds number decreases the discharge coefficient. Once the flow breaks away from the wall, there is no Reynolds number dependence of C_D.[16]

For well-designed ports (e.g., Fig. 6-13) the discharge coefficient of the port and valve assembly need be no lower than that of the isolated valve (except when the port is used to generate swirl). However, if the cross-sectional area of the port is not sufficient or the radius of the surface at the inside of the bend is too small, a significant reduction in C_D for the assembly can result.[16]

At high engine speeds, unless the inlet valve is of sufficient size, the inlet flow during part of the induction process can become choked (i.e., reach sonic velocity at the minimum valve flow area). Choking substantially reduces volumetric efficiency. Various definitions of inlet Mach number have been used to identify the onset of choking. Taylor and coworkers[7] correlated volumetric efficiencies measured on a range of engine and inlet valve designs with an *inlet Mach index Z* formed from an average gas velocity through the inlet valve:

$$Z = \frac{A_p \bar{S}_p}{C_i A_i a} \tag{6.14}$$

where A_i is the nominal inlet valve area $(\pi D_v^2/4)$, C_i is a mean valve discharge coefficient based on the area A_i, and a is the sound speed. From the method used to determine C_i, it is apparent that $C_i A_i$ is the average effective open area of the valve (it is the average value of $C_D \pi D_v L_v$). Z corresponds closely, therefore, to the mean Mach number in the inlet valve throat. Taylor's correlations show that η_v decreases rapidly for $Z \geq 0.5$. An alternative equivalent approach to this problem has been developed, based on the average flow velocity through the valve during the period the valve is open.[19] A *mean inlet Mach number* was defined:

$$\bar{M}_i = \frac{\bar{v}_i}{a} \tag{6.15}$$

where $\bar{v}_i$ is the mean inlet flow velocity during the valve open period. $\bar{M}_i$ is related to Z via

$$\bar{M}_i = \frac{Z(\eta_v/100)180}{\theta_{\text{IVC}} - \theta_{\text{IVO}}} \tag{6.16}$$

This mean inlet Mach number correlates volumetric efficiency characteristics better than the Mach index. For a series of modern small four-cylinder engines, when $\bar{M}_i$ approaches 0.5 the volumetric efficiency decreases rapidly. This is due to the flow becoming choked during part of the intake process. This relationship can be used to size the inlet valve for the desired volumetric efficiency at maximum engine speed. Also, if the inlet valve is closed too early, volumetric efficiency will decrease gradually with increasing $\bar{M}_i$, for $\bar{M}_i < 0.5$, even if the valve open area is sufficiently large.[19]

EXHAUST VALVES. In studies of the flow from the cylinder through an exhaust poppet valve, different flow regimes at low and high lift occur, as shown in Fig. 6-17. Values of C_D based on the valve curtain area, for several different exhaust valve and port combinations, are given in Fig. 6-18. A sharp-cornered isolated poppet valve (i.e., straight pipe downstream, no port) gives the best performance.

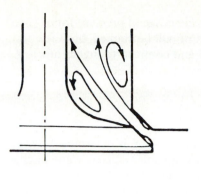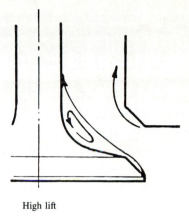

Low lift   High lift

FIGURE 6-17
Flow pattern through exhaust valve at low and high lift.[16]

At high lifts, $L_v/D_v \geq 0.2$, the breakaway of the flow reduces the discharge coefficient. (At $L_v/D_v = 0.25$ the effective area is about 90 percent of the minimum geometric area. For $L_v/D_v < 0.2$ it is about 95 percent.[16]) The port design significantly affects C_D at higher valve lifts, as indicated by the data from four port designs in Fig. 6-18. Good designs can approach the performance of isolated

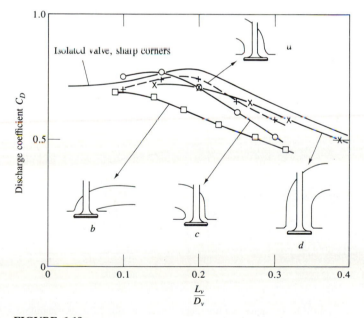

FIGURE 6-18
Discharge coefficient as function of valve lift for several exhaust valve and port designs.[16] a,[20] b,[15] c,[20] d.[21]

valves, however. Exhaust valves operate over a wide range of pressure ratios (1 to 5). For pressure ratios greater than about 2 the flow will be choked, but the effect of pressure ratio on discharge coefficient is small and confined to higher lifts (e.g., ± 5 percent at $L_v/D_v = 0.3$).[15]

6.4 RESIDUAL GAS FRACTION

The residual gas fraction in the cylinder during compression is determined by the exhaust and inlet processes. Its magnitude affects volumetric efficiency and engine performance directly, and efficiency and emissions through its effect on working-fluid thermodynamic properties. The residual gas fraction is primarily a function of inlet and exhaust pressures, speed, compression ratio, valve timing, and exhaust system dynamics.

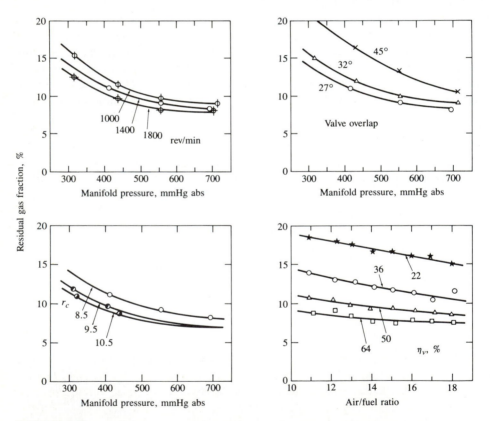

FIGURE 6-19
Residual gas fraction for 2-dm³ four-cylinder spark-ignition engine as a function of intake manifold pressure for a range of speeds, compression ratios, and valve overlaps: also as a function of air/fuel ratio for a range of volumetric efficiencies. Operating conditions, unless noted: speed = 1400 rev/min, $A/F = 14.5$, spark timing set to give 0.95 maximum torque, compression ratio = 8.5.[22]

The residual gas mass fraction x_r (or burned gas fraction if EGR is used) is usually determined by measuring the CO_2 concentration in a sample of gas extracted from the cylinder during the compression stroke. Then

$$x_r = \frac{(\tilde{x}_{CO_2})_C}{(\tilde{x}_{CO_2})_e} \qquad (6.17)$$

where the subscripts C and e denote compression and exhaust, and $\tilde{x}_{CO_2}$ are mole fractions in the wet gas. Usually CO_2 volume or mole fractions are measured in dry gas streams (see Sec. 4.9). A correction factor K,

$$K = \frac{(\tilde{x}_i)_{\text{wet}}}{(\tilde{x}_i)_{\text{dry}}} = \frac{1}{1 + 0.5[y(\tilde{x}^*_{CO_2} + x^*_{CO}) - 0.74\tilde{x}^*_{CO}]} \qquad (6.18)$$

where y is the molar hydrogen/carbon ratio of the fuel and $\tilde{x}^*_{CO_2}$, $\tilde{x}^*_{CO}$ are dry mole fractions, can be used to convert the dry mole fraction measurements.

Residual gas measurements in a spark-ignition engine are given in Fig. 6-19, which shows the effect of changes in speed, valve overlap, compression ratio, and air/fuel ratio for a range of inlet manifold pressures for a 2-dm³, 88.5-mm bore, four-cylinder engine.[22] The effect of variations in spark timing was negligible. Inlet pressure, speed, and valve overlap are the most important variables, though the exhaust pressure also affects the residual fraction.[23] Normal settings for inlet valve opening (about 15° before TC) and exhaust valve closing (about 12° after TC) provide sufficient overlap for good scavenging, but avoid excessive backflow from the exhaust port into the cylinder.

Residual gas fractions in diesel engines are substantially lower than in SI engines because inlet and exhaust pressures are comparable in magnitude and the compression ratio is 2 to 3 times as large. Also, a substantial fraction of the residual gas is air.

6.5 EXHAUST GAS FLOW RATE AND TEMPERATURE VARIATION

The exhaust gas mass flow rate and the properties of the exhaust gas vary significantly during the exhaust process. The origin of this variation for an ideal exhaust process is evident from Fig. 5-3. The thermodynamic state (pressure, temperature, etc.) of the gas in the cylinder varies continually during the exhaust blowdown phase, until the cylinder pressure closely approaches the exhaust manifold pressure. In the real exhaust process, the exhaust valve restricts the flow out of the cylinder, the valve lift varies with time, and the cylinder volume changes during the blowdown process, but the principles remain the same.

Measurements have been made of the variation in mass flow rate through the exhaust valve and gas temperature at the exhaust port exit during the exhaust process of a spark-ignition engine.[24] Figure 6-20 shows the instantaneous mass flow rate data at three different engine speeds. The blowdown and displacement

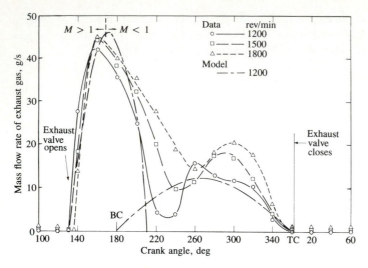

FIGURE 6-20

Instantaneous mass flow rate of exhaust gas through the valve versus crank angle: equivalence ratio = 1.2, wide-open throttle, compression ratio = 7. Dash-dot line is one-dimensional compressible flow model for blowdown and incompressible displacement model for exhaust stroke.[24]

phases of the exhaust process are evident. Simple quasi-steady models of these phases give good agreement with the data at lower engine speeds. The blowdown model shown applies orifice flow equations to the flow across the exhaust valve using the measured cylinder pressure and estimated gas temperature for upstream stagnation conditions. Equation (C.9) is used when the pressure ratio across the valve exceeds the critical value. Equation (C.8) is used when the pressure ratio is less than the critical value. The displacement model assumes the gas in the cylinder is incompressible as the piston moves through the exhaust stroke. As engine speed increases, the crank angle duration of the blowdown phase increases. There is evidence of dynamic effects occurring at the transition between the two phases. The peak mass flow rate during blowdown does not vary substantially with speed since the flow is choked. The mass flow rate at the time of maximum piston speed during displacement scales approximately with piston speed. As the inlet manifold pressure is reduced below the wide-open throttle value, the proportion of the charge which exits the cylinder during the blowdown phase decreases but the mass flow rate during displacement remains essentially constant.

The exhaust gas temperature varies substantially through the exhaust process, and decreases due to heat loss as the gas flows past the exhaust valve and through the exhaust system.

Figure 6-21 shows the measured cylinder pressure, calculated cylinder gas temperature and exhaust mass flow rate, and measured gas temperature at the exhaust port exit for a single-cylinder spark-ignition engine at mid-load and low speed.[25] The average cylinder-gas temperature falls rapidly during blowdown, and continues to fall during the exhaust stroke due to heat transfer to the cylin-

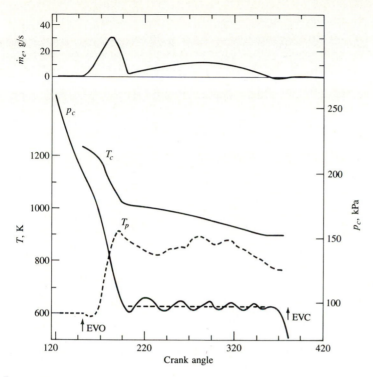

FIGURE 6-21
Measured cylinder pressure p_c, calculated cylinder-gas temperature T_c, exhaust mass flow rate $\dot{m}_e$, and measured gas temperature at exhaust port exit T_p, for single-cylinder spark-ignition engine. Speed = 1000 rev/min, imep = 414 kPa, equivalence ratio = 1.2, spark timing − 10° BTC, r_c − 7.2.[25]

der walls. The gas temperature at the port exit at the start of the exhaust flow pulse is a mixture of hotter gas which has just left the cylinder and cooler gas which left the cylinder at the end of the previous exhaust process and has been stationary in the exhaust port while the valve has been closed. The port exit temperature has a minimum where the transition from blowdown flow to displacement occurs, and the gas comes momentarily to rest and loses a substantial fraction of its thermal energy to the exhaust port walls.

Figure 6-22 shows the effect of varying load and speed on exhaust port exit temperatures. Increasing load $(A \rightarrow B \rightarrow C)$ increases the mass and temperature in the blowdown pulse. Increasing speed $(B \rightarrow D)$ raises the gas temperature throughout the exhaust process. These effects are the result of variations in the relative importance of heat transfer in the cylinder and heat transfer to the exhaust valve and port. The time available for heat transfer, which depends on engine speed and exhaust gas flow rate, is the most critical factor. The exhaust temperature variation with equivalence ratio follows from the variation in expansion stroke temperatures, with maximum values at $\phi = 1.0$ and lower values for leaner and richer mixtures.[24] Diesel engine exhaust temperatures are significantly

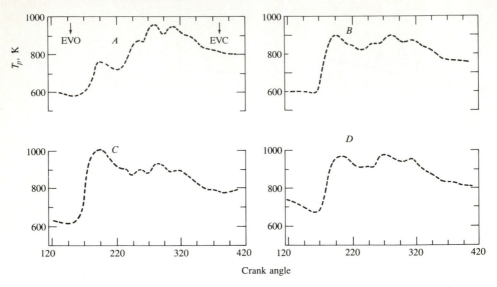

FIGURE 6-22
Measured gas temperature at exhaust port exit as a function of crank angle, single-cylinder spark-ignition engine, for different loads and speeds. Curve A: imep = 267 kPa, 1000 rev/min; curve B: imep = 414 kPa, 1000 rev/min; curve C: imep = 621 kPa, 1000 rev/min; curve D: imep = 414 kPa, 1600 rev/min. Equivalence ratio = 1.2, spark timing = $10°$ BTC, compression ratio = 7.2.[25]

lower than spark-ignition engine exhaust temperatures because of the lean oper-ating equivalence ratio and their higher expansion ratio during the power stroke.

The average exhaust gas temperature is an important quantity for deter-mining the performance of turbochargers, catalytic converters, and particulate traps. The time-averaged exhaust temperature does not correspond to the average energy of the exhaust gas because the flow rate varies substantially. An enthalpy-averaged temperature

$$\bar{T}_h = \left(\int_{\text{EVO}}^{\text{EVC}} \dot{m} c_p T_g \, d\theta \right) \Big/ \left(\int_{\text{EVO}}^{\text{EVC}} \dot{m} c_p \, d\theta \right) \tag{6.19}$$

is the best indicator of exhaust thermal energy. Average exhaust gas temperatures are usually measured with a thermocouple. Thermocouple-averaged temperatures are close to time-averaged temperatures. Mass-averaged exhaust temperatures (which are close to $\bar{T}_h$ if c_p variations are small) for a spark-ignition engine at the exhaust port exit are about 100 K higher than time-averaged or thermocouple-determined temperatures. Mass-average temperatures in the cylinder during the exhaust process are about 200 to 300 K higher than mass-averaged port tem-peratures. All these temperatures increase with increasing speed, load, and spark retard, with speed being the variable with the largest impact.[26]

6.6 SCAVENGING IN TWO-STROKE CYCLE ENGINES

6.6.1 Two-Stroke Engine Configurations

In two-stroke cycle engines, each outward stroke of the piston is a power stroke. To achieve this operating cycle, the fresh charge must be supplied to the engine cylinder at a high-enough pressure to displace the burned gases from the previous cycle. Raising the pressure of the intake mixture is done in a separate pump or blower or compressor. The operation of clearing the cylinder of burned gases and filling it with fresh mixture (or air) the combined intake and exhaust process—is called *scavenging*. However, air capacity is just as important as in the four-stroke cycle; usually, a greater air mass flow rate must be achieved to obtain the same output power. Figures 1-12, and 1-5 and 1-24 show sectioned drawings of a two-stroke spark-ignition engine and two two-stroke diesels.

The different categories of two-stroke cycle scavenging flows and the port (and valve) arrangements that produce them are illustrated in Figs. 6-23 and 6-24. Scavenging arrangements are classified into: (*a*) *cross-scavenged*, (*b*) *loop-scavenged*, and (*c*) *uniflow-scavenged configurations*. The location and orientation of the scavenging ports control the scavenging process, and the most common arrangements are indicated. Cross- and loop-scavenging systems use exhaust and inlet ports in the cylinder wall, uncovered by the piston as it approaches BC.[27] The uniflow system may use inlet ports with exhaust valves in the cylinder head,

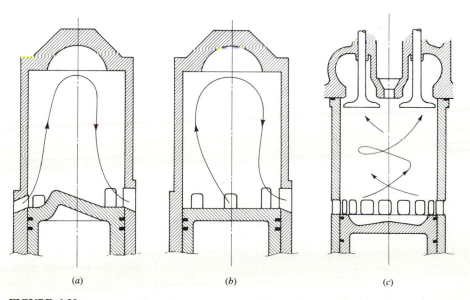

(a) $\qquad\qquad$ (b) $\qquad\qquad$ (c)

FIGURE 6-23
(*a*) Cross-scavenged, (*b*) loop-scavenged, and (*c*) uniflow-scavenged two-stroke cycle flow configurations.

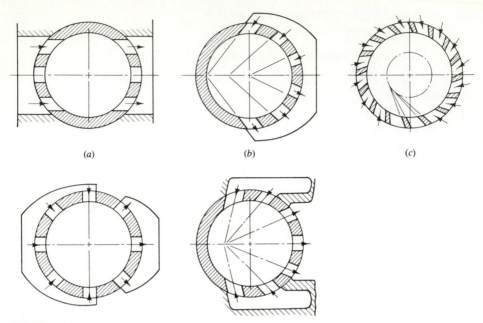

FIGURE 6-24
Common porting arrangements that go with (a) cross-scavenged, (b) loop-scavenged, and (c) uniflow-scavenged configurations.

or inlet and exhaust ports with opposed pistons. Despite the different flow patterns obtained with each cylinder geometry, the general operating principles are similar. Air in a diesel, or fuel-air mixture in a spark-ignition engine, must be supplied to the inlet ports at a pressure higher than the exhaust system pressure.

Figure 6-25 illustrates the principles of the scavenging process for a uniflow engine design. Between 100 and 110° after TC, the exhaust valve opens and a blowdown discharge process commences. Initially, the pressure ratio across the exhaust valve exceeds the critical value (see App. C) and the flow at the valve will be sonic: as the cylinder pressure decreases, the pressure ratio drops below the critical value. The discharge period up to the time of the scavenging port opening is called the blowdown (or free exhaust) period. The scavenging ports open between 60 and 40° before BC when the cylinder pressure slightly exceeds the scavenging pump pressure. Because the burned gas flow is toward the exhaust valves, which now have a large open area, the exhaust flow continues and no backflow occurs. When the cylinder pressure falls below the inlet pressure, air enters the cylinder and the scavenging process starts. This flow continues as long as the inlet ports are open and the inlet total pressure exceeds the pressure in the cylinder. As the cylinder pressure rises above the exhaust pressure, the fresh charge flowing into the cylinder displaces the burned gases: a part of the fresh charge mixes with the burned gases and is expelled with them. The exhaust valves usually close after the inlet ports close. Since the flow in the cylinder is toward the exhaust valve, additional scavenging is obtained. Figure 1-16 illustrates the

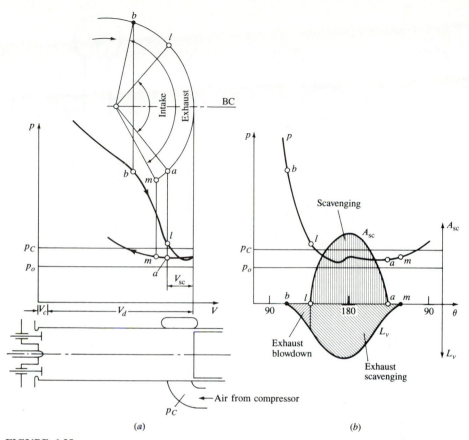

FIGURE 6-25
Gas exchange process in two-stroke cycle uniflow-scavenged diesel engine: (a) valve and port timing and pressure-volume diagram; (b) pressure, scavenging port open area A_{sc}, and exhaust valve lift L_v as functions of crank angle.[1]

similar sequence of events for a loop-scavenged engine. Proper flow patterns for the fresh charge are extremely important for good scavenging and charging of the cylinder.

Common methods for supercharging or pressurizing the fresh charge are shown in Fig. 6-26. In large two-stroke cycle engines, more complex combinations of these approaches are often used, as shown in Fig. 1-24.

6.6.2 Scavenging Parameters and Models

The following overall parameters are used to describe the scavenging process.[13]
The *delivery ratio* Λ:

$$\Lambda = \frac{\text{mass of delivered air (or mixture) per cycle}}{\text{reference mass}} \qquad (6.20)$$

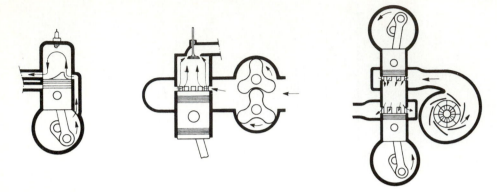

FIGURE 6-26
Common methods of pressurizing the fresh charge in two-stroke cycle engines: left, crankcase compression; center, roots blower; right, centrifugal compressor.[7]

compares the actual scavenging air mass (or mixture mass) to that required in an ideal charging process.† The reference mass is defined as displaced volume × ambient air (or mixture) density. Ambient air (or mixture) density is determined at atmospheric conditions or at intake conditions. This definition is useful for experimental purposes. For analytical work, it is often convenient to use the trapped cylinder mass m_{tr} as the reference mass.

The *trapping efficiency* η_{tr}:

$$\eta_{tr} = \frac{\text{mass of delivered air (or mixture) retained}}{\text{mass of delivered air (or mixture)}} \tag{6.21}$$

indicates what fraction of the air (or mixture) supplied to the cylinder is retained in the cylinder.

The *scavenging efficiency* η_{sc}:

$$\eta_{sc} = \frac{\text{mass of delivered air (or mixture) retained}}{\text{mass of trapped cylinder charge}} \tag{6.22}$$

indicates to what extent the residual gases in the cylinder have been replaced with fresh air.

The *purity* of the charge:

$$\text{Purity} = \frac{\text{mass of air in trapped cylinder charge}}{\text{mass of trapped cylinder charge}} \tag{6.23}$$

indicates the degree of dilution, with burned gases, of the unburned mixture in the cylinder.

---

† If scavenging is done with fuel-air mixture, as in spark-ignition engines, then mixture mass is used instead of air mass.

The *charging efficiency* η_{ch}:

$$\eta_{ch} = \frac{\text{mass of delivered air (or mixture) retained}}{\text{displaced volume} \times \text{ambient density}} \tag{6.24}$$

indicates how effectively the cylinder volume has been filled with fresh air (or mixture). Charging efficiency, trapping efficiency, and delivery ratio are related by

$$\eta_{ch} = \Lambda\eta_{tr} \tag{6.25}$$

When the reference mass in the definition of delivery ratio is the trapped cylinder mass m_{tr} (or closely approximated by it) then

$$\eta_{sc} = \Lambda\eta_{tr} \tag{6.26}$$

In real scavenging processes, mixing occurs as the fresh charge displaces the burned gases and some of the fresh charge may be expelled. Two limiting ideal models of this process are: (1) perfect displacement and (2) complete mixing. Perfect displacement or scavenging would occur if the burned gases were pushed out by the fresh gases without any mixing. Complete mixing occurs if entering fresh mixture mixes instantaneously and uniformly with the cylinder contents.

For *perfect displacement* (with m_{tr} as the reference mass in the delivery ratio),

$$\begin{aligned} \eta_{sc} = \Lambda \quad &\text{and} \quad \eta_{tr} = 1 \quad &&\text{for } \Lambda \leqslant 1 \\ \eta_{sc} = 1 \quad &\text{and} \quad \eta_{tr} = \Lambda^{-1} \quad &&\text{for } \Lambda > 1 \end{aligned} \tag{6.27}$$

For the *complete mixing* limit, consider the scavenging process as a quasi-steady flow process. Between time t and $t + dt$, a mass element dm_{ad} of air is delivered to the cylinder and is uniformly mixed throughout the cylinder volume. An equal amount of fluid, with the same proportions of air and burned gas as the cylinder contents at time t, leaves the cylinder during this time interval. Thus the mass of air delivered between t and $t + dt$ which is retained, dm_{ar}, is given by

$$dm_{ar} = dm_{ad}\left(1 - \frac{m_{ar}}{m_{tr}}\right)$$

Assuming m_{tr} is constant, this integrates over the duration of the scavenging process to give

$$\frac{m_{ar}}{m_{tr}} = 1 - \exp\left(\frac{m_{ad}}{m_{tr}}\right) \tag{6.28}$$

Thus, for *complete mixing*, with the above definitions,

$$\eta_{sc} = 1 - e^{-\Lambda}$$
$$\eta_{tr} = \frac{1}{\Lambda}(1 - e^{-\Lambda}) \tag{6.29}$$

Figure 6-27 shows η_{sc} and η_{tr} for the perfect displacement and complete mixing assumptions as a function of Λ, the delivery ratio.

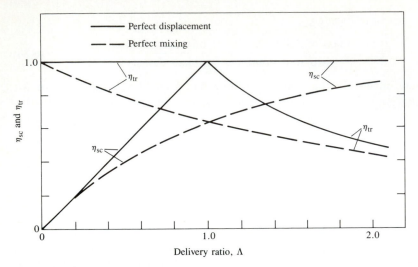

FIGURE 6-27
Scavenging efficiency η_{sc} and trapping efficiency η_{tr} versus delivery ratio Λ for perfect displacement and complete mixing models.

An additional possibility is the direct flow of fresh mixture through the cylinder into the exhaust without entraining burned gases. This is called *short-circuiting*; it is obviously undesirable since some fresh air or mixture is wasted. There is no simple model for this process. When short-circuiting occurs, lower scavenging efficiencies result even though the volume occupied by the short-circuiting flow through the cylinder does displace an equal volume of the burned gases. Another phenomenon which reduces scavenging efficiency is the formation of *pockets* or *dead zones* in the cylinder volume where burned gases can become trapped and escape displacement or entrainment by the fresh scavenging flow. These unscavenged zones are most likely to occur in regions of the cylinder that remain secluded from the main fresh mixture flow path.

6.6.3 Actual Scavenging Processes

Several methods have been developed for determining what occurs in actual cylinder scavenging processes.[27] Accurate measurement of scavenging efficiency is difficult due to the problem of measuring the trapped air mass. Estimation of η_{sc} from indicated mean effective pressure and from gas sampling are the most reliable methods.[7] Flow visualization experiments[28–30] in liquid analogs of the cylinder and flow velocity mapping techniques[31] have proved useful in providing a qualitative picture of the scavenging flow field and identifying problems such as short-circuiting and dead volumes.

Flow visualization studies indicate the key features of the scavenging process. Figure 6-28 shows a sequence of frames from a movie of one liquid scavenging another in a model of a large two-stroke cycle loop-scavenged

FIGURE 6-28
Photos of one fluid scavenging another in liquid analog experiments in model loop-scavenged engine cylinder. Top two rows: View perpendicular to scavenging loop. Bottom row: Orthogonal views. Dark denser fluid displacing light less dense fluid.[29]

diesel.[29] The physical variables were scaled to maintain the same values of the appropriate dimensionless numbers for the liquid analog flow and the real engine flow. The density of the liquid representing air (which is dark) was twice the density of the liquid representing burned gas (which is clear). Early in the scavenging process, the fresh air jets penetrate into the burned gas and displace it first toward the cylinder head and then toward the exhaust ports (the schematic gives the location of the ports). During this initial phase, the outflowing gas contains no air; pure displacement of the burned gas from the cylinder is being achieved. Then short-circuiting losses start to occur, due to the damming-up or buildup of fresh air on the cylinder wall opposite the exhaust ports. The short-circuiting fluid flows directly between the scavenge ports and the exhaust ports above them. Since this damming-up of the inflowing fresh air back toward the exhaust ports continues, short-circuiting losses will also continue. While the scavenging front remains distinct as it traverses the cylinder, its turbulent character indicates that mixing between burned gas and air across the front is taking place. For both these reasons (short-circuiting and short-range mixing), the outflowing gas, once the "displacement" phase is over, contains an increasing amount of fresh air.

Outflowing fluid composition measurements from this model study of a Sulzer two-stroke loop-scavenged diesel engine confirm this sequence of events. At 24 crank angle degrees after the onset of scavenging, fresh air due to short-circuiting was detected in the exhaust. At the time the displacement front reached the exhaust port (65° after the onset of scavenging), loss of fresh air due to scavenging amounted to 13 percent of the scavenge air flow. The actual plot of degree of purity (or η_{sc}) versus delivery ratio (Λ) closely followed the perfect displacement line for $\Lambda < 0.4$. For $\Lambda > 0.4$, the shape of the actual curve was similar in shape to the complete mixing curve.

Engine tests confirm these results from model studies. Initially, the exhausted gas contains no fresh air or mixture; only burned gas is being displaced from the cylinder. However, within the cylinder both displacement *and* mixing at the interface between burned gas and fresh gas are occurring. The departure from perfect scavenging behavior is evident when fresh mixture first appears in the exhaust. For loop-scavenged engines this is typically when $\Lambda \approx 0.4$. For uniflow scavenging this perfect scavenging phase lasts somewhat longer; for cross-scavenging it is over sooner (because the short-circuiting path is shorter).

The mixing that occurs is short-range mixing, not mixing throughout the cylinder volume. The jets of scavenging mixture, on entering the cylinder, mix readily with gases in the immediate neighborhood of the jet efflux. More efficient scavenging—i.e., less mixing—is obtained by reducing the size of the inlet ports while increasing their number.[32] It is important that the jets from the inlet ports slow down significantly once they enter the cylinder. Otherwise the scavenging front will reach the exhaust ports or valves too early. The jets are frequently directed to impinge on each other or against the cylinder wall. Swirl in uniflow-scavenged systems may be used to obtain an equivalent result.

The most desirable loop-scavenging flow is illustrated in Fig. 6-29. The

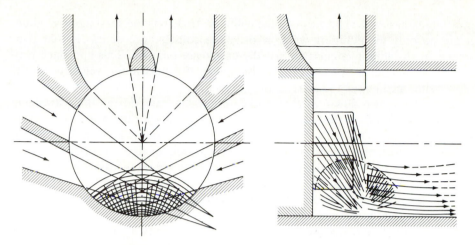

FIGURE 6-29
Desirable air flow in loop-scavenged engine: air from the entering jets impinges on far cylinder wall
and flows toward the cylinder head.[31]

scavenging jets enter symmetrically with sufficient velocity to fill up about half
the cylinder cross section, and thereafter flow at lower velocity along the cylinder
wall toward the cylinder head. By proper direction of the scavenging jets it is
possible to achieve almost no outflow in the direction of the exhaust from the
cross-hatched stagnation zone on the opposite cylinder wall. In fact, measurement of the velocity profile in this region is a good indicator of the effectiveness
of the scavenging flow. If the flow along the cylinder wall toward the head is
stable, i.e., if its maximum velocity occurs near the wall and the velocity is near
zero on the plane perpendicular to the axis of symmetry of the ports (which
passes through the cylinder axis), the scavenging flow will follow the desired path.
If there are "tongues" of scavenging flow toward the exhaust port, either in the
center of the cylinder or along the walls, then significant short-circuiting will
occur.[31]

   In uniflow-scavenged configurations, the inlet ports are evenly spaced
around the full circumference of the cylinder and are usually directed so that the
scavenging jets create a swirling flow within the cylinder (see Fig. 6-24). Results of
measurements of scavenging front location in rig flow tests of a valved uniflow
two-stroke diesel cylinder, as the inlet port angle was varied to give a wide range
of swirl, showed that inlet jets directed tangentially to a circle of half the cylinder
radius gave the most stable scavenging front profile over a wide range of conditions.[33]

   Though the scavenging processes in spark-ignition and diesel two-stroke
engines are similar, these two types of engine operate with quite different delivery
ratios. In mixture-scavenged spark-ignition engines, any significant expulsion of
fresh fluid with the burned gas results in a significant loss of fuel and causes high
hydrocarbon emissions as well as loss of the energy expended in pumping the

flow which passes straight through the cylinder. In diesels the scavenging medium is air, so only the pumping work is lost. One consequence of this is that two-stroke spark-ignition engines are usually crankcase pumped. This approach provides the maximum pressure and thus also the maximum velocity in the scavenging medium at the start of the scavenging process just after the cylinder pressure has blown down; as the crankcase pressure falls during the scavenging process, the motion of the scavenging front within the cylinder also slows down. Figure 6-30 shows the delivery ratio and trapping, charging, and scavenging efficiencies of two crankcase-scavenged spark-ignition engines as a function of engine speed. These quantities depend significantly on intake and exhaust port design and open period and the exhaust system configuration.[34-36] For two-stroke cycle spark-ignition engines, which use crankcase pumping, delivery ratios vary between about 0.5 and 0.8.

Figure 6-31 shows scavenging data typical of large two-stroke diesels.[37] The purity (mass of air in trapped cylinder charge/mass of trapped cylinder charge) is shown as a function of the delivery ratio. The different scavenging configurations have different degrees of effectiveness, with uniflow scavenging being the most efficient. These diesel engines normally operate with delivery ratios in the range 1.2 to 1.4.

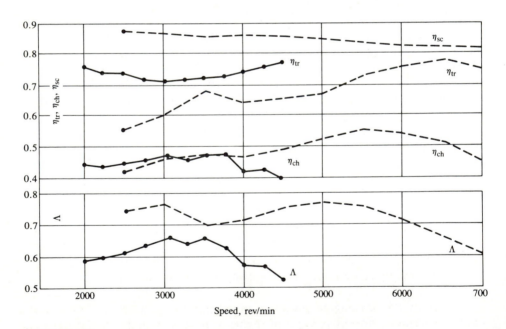

FIGURE 6-30
Delivery ratio Λ, trapping efficiency η_{tr}, charging efficiency η_{ch}, and scavenging efficiency η_{sc}, at full load, as functions of speed for two single-cylinder two-stroke cycle spark-ignition engines. Solid line is 147 cm^3 displacement engine.[34] Dashed line is loop-scavenged 246 cm^3 displacement engine.[35]

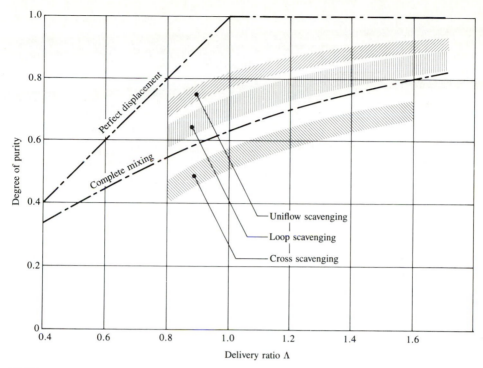

FIGURE 6-31
Purity as a function of delivery ratio Λ for different types of large marine two-stroke diesel engines.[37]

6.7 FLOW THROUGH PORTS

The importance of the intake and exhaust ports to the proper functioning of the two-stroke cycle scavenging process is clear from the discussion in Sec. 6.6. The crank angle at which the ports open, the size, number, geometry, and location of the ports around the cylinder circumference, and the direction and velocity of the jets issuing from the ports into the cylinder all affect the scavenging flow. A summary of the information available on flow through piston-controlled ports can be found in Annand and Roe.[16] Both the flow resistance of the inlet and exhaust port configurations, as well as the details of the flow pattern produced by the port system inside the cylinder during scavenging, are important. Figure 6-32 defines the essential geometrical characteristics of inlet ports. Rectangular ports make best use of the cylinder wall area and give precise timing control. Ports can be tapered, and may have axial and tangential inclination as shown.

Figure 6-33 illustrates the flow patterns expected downstream of piston-controlled inlet ports. For small openings, the flow remains attached to the port walls. For fully open ports with sharp corners the flow detaches at the upstream corners. Both a rounded entry and converging taper to the port help prevent flow detachment within the port. Discharge coefficients for ports have been measured as a function of the open fraction of the port, the pressure ratio across the port,

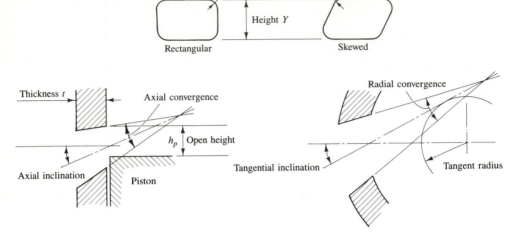

FIGURE 6-32
Parameters which define geometry of inlet ports.[16]

and port geometry and inclination (see Ref. 16 for a detailed summary). The most appropriate reference area for evaluating the discharge coefficient is the open area of the port (see Fig. 6-32). For the open height h_p less than $(Y - r)$ but greater than r this is

$$A_R = Xh_p - 0.43r^2 \tag{6.30}$$

where Y is the port height, X the port width, and r the corner radius. For $h_p = Y$, the reference area is

$$A_R = XY - 0.86r^2 \tag{6.31}$$

The effect of variations in geometry and operating conditions on the discharge coefficient C_D can usually be interpreted by reference to the flow patterns illustrated in Fig. 6-33. The effects of inlet port open fraction and port geometry on C_D are shown in Fig. 6-34: geometry effects are most significant at small and large open fractions.[20] C_D varies with pressure ratio, increasing as the pressure

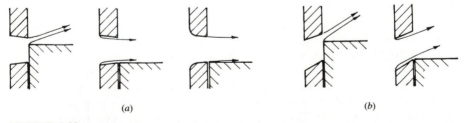

(a)   (b)

FIGURE 6-33
Flow pattern through piston-controlled inlet ports: (a) port axis perpendicular to wall; small opening and large opening with sharp and rounded entry; (b) port axis inclined.[16]

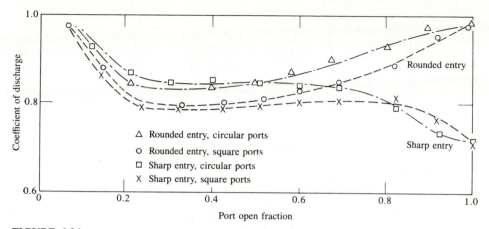

FIGURE 6-34
Discharge coefficients as a function of port open fraction (uncovered height/port height) for different inlet port designs. Pressure ratio across port = 2.35.[20]

ratio increases. Empirical relations that predict this variation with pressure ratio have been developed.[38]

Tangentially inclined inlet ports are used when swirl is desired to improve scavenging or when jet focusing or impingement within the cylinder off the cylinder axis is required (see Sec. 6.6.3). The discharge coefficient decreases as the jet tangential inclination increases. The jet angle and the port angle can deviate significantly from each other depending on the details of the port design and the open fraction.[31]

In piston-controlled exhaust ports, the angle of the jet from a thin-walled exhaust port increases as indicated in Fig. 6-35.[31] In thick ports, the walls are

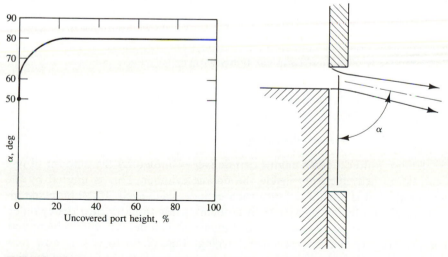

FIGURE 6-35
Angle of jet exiting exhaust port as a function of open port height.[31]

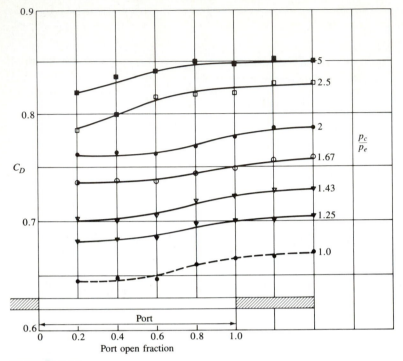

FIGURE 6-36
Discharge coefficient of a single rectangular exhaust port (7.6 mm wide × 12.7 mm high) in the wall of a 51-mm bore cylinder as a function of open fraction and pressure ratio. Steady-flow rig tests at 21°C. p_c = cylinder pressure, p_e = exhaust system pressure.[39]

usually tapered to allow the outward flow to diffuse. The pressure ratio across the exhaust ports varies substantially during the exhaust process. The pressure ratio has a significant effect on the exhaust port discharge coefficient, as shown in Fig. 6-36. The changes in exit jet angle and separation point explain the effects of increasing open fraction and pressure ratio. The discharge coefficient also increases modestly with increasing gas temperature.[39]

6.8 SUPERCHARGING AND TURBOCHARGING

6.8.1 Methods of Power Boosting

The maximum power a given engine can deliver is limited by the amount of fuel that can be burned efficiently inside the engine cylinder. This is limited by the amount of air that is introduced into each cylinder each cycle. If the inducted air is compressed to a higher density than ambient, prior to entry into the cylinder, the maximum power an engine of fixed dimensions can deliver will be increased. This is the primary purpose of supercharging; Eqs. (2.39) to (2.41) show how power, torque, and mean effective pressure are proportional to inlet air density.

The term *supercharging* refers to increasing the air (or mixture) density by increasing its pressure prior to entering the engine cylinder. Three basic methods are used to accomplish this. The first is *mechanical supercharging* where a separate pump or blower or compressor, usually driven by power taken from the engine, provides the compressed air. The second method is *turbocharging*, where a turbocharger—a compressor and turbine on a single shaft—is used to boost the inlet air (or mixture) density. Energy available in the engine's exhaust stream is used to drive the turbocharger turbine which drives the turbocharger compressor which raises the inlet fluid density prior to entry to each engine cylinder. The third method—*pressure wave supercharging*—uses wave action in the intake and exhaust systems to compress the intake mixture. The use of intake and exhaust manifold tuning to increase volumetric efficiency (see Sec. 6.2.2) is one example of this method of increasing air density. An example of a pressure wave supercharging device is the Comprex, which uses the pressure available in the exhaust gas stream to compress the inlet mixture stream by direct contact of the fluids in narrow flow channels (see Sec. 6.8.5). Figure 6-37 shows typical arrangements of the different supercharging and turbocharging systems. The most common arrangements use a mechanical supercharger (Fig. 6-37a) or turbocharger (Fig. 6-37b). Combinations of an engine-driven compressor and a turbocharger (Fig. 6-37c) are used (e.g., in large marine engines; Fig. 1-24). Two-stage turbocharging (Fig. 6-37d) is one viable approach for providing very high boost pressures (4 to 7 atm) to obtain higher engine brake mean effective pressures. Turbocompounding, i.e., use of a second turbine in the exhaust directly geared to the engine drive shaft (Fig. 6-37e), is an alternative method of increasing engine power (and efficiency). Charge cooling with a heat exchanger (often called an aftercooler or intercooler) after compression, prior to entry to the cylinder, can be used to increase further the air or mixture density as shown in Fig. 6-37f.

Supercharging is used in four-stroke cycle engines to boost the power per unit displaced volume. Some form of supercharging is necessary in two-stroke cycle engines to raise the fresh air (or mixture) pressure above the exhaust pressure so that the cylinder can be scavenged effectively. With additional boost in two-stroke cycle engines, the power density is also raised. This section reviews the operating characteristics of the blowers, compressors, turbines, and wave-compression devices used to increase inlet air or mixture density or convert exhaust-gas availability to work. The operating characteristics of supercharged and turbocharged engine systems are discussed in Chap. 15.

6.8.2 Basic Relationships

Expressions for the work required to drive a blower or compressor and the work produced by a turbine are obtained from the first and second laws of thermodynamics. The first law, in the form of the steady flow energy equation, applied to a control volume around the turbomachinery component is

$$\dot{Q} - \dot{W} = \dot{m}\left[\left(h + \frac{C^2}{2} + gz\right)_{\text{out}} - \left(h + \frac{C^2}{2} + gz\right)_{\text{in}}\right] \qquad (6.32)$$

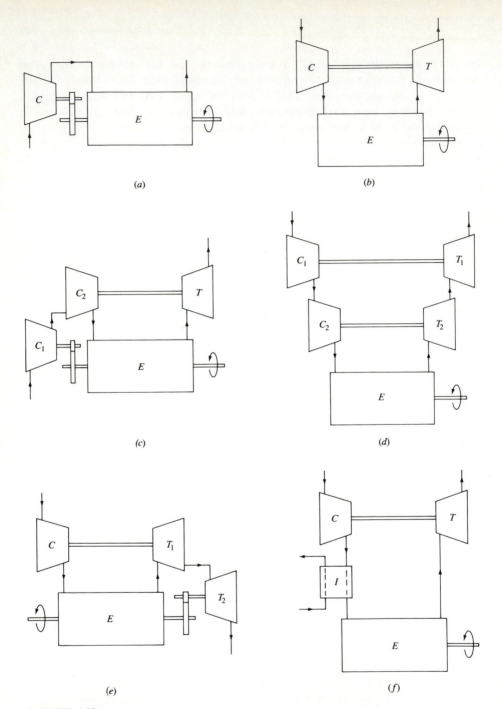

FIGURE 6-37
Supercharging and turbocharging configurations: (*a*) mechanical supercharging; (*b*) turbocharging; (*c*) engine-driven compressor and turbocharger; (*d*) two-stage turbocharging; (*e*) turbocharging with turbocompounding; (*f*) turbocharger with intercooler. C Compressor, E Engine, I Inter-cooler, T Turbine.

250

where $\dot{Q}$ is the heat-transfer rate into the control volume, $\dot{W}$ is the shaft work-transfer rate out of the control volume, $\dot{m}$ is the mass flow, h is the specific enthalpy, $C^2/2$ is the specific kinetic energy, and gz is the specific potential energy (which is not important and can be omitted).

A *stagnation* or *total enthalpy*, h_0, can be defined as

$$h_0 = h + \frac{C^2}{2} \tag{6.33}$$

For an ideal gas, with constant specific heats, a *stagnation* or *total temperature* follows from Eq. (6.33):

$$T_0 = T + \frac{C^2}{2c_p} \tag{6.34}$$

A *stagnation* or *total pressure* is also defined: it is the pressure attained if the gas is isentropically brought to rest:

$$p_0 = p\left(\frac{T_0}{T}\right)^{\gamma/(\gamma-1)} \tag{6.35}$$

$\dot{Q}$ in Eq. (6.32) for pumps, blowers, compressors, and turbines is usually small enough to be neglected. Equation (6.32) then gives the work-transfer rate as

$$-\dot{W} = \dot{m}(h_{0,\,\text{out}} - h_{0,\,\text{in}}) \tag{6.36}$$

A component efficiency is used to relate the actual work-transfer rate to the work-transfer rate required (or produced) by an equivalent reversible adiabatic device operating between the same pressures. The second law is then used to determine this reversible adiabatic work-transfer rate, which is that occurring in an isentropic process.

For a compressor, the *compressor isentropic efficiency* η_C is

$$\eta_C = \frac{\text{reversible power requirement}}{\text{actual power requirement}} \tag{6.37}$$

Figure 6-38 shows the end states of the gas passing through a compressor on an *h-s* diagram. Both static (p_1, p_2) and stagnation (p_{01}, p_{02}) constant-pressure lines are shown. The total-to-total isentropic efficiency is, from Eq. (6.37),

$$\eta_{\text{CTT}} = \frac{h_{02s} - h_{01}}{h_{02} - h_{01}} \tag{6.38}$$

which, since c_p is essentially constant for air, or fuel-air mixture, becomes

$$\eta_{\text{CTT}} = \frac{T_{02s} - T_{01}}{T_{02} - T_{01}} \tag{6.39}$$

Since the process 01 to 02s is isentropic,

$$T_{02s} = T_{01}\left(\frac{p_{02}}{p_{01}}\right)^{(\gamma-1)/\gamma}$$

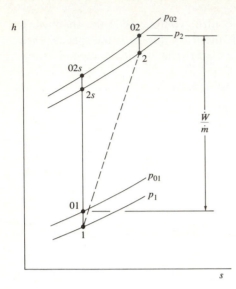

FIGURE 6-38
Enthalpy-entropy diagram for compressor. Inlet state 01, exit state 2; equivalent isentropic compressor exit state 2s.

Equation (6.39) becomes

$$\eta_{CTT} = \frac{(p_{02}/p_{01})^{(\gamma-1)/\gamma} - 1}{(T_{02}/T_{01}) - 1} \tag{6.40}$$

In deriving Eq. (6.40) it has been tacitly assumed that the kinetic energy pressure head $(p_{02} - p_2)$ can be recovered. In internal combustion engine applications the compressor feeds the engine via a large manifold, and much of this kinetic energy will be dissipated. The blower or compressor should be designed for effective recovery of this kinetic energy before the exit duct. Since the kinetic energy of the gas leaving the compressor is not usually recovered, a more realistic definition of efficiency is based on exit static conditions:[40]

$$\eta_{CTS} = \frac{T_{2s} - T_{01}}{T_{02} - T_{01}} = \frac{(p_2/p_{01})^{(\gamma-1)/\gamma} - 1}{(T_{02}/T_{01}) - 1} \tag{6.41}$$

This is termed the total-to-static efficiency. The basis on which the efficiency is calculated should always be clearly stated.

The work-transfer rate or power required to drive the compressor is obtained by combining Eq. (6.36), the ideal gas model, and Eq. (6.40):

$$-\dot{W}_C = \dot{m}_i c_{p,i}(T_{02} - T_{01}) = \frac{\dot{m}_i c_{p,i} T_{01}}{\eta_{CTT}} \left[\left(\frac{p_{02}}{p_{01}} \right)^{(\gamma-1)/\gamma} - 1 \right] \tag{6.42}$$

where the subscript i denotes inlet mixture properties. If η_{CTS} is used to define the compressor performance, then p_2 replaces p_{02} in Eq. (6.42). Equation (6.42) gives the thermodynamic power requirement. There will also be mechanical losses in the blower or compressor. Thus the power required to drive the device, $-\dot{W}_{C,D}$, will be

$$-\dot{W}_{C,D} = -\frac{\dot{W}_C}{\eta_m} \tag{6.43}$$

where η_m is the blower or compressor mechanical efficiency.

Figure 6-39 shows the gas states at inlet and exit to a turbine on an h-s diagram. State 03 is the inlet stagnation state; 4 and 04 are the exit static and stagnation states, respectively. States 4s and 04s define the static and stagnation exit states of the equivalent reversible adiabatic turbine. The *turbine isentropic efficiency* is defined as

$$\eta_T = \frac{\text{actual power output}}{\text{reversible power output}} \tag{6.44}$$

Thus, the total-to-total turbine efficiency is

$$\eta_{TTT} = \frac{h_{03} - h_{04}}{h_{03} - h_{04s}} \tag{6.45}$$

If the exhaust gas is modeled as an ideal gas with constant specific heats, then Eq. (6.45) can be written

$$\eta_{TTT} = \frac{T_{03} - T_{04}}{T_{03} - T_{04s}} = \frac{1 - (T_{04}/T_{03})}{1 - (p_{04}/p_{03})^{(\gamma-1)/\gamma}} \tag{6.46}$$

Note that for exhaust gas over the temperature range of interest, c_p may vary significantly with temperature (see Figs. 4-10 and 4-17).

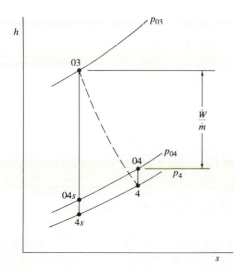

FIGURE 6-39
Enthalpy-entropy diagram for a turbine. Inlet state 03, exit state 4; equivalent isentropic turbine exit state 4s.

Since the kinetic energy at the exit of a turbocharger turbine is usually wasted, a total-to-static turbine isentropic efficiency, where the reversible adiabatic power output is that obtained between inlet stagnation conditions and the exit static pressure, is more realistic:[40]

$$\eta_{TTS} = \frac{h_{03} - h_{04}}{h_{03} - h_{4s}} = \frac{T_{03} - T_{04}}{T_{03} - T_{4s}} = \frac{1 - (T_{04}/T_{03})}{1 - (p_4/p_{03})^{(\gamma - 1)/\gamma}} \tag{6.47}$$

The power delivered by the turbine is given by [Eqs. (6.36) and (6.46)]

$$\dot{W}_T = \dot{m}_e(h_{03} - h_{04}) = \dot{m}_e c_{p,e}(T_{03} - T_{04}) = \dot{m}_e c_{p,e} \eta_{TTT} T_{03}\left[1 - \left(\frac{p_{04}}{p_{03}}\right)^{(\gamma_e - 1)/\gamma_e}\right] \tag{6.48}$$

where the subscript e denotes exhaust gas properties. If the total-to-static turbine efficiency (η_{TTS}) is used in the relation for $\dot{W}_T$, then p_4 replaces p_{04} in Eq. (6.48). With a turbocharger, the turbine is mechanically linked to the compressor. Hence, at constant turbocharger speed,

$$- \dot{W}_C = \eta_m \dot{W}_T \tag{6.49}$$

where η_m is the mechanical efficiency of the turbocharger. The mechanical losses are mainly bearing friction losses. The mechanical efficiency is usually combined with the turbine efficiency since these losses are difficult to separate out.

It is advantageous if the operating characteristics of blowers, compressors, and turbines can be expressed in a manner that allows easy comparison between different designs and sizes of devices. This can be done by describing the performance characteristics in terms of dimensionless numbers.[40] The most important dependent variables are: mass flow rate $\dot{m}$, component isentropic efficiency η, and temperature difference across the device ΔT_0. Each of these are a function of the independent variables: $p_{0,\text{in}}$, $p_{0,\text{out}}$ (or p_{out}), $T_{0,\text{in}}$, N(speed), D(characteristic dimension), R(gas constant), γ (c_p/c_v),and μ(viscosity); i.e.,

$$\dot{m}, \eta, \Delta T_0 = f(p_{0,\text{in}}, p_{0,\text{out}}, T_{0,\text{in}}, N, D, R, \gamma, \mu) \tag{6.50}$$

By dimensional analysis, these eight independent variables can be reduced to four dimensionless groups:

$$\frac{\dot{m}\sqrt{RT_{0,\text{in}}}}{p_{0,\text{in}} D^2}, \eta, \frac{\Delta T_0}{T_{0,\text{in}}} = f\left(\frac{ND}{\sqrt{RT_{0,\text{in}}}}, \frac{p_{0,\text{out}}}{p_{0,\text{in}}}, \frac{\dot{m}}{\mu D}, \gamma\right) \tag{6.51}$$

The Reynolds number, $\dot{m}/(\mu D)$, has little effect on performance and γ is fixed by the gas. Therefore these variables can be omitted and Eq. (6.51) becomes

$$\frac{\dot{m}\sqrt{RT_{0,\text{in}}}}{p_{0,\text{in}} D^2}, \eta, \frac{\Delta T_0}{T_{0,\text{in}}} = f\left(\frac{ND}{\sqrt{RT_{0,\text{in}}}}, \frac{p_{0,\text{out}}}{p_{0,\text{in}}}\right) \tag{6.52}$$

For a particular device, the dimensions are fixed and the value of R is fixed. So it has become the convention to plot

$$\frac{\dot{m}\sqrt{T_{0,\,in}}}{p_{0,\,in}},\ \eta,\ \frac{\Delta T_0}{T_{0,\,in}} = f\left(\frac{N}{\sqrt{T_{0,\,in}}},\ \frac{p_{0,\,out}}{p_{0,\,in}}\right) \tag{6.53}$$

$\dot{m}\sqrt{T_{0,\,in}}/p_{0,\,in}$ is referred to as the corrected mass flow; $N/\sqrt{T_{0,\,in}}$ is referred to as the corrected speed. The disadvantage of this convention of removing D and R is that the groups of variables are no longer dimensionless, and performance plots or maps relate to a specific machine.

Compressor characteristics are usually plotted in terms of the pressure ratio (p_{02}/p_{01}) or (p_2/p_{01}) against the corrected mass flow $(\dot{m}\sqrt{T_{01}}/p_{01})$ along lines of constant corrected speed $(N/\sqrt{T_{01}})$. Contours of constant efficiency are superposed. Similar plots are used for turbines: p_{03}/p_4 against $\dot{m}\sqrt{T_{03}}/p_{03}$ along lines of constant $N/\sqrt{T_{03}}$. Since these occupy a narrow region of the turbine performance map, other plots are often used (see Sec. 6.8.4).

6.8.3 Compressors

Practical mechanical supercharging devices can be classified into: (1) sliding vane compressors, (2) rotary compressors, and (3) centrifugal compressors. The first two types are positive displacement compressors; the last type is an aerodynamic compressor. Four different types of positive displacement compressors are illustrated in Fig. 6-40.

In the sliding vane compressor (Fig. 6-40a), deep slots are cut into the rotor to accommodate thin vanes which are free to move radially. The rotor is mounted eccentrically in the housing. As the rotor rotates, the centrifugal forces acting on the vanes force them outward against the housing, thereby dividing the crescent-shaped space into several compartments. Ambient air is drawn through the intake port into each compartment as its volume increases to a maximum. The trapped air is compressed as the compartment volume decreases, and is then discharged through the outlet port. The flow capacity of the sliding vane compressor depends on the maximum induction volume which is determined by the housing cylinder bore, rotor diameter and length, eccentricity, number of vanes, dimensions of the inlet and outlet ports. The actual flow rate and pressure rise at constant speed will be reduced by leakage. Also, heat transfer from the moving vanes and rotor and stator surfaces will reduce compression efficiency unless cooling is used to remove the thermal energy generated by friction between the vanes, and the rotor and stator. The volumetric efficiency can vary between 0.6 and 0.9 depending on the size of the machine, the quality of the design, and the method of lubrication and cooling employed. The displaced volume V_D is given by

$$V_D = \pi \varepsilon l(2r + \varepsilon) \tag{6.54}$$

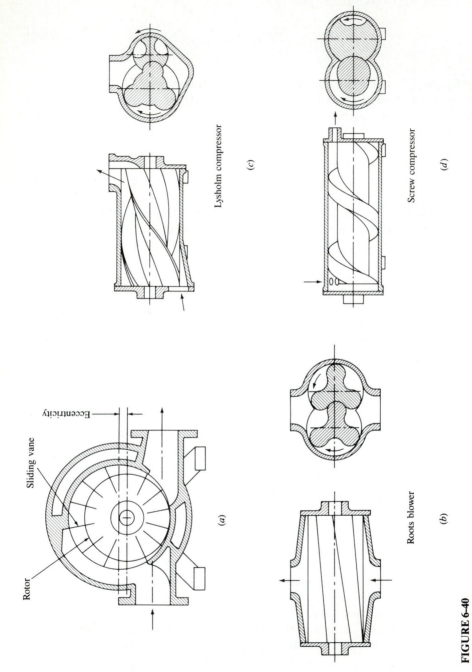

Lysholm compressor

(c)

Screw compressor

(d)

(a)

Sliding vane

Eccentricity

Rotor

Roots blower

(b)

FIGURE 6-40

Positive displacement compressors: (a) sliding vane compressor; (b) roots blower; (c) Lysholm compressor; (d) screw compressor.[41]

where r is the rotor radius, ε the eccentricity, and l the axial length of the compressor. The mass flow rate parameter is

$$\frac{\dot{m}\sqrt{T_0/T_{std}}}{p_0/p_{std}} = \text{constant} \times \rho_i \eta_v N \varepsilon l (2r + \varepsilon) \tag{6.55}$$

where η_v is the device volumetric efficiency, N its speed, and the subscripts i, 0 and std refer to inlet, inlet stagnation, and standard atmospheric conditions, respectively. Figure 6-41 shows the performance characteristics of a typical sliding vane compressor. The mass flow rate at constant speed depends on the pressure ratio only through its (weak) effect on volumetric efficiency. The isentropic efficiency is relatively low.[41]

An alternative positive displacement supercharger is the roots blower (Fig. 6-40b). The two rotors are connected by gears. The working principles are as follows. Air trapped in the recesses between the rotor lobes and the housing is carried toward the delivery port without significant change in volume. As these recesses open to the delivery line, since the suction side is closed, the trapped air is suddenly compressed by the backflow from the higher-pressure delivery line. This intermittent delivery produces nonuniform torque on the rotor and pressure pulses in the delivery line. Roots blowers are most suitable for small pressure ratios (about 1.2). The volumetric efficiency depends on the running clearances, rotor length, rotational speed, and pressure ratio. In the three-lobe machines shown (two lobes are sometimes used) the volume of each recess V_R is

$$V_R = 0.546R^2 l$$

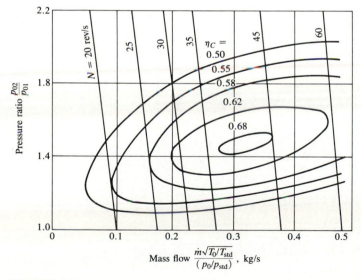

FIGURE 6-41
Performance map for sliding vane compressor.[41]

where R is the rotor radius and l the blower length. The mass flow parameter is

$$\frac{\dot{m}\sqrt{T_0/T_{std}}}{p_0/p_{std}} = \text{constant} \times \rho_i \eta_v N R^2 l \qquad (6.56)$$

A performance map of a typical small roots blower is shown in Fig. 6-42. It is similar in character to that of the sliding vane compressor. At constant speed, the flow rate depends on increasing pressure ratio only through the resulting decrease in volumetric efficiency (Eq. 6.56).[41] The advantage of the roots blower is that its performance range is not limited by surge and choking as is the centrifugal compressor (see below). Its disadvantages are its high noise level, poor efficiency, and large size.[42]

Screw compressors (Fig. 6-40c and d) must be precision machined to achieve close tolerances between rotating and stationary elements for satisfactory operation. They run at speeds between 3000 and 30,000 rev/min. It is usually necessary to cool the rotors internally. High values of volumetric and isentropic efficiency are claimed.[41]

A *centrifugal compressor* is primarily used to boost inlet air or mixture density coupled with an exhaust-driven turbine in a turbocharger. It is a single-stage radial flow device, well suited to the high mass flow rates at the relatively low pressure ratios (up to about 3.5) required by the engine. To operate efficiently it must rotate at high angular speed. It is therefore much better suited to direct coupling with the exhaust-driven turbine of the turbocharger than to mechanical coupling through a gearbox to the engine for mechanical supercharging.

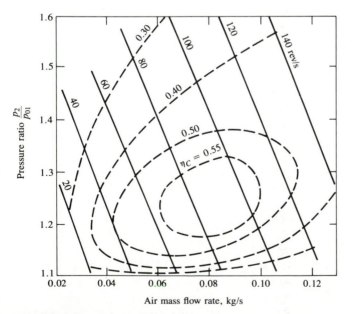

FIGURE 6-42
Performance map at standard inlet conditions for roots blower.[42]

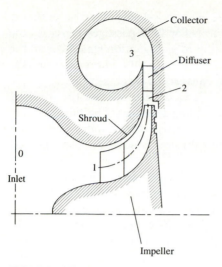

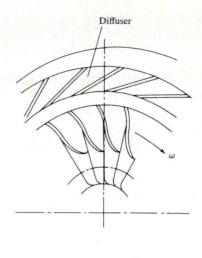

FIGURE 6-43
Schematic of centrifugal compressor.[40]

The centrifugal compressor consists of a stationary inlet casing, a rotating bladed impellor, a stationary diffuser (with or without vanes), and a collector or volute to bring the compressed air leaving the diffuser to the engine intake system (see Fig. 6-43). Figure 6-44 indicates, on an *h-s* diagram, how each component contributes to the overall pressure rise across the compressor. Air at stagnation

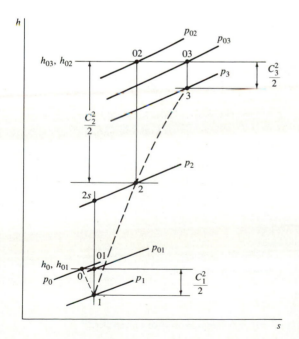

FIGURE 6-44
Enthalpy-entropy diagram for flow through centrifugal compressor.

state 0 is accelerated in the inlet to pressure p_1 and velocity C_1. The enthalpy change 01 to 1 is $C_1^2/2$. Compression in the impeller flow passages increases the pressure to p_2 and velocity to C_2, corresponding to a stagnation state 02 if all the exit kinetic energy were recovered. The isentropic equivalent compression process has an exit static state 2s. The diffuser, 2 to 3, converts as much as practical of the air kinetic energy at exit to the impeller $(C_2^2/2)$ to a pressure rise $(p_3 - p_2)$ by slowing down the gas in carefully shaped expanding passages. The final state, in the collector, has static pressure p_3, low kinetic energy $C_3^2/2$, and a stagnation pressure p_{03} which is less than p_{02} since the diffusion process is incomplete as well as irreversible.[40]

The work transfer to the gas occurs in the impeller. It can be related to the change in gas angular momentum via the velocity components at the impeller entry and exit, which are shown in Fig. 6-45. Here C_1 and C_2 are the absolute gas velocities, U_1 and U_2 are the tangential blade velocities, and w_1 and w_2 are the gas velocities relative to the impeller all at inlet (1) and exit (2), respectively. The torque T exerted on the gas by the impeller equals the rate of change of angular momentum:

$$T = \dot{m}(r_2 C_{\theta_2} - r_1 C_{\theta_1}) \tag{6.57}$$

The rate of work transfer to the gas is given by

$$-\dot{W}_C = T\omega = \dot{m}\omega(r_2 C_{\theta_2} - r_1 C_{\theta_1}) = \dot{m}(U_2 C_{\theta_2} - U_1 C_{\theta_1}) \tag{6.58}$$

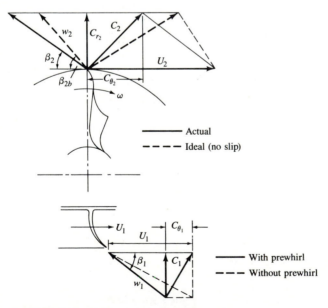

—— Actual
---- Ideal (no slip)

—— With prewhirl
---- Without prewhirl

FIGURE 6-45
Velocity diagrams at inlet (1) and exit (2) to centrifugal compressor rotor or impeller.[40]

This is often called the Euler equation. Normally in compressors the inlet flow is axial so $C_{\theta_1} = 0$. Thus Eq. (6.58) can be written:

$$-\frac{\dot{W}_c}{\dot{m}} = U_2 C_{\theta_2} = U_2\left(1 - \frac{C_{r_2}}{U_2}\cot\beta_2\right) \qquad (6.59)$$

where β_2 is the backsweep angle. In the ideal case with no slip, β_2 is the blade angle, β_{2b}. In practice, there is slip and β_2 is less than β_{2b}. Many compressors have radial vanes (i.e., $\beta_{2b} = 90°$). A recent trend is backswept vanes ($\beta_{2b} < 90°$) which give higher efficiency. Since work transfer to the gas occurs only in the impeller, the work-transfer rate given by Eq. (6.59) equals the change in stagnation enthalpy $(h_{03} - h_{01})$ in Fig. 6-44 [see Eq. (6.36)].

The operating characteristics of the centrifugal compressor are usually described by a *performance map*. This shows lines of constant compressor efficiency η_C, and constant corrected speed $N/\sqrt{T_{0,\,in}}$, on a plot of pressure ratio $p_{0,\,out}/p_{0,\,in}$ against corrected mass flow $\dot{m}\sqrt{T_{0,\,in}}/p_{0,\,in}$ [see Eq. (6.53)]. Figure 6-46 indicates the form of such a map. The stable operating range in the center of the map is separated from an unstable region on the left by the *surge line*. When the mass flow is reduced at a constant pressure ratio, local flow reversal eventually occurs in the boundary layer. Further reductions in mass flow cause the flow to reverse completely, causing a drop in pressure. This relieves the adverse pressure gradient. The flow reestablishes itself, builds up again, and the process repeats. Compressors should not be operated in this unstable regime. The

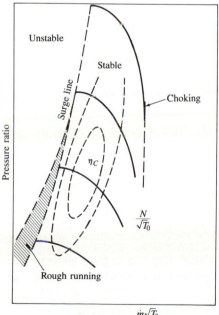

FIGURE 6-46

Schematic of compressor operating map showing stable operating range.[40]

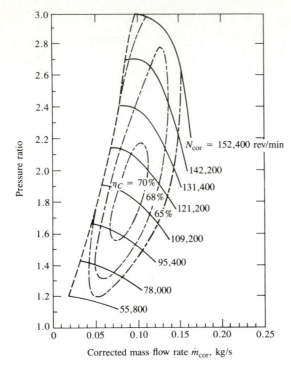

FIGURE 6-47
Centrifugal compressor operating map. Lines of constant corrected speed and compressor efficiency are plotted on a graph of pressure ratio against corrected mass flow.[43]

stable operating regime is limited on the right by choking. The velocities increase as $\dot{m}$ increases, and eventually the flow becomes sonic in the limiting area of the machine. Extra mass flow through the compressor can only be obtained by higher speed. When the diffuser is choked, compressor speed may rise substantially with only a limited increase in the mass flow rate.[40]

Figure 6-47 shows an actual turbocharger compressor performance map. In practice, the map variables corrected speed and mass flow rate are usually defined as[44]

$$N_{cor} = N \left(\frac{T_{ref}}{T_{0, in}} \right)^{1/2}$$

$$\dot{m}_{cor} = \dot{m} \left(\frac{T_{0, in}}{T_{ref}} \right)^{1/2} \left(\frac{p_{ref}}{p_{0, in}} \right)$$

(6.60)

where T_{ref} and p_{ref} are standard atmospheric temperature and pressure, respectively. Though the details of different compressor maps vary, their general characteristics are similar. The high efficiency region runs parallel to the surge line (and close to it for vaneless diffusers). A wide flow *range* for the compressor (see Fig. 6-46) is important in turbochargers used for transportation applications.

6.8.4 Turbines

The turbocharger turbine is driven by the energy available in the engine exhaust. The ideal energy available is shown in Fig. 6-48. It consists of the blowdown work transfer produced by expanding the gas in the cylinder at exhaust valve opening to atmospheric pressure (area *abc*) and (for the four-stroke cycle engine) the work done by the piston displacing the gases remaining in the cylinder after blowdown (area *cdef*).

The reciprocating internal combustion engine is inherently an unsteady pulsating flow device. Turbines can be designed to accept such an unsteady flow, but they operate more efficiently under steady flow conditions. In practice, two approaches for recovering a fraction of the available exhaust energy are commonly used: constant-pressure turbocharging and pulse turbocharging. In *constant-pressure turbocharging*, an exhaust manifold of sufficiently large volume to damp out the mass flow and pressure pulses is used so that the flow to the turbine is essentially steady. The disadvantage of this approach is that it does not make full use of the high kinetic energy of the gases leaving the exhaust port; the losses inherent in the mixing of this high-velocity gas with a large volume of low-velocity gas cannot be recovered. With *pulse turbocharging*, short small-cross-section pipes connect each exhaust port to the turbine so that much of the kinetic energy associated with the exhaust blowdown can be utilized. By suitably grouping the different cylinder exhaust ports so that the exhaust pulses are sequential and have minimum overlap, the flow unsteadiness can be held to an acceptable level. The turbine must be specifically designed for this pulsating flow to achieve adequate efficiencies. The combination of increased energy available at the turbine, with reasonable turbine efficiencies, results in the pulse system being more commonly used for larger diesels.[40] For automotive engines, constant-pressure turbocharging is used.

Two types of turbines are used in turbochargers: radial and axial flow turbines. The radial flow turbine is similar in appearance to the centrifugal compressor; however, the flow is radially inward not outward. Radial flow turbines are

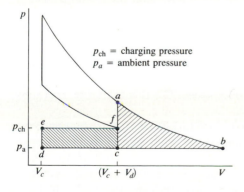

FIGURE 6-48
Constant-volume cycle p-V diagram showing available exhaust energy.

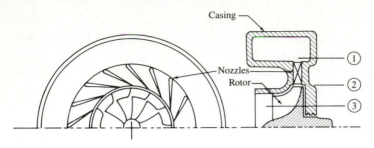

FIGURE 6-49
Schematic of radial flow turbine.

normally used in automotive or truck applications. Larger engines—locomotive, stationary, or marine—use axial flow turbines.

A drawing of a radial flow turbine is shown in Fig. 6-49. It consists of an inlet casing or scroll, a set of inlet nozzles (often omitted with small turbines), and the turbine rotor or wheel. The function of each component is evident from the *h-s* diagram and velocity triangles in Fig. 6-50. The nozzles (01-2) accelerate the flow, with modest loss in stagnation pressure. The drop in stagnation enthalpy, and hence the work transfer, occurs solely in the rotor passages, 2-3: hence, the rotor is designed for minimum kinetic energy $C_3^2/2$ at exit. The velocity triangles at inlet and exit relate the work transfer to the change in angular momentum via the Euler equation:

$$\dot{W}_T = T\omega = \dot{m}\omega(r_2 C_{\theta_2} - r_3 C_{\theta_3}) = \dot{m}(U_2 C_{\theta_2} - U_3 C_{\theta_3}) \qquad (6.61)$$

where T is the torque and ω the rotor angular speed. For maximum work transfer the exit velocity should be axial. The work-transfer rate relates to the change in stagnation enthalpy via

$$\dot{W}_T = \dot{m}(h_{02} - h_{03}) = \dot{m}(h_{01} - h_{03}) \qquad (6.62)$$

The turbine isentropic efficiency is given by Eqs. (6.44) to (6.47).

Many different types of plots have been used to define radial flow turbine characteristics. Figure 6-51 shows lines of constant corrected speed and efficiency on a plot of pressure ratio versus corrected mass flow rate. As flow rate increases at a given speed, it asymptotically approaches a limit corresponding to the flow becoming choked in the stator nozzle blades or the rotor. For turbines, efficiency is usually presented on a different diagram because the operating regime in Fig. 6-51 is narrow. Figure 6-52 shows an alternative plot for a radial turbine: corrected mass flow rate against corrected rotor speed. On this map, the operating regime appears broader.

A schematic of a turbocharger axial flow turbine is shown in Fig. 6-53. Usually a single stage is sufficient to expand the exhaust gas efficiently through the pressure ratios associated with engine turbocharging. This turbine consists of an annular flow passage, a single row of nozzles or stator blades, and a rotating blade ring. The changes in gas state across each component are similar to those of the radial turbine shown in the *h-s* diagram of Fig. 6-50. The velocity triangles at entry and exit to the rotor, shown in Fig. 6-54, relate the work transfer from the gas to the rotor to the change in angular momentum:

$$\dot{W}_T = \omega T = \dot{m}\omega(r_2 \, C_{\theta_2} + r_3 \, C_{\theta_3})$$

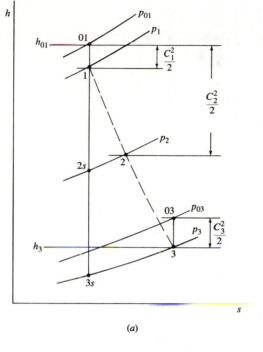

(a)

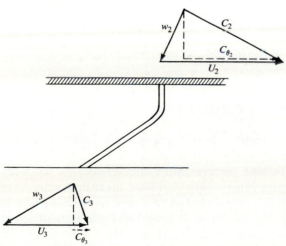

(b)

FIGURE 6-50
(a) Enthalpy-entropy diagram for radial turbine. (b) Velocity diagrams at turbine rotor inlet (2) and exit (3).

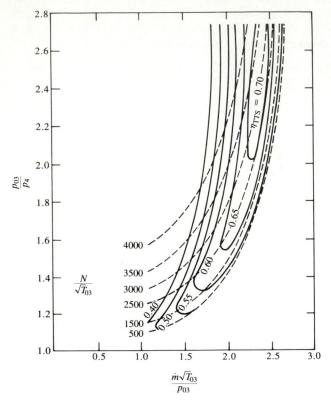

FIGURE 6-51
Radial turbine performance map showing lines of constant corrected speed and efficiency on a plot of pressure ratio versus corrected mass flow rate. T_{03} = turbine inlet temperature (K), p_{03} = turbine inlet pressure (bar), p_4 = turbine exit pressure (bar), $\dot{m}$ = mass flow rate (kg/s), N = speed (rev/min).[40]

Since the mid-radius r_2 usually equals the mid-radius r_3,

$$\dot{W}_T = \dot{m}U(C_{\theta_2} + C_{\theta_3}) = \dot{m}U(C_2 \sin \alpha_2 + C_3 \sin \alpha_3)$$
$$= \dot{m}U(C_{z_2} \tan \beta_2 + C_{z_3} \tan \beta_3) \tag{6.63}$$

Equation (6.62) relates the work-transfer rate to the stagnation enthalpy change as in the radial turbine.

Figure 6-55 shows axial turbine performance characteristics on the standard dimensionless plot of pressure ratio versus corrected mass flow rate. Here the constant speed lines converge to a single choked flow limit as the mass flow is increased. In the radial turbine, the variation in centrifugal effects with speed cause a noticeable spread in the constant speed lines (Fig. 6-51).

An alternative performance plot for turbines is efficiency versus *blade speed ratio*. This ratio is the blade speed U (at its mean height) for an axial flow turbine

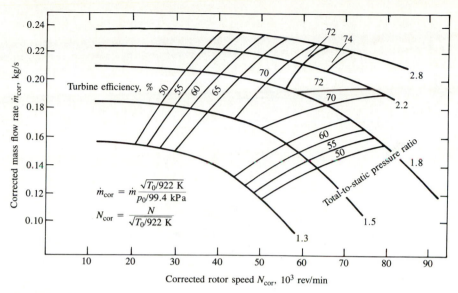

FIGURE 6-52
Alternative radial turbine performance map: corrected mass flow rate is plotted against corrected rotor speed.[45]

or the wheel tip speed for a radial flow turbine, divided by the velocity equivalent of the isentropic enthalpy drop across the turbine stage, C_s; i.e.,

$$\text{Blade speed ratio} = \frac{U}{C_s}$$

where

$$C_s = [2(h_{03} - h_{4s})]^{1/2} \tag{6.64}$$

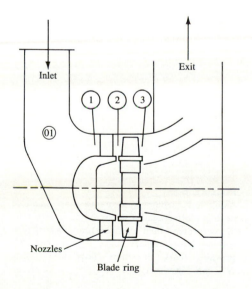

FIGURE 6-53
Schematic of single-stage axial flow turbine.

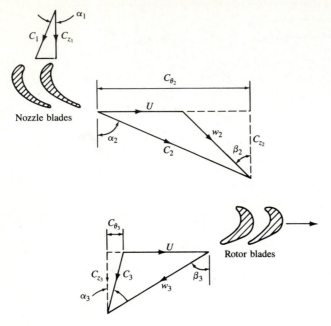

FIGURE 6-54
Velocity diagrams at entry (2) and exit (3) to axial flow turbine blade ring.[40]

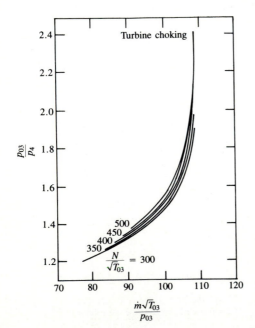

FIGURE 6-55
Axial flow turbine performance map: pressure ratio is plotted against corrected mass flow rate. T_{03} = turbine inlet temperature (K), p_{03} = turbine inlet pressure (bar), p_4 = turbine exit pressure (bar), $\dot{m}$ = mass flow rate (kg/s), N = speed (rev/min).[40]

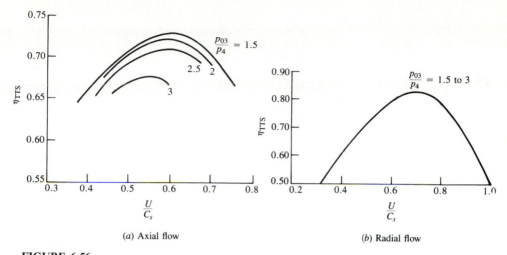

FIGURE 6-56
Plot of turbine total-to-static efficiency versus blade speed ratio U/C_s for (a) axial flow and (b) radial flow turbines.[40]

This method of displaying performance is useful for matching compressor and turbine wheel size for operation of the turbine at optimum efficiency. Figure 6-56 shows such plots for an axial and radial flow turbine. The peak efficiency can occur for $0.4 < U/C_s < 0.8$, depending on turbine design and application.[40]

For a given turbocharger, the compressor and turbine characteristics are linked. Since the compressor and turbine are on a common shaft with speed N:

$$\frac{N}{\sqrt{T_{01}}} = \frac{N}{\sqrt{T_{03}}}\left(\frac{T_{03}}{T_{01}}\right)^{1/2} \tag{6.65}$$

For $\dot{m}_C = \dot{m}_T = \dot{m}$ (if $\dot{m}_C[1 + (F/A)] = \dot{m}_T$, the equation is easily modified):

$$\frac{\dot{m}\sqrt{T_{01}}}{p_{01}} = \frac{\dot{m}\sqrt{T_{03}}}{p_{03}}\left(\frac{p_{03}}{p_{01}}\right)\left(\frac{T_{01}}{T_{03}}\right)^{1/2} \tag{6.66}$$

Since the compressor and turbine powers are equal in magnitude:

$$h_{02} - h_{01} = \eta_m(h_{03} - h_{04}) \tag{6.67}$$

or, with an ideal gas model,

$$c_{p,C}(T_{02} - T_{01}) = \eta_m c_{p,T}(T_{03} - T_{04}) \tag{6.68}$$

Equation (6.68), with Eqs. (6.40) and (6.46), gives

$$\left(\frac{p_{02}}{p_{01}}\right)^{(\gamma_C - 1)/\gamma_C} - 1 = \eta_C \eta_T \eta_m \frac{c_{p,T}}{c_{p,C}}\left[1 - \left(\frac{p_4}{p_{03}}\right)^{(\gamma_T - 1)/\gamma_T}\right]\frac{T_{03}}{T_{01}} \tag{6.69}$$

Assuming that the turbine exit pressure p_4 equals atmospheric pressure p_{01}, the equilibrium or steady-state running lines for constant values of T_{03}/T_{01} can be

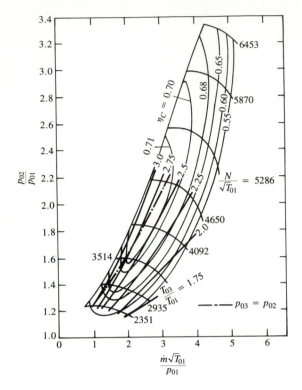

FIGURE 6-57
Steady-state turbocharger operating lines plotted as constant T_{03}/T_{01} lines on compressor map. Turbine characteristics defined by Fig. 6-51. p_{01} = compressor inlet pressure (bar), p_{02} = compressor exit pressure (bar), T_{01} = compressor inlet temperature (K), T_{03} = turbine inlet temperature (K), $\dot{m}$ = mass flow rate (kg/s), N = speed (rev/min).[40]

determined. Figure 6-57 shows an example of such a set of turbocharger characteristics, plotted on a turbocharger compressor map for a radial turbine with characteristics similar to Fig. 6-51. The dash-dot-dash line is for $p_{02} = p_{03}$. To the right of this line, $p_{03} > p_{02}$; to the left of this line $p_{02} > p_{03}$.[40]

The problem of overspeeding the turbocharger and generating very high cylinder pressures often requires that some of the exhaust be bypassed around the turbine. The bypass valve or *wastegate* is usually built into the turbocharger casing. It consists of a spring-loaded valve acting in response to the inlet manifold pressure on a controlling diaphragm. When the wastegate is open, only a portion of the exhaust gases will flow through the turbine and generate power; the remainder passes directly into the exhaust system downstream of the turbine.

6.8.5 Wave-Compression Devices

Pressure wave superchargers make use of the fact that if two fluids having different pressures are brought into direct contact in long narrow channels, equilization of pressure occurs faster than mixing. One such device, the Comprex, has been developed for internal combustion engine supercharging which operates using this principle.[46] It is shown schematically in Fig. 6-58. The working channels of the Comprex are arranged on a rotor or cell wheel (*b*) which is rotated

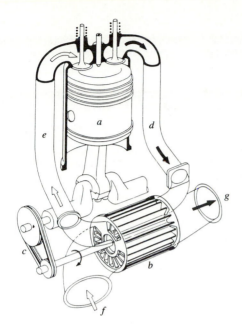

FIGURE 6-58
Schematic of Comprex supercharger.[47] *a* Engine, *b* Cell wheel or rotor, *c* Belt drive, *d* High-pressure exhaust gas (G-HP), *e* High-pressure air (A-HP), *f* Low-pressure air (A-LP), *g* Low-pressure exhaust gas (G-LP)

between two castings by a belt driven from the crankshaft (*c*). There is no contact between the rotor and the casing, but the gaps are kept small to minimize leakage. The belt drive merely overcomes friction and maintains the rotor at a speed proportional to engine speed (usually 4 or 5 times faster): it provides no compression work. One casing (the air casing) contains the passage which brings low-pressure air (*f*) to one set of ports and high-pressure air (*e*) from another set of ports in the rotor-side inner casing. The other casing (the gas casing) connects the high-pressure engine exhaust gas (*d*) to one set of ports at the other end of the rotor, and connects a second set of ports to the exhaust system (*g*). Fluid can flow into and out of the rotor channels through these ports. The exhaust gas inlet port is made small enough to cause a significant pressure rise in the exhaust manifold (e.g., 2 atm) when the engine is operated at its rated power. The pressure wave process does not depend on the pressure and flow fluctuations within the manifold caused by individual cylinder exhaust events: its operation can be explained assuming constant pressure at each set of ports. As the rotor makes one revolution, the ends of each channel are alternatively closed, or are open to a flow passage. By appropriate arrangement of these passages and selection of the geometry and location of the ports, an efficient energy transfer between the engine exhaust gases and the fresh charge can be realized.[46]

The wave-compression process in the Comprex can be explained in more detail with the aid of Fig. 6-59, where the rotational motion of the channels has been unrolled. Consider the channel starting at the top; it is closed at both ends and contains air at atmospheric pressure. As it opens at the upper edge of the high-pressure gas (G-HP) duct, a compression or shock wave (1) propagates from

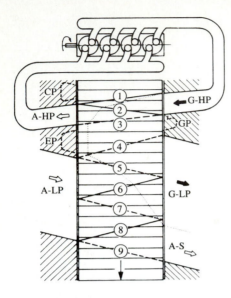

FIGURE 6-59
Unrolled view of the Comprex pressure-wave process.[47] A Air, G Gas, S Scavenging, HP High pressure, LP Low pressure; CP, EP, GP are pockets.

the right end of the channel toward the left, compressing the air through which it passes. The compressed air behind the wave occupies less space so the high-pressure exhaust gas moves into the channel as indicated by the dotted line. This line is the boundary between the two fluids. As this wave (1) reaches the left end, the channel is opened and compressed air flows into the engine inlet duct (A-HP). The inlet duct is shaped to provide the same mass flow at lower velocity: this deceleration of the air produces a second compression wave (2) which propagates back into the channel. As a result the compressed air leaving the cell on the left has a higher pressure than the driving gas on the right. As this wave (2) arrives at the right-hand side, the high-pressure gas (G-HP) channel closes. An expansion wave (3) then propagates back to the left, separating the now motionless and partly expanded fluid on the right from still-moving fluid on the left. When this wave (3) reaches the left-hand end, A-HP is closed and all the gases in the channel are at rest. Note that the first gas particles (dotted line) have not quite reached the air end of the channel: a cushion of air remains to prevent break-through.

The cell's contents are still at a higher pressure than the low pressure in the exhaust gas duct. When the right-hand end of the cell reaches this duct, the cell's contents expand into the exhaust. This motion is transferred through the channel by an expansion wave (4) which propagates to the left at sonic speed. When this wave reaches the left-hand end, the cell opens to the low-pressure air duct (A-LP) and fresh air is drawn into the cell. The flow to the right continues, but with decreasing speed due to wave action (5, 6, 7, 8) and pressure losses at each end of the cell. When the dotted line—the interface between air and the exhaust gas—reaches the right end of the cell, all the driving gas has left. The cell is then purged by the scavenging air flow (A-S) and filled with fresh air at atmospheric pressure. At wave (9), the cell is closed at both ends, restoring it to its initial state.[47]

The speed of these pressure waves is the local sound speed and is a function of local gas temperature only. Thus, the above process will only work properly for a given exhaust gas temperature at a particular cell speed. The operating range is extended by the use of "pockets" as shown in Fig. 6-59. The pockets prevent the reflection of sound waves from a closed channel end which would cause a substantial change in flow velocity in the channel. These pockets, marked CP and EP on the air side and GP on the exhaust gas side, allow flow from one channel to adjacent channels via the pocket if the wave action requires it. Thus the device can be tuned for full-load medium-speed operation and still give acceptable performance at other loads and speeds because the pockets allow the particle paths to change without major losses.[46]

Figure 6-60 shows the apparent compressor performance map of a Comprex when connected to a small three-cylinder diesel engine. Note that the map depends on the engine to which the device is coupled because the exhaust gas expansion process and fresh air compression process occur within the same rotor. The volume flow rate is the net air: it is the total air flow into the device less the scavenging air flow. The values of isentropic efficiency [defined by Eq. (6.39)] are comparable to those of mechanical and aerodynamic compressors.

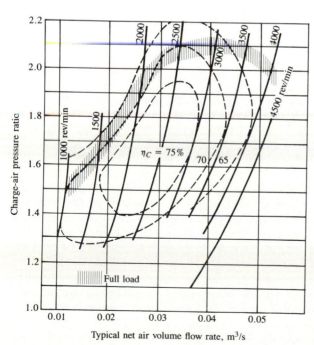

Typical net air volume flow rate, m³/s

FIGURE 6-60

Apparent compressor map of Comprex connected to a 1.2-dm³ diesel engine: charge-air pressure ratio plotted versus net air volume flow rate (total air flow less scavenging air flow).[46]

PROBLEMS

6.1. A conventional spark-ignition engine operating with gasoline will not run smoothly (due to incomplete combustion) with an equivalence ratio leaner than about $\phi = 0.8$. It is desirable to extend the smooth operating limit of the engine to leaner equivalence ratios so that at part-throttle operation (with intake pressure less than 1 atmosphere) the pumping work is reduced. Leaner than normal operation can be achieved by adding hydrogen gas (H_2) to the mixture in the intake system. The addition of H_2 makes the fuel-air mixture easier to burn.

(a) The fuel composition with "mixed" fuel operation is $H_2 + C_8H_{18}$—one mole of hydrogen to every mole of gasoline, which is assumed the same as isooctane. What is the stoichiometric air/fuel ratio for the "mixed" fuel?

(b) The lower heating value of H_2 is 120 MJ/kg and for isooctane is 44.4 MJ/kg. What is the heating value per kilogram of fuel mixture?

(c) Engine operation with isooctane and the mixed ($H_2 + C_8H_{18}$) fuel is compared in a particular engine at a part-load condition (brake mean effective pressure of 275 kPa and 1400 rev/min). You are given the following information about the engine operation:

Fuel	C_8H_{18}	$H_2 + C_8H_{18}$
Equivalence ratio	0.8	0.5
Gross indicated fuel conversion efficiency	0.35	0.4
Mechanical rubbing friction mep	138 kPa	138 kPa
Inlet manifold pressure	46 kPa	?
Pumping mep	55 kPa	?

Estimate approximately the inlet manifold pressure and the pumping mean effective pressure with ($H_2 + C_8H_{18}$) fuel. Explain your method and assumptions clearly. Note that mechanical efficiency η_m is defined as

$$\eta_m = \frac{bmep}{imep_g} = \frac{bmep}{bmep + rfmep + pmep}$$

6.2. Hydrogen is a possible future fuel for spark-ignition engines. The lower heating value of hydrogen is 120 MJ/kg and for gasoline (C_8H_{14}) is 44 MJ/kg. The stoichiometric air/fuel ratio for hydrogen is 34.3 and for gasoline is 14.4. A disadvantage of hydrogen fuel in the SI engine is that the partial pressure of hydrogen in the H_2-air mixture reduces the engine's volumetric efficiency, which is proportional to the partial pressure of air. Find the partial pressure of air in the intake manifold downstream of the hydrogen fuel-injection location at wide-open throttle when the total intake manifold pressure is 1 atmosphere; the equivalence ratio is 1.0. Then estimate the ratio of the fuel energy per unit time entering a hydrogen-fueled engine operating with a stoichiometric mixture to the fuel energy per unit time entering an identical gasoline-fueled engine operating at the same speed with a stoichiometric mixture. (Note that the "fuel energy" per unit mass of *fuel* is the fuel's heating value.)

6.3. Sketch (a) shows an ideal cycle p-V diagram for a conventional throttled spark-ignition engine, 1-2-3-4-5-6-7-1. The gas properties c_v, c_p, γ, R throughout the cycle are constant. The mass of gas in the cylinder is m. The exhaust pressure is p_e.

Sketch (b) shows an ideal cycle p-V diagram 1-2-3-4-5-6-8-1 for a spark-ignition engine with novel inlet valve timing. The inlet manifold is unthrottled; it has essentially the same pressure as the exhaust. To reduce the mass inducted at part

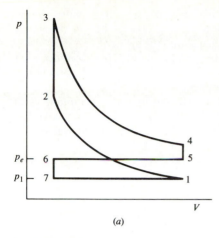
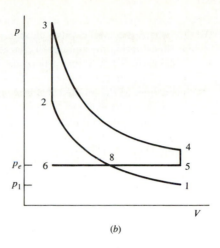

(a)   (b)

FIGURE P6-3

load, the inlet valve is closed rapidly *partway through the intake stroke* at point 8. The gas in the cylinder at inlet valve closing at 8 is then expanded isentropically to 1 with the inlet valve closed. The pressure p_1 at the start of compression is the same for both cycles.

(a) Indicate on p-V diagrams the area that corresponds to the pumping work per cycle for cycles (a) and (b). Which area is greater?

(b) Derive expressions for the pumping work per cycle W_p in terms of m, c_v, γ, T_1, (p_e/p_1), and the compression ratio r_c for cycles (a) and (b). Be consistent about the signs of the work transfers to and from the gas.

(c) For $\gamma = 1.3$, $r_c = 8$, and $(p_e/p_1) = 2$ find the ratio $W_p(b)/W_p(a)$, assuming the values of T_1 and m are the same in both cases.

6.4. For four-stroke cycle engines, the inlet and exhaust valve opening and closing crank angles are typically: IVO 15° BTC; IVC 50° ABC; EVO 55° BBC; EVC 10° ATC. Explain why these valve timings improve engine breathing relative to valve opening and closing at the beginnings and ends of the intake and exhaust strokes. Are there additional design issues that are important?

6.5. Estimate approximately the pressure drop across the inlet valve about halfway through the intake stroke and across the exhaust valve halfway through the exhaust stroke, when the piston speed is at its maximum for a typical four-stroke cycle spark-ignition engine with $B = L = 85$ mm at 2500 and 5000 rev/min at WOT. Assume appropriate values for any valve and port geometric details required, and for the gas composition and state.

6.6. Using the data in Fig. 6-21, estimate the fraction of the original mass left in the cylinder: (a) at the end of the blowdown process and (b) at the end of the exhaust stroke.

6.7. Compare the engine residual gas fraction data in Fig. 6-19 with ideal cycle estimates of residual gas fraction as follows. Using Eq. (5.47) plot the fuel-air cycle residual mass fraction x_r against p_i/p_e for $r_c = 8.5$ on the same graph as the engine data in Fig. 6-19 at 1400 rev/min and 27° valve overlap. Assume $T_r = 1400$ K and $(\gamma - 1)/\gamma = 0.24$ in Eq. (5.47). Suggest an explanation for any significant difference.

6.8. One concept that would increase SI engine efficiency is early intake valve closing (EIVC) where the intake valve closes *before* the piston reaches BC on the intake stroke, thus limiting the amount of charge inducted into the cylinder.

(a) Explain why EIVC improves engine efficiency at part load. (*Hint*: consider what must happen to the inlet manifold pressure in order to maintain constant mass in the cylinder as the intake valve is closed sooner.)

(b) This part load reduction in charge could be achieved by using late intake valve closing where the intake valve is not closed until the compression stroke has pushed some of the cylinder gases back out into the intake manifold. Based on a comparison of p-V diagrams, is this method inferior to EIVC?

6.9. An eight-cylinder turbocharged aftercooled four-stroke cycle diesel engine operates with an inlet pressure of 1.8 atmospheres at its maximum rated power at 2000 rev/min. $B = 128$ mm, $L = 140$ mm, η_v (based on inlet manifold conditions of 1.8 atm and 325 K after the aftercooler) = 0.9. The compressor isentropic efficiency is 0.7.

(a) Calculate the power required to drive the turbocharger compressor.

(b) If the exhaust gas temperature is 650°C and the turbocharger isentropic efficiency is 0.65, estimate the pressure at turbine inlet. The turbine exhausts to the atmosphere.

6.10. The charging efficiency of two-stroke cycle diesel engines can be estimated from measurement of the concentration of O_2 and CO_2 in the burned gases within the cylinder, or in the exhaust blowdown pulse prior to any mixing with fresh air. The engine bore = 125 mm, stroke = 150 mm, compression ratio = 15. The fuel flow rate at 1800 rev/min is 1.6 g/s per cylinder. The conditions used to evaluate the air density for the reference mass are 300 K and 1 atm. The molar concentrations (dry) of CO_2 and O_2 in the in-cylinder burned gases are 7.2 and 10.4 percent (see Fig. 4-22). The scavenging air flow rate is 80 g/s. Evaluate (a) the charging efficiency, (b) the delivery ratio, and (c) the trapping efficiency (assuming the trapped mass equals the reference mass).

REFERENCES

1. Khovakh, M.: *Motor Vehicle Engines*, English Translation, Mir Publishers, Moscow, 1976.
2. Matsuoka, S., Tasaka, H., and Tsuruta, J.: "The Evaporation of Fuel and Its Effect on Volumetric Efficiency," JARI technical memorandum no. 2, pp. 17–22, 1971.
3. Takizawa, M., Uno, T., Oue, T., and Yura, T.: "A Study of Gas Exchange Process Simulation of an Automotive Multi-Cylinder Internal Combustion Engine," SAE paper 820410, *SAE Trans.*, vol. 91, 1982.
4. Kay, I. W.: "Manifold Fuel Film Effects in an SI Engine," SAE paper 780944, 1978.
5. Ohata, A., and Ishida, Y.: "Dynamic Inlet Pressure and Volumetric Efficiency of Four Cycle Four Cylinder Engine," SAE paper 820407, *SAE Trans.*, vol. 91, 1982.
6. Benson, R. S., and Whitehouse, N. D.: *Internal Combustion Engines*, vol. 2, Pergamon Press, 1979.
7. Taylor, C. F.: *The Internal-Combustion Engine in Theory and Practice*, vol. 1, 2d ed., revised, MIT Press, Cambridge, Mass., 1985.
8. Hofbauer, P., and Sator, K.: "Advanced Automotive Power Systems, Part 2: A Diesel for a Subcompact Car," SAE paper 770113, *SAE Trans.*, vol. 86, 1977.
9. Armstrong, D. L., and Stirrat, G. F.: "Ford's 1982 3.8L V6 Engine," SAE paper 820112, *SAE Trans.*, vol. 91, 1982.
10. Chapman, M., Novak, J. M., and Stein, R. A.: "Numerical Modeling of Inlet and Exhaust Flows in Multi-Cylinder Internal Combustion Engines," in *Flows in Internal Combustion Engines*, Winter Annual Meeting, ASME, New York, 1982.

11. Barnes-Moss, H. W.: "A Designers Viewpoint," in *Passenger Car Engines, Conference Proceedings*, pp. 133–147, Institution of Mechanical Engineers, London, 1975.

12. Asmus, T. W.: "Valve Events and Engine Operation," SAE paper 820749, *SAE Trans.*, Vol. 91, 1982.

13. SAE Recommended Practice, "Engine Terminology and Nomenclature—General," in *SAE Handbook*, J604d.

14. Kastner, L. J., Williams, T. J., and White, J. B.: "Poppet Inlet Valve Characteristics and Their Influence on the Induction Process," *Proc. Instn Mech. Engrs*, vol. 178, pt. 1, no. 36, pp. 955–978, 1963–1964.

15. Woods, W. A., and Khan, S. R.: "An Experimental Study of Flow through Poppet Valves," *Proc. Instn Mech. Engrs*, vol. 180, pt. 3N, pp. 32–41, 1965–1966.

16. Annand, W. J. D., and Roe, G. E.: *Gas Flow in the Internal Combustion Engine*, Haessner Publishing, Newfoundland, N.J., 1974.

17. Tanaka, K.: "Air Flow through Exhaust Valve of Conical Seat," *Int. Congr. Appl. Mech.*, vol. 1, pp. 287–295, 1931.

18. Bicen, A. F., and Whitelaw, J. H.: "Steady and Unsteady Air Flow through an Intake Valve of a Reciprocating Engine," in *Flows in Internal Combustion Engines—II*, FED-vol. 20, Winter Annual Meeting, ASME, 1984.

19. Fukutani, I., and Watanabe, E.: "An Analysis of the Volumetric Efficiency Characteristics of 4-Stroke Cycle Engines Using the Mean Inlet Mach Number Mim," SAE paper 790484, *SAE Trans.*, vol. 88, 1979.

20. Wallace, W. B.: "High-Output Medium-Speed Diesel Engine Air and Exhaust System Flow Losses," *Proc. Instn Mech. Engrs*, vol. 182, pt. 3D, pp. 134–144, 1967–1968.

21. Cole, B. N., and Mills, B.: "The Theory of Sudden Enlargements Applied to Poppet Exhaust-Valve, with Special Reference to Exhaust-Pulse Scavenging," *Proc. Instn Mech. Engrs*, pt. 1B, pp. 364–378, 1953.

22. Toda, T., Nohira, H., and Kobashi, K.: "Evaluation of Burned Gas Ratio (BGR) as a Predominant Factor to NO_x," SAE paper 760765, *SAE Trans.*, vol. 85, 1976.

23. Benson, J. D., and Stebar, R. F.: "Effects of Charge Dilution on Nitric Oxide Emission from a Single-Cylinder Engine," SAE paper 710008, *SAE Trans.*, vol. 80, 1971.

24. Tabaczynski, R. J., Heywood, J. B., and Keck, J. C.: "Time-Resolved Measurements of Hydrocarbon Mass Flow Rate in the Exhaust of a Spark-Ignition Engine," SAE paper 720112, *SAE Trans.*, vol. 81, 1972.

25. Caton, J. A., and Heywood, J. B.: "An Experimental and Analytical Study of Heat Transfer in an Engine Exhaust Port," *Int. J. Heat Mass Transfer*, vol. 24, no. 4, pp. 581–595, 1981.

26. Caton, J. A.: "Comparisons of Thermocouple, Time-Averaged and Mass-Averaged Exhaust Gas Temperatures for a Spark-Ignited Engine," SAE paper 820050, 1982.

27. Phatak, R. G.: "A New Method of Analyzing Two-Stroke Cycle Engine Gas Flow Patterns," SAE paper 790487, *SAE Trans.*, vol. 88, 1979.

28. Rizk, W.: "Experimental Studies of the Mixing Processes and Flow Configurations in Two-Cycle Engine Scavenging," *Proc. Instn Mech. Engrs*, vol. 172, pp. 417–437, 1958.

29. Dedeoglu, N.: "Scavenging Model Solves Problems in Gas Burning Engine," SAE paper 710579, *SAE Trans.*, vol. 80, 1971.

30. Sher, E.: "Investigating the Gas Exchange Process of a Two-Stroke Cycle Engine with a Flow Visualization Rig," *Israel J. Technol.*, vol. 20, pp. 127–136, 1982.

31. Jante, A.: "Scavenging and Other Problems of Two-Stroke Cycle Spark-Ignition Engines," SAE paper 680468, *SAE Trans.*, vol. 77, 1968.

32. Kannappan, A.: "Cumulative Sampling Technique for Investigating the Scavenging Process in Two-Stroke Engine," ASME paper 74-DGP-11, 1974.

33. Ohigashi, S., Kashiwada, Y., and Achiwa, J.: "Scavenging the 2-Stroke Diesel Engine," *Bull. JSME*, vol. 3, no. 9, pp. 130–136, 1960.

34. Huber, E. W.: "Measuring the Trapping Efficiency of Internal Combustion Engines through Continuous Exhaust Gas Analysis," SAE paper 710144, *SAE Trans.*, vol. 80, 1971.

35. Blair, G. P., and Kenny, R. G.: "Further Developments in Scavenging Analysis for Two-Cycle Engines," SAE paper 800038, *SAE Trans.*, vol. 89, 1980.

36. Baudequin, F., and Rochelle, P.: "Some Scavenging Models for Two-Stroke Engines," *Proc. Instn Mech. Engrs, Automobile Division*, vol. 194, no. 22, pp. 203–210, 1980.
37. Gyssler, G.: "Problems Associated with Turbocharging Large Two-Stroke Diesel Engines," *Proc. CIMAC*, paper B.16, 1965.
38. Annand, W. J. D.: "Compressible Flow through Square-Edged Orifices: An Empirical Approximation for Computer Calculations," *J. Mech. Engng Sci.*, vol. 8, p. 448, 1966.
39. Benson, R. S.: "Experiments on a Piston Controlled Port," *The Engineer*, vol. 210, pp. 875–880, 1960.
40. Watson, N., and Janota, M. S.: *Turbocharging the Internal Combustion Engine*, Wiley-Interscience Publications, John Wiley, New York, 1982.
41. Bhinder, F. S.: "Supercharging Compressors—Problems and Potential of the Various Alternatives," SAE paper 840243, 1984.
42. Bhinder, F. S.: "Some Fundamental Considerations Concerning the Pressure Charging of Small Diesel Engines," SAE paper 830145, 1983.
43. Brandstetter, W., and Dziggel, R.: "The 4- and 5-Cylinder Turbocharged Diesel Engines for Volkswagen and Audi," SAE paper 820441, *SAE Trans.*, vol. 91, 1982.
44. SAE Recommended Practice, "Turbocharger Nomenclature and Terminology," in *SAE Handbook*, J922.
45. Flynn, P. F.: "Turbocharging Four-Cycle Diesel Engines," SAE paper 790314, *SAE Trans.*, vol. 88, 1979.
46. Gyarmathy, G.: "How Does the Comprex Pressure-Wave Supercharger Work?" SAE paper 830234, 1983.
47. Kollbrunner, T. A.: "Comprex Supercharging for Passenger Diesel Car Engines," SAE paper 800884, *SAE Trans.*, vol. 89, 1980.

CHAPTER
7

SI ENGINE FUEL METERING AND MANIFOLD PHENOMENA

7.1 SPARK-IGNITION ENGINE MIXTURE REQUIREMENTS

The task of the engine induction and fuel systems is to prepare from ambient air and fuel in the tank an air-fuel mixture that satisfies the requirements of the engine over its entire operating regime. In principle, the optimum air/fuel ratio for a spark-ignition engine is that which gives the required power output with the lowest fuel consumption, consistent with smooth and reliable operation. In practice, the constraints of emissions control may dictate a different air/fuel ratio, and may also require the recycling of a fraction of the exhaust gases (EGR) into the intake system. The relative proportions of fuel and air that provide the lowest fuel consumption, smooth reliable operation, and satisfy the emissions requirements, at the required power level, depend on engine speed and load. Mixture requirements and preparation are usually discussed in terms of the air/fuel ratio or fuel/air ratio (see Sec. 2.9) and percent EGR [see Eq. (4.2)]. While the fuel metering system is designed to provide the appropriate fuel flow for the *actual* air flow at each speed and load, the relative proportions of fuel and air can be stated more generally in terms of the fuel/air equivalence ratio ϕ, which is the actual fuel/air ratio normalized by dividing by the stoichiometric fuel/air ratio [Eq.

(3.8)]. The combustion characteristics of fuel-air mixtures and the properties of combustion products, which govern engine performance, efficiency, and emissions, correlate best for a wide range of fuels relative to the stoichiometric mixture proportions. Where appropriate, therefore, the equivalence ratio will be used as the defining parameter. A typical value for the stoichiometric air/fuel ratio of gasoline is 14.6.† Thus, for gasoline,

$$\phi \approx \frac{14.6}{A/F} \tag{7.1}$$

The effects of equivalence ratio variations on engine combustion, emissions, and performance are discussed more fully in Chaps. 9, 11, and 15. A brief summary is sufficient here. Mixture requirements are different for full-load (wide-open throttle) and for part-load operation. At the former operating condition, complete utilization of the *inducted air* to obtain maximum power for a given displaced volume is the critical issue. Where less than the maximum power at a given speed is required, efficient utilization of the *fuel* is the critical issue. At wide-open throttle, maximum power for a given volumetric efficiency is obtained with rich-of-stoichiometric mixtures, $\phi \approx 1.1$ (see the discussion of the fuel-air cycle results in Sec. 5.5.3). Mixtures that are richer still are sometimes used to increase volumetric efficiency by increasing the amount of charge cooling that accompanies fuel vaporization [see Eq. (6.5)], thereby increasing the inducted air density.

At part-load (or part-throttle) operating conditions, it is advantageous to dilute the fuel-air mixture, either with excess air or with recycled exhaust gas. This dilution improves the fuel conversion efficiency for three reasons:[1] (1) the expansion stroke work for a given expansion ratio is increased as a result of the change in thermodynamic properties of the burned gases—see Secs. 5.5.3 and 5.7.4; (2) for a given mean effective pressure, the intake pressure increases with increasing dilution, so pumping work decreases—see Fig. 5-10; (3) the heat losses to the walls are reduced because the burned gas temperatures are lower. In the absence of strict engine NO_x emission requirements, excess air is the obvious diluent, and at part throttle engines have traditionally operated lean. When tight control of NO_x, HC, and CO emissions is required, operation of the engine with a stoichiometric mixture is advantageous so that a three-way catalyst‡ can be used to clean up the exhaust. The appropriate diluent is then recycled exhaust gases which significantly reduces NO_x emissions from the engine itself. The amount of diluent that the engine will tolerate at any given speed and load depends on the details of the engine's combustion process. Increasing excess air

† Typical value only. Most gasolines have $(A/F)_s$ in the range 14.4 to 14.7. $(A/F)_s$ could lie between 14.1 and 15.2.

‡ A three-way catalyst system, when operated with a close-to-stoichiometric mixture, achieves substantial reductions in NO_x, CO, and HC emissions simultaneously; see Sec. 11.6.2.

or the amount of recycled exhaust slows down the combustion process and increases its variability from cycle to cycle. A certain minimum combustion repeatability or stability level is required to maintain smooth engine operation. Deterioration in combustion stability therefore limits the amount of dilution an engine can tolerate. As load decreases, less dilution of the *fresh* mixture can be tolerated because the internal dilution of the mixture with residual gas increases (see Sec. 6.4). At idle conditions, the fresh mixture will not usually tolerate any EGR and may need to be stoichiometric or fuel-rich to obtain adequate combustion stability.

Mixture composition requirements over the engine load and speed range are illustrated schematically for the two approaches outlined above in Fig. 7-1. If stoichiometric operation and EGR are not required for emissions control, as load increases the mixture is leaned out from a fuel-rich or close-to-stoichiometric composition at very light load. As wide-open throttle operation is approached at each engine speed, the mixture is steadily enriched to rich-of-stoichiometric at the maximum bmep point. With the stoichiometric operating conditions required for three-way-catalyst-equipped engines, when EGR is used, the percentage of recycled exhaust increases from zero at light load to a maximum at mid-load, and then decreases to zero as wide-open throttle conditions are approached so maximum bmep can be obtained. Combinations of these strategies are possible. For example, lean operation at light load can be used for best efficiency, and

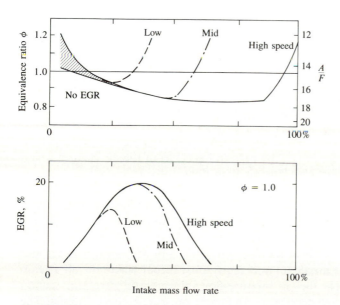

FIGURE 7-1
Typical mixture requirements for two common operating strategies. Top diagram shows equivalence ratio variation with intake mass flow rate (percent of maximum flow at rated speed) at constant low, mid, and high engine speeds. Bottom diagram shows recycled exhaust (EGR) schedule as a function of intake flow rate, for low, mid, and high speeds for stoichiometric operation.

stoichiometric mixtures (with a three-way catalyst) and/or EGR can be used at mid loads to control NO_x emissions.

In practical spark-ignition engine induction systems, the fuel and air distribution between engine cylinders is not uniform (and also varies in each individual cylinder on a cycle-by-cycle basis). A spread of one or more air/fuel ratios between the leanest and richest cylinders over the engine's load and speed range is not uncommon in engines with conventional carburetors. The average mixture composition must be chosen to avoid excessive combustion variability in the leanest operating cylinder. Thus, as the spread in mixture nonuniformity increases, the mean equivalence ratio must be moved toward stoichiometric and away from the equivalence ratio which gives minimum fuel consumption.

7.2 CARBURETORS

7.2.1 Carburetor Fundamentals

A carburetor has been the most common device used to control the fuel flow into the intake manifold and distribute the fuel across the air stream. In a carburetor, the air flows through a converging-diverging nozzle called a venturi. The pressure difference set up between the carburetor inlet and the throat of the nozzle (which depends on the air flow rate) is used to meter the appropriate fuel flow for that air flow. The fuel enters the air stream through the fuel discharge tube or ports in the carburetor body and is atomized and convected by the air stream past the throttle plate and into the intake manifold. Fuel evaporation starts within the carburetor and continues in the manifold as fuel droplets move with the air flow and as liquid fuel flows over the throttle and along the manifold walls. A modern carburetor which meters the appropriate fuel flow into the air stream over the complete engine operating range is a highly developed and complex device. There are many types of carburetors; they share the same basic concepts which we will now examine.

Figure 7-2 shows the essential components of an elementary carburetor. The air enters the intake section of the carburetor (1) from the air cleaner which removes suspended dust particles. The air then flows into the carburetor venturi (a converging-diverging nozzle) (2) where the air velocity increases and the pressure decreases. The fuel level is maintained at a constant height in the float chamber (3) which is connected via an air duct (4) to the carburetor intake section (1). The fuel flows through the main jet (a calibrated orifice) (5) as a result of the pressure difference between the float chamber and the venturi throat and through the fuel discharge nozzle (6) into the venturi throat where the air stream atomizes the liquid fuel. The fuel-air mixture flows through the diverging section of the venturi where the flow decelerates and some pressure recovery occurs. The flow then passes the throttle valve (7) and enters the intake manifold.

Note that the flow may be unsteady even when engine load and speed are constant, due to the periodic filling of each of the engine cylinders which draws air through the carburetor venturi. The induction time, $1/(2N)$ (20 ms at 1500

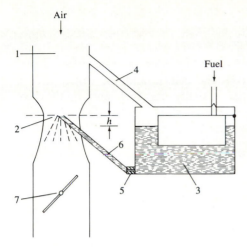

Air

Fuel

FIGURE 7-2

Schematic of elementary carburetor.

1 Inlet section
2 Venturi throat
3 Float chamber
4 Pressure equalizing passage
5 Calibrated orifice
6 Fuel discharge tube
7 Throttle plate

rev/min), is the characteristic time of this periodic cylinder filling process. Generally, the characteristic times of changes in throttle setting are longer; it takes several engine operating cycles to reestablish steady-state engine operation after a sudden change in throttle position.[2] It is usually assumed that the flow processes in the carburetor can be modeled as quasi steady.

FLOW THROUGH THE VENTURI. Equation (C.8) in App. C relates the mass flow rate of a gas through a flow restriction to the upstream stagnation pressure and temperature, and the pressure at the throat. For the carburetor venturi:

$$\dot{m}_a = \frac{C_{DT} A_T p_0}{\sqrt{RT_0}} \left(\frac{p_T}{p_0}\right)^{1/\gamma} \left\{\frac{2\gamma}{\gamma - 1}\left[1 - \left(\frac{p_T}{p_0}\right)^{(\gamma - 1)/\gamma}\right]\right\}^{1/2} \tag{7.2}$$

where C_{DT} and A_T are the discharge coefficient and area of the venturi throat, respectively. If we assume the velocity at the carburetor inlet can be neglected, Eq. (7.2) can be rearranged in terms of the pressure drop from upstream conditions to the venturi throat for the air stream, $\Delta p_a = p_0 - p_T$, as

$$\dot{m}_a = C_{DT} A_T (2\rho_{a0} \Delta p_a)^{1/2} \Phi \tag{7.3}$$

where

$$\Phi = \left[\left(\frac{\gamma}{\gamma - 1}\right) \frac{(p_T/p_0)^{2/\gamma} - (p_T/p_0)^{(\gamma + 1)/\gamma}}{1 - (p_T/p_0)}\right]^{1/2} \tag{7.4}$$

and accounts for the effects of compressibility. Figure C-3 shows the value of Φ as a function of pressure drop. For the normal carburetor operating range, where $\Delta p_a/p_0 \le 0.1$, the effects of compressibility which reduce Φ below 1.0 are small.

FLOW THROUGH THE FUEL ORIFICE. Since the fuel is a liquid and therefore essentially incompressible, the fuel flow rate through the main fuel jet is given by

Eq. (C.2) in App. C as

$$\dot{m}_f = C_{D_o} A_o (2\rho_f \Delta p_f)^{1/2} \tag{7.5}$$

where C_{D_o} and A_o are the discharge coefficient and area of the orifice, respectively, Δp_f is the pressure difference across the orifice, and the orifice area is assumed much less than the passage area. Usually, the fuel level in the float chamber is held below the fuel discharge nozzle, as shown in Fig. 7-2, to prevent fuel spillage when the engine is inclined to the horizontal (e.g., in a vehicle on a slope). Thus,

$$\Delta p_f = \Delta p_a - \rho_f gh$$

where h is typically of order 10 mm.

The discharge coefficient C_{D_o} in Eq. (7.5) represents the effect of all deviations from the ideal one-dimensional isentropic flow. It is influenced by many factors of which the most important are the following: (1) fluid mass flow rate; (2) orifice length/diameter ratio; (3) orifice/approach-area ratio; (4) orifice surface area; (5) orifice surface roughness; (6) orifice inlet and exit chamfers; (7) fluid specific gravity; (8) fluid viscosity; and (9) fluid surface tension. The use of the orifice Reynolds number, $\mathrm{Re}_o = \rho V D_o / \mu$, as a correlating parameter for the discharge coefficient accounts for effects of mass flow rate, fluid density and viscosity, and length scale to a good first approximation. The discharge coefficient of a typical carburetor main fuel-metering system orifice increases smoothly with increasing Re_o.[3]

CARBURETOR PERFORMANCE. The air/fuel ratio delivered by this carburetor is given by

$$\left(\frac{A}{F}\right) = \frac{\dot{m}_a}{\dot{m}_f} = \left(\frac{C_{DT}}{C_{D_o}}\right)\left(\frac{A_T}{A_o}\right)\left(\frac{\rho_{ao}}{\rho_f}\right)^{1/2}\left(\frac{\Delta p_a}{\Delta p_a - \rho_f gh}\right)^{1/2} \Phi \tag{7.6}$$

and the equivalence ratio $\phi \; [=(A/F)_s/(A/F)]$ by

$$\phi = \frac{(A/F)_s}{\Phi}\left(\frac{C_{D_o}}{C_{DT}}\right)\left(\frac{A_o}{A_T}\right)\left(\frac{\rho_f}{\rho_{ao}}\right)^{1/2}\left(1 - \frac{\rho_f gh}{\Delta p_a}\right)^{1/2} \tag{7.7}$$

where $(A/F)_s$ is the stoichiometric air/fuel ratio. The terms A_o, A_T, ρ_f, and ρ_{ao} are all constant for a given carburetor, fuel, and ambient conditions. Also, except for very low flows, $\rho_f gh \ll \Delta p_a$. The discharge coefficients, C_{D_o} and C_{DT}, and Φ vary with flow rates, however. Hence, the equivalence ratio of the mixture delivered by an elementary carburetor is not constant.

Figure 7-3 illustrates the performance of the elementary carburetor. The top set of curves shows how Φ, C_{DT}, and C_{D_o} typically vary with the venturi pressure drop.[4] Note that for $\Delta p_a \le \rho_f gh$ there is no fuel flow. Once fuel starts to flow, as a consequence of these variations the fuel flow rate increases more rapidly than the air flow rate. The carburetor delivers a mixture of increasing fuel/air equivalence ratio as the flow rate increases. Suppose the venturi and orifice are sized to

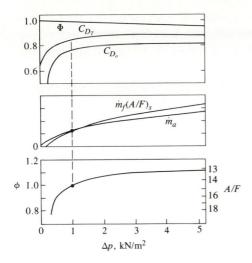

FIGURE 7-3
Performance of elementary carburetor: variation of C_{D_T}, C_{D_o}, Φ, $\dot{m}_f$ $(A/F)_s$, $\dot{m}_a$, and equivalence ratio ϕ with venturi pressure drop.

give a stoichiometric mixture at an air flow rate corresponding to 1 kN/m² venturi pressure drop (middle graph in Fig. 7-3). At higher air flow rates, the carburetor will deliver a fuel-rich mixture; at very high flow rates it will eventually deliver an essentially constant equivalence ratio. At lower air flow rates, the mixture delivered leans out rapidly. Thus, the elementary carburetor cannot provide the variation in mixture ratio which the engine requires over the complete load range at any given speed (see Fig. 7-1).

The deficiencies of the elementary carburetor can be summarized as follows:

1. At low loads the mixture becomes leaner; the engine requires the mixture to be enriched at low loads.
2. At intermediate loads, the mixture equivalence ratio increases slightly as the air flow increases. The engine requires an almost constant equivalence ratio.
3. As the air flow approaches the maximum wide-open throttle value, the equivalence ratio remains essentially constant. However, the mixture equivalence ratio should increase to 1.1 or greater to provide maximum engine power.
4. The elementary carburetor cannot compensate for transient phenomena in the intake manifold. Nor can it enrich the mixture during engine starting and warm-up.
5. The elementary carburetor cannot adjust to changes in ambient air density (due primarily to changes in altitude).

7.2.2   Modern Carburetor Design

The changes required in the elementary carburetor so that it provides the equivalence ratio versus air flow distribution shown in Fig. 7-1 are:

1. The *main metering system* must be compensated to provide essentially constant lean or stoichiometric mixtures over the 20 to 80 percent air flow range.
2. An *idle system* must be added to meter the fuel flow at idle and light loads.
3. An *enrichment system* must be added so the engine can provide its maximum power as wide-open throttle is approached.
4. An *accelerator pump* which injects additional fuel when the throttle is opened rapidly is required to maintain constant the equivalence ratio delivered to the engine cylinder.
5. A *choke* must be added to enrich the mixture during engine starting and warm-up to ensure a combustible mixture within each cylinder at the time of ignition.
6. *Altitude compensation* is required to adjust the fuel flow to changes in air density.

In addition, it is necessary to increase the magnitude of the pressure drop available for controlling the fuel flow. Two common methods used to achieve this are the following.

BOOST VENTURIS. The carburetor venturi should give as large a vacuum at the throat as possible at maximum air flow, within the constraints of a low pressure loss across the complete venturi and diffuser. In a single venturi, as the diameter of the throat is decreased at a given air flow to increase the flow velocity and hence the metering signal at the throat, the pressure loss increases. A higher vacuum signal at the venturi throat and higher velocities for improved atomization can be obtained without increasing the overall pressure loss through the use of multiple venturis. Figure 7-4 shows the geometry and the pressure distribution in a typical double-venturi system. A boost venturi is positioned upstream of the throat of the larger main venturi, with its discharge at the location of maximum velocity in the main venturi. Only a fraction of the air flows through the boost venturi. Since the pressure at the boost venturi *exit* equals the pressure

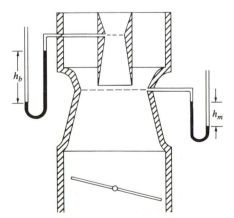

FIGURE 7-4
Schematic of carburetor double-venturi system.

at the main venturi *throat*, a higher vacuum $\Delta p_b = \rho_m g h_b$ is obtained at the boost venturi throat which can be used to obtain more precise metering of the fuel (ρ_m is the manometer fluid density). Best results are obtained with the boost venturi exit slightly upstream (≈ 5 mm) of the main venturi throat. Because only a fraction of the total air flow goes through the boost venturi, the use of multiple venturis makes it possible to obtain a high velocity air stream (up to 200 m/s) where the fuel is introduced at the boost venturi throat, and adequate vacuum, and to reduce the pressure loss across the total venturi system, without increasing the height of the carburetor. The fuel is better atomized in the smaller boost venturi with its higher air velocity, and since this air and fuel mixture is discharged centrally into the surrounding venturi, a more homogeneous mixture results. The vacuum developed at the venturi throat of a typical double-venturi system is about twice the theoretical vacuum of a single venturi of the same flow area.[5] A triple-venturi system can be used to give further increases in metering signal. The overall discharge coefficient of a multiple-venturi carburetor is lower than a single-venturi carburetor of equal cross-sectional area. The throat area of the main venturi in a multiple-venturi system is usually increased, therefore, above the single-venturi size to compensate for this. Some decrease in air stream velocity is tolerated to maintain a high discharge coefficient (and hence a high volumetric efficiency).[6]

MULTIPLE-BARREL CARBURETORS. Use of carburetors with venturi systems in parallel is a common way of maintaining an adequate part-load metering signal, high volumetric efficiency at wide-open throttle, and minimum carburetor height as engine size and maximum air flow increases. As venturi size in a single-barrel carburetor is increased to provide a higher engine air flow at maximum power, the venturi length increases and the metering signal generated at low flows decreases. Maximum wide-open throttle air flow is some 30 to 70 times the idle air flow (the value depending on engine displacement). Two-barrel carburetors usually consist of two single-barrel carburetors mounted in parallel. Four-barrel carburetors consist of a pair of two-barrel carburetors in parallel, with throttle plates compounded on two shafts. Air flows through the primary barrel(s) at low and intermediate engine loads. At higher loads, the throttle plate(s) on the secondary barrel(s) (usually of larger cross-sectional area) start to open when the air flow exceeds about 50 percent of the maximum engine air flow.

There are many different designs of complete carburetors. The operating principles of the methods most commonly used to achieve the above listed modifications will now be reviewed. Figure 7-5 shows a schematic of a conventional modern carburetor and the names of the various components and fuel passages.

COMPENSATION OF MAIN METERING SYSTEM. Figure 7-6 shows a main fuel-metering system with air-bleed compensation. As the pressure at the venturi throat decreases, air is bled through an orifice (or series of orifices) into the main fuel well. This flow reduces the pressure difference across the main fuel-metering orifice which no longer experiences the full venturi vacuum. The mixing of bleed

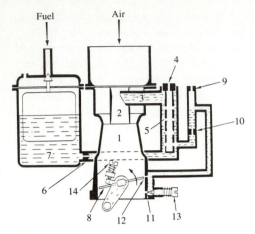

FIGURE 7-5
Schematic of modern carburetor.

1 Main venturi	8 Throttle plate
2 Boost venturi	9 Idle air-bleed orifice
3 Main metering spray tube or nozzle	10 Idle fuel orifice
4 Air-bleed orifice	11 Idle mixture orifice
5 Emulsion tube or well	12 Transition orifice
6 Main fuel-metering orifice	13 Idle mixture adjusting screw
7 Float chamber	14 Idle throttle setting adjusting screw

Fuel enters the air stream from the main metering system through (3). At idle, fuel enters air at (11). During transition, fuel enters at (11), (12), and (3). (*Courtesy S.p.A.E. Weber.*)

air with the fuel forms an emulsion which atomizes more uniformly to smaller drop sizes on injection at the venturi throat. The schematic in Fig. 7-6 illustrates the operating principle. When the engine is not running, the fuel is at the same level in the float bowl and in the main well. With the engine running, as the throttle plate is opened, the air flow and the vacuum in the venturi throat increase. For $\Delta p_v (= p_0 - p_v) < \rho_f g h_1$, there is no fuel flow from the main metering system. For $\rho_f g h_1 < \Delta p_v < \rho_f g h_2$, only fuel flows through the main well and nozzle, and the system operates just like an elementary carburetor. For $\Delta p_v > \rho_f g h_2$, air enters the main well together with fuel. The amount of air entering the well is controlled by the size of the main air-bleed orifice. The amount of air is

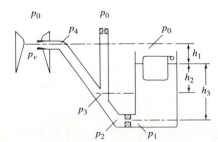

FIGURE 7-6
Schematic of main metering system with air-bleed compensation.

small and does not significantly affect the composition of the mixture. The air-bleed mass flow rate is given by

$$\dot{m}_{a_b} = C_{D_b} A_b [2(p_0 - p_3)\rho_a]^{1/2} \tag{7.8}$$

where C_{D_b} and A_b are the discharge coefficient and the area of the air-bleed orifice. The fuel mass flow rate through the fuel orifice is given by

$$\dot{m}_f = C_{D_o} A_o [2(p_1 - p_2)\rho_f]^{1/2} \tag{7.9}$$

where

$$p_1 = p_0 + \rho_f g h_3 \quad \text{and} \quad p_2 = p_3 + \rho_f g(h_3 - h_2)$$

The density of the emulsion ρ_{em} in the main well and nozzle is usually approximated by

$$\rho_{em} = \frac{\dot{m}_{a_b} + \dot{m}_f}{\dot{m}_{a_b}/\rho_a + \dot{m}_f/\rho_f} \tag{7.10}$$

Since typical values are $\rho_f = 730$ kg/m^3 and $\rho_a = 1.14$ kg/m^3, usually $\rho_f \gg \rho_{em} \gg \rho_a$. Thus, as the air-bleed flow rate increases, the height of the column of emulsion becomes less significant. However, the decrease in emulsion density due to increasing air bleed increases the flow velocity, which results in a significant pressure drop across the main nozzle. This pressure drop depends on nozzle length and diameter, fuel flow rate, bleed air flow rate, relative velocity between fuel and bleed air, and fuel properties. It is determined empirically, and has been found to correlate with ρ_{em} [as defined by Eq. (7.10)].[2, 6] The pressure loss at the main discharge nozzle with two-phase flow can be several times the pressure loss with single-phase flow.

Figure 7-7 illustrates the behavior of the system shown in Fig. 7-6: $\dot{m}_a$, $\dot{m}_f$, and the fuel/air equivalence ratio ϕ are plotted against Δp_v. Once the bleed system is operating ($\Delta p_v > \rho_f g h_2$) the fuel flow rate is reduced below its equivalent elementary carburetor value (the $A_b = 0$ line). As the bleed orifice area is increased, in the limit of large A_b and neglecting the pressure losses in the main nozzle, the fuel flow rate remains constant ($A_b \to \infty$). An appropriate choice of bleed orifice area A_b will provide the desired equivalence ratio versus pressure drop or air flow characteristic.

Additional control flexibility is obtained in practice through use of a second orifice, or of a series of holes in the main well or emulsion tube as shown in Fig. 7-5. Main metering systems with controlled air bleed provide reliable and stable control of mixture composition at part throttle engine operation. They are simple, have considerable design flexibility, and atomize the fuel effectively. In some carburetor designs, an additional compensation system consisting of a tapered rod or needle in the main metering orifice is used. The effective open area of the main metering orifice, and hence the fuel flow rate, can thus be directly related to throttle position (and manifold vacuum).

A wide range of two-phase flow patterns can be generated by bleeding an air flow into a liquid flow. Fundamental studies of the generation and flow of

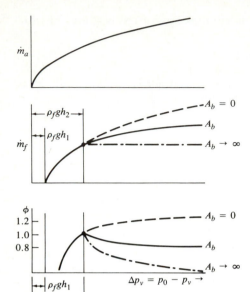

FIGURE 7-7
Metering characteristics of system with air-bleed compensation: mass flow rate of air $\dot{m}_a$, mass flow rate of fuel $\dot{m}_f$, and equivalence ratio ϕ as functions of venturi pressure drop for different air-bleed orifice areas A_b.

two-phase mixtures in small diameter tubes with bleed holes similar to those used in carburetors have been carried out.[7] For a given pipe and bleed hole size, the type of flow pattern set up depends on the flow rates of the two phases.

IDLE SYSTEM. The idle system is required because at low air flows through the carburetor insufficient vacuum is created at the venturi throat to draw fuel into the air stream. However, at idle and light loads, the manifold vacuum is high, with the pressure drop occurring across the almost-closed throttle plate. This low manifold pressure at idle is exploited for the idle fuel system by connecting the main fuel well to an orifice in the throttle body downstream of the throttle plate. Figure 7-5 shows the essential features of an idle system. The main well (5) is connected through one or more orifices (10), past one or more idle air-bleed orifices (or holes) (9), past an idle mixture adjusting screw (13), to the idle discharge port (11) in the throttle body. Emulsifying air is admitted into the idle system [at (9) and (12)] to reduce the pressure drop across the idle port and permit larger-sized ports (which are easier to manufacture) to be used. Satisfactory engine operation at idle is obtained empirically by means of the idle throttle position stop adjustment (14) and the idle mixture adjustment (13). As the throttle is opened from its idle position, the idle metering system performs a transitional function. One or more holes (12) located above the idle discharge port (11) assist as air bleeds when the throttle is at or near its idle position. As the throttle plate opens and the air flow increases, these additional discharge holes are exposed to the manifold vacuum. Additional fuel is forced out of these holes into the air stream to provide the appropriate mixture ratio. As the throttle plate is opened further, the main fuel metering system starts to supply fuel also. Because the two systems are coupled, they interact and the main system behavior in this *transition region* is modified by the fuel flow through the idle system. The total combined fuel flow provides a rich (or close-to-stoichiometric) mixture at idle, a progressive leaning of the mixture as air flow increases, and eventually (as the main system takes over full control of the fuel flow rate) an approximately constant mixture composition.

POWER ENRICHMENT SYSTEM. This system delivers additional fuel to enrich the mixture as wide-open throttle is approached so the engine can deliver its maximum power. The additional fuel is normally introduced via a submerged valve which communicates directly with the main discharge nozzle, bypassing the metering orifice. The valve, which is spring loaded, is operated either mechanically through a linkage with the throttle plate (opening as the throttle approaches its wide-open position) or pneumatically (using manifold vacuum).

ACCELERATOR PUMP. When the throttle plate is opened rapidly, the fuel-air mixture flowing into the engine cylinder leans out temporarily. The primary reason for this is the time lag between fuel flow into the air stream at the carburetor and the fuel flow past the inlet valve (see Sec. 7.6.3). While much of the fuel flow into the cylinder is fuel vapor or small fuel droplets carried by the air stream, a fraction of the fuel flows onto the manifold and port walls and forms a liquid film. The fuel which impacts on the walls evaporates more slowly than fuel carried by the air stream and introduces a lag between the air/fuel ratio produced at the carburetor and the air/fuel ratio delivered to the cylinder. An accelerator pump is used as the throttle plate is opened rapidly to supply additional fuel into the air stream at the carburetor to compensate for this leaning effect. Typically, fuel is supplied to the accelerator pump chamber from the float chamber via a small hole in the bottom of the fuel bowl, past a check valve. A pump diaphragm and stem is actuated by a rod attached to the throttle plate lever. When the throttle is opened to increase air flow, the rod-driven diaphragm will increase the fuel pressure which shuts the valve and discharges fuel past a discharge check valve or weight in the discharge passage, through the accelerator pump discharge nozzle(s), and into the air stream. A calibrated orifice controls the fuel flow. A spring connects the rod and diaphragm to extend the fuel discharge over the appropriate time period and to reduce the mechanical strain on the linkage.

CHOKE. When a cold engine is started, especially at low ambient temperatures, the following factors introduce additional special requirements for the complete carburetor:

1. Because the starter-cranked engine turns slowly (70 to 150 rev/min) the intake manifold vacuum developed during engine start-up is low.
2. This low manifold vacuum draws a lower-than-normal fuel flow from the carburetor idle system.

3. Because of the low manifold temperature and vacuum, fuel evaporation in the carburetor, manifold, and inlet port is much reduced.

Thus, during cranking, the mixture which reaches the engine cylinder would be too lean to ignite. Until normal manifold conditions are established, fuel distribution is also impaired. To overcome these deficiencies and ensure prompt starts and smooth operation during engine warm-up, the carburetor must supply a fuel-rich mixture. This is obtained with a choke. Once normal manifold conditions are established, the choke must be excluded. The primary element of a typical choke system is a plate, upstream of the carburetor, which can close off the intake system. At engine start-up, the choke plate is closed to restrict the air flow into the carburetor barrel. This causes almost full manifold vacuum within the venturi which draws a large fuel flow through the main orifice. When the engine starts, the choke is partly opened to admit the necessary air flow and reduce the vacuum in the venturi to avoid flooding the intake with fuel. As the engine warms up, the choke is opened either manually or automatically with a thermostatic control. For normal engine operation the choke plate is fully open and does not influence carburetor performance. A manifold vacuum control is often used to close the choke plate partially if the engine is accelerated during warm-up. During engine warm-up the idle speed is increased to prevent engine stalling. A fast idle cam is rotated into position by the automatic choke lever.

ALTITUDE COMPENSATION. An inherent characteristic of the conventional float type carburetor is that it meters fuel flow in proportion to the air *volume* flow rate. Air density changes with ambient pressure and temperature, with changes due to changes in pressure with altitude being most significant. For example, at 1500 m above sea level, mean atmospheric pressure is 634 mmHg or 83.4 percent of the mean sea-level value. While ambient temperature variations, winter to summer, can produce changes of comparable magnitude, the temperature of the air entering the carburetor for warmed-up engine operation is controlled to within much closer tolerances by drawing an appropriate fraction of the air from around the exhaust manifold.

Equation (7.6) shows how the air/fuel ratio delivered by the main metering system will vary with inlet air conditions. The primary dependence is through the $\sqrt{\rho_{a0}}$ term; depending on what is held constant (e.g., throttle setting or air mass flow rate) there may be an additional, much smaller dependence through Φ and Δp_a (see Ref. 5). To a good approximation, the enrichment E with increasing altitude z is given by

$$1 + E = \frac{(F/A)_z}{(F/A)_0} = \left(\frac{\rho_{a0}}{\rho_{az}}\right)^{1/2} \tag{7.11}$$

For $z = 1500$ m, $E = 9.5$ percent; thus, a cruise equivalence ratio of 0.9 or $(A/F) = 16.2$ would be enriched to close to stoichiometric.

The effects of increase in altitude on the carburetor flow curve shown in Fig. 7-1 are: (1) to enrich the entire part-throttle portion of the curve and (2) to bring in the power-enrichment system at a lower air flow rate due to decreased manifold vacuum. To reduce the impact to changes in altitude on engine emissions of CO and HC, modern carburetors are altitude compensated. A number of methods can be used to compensate for changes in ambient pressure with altitude:

1. *Venturi bypass method.* To keep the air volume flow rate through the venturi equal to what it was at sea-level atmospheric pressure (calibration condition), a bypass circuit around the venturi for the additional volume flow is provided.
2. *Auxiliary jet method.* An auxiliary fuel metering orifice with a pressure-controlled tapered metering rod connects the fuel bowl to the main well in parallel with the main metering orifice.
3. *Fuel bowl back-suction method.* As altitude increases, an aneroid bellows moves a tapered rod from an orifice near the venturi throat, admitting to the bowl an increasing amount of the vacuum signal developed at the throat.
4. *Compensated air-bleed method.* The orifices in the bleed circuits to each carburetor system are fitted with tapered metering pins actuated by a single aneroid bellows.[8]

TRANSIENT EFFECTS. The pulsating and transient nature of the flow through a carburetor during actual engine operation is illustrated by the data shown in Fig. 7-8.[2] The changes in pressure with time in the intake manifold and at the boost venturi throat of a standard two-barrel carburetor installed on a production V-8 engine are shown as the throttle is opened from light load (22°) to wide-open throttle at 1000 rev/min. Note the rapid increase in boost venturi suction as the throttle is suddenly opened. This results from the sudden large increase in the air flow rate and corresponding increase in air velocity within the boost venturi. Note also that the pressure fluctuations decay rapidly, and within a few engine revolutions have stabilized at the periodic values associated with the new throttle angle. At wide-open throttle, the pulsating nature of the flow as each

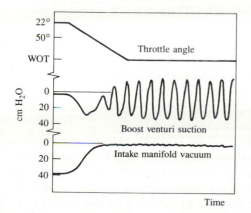

FIGURE 7-8
Throttle angle, boost venturi suction, and intake manifold vacuum variation with time as throttle is opened from light load (22°) to wide-open throttle at 1000 rev/min. Standard two-barrel carburetor and production V-8 engine.[2]

cylinder draws in its charge is evident. The pressure drop across the main meter-ing jet also fluctuates. The pulsations in the venturi air flow (and hence fuel flow) due to the filling of each cylinder in turn are negligibly small at small throttle angles and increase to a maximum at wide-open throttle. At small throttle open-ings, the choked flow at the throttle plate prevents the manifold pressure fluctua-tions from propagating upstream into the venturi. The effective time-averaged boost venturi suction is greater for the pulsating flow case than for the steady flow case. If the ratio of the effective metering signal for a pulse cycle to that for steady air flow at the same average mass flow is denoted as $1 + \Omega$, where Ω is the pulsation factor, then Ω is related to the amplitude and frequency of pressure waves within the intake manifold as well as the damping effect of the throttle plate. An empirical equation for Ω is

$$\Omega = \frac{\text{constant} \times (1 - M)p_m n_R}{N n_{c/b}} \tag{7.12}$$

where M is the throttle plate Mach number, p_m the manifold pressure, n_R the number of revolutions per power stroke, N the crank speed, and $n_{c/b}$ the number of cylinders per barrel. The value of the constant depends on carburetor and engine geometry. For p_m in kilonewtons per square meter and N in revolutions per minute a typical value for the constant is 7.3.[2] Thus, at wide-open throttle at 1500 rev/min, Ω has a value of about 0.2. The transient behavior of the air and fuel flows in the manifold are discussed more fully in Sec. 7.6.

7.3   FUEL-INJECTION SYSTEMS

7.3.1   Multipoint Port Injection

The fuel-injection systems for conventional spark-ignition engines inject the fuel into the engine intake system. This section reviews systems where the fuel is injected into the intake port of each engine cylinder. Thus these systems require one injector per cylinder (plus, in some systems, one or more injectors to supple-ment the fuel flow during starting and warm-up). There are both mechanical injection systems and electronically controlled injection systems. The advantages of port fuel injection are increased power and torque through improved volu-metric efficiency and more uniform fuel distribution, more rapid engine response to changes in throttle position, and more precise control of the equivalence ratio during cold-start and engine warm-up. Fuel injection allows the amount of fuel injected per cycle, for each cylinder, to be varied in response to inputs derived from sensors which define actual engine operating conditions. Two basic approaches have been developed; the major difference between the two is the method used to determine the air flow rate.

Figure 7-9 shows a schematic of a *speed-density* system, where engine speed and manifold pressure and air temperature are used to calculate the engine air flow. The electrically driven fuel pump delivers the fuel through a filter to the fuel line. A pressure regulator maintains the pressure in the line at a fixed value (e.g.,

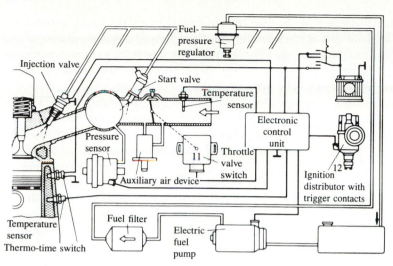

FIGURE 7-9
Speed-density electronic multipoint port fuel-injection system: Bosch D-Jetronic System.[9] (*Courtesy Robert Bosch GmbH and SAE.*)

270 kN/m², 39 lb/in², usually relative to manifold pressure to maintain a constant fuel pressure drop across the injectors). Branch lines lead to each injector; the excess fuel returns to the tank via a second line. The inducted air flows through the air filter, past the throttle plate to the intake manifold. Separate runners and branches lead to each inlet port and engine cylinder. An electromagnetically actuated fuel-injection valve (see Fig. 7-10) is located either in the intake manifold tube or the intake port of each cylinder. The major components of the injector are the valve housing, the injector spindle, the magnetic plunger to which the spindle is connected, the helical spring, and the solenoid coil. When the solenoid is not excited, the solenoid plunger of the magnetic circuit is forced, with its seal, against the valve seat by the helical spring and closes the fuel passage. When the solenoid coil is excited, the plunger is attracted and lifts the spindle about

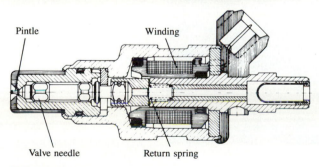

FIGURE 7-10
Cross section of fuel injector.[10]

0.15 mm so that fuel can flow through the calibrated annular passage around the valve stem. The front end of the injector spindle is shaped as an atomizing pintle with a ground top to atomize the injected fuel. The relatively narrow spray cone of the injector, shown in the photo in Fig. 7-11, minimizes the intake manifold wall wetting with liquid fuel. The mass of fuel injected per injection is controlled by varying the duration of the current pulse that excites the solenoid coil. Typical injection times for automobile applications range from about 1.5 to 10 ms.[11]

The appropriate coil excitation pulse duration or width is set by the electronic control unit (ECU). In the speed-density system, the primary inputs to the ECU are the outputs from the manifold pressure sensor, the engine speed sensor (usually integral with the distributor), and the temperature sensors installed in the intake manifold to monitor air temperature and engine block to monitor the water-jacket temperature—the latter being used to indicate fuel-enrichment requirements during cold-start and warm-up. For warm-engine operation, the mass of air per cylinder per cycle m_a is given by

$$m_a = \eta_v(N)\rho_a(T_i, p_i)V_d = \frac{\eta_v V_d p_i}{R_a T_i} \tag{7.13}$$

where η_v is the volumetric efficiency, N is engine speed, ρ_a is the inlet air density, and V_d is the displaced volume per cylinder. The electronic control unit forms the pulse which excites the injector solenoids. The pulse width depends primarily on the manifold pressure; it also depends on the variation in volumetric efficiency η_v with speed N and variations in air density due to variations in air temperature. The control unit also initiates mixture enrichment during cold-engine operation and during accelerations that are detected by the throttle sensor.

FIGURE 7-11
Short time-exposure photograph of liquid fuel spray from Bosch-type injector. (*Courtesy Robert Bosch GmbH.*)

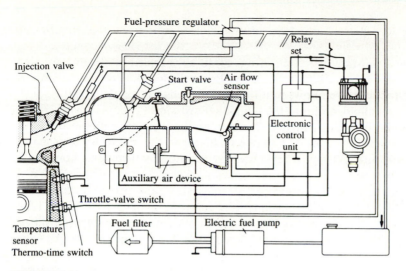

FIGURE 7-12
Electronic multipoint port fuel-injection system with air-flow meter: Bosch L-Jetronic system.[9] (*Courtesy Robert Bosch GmbH and SAE.*)

Figure 7-12 shows an alternative EFI system (the Bosch L-Jetronic) which uses an air-flow meter to measure air flow directly. The air-flow meter is placed upstream of the throttle. The meter shown measures the force exerted on a plate as it is displaced by the flowing air stream; it provides a voltage proportional to the air flow rate. An alternative air-flow measuring approach is to use a hot-wire air mass flow meter.[10] The advantages of direct air-flow measurement are:[12] (1) automatic compensation for tolerances, combustion chamber deposit buildup, wear and changes in valve adjustments; (2) the dependence of volumetric efficiency on speed and exhaust backpressure is automatically accounted for; (3) less acceleration enrichment is required because the air-flow signal precedes the filling of the cylinders; (4) improved idling stability; and (5) lack of sensitivity of the system to EGR since only the fresh air flow is measured.

The mass of air inducted per cycle to each cylinder, m_a, varies as

$$m_a \propto \frac{\dot{m}_a}{N} \tag{7.14}$$

Thus the primary signals for the electronic control unit are air flow and engine speed. The pulse width is inversely proportional to speed and directly proportional to air flow. The engine block temperature sensor, starter switch, and throttle valve switch provide input signals for the necessary adjustments for cold-start, warm-up, idling, and wide-open throttle enrichment.

For cold-start enrichment, one (or more) cold-start injector valve is used to inject additional fuel into the intake manifold (see Figs. 7-9 and 7-12). Since short opening and closing times are not important, this valve can be designed to provide extremely fine atomization of the fuel to minimize the enrichment required.

Mechanical, air-flow-based metering, continuous injection systems are also used. Figure 7-13 shows a schematic of the Bosch K-Jetronic system.[9, 10] Air is drawn through the air filter, flows past the air-flow sensor, past the throttle valve, into the intake manifold, and into the individual cylinders. The fuel is sucked out of the tank by a roller-cell pump and fed through the fuel accumulator and filter to the fuel distributor. A primary pressure regulator in the fuel distributor maintains the fuel pressure constant. Excess fuel not required by the engine flows back to the tank. The mixture-control unit consists of the air-flow sensor and fuel distributor. It is the most important part of the system, and provides the desired metering of fuel to the individual cylinders by controlling the cross section of the metering slits in the fuel distributor. Downstream of each of these metering slits is a differential pressure valve which for different flow rates maintains the pressure drop at the slits constant.

Fuel-injection systems offer several options regarding the timing and location of each injection relative to the intake event.[10] The K-Jetronic mechanical injection system injects fuel continuously in front of the intake valves with the spray directed toward the valves. Thus about three-quarters of the fuel required for any engine cycle is stored temporarily in front of the intake valve, and one-quarter enters the cylinder directly during the intake process.

With electronically controlled injection systems, the fuel is injected intermittently toward the intake valves. The fuel-injection pulse width to provide the appropriate mass of fuel for each cylinder air charge varies from about 1.5 to 10 ms over the engine load and speed range. In crank angle degrees this varies

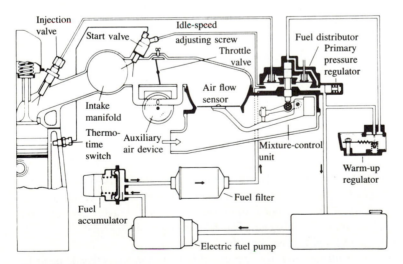

FIGURE 7-13
Mechanical multipoint port fuel-injection system: Bosch K-Jetronic system.[9] (*Courtesy Robert Bosch GmbH and SAE.*)

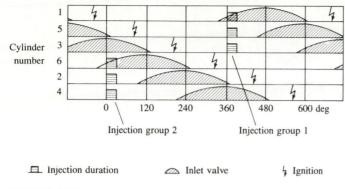

Injection group 2   Injection group 1

⊑ Injection duration   ◿ Inlet valve   ↳ Ignition

FIGURE 7-14
Injection pulse diagram for D-Jetronic system in six-cylinder engine.[10]

from about 10° at light load and low speed to about 300° at maximum speed and load. Thus the pulse width varies from being much less than to greater than the duration of the intake stroke. To reduce the complexity of the electronic control unit, groups of injectors are often operated simultaneously. In the Bosch L-Jetronic system, all injectors are operated simultaneously. To achieve adequate mixture uniformity, given the short pulse width relative to the intake process over much of the engine load-speed range, fuel is injected twice per cycle; each injection contributes half the fuel quantity required by each cylinder per cycle. (This approach is called simultaneous double-firing.) In the speed-density system, the injectors are usually divided into groups, each group being pulsed simultaneously. For example, for a six-cylinder engine, two groups of three injectors may be used. Injection for each group is timed to occur while the inlet valves are closed or just starting to open, as shown in Fig. 7-14. The other group of injectors inject one crank revolution later. Sequential injection timing, where the phasing of *each* injection pulse relative to its intake valve lift profile is the same, is another option. Engine performance and emissions do change as the timing of the start of injection relative to inlet valve opening is varied. Injection with valve lift at its maximum, or decreasing, is least desirable.[10]

7.3.2 Single-Point Throttle-Body Injection

Single-point fuel-injection systems, where one or two electronically controlled injectors meter the fuel into the air flow directly above the throttle body, are also used. They provide straightforward electronic control of fuel metering at lower cost than multipoint port injection systems. However, as with carburetor systems, the problems associated with slower transport of fuel than the air from upstream of the throttle plate to the cylinder must now be overcome (see Sec. 7.6). Figure 7-15 shows a cutaway of one such system.[13] Two injectors, each in a separate air-flow passage with its own throttle plate, meter the fuel in response to calibrations of air flow based on intake manifold pressure, air temperature, and

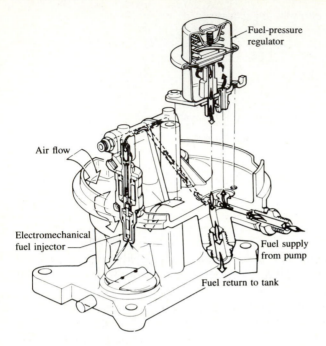

FIGURE 7-15
Cutaway drawing of a two-injector throttle-body electronic fuel-injection system.[13]

engine speed using the speed-density EFI logic described in the previous section. Injectors are fired alternatively or simultaneously, depending on load and speed and control logic used. Under alternative firing, each injection pulse corresponds to one cylinder filling. Smoothing of the fuel-injection pulses over time is achieved by proper placement of the fuel injector assembly above the throttle bore and plate. The walls and plate accumulate liquid fuel which flows in a sheet toward the annular throttle opening (see Sec. 7.6.3). The high air velocity created

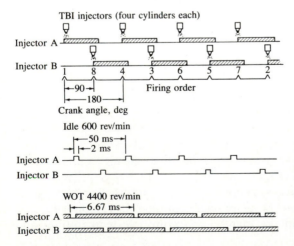

FIGURE 7-16
Injector fuel delivery schedule for two-injector throttle-body injection system for eight-cylinder engine with dual plane intake manifold. Each injection nozzle feeds one plane of the manifold and its four cylinders.[14]

by the pressure drop across the throttle shears and atomizes the liquid sheet. Vigorous mixing of fuel and air then occurs, especially at part throttle, and provides good mixture uniformity and distribution between cylinders. Injector fuel delivery scheduling is illustrated in Fig. 7-16 for an eight-cylinder engine for a throttle-body fuel-injection system.[14]

7.4 FEEDBACK SYSTEMS

It is possible to reduce engine emissions of the three pollutants—unburned hydrocarbons, carbon monoxide, and oxides of nitrogen—with a single catalyst in the exhaust system if the engine is operated very close to the stoichiometric air/fuel ratio. Such systems (called three-way catalyst systems) are described in more detail in Sec. 11.6.2. The engine operating air/fuel ratio is maintained close to stoichiometric through the use of a sensor in the exhaust system, which provides a voltage signal dependent on the oxygen concentration in the exhaust gas stream. This signal is the input to a feedback system which controls the fuel feed to the intake.

The sensor [called an oxygen sensor or lambda sensor—λ being the symbol used for the relative air/fuel ratio, Eq. (3.9)] is an oxygen concentration cell with a solid electrolyte through which the current is carried by oxygen ions. The electrolyte is yttria (Y_2O_3)-stabilized zirconia (ZrO_2) ceramic which separates two gas chambers (the exhaust manifold and the atmosphere) which have different oxygen partial pressures. The cell can be represented as a series of interfaces as follows:

Exhaust	Metal	Ceramic	Metal	Air
p'_{O_2}	M_e	ZrO_2, Y_2O_3	M_e	p''_{O_2}

p''_{O_2} is the oxygen partial pressure of the air (≈ 20 kN/m^2) and p'_{O_2} is the equilibrium oxygen partial pressure in the exhaust gases. An electrochemical reaction takes place at the metal electrodes:

$$O_2 + 4M_e \rightleftharpoons 2O^{2-}$$

and the oxygen ions transport the current across the cell. The open-circuit output voltage of the cell V_o can be related to the oxygen partial pressures p'_{O_2} and p''_{O_2} through the Nernst equation:

$$V_o = \frac{RT}{4F} \ln \left(\frac{p''_{O_2}}{p'_{O_2}} \right) \tag{7.15}$$

where F is the Faraday constant. Equilibrium is established in the exhaust gases by the catalytic activity of the platinum metal electrodes. The oxygen partial pressure in equilibrated exhaust gases decreases by many orders of magnitude as the equivalence ratio changes from 0.99 to 1.01, as shown in Fig. 7-17a. Thus the sensor output voltage increases rapidly in this transition from a lean to a rich

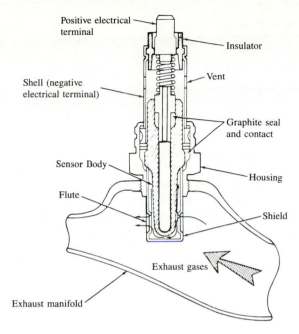

Positive electrical terminal

Insulator

Shell (negative electrical terminal)

Vent

Graphite seal and contact

Sensor Body

Housing

Flute

Shield

Exhaust gases

Exhaust manifold

FIGURE 7-18
Cross-section drawing of exhaust oxygen sensor.[16]

output varies the fuel quantity linearly in the opposite direction to the sign of the comparator signal. There is a time lag τ_L in the loop composed of the transport time of fuel-air mixture from the point of fuel admission in the intake system to the sensor location in the exhaust, and the sensor and control system time delay. Because of this time lag, the controller continues to influence the fuel flow rate in the same direction, although the reference point $\phi = 1.0$ has been passed, as shown in Fig. 7-19b. Thus, oscillations in the equivalence ratio delivered to the engine exist even under steady-state conditions of closed-loop control. This behavior of the control system is called the *limit cycle*. The frequency f of this limit cycle is given by

$$f_{LC} = \frac{1}{4\tau_L} \tag{7.16}$$

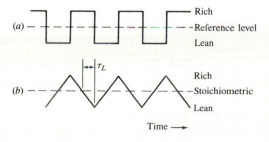

(a)

Rich

Reference level

Lean

τ_L

(b)

Rich

Stoichiometric

Lean

Time

FIGURE 7-19
Operation of electronic control unit for closed-loop feedback: (a) sensor signal compared with reference level; (b) controller output voltage—the integrated comparator output.[12]

and the change in equivalence ratio peak-to-peak is

$$\Delta\phi = 2K\tau_L \tag{7.17}$$

where K is the integrator gain (in equivalence ratio units per unit time).

Depending on the details of the three-way catalyst used for cleanup of all three pollutants (CO, HC, and NO_x) in the exhaust, the optimum average equivalence ratio may not be precisely the stoichiometric value. Furthermore, the reference voltage for maximum sensor durability may not correspond exactly to the stoichiometric point or the desired catalyst mean operating point. While a small shift ($\sim \pm 1$ percent) in operating point from the stoichiometric can be obtained by varying the reference voltage level, larger shifts are obtained by modifying the control loop to provide a steady-state bias. One method of providing a bias—asymmetrical gain rate biasing[17]—uses two separate integrator circuits with different gain rates K^+ and K^- to integrate the comparator output, depending on whether the comparator output is positive (rich) or negative (lean). An alternative biasing technique incorporates an additional delay time τ_D so that the controller output continues decreasing (or increasing) even though the sensor signal has switched from the high to the low level (or vice versa). By introducing this additional delay only on the negative slope of the sensor signal, a net lean bias is produced. Introducing the additional delay on the positive slope of the sensor signal produces a net rich bias.[12]

Note that the sensor only operates at elevated temperatures. During engine start-up and warm-up, the feedback system does not operate and conventional controls are required to obtain the appropriate fuel-air mixture for satisfactory engine operation.

7.5 FLOW PAST THROTTLE PLATE

Except at or close to wide-open throttle, the throttle provides the minimum flow area in the entire intake system. Under typical road-load conditions, more than 90 percent of the total pressure loss occurs across the throttle plate. The minimum-to-maximum flow area ratio is large—typically of order 100. Throttle

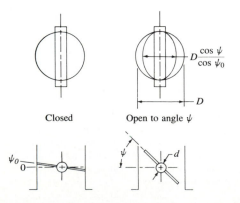

Closed   Open to angle ψ

FIGURE 7-20
Throttle plate geometry.[2]

(a) 20° throttle plate angle

(b) 30° throttle plate angle

(c) 45° throttle plate angle

(d) 60° throttle plate angle

FIGURE 7-21
Photographs of flow in two-dimensional hydraulic analog of carburetor venturi, throttle plate, and manifold plenum floor at different throttle plate angles.[18]

plate geometry and parameters are illustrated in Fig. 7-20. A throttle plate of conventional design such as Fig. 7-20 creates a three-dimensional flow field. At part-throttle operating conditions the throttle plate angle is in the 20 to 45° range and jets issue from the "crescent moon"-shaped open areas on either side of the throttle plate. The jets are primarily two dimensional. Figure 7-21 shows photographs taken of a two-dimensional hydraulic analog of a typical carburetor venturi and throttle plate in steady flow at different throttle angles. The path lines traced by the particles in the flow indicate the relative magnitude of the flow velocity.[18] The flow accelerates through the carburetor venturi (separation occurs at the corners of the entrance section); it then divides on either side of the throttle plate. There is a stagnation point on the upstream side of the throttle plate. The wake of the throttle plate contains two vortices which rotate in opposite directions. The jets on either side of the wake (at part throttle) are at or near sonic velocity. There is little or no mixing between the two jets. Thus, if maldistribution of the fuel-air mixture occurs above the throttle plate, it is not corrected immediately below the throttle plate.

In analyzing the flow through the throttle plate, the following factors should be considered:[2, 19, 20]

1. The throttle plate shaft is usually of sufficient size to affect the throttle open area.
2. To prevent binding in the throttle bore, the throttle plate is usually completely closed at some nonzero angle (5, 10, or 15°).
3. The discharge coefficient of the throttle plate is less than that of a smooth converging-diverging nozzle, and varies with throttle angle, pressure ratio, and throttle plate Reynolds number.
4. Due to the manufacturing tolerances involved, there is usually some minimum leakage area even when the throttle plate is closed against the throttle bore. This leakage area can be significant at small throttle openings.
5. The measured pressure drop across the throttle depends (± 10 percent) on the circumferential location of the downstream pressure tap.
6. The pressure loss across the throttle plate under the actual flow conditions (which are unsteady even when the engine speed and load are constant, see Fig. 7-8) may be less than under steady flow conditions.

The throttle plate open area A_{th}, as a function of angle ψ for the geometry in Fig. 7-20, is given by[2]

$$\frac{4A_{th}}{\pi D^2} = \left(1 - \frac{\cos \psi}{\cos \psi_0}\right) + \frac{2}{\pi}\left[\frac{a}{\cos \psi}(\cos^2 \psi - a^2 \cos^2 \psi_0)^{1/2}\right.$$
$$\left. - \frac{\cos \psi}{\cos \psi_0}\sin^{-1}\left(\frac{a \cos \psi_0}{\cos \psi}\right) - a(1 - a^2)^{1/2} + \sin^{-1} a\right] \quad (7.18)$$

where $a = d/D$, d is the throttle shaft diameter, D is the throttle bore diameter, and ψ_0 is the throttle plate angle when tightly closed against the throttle bore. When $\psi = \cos^{-1}(a \cos \psi_0)$, the throttle open area reaches its maximum value ($\approx \pi D^2/4 - dD$). The throttle plate discharge coefficient (which varies with A_{th}), and minimum leakage area, must be determined experimentally.

The mass flow rate through the throttle valve can be calculated from standard orifice equations for compressible fluid flow [see App. C, Eqs. (C-8) and

(C-9)]. For pressure ratios across the throttle less than the critical value ($p_T/p_0 = 0.528$), the mass flow rate is given by

$$\dot{m}_{th} = \frac{C_D A_{th} p_0}{\sqrt{R T_0}} \left(\frac{p_T}{p_0}\right)^{1/\gamma} \left\{\frac{2\gamma}{\gamma - 1}\left[1 - \left(\frac{p_T}{p_0}\right)^{(\gamma-1)/\gamma}\right]\right\}^{1/2} \tag{7.19}$$

where A_{th} is the throttle plate open area [Eq. (7.18)], p_0 and T_0 are the upstream pressure and temperature, p_T is the pressure downstream of the throttle plate (assumed equal to the pressure at the minimum area: i.e., no pressure recovery occurs), and C_D is the discharge coefficient (determined experimentally). For pressure ratios greater than the critical ratio, when the flow at the throttle plate is choked,

$$\dot{m}_{th} = \frac{C_D A_{th} p_0}{\sqrt{R T_0}} \gamma^{1/2} \left(\frac{2}{\gamma + 1}\right)^{(\gamma+1)/2(\gamma-1)} \tag{7.20}$$

The relation between air flow rate, throttle angle, intake manifold pressure, and engine speed for a two-barrel carburetor and a 4.7-dm³ (288-in³) displacement eight-cylinder production engine is shown in Fig. 7-22. While the lines are predictions from a quasi-steady computer simulation, the agreement with data is excellent. The figure shows that for an intake manifold pressure below the critical

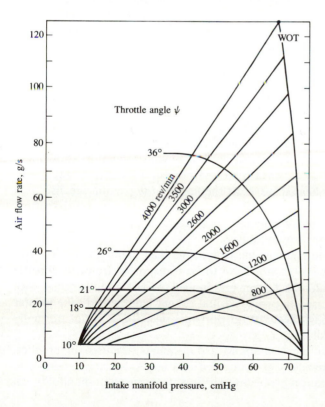

FIGURE 7-22
Variation in air flow rate past a throttle, with inlet manifold pressure, throttle angle, and engine speed. 4.7-dm³ displacement eight-cylinder engine.[2]

value $(0.528 \times p_{atm} = 53.5 \text{ kN/m}^2 = 40.1 \text{ cmHg})$ the air flow rate at a given throttle position is independent of manifold pressure and engine speed because the flow at the throttle plate is choked.[2]

7.6 FLOW IN INTAKE MANIFOLDS

7.6.1 Design Requirements

The details of the air and fuel flow in intake manifolds are extremely complex. The combination of pulsating flow into each cylinder, different geometry flow paths from the plenum beneath the throttle through each runner and branch of the manifold to each inlet port, liquid fuel atomization, vaporization and transport phenomena, and the mixing of EGR with the fresh mixture under steady-state engine operating conditions are difficult enough areas to untangle. During engine transients, when the throttle position is changed, the fact that the processes which govern the air and the fuel flow to the cylinder are substantially different introduces additional problems. This section reviews our current understanding of these phenomena.

Intake manifolds consist typically of a plenum, to the inlet of which bolts the throttle body, with individual runners feeding branches which lead to each cylinder (or the plenum can feed the branches directly). Important design criteria are: low air flow resistance; good distribution of air and fuel between cylinders; runner and branch lengths that take advantage of ram and tuning effects; sufficient (but not excessive) heating to ensure adequate fuel vaporization with carbureted or throttle-body injected engines. Some compromises are necessary; e.g., runner and branch sizes must be large enough to permit adequate flow without allowing the air velocity to become too low to transport the fuel droplets. Some of these design choices are illustrated in Fig. 7-23 which shows an inlet manifold and carburetor arrangement for a modern four-cylinder 1.8-dm³ engine. In this design the four branches that link the plenum beneath the carburetor and throttle with the inlet ports are similar in length and geometry, to provide closely comparable flow paths. This manifold is heated by engine coolant as shown and uses an electrically heated grid beneath the carburetor to promote rapid fuel evaporation.[21] Exhaust gas heated stoves at the floor of the plenum are also used in some intake manifolds to achieve adequate fuel vaporization and distribution. Note that with EGR, the intake manifold may contain passages to bring the exhaust gas to the plenum or throttle body.

With port fuel-injection systems, the task of the inlet manifold is to control the air (and EGR) flow. Fuel does not have to be transported from the throttle body through the entire manifold. Larger and longer runners and branches, with larger angle bends, can be used to provide equal runner lengths and take greater advantage of ram and tuning effects. With port fuel injection it is not normally necessary to heat the manifold.

A large number of different manifold arrangements are used in practice. Different cylinder arrangements (e.g., four, V-six, in-line-six, etc.) provide quite different air and fuel distribution problems. Air-flow phenomena in manifolds can

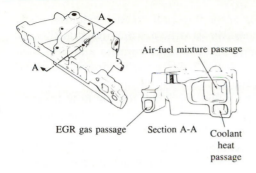

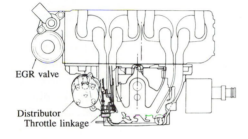

FIGURE 7-23
Inlet manifold for four-cylinder 1.8-dm³ displacement spark-ignition engine.[21]

be considered as unaffected by the fuel flow. The reverse is definitely not the case: the fuel flow—liquid and vapor—depends strongly on the air flow. These two topics will therefore be reviewed in sequence.

7.6.2 Air-Flow Phenomena

The air flow out of the manifold occurs in a series of pulses, one going to each cylinder. Each pulse is approximately sinusoidal in shape. For four- and eight-cylinder engines, these flow pulses sequence such that the outflow is essentially zero between pulses. For six-cylinder arrangements the pulses will overlap. When the engine is throttled, backflow from the cylinder into the intake manifold occurs during the early part of the intake process until the cylinder pressure falls below the manifold pressure. Backflow can also occur early in the compression stroke before the inlet valve closes, due to rising cylinder pressure. The flow at the throttle will fluctuate as a consequence of the pulsed flow out of the manifold into the cylinders. At high intake vacuum, the flow will be continuously inward at the throttle and flow pulsations will be small. When the outflow to the cylinder which is undergoing its intake stroke is greater than the flow through the throttle, the cylinder will draw mixture from the rest of the intake manifold. During the portion of the intake stroke when the flow into the cylinder is lower than the flow through the throttle, mixture will flow back into the rest of the manifold. At wide-open throttle when the flow restriction at the throttle is a minimum, flow pulsations at the throttle location will be much more pronounced.[19]

The air flows to each cylinder of a multicylinder engine, even under steady operating conditions, are not identical. This is due to differences in runner and branch length and other geometric details of the flow path to each cylinder. Also, as each cylinder's intake flow commences, air is drawn from the branch and runner leading to the cylinder, the plenum, and the other runners and branches feeding the plenum, as well as past the throttle. This "drawing down" of other parts of the intake manifold depends on the arrangement of the plenum, runners, and branches, and the firing order of the cylinders. Thus the air flow to each individual cylinder is affected by the details of its own branch, how that branch connects to the rest of the intake manifold, and the cylinder firing order.[22] The differences between the air flows to individual cylinders have been measured. Variations in the average air mass flow rate to each individual cylinder of up to about 5 percent above and below the average are quite common. Larger peak-to-peak variations (± 15 percent) have been measured. The extent of each cylinder's difference from the average flow varied significantly as engine speed and load were varied.[23, 24]

Typical quantities that characterize manifold air flow are given in Table 7.1 for a four- and an eight-cylinder spark-ignition engine. The volume of mixture pulled into each cylinder per cycle is about the same as the volume of one direct flow path between the throttle plate and inlet valve. Thus, one stroke loads the manifold, the next one pulls the charge into the cylinder.

An additional phenomenon becomes important when engine load is changed by opening or closing the throttle: the mass of air in the induction system volume takes a finite time to adjust to the new engine operating conditions. For example, as the throttle is opened the air flow into the manifold increases as the throttle open area increases. However, due to the finite volume of

TABLE 7.1
Parameters that characterize manifold air flow

Engine geometry	I-4†	V-8‡
Typical flow-path distance between throttle bore and intake valve, cm	33	30
Average intake-passage flow area, cm^2	9.4	16
Volume of one direct flow path from throttle bore to intake valve, cm^3	300	500

Range of speeds, etc.	Maximum	Minimum
Crankshaft, rev/min	5000	650
Peak air velocity in manifold branch, m/s	130†, 100‡	15
Peak Reynolds number in manifold branch	4×10^5	5×10^4
Duration of individual cylinder intake process, ms	6	46

† 1.8-dm^3 four-cylinder in-line SI engine.[21]
‡ 5.6-dm^3 V eight-cylinder SI engine.[25]

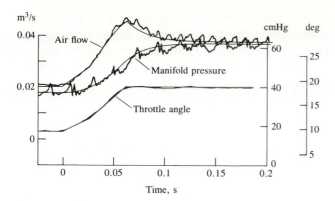

FIGURE 7-24
Throttle angle, intake manifold pressure, and air flow rate past the throttle versus time for 10° part-load throttle opening. 5-dm³ V-8 engine.[25]

the manifold, the pressure level in the manifold increases more slowly than would be the case if steady-state conditions prevailed at each throttle position. Thus, the pressure difference across the throttle is larger than it would be under steady flow conditions and the throttle air flow overshoots its steady-state value. The air flow into each cylinder depends on the pressure in the manifold, so this lags the throttle air flow. This transient air-flow phenomenon affects fuel metering. For throttle-body injection or a carburetor, fuel flow should be related to throttle air flow. For port fuel injection, fuel flow should be related to cylinder air flow. Actual results for the air flow rate and manifold pressure in response to an opening of the throttle (increase in throttle angle) are shown in Fig. 7-24. The overshoot in throttle air flow and lag in manifold pressure as the throttle angle is increased are evident. Opposite effects will occur for a decrease in throttle angle.

AIR-FLOW MODELS. Several models of the flow in an intake manifold have been proposed.[26, 27] One simple manifold model that describes many of the above phenomena is the *plenum* or *filling and emptying* model. It is based on the assumption that at any given time the manifold pressure is uniform. The continuity equation for air flow into and out of the intake manifold is

$$\frac{dm_{a, m}}{dt} = \dot{m}_{a, \text{th}} - \sum \dot{m}_{a, \text{cyl}} \tag{7.21}$$

where $m_{a, m}$ is the mass of air in the manifold, and $\dot{m}_{a, \text{th}}$ and $\dot{m}_{a, \text{cyl}}$ are the air mass flow rates past the throttle and into each cylinder, respectively. The flow rate past the throttle is given by Eq. (7.19) or (7.20). For manifold pressures sufficiently low to choke the flow past the throttle plate, the flow rate is independent of manifold pressure. The mass flow rate to the engine cylinders can be modeled at several levels of accuracy. The air flow through the valve to each cylinder can be computed from the valve area, discharge coefficient, and pressure difference across the valve; or a sine wave function can be assumed. In the general case, Eq. (7.21) must be combined with the first law for an open system (see Sec. 14.2.2). For calculating the manifold response to a change in load or throttle setting, simplifying assumptions can be made. A quasi-steady approximation for the cylinder air flow:

$$\sum \dot{m}_{a,\,\text{cyl}} = \frac{\eta_v \rho_{a,\,m} V_d N}{2}$$

is usually adequate, and the air temperature can be assumed constant.[25] Then, using the ideal gas law for the manifold, $p_m V_m = m_{a,\,m} R_a T_m$, Eq. (7.21) can be written as

$$\frac{dp_m}{dt} + \frac{\eta_v V_d N}{2V_m} p_m = \dot{m}_{a,\,\text{th}} \frac{R T_m}{V_m} \tag{7.22}$$

Both η_v and $\dot{m}_{a,\,\text{th}}$ have some dependence on p_m [e.g., see Eq. (6.2)]. In the absence of this weak dependence, Eq. (7.22) would be a first-order equation for p_m with a time constant $\tau = 2V_m/(\eta_v V_d N) \approx V_m/\dot{V}_{\text{cyl}}$, which is 2 to 4 times the intake stroke duration. The smooth curves in Fig. 7-24 are predictions made with Eq. (7.22) and show good agreement with the data. The plenum model is useful for investigating manifold pressure variations that result from load changes. It provides no information concerning pressure variations associated with momentum effects.

Helmholtz resonator models for the intake system have been proposed. This type of model can predict the resonant frequencies of the combined intake and engine cylinder system, and hence the engine speeds at which increases in air flow due to intake tuning occur. It does not predict the magnitude of the increase in volumetric efficiency. The Helmholtz resonator theory analyzes what happens during one inlet stroke, as the air in the manifold pipe is acted on by a forcing function produced by the piston motion. As the piston moves downward during the intake stroke, a reduced pressure occurs at the inlet valve relative to the pressure at the open end of the inlet pipe. A rarefraction wave travels down the intake pipe to the open end and is reflected as a compression wave. A positive tuning effect occurs when the compression wave arrives at the inlet valve as the valve is closing.[27] A single-cylinder engine modeled as a Helmholtz resonator is shown in Fig. 7-25a. The effective resonator volume V_{eff} is chosen to be one-half of the displaced volume plus the clearance volume; the piston velocity is then close to its maximum and the pressure in the inlet system close to its minimum. The tuning peak occurs when the natural frequency of the cylinder volume coupled to the pipe is about twice the piston frequency. For a single-cylinder, fed by a single pipe open to the atmosphere, the resonant tuning speed N_t is given by

$$N_t(\text{rev/min}) = \frac{955}{K} a\left(\frac{A}{lV_{\text{eff}}}\right)^{1/2} \tag{7.23}$$

where a is the sound speed (m/s), A the effective cross-sectional area of the inlet system (cm²), l the effective length of the inlet system (cm), K a constant equal to about 2 for most engines, and $V_{\text{eff}} = V_d(r_c + 1)/[2(r_c - 1)]$ (cm³).[28]

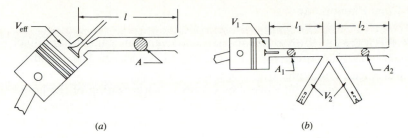

FIGURE 7-25
Helmholtz resonator models for (a) single-cylinder engine and (b) multicylinder engine.[27]

The Helmholtz theory for multicylinder engines treats the pipes of cylinders not undergoing induction as an additional volume. The two pipes, (l_1, A_1) and (l_2, A_2), and two volumes, V_1 and V_2, in Fig. 7-25b form a vibrating system with two degrees of freedom and two resonant frequencies. The following equation, based on an electrical analog (in which capacitors represent volumes and inductors pipes), gives the two frequencies at which the manifold shown in Fig. 7-25b would be tuned:[28]

$$f_{\mp} = \frac{1}{2\pi} \left\{ \frac{(\alpha\beta + \alpha + 1) \mp [(\alpha\beta + \alpha + 1)^2 - 4\alpha\beta]^{1/2}}{2\alpha\beta L_1 C_1} \right\}^{1/2} \tag{7.24}$$

where $\alpha = L_2/L_1$, $\beta = C_2/C_1$, $C_1 = V_1$, $C_2 = V_2$, $L_1 = (l/A)_1$, $L_2 = (l/A)_2$, and $V_{\text{eff}} = V_1$. The Helmholtz theory predicts the engine speeds at which positive tuning resonances occur with reasonable accuracy.[27]

The dynamics of the flow in multicylinder intake (and exhaust) systems can be modeled most completely using one-dimensional unsteady compressible flow equations. The standard method of solution of the governing equations has been the method of characteristics (see Benson[29]). Recently, finite difference techniques which are more efficient have been used.[30] The assumptions usually made in this type of analysis are:

1. The intake (or exhaust) system can be modeled as a combination of pipes, junctions, and plenums.
2. Flow in the pipes is one dimensional and no axial heat conduction occurs.
3. States in the engine cylinders and plenums are uniform in space.
4. Boundary conditions are considered quasi steady.
5. Coefficients of discharge, heat transfer, pipe friction, and bend losses for steady flow are valid for unsteady flow.
6. The gases can be modeled as ideal gases.

This approach to intake and exhaust flow analysis is discussed more fully in Sec. 14.3.4.

7.6.3   Fuel-Flow Phenomena

TRANSPORT PROCESSES. With conventional spark-ignition engine liquid-fuel metering systems, the fuel enters the air stream as a liquid jet. The liquid jet atomizes into droplets. These mix with the air and also deposit on the walls of the intake system components. The droplets vaporize; vaporization of the liquid fuel on the walls occurs. The flow of liquid fuel along the walls can be significant. The transport of fuel as vapor, droplets, and liquid streams or films can all be important. The fuel transport processes in the intake system are obviously extremely complex.

The details of the fuel transport process are different for multipoint fuel-injection systems than for carburetor and throttle-body injection systems. For the latter systems, fuel must be transported past the throttle plate and through the complete intake manifold. For the former systems, the liquid fuel is injected in the inlet port, toward the back of the intake valve. For all these practical fuel meter-ing systems, the quality of the mixture entering the engine is imperfect. The fuel, air, recycled exhaust, mixture is not homogeneous; the fuel may not be fully vaporized as it enters the engine. The charge going to each cylinder is not usually uniform in fuel/air ratio throughout its volume, and the distribution of fuel between the different engine cylinders is not exactly equal. During engine tran-sients, when engine fuel and air requirements and manifold conditions change, it is obvious that the above fuel transport processes will not all vary with time in the same way. Thus, in addition to the transient non-quasi-steady air-flow phe-nomena described above, there are transient fuel-flow phenomena. These have to be compensated for in the fuel metering strategy.

Since gasoline, the standard spark-ignition engine fuel, is a mixture of a large number of individual hydrocarbons it has a boiling temperature range rather than a single boiling point. Typically, this range is 30 to 200°C. Individual hydrocarbons have the saturation pressure-temperature relationships of a pure substance. The lower the molecular weight, the higher will be the saturated vapor pressure at a given temperature. The boiling point of hydrocarbons depends pri-marily on their molecular weight: the vapor pressure also depends on molecular structure. The equilibrium state of a hydrocarbon-air mixture depends therefore on the vapor pressure of the hydrocarbon at the given temperature, the relative amounts of the hydrocarbon and air, and the total pressure of the mixture. The equilibrium fraction of fuel evaporated at a given temperature and pressure can be calculated from Bridgeman charts[31] and the distillation characteristics of the fuel (defined by the ASTM distillation curve[32]). Figure 7-26a shows the effect of mixture temperature on percent of indolene fuel (a specific gasoline) evaporated at equilibrium at atmospheric pressure. Figure 7-26b shows the effect of reduced manifold pressure on the amount evaporated.[18] While insufficient time is usually available in the manifold to establish equilibrium, the trends shown are indicative of what happens in practice: lower pressures increase the relative amount of fuel vaporized and charge heating is usually required to vaporize a substantial frac-tion of the fuel.

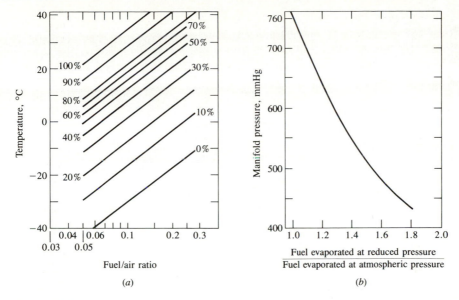

FIGURE 7-26
(a) Percentage of indolene fuel evaporated at equilibrium at 1 atmosphere pressure. (b) Effect of pressure on amount of indolene fuel evaporated.[18]

For carbureted and throttle-body injection systems, the fuel path is the following. Until the throttle plate is close to fully opened, most of the fuel metered into the air stream impacts on the throttle plate and throttle-body walls. Only a modest fraction of the fuel vaporizes upstream of the throttle. The liquid is re-entrained as the air flows at high velocity past the throttle plate. The fuel does not usually divide equally on either side of the throttle plate axis. The air undergoes a 90° bend in the plenum beneath the throttle; much of the fuel which has not evaporated is impacted on the manifold floor. Observations of fuel behavior in intake manifolds with viewing ports or transparent sections show that there is substantial liquid fuel on the walls with carburetor fuel metering systems. Figure 7-27 shows the engine conditions under which liquid fuel was observed on the floor of the manifold plenum beneath the throttle plate and in the manifold runners, in a standard four-cylinder production engine.[23] This manifold was heated by engine coolant at 90°C. The greatest amount of liquid was present at high engine loads and low speeds. Heating the manifold to a higher temperature with steam at 115°C resulted in a substantial reduction in the amount of liquid: there was no extensive puddling on the plenum floor, liquid films or rivulets were observed in a zone bounded by 120 mmHg vacuum and 2500 rev/min, and there were no films or rivulets in the manifold runner. Depending on engine operating conditions, transport of fuel as a liquid film or rivulet in the manifold and vaporization from these liquid fuel films and rivulets and subsequent transport as vapor may occur.

Vaporized fuel and liquid droplets which remain suspended in the air

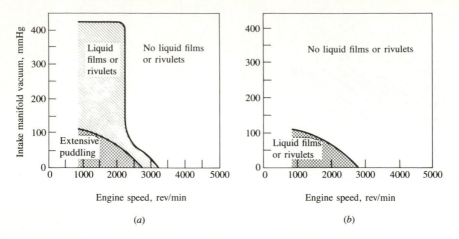

FIGURE 7-27
Regions of engine load and speed range where extensive pools or puddles, liquid films, or rivulets were observed: (*a*) on manifold plenum floor and (*b*) in manifold runner. Four-cylinder automobile engine. Manifold heated by coolant at 90°C.[23]

stream will be transported with the air stream. However, droplet deposition on the manifold walls may occur due to gravitational settling and to inertial effects as the flow goes round bends in the manifold.

The fuel transport processes for port fuel-injection systems are different and will depend significantly on the timing and duration of the injection pulse. Fuel is injected onto the back of the inlet valve (and surrounding port wall), usually while the valve is closed or only partly open. Vaporization of liquid fuel off the valve and walls occurs, enhanced by the backflow of hot residual gases from the cylinder (especially at part load). There is evidence that, even under fully warmed-up engine conditions, some fuel is carried as liquid drops into the cylinder.[33]

FUEL DROPLET BEHAVIOR. With carburetor and throttle-body injection systems, the liquid fuel atomizes as it enters the air stream. In the carburetor venturi this occurs as the fuel-air emulsion from the fuel jet(s) enters the high-velocity (> 100 m/s) air stream. With an injector, the velocity of the liquid jet as it exits the nozzle is high enough to shatter the flowing liquid, and its interaction with the coaxial air flow further atomizes the fuel. Typical droplet-size distributions are not well defined; size would vary over the load and speed range. Droplet diameters in the 25 to 100 μm range are usually assumed to be representative: larger drops are also produced. The liquid fuel drops are accelerated by the surrounding air stream and start to vaporize. Vaporization rates have been calculated using established formulas for heat and mass transfer between a droplet and a surrounding flowing gas stream (see Ref. 34 for a review of methods of calculating droplet vaporization rates). Calculations of fuel vaporization in a carburetor venturi and upstream of the throttle plate show that the temperature of the liquid fuel droplets decreases rapidly (by up to about 30°C[35]), and the fraction of the fuel vaporized is small (in the 2 to 15 percent range[35, 36]).

Liquid fuel drops, due to their density being many times that of the air, will not exactly follow the air flow. Droplet impaction on the walls may occur as the flow changes direction, and the greater inertia of the droplets causes them to move across the streamlines to the outer wall. Deposition on the manifold floor due to gravity may also occur. The equation of motion for an individual droplet in a flowing gas stream is

$$(\tfrac{1}{6}\pi D_d^3 \rho_f)\mathbf{a} = m_d \mathbf{g} - \tfrac{1}{2}(\mathbf{v}_d - \mathbf{v}_g)|\mathbf{v}_d - \mathbf{v}_g|\rho_g C_D \frac{\pi D_d^2}{4} \tag{7.25}$$

where D_d is the droplet diameter, ρ_f and ρ_g are liquid and gas densities, $\mathbf{v}_d$ and $\mathbf{v}_g$ are the droplet and gas velocities, $\mathbf{a}$ is the droplet acceleration, $\mathbf{g}$ acceleration due to gravity, and C_D is the drag coefficient. For $6 < \text{Re} < 500$ the drag coefficient of an evaporating droplet is a strong function of the Reynolds number, Re: e.g.,

$$C_D = 27 \, \text{Re}^{-0.84} \tag{7.26}$$

where $\text{Re} = (\rho_g D_d |\mathbf{v}_d - \mathbf{v}_g|/\mu_g)$.

Studies of droplet impaction and evaporation using the above equations and typical manifold conditions and geometries indicate the following.[26, 35, 37] For 90° bends, drops of less than 10 μm diameter are essentially carried by the gas stream (<10 percent impaction); almost all droplets larger than 25 μm impact on the walls. Droplet sizes produced first in the carburetor venturi or fuel injector spray and then by secondary atomization as liquid fuel is entrained from the throttle plate and throttle-body walls depend on the local gas velocity: higher local relative velocities between the gas and liquid produce smaller drop sizes. Approximate estimates which combine the two phenomena outlined above show that at low engine air flow rates, almost all of the fuel will impact first on the throttle plate and then on the manifold floor as the flow turns 90° into the manifold runners. At high air flows, because the drops are smaller, a substantial fraction of the drops may stay entrained in the air flow. Secondary atomization at the throttle at part-load operating conditions is important to the fuel transport process: the very high air velocities at the edge of the throttle plate produce droplets of order or less than 10 μm diameter. However, coalescence and deposition on the walls and subsequent reentrainment probably increase the mean droplet size. In the manifold, gravitational settling of large (>100 μm) droplets would occur at low air flow rates,[38] but these drops are also likely to impact the walls due to their inertia as the flow is turned.

Estimates of droplet evaporation rates in the manifold indicate the following. With a representative residence time in the manifold of about one crank revolution (10 ms at 6000 rev/min, 100 ms at 600 rev/min), only drops of size less than about 10 μm will evaporate at the maximum speed; 100 μm droplets will not vaporize fully at any speed. Most of these large droplets impact on the walls, anyway. Drops small enough to be carried by the air stream are likely to vaporize in the manifold.[26]

FUEL-FILM BEHAVIOR. The fuel which impacts on the wall will also vaporize and, depending on where in the manifold deposition occurs and the local manifold geometry, may be transported along the manifold as a liquid film or rivulet. If the vaporization rate off the wall is sufficiently high, then a liquid film will not build up. Any liquid film or pool on the manifold floor or walls is important because it introduces additional fuel transport processes—deposition, liquid transport, and evaporation—which together have a much longer time constant than the air transport process. Thus changes in the air and the fuel flow into each engine cylinder, during a change in engine load, will not occur in phase with each other unless compensation is made for the slower fuel transport.

Several models of the behavior of liquid-fuel wall-films have been developed. One approach analyzes a liquid puddle on the floor of the manifold plenum.[38] Metered fuel enters the puddle; fuel leaves primarily through vaporization. The equation for rate of change of mass of fuel in the puddle is

$$\dot{m}_{f,\,p} = \dot{m}_{f,\,\text{in}} - \dot{m}_{f,\,\text{out}} = x\dot{m}_{f,\,m} - \frac{m_{f,\,p}}{\tau} \tag{7.27}$$

where $m_{f,\,p}$ is the mass of fuel in the puddle, $\dot{m}_{f,\,m}$ is the metered fuel flow rate, and x is the fraction of the metered flow that enters the puddle. It is assumed that the reentrainment/evaporation rate is proportional to the mass of fuel in the puddle divided by the characteristic time τ of the reentrainment/evaporation process. The puddle behavior predicted by this model in response to a step increase in engine load is shown in Fig. 7-28a. Because only part $(1 - x)$ of the fuel flows directly with the air, as the throttle is opened rapidly a lean air/fuel ratio excursion is predicted. Figure 7-28b shows that this behavior (without any metering compensation) is observed in practice. Estimates of the volume of fuel in the puddle (for a 5-liter V-8 engine) are of order 1000 mm³, and increase with

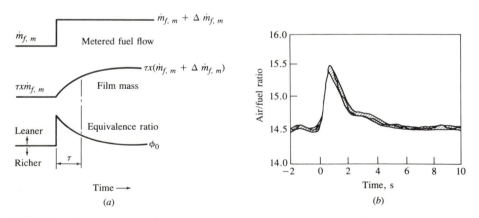

FIGURE 7-28
(a) Predicted behavior of the fuel film for an uncompensated step change in engine operating conditions. (b) Observed variation in air/fuel ratio for uncompensated throttle opening at 1600 rev/min which increased manifold pressure from 48 to 61 cmHg.[38]

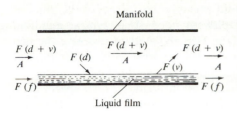

Manifold

Liquid film

FIGURE 7-29
Schematic of fuel flow paths in the manifold when liquid film flows along the manifold runner floor.
A   Air
F   Fuel
f   Liquid fuel film
d   Liquid fuel droplets
v   Fuel vapor

increasing load and speed. The time constant is of order 2 seconds for a fully warmed-up engine; it varies with engine operating conditions and is especially sensitive to intake manifold temperature. Such models have been used primarily to develop fuel metering strategies which compensate for the fuel transport lag.[38]

An alternative model, for liquid film flow in the manifold runner and branch, has been developed.[37] Fuel is deposited on the manifold walls and forms a film which flows toward the cylinder due to the shear force at the gas/liquid interface as shown in Fig. 7-29. Vaporization from the film also occurs. An analysis of the dynamics of the fuel film leads to expressions for steady-state film velocity and thickness. As air and metered fuel flows change due to a throttle position change, the characteristic time for reestablishing steady state is $l/(2u_f)$, where l is the manifold length and u_f the average film velocity. This characteristic response time is of order 1 second for typical manifold conditions, in approximate agreement with values obtained from transient engine experiments.

A more extensive analysis of both fuel droplet and film evaporation in a complete carburetor, throttle, manifold system,[35] with a multicomponent model for gasoline based on its distillation curve, indicates the following phenomena are important. Secondary atomization of the liquid fuel at the throttle, which produces the smallest droplet sizes when the throttle open angle is small, significantly increases the fraction of fuel evaporation in the manifold. Increasing inlet air temperature increases the fraction of fuel vaporized; this effect is larger at lower loads since secondary atomization under these conditions increases the liquid fuel surface area significantly. Heating the wall, which heats the liquid film on the wall directly, provides a greater increase in fraction evaporated than does equivalent heating of the air flow upstream of the carburetor. Due to the multi-component nature of the fuel, the residual liquid fuel composition changes significantly as fuel is transported from the carburetor to the manifold exit. Of the full boiling range liquid composition at entry, all the light ends, most of the mid-range components, but only a modest amount of the high boiling point fraction have evaporated at the manifold exit. The predicted fuel fraction evaporated ranged from 40 to 60 percent for the conditions examined. One set of measurements of the fraction of fuel vaporized in the manifold of a warmed-up four-cylinder engine showed that 70 to 80 percent of the fuel had vaporized, confirming that under these operating conditions "most" but not necessarily "all" the fuel enters the cylinder in vapor form.[39]

The engine operating range where fuel puddling, fuel films, and rivulets are observed (see Fig. 7-27) can now be explained. At light load, secondary atomization at the throttle and the lower manifold pressure would reduce the amount of liquid fuel impinging on the manifold plenum floor. Also, typical manifold heating at light load substantially exceeds the heat required to vaporize the fuel completely,[40] and manifold floor temperatures are of order 15°C higher than at full load. All the above is consistent with less liquid on the floor and none in the runners at light load, compared to what occurs at full load. At high speed, drop sizes produced in the carburetor are much smaller, so impingement on the walls is much reduced.

The fuel flow to each cylinder per cycle is not exactly the same. There is a *geometric variation* where fuel is not divided equally among individual cylinders. There is also a *time variation* under steady-state engine conditions where the air/fuel ratio in a given cylinder varies cycle-by-cycle.[41] Data on time-averaged air/fuel ratios in each cylinder of multicylinder engines show that the extent of the maldistribution varies from engine to engine, and for a particular engine varies over the load and speed range. Spreads in the equivalence ratio (maximum to minimum) of about 5 percent of the mean value are typical at light load for carbureted engines. Largest variations between cylinders usually occur at wide-open throttle. WOT spreads in the equivalence ratio of about 15 percent of the mean appear to be typical, again for carbureted engines, while spreads as high as 20 to 30 percent are not uncommon at particular speeds for some engines.[23, 40] Time variations are less well defined; the limited data available suggest they could be of comparable magnitude.[41]

With multipoint port fuel-injection systems, the fuel transport processes are substantially different and are not well understood. Air-flow phenomena are comparable to those with carbureted or throttle-body injection systems. However, manifold design can be optimized for air flow alone since fuel transport from the throttle through the manifold is no longer a design constraint. Because the manufacture and operation of individual fuel injectors are not identical, there is still some variation in fuel mass injected cylinder-to-cylinder and cycle-to-cycle. Since individual cylinder air flows depend on the design of the manifold, whereas the amount of fuel injected does not, uniform air distribution is especially important with port injection systems. The fuel vaporization and transport processes will depend on the duration of injection and the timing of injection pulse(s) relative to the intake valve-lift profile. Some of the injected fuel will impinge on the port walls, valve stem, and backside of the valve, especially when injection toward a closed valve occurs. Backflow of hot residual gases at part-load operation will have a substantial effect on fuel vaporization. Compensation for fuel lag during transient engine operation is still required; sudden throttle openings are accompanied by a "lean spike" in the mixture delivered to the engine, comparable to though smaller than that shown in Fig. 7-28 for a throttle-body fuel-injection system. Thus wall wetting, evaporation off the wall, and liquid flow along the wall are all likely to be important with port fuel-injection systems also.

With port fuel-injection systems, liquid fuel enters the cylinder and droplets are present during intake and compression. Limited measurements have been made of the distribution, size, and number density of these fuel droplets. During intake, the droplet number density in the clearance volume increased to a maximum at the end of injection (the injection lasted from 45 to 153° ATC) and then decreased due to evaporation during compression to a very small value at the time of ignition. Average droplet size during intake was 10 to 20 μm in diameter; it increased during compression as the smaller drops in the distribution evaporated. At the conditions tested, some 10 to 20 percent of the fuel was in droplet form at the end of injection. At ignition, the surviving droplets contained a negligible fraction of the fuel. During injection, the distribution of droplets across the clearance volume was nonuniform. It became much more uniform with time, after injection ended.[33]

PROBLEMS

7.1. The equivalence ratio in a conventional spark-ignition engine varies from no load (idle) to full load, at a fixed engine speed, as shown at the top of Fig. P7-1. (By load is meant the percentage of the maximum brake torque at that speed.) Also shown is the variation in total friction (pumping plus mechanical rubbing plus accessory friction). Using formats similar to those shown, draw *carefully proportioned qualitative* graphs of the following parameters versus load (0 to 100 percent):

Combustion efficiency, η_c
Gross indicated fuel conversion efficiency, $\eta_{f,\mathrm{ig}}$
Gross indicated mean effective pressure, imep_g
Brake mean effective pressure, bmep
Mechanical efficiency, η_m

Indicate clearly where the maximum occurs if there is one, and where the value is zero or unity or some other obvious value, if appropriate. Provide a *brief justification* for the shape of the curves you draw.

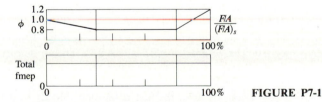

FIGURE P7-1

7.2. The four-cylinder spark-ignition engine shown in the figure uses an oxygen sensor in the exhaust system to determine whether the exhaust gas composition is lean or rich of the stoichiometric point, and a throttle-body injection system with feedback to maintain engine operation close to stoichiometric. However, since there is a time delay between a change in the fuel/air ratio at the injector location and the detection of that change by the sensor (corresponding to the flow time between the injector and the sensor), the control system shown results in oscillations in fuel/air ratio about the stoichiometric point.

(a) Estimate the average flow time between the injector and the sensor at an engine speed of 2000 rev/min.

(b) The sensor and control unit provide a voltage V of $+V_s$ volts when the fuel/air equivalence ratio ϕ is less than one and a voltage of $-V_s$ volts when ϕ is greater than one. The feedback injection system provides a fuel/air ratio (F/A) given by

$$\left(\frac{F}{A}\right) = \left(\frac{F}{A}\right)_{t=0} (1 + CVt)$$

where t is the time (in seconds) after the voltage signal last changed sign, $(F/A)_{t=0}$ is the fuel/air ratio at the injector at $t = 0$, and C is a constant. Develop carefully proportioned quantitative sketches of the variation in the fuel/air ratio at the injector and at the exhaust sensor, with time, showing the phase relation between the two curves. Explain briefly how you developed these graphs.

(c) Find the value of the constant C, in volts^{-1}-seconds^{-1} (the feedback system gain), such that (F/A) variations about the stoichiometric value do not exceed ± 10 percent for $V_s = 1$ V.

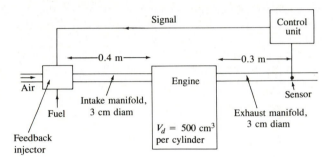

FIGURE P7-2

7.3. In many spark-ignition engines, liquid fuel is added to the inlet air upstream of the inlet manifold above the throttle. The inlet manifold is heated to ensure that under steady-state conditions the fuel is vaporized before the mixture enters the cylinder.

(a) At normal wide-open throttle operating conditions, in a four-stroke cycle 1.6-dm^3 displacement four-cylinder engine, at 2500 rev/min, the temperature of the air entering the carburetor is 40°C. The heat of vaporization of the fuel is 350 kJ/kg and the rate of heat transfer to the intake mixture is 1.4 kW. Estimate the temperature of the inlet mixture as it passes through the inlet valve, assuming that the fuel is fully vaporized. The volumetric efficiency is 0.85. The air density is 1.06 kg/m^3 and c_p for air is 1 kJ/kg·K. You may neglect the effects of the heat capacity of the liquid and vapor fuel.

(b) With port electronic fuel-injection systems, the fuel is injected directly into the intake port. The intake manifold is no longer heated. However, the fuel is only partly vaporized prior to entering the cylinder. Estimate the mixture temperature as it passes through the inlet valve with the EFI system, assuming that the air temperature entering the intake manifold is still 40°C and 50 percent of the fuel is vaporized.

(c) Estimate the ratio of the maximum indicated power obtained at these conditions with this engine with a carburetor, to the maximum power obtained with port fuel injection. Assume that the inlet valve is the dominant restriction in the flow into the engine and that the pressure ratio across the inlet valve is the same for both carbureted and port-injection fueled engines. The intake mixture pressure and equivalence ratio remain the same in both these cases.

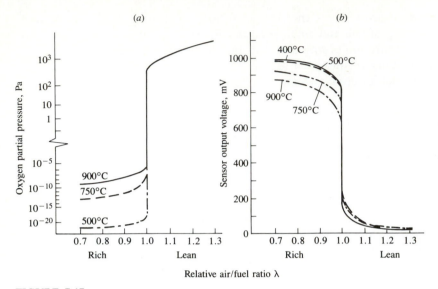

FIGURE 7-17
Oxygen-sensor characteristics. Variation as a function of relative air/fuel ratio and temperature of: (a) oxygen partial pressure in equilibrated combustion products; (b) sensor output voltage.[15]

mixture at the stoichiometric point, as shown in Fig. 7-17b. Since this transition is not temperature dependent, it is well suited as a sensor signal for a feedback system.[15]

Figure 7-18 shows a cross-section drawing of a lambda sensor, screwed into the wall of the exhaust manifold. This location provides rapid warm-up of the sensor following engine start-up. It also gives the shortest flow time from the fuel injector or carburetor location to the sensor—a delay time which is important in the operation of the feedback system. The sensor body is made of ZrO_2 ceramic stabilized with Y_2O_3 to give adequate low-temperature electrical conductivity. The inner and outer electrodes are 10-μm thick porous platinum layers to provide the required catalytic equilibration. The outer electrode which is exposed to the exhaust gases is protected against corrosion and erosion by a 100-μm spinal coat and a slotted shield. Air passes to the inner electrode through holes in the protective sleeve. The shield, protective sleeve, and housing are made from heat- and corrosion-resistant steel alloys. Such sensors were first developed for air/fuel ratio control at close to the stoichiometric value. Use of a similar sensor to control air/fuel ratios at lean-of-stoichiometric values during part-throttle engine operation is also feasible.

For closed-loop feedback control at close-to-stoichiometric, use is made of the sensor's low-voltage output for lean mixtures and a high-voltage output for rich mixtures. A control voltage reference level is chosen at about the mid-point of the steep transition in Fig. 7-17b. In the electronic control unit the sensor signal is compared to the reference voltage in the comparator as shown in Fig. 7-19a. The comparator output is then integrated in the integral controller whose

7.4. Port fuel-injection systems are replacing carburetors in automobile spark-ignition engines. List the major advantages and any disadvantages of fuel metering with port fuel injection relative to carburetion.

7.5. With multipoint port fuel injection and single-point injection systems, the fuel flow rate is controlled by the injection pulse duration. If each injector operates continuously at the maximum rated power point (wide-open throttle, $A/F = 12$, 5500 rev/min) of an automobile spark-ignition engine, estimate approximately the injection pulse duration (in crank angle degrees) for the same engine at idle. Idle conditions are: 700 rev/min, 0.3 atm inlet manifold pressure, stoichiometric mixture.

7.6. The fuel-air cycle results indicate that the maximum imep is obtained with gasoline-air mixtures at equivalence ratios of about 1.0. In practice, the maximum wide-open throttle power of a spark-ignition engine at a given speed is obtained with an air/fuel ratio of about 12. The vaporization of the additional gasoline lowers the temperature of the inlet air and the richer mixture has a lower ratio of specific heats γ_u during compression. Estimate approximately the change in mixture temperature due to vaporization of the additional fuel used to decrease A/F from 14.6 (an equivalence ratio of 1.0) to 12.2 in the intake system, and the combined effect of vaporization and lower γ_u on the unburned mixture temperature at WOT when the cylinder pressure is at its peak of 40 atm. (The principal effect of the richer mixture is its impact on knock.)

7.7. (a) Plot dimensionless throttle plate open area $4A_{th}/(\pi D^2)$ as a function of throttle plate angle ψ. Assume $\psi_0 = 10°$, D (throttle bore diameter) = 57 mm, d (throttle shaft diameter) = 10.4 mm. What is the *throttle plate* area?

(b) Estimate the average velocity of the air flowing through the throttle plate open area for $\psi = 26°$ at 3000 rev/min and $\psi = 36°$ at 2000 rev/min. Use the relationship between ψ, engine speed, and inlet manifold pressure given in Fig. 7-22. Assume a discharge coefficient $C_D = 0.8$.

(c) For the throttle of part (a), estimate and plot the total force on the throttle plate and shaft, and the force parallel and perpendicular to the throttle bore axis (i.e., in the mean flow direction and normal to that direction) as a function of throttle angle at 2000 rev/min. Again use Fig. 7-22 for the relationship between ψ and inlet manifold pressure.

7.8. For the engine and intake manifold shown in Fig. 7-23, estimate the ratio of the intake manifold runner cross-sectional area to $(\pi B^2/4)$, the ratio of the length of the flow path from the intake manifold entrance to the inlet valve seat to the bore, the ratio of the volume of each inlet port to each cylinder's displaced volume, and the ratio of the volume of each intake manifold runner to each cylinder's displaced volume. The cylinder bore is 89 mm.

REFERENCES

1. Nakajima, Y., Sugihara, K., Takagi, Y., and Muranaka, S.: "Effects of Exhaust Gas Recirculation on Fuel Consumption," in *Proceedings of Institution of Mechanical Engineers, Automobile Division*, vol. 195, no. 30, pp. 369–376, 1981.
2. Harrington, D. L., and Bolt, J. A.: "Analysis and Digital Simulation of Carburetor Metering," SAE paper 700082, *SAE Trans.*, vol. 79, 1970.
3. Bolt, J. A., Derezinski, S. J., and Harrington, D. L.: "Influence of Fuel Properties on Metering in Carburetors," SAE paper 710207, *SAE Trans.*, vol. 80, 1971.

4. Khovakh, M.: *Motor Vehicle Engines* (English translation), Mir Publishers, Moscow, 1976.
5. Bolt, J. A., and Boerma, M. J.: "Influence of Air Pressure and Temperature on Carburetor Metering," SAE paper 660119, 1966.
6. Shinoda, K., Koide, H., and Yii, A.: "Analysis and Experiments on Carburetor Metering at the Transition Region to the Main System," SAE paper 710206, *SAE Trans.*, vol. 80, 1971.
7. Oya, T.: "Upward Liquid Flow in a Small Tube into which Air Streams," *Bull. JSME*, vol. 14, no. 78, pp. 1320–1329, 1971.
8. Wrausmann, R. C., and Smith, R. J.: "An Approach to Altitude Compensation of the Carburetor," SAE paper 760286, 1976.
9. Bosch, *Automotive Handbook*, 1st English ed., Robert Bosch GmbH, Stuttgart, 1978.
10. Glöckler, O., Knapp, H., and Manger, H.: "Present Status and Future Development of Gasoline Fuel Injection Systems for Passenger Cars," SAE paper 800467, 1980.
11. Greiner, M., Romann, P., and Steinbrenner, U.: "BOSCH Fuel Injectors—New Developments," SAE paper 870124, 1987.
12. Gorille, I., Rittmannsberger, N., and Werner, P.: "Bosch Electronic Fuel Injection with Closed Loop Control," SAE paper 750368, *SAE Trans.*, vol. 84, 1975.
13. Czadzeck, G. H.: "Ford's 1980 Central Fuel Injection System," SAE paper 790742, 1979.
14. Bowler, L. L.: "Throttle Body Fuel Injection (TBI)—An Integrated Engine Control System," SAE paper 800164, *SAE Trans.*, vol. 89, 1980.
15. Hamann, E., Manger, H., and Steinke, L.: "Lambda-Sensor with Y_2O_3-Stabilized ZrO_2-Ceramic for Application in Automotive Emission Control Systems," SAE paper 770401, *SAE Trans.*, vol. 86, 1977.
16. Seiter, R. E., and Clark, R. J.: "Ford Three-Way Catalyst and Feedback Fuel Control System," SAE paper 780203, *SAE Trans.*, vol. 87, 1978.
17. Camp, J., and Rachel, T.: "Closed-Loop Electronic Fuel and Air Control of Internal Combustion Engines," SAE paper 750369, 1975.
18. Liimatta, D. R., Hurt, R. F., Deller, R. W., and Hull, W. L.: "Effects of Mixture Distribution on Exhaust Emissions as Indicated by Engine Data and the Hydraulic Analogy," SAE paper 710618, *SAE Trans.*, vol. 80, 1971.
19. Benson, R. S., Baruah, P. C., and Sierens, I. R.: "Steady and Non-steady Flow in a Simple-Carburetor," in *Proceedings of Institution of Mechanical Engineers*, vol. 188, no. 53/74, pp. 537–548, 1974.
20. Woods, W. A., and Goh, G. K.: "Compressible Flow through a Butterfly Throttle Valve in a Pipe," in *Proceedings of Institution of Mechanical Engineers*, vol. 193, no. 10, pp. 237–244, 1979.
21. Walker, J. W.: "The GM 1.8 Liter L-4 Gasoline Engine Designed by Chevrolet," SAE paper 820111, *SAE Trans.*, vol. 91, 1982.
22. Chapman, M.: "Two Dimensional Numerical Simulation of Inlet Manifold Flow in a Four Cylinder Internal Combustion Engine," SAE paper 790244, 1979.
23. Kay, I. W.: "Manifold Fuel Film Effects in an SI Engine," SAE paper 780944, 1978.
24. Brandstetter, W. R., and Carr, M. J.: "Measurement of Air Distribution in a Multicylinder Engine by Means of a Mass Flow Probe," SAE paper 730494, 1973.
25. Aquino, C. F.: "Transient *A/F* Control Characteristics of the 5 Liter Central Fuel Injection Engine," SAE 810494, *SAE Trans.*, vol. 90, 1981.
26. Trayser, D. A., Creswick, F. A., Giesike, J. A., Hazard, H. R., Weller, A. E., and Locklin, D. W.: "A Study of the Influence of Fuel Atomization, Vaporization, and Mixing Processes on Pollutant Emissions from Motor-Vehicle Powerplants," Battelle Memorial Institute, Columbus, Ohio, 1969.
27. Tabaczysnki, R. J.: "Effects of Inlet and Exhaust System Design on Engine Performance," SAE paper 821577, 1982.
28. Engelman, H. W.: "Design of a Tuned Intake Manifold," ASME paper 73-WA/DGP-2, 1973.
29. Benson, R. S.: in J. H. Horlock and D. E. Winterbone (eds.), *The Thermodynamics and Gas Dynamics of Internal Combustion Engines*, vol. 1, Clarendon Press, Oxford, 1982.
30. Chapman, M., Novak, J. M., and Stein, R. A.: "Numerical Modeling of Inlet and Exhaust Flows in Multi-cylinder Internal Combustion Engines," in *Flows in Internal Combustion Engines*, ASME Winter Annual Meeting, Nov. 14–19, 1982, ASME, New York.

31. Bridgeman, O. C.: "Equilibrium Volatility of Motor Fuels from the Standpoint of Their Use in Internal Combustion Engines," National Bureau of Standards research paper 694, 1934.
32. ASTM Standard Method: "Distillation of Petroleum Products," ANSI/ASTM D86 (1P 123/68).
33. Peters, B. D.: "Laser-Video Imaging and Measurement of Fuel Droplets in a Spark-Ignition Engine," in *Proceedings of Conference on Combustion in Engineering*, Oxford, Apr. 11–14, 1983, Institution of Mechanical Engineers, 1983.
34. Sirignano, W. A.: "Fuel Droplet Vaporization and Spray Combustion Theory," *Prog. Energy and Combust. Sci.*, vol. 9, pp. 291–322, 1983.
35. Boam, D. J., and Finlay, I. C.: "A Computer Model of Fuel Evaporation in the Intake System of a Carbureted Petrol Engine," Conference on *Fuel Economy and Emissions of Lean Burn Engines*, London, June 12–14, 1979, paper C89/79, Institution of Mechanical Engineers, 1979.
36. Yun, H. J., and Lo, R. S.: "Theoretical Studies of Fuel Droplet Evaporation and Transportation in a Carburetor Venturi," SAE paper 760289, 1976.
37. Servati, H. B., and Yuen, W. W.: "Deposition of Fuel Droplets in Horizontal Intake Manifolds and the Behavior of Fuel Film Flow on Its Walls," SAE paper 840239, *SAE Trans.*, vol. 93, 1984.
38. Hires, S. D., and Overington, M. T.: "Transient Mixture Strength Excursions—An Investigation of Their Causes and the Development of a Constant Mixture Strength Fueling Strategy," SAE paper 810495, *SAE Trans.*, vol. 90, 1981.
39. Collins, M. H.: "A Technique to Characterize Quantitatively the Air/Fuel Mixture in the Inlet Manifold of a Gasoline Engine," SAE paper 690515, *SAE Trans.*, vol. 78, 1969.
40. Blackmore, D. R., and Thomas, A.: *Fuel Economy of the Gasoline Engine*, John Wiley, 1977.
41. Yu, H. T. C.: "Fuel Distribution Studies—A New Look at an Old Problem," *SAE Trans.*, vol. 71, pp. 596–613, 1963.

Gas motion within the engine cylinder is one of the major factors that controls the combustion process in spark-ignition engines and the fuel-air mixing and combustion processes in diesel engines. It also has a significant impact on heat transfer. Both the bulk gas motion and the turbulence characteristics of the flow are important. The initial in-cylinder flow pattern is set up by the intake process. It may then be substantially modified during compression. This chapter reviews the important features of gas motion within the cylinder set up by flows into and out of the cylinder through valves or ports, and by the motion of the piston.

8.1 INTAKE JET FLOW

The engine intake process governs many important aspects of the flow within the cylinder. In four-stroke cycle engines, the inlet valve is the minimum area for the flow (see Sec. 6.3) so gas velocities at the valve are the highest velocities set up during the intake process. The gas issues from the valve opening into the cylinder as a conical jet and the radial and axial velocities in the jet are about 10 times the mean piston speed. Figure 8-1 shows the radial and axial velocity components close to the valve exit, measured during the intake process, in a motored model engine with transparent walls and single valve located on the cylinder axis, using laser doppler anemometry (see next section).[1] The jet separates from the valve

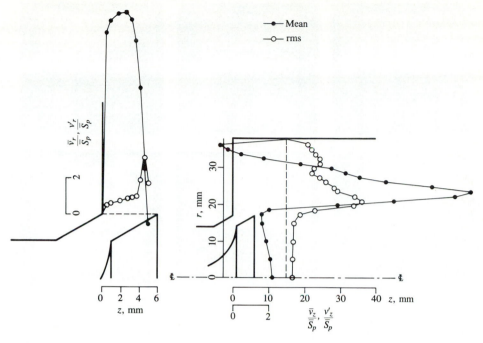

FIGURE 8-1
Radial mean velocity $\bar{v}_r$ and root mean square (rms) velocity fluctuations v'_r at the valve exit plane, and axial mean velocity $\bar{v}_z$ and rms velocity fluctuation v'_z 15 mm below the cylinder head, at 36° ATC in model engine operated at 200 rev/min. Valve lift = 6 mm. Velocities normalized by mean piston speed.[1]

seat and lip, producing shear layers with large velocity gradients which generate turbulence. This separation of the jet sets up recirculation regions beneath the valve head and in the corner between the cylinder wall and cylinder head.

The motion of the intake jet within the cylinder is shown in the schlieren photographs in Fig. 8-2 taken in a transparent engine. This engine had a square cross-section cylinder made up of two quartz walls and two steel walls, to permit easy optical access. The schlieren technique makes regions with density gradients in the flow show up as lighter or darker regions on the film.[2] The engine was throttled to one-half an atmosphere intake pressure, so the jet starts after the intake stroke has commenced, at 35° ATC, following backflow of residual into the intake manifold. The front of the intake jet can be seen propagating from the valve to the cylinder wall at several times the mean piston speed. Once the jet reaches the wall ($\theta > 41°$ ATC), the wall deflects the major portion of the jet downward toward the piston; however, a substantial fraction flows upward toward the cylinder head. The highly turbulent nature of the jet is evident.

The interaction of the intake jet with the wall produces large-scale rotating flow patterns within the cylinder volume. These are easiest to visualize where the engine geometry has been simplified so the flow is axisymmetric. The photograph

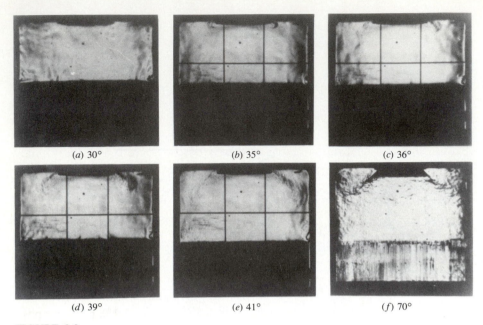

(a) 30° (b) 35° (c) 36°

(d) 39° (e) 41° (f) 70°

FIGURE 8-2
Sequence of schlieren photographs of intake jet as it develops during intake stroke. Numbers are crank angle degrees after TC.[2]

in Fig. 8-3 of a water analog of an engine intake flow was taken in a transparent model of an engine cylinder and piston. The valve is located in the center of the cylinder head, and the flow into the valve is along the cylinder axis. The experimental parameters have been scaled so that the appropriate dimensionless numbers which govern the flow, the Reynolds and Strouhal numbers, were maintained equal to typical engine values. The photograph shows the major features

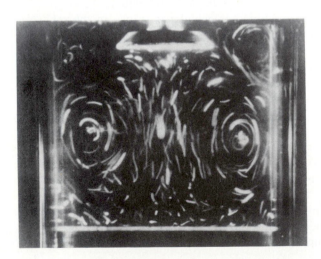

FIGURE 8-3
Large-scale rotating flow pattern set up within the cylinder by the intake jet. Photograph of streak lines in water flow into model engine with axisymmetric valve.[3]

of the intake generated flow in a thin illuminated plane through the cylinder axis. The streaks are records of the paths of tracer particles in the flow during the period the camera shutter is open. The bulk of the cylinder as the piston moves down is filled with a large ring vortex, whose center moves downward and remains about halfway between the piston and the head. The upper corner of the cylinder contains a smaller vortex, rotating in the opposite direction. These vortices persist until about the end of the intake stroke, when they became unstable and break up.[3]

With inlet valve location and inlet port geometry more typical of normal engine practice, the intake generated flow is more complex. However, the presence of large-scale rotating flow patterns can still be discerned. Figure 8-4a shows the effect of off-axis valve location (with the flow into the valve still parallel to the cylinder and valve axis). During the first half of the inlet stroke, at least, a flow pattern similar in character to that in Fig. 8-3 is evident. The vortices are now displaced to one side, however, and the planes of their axes of rotation are no longer perpendicular to the cylinder axis but are tipped at an angle to it. The vortices become unstable and break up earlier in the intake stroke than was the case with the axisymmetric flow.[3] With an offset valve and a normal inlet port configuration which turns the flow through 50 to 70° (see Fig. 6-13), photographs

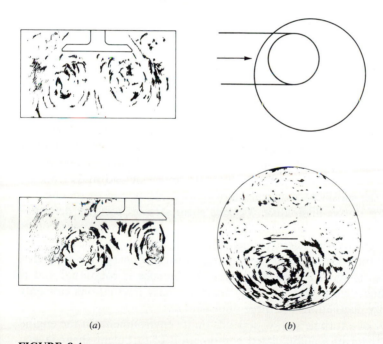

(a)　　　　　　　　　　　　　　　　　(b)

FIGURE 8-4
Sketches from: (a) streak photographs of in-cylinder intake generated flow in water analog of intake process in model engine with offset inlet valve, at 90° ATC;[3] (b) streak photographs of flow in diametral plane; 30 mm below cylinder head, with intake port and valve geometry shown, with steady water flow into cylinder. Valve lift = 4 mm.[4]

of the flow pattern in a diametral plane show an additional large-scale rotation. Figure 8-4*b* shows the flow pattern observed in a water-flow model of the cylinder in a plane 30 mm (one-third of the bore) from the cylinder head, with a standard inlet port design. The direction of flow with this vortex pair is toward the left across the center of the cylinder. This flow pattern occurs because the cylinder wall closest to the valve impedes the flow out of the valve and forces the flow on either side of the plane passing through the valve and cylinder axes to circulate around the cylinder in opposite directions. The upper vortex follows the flow direction of the port and becomes larger still as the valve lift increases. The details of this aspect of the intake flow depend on the port design, valve stem orientation, and the valve lift.[4] With suitable port and/or cylinder head design, it is possible to develop a single vortex flow within the bulk of the cylinder. The production and characteristics of such "swirling" flows are reviewed in Sec. 8.3.

In summary, the jet-like character of the intake flow, interacting with the cylinder walls and moving piston, creates large-scale rotating flow patterns within the cylinder. The details of these flows are strongly dependent on the inlet port, valve, and cylinder head geometry. These flows appear to become unstable, either during the intake or the compression process, and break down into three-dimensional turbulent motions. Recirculating flows of this type are usually sensitive to small variations in the flow: hence there are probably substantial cycle-by-cycle flow variations.[5]

8.2 MEAN VELOCITY AND TURBULENCE CHARACTERISTICS

8.2.1 Definitions

The flow processes in the engine cylinder are turbulent. In turbulent flows, the rates of transfer and mixing are several times greater than the rates due to molecular diffusion. This turbulent "diffusion" results from the local fluctuations in the flow field. It leads to increased rates of momentum and heat and mass transfer, and is essential to the satisfactory operation of spark-ignition and diesel engines. Turbulent flows are always dissipative. Viscous shear stresses perform deformation work on the fluid which increases its internal energy at the expense of its turbulence kinetic energy. So energy is required to generate turbulence: if no energy is supplied, turbulence decays. A common source of energy for turbulent velocity fluctuations is shear in the mean flow. Turbulence is rotational and is characterized by high fluctuating vorticity: these vorticity fluctuations can only persist if the velocity fluctuations are three dimensional.[6]

The character of a turbulent flow depends on its environment. In the engine cylinder, the flow involves a complicated combination of turbulent shear layers, recirculating regions, and boundary layers. The flow is unsteady and may exhibit substantial cycle-to-cycle fluctuations. Both large-scale and small-scale turbulent motions are important factors governing the overall behavior of the flow.[5]

An important characteristic of a turbulent flow is its irregularity or randomness. Statistical methods are therefore used to define such a flow field. The quantities normally used are: the mean velocity, the fluctuating velocity about the mean, and several length and time scales. In a steady turbulent flow situation, the instantaneous local fluid velocity U (in a specific direction) is written:

$$U(t) = \bar{U} + u(t) \tag{8.1}$$

For steady flow, the mean velocity $\bar{U}$ is the time average of $U(t)$:

$$\bar{U} = \lim_{\tau \to \infty} \frac{1}{\tau} \int_{t_0}^{t_0 + \tau} U(t) \, dt \tag{8.2}$$

The fluctuating velocity component u is defined by its root mean square value, the turbulence intensity, u':

$$u' = \lim_{\tau \to \infty} \left(\frac{1}{\tau} \int_{t_0}^{t_0 + \tau} u^2 \, dt \right)^{1/2} \tag{8.3a}$$

Alternatively,

$$u' = \lim_{\tau \to \infty} \left[\frac{1}{\tau} \int_{t_0}^{t_0 + \tau} (U^2 - \bar{U}^2) \, dt \right]^{1/2} \tag{8.3b}$$

since the time average of $(u\bar{U})$ is zero.

In engines, the application of these turbulence concepts is complicated by the fact that the flow pattern changes during the engine cycle. Also, while the overall features of the flow repeat each cycle, the details do not because the mean flow can vary significantly from one engine cycle to the next. There are both cycle-to-cycle variations in the mean or bulk flow at any point in the cycle, as well as turbulent fluctuations about that specific cycle's mean flow.

One approach used in quasi-periodic flows such as that which occurs in the engine cylinder is *ensemble-averaging* or *phase-averaging*. Usually, velocity measurements are made over many engine cycles, and over a range of crank angles. The instantaneous velocity at a specific crank angle position θ in a particular cycle i can be written as

$$U(\theta, i) = \bar{U}(\theta, i) + u(\theta, i) \tag{8.4}$$

The ensemble- or phase-averaged velocity, $\bar{U}(\theta)$, is defined as the average of values at a specific phase or crank angle in the basic cycle. Figure 8-5 shows this approach applied schematically to the velocity variation during a two-stroke engine cycle, with small and large cycle-to-cycle variations. The ensemble-averaged velocity is the average over a large number of measurements taken at the same crank angle (two such points are indicated by dots):

$$\bar{U}_{EA}(\theta) = \frac{1}{N_c} \sum_{i=1}^{N_c} U(\theta, i) \tag{8.5}$$

where N_c is the number of cycles for which data are available. By repeating this

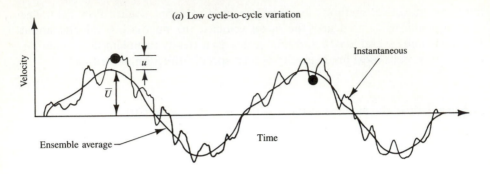

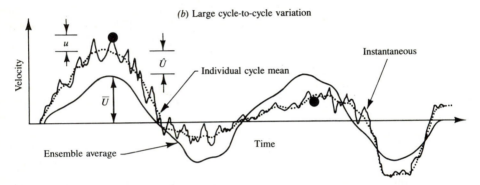

FIGURE 8-5
Schematic of velocity variation with crank angle at a fixed location in the cylinder during two consecutive cycles of an engine. Dots indicate measurements of instantaneous velocity at the same crank
angle. Ensemble- or phase-averaged velocity obtained by averaging over a large number of such
measurements shown as solid smooth line. Top graph: low cycle-to-cycle flow variations. Here the
individual-cycle mean velocity and ensemble-averaged velocity are closely comparable. Bottom
graph: large cycle-to-cycle variations. Here the individual-cycle mean velocity (dotted line) is different
from the ensemble-averaged mean by $\hat{U}$. The turbulent fluctuation u is then defined in relation to the
individual-cycle mean.[5]

process at many crank angle locations, the ensemble-averaged velocity profile
over the complete cycle is obtained.

The ensemble-averaged mean velocity is only a function of crank angle
since the cyclic variation has been averaged out. The difference between the mean
velocity in a particular cycle and the ensemble-averaged mean velocity over many
cycles is defined as the cycle-by-cycle variation in mean velocity:

$$\hat{U}(\theta, i) = \bar{U}(\theta, i) - \bar{U}_{EA}(\theta) \qquad (8.6)$$

Thus the instantaneous velocity, given by Eq. (8.4), can be split into three components:[7]

$$U(\theta, i) = \bar{U}_{EA}(\theta) + \hat{U}(\theta, i) + u(\theta, i) \qquad (8.7)$$

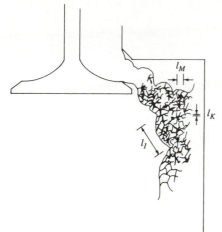

FIGURE 8-6
Schematic of jet created by flow through the intake
valve indicating its turbulent structure.[5, 6]

Figure 8-5 illustrates this breakdown of the instantaneous velocity into an
ensemble-averaged component, an individual-cycle mean velocity, and a component which randomly fluctuates in time at a particular point in space in a
single cycle. This last component is the conventional definition of the turbulent
velocity fluctuation. Whether this differs significantly from the fluctuations about
the ensemble-averaged velocity depends on whether the cycle-to-cycle fluctuations are small or large. The figure indicates these two extremes.†

In turbulent flows, a number of length scales exist that characterize different
aspects of the flow behavior. The largest eddies in the flow are limited in size by
the system boundaries. The smallest scales of the turbulent motion are limited by
molecular diffusion. The important length scales are illustrated by the schematic
of the jet issuing into the cylinder from the intake valve in Fig. 8-6. The eddies
responsible for most of the turbulence production during intake are the large
eddies in the conical inlet jet flow. These are roughly equal in size to the local jet
thickness. This scale is called the *integral scale*, l_I: it is a measure of the largest
scale structure of the flow field. Velocity measurements made at two points
separated by a distance x significantly less than l_I will correlate with each other;
with $x \gg l_I$, no correlation will exist. The integral length scale is, therefore,
defined as the integral of the autocorrelation coefficient of the fluctuating velocity
at two adjacent points in the flow with respect to the variable distance between

---

† There is considerable debate as to whether the fluctuating components of the velocity $U(\theta, i)$ defined
by Eq. (8.7) (cycle fluctuations in the mean velocity and fluctuations in time about the individual cycle
mean) are physically distinct phenomena. The high-frequency fluctuations in velocity are often defined
as "turbulence." The low-frequency fluctuations are generally attributed to the variations in the mean
flow between individual cycles, a phenomenon that is well established. Whether this distinction is
valid has yet to be resolved.

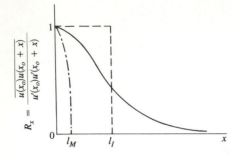

FIGURE 8-7
Spatial velocity autocorrelation R_x as a function of x, defining the integral length scale l_I and the micro length scale l_M.

the points, as shown in Fig. 8-7: i.e.,

$$l_I = \int_0^\infty R_x \, dx \qquad (8.8a)$$

where

$$R_x = \frac{1}{N_m - 1} \sum_{i=1}^{N_m} \frac{\overline{u(x_0)u(x_0 + x)}}{\overline{u'(x_0)u'(x_0 + x)}} \qquad (8.8b)$$

This technique for determining the integral scale requires simultaneous measurements at two points. Due to the difficulty of applying such a technique in engines, most efforts to determine length scales have first employed correlations to determine the *integral time scale*, τ_I. The integral time scale of turbulence is defined as a correlation between two velocities at a fixed point in space, but separated in time:

$$\tau_I = \int_0^\infty R_t \, dt \qquad (8.9a)$$

where

$$R_t = \frac{1}{N_m - 1} \sum_{i=1}^{N_m} \frac{\overline{u(t_0)u(t_0 + t)}}{\overline{u'(t_0)u'(t_0 + t)}} \qquad (8.9b)$$

and N_m is the number of measurements. Under conditions where the turbulence pattern is convected past the observation point without significant distortion and the turbulence is relatively weak, the integral length and time scales are related by

$$l_I = \bar{U}\tau_I \qquad (8.10)$$

In flows where the large-scale structures are convected, τ_I is a measure of the time it takes a large eddy to pass a point. In flows without mean motion, the integral time scale is an indication of the lifetime of an eddy.[5, 8]

Superposed on this large-scale flow is a range of eddies of smaller and smaller size, fed by the continual breakdown of larger eddies. Since the smaller eddies respond more rapidly to changes in local flow pattern, they are more likely to be isotropic (have no preferred direction) than are the large eddies, and have a structure like that of other turbulent flows. The dissipation of turbulence energy takes place in the smallest structures. At this smallest scale of the turbulent motion, called the *Kolmogorov scale*, molecular viscosity acts to dissipate small-scale kinetic energy into heat. If ε is the energy dissipation rate per unit mass and v the kinematic viscosity, Kolmogorov length and time scales are defined by

$$l_K = \left(\frac{v^3}{\varepsilon}\right)^{1/4} \qquad \tau_K = \left(\frac{v}{\varepsilon}\right)^{1/2} \tag{8.11}$$

The Kolmogorov length scale indicates the size of the smallest eddies. The Kolmogorov time scale characterizes the momentum-diffusion of these smallest structures.

A third scale is useful in characterizing a turbulent flow. It is called the *microscale* (or Taylor microscale). The micro length scale l_M is defined by relating the fluctuating strain rate of the turbulent flow field to the turbulence intensity:

$$\frac{\partial u}{\partial x} \approx \frac{u'}{l_M} \tag{8.12}$$

It can be determined from the curvature of the spatial correlation curve at the origin, as shown in Fig. 8-7.[5, 6] More commonly, the micro time scale τ_M is determined from the temporal autocorrelation function of Eq. (8.9):

$$\tau_M^2 = -\frac{2}{(\partial^2 R_t/\partial t^2)_{t_0}}$$

For turbulence which is homogeneous (has no spatial gradients) and is isotropic (has no preferred direction), the microscales l_M and τ_M are related by

$$l_M = \bar{U}\tau_M \tag{8.13}$$

These different scales are related as follows. The turbulent kinetic energy per unit mass in the large-scale eddies is proportional to u'^2. Large eddies lose a substantial fraction of this energy in one "turnover" time l_I/u'. In an equilibrium situation the rate of energy supply equals the rate of dissipation:

$$\varepsilon \approx \frac{u'^3}{l_I}$$

Thus,

$$\frac{l_K}{l_I} \approx \left(\frac{u'l_I}{v}\right)^{-3/4} = \text{Re}_T^{-3/4} \tag{8.14}$$

where Re_T is the turbulent Reynolds number, $u'l_I/v$.

Within the restrictions of homogeneous and isotropic turbulence, an energy budget can be used to relate l_I and l_M:[6]

$$\varepsilon = \frac{Au'^3}{l_I} = \frac{15vu'^2}{l_M^2}$$

where A is a constant of order 1. Thus,

$$\frac{l_M}{l_I} = \left(\frac{15}{A}\right)^{1/2} \text{Re}_T^{-1/2} \tag{8.15}$$

These restrictions are not usually satisfied within the engine cylinder during intake. They are approximately satisfied at the end of compression.

8.2.2 Application to Engine Velocity Data

As has been explained above, it is necessary to analyze velocity data on an individual cycle basis as well as using ensemble-averaging techniques. The basic definitions for obtaining velocities which characterize the flow will now be developed. The ensemble-averaged velocity $\bar{U}_{EA}$ has already been defined by Eq. (8.5). The ensemble-averaged fluctuation intensity $u'_{F, EA}$ is given by

$$u'_{F, EA}(\theta) = \left\{\frac{1}{N_c} \sum_{i=1}^{N_c} [u(\theta, i)]^2\right\}^{1/2} = \left\{\frac{1}{N_c} \sum_{i=1}^{N_c} [U(\theta, i)^2 - \bar{U}_{EA}(\theta)^2]\right\}^{1/2} \tag{8.16}$$

It includes all fluctuations about the ensemble-averaged mean velocity.

Use of Eqs. (8.5) and (8.16) requires values for U and u at each specific crank angle under consideration. While some measurement techniques (e.g., hot-wire anemometry) provide this, the preferred velocity measurement method (laser doppler anemometry) provides an intermittent signal. With laser doppler anemometry (LDA), interference fringes are produced within the small volume created by the intersection of two laser beams within the flow field. When a small particle passes through this volume, it scatters light at a frequency proportional to the particle velocity. By seeding the flow with particles small enough to be carried without slip by the flow and collecting this scattered light, the flow velocity is determined.[9] A signal is only produced when a particle moves through the measurement volume; thus one collects data as velocity crank angle pairs. It is necessary, therefore, to perform the ensemble-averaging over a finite crank angle window $\Delta\theta$ around the specific crank angle of interest, $\bar{\theta}$. The ensemble-averaged velocity equation becomes

$$\bar{U}_{EA}(\bar{\theta}) = \frac{1}{N_t} \sum_{i=1}^{N_c} \sum_{j=1}^{N_i} U_{i, j}\left(\bar{\theta} \pm \frac{\Delta\theta}{2}\right) \tag{8.17}$$

where N_i is the number of velocity measurements recorded in the window during the ith cycle, N_c is the number of cycles, and N_t is the total number of measurements.† The corresponding equation for the ensemble-averaged root mean square

---

† This need to ensemble-average over a finite crank angle window introduces an error called crank angle broadening, due to change in the mean velocity across the window. This error depends on the velocity gradient, and can be made negligibly small by suitable choice of window size.[9–11]

velocity fluctuation is

$$u'_{F,\,\mathrm{EA}}(\bar{\theta}) = \left\{ \frac{1}{N_t} \sum_{i=1}^{N_c} \sum_{j=1}^{N_i} \left[u_{i,\,j} \left(\bar{\theta} \pm \frac{\Delta\theta}{2} \right) \right]^2 \right\}^{1/2} \tag{8.18}$$

where

$$u_{i,\,j} = U_{i,\,j} - \bar{U}_{\mathrm{EA}} \tag{8.19}$$

As has already been explained, this definition of fluctuation intensity [the ensemble-averaged rms velocity fluctuation, Eq. (8.18)] includes cyclic variations in the mean flow as well as the turbulent fluctuations about each cycle's mean flow.[7] It is necessary to determine the mean and fluctuating velocities on an individual-cycle basis to characterize the flow field more completely. The critical part of this process is defining the mean velocity at a specific crank angle (or within a small window centered about that crank angle) in each cycle. Several methods have been used to determine this individual-cycle mean velocity (e.g., moving window, low-pass filtering, data smoothing, conditional sampling; see Ref. 7 for a summary). A high data rate is required.

In this individual-cycle velocity analysis the individual-cycle time-averaged or mean velocity $\bar{U}(\bar{\theta}, i)$ is first determined.[7, 12] The ensemble average of this mean velocity

$$\bar{U}_{\mathrm{EA}}(\bar{\theta}) = \frac{1}{N_c} \sum_{i=1}^{N_c} \bar{U}\left(\bar{\theta} \pm \frac{\Delta\theta}{2}, i \right) \tag{8.20}$$

is identical to the ensemble-averaged value given by Eq. (8.17). The root mean square fluctuation in individual-cycle mean velocity can then be determined from

$$U_{\mathrm{RMS}}(\bar{\theta}) = \left\{ \frac{1}{N_c} \sum_{i=1}^{N_c} \left[\bar{U}\left(\bar{\theta} \pm \frac{\Delta\theta}{2}, i \right) - \bar{U}_{\mathrm{EA}}(\bar{\theta}) \right]^2 \right\}^{1/2} \tag{8.21}$$

This indicates the magnitude of the cyclic fluctuations in the mean motion. The instantaneous velocity fluctuation from the mean velocity, within a specified window $\Delta\theta$ at a particular crank angle $\bar{\theta}$, is obtained from Eq. (8.4). This instantaneous velocity fluctuation is ensemble-averaged, because it varies substantially cycle-by-cycle and because the amount of data is usually insufficient to give reliable individual-cycle results:

$$u'_{T,\,\mathrm{EA}}(\bar{\theta}) = \left\{ \frac{1}{N_c} \sum_{i=1}^{N_c} \left[U\left(\bar{\theta} \pm \frac{\Delta\theta}{2}, i \right) - \bar{U}\left(\bar{\theta} \pm \frac{\Delta\theta}{2}, i \right) \right]^2 \right\}^{1/2} \tag{8.22}$$

This quantity is the ensemble-averaged turbulence intensity.

Several different techniques have been used to measure gas velocities within the engine cylinder (see Refs. 13 and 14 for brief reviews and references). The technique which provides most complete and accurate data is laser doppler anemometry.[9] Sample results obtained with this technique will now be reviewed to illustrate the major features of the in-cylinder gas motion. The available results must be interpreted with caution since they have been obtained in special engines where the geometry and flow have been modified to make the experiments and their interpretation easier. Also, the flow within the cylinder is three dimensional in nature. It takes measurements at many points within the flow field and the use of a flow visualization technique to characterize the flow adequately.

Figure 8-8 shows ensemble-averaged velocities throughout the engine cycle at two measurement locations in a special L-head engine designed to generate a swirling flow within the cylinder. The engine was motored at 300 rev/min, giving a mean piston speed of 0.76 m/s. Figure 8-8b shows the mean velocity in the path of the swirling intake flow within the clearance volume, in the swirl direction. High velocities are generated during the intake process, rising to a maximum and then decreasing in response to the piston motion (see Fig. 2-2). During the compression stroke, the velocity continues to decrease but at a much slower rate. This is a motored engine cycle. A comparison of intake and compression velocities with an equivalent firing cycle showed close agreement.[15] The expansion and

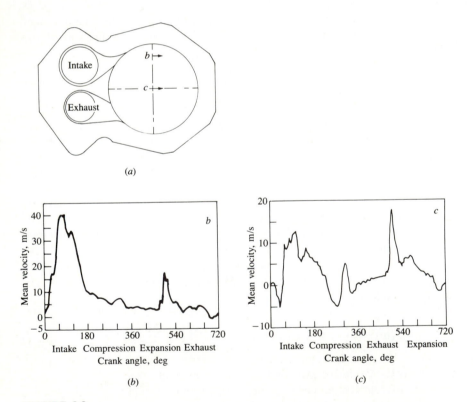

(a)

(b)

(c)

FIGURE 8-8
Ensemble-averaged velocities throughout the engine cycle in motored four-stroke L-head engine: 300 rev/min, mean piston speed 0.76 m/s. (a) Engine schematic showing measurement locations and velocity directions; (b) velocity at b in intake flow path; (c) velocity at c on cylinder axis.[11]

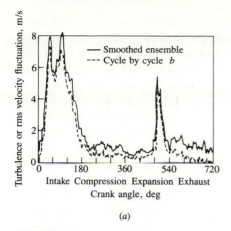

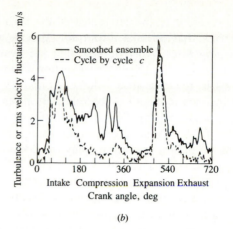

(a)

(b)

FIGURE 8-9
Ensemble-averaged rms velocity fluctuation and ensemble-averaged individual-cycle turbulence inten-sity as a function of crank angle: (a) at location b in Fig. 8-8a; (b) at location c in Fig. 8-8a.[11]

exhaust stroke velocities are not typical of firing engine behavior, however.†
Figure 8-8c shows the mean velocity in the clearance volume in the same direc-tion but on the cylinder axis. At this location, positive and negative flow veloc-ities were measured. Since this location is out of the path of the intake generated flow, velocities during the intake stroke are much lower. The nonhomogeneous character of this particular ensemble-mean flow is evident.

Figure 8-9 shows the ensemble-averaged rms velocity fluctuation (which includes contributions from cycle-by-cycle variations in the mean flow and turbulence) and the ensemble-averaged individual-cycle turbulence intensity at these same two locations and directions. The difference between the two curves in each graph is an indication of the cycle-by-cycle variation in the mean flow [see Eq. (8.7)]. During the intake process, within the directed intake flow pattern, the cycle-by-cycle variation in the mean flow is small in comparison to the high turbulence levels created by the intake flow. Outside this directed flow region, again during intake, this cycle-by-cycle contribution is more significant relative to the turbulence. During compression, the cycle-by-cycle mean flow variation is comparable in magnitude to the ensemble-averaged turbulence intensity. It is therefore highly significant.

Two important questions regarding the turbulence in the engine cylinder are whether it is homogeneous (i.e., uniform) and whether it is isotropic (i.e., independent of direction). The data already presented in Figs. 8-8 and 8-9 show that during intake the flow is far from homogeneous. Nor is it isotropic.[11]

---

† The increase in velocity when the exhaust valve opens is due to the flow of gas *into* the cylinder because, due primarily to heat losses, the cylinder pressure is then below 1 atm.

However, it is the character of the turbulence at the end of the compression process that is most important: that is what controls the fuel-air mixing and burning rates. Only limited data are available which relate to these questions. With open disc-shaped combustion chambers, measurements at different locations in the clearance volume around TC at the end of compression show the turbulence intensity to be relatively homogeneous. In the absence of an intake generated swirling flow, the turbulence intensity was also essentially isotropic near TC.[16] These specific results support the more general conclusion that the inlet boundary conditions play the dominant role in the generation of the mean flow and turbulence fields during the intake stroke. However, in the absence of swirl, this intake generated flow structure has almost disappeared by the time the compression process commences. The turbulence levels follow this trend in the mean flow, with the rapid decay process lasting until intake valve closing. Later in the compression process the turbulence becomes essentially homogeneous.[17]

When a swirling flow is generated during intake, an almost solid-body rotating flow develops which remains stable for much longer than the inlet jet generated rotating flows illustrated in Fig. 8-3. With simple disc-shaped chambers, the turbulence still appears to become almost homogeneous at the end of compression. With swirl and bowl-in-piston geometry chambers, however, the flow is more complex (see Sec. 8.3).

The flow through the intake valve or port is responsible for many features of the in-cylinder motion. The flow velocity through the valve is proportional to the piston speed [see Eq. (6.10) for pseudo valve flow velocity, and Eq. 2.10)]. It would be expected therefore that in-cylinder flow velocities at different engine speeds would scale with mean piston speed [Eq. (2.11)]. Figure 8-10 shows ensemble-averaged mean and rms velocity fluctuations, normalized by the mean piston speed through the cycle at three different engine speeds. The measurement location is in the path of the intake generated swirling flow (point b in Fig. 8-8a). All the curves have approximately the same shape and magnitude, indicating the

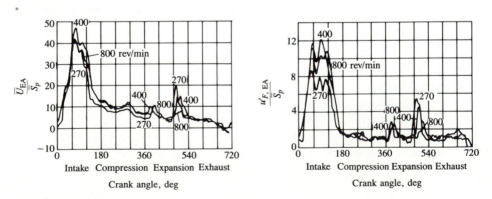

FIGURE 8-10
Ensemble-averaged mean and rms velocity fluctuations, normalized by mean piston speed, throughout the engine cycle for three engine speeds. Location b in Fig. 8-8a.[11]

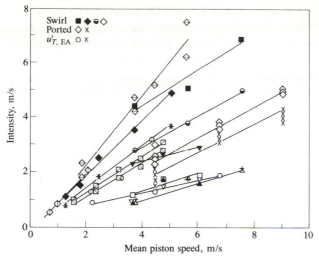

FIGURE 8-11
Individual-cycle turbulence intensity $u'_{T,\,EA}$ (OX) and ensemble-averaged rms fluctuation velocity (remaining symbols) at TC at the end of compression, for a number of different flow configurations and chamber geometries as a function of mean piston speed.[16] Two data sets for two-stroke ported engines. Four data sets with intake generated swirl.

appropriateness of this velocity scaling.† Other results support this conclusion, though in the absence of an ordered mean motion such as swirl when the ensemble-averaged mean velocities at the end of compression are low, this scaling for the mean velocity does not always hold.[16] Figure 8-11 shows a compilation of ensemble-averaged rms fluctuation velocity or ensemble-averaged individual-cycle turbulence intensity results at TC at the end of compression, from 13 different flow configurations and combustion chamber geometries. Two of these sets of results are from two-stroke cycle ported configurations. The measured fluctuating velocities or turbulence intensities are plotted against mean piston speed. The linear relationship holds well. There is a substantial variation in the proportionality constant, in part because in most of these studies (identified in the figure) the ensemble-averaged rms fluctuation velocity was the quantity measured. Since this includes the cycle-by-cycle fluctuation in the mean velocity, it is larger (by up to a factor of 2) than the average turbulence intensity $u'_{T,\,EA}$.

A consensus conclusion is emerging from these studies that the turbulence intensity at top-center, with open combustion chambers in the absence of swirl, has a maximum value equal to about half the mean piston speed:[16, 18]

$$u'_{T,\,EA}(TC) \approx 0.5\bar{S}_p \tag{8.23}$$

---

† Note that because of the valve and combustion chamber of this particular engine, the ratio of $\bar{U}$ to $\bar{S}_p$ is higher than is typical of normal engine geometries.

The available data show that the turbulence intensity at TC with swirl is usually higher than without swirl[16] (see the four data sets with swirl in Fig. 8-11). Some data, however, indicate that the rms fluctuation intensity with swirl may be lower.[18] The ensemble-averaged cyclic variation in individual-cycle mean velocity at the end of compression also scales with mean piston speed. This quantity can be comparable in magnitude to the turbulence intensity. It usually decreases when a swirling flow is generated within the cylinder during the intake process.[11, 16]

During the compression stroke, and also during combustion while the cylinder pressure continues to rise, the unburned mixture is compressed. Turbulent flow properties are changed significantly by the large and rapidly imposed distortions that result from this compression. Such distortions, in the absence of dissipation, would conserve the angular momentum of the flow: rapid compression would lead to an increase in vorticity and turbulence intensity. There is evidence that, with certain types of in-cylinder flow pattern, an increase in turbulence intensity resulting from piston motion and combustion does occur toward the end of the compression process. The compression of large-scale rotating flows can cause this increase due to the increasing angular velocity required to conserve angular momentum resulting in a growth in turbulence generation by shear.[19]

Limited results are available which characterize the turbulence time and length scales in automobile engine flows. During the intake process, the integral length scale is of the order of the intake jet diameter, which is of the order of the valve lift ($\lesssim 10$ mm in automobile-size engines). During compression the flow relaxes to the shape of the combusion chamber. The integral time scale at the end of compression decreases with increasing engine speed. It is of order 1 ms at engine speeds of about 1000 rev/min. The integral length scale at the end of compression is believed to scale with the clearance height and varies little with engine speed. It decreases as the piston approaches TC to about 2 mm ($0.2 \times$ clearance height). The micro time scale at the end of compression is of order 0.1 ms at 1000 rev/min, and decreases as engine speed increases (again in automobile-size engine cylinders). Micro length scales are of order 1 mm at the end of compression and vary little with engine speed. Kolmogorov length scales at the end of compression are of order 10^{-2} mm.[8, 20, 21]

8.3 SWIRL

Swirl is usually defined as organized rotation of the charge about the cylinder axis. Swirl is created by bringing the intake flow into the cylinder with an initial angular momentum. While some decay in swirl due to friction occurs during the engine cycle, intake generated swirl usually persists through the compression, combustion, and expansion processes. In engine designs with bowl-in-piston combustion chambers, the rotational motion set up during intake is substantially modified during compression. Swirl is used in diesels and some stratified-charge engine concepts to promote more rapid mixing between the inducted air charge and the injected fuel. Swirl is also used to speed up the combustion process in spark-ignition engines. In two-stroke engines it is used to improve scavenging. In some designs of prechamber engines, organized rotation about the prechamber axis is also called swirl. In prechamber engines where swirl within the precombustion chamber is important, the flow into the prechamber during the compression process creates the rotating flow. Prechamber flows are discussed in Sec. 8.5.

8.3.1 Swirl Measurement

The nature of the swirling flow in an actual operating engine is extremely difficult to determine. Accordingly, steady flow tests are often used to characterize the swirl. Air is blown steadily through the inlet port and valve assembly in the cylinder head into an appropriately located equivalent of the cylinder. A common technique for characterizing the swirl within the cylinder has been to use a light paddle wheel, pivoted on the cylinder centerline (with low friction bearings), mounted between 1 and 1.5 bore diameters down the cylinder. The paddle wheel diameter is close to the cylinder bore. The rotation rate of the paddle wheel is used as a measure of the air swirl. Since this rotation rate depends on the location of the wheel and its design, and the details of the swirling flow, this technique is being superseded by the impulse swirl meter shown in Fig. 8-12. A honeycomb flow straightener replaces the paddle wheel: it measures the total torque exerted by the swirling flow. This torque equals the flux of angular

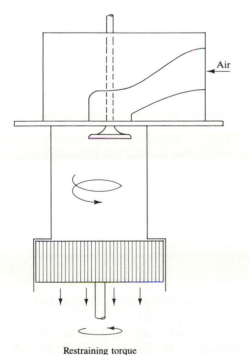

Restraining torque

FIGURE 8-12
Schematic of steady-flow impulse torque swirl meter.[22]

momentum through the plane coinciding with the flow-straightener upstream face.

For each of these approaches, a *swirl coefficient* is defined which essentially compares the flow's angular momentum with its axial momentum. For the paddle wheel, the swirl coefficient C_s is defined by

$$C_s = \frac{\omega_p B}{v_0} \tag{8.24}$$

where ω_p is the paddle wheel angular velocity ($=2\pi N_p$, where N_p is the rotational speed) and the bore B has been used as the characteristic dimension. The characteristic velocity, v_0, is derived from the pressure drop across the valve using an incompressible flow equation:

$$v_0 = \left[\frac{2(p_0 - p_c)}{\rho}\right]^{1/2} \tag{8.25}$$

or a compressible flow equation:

$$v_0 = \left\{\frac{2\gamma}{(\gamma - 1)} \frac{p_0}{\rho_0}\left[1 - \left(\frac{p_c}{p_0}\right)^{(\gamma - 1)/\gamma}\right]\right\}^{1/2} \tag{8.26}$$

where the subscripts 0 and c refer to upstream stagnation and cylinder values, respectively. The difference between Eqs. (8.25) and (8.26) is usually small. With the impulse torque meter, characteristic velocity and length scales must also be introduced. Several swirl parameters have been defined.[22, 23] The simplest is

$$C_s = \frac{8T}{\dot{m} v_0 B} \tag{8.27}$$

where T is the torque and $\dot{m}$ the air mass flow rate. The velocity v_0, defined by Eq. (8.25) or Eq. (8.26), and the bore have again been used to obtain a dimensionless coefficient. Note that for solid-body rotation of the fluid within the cylinder at the paddle wheel speed ω_p, Eqs. (8.24) and (8.27) give identical swirl coefficients. In practice, because the swirling flow is not solid-body rotation and because the paddle wheel usually lags the flow due to slip, the impulse torque meter gives higher swirl coefficients.[23] When swirl measurements are made in an operating engine, a *swirl ratio* is normally used to define the swirl. It is defined as the angular velocity of a solid-body rotating flow ω_s, which has equal angular momentum to the actual flow, divided by the crankshaft angular rotational speed:

$$R_s = \frac{\omega_s}{2\pi N} \tag{8.28}$$

During the induction stroke in an engine the flow and the valve open area, and consequently the angular momentum flux into the cylinder, vary with crank angle. Whereas in rig tests the flow and valve open area are fixed and the angular momentum passes down the cylinder continuously, in the engine intake process the momentum produced under corresponding conditions of flow and valve lift remains in the cylinder. Steady-state impulse torque-meter flow rig data can be used to estimate engine swirl in the following manner.[23] Assuming that the port and valve retain the same characteristics under the transient conditions of the engine as on the steady-flow rig, the equivalent solid-body angular velocity ω_s at the end of the intake process is given by

$$\omega_s = \frac{8}{B^2} \left(\int_{\theta_1}^{\theta_2} T \, d\theta \right) \Big/ \left(\int_{\theta_1}^{\theta_2} \dot{m} \, d\theta \right)$$

where θ_1 and θ_2 are crank angles at the start and end of the intake process and the torque T and mass flow rate $\dot{m}$ are evaluated at the valve lift corresponding to the local crank angle. Using Eq. (8.27) for T, Eq. (6.11) for $\dot{m}$, assuming v_0 and ρ are constant throughout the intake process, and introducing volumetric efficiency η_v based on intake manifold conditions via Eq. (2.27), it can be shown that

$$R_s = \frac{\omega_s}{2\pi N} = \pi\eta_v \, BL \left[\int_{\theta_1}^{\theta_2} (A_v C_D) C_s \, d\theta \right] \Big/ \left[\int_{\theta_1}^{\theta_2} (A_v C_D) \, d\theta \right]^2 \qquad (8.29)$$

where $A_v C_D$ is the effective valve open area at each crank angle. Note that the crank angle in Eq. (8.29) should be in radians. Except for its (weak) dependence on η_v, Eq. (8.29) gives R_s independent of operating conditions directly from rig test results and engine geometry.

The relationship between steady-flow rig tests (which are extensively used because of their simplicity) and actual engine swirl patterns is not fully understood. Steady-flow tests adequately describe the swirl generating characteristics of the intake port and valve (at fixed valve lift) and are used extensively for this purpose. However, the swirling flow set up in the cylinder during intake can change significantly during compression.

8.3.2 Swirl Generation during Induction

Two general approaches are used to create swirl during the induction process. In one, the flow is discharged into the cylinder tangentially toward the cylinder wall, where it is deflected sideways and downward in a swirling motion. In the other, the swirl is largely generated within the inlet port: the flow is forced to rotate about the valve axis *before* it enters the cylinder. The former type of motion is achieved by forcing the flow distribution around the circumference of the inlet valve to be nonuniform, so that the inlet flow has a substantial net angular momentum about the cylinder axis. The directed port and deflector wall port in Fig. 8-13 are two common ways of achieving this result. The directed port brings the flow toward the valve opening in the desired tangential direction. Its passage is straight, which due to other cylinder head requirements restricts the flow area and results in a relatively low discharge coefficient. The deflector wall port uses the port inner side wall to force the flow preferentially through the outer periphery of the valve opening, in a tangential direction. Since only one wall is used to obtain a directional effect, the port areas are less restrictive.

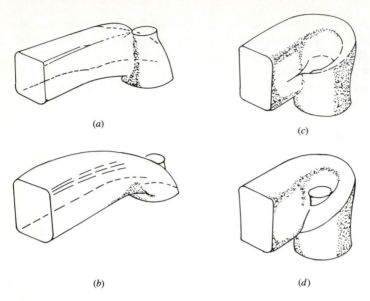

FIGURE 8-13
Different types of swirl-generating inlet ports: (a) deflector wall; (b) directed; (c) shallow ramp helical; (d) steep ramp helical.[24]

Flow rotation about the cylinder axis can also be generated by masking off or shrouding part of the peripheral inlet valve open area, as shown in Fig. 8-14. Use is often made of a mask or shroud on the valve in research engines because changes can readily be made. In production engines, the added cost and weight, problems of distortion, the need to prevent valve rotation, and reduced volumetric efficiency make masking the valve an unattractive approach. The more practical alternative of building a mask on the cylinder head around part of the inlet valve periphery is used in production spark-ignition engines to generate swirl. It can easily be incorporated in the cylinder head casting process.

The second broad approach is to generate swirl within the port, about the valve axis, prior to the flow entering the cylinder. Two examples of such *helical ports* are shown in Fig. 8-13. Usually, with helical ports, a higher flow discharge coefficient at equivalent levels of swirl is obtained, since the whole periphery of the valve open area can be fully utilized. A higher volumetric efficiency results. Also, helical ports are less sensitive to position displacements, such as can occur in casting, since the swirl generated depends mainly on the port geometry above the valve and not the position of the port relative to the cylinder axis.

Figure 8-15 compares steady-state swirl-rig measurements of examples of the ports in Fig. 8-13. The rig swirl number increases with increasing valve lift, reflecting the increasing impact of the port shape and decreasing impact of the flow restriction between the valve head and seat. Helical ports normally impart more angular momentum at medium lifts than do directed ports.[23, 25] The swirl

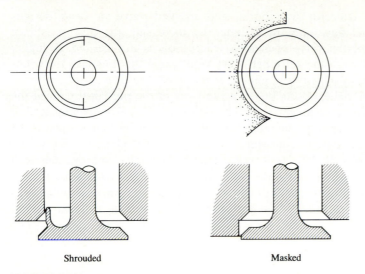

Shrouded   Masked

FIGURE 8-14
Shrouded inlet valve and masked cylinder head approaches for producing net in-cylinder angular momentum.

ratios for these ports calculated from this rig data using Eqs. (8.27) and (8.29) are: 2.5 for the directed port, 2.9 for the shallow ramp helical, and 2.6 for the steep ramp helical. Vane swirl-meter swirl ratios were about 30 percent less. These impulse-swirl-meter derived engine swirl ratios are within about 20 percent of the solid-body rotation rate which has equal angular momentum to that of the cylinder charge determined from tangential velocity measurements made within the cylinder of an operating engine with the same port, at the end of the induction process.[23]

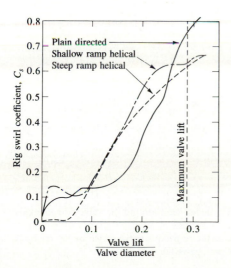

FIGURE 8-15
Steady-state torque meter swirl measurements of directed, shallow ramp, and steep ramp helical ports as a function of inlet valve lift/diameter ratio.[23]

Directed and deflector wall ports, and masked valve or head designs produce a tangential flow into the cylinder by increasing the flow resistance through that part of the valve open area where flow is not desired. A highly nonuniform flow through the valve periphery results and the flow into the cylinder has a substantial v_θ velocity component in the same direction about the cylinder axis. In contrast, helical ports produce the swirl in the port upstream of the valve, and the velocity components v_r, and v_z through the valve opening, and v_θ *about the valve axis* are approximately uniform around the valve open area. Figure 8-16 shows velocity data measured at the valve exit plane in steady-flow rig tests with examples of these two types of port. The valve and cylinder wall locations are shown. In Fig. 8-16a, the deflector wall of the tangentially oriented port effectively prevents any significant flow around half the valve periphery. In contrast, in Fig. 8-16b with the helical port, the air flows into the cylinder around

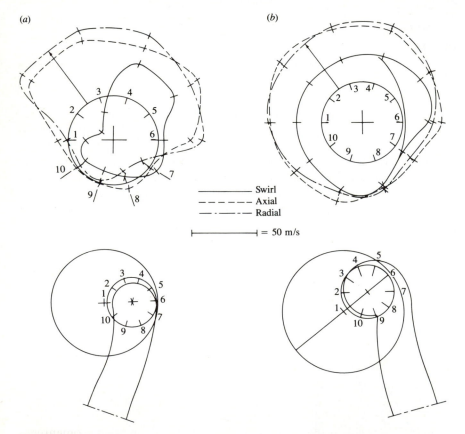

FIGURE 8-16
Swirl, axial, and radial velocities measured 2 mm from cylinder head around the valve circumference for (a) tangential deflector-wall port and (b) helical port; magnitude of velocity is given by the distance along a radial line (from valve axis), from valve outline to the respective curve scaled by the reference length (examples of radial velocity indicated by two arrows); valve lift = 12.8 mm.[26, 27]

the full valve open area. The radial and axial velocities are essentially uniform around the valve periphery. The swirl velocity about the valve axis (anticlockwise when viewed from above) for this helical port is relatively uniform and is about half the magnitude of the radial and axial velocities.

The swirling air flow within the cylinder of an operating engine is not uniform. The velocities generated at the valve at each point in the induction process depend on the valve open area and piston velocity. The velocities are highest during the first half of the intake process as indicated in Fig. 6-15. Thus, the swirl velocities generated during this portion of the induction stroke are higher than the swirl generated during the latter half of the stroke: there is swirl stratification. Also, the flow pattern close to the cylinder head during induction is comparatively disorganized, and not usually close to a solid-body rotation. It consists of a system of vortices, created by the high-velocity tangential or spiraling intake jet. Further down the cylinder, the flow pattern is closer to solid-body rotation with the swirl velocity increasing with increasing radius.[23, 24] This more ordered flow directly above the piston produces higher swirl velocities in that region of the cylinder. As the piston velocity decreases during intake, the swirl pattern redistributes, with swirl speeds close to the piston decreasing and swirl speeds in the center of the cylinder increasing.[27] Note that the axis of rotation of the in-cylinder gases may not exactly coincide with the cylinder axis.

8.3.3 Swirl Modification within the Cylinder

The angular momentum of the air which enters the cylinder at each crank angle during induction decays throughout the rest of the intake process and during the compression process due to friction at the walls and turbulent dissipation within the fluid. Typically one-quarter to one-third of the initial moment of momentum about the cylinder axis will be lost by top-center at the end of compression. However, swirl velocities in the charge can be substantially increased during compression by suitable design of the combustion chamber. In many designs of direct-injection diesel, air swirl is used to obtain much more rapid mixing between the fuel injected into the cylinder and the air than would occur in the absence of swirl. The tangential velocity of the swirling air flow set up inside the cylinder during induction is substantially increased by forcing most of the air into a compact bowl-in-piston combustion chamber, usually centered on the cylinder axis, as the piston approaches its top-center position. Neglecting the effects of friction, angular momentum is conserved, and as the moment of inertia of the air is decreased its angular velocity must increase.

However, the total angular momentum of the charge within the cylinder does decay due to friction at the chamber walls. The angular momentum of the cylinder charge Γ_c changes with time according to the moment of momentum conservation equation:

$$\frac{d\Gamma_c}{dt} = J_i - T_f \tag{8.30}$$

where J_i is the flux of angular momentum into the cylinder and T_f is the torque due to wall friction. At each point in the intake process J_i is given by

$$J_i = \int_{A_v} \rho r v_\theta \, \mathbf{v} \cdot d\mathbf{A}_v \tag{8.31}$$

where $d\mathbf{A}_v$ is an element of the valve open area, as defined in Fig. 8-17. While the angular momentum entering the cylinder during the intake process is

$$\Gamma_{c,\,i} = \int_{tivo}^{tivc} \int_{A_v} \rho r v_\theta \cdot d\mathbf{A}_v \, dt$$

the actual angular momentum within the cylinder at the end of induction will be less, due to wall friction during the intake process. Friction continues through the compression process so the total charge angular momentum at the end of compression is further reduced.

There is friction on the cylinder wall, cylinder head, and piston crown (including any combustion chamber within the crown). This friction can be estimated with sufficient accuracy using friction formulas developed for flow over a flat plate, with suitable definition of characteristic length and velocity scales. Friction on the cylinder wall can be estimated from the wall shear stress:

$$\tau = \tfrac{1}{2}\rho \left(\frac{\omega_s B}{2} \right)^2 C_F \tag{8.32}$$

where ω_s is the equivalent solid-body swirl. The friction factor C_F is given by the flat plate formula:

$$C_F = 0.037\lambda(\mathrm{Re}_B)^{-0.2} \tag{8.33}$$

where λ is an empirical constant introduced to allow for differences between the flat plate and cylinder wall ($\lambda \approx 1.5$)[28] and Re_B is the equivalent of the flat plate Reynolds number [$\mathrm{Re}_B = \rho(B\omega_s/2)(\pi B)/\mu$]. Friction on the cylindrical walls of a piston cup or bowl can be obtained from the above expressions with D_B, the bowl diameter, replacing the bore.

Friction on the cylinder head, piston crown, and piston bowl floor can be estimated from expressions similar to Eqs. (8.32) and (8.33). However, since the

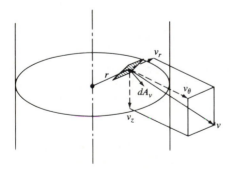

FIGURE 8-17
Definition of symbols in equation for angular momentum flux into the cylinder [Eq. (8.31)].

tangential velocity v_θ at the wall varies with radius, the shear stress should be evaluated at each radius and integrated over the surface: e.g.,[29]

$$\tau(r) = C_1 \tfrac{1}{2}\rho [v_\theta(r)]^2 \, \text{Re}^{-0.2} \tag{8.34}$$

with
$$\text{Re} = \frac{\rho v_\theta(r) r}{\mu}$$

where C_1 is an empirical constant (≈ 0.055). An alternative approximate approach is to evaluate these components of the wall shear stress at the mean radius.[28]

Next, consider the effects on swirl of radially inward displacement of the air charge during compression. The most common example of this phenomenon occurs with the bowl-in-piston combustion chamber design of medium- and high-speed direct-injection diesels (see Sec. 10.2.1). However, in spark-ignition engines where swirl is used to increase the burning rate, the shape of the combustion chamber close to top-center can also force radially inward motion of the charge. For a given swirling in-cylinder flow at the end of induction and neglecting the effects of friction, as the moment of inertia of the air about the cylinder axis is decreased the air's angular velocity must increase to conserve angular momentum. For example, for solid-body rotation of the cylinder air charge of mass m_c, the initial angular momentum $\Gamma_{c,i}$ and solid-body rotation $\omega_{s,i}$ are related at bottom-center by

$$\Gamma_{c,i} = I_c \omega_{s,i}$$

where I_c is the moment of inertia of the charge about the cylinder axis. For a disc-shaped combustion chamber, $I_o = m_a B^2/8$ and is constant. For a bowl-in-piston combustion chamber,

$$I_c = \frac{m_c B^2}{8} \frac{[(z/h_B) + (D_B/B)^4]}{[(z/h_B) + (D_B/B)^2]} \tag{8.35}$$

where D_B and h_B are the diameter and depth of the bowl, respectively, and z is the distance of the piston crown from the cylinder head. At TC crank position, $z \approx 0$ and $I_c \approx m_c D_B^2/8$. At the end of induction, $I_c \approx m_c B^2/8$. Thus, in the absence of friction ω_s would increase by $(B/D_B)^2$, usually a factor of about 4.

In an operating engine with this bowl-in-piston chamber design, the observed increase in swirl in the bowl is less; it is usually about a factor of 2 to 3.[23, 25] This is because of wall friction, dissipation in the fluid due to turbulence and velocity gradients, and the fact that a fraction of the fluid remains in the clearance height above the piston crown. The loss in angular momentum due to these effects will vary with geometric details, initial swirl flow pattern, and engine speed.

Swirl velocity distributions in the cylinder at the end of induction show the tangential velocity increasing with radius, except close to the cylinder wall where friction causes the velocity to decrease. While the velocity distribution is not that of a solid-body rotation, depending on port design and operating conditions it is often close to solid-body rotation.[23, 25] Departures from the solid-body velocity distribution are greater at higher engine speeds, suggesting that the flow pattern in the cylinder at this point in the cycle is still developing with time.[23, 30] In the absence of radially inward gas displacement during compression, the flow pattern continues to develop toward a solid-body distribution throughout the compression stroke.[25] Swirl ratios of 3 to 5 at top-center can be achieved with the ports shown in Fig. 8-13, with flat-topped pistons (i.e., in the absence of any swirl amplification during compression).[23, 25]

With combustion chambers where the chamber radius is less than the cylinder bore, such as the bowl in piston, the tangential velocity distribution with radius will change during compression. Even if the solid-body rotation assumption is reasonable at the end of induction, the profile will distort as gas moves into the piston bowl. Neglecting the effects of friction, the angular momentum of each fluid element will remain constant as it moves radially inward. Thus the increase in tangential velocity of each fluid element as it moves radially inward is proportional to the change in the reciprocal of its radius. Measurements of the swirl velocity distribution within the cylinder of bowl-in-piston engine designs support this description. The rate of displacement of gas into the bowl depends on the bowl volume V_B, cylinder volume V, and piston speed S_p, at that particular piston position:

$$\frac{dm_B}{dt} = \frac{m_c}{L}\left(\frac{V_B}{V}\right)\left(\frac{V_d}{V}\right)S_p$$

The gas velocity into the bowl will therefore increase rapidly toward the end of the compression stroke and reach a maximum just before TC (see Sec. 8.4 where this radial "squish" motion is discussed more fully). Thus, there is a rapid increase in v_θ in the bowl as the crank angle approaches TC. The lower layers of the bowl rotate slower than the upper layers because that gas entered the bowl earlier in the compression process.[23, 25]

Velocity measurements illustrating the development of this radial distribution in tangential velocity are shown in Fig. 8-18. These measurements were made by analysing the motion of burning carbon particles in the cylinder of an operating diesel engine from movies of the combustion process. The figure shows the engine geometry and the data compared with a model based on gas displacement and conservation of angular momentum in each element of the charge as it is displaced inward. Different swirl velocity profiles exist within and outside the bowl as the piston approaches TC. Swirl velocities within the bowl increase as TC is approached, roughly as predicted by the ideal model. Outside the bowl, the swirl velocity decreases with increasing radius due to the combined effects of friction and inward gas displacement as the clearance height decreases.

Swirl ratios in bowl-in-piston engine designs of up to about 15 can be achieved with $D_B \approx 0.5B$, at top-center. Amplification factors relative to flat-topped piston swirl are typically about 2.5 to 3, some 30 percent lower than the ideal factor of $(B/D_B)^2$ given by Eq. (8.35) as $z \to 0$. This difference is due to the mass remaining within the clearance height which does not enter the bowl, and

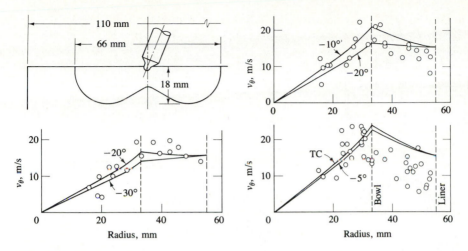

FIGURE 8-18
Velocity measurements as a function of radius across the combustion chamber of a firing, bowl-in-piston, direct-injection diesel engine. Schematic shows the chamber geometry. Solid lines are calculations based on the assumption of constant angular momentum for fluid elements as they move radially inward.[31]

the effects of wall friction (enhanced by the higher gas velocities in the bowl). Sometimes the bowl axis is offset from the cylinder axis and some additional loss in swirl amplification results.[25]

The effect of swirl generation during induction on velocity fluctuations in the combustion chamber at the end of compression has been examined.[32] The turbulence intensity with swirl was higher than without swirl (with the same chamber geometry). Integral scales of the turbulence were smaller with swirl than without. Cyclic fluctuations in the mean velocity are, apparently, reduced by swirl. Also, some studies show that the ensemble-averaged fluctuation intensity goes down when swirl is introduced.[18] There is evidence that swirl makes the turbulence intensity more homogeneous.[30]

8.4 SQUISH

Squish is the name given to the radially inward or transverse gas motion that occurs toward the end of the compression stroke when a portion of the piston face and cylinder head approach each other closely. Figure 8-19 shows how gas is thereby displaced into the combustion chamber. Figure 8-19a shows a typical wedge-shaped SI engine combustion chamber and Fig. 8-19b shows a bowl-in-piston diesel combustion chamber. The amount of squish is often defined by the *percentage squish area*: i.e., the percentage of the piston area, $\pi B^2/4$, which closely approaches the cylinder head (the shaded areas in Fig. 8-19). Squish-generated gas motion results from using a compact combustion chamber geometry.

A theoretical squish velocity can be calculated from the instantaneous dis-

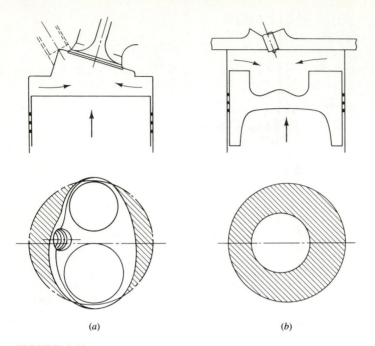

(a)   (b)

FIGURE 8-19
Schematics of how piston motion generates squish: (a) wedge-shaped SI engine combustion chamber;
(b) bowl-in-piston direct-injection diesel combustion chamber.

placement of gas across the inner edge of the squish region (across the dashed lines in the drawings in Fig. 8-20a and b), required to satisfy mass conservation. Ignoring the effects of gas dynamics (nonuniform pressure), friction, leakage past the piston rings, and heat transfer, expressions for the squish velocity are:

1. *Bowl-in-piston chamber* (Fig. 8-20a):[33]

$$\frac{v_{sq}}{S_p} = \frac{D_B}{4z}\left[\left(\frac{B}{D_B}\right)^2 - 1\right]\frac{V_B}{A_c z + V_B} \tag{8.36}$$

where V_B is the volume of the piston bowl, A_c is the cross-sectional area of the cylinder ($\pi B^2/4$), S_p is the instantaneous piston speed [Eq. (2.11)], and z is the distance between the piston crown top and the cylinder head ($z = c + Z$, where $Z = l + a - s$; see Fig. 2-1).

2. *Simple wedge chamber* (Fig. 8-20b):[34]

$$\frac{v_{sq}}{S_p} = \frac{A_s}{b(Z + c)}\left(1 - \frac{Z + c}{C + Z}\right) \tag{8.37}$$

where A_s is the squish area, b is the width of the squish region, and C is $Z/(r_c - 1)$ evaluated at the end of induction.

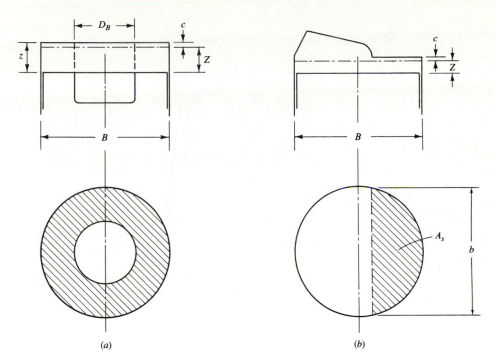

FIGURE 8-20
(a) Schematic of axisymmetric bowl-in-piston chamber for Eq. (8.36). (b) Schematic of wedge chamber with transverse squish for Eq. (8.37).

The theoretical squish velocity for a bowl-in-piston engine normalized by the mean piston speed $\bar{S}_p$ is shown in Fig. 8-21 for different ratios of D_B/B and clearance heights c. The maximum squish velocity occurs at about $10°$ before TC. After TC, v_{sq} is negative; a reverse squish motion occurs as gas flows out of the bowl into the clearance height region. Under motored conditions this is equal to the forward motion.

These models omit the effects of gas inertia, friction, gas leakage past the piston rings, and heat transfer. Gas inertia and friction effects have been shown to be small. The effects of gas leakage past the piston rings and of heat transfer are more significant. The squish velocity decrement Δv_L due to leakage is proportional to the mean piston speed and a dimensionless leakage number:

$$N_L = A_{E,\,L} \frac{\sqrt{\gamma R T_{\text{IVC}}}}{N V_d} \qquad (8.38)$$

where $A_{E,\,L}$ is the effective leakage area and T_{IVC} is the temperature of the cylinder gases at inlet valve closing. Leakage was modeled as a choked flow through the effective leakage area. Values of $\Delta v_L/v_{sq}$ are shown in Fig. 8-22. The effect of leakage on v_{sq} is small for normal gas leakage rates. A decrement on squish

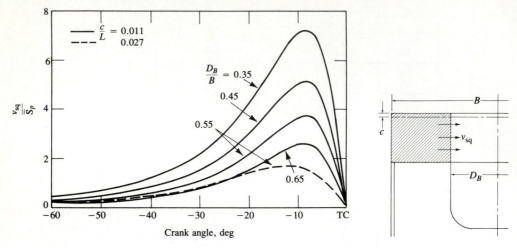

FIGURE 8-21
Theoretical squish velocity divided by mean piston speed for bowl-in-piston chambers, for different D_B/B and c/L (clearance height/stroke). $B/L = 0.914$, $V_B/V_d = 0.056$, connecting rod length/crank radius = 3.76.[35]

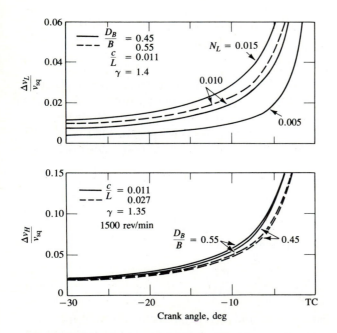

FIGURE 8-22
Values of squish velocity decrement due to leakage Δv_L and heat transfer Δv_H, normalized by the ideal squish velocity, as a function of crank angle.[35]

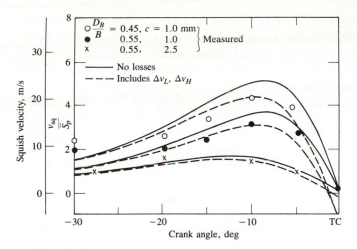

FIGURE 8-23
Comparison of measured squish velocities in bowl-in-piston combustion chambers, with different bowl diameter/bore ratios and clearance heights, to calculated ideal squish velocity (solid lines) and calculations corrected for leakage and heat transfer (dashed lines). Bore = 85 mm, stroke = 93 mm, 1500 rev/min.[35]

velocity due to heat transfer, Δv_H, has also been derived, using standard engine heat-transfer correlations (see Sec. 12.4). Values of $\Delta v_H/v_{sq}$ are also shown in Fig. 8-22. Again the effects are small in the region of maximum squish, though they become more important as the squish velocity decreases from its maximum value as the piston approaches TC.

Velocity measurements in engines provide good support for the above theory. The ideal theory adequately predicts the dependence on engine speed.[36] With appropriate corrections for leakage and heat-transfer effects, the above theory predicts the effects of the bowl diameter/bore ratio and clearance height on squish velocity (see Fig. 8-23). The change in direction of the radial motion as the piston moves through TC has been demonstrated under motored engine conditions. Under firing conditions, the combustion generated gas expansion in the open portion of the combustion chamber substantially increases the magnitude of the reverse squish motion after TC.[37]

8.5 PRECHAMBER ENGINE FLOWS

Small high-speed diesel engines use an auxiliary combustion chamber, or prechamber, to achieve adequate fuel-air mixing rates. The prechamber is connected to the main combustion chamber above the piston via a nozzle, passageway, or one or more orifices. Flow of air through this restriction into the prechamber during the compression process sets up high velocities in the prechamber at the time the fuel-injection process commences. This results in the required high fuel-air mixing rates. Figures 1-21 and 10-2 show examples of these prechamber or indirect-injection diesels. The two most common designs of auxiliary chamber are: the swirl chamber (Fig. 10-2a), where the flow through the passageway enters the chamber tangentially producing rapid rotation within the chamber, and the prechamber (Fig. 10-2b) with one or more connecting orifices designed to produce a highly turbulent flow but no ordered motion within the chamber. Auxiliary chambers are sometimes used in spark-ignition engines. The torch-ignition three-valve stratified-charge engine (Fig. 1-27) is one such concept. The prechamber is used to create a rich mixture in the vicinity of the spark plug to promote rapid flame development. An alternative concept uses the prechamber around the spark plug to generate turbulence to enhance the early stages of combustion, but has no mixture stratification.

The most critical phase of flow into the prechamber occurs towards the end of compression. While this flow is driven by a pressure difference between the main chamber above the piston and the auxiliary chamber, this pressure difference is small, and the mass flow rate and velocity at the nozzle, orifice, or passageway can be estimated using a simple gas displacement model. Assuming that the gas density throughout the cylinder is uniform (an adequate assumption toward the end of compression—the most critical period), the mass in the prechamber m_P is given by $m_c(V_P/V)$, where m_c is the cylinder mass, V the cylinder volume, and V_P the prechamber volume. The mass flow rate through the throat of the restriction is, therefore,

$$\dot{m} = \frac{dm_P}{dt} = -\frac{m_c V_P}{V^2}\frac{dV}{dt} \tag{8.39}$$

Using the relations $dV/dt = -(\pi B^2/4)S_p$ where S_p is the instantaneous piston speed, $V_d = \pi B^2 L/4$, and $\bar{S}_p = 2NL$, Eq. (8.39) can be written as

$$\frac{\dot{m}}{m_c N} = 2(r_c - 1)\left(\frac{V_P}{V_c}\right)\left(\frac{S_p}{\bar{S}_p}\right)\left(\frac{V_c}{V}\right)^2 \tag{8.40}$$

where V_c is the clearance volume, $S_p/\bar{S}_p$ is given by Eq. (2.11), and V/V_c is given by Eq. (2.6). The gas velocity at the throat v_T can be obtained from $\dot{m}$ via the relation $\rho v_T A_T = \dot{m}$, the density $\rho = m_c/V$, and Eq. (8.40):

$$\frac{v_T}{\bar{S}_p} = \left(\frac{V_P}{V_c}\right)\left(\frac{\pi B^2/4}{A_T}\right)\left(\frac{S_p}{\bar{S}_p}\right)\left(\frac{V_c}{V}\right) \tag{8.41}$$

where A_T is the effective cross-sectional area of the throat. The variation of $\dot{m}/(m_c N)$ and $v_T/\bar{S}_p$ with crank angle during the compression process for values of r_c, V_P/V_c, and $A_T/(\pi B^2/4)$ typical of a swirl prechamber diesel are shown in Fig. 8-24. The velocity reaches its peak value about 20° before TC: very high gas velocities, an order of magnitude or more larger than the mean piston speed, can be achieved depending on the relative effective throat area. Note that as the piston approaches TC, first the nozzle velocity and then the mass flow rate decrease to zero. After TC, in the absence of combustion, an equivalent flow in the reverse direction out of prechamber would occur. Combustion in the pre-

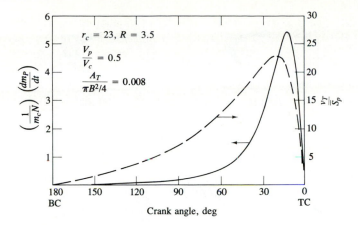

FIGURE 8-24
Velocity and mass flow rate at the prechamber nozzle throat, during compression, for a typical small swirl-prechamber automotive diesel.

chamber diesel usually starts just before TC, and the pressure in the prechamber then rises significantly above the main chamber pressure. The outflow from the prechamber is then governed by the development of the combustion process, and the above simple gas displacement model no longer describes the flow. This combustion generated prechamber gas motion is discussed in Sec. 14.4.4.

In prechamber stratified-charge engines, the flow of gas into the prechamber during compression is critical to the creation of an appropriate mixture in the prechamber at the crank angle when the mixture is ignited. In the concept shown in Fig. 1-27, a very rich fuel-air mixture is fed directly to the prechamber during intake via the prechamber intake valve, while a lean mixture is fed to the main chamber via the main intake valve. During compression, the flow into the prechamber reduces the prechamber equivalence ratio to a close-to-stoichiometric value at the time of ignition. Figure 8-25 shows a gas displacement calculation of this process and relevant data; the prechamber equivalence ratio, initially greater

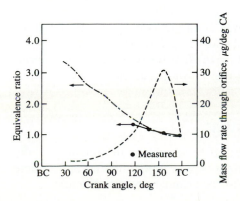

FIGURE 8-25
Effect of gas flow into the prechamber during compression on the prechamber equivalence ratio in a three-valve prechamber stratified-charge engine. Calculations based on gas displacement model.[38]

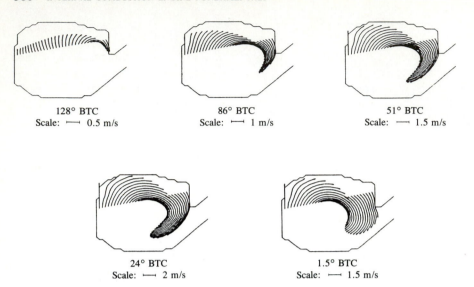

128° BTC
Scale: ⊢━━ 0.5 m/s

86° BTC
Scale: ⊢━━ 1 m/s

51° BTC
Scale: ⊢━━ 1.5 m/s

24° BTC
Scale: ⊢━━ 2 m/s

1.5° BTC
Scale: ⊢━━ 1.5 m/s

FIGURE 8-26
Calculations of developing flow field in (two-dimensional) swirl prechamber during compression process. Lines are instantaneous flow streamlines, analogous to streak photographs of flow field.[41]

than 3, is leaned out to unity as mass flows through the orifice into the prechamber (whose volume is 8.75 percent of the clearance volume).[38] Charts for estimating the final equivalence ratio, based on gas displacement, for this prechamber concept are available.[39]

The velocity field set up inside the prechamber during compression is strongly dependent on the details of the nozzle and prechamber geometry. Velocities vary linearly with mean piston speed.[40] In swirl prechambers, the nozzle flow sets up a vortex within the chamber. Figure 8-26 shows calculations of this developing flow field; instantaneous flow streamlines have been drawn in, with the length of the streamlines indicating how the particles of fluid move relative to each other.[41] The velocities increase with increasing crank angle as the compression process proceeds, and reach a maximum at about 20° before TC. Then, as the piston approaches TC and the flow through the passageway decreases to zero, the vortex in the swirl chamber expands to fill the entire chamber and mean velocities decay. Very high swirl rates can be achieved just before TC: local swirl ratios of up to 60 at intermediate radii and up to 20 at the outer radius have been measured. These high swirl rates produce large centrifugal accelerations.

8.6 CREVICE FLOWS AND BLOWBY

The engine combustion chamber is connected to several small volumes usually called *crevices* because of their narrow entrances. Gas flows into and out of these volumes during the engine operating cycle as the cylinder pressure changes.

The largest crevices are the volumes between the piston, piston rings, and cylinder wall. Some gas flows out of these regions into the crankcase; it is called *blowby*. Other crevice volumes in production engines are the threads around the spark plug, the space around the plug center electrode, the gap around the fuel injector, crevices between the intake and exhaust valve heads and cylinder head, and the head gasket cutout. Table 8.1 shows the size and relative importance of these crevice regions in one cylinder of a production V-6 spark-ignition engine determined from measurements of cold-engine components. Total crevice volume is a few percent of the clearance volume, and the piston and ring crevices are the dominant contributors. When the engine is warmed up, dimensions and crevice volumes will change.

The important crevice processes occurring during the engine cycle are the following. As the cylinder pressure rises during compression, unburned mixture or air is forced into each crevice region. Since these volumes are thin they have a large surface/volume ratio; the gas flowing into the crevice cools by heat transfer to close to the wall temperature. During combustion while the pressure continues to rise, unburned mixture or air, depending on engine type, continues to flow into these crevice volumes. After flame arrival at the crevice entrance, burned gases will flow into each crevice until the cylinder pressure starts to decrease. Once the crevice gas pressure is higher than the cylinder pressure, gas flows back from each crevice into the cylinder.

The volumes between the piston, piston rings, and cylinder wall are shown schematically in Fig. 8-27. These crevices consist of a series of volumes (numbered 1 to 5) connected by flow restrictions such as the ring side clearance and ring gap. The geometry changes as each ring moves up and down in its ring groove, sealing either at the top or bottom ring surface. The gas flow, pressure distribution, and ring motion are therefore coupled. Figure 8-28 illustrates this behavior: pressure distributions, ring motion, and mass flow of gas into and out

TABLE 8.1
V-6 engine crevice data†[42]

	cm^3	%
Displaced volume per cylinder	632	
Clearance volume per cylinder	89	100
Volume above first ring	0.93	1.05
Volume behind first ring	0.47	0.52
Volume between rings	0.68	0.77
Volume behind second ring	0.47	0.52
Total ring crevice volume	2.55	2.9
Spark plug thread crevice	0.25	0.28
Head gasket crevice	0.3	0.34
Total crevice volume	3.1	3.5

† Determined for cold engine.

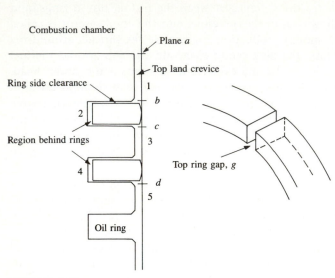

FIGURE 8-27
Schematic of piston and ring assembly in automotive spark-ignition engine.

of the regions defined by planes a, b, c, d, and through the ring gap g are plotted versus crank angle through compression and expansion. These results come from an analysis of these regions as volumes connected by passageways, with a pre-scribed cylinder pressure versus crank angle profile coupled with a dynamic model for ring motion, and assuming that the gas temperature equals the wall temperature.[42] During compression and combustion, the rings are forced to the groove lower surfaces and mass flows *into* all the volumes in this total crevice region. The pressure above and behind the first ring is essentially the same as the cylinder pressure; there is a substantial pressure drop across each ring, however. Once the cylinder pressure starts to decrease (after 15° ATC) gas flows out of regions 1 and 2 in Fig. 8-27 into the cylinder, but continues to flow *into* regions 3, 4, and 5 until the pressure in the cylinder falls below the pressure beneath the top ring. The top ring then shifts to seal with the upper grove surface and gas flows out of regions 2, 3, and 4 (which now have the same pressure), both into the cylinder and as blowby into the crackcase. Some 5 to 10 percent of the total cylinder charge is trapped in these regions at the time of peak cylinder pressure. Most of this gas returns to the cylinder; about 1 percent goes to the crankcase as blowby. The gas flow back into the cylinder continues throughout the expansion process. In spark-ignition engines this phenomenon is a major contributor to unburned hydrocarbon emissions (see Sec. 11.4.3). In all engines it results in a loss of power and efficiency.

There is substantial experimental evidence to support the above description of flow in the piston ring crevice region. In a special square-cross-section flow visualization engine, both the low-velocity gas expansion out of the volume above the first ring after the time of peak pressure and the jet-like flows through

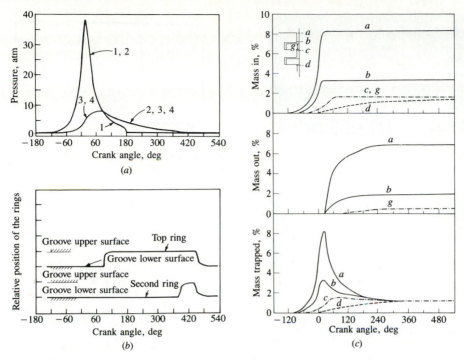

FIGURE 8-28

(a) Pressures in the combustion chamber (1), in region behind top ring (2), in region between rings (3), and behind second ring (4); (b) relative position of top and second rings; (c) percentage of total cylinder mass that flows into and out of the different crevice regions across planes a, b, c, and d and through the ring gap g in Fig. 8-27, and the percentage of mass trapped beneath these planes, as a function of crank angle. Automotive spark-ignition engine at wide-open throttle and 2000 rev/min.[42]

the top ring gap later in the expansion process when the pressure difference across the ring changes sign have been observed. Figure 8-29 shows these flows with explanatory schematics.

Blowby is defined as the gas that flows from the combustion chamber past the piston rings and into the crankcase. It is forced through any leakage paths afforded by the piston-bore-ring assembly in response to combustion chamber pressure. If there is good contact between the compression rings and the bore, and the rings and the bottom of the grooves, then the only leakage path of consequence is the ring gap. Blowby of gases from the cylinder to the crankcase removes gas from these crevice regions and thereby prevents some of the crevice gases from returning to the cylinder. Crankcase blowby gases used to be vented directly to the atmosphere and constituted a significant source of HC emissions. The crankcase is now vented to the engine intake system and the blowby gases are recycled. Blowby at a given speed and load is controlled primarily by the greatest flow resistance in the flow path between the cylinder and the crankcase. This is the smallest of the compression ring ring-gap areas. Figure 8-30 shows how measured blowby flow rates increase linearly with the smallest gap area.[43]

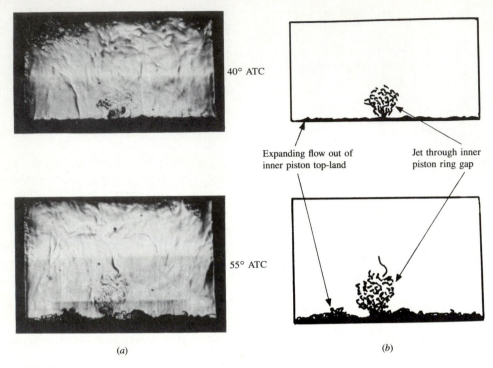

FIGURE 8-29
Schlieren photographs of the flow out of the piston–cylinder wall crevices during the expansion stroke. A production piston was inserted into the square cross-section piston of the visualization engine. Gas flows at low velocity out of the crevice entrance all around the production piston circumference once the cylinder pressure starts decreasing early in the expansion stroke. Gas flows out of the ring gap as a jet once the pressure above the ring falls below the pressure beneath the ring.[42]

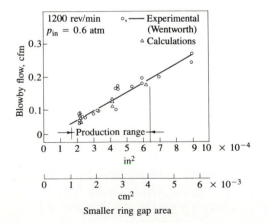

FIGURE 8-30
Measured blowby for one cylinder of an automobile spark-ignition engine as a function of the smallest ring gap area, compared with blowby calculations based on flow model described in text.[42, 43]

Calculations of blowby based on the model described earlier are in good agreement.[42] Extrapolation back to the zero gap area gives nearly zero blowby. Note, however, that if the bore finish is rough, or if the rings do not contact the bore all around, or if the compression rings lift off the bottom of the groove, this linear relationship may no longer hold.

8.7 FLOWS GENERATED BY PISTON–CYLINDER WALL INTERACTION

Because a boundary layer exists on the cylinder wall, the motion of the piston generates unusual flow patterns in the corner formed by the cylinder wall and the piston face. When the piston is moving away from top-center a sink-type flow occurs. When the piston moves toward top-center a vortex flow is generated. Figure 8-31 shows schematics of these flows (in a coordinate frame with the piston face at rest). The vortex flow has been studied because of its effect on gas motion at the time of ignition and because it has been suggested as a mechanism for removing hydrocarbons off the cylinder wall during the exhaust stroke (see Sec. 11.4.3).

The vortex flow has been studied in cylinders with water as the fluid over the range of Reynolds numbers typical of engine operation.[44, 45] Laminar, transition, and turbulent flow regimes have been identified. It has been shown that a quasi-steady flow assumption is valid and that

$$\frac{A_V}{L^2} = f\left(\frac{v_w L}{v}\right)$$

where A_V is the vortex area (area inside the dashed line in Fig. 8-31), L is the stroke, v_w is the wall velocity in piston stationary coordinates ($v_w = S_p$ in the engine), v is the kinematic viscosity, and ($v_w L/v$) is a Reynolds number.

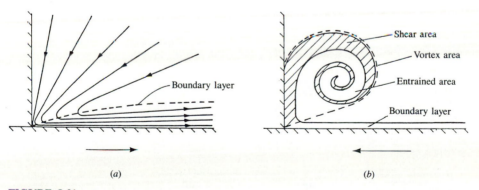

(a)   (b)

FIGURE 8-31
Schematics of the flow pattern set up in the piston face–cylinder wall corner, in piston-stationary coordinates, due to the boundary layer on the cylinder wall. Piston crown on left; cylinder wall at bottom. (a) Sink flow set up during intake and expansion; (b) vortex flow set up during compression and exhaust.[44] Arrow shows cylinder wall velocity relative to piston.

For the laminar flow regime, a good assumption is that A_V is proportional to the shear area in the vortex (shown cross-hatched), which equals the boundary-layer area; this can be estimated from boundary-layer theory. In the turbulent flow regime, an entrainment theory was used, which assumed that the rate of change of vortex area was proportional to the product of the exposed perimeter of the vortex and the velocity difference between the vortex and the stationary fluid ($\approx v_w$). The relevant relationships are:

For $(v_w L/v) \leq 2 \times 10^4$:
$$\frac{A_V}{L^2} = \left(\frac{v_w L}{v}\right)^{-1/2}$$
(8.42a)

For $(v_w L/v) \geq 2 \times 10^4$:
$$\frac{A_V}{L^2} = 0.006$$
(8.42b)

Figure 8-32 shows these two theories correlated against hydraulic analog data.

These theories are for constant values of v. During compression, v decreases substantially as the gas temperature and pressure increase (v decreases by a factor of 4 for a compression ratio of 8). This will decrease the size of the vortex until the turbulent regime is reached. During the exhaust stroke following blowdown, v will remain approximately constant as the pressure and temperature do not change significantly. Typical parameter values at 1500 rev/min are: $\bar{S}_p = 5$ m/s, $L = 0.1$ m; average values of v are 1.2×10^{-5} and 1.4×10^{-4} m²/s for compression and exhaust stroke, respectively. Hence a Reynolds number for the compression stroke is 4×10^4, $A_V/L^2 \approx 0.006$, and the vortex diameter $d_V \approx 0.09L$. For the exhaust stroke, the Reynolds number is 4×10^3, $A_V/L^2 \approx 0.015$, and $d_V \approx 0.14L$. Thus the vortex dimensions at the end of the upward stroke of the piston are comparable to the engine clearance height.

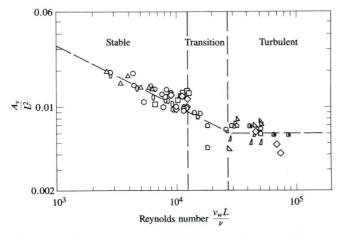

FIGURE 8-32
Ratio of area of vortex in piston face–cylinder wall corner to square of stroke, as a function of Reynolds number based on piston velocity, for piston moving toward the cylinder head.[44]

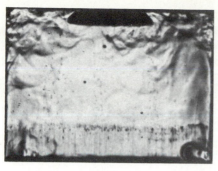

60° BTC

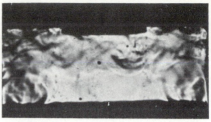

20° BTC

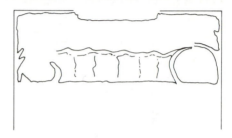

FIGURE 8-33
Schlieren photographs of in-cylinder flow during later stages of exhaust stroke. Growing vortex in the piston face–cylinder wall corner and turbulent outflow toward the valve are apparent at 60° BTC. At 20° BTC, the vortex has grown to of order 0.2B diameter.[42]

This vortex flow has been observed in an operating engine. Figure 8-33 shows schlieren photographs taken during the exhaust stroke in a special square-cross-section flow visualization spark-ignition engine. The accompanying schematic identifies the vortex structure which is visible in the photo because the cool boundary-layer gas is being scraped off the cylinder wall by the upward-moving piston and "rolled up." The vortex diameter as the piston approaches TC is about 20 percent of the bore.

PROBLEMS

8.1. (a) Estimate the ratio of the maximum gas velocity in the center of the hollow cone inlet jet to the mean piston speed from the data in Fig. 8-1.

    (b) Compare this ratio with the ratio of inlet valve pseudo flow velocity determined from Fig. 6-15 to the mean piston speed at the same crank angle. The engine is that of Fig. 1-4.

    (c) Are the engine velocity data in (a) consistent with the velocity calculated from the simple piston displacement model of (b)? Explain.

8.2. Given the relationship between turbulence intensity and mean piston speed [Eq. (8.23)] and that the turbulence integral scale is $\approx 0.2 \times$ clearance height, use Eqs. (8.14) and (8.15) to estimate the following quantities for a spark-ignition engine with bore = stroke = 86 mm, $r_c = 9$, at 1000 and 5000 rev/min and wide-open throttle:

    (a) Mean and maximum piston speed, maximum gas velocity through the inlet valve (see Prob. 8.1)

    (b) Turbulence intensity, integral length scale, micro length scale, and Kolmogorov length scale, all at TC

8.3. The swirl ratio at the end of induction at 2000 rev/min in a direct-injection diesel engine of bore = stroke = 100 mm is 4.0. What is the average tangential velocity (evaluated at the inlet valve–axis radial location) required to give this swirl ratio? What is the ratio of this velocity to the mean piston speed and to the mean flow velocity through the inlet valve estimated from the average valve open area and open time?

8.4. (a) Derive a relationship for the depth (or height) h_B of a disc-shaped bowl-in-piston direct-injection diesel engine combustion chamber in terms of compression ratio r_c, bore B, stroke L, bowl diameter D_B, and top-center cylinder-head to piston-crown clearance c. For $B = L = 100$ mm, $r_c = 16$, $D_B = 0.5B$, $c = 1$ mm find the fraction of the air charge within the bowl at TC.

(b) If the swirl ratio at the end of induction at 2500 rev/min is 3 find the swirl ratio and average angular velocity in the bowl-in-piston chamber of dimensions given above. Assume the swirling flow is always a solid-body rotation. Compare the tangential velocity at the bowl edge with the mean piston speed. Neglect any friction effects.

(c) What would the swirl ratio be if the top-center clearance height was zero?

8.5. Using Eq. (8.37) and Fig. 8-20b plot the squish velocity divided by the mean piston speed at 10° BTC (the approximate location of the maximum) as a function of squish area expressed as a percentage of the cylinder cross section, $A_s/(\pi B^2/4) \times 100$, from 50 to 0 percent. $r_c = 10$, $c/B = 0.01$, $B/L = 1$, $R = l/a = 3.5$.

8.6. Figure 8-24 shows the velocity at the prechamber nozzle throat during compression for dimensions typical of a small swirl chamber indirect-injection diesel. Assuming that the swirl chamber shape is a disc of height equal to the diameter, that the nozzle throat is at $0.8 \times$ prechamber radius, and that the flow enters the prechamber tangentially, estimate the swirl ratio based on the total angular momentum about the swirl chamber axis in the prechamber at top-center. Assume $B = L$; neglect friction.

8.7. The total crevice volume in an automobile spark-ignition engine is about 3 percent of the clearance volume. If the gas in these crevice regions is close to the wall temperature (450 K) and at the cylinder pressure, estimate the fraction of the cylinder mass within these crevice regions at these crank angles: inlet valve closing (50° ABC), spark discharge (30° BTC), maximum cylinder pressure (15° ATC), exhaust valve opening (60° BBC), TC of the exhaust stroke. Use the information in Fig. 1-8 for your input data, and assume the inlet pressure is 0.67 atm.

REFERENCES

1. Bicen, A. F., Vafidis, C., and Whitelaw, J. H.: "Steady and Unsteady Airflow through the Intake Valve of a Reciprocating Engine," *ASME Trans., J. Fluids Engng*, vol. 107, pp. 413–420, 1985.
2. Namazian, M., Hansen, S. P., Lyford-Pike, E. J., Sanchez-Barsse, J., Heywood, J. B., and Rife, J.: "Schlieren Visualization of the Flow and Density Fields in the Cylinder of a Spark-Ignition Engine," SAE paper 800044, *SAE Trans.*, vol. 89, 1980.
3. Ekchian, A., and Hoult, D. P.: "Flow Visualization Study of the Intake Process of an Internal Combustion Engine," SAE paper 790095, *SAE Trans.*, vol. 88, 1979.
4. Hirotomi, T., Nagayama, I., Kobayashi, S., and Yamamasu, M.: "Study of Induction Swirl in a Spark Ignition Engine," SAE paper 810496, *SAE Trans.*, vol. 90, 1981.
5. Reynolds, W. C.: "Modeling of Fluid Motions in Engines—An Introductory Overview," in J. N. Mattavi and C. A. Amann (eds.), *Combustion Modelling in Reciprocating Engines*, pp. 69–124, Plenum Press, 1980.
6. Tennekes, H., and Lumley, J. L.: *A First Course in Turbulence*, MIT Press, 1972.

7. Rask, R. B.: "Laser Doppler Anemometer Measurements of Mean Velocity and Turbulence in Internal Combustion Engines," ICALEO '84 Conference Proceedings, vols. 45 and 47, *Inspection, Measurement and Control and Laser Diagnostics and Photochemistry*, Laser Institute of America, Boston, November 1984.

8. Tabaczynski, R. J.: "Turbulence and Turbulent Combustion in Spark-Ignition Engines," *Prog. Energy Combust. Sci.*, vol. 2, pp. 143–165, 1976.

9. Witze, P. O.: "A Critical Comparison of Hot-Wire Anemometry and Laser Doppler Velocimetry for I.C. Engine Applications," SAE paper 800132, *SAE Trans.*, vol. 89, 1980.

10. Witze, P. O., Martin, J. K., and Borgnakke, C.: "Conditionally-Sampled Velocity and Turbulence Measurements in a Spark Ignition Engine," *Combust. Sci. Technol.*, vol. 36, pp. 301–317, 1984.

11. Rask, R. B.: "Comparison of Window, Smoothed-Ensemble, and Cycle-by-Cycle Data Reduction Techniques for Laser Doppler Anemometer Measurements of In-Cylinder Velocity," in T. Morel, R. P. Lohmann, and J. M. Rackley (eds.), *Fluid Mechanics of Combustion Systems*, pp. 11–20, ASME, New York, 1981.

12. Liou, T-M., and Santavicca, D. A.: "Cycle Resolved LDV Measurements in a Motored IC Engine," *ASME Trans., J. Fluids Engng*, vol. 107, pp. 232–240, 1985.

13. Amann, C. A.: "Classical Combustion Diagnostics for Engine Research," SAE paper 850395, in *Engine Combustion Analysis: New Approaches*, P-156, SAE, 1985.

14. Dyer, T. M.: "New Experimental Techniques for In-Cylinder Engine Studies," SAE paper 850396, in *Engine Combustion Analysis: New Approaches*, P-156, SAE, 1985.

15. Rask, R. B.: "Laser Doppler Anemometer Measurements in an Internal Combustion Engine," SAE paper 790094, *SAE Trans.*, vol. 88, 1979.

16. Liou, T.-M., Hall, M., Santavicca, D. A., and Bracco, F. V.: "Laser Dopper Velocimetry Measurements in Valved and Ported Engines," SAE paper 840375, *SAE Trans.*, vol. 93, 1984.

17. Arcoumanis, C., and Whitelaw, J. H.: "Fluid Mechanics of Internal Combustion Engines: A Review," *Proc. Instn Mech. Engrs.*, vol. 201, pp. 57–74, 1987.

18. Bopp, S., Vafidis, C., and Whitelaw, J. H.: "The Effect of Engine Speed on the TDC Flowfield in a Motored Reciprocating Engine," SAE paper 860023, 1986.

19. Wong, V. W., and Hoult, D. P.: "Rapid Distortion Theory Applied to Turbulent Combustion," SAE paper 790357, *SAE Trans.*, vol. 88, 1979.

20. Fraser, R. A., Felton, P. G., and Bracco, F. V.,: "Preliminary Turbulence Length Scale Measurements in a Motored IC Engine," SAE paper 860021, 1986.

21. Ikegami, M., Shioji, M., and Nishimoto, K.: "Turbulence Intensity and Spatial Integral Scale during Compression and Expansion Strokes in a Four-cycle Reciprocating Engine," SAE paper 870372, 1987.

22. Uzkan, T., Borgnakke, C., and Morel, T.: "Characterization of Flow Produced by a High-Swirl Inlet Port," SAE paper 830266, 1983.

23. Monaghan, M. L., and Pettifer, H. F.: "Air Motion and Its Effects on Diesel Performance and Emissions," SAE paper 810255, in *Diesel Combustion and Emissions*, pt. 2, SP-484, *SAE Trans.*, vol. 90, 1981.

24. Tindal, M. J., Williams, T. J., and Aldoory, M.: "The Effect of Inlet Port Design on Cylinder Gas Motion in Direct Injection Diesel Engines," in *Flows in Internal Combustion Engines*, pp. 101–111, ASME, New York, 1982.

25. Brandl, F., Reverencic, I., Cartellieri, W., and Dent, J. C.: "Turbulent Air Flow in the Combustion Bowl of a D.I. Diesel Engine and Its Effect on Engine Performance," SAE paper 790040, *SAE Trans.*, vol. 88, 1979.

26. Brandstätter, W., Johns, R. J. R., and Wigley, G.: "Calculation of Flow Produced by a Tangential Inlet Port," in *International Symposium on Flows in Internal Combustion Engines—III*, FED vol. 28, pp. 135–148, ASME, New York, 1985.

27. Brandstätter, W., Johns, R. J. R., and Wigley, G.: "The Effect of Inlet Port Geometry on In-Cylinder Flow Structure," SAE paper 850499, 1985.

28. Davis, G. C., and Kent, J. C.: "Comparison of Model Calculations and Experimental Measurements of the Bulk Cylinder Flow Processes in a Motored PROCO Engine," SAE paper 790290, 1979.

29. Borgnakke, C., Davis, G. C., and Tabaczynski, R. J.: "Predictions of In-Cylinder Swirl Velocity and Turbulence Intensity for an Open Chamber Cup in Piston Engine," SAE paper 810224, *SAE Trans.*, vol. 90, 1981.

30. Arcoumanis, C., Bicen, A. F., and Whitelaw, J. H.: "Squish and Swirl-Squish Interaction in Motored Model Engines," *Trans. ASME, J. Fluids Engng*, vol. 105, pp. 105–112, 1983.

31. Ikegami, M., Mitsuda, T., Kawatchi, K., and Fujikawa, T.: "Air Motion and Combustion in Direct Injection Diesel Engines," JARI technical memorandum no. 2, pp. 231–245, 1971.

32. Liou, T.-M., and Santavicca, D. A.: "Cycle Resolved Turbulence Measurements in a Ported Engine With and Without Swirl," SAE paper 830419, *SAE Trans.*, vol. 92, 1983.

33. Fitzgeorge, D., and Allison, J. L.: "Air Swirl in a Road-Vehicle Diesel Engine," *Proc. Instn Mech. Engrs* (A.D.), no. 4, pp. 151–168, 1962–1963.

34. Lichty, L. C.: *Combustion Engine Processes*, McGraw-Hill, 1967.

35. Shimamoto, Y., and Akiyama, K.: "A Study of Squish in Open Combustion Chambers of a Diesel Engine," *Bull. JSME*, vol. 13, no. 63, pp. 1096–1103, 1970.

36. Dent, J. C., and Derham, J. A.: "Air Motion in a Four-Stroke Direct Injection Diesel Engine," *Proc. Instn Mech. Engrs*, vol. 188, 21/74, pp. 269–280, 1974.

37. Asanuma, T., and Obokata, T.: "Gas Velocity Measurements of a Motored and Firing Engine by Laser Anemometry," SAE paper 790096, *SAE Trans.*, vol. 88, 1979.

38. Asanuma, T., Babu, M. K. G., and Yagi, S.: "Simulation of Thermodynamic Cycle of Three-Valve Stratified Charge Engine," SAE paper 780319, *SAE Trans.*, vol. 87, 1978.

39. Hires, S. D., Ekchian, A., Heywood, J. B., Tabaczynski, R. J., and Wall, J. C.: "Performance and NO_x Emissions Modelling of a Jet Ignition Prechamber Stratified Charge Engine," SAE paper 760161, *SAE Trans.*, vol. 85, 1976.

40. Zimmerman, D. R.: "Laser Anemometer Measurements of the Air Motion in the Prechamber of an Automotive Diesel Engine," SAE paper 830452, 1983.

41. Meintjes, K., and Alkidas, A. C.: "An Experimental and Computational Investigation of the Flow in Diesel Prechambers," SAE paper 820275, *SAE Trans.*, vol. 91, in *Diesel Engine Combustion, Emissions, and Particulates*, P-107, SAE, 1982.

42. Namazian, M., and Heywood, J. B.: "Flow in the Piston-Cylinder-Ring Crevices of a Spark-Ignition Engine: Effect on Hydrocarbon Emissions, Efficiency and Power," SAE paper 820088, *SAE Trans.*, vol. 91, 1982.

43. Wentworth, J. T.: "Piston and Ring Variables Affect Exhaust Hydrocarbon Emissions," SAE paper 680109, *SAE Trans.*, vol. 77, 1968.

44. Tabaczynski, R. J., Hoult, D. P., and Keck, J. C.: "High Reynolds Number Flow in a Moving Corner," *J. Fluid Mech.*, vol. 42, pp. 249–255, 1970.

45. Daneshyar, H. F., Fuller, D. E., and Deckker, B. E. L.: "Vortex Motion Induced by the Piston of an Internal Combustion Engine," *Int. J. Mech. Sci.*, vol. 15, pp. 381–390, 1973.

CHAPTER

9

COMBUSTION IN SPARK-IGNITION ENGINES

9.1 ESSENTIAL FEATURES OF PROCESS

In a conventional spark-ignition engine the fuel and air are mixed together in the intake system, inducted through the intake valve into the cylinder, where mixing with residual gas takes place, and then compressed. Under normal operating conditions, combustion is initiated towards the end of the compression stroke at the spark plug by an electric discharge. Following inflammation, a turbulent flame develops, propagates through this essentially premixed fuel, air, burned gas mixture until it reaches the combustion chamber walls, and then extinguishes. Photographs of this process taken in operating engines illustrate its essential features. Figure 9-1 (color plate) shows a sequence of frames from a high-speed color movie of the combustion process in a special single-cylinder engine with a glass piston crown.[1] The spark discharge is at $-30°$. The flame first becomes visible in the photos at about $-24°$. The flame, approximately circular in outline in this

---

FIGURE 9-1 (On color plate opposite p. 498)
Color photographs from high-speed movie of spark-ignition engine combustion process, taken through glass piston crown. Ignition timing 30° BTC, light load, 1430 rev/min, $(A/F) = 19$.[1]

371

view through the piston, then propagates outward from the spark plug location. The blue light from the flame is emitted most strongly from the front. The irregular shape of the turbulent flame front is apparent. At TC the flame diameter is about two-thirds of the cylinder bore. The flame reaches the cylinder wall farthest from the spark plug about 15° ATC, but combustion continues around parts of the chamber periphery for another 10°. At about 10° ATC, additional radiation—initially white, turning to pinky-orange—centered at the spark plug location is evident. This afterglow comes from the gases behind the flame which burned earlier in the combustion process, as these are compressed to the highest temperatures attained within the cylinder (at about 15° ATC) while the rest of the charge burns.[2, 3]

Additional features of the combustion process are evident from the data in Fig. 9-2, taken from several consecutive cycles of an operating spark-ignition engine. The cylinder pressure, fraction of the charge mass which has burned (determined from the pressure data, see Sec. 9.2), and fraction of the cylinder volume enflamed by the front (determined from photographs like Fig. 9-1) are shown, all as a function of crank angle.[4] Following spark discharge, there is a period during which the energy release from the developing flame is too small for the pressure rise due to combustion to be discerned. As the flame continues to grow and propagate across the combustion chamber, the pressure then steadily rises above the value it would have in the absence of combustion. The pressure reaches a maximum after TC but before the cylinder charge is fully burned, and then decreases as the cylinder volume continues to increase during the remainder of the expansion stroke.

The flame development and subsequent propagation obviously vary, cycle-by-cycle, since the shape of the pressure, volume fraction enflamed, and mass fraction burned curves for each cycle differ significantly. This is because flame growth depends on local mixture motion and composition. These quantities vary in successive cycles in any given cylinder and may vary cylinder-to-cylinder. Especially significant are mixture motion and composition in the vicinity of the spark plug at the time of spark discharge since these govern the early stages of flame development. Cycle-by-cycle and cylinder-to-cylinder variations in combustion are important because the extreme cycles limit the operating regime of the engine (see Sec. 9.4.1).

Note that the volume fraction enflamed curves rise more steeply than the mass fraction burned curves. In large part, this is because the density of the unburned mixture ahead of the flame is about four times the density of the burned gases behind the flame. Also, there is some unburned mixture behind the visible front to the flame: even when the entire combustion chamber is fully enflamed, some 25 percent of the mass has still to burn. From this description it is plausible to divide the combustion process into four distinct phases: (1) spark ignition; (2) early flame development; (3) flame propagation; and (4) flame termination. Our understanding of each of these phases will be developed in the remainder of this chapter.

The combustion event must be properly located relative to top-center to

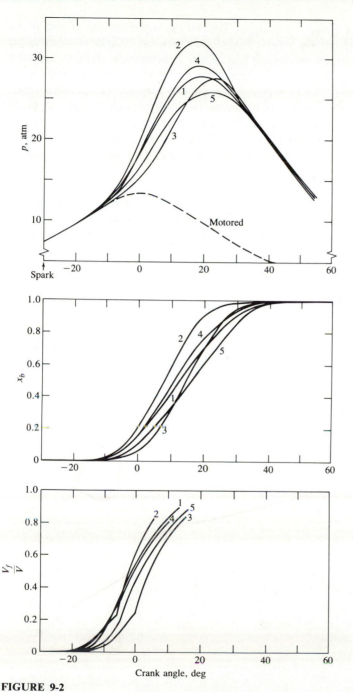

FIGURE 9-2
Cylinder pressure, mass fraction burned, and volume fraction enflamed for five consecutive cycles in a spark-ignition engine as a function of crank angle. Ignition timing 30° BTC, wide-open throttle, 1044 rev/min, $\phi = 0.98$.[4]

obtain the maximum power or torque. The combined duration of the flame development and propagation process is typically between 30 and 90 crank angle degrees. Combustion starts before the end of the compression stroke, continues through the early part of the expansion stroke, and ends after the point in the cycle at which the peak cylinder pressure occurs. The pressure versus crank angle curves shown in Fig. 9-3a allow us to understand why engine torque (at given engine speed and intake manifold conditions) varies as spark timing is varied relative to TC. If the start of the combustion process is progressively advanced before TC, the compression stroke work transfer (which is *from* the piston *to* the cylinder gases) increases. If the end of the combustion process is progressively delayed by retarding the spark timing, the peak cylinder pressure occurs later in the expansion stroke and is reduced in magnitude. These changes reduce the expansion stroke work transfer *from* the cylinder gases *to* the piston. The optimum timing which gives the maximum brake torque—called *maximum brake torque, or MBT, timing*—occurs when the magnitudes of these two opposing trends just offset each other. Timing which is advanced or retarded from this optimum gives lower torque. The optimum spark setting will depend on the rate of flame development and propagation, the length of the flame travel path across the combustion chamber, and the details of the flame termination process after it reaches the wall. These depend on engine design and operating conditions, and the properties of the fuel, air, burned gas mixture. Figure 9-3b shows the effect of variations in spark timing on brake torque for a typical spark-ignition engine. The maximum is quite flat.

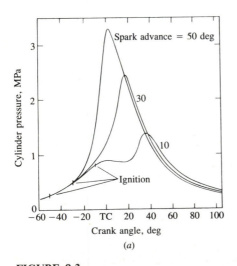

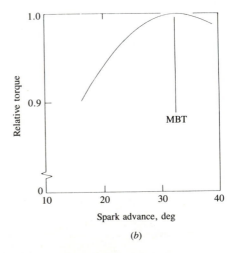

(a)   (b)

FIGURE 9-3
(a) Cylinder pressure versus crank angle for overadvanced spark timing (50°), MBT timing (30°), and retarded timing (10°). (b) Effect of spark advance on brake torque at constant speed and (A/F), at wide-open throttle. MBT is maximum brake torque timing.[5]

Empirical rules for relating the mass burning profile and maximum cylinder pressure to crank angle at MBT timing are often used. For example, with optimum spark timing: (1) the maximum pressure occurs at about 16° after TC; (2) half the charge is burned at about 10° after TC. In practice, the spark is often retarded to give a 1 or 2 percent reduction in brake torque from the maximum value, to permit a more precise definition of timing relative to the optimum.

So far we have described normal combustion in which the spark-ignited flame moves *steadily* across the combustion chamber until the charge is fully consumed. However, several factors—e.g., fuel composition, certain engine design and operating parameters, and combustion chamber deposits—may prevent this normal combustion process from occurring. Two types of abnormal combustion have been identified: knock and surface ignition.

Knock is the most important abnormal combustion phenomenon. Its name comes from the noise that results from the autoignition of a portion of the fuel, air, residual gas mixture ahead of the advancing flame. As the flame propagates across the combustion chamber, the unburned mixture ahead of the flame—called the *end gas*—is compressed, causing its pressure, temperature, and density to increase. Some of the end-gas fuel-air mixture may undergo chemical reactions prior to normal combustion. The products of these reactions may then autoignite: i.e., spontaneously and rapidly release a large part or all of their chemical energy. When this happens, the end gas burns very rapidly, releasing its energy at a rate 5 to 25 times that characteristic of normal combustion. This causes high-frequency pressure oscillations inside the cylinder that produce the sharp metallic noise called knock.

The presence or absence of knock reflects the outcome of a race between the advancing flame front and the precombustion reactions in the unburned end gas. Knock will not occur if the flame front consumes the end gas before these reactions have time to cause the fuel-air mixture to autoignite. Knock will occur if the precombustion reactions produce autoignition before the flame front arrives.

The other important abnormal combustion phenomenon is *surface ignition*. Surface ignition is ignition of the fuel-air charge by overheated valves or spark plugs, by glowing combustion-chamber deposits, or by any other hot spot in the engine combustion chamber: it is ignition by any source other than normal spark ignition. It may occur before the spark plug ignites the charge (preignition) or after normal ignition (postignition). It may produce a single flame or many flames. Uncontrolled combustion is most evident and its effects most severe when it results from preignition. However, even when surface ignition occurs after the spark plug fires (postignition), the spark discharge no longer has complete control of the combustion process.

Surface ignition may result in knock. Knock which occurs following normal spark ignition is called *spark knock* to distinguish it from knock which has been preceded by surface ignition. Abnormal combustion phenomena are reviewed in more detail in Sec. 9.6.

9.2 THERMODYNAMIC ANALYSIS OF SI ENGINE COMBUSTION

9.2.1 Burned and Unburned Mixture States

Because combustion occurs through a flame propagation process, the changes in state and the motion of the unburned and burned gas are much more complex than the ideal cycle analysis in Chapter 5 suggests. The gas pressure, temperature, and density change as a result of changes in volume due to piston motion. During combustion, the cylinder pressure increases due to the release of the fuel's chemical energy. As each element of fuel-air mixture burns, its density decreases by about a factor of four. This combustion-produced gas expansion compresses the unburned mixture ahead of the flame and displaces it toward the combustion chamber walls. The combustion-produced gas expansion also compresses those parts of the charge which have already burned, and displaces them back toward the spark plug. During the combustion process, the unburned gas elements move away from the spark plug; following combustion, individual gas elements move back toward the spark plug. Further, elements of the unburned mixture which burn at different times have different pressures and temperatures just prior to combustion, and therefore end up at different states after combustion. The thermodynamic state and composition of the burned gas is, therefore, non-uniform. A first law analysis of the spark-ignition engine combustion process enables us to quantify these gas states.

Consider the schematic of the engine cylinder while combustion is in progress, shown in Fig. 9-4. Work transfer occurs between the cylinder gases and the piston (*to the gas* before TC; *to the piston* after TC). Heat transfer occurs to the chamber walls, primarily from the burned gases. At the temperatures and pressures typical of spark-ignition engines it is a reasonable approximation to assume that the volume of the reaction zone *where combustion is actually occurring* is a negligible fraction of the chamber volume even though the thickness of the turbulent flame may not be negligible compared with the chamber dimensions (see Sec. 9.3.2). With normal engine operation, at any point in time or crank angle, the pressure throughout the cylinder is close to uniform. The condi-

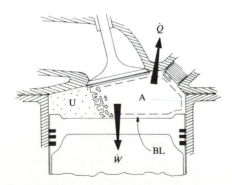

FIGURE 9-4
Schematic of flame in the engine cylinder during combustion: unburned gas (U) to left of flame, burned gas to right. A denotes adiabatic burned-gas core, BL denotes thermal boundary layer in burned gas, $\dot{W}$ is work-transfer rate to piston, $\dot{Q}$ is heat-transfer rate to chamber walls.

tions in the burned and unburned gas are then determined by conservation of mass:

$$\frac{V}{m} = \int_0^{x_b} v_b \, dx + \int_{x_b}^1 v_u \, dx \qquad (9.1)$$

and conservation of energy:

$$\frac{U_0 - W - Q}{m} = \int_0^{x_b} u_b \, dx + \int_{x_b}^1 u_u \, dx \qquad (9.2)$$

where V is the cylinder volume, m is the mass of the cylinder contents, v is the specific volume, x_b is the mass fraction burned, U_0 is the internal energy of the cylinder contents at some reference point θ_0, u is the specific internal energy, W is the work done on the piston, and Q is the heat transfer to the walls. The subscripts u and b denote unburned and burned gas properties, respectively. The work and heat transfers are

$$W = \int_{V_0}^V p \, dV' \qquad Q = \int_{\theta_0}^\theta \left(\frac{\dot{Q}}{360N}\right) d\theta \qquad (9.3)$$

where $\dot{Q}$ is the instantaneous heat-transfer rate to the chamber walls.

To proceed further, models for the thermodynamic properties of the burned and unburned gases are required. Several categories of models are described in Chap. 4. Accurate calculations of the state of the cylinder gases require an equilibrium model (or good approximation to it) for the burned gas and an ideal gas mixture model (of frozen composition) for the unburned gas (see Table 4.2). However, useful illustrative results can be obtained by assuming that the burned and unburned gases are different ideal gases, each with constant specific heats;[6] i.e.,

$$pv_b = R_b T_b \qquad u_b = c_{v,b} T_b + h_{f,b} \qquad (9.4)$$
$$pv_u = R_u T_u \qquad u_u = c_{v,u} T_u + h_{f,u} \qquad (9.5)$$

Combining Eqs. (9.1) to (9.5) gives

$$\frac{pV}{m} = x_b R_b \bar{T}_b + (1 - x_b) R_u \bar{T}_u \qquad (9.6)$$

and

$$\frac{U_0 - W - Q}{m} = x_b(c_{v,b} \bar{T}_b + h_{f,b}) + (1 - x_b)(c_{v,u} \bar{T}_u + h_{f,u}) \qquad (9.7)$$

where

$$\bar{T}_b = \frac{1}{x_b} \int_0^{x_b} T_b \, dx \qquad \bar{T}_u = \frac{1}{1 - x_b} \int_{x_b}^1 T_u \, dx$$

are the mean temperatures of the burned and unburned gases. Equations (9.6) and (9.7) may now be solved to obtain

$$x_b = \frac{pV - p_0 V_0 + (\gamma_b - 1)(W + Q) + (\gamma_b - \gamma_u)mc_{v,u}(\bar{T}_u - T_0)}{m[(\gamma_b - 1)(h_{f,u} - h_{f,b}) + (\gamma_b - \gamma_u)c_{v,u}\bar{T}_u]} \tag{9.8}$$

and

$$\bar{T}_b = \frac{R_u}{R_b}\bar{T}_u + \frac{pV - mR_u\bar{T}_u}{mR_b x_b} \tag{9.9}$$

If we now assume the unburned gas is initially uniform and undergoes isentropic compression, then

$$\frac{\bar{T}_u}{T_0} = \left(\frac{p}{p_0}\right)^{(\gamma_u - 1)/\gamma_u} \tag{9.10}$$

This equation, with Eqs. (9.8) and (9.9) enables determination of both x_b and $\bar{T}_b$ from the thermodynamic properties of the burned and unburned gases, and known values of p, V, m, and $\dot{Q}$. Alternatively, if x_b is known then p can be determined. Mass fraction burned and cylinder gas pressure are uniquely related.

While Eq. (9.9) defines a mean burned gas temperature, the burned gas is not uniform. Mixture which burns early in the combustion process is further compressed after combustion as the remainder of the charge is burned. Mixture which burns late in the combustion process is compressed prior to combustion and, therefore, ends up at a different final state. A temperature gradient exists across the burned gas with the earlier burning portions at the higher temperature.[7, 8] Two limiting models bracket what occurs in practice: (1) a *fully mixed* model, where it is assumed that each element of mixture which burns mixes instantaneously with the already burned gases (which therefore have a uniform temperature), and (2) an *unmixed* model, where it is assumed that no mixing occurs between gas elements which burn at different times.

In the fully mixed model the burned gas is uniform, $T_b = \bar{T}_b$, and the equations given above fully define the state of the cylinder contents. In the unmixed model, the assumption is made that no mixing occurs between gas elements that burn at different times, and each burned gas element is therefore isentropically compressed (and eventually expanded) after combustion.† Thus:

$$\frac{T_b(x'_b, x_b)}{T_b(x'_b)} = \left[\frac{p(x_b)}{p(x'_b)}\right]^{(\gamma_b - 1)/\gamma_b} \tag{9.11}$$

---

† This model applies to burned gas regions of the chamber *away* from the walls. Heat transfer to the walls results in a thermal boundary layer on the walls which grows with time. The gas in the boundary layer is not isentropically compressed and expanded.

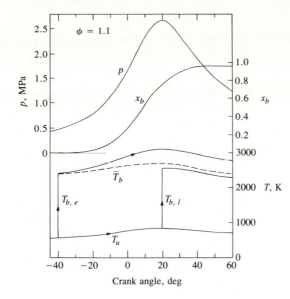

FIGURE 9-5
Cylinder pressure, mass fraction burned, and gas temperatures as functions of crank angle during combustion. T_u is unburned gas temperature, T_b is burned gas temperature, the subscripts e and l denote early and late burning gas elements, and $\bar{T}_b$ is the mean burned gas temperature.[9] (*Reprinted with permission. Copyright 1973, American Chemical Society.*)

where $T_b(x_b', x_b)$ is the temperature of the element which burned at the pressure $p(x_b')$ when the pressure is $p(x_b)$, and

$$T_b(x_b') = \frac{h_{f,u} - h_{f,b} + c_{p,u}\,T_u(x_b')}{c_{p,b}} \qquad (9.12)$$

is the temperature resulting from isenthalpic combustion of the unburned gas at $T_u(x_b')$, $p(x_b')$. An example of the temperature distribution computed with this model is shown in Fig. 9-5. A mixture element that burns right at the start of the combustion process reaches, in the absence of mixing, a peak temperature after combustion about 400 K higher than an element that burns toward the end of the combustion process. The mean burned gas temperature is closer to the lower of these temperatures. These two models approximate respectively to situations where the time scale that characterizes the turbulent mixing process in the burned gases is (1) much less than the overall burning time (for the fully mixed model) or (2) much longer than the overall burning time (for the unmixed model). The real situation lies in between.

Measurements of burned gas temperatures have been made in engines using spectroscopic techniques through quartz windows in the cylinder head. Examples of measured temperatures are shown in Fig. 9-6. The solid lines marked A, B, and C are the burned gas temperatures measured by Rassweiler and Withrow[7] using the sodium line reversal technique in an L-head engine, for the spark plug end (A), the middle (B), and the opposite end (C) of the chamber, respectively. Curves labeled W_2 and W_3 were measured by Lavoie[8] through two different windows, W_2 and W_3 (with W_2 closer to the spark), again in an L-head engine. Each set of experimental temperatures shows a temperature gradient across the burned gas comparable to that predicted, and the two sets have similar shapes.

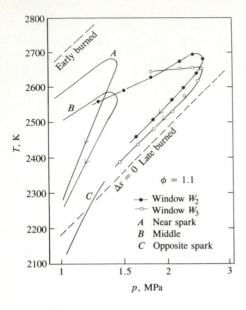

FIGURE 9-6
Burned gas temperatures measured using spectroscopic techniques through windows in the cylinder head, as a function of cylinder pressure. Temperatures measured closer to spark plug have higher values. Dashed lines show isentropic behavior.[7, 8]

In the unmixed model, the temperature of each burned gas element follows a different isentropic line as it is first compressed as p increases to p_{max} and then expanded as the pressure falls after p_{max}. The measured temperature curves in Fig. 9-6 do not follow the calculated isentropes because of gas motion past the observation ports. As has already been mentioned, the expansion of a gas element which occurs during combustion compresses the gas ahead of the flame and moves it away from the spark plug. At the same time, previously burned gas is compressed and moved back toward the spark plug. Defining this motion in an engine requires sophisticated flow models, because the combustion chamber shape is rarely symmetrical, the spark plug is not usually centrally located, and often there is a bulk gas motion at the time combustion is initiated. However, the gas motion in a spherical or cylindrical combustion bomb with central ignition which can readily be computed illustrates the features of the combustion-induced motion in an engine. Figure 9-7 shows calculated particle trajectories for a stoichiometric methane-air mixture, initially at ambient conditions, as a laminar

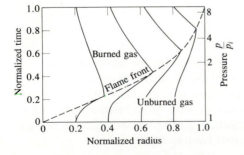

FIGURE 9-7
Particle trajectories in unburned and burned gas as flame propagates outward at constant velocity from the center of a spherical combustion bomb. Stoichiometric methane-air mixture initially at 1 atm and 300 K.

flame with a constant burning velocity propagates outward from the center of a spherical container. Applying this gas motion model to an engine, it can be concluded that a window in the cylinder head initially views earlier burned gas (of higher temperature and entropy) and that as more of the charge burns, the window views later burned gas of progressively lower entropy. The experimental curves fit this description: they cross the constant entropy lines toward lower entropy. Note that the gradient in temperature persists well into the expansion stroke, indicating that the "unmixed" model is closer to reality than the "fully mixed" model.

   More accurate calculations relating the mass fraction burned, gas pressure, and gas temperature distribution are often required. Note that the accuracy of such calculations depends on the accuracy with which the time-varying heat loss to the chamber walls can be estimated (see Sec. 12.4.3) and whether flows into and out of crevice regions are significant (see Sec. 8.6), as well as the accuracy of the models used to describe the thermodynamic properties of the gases. Appropriate more accurate models for the thermodynamic properties are: an equilibrium model for the burned gas, and specific heat models which vary with temperature for each of the components of the unburned mixture (see Secs. 4.1 and 4.7). In the absence of significant crevice effects, Eqs. (9.1) and (9.2) can be written as

$$\frac{V}{m} = \bar{v}_b x_b + \bar{v}_u(1 - x_b) \tag{9.13}$$

$$\frac{U_0 - W - Q}{m} = \bar{u}_b x_b + \bar{u}_u(1 - x_b) \tag{9.14}$$

where

$$\bar{v}_b = \frac{1}{x_b}\int_0^{x_b} v_b \, dx \qquad \text{and} \qquad \bar{v}_u = \frac{1}{1 - x_b}\int_{x_b}^1 v_u \, dx$$

and similar definitions hold for $\bar{u}_b$ and $\bar{u}_u$. For a given equivalence ratio, fuel and burned gas fraction:

$$h_u = h_u(T_u) \qquad h_b = h_b(T_b, p) \tag{9.15a, b}$$

$$pv_u = \left(\frac{\tilde{R}}{M_u}\right)T_u \qquad pv_b = \left(\frac{\tilde{R}}{M_b}\right)T_b \tag{9.16a, b}$$

and
$$\bar{u}_u = \bar{h}_u - p\bar{v}_u \qquad \bar{u}_b = \bar{h}_b - p\bar{v}_b \tag{9.17a, b}$$

To simplify the calculations, it is convenient to assume that, for the burned gas, $\bar{u}_b = u_b(\bar{T}_b, p)$ and $\bar{v}_b = v_b(\bar{T}_b, p)$. This corresponds to the fully mixed assumption described above. The effect of neglecting the temperature distribution in the calculation of mass fraction burned is small. In addition, the heat losses from the unburned gas can usually be neglected; the unburned gas is then compressed isentropically. $\bar{T}_u$ is specified for some initial state of the unburned gas (where

$x_b = 0$) by p_0, V_0, M_u, and the mass of charge m. Then, since for any isentropic process

$$\left(\frac{\partial T}{\partial p}\right)_s = \frac{v - (\partial h/\partial p)_T}{(\partial h/\partial T)_p}$$

(9.18)

$\bar{T}_u$ can be determined.

Equations (9.13) to (9.18) constitute a set of nine equations for the nine unknowns $\bar{v}_u$, $\bar{v}_b$, $\bar{u}_u$, $\bar{u}_b$, $\bar{h}_u$, $\bar{h}_b$, $\bar{T}_u$, $\bar{T}_b$, and x_b or p. One convenient solution method is to eliminate x_b from Eqs. (9.13) and (9.14) to obtain

$$\frac{(V/m) - \bar{v}_u}{\bar{v}_b - \bar{v}_u} - \frac{(U/m) - \bar{u}_u}{\bar{u}_b - \bar{u}_u} = f(\bar{T}_b, \bar{T}_u) = 0$$

(9.19)

where $U = U_0 - W - Q$. $\bar{T}_u$ can be determined from Eq. (9.18). Equation (9.19) can then be solved using an appropriate iterative technique for $\bar{T}_b$, and x_b can be obtained from Eq. (9.13). An alternative formulation based on the rate of change of pressure $dp/d\theta$ and equations for $d\bar{T}_u/d\theta$, $d\bar{T}_b/d\theta$, $dm_b/d\theta$, and $dV_b/d\theta$ can be found in Ref. 10. Some examples of mass fraction burned curves obtained from measured pressure data, with gasoline and methanol fuels, are shown in Fig. 9-8. With accurate pressure versus crank angle records, values of final mass fraction burned should be close to but lower than unity, usually in the range 0.93 to 0.98: the difference from unity is the combustion inefficiency for lean mixtures (see Fig. 3-9) and incomplete oxygen utilization for rich mixtures (see Fig. 4-20).

More accurate burned gas temperature calculations need to account for the presence of a thermal boundary layer (of order 1 mm thick) around the combustion chamber walls (see Sec. 12.6.5). The burned gas region in Fig. 9-4 can be

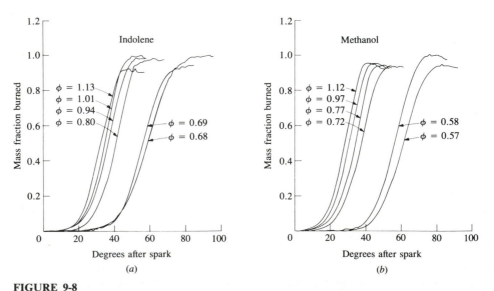

FIGURE 9-8
Mass fraction burned curves determined from measured cylinder pressure data using two-zone combustion model: (a) gasoline; (b) methanol. ϕ = fuel/air equivalence ratio.[11]

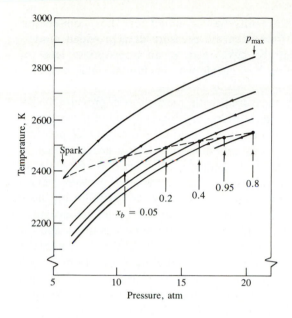

FIGURE 9-9
Calculated temperature distribution in the adiabatic core of the burned gas zone for the unmixed model assuming thermodynamic equilibrium. $\phi = 1.0$. Dashed line is temperature of each element just after it burns.

divided into an adiabatic core and a boundary layer that grows in thickness with time. In the adiabatic core, in the absence of mixing between gas elements that burn at different times, burned gas is compressed and then expanded isentropically. The burned gas temperature distribution can be calculated as follows. Given the pressure versus crank angle data, the unburned mixture state can be determined using Eq. (9.18) above. Each small element of unburned mixture burns in a constant-enthalpy constant-pressure process. So the burned state of an element of unburned charge, which burns at $p = p_i$, can be obtained from the relation

$$h_b(T_{b,i}, p_i) = h_u(T_{u,i}, p_i)$$

After combustion, this element which burned at $p = p_i$ is compressed and expanded along the isentropic:

$$s_b(T_b, p) = s_b(T_{b,i}, p_i)$$

An example of the temperature distribution computed in this manner for this unmixed model in the burned gas adiabatic core is shown in Fig. 9-9. The element ignited by the spark is compressed to the highest peak temperature at p_{max}. The temperature difference across the bulk of the charge ($0.05 < x_b < 0.95$) is about 200 K.

9.2.2 Analysis of Cylinder Pressure Data

Cylinder pressure changes with crank angle as a result of cylinder volume change, combustion, heat transfer to the chamber walls, flow into and out of crevice regions, and leakage. The first two of these effects are the largest. The effect of volume change on the pressure can readily be accounted for; thus, combustion rate information can be obtained from accurate pressure data provided models for the remaining phenomena can be developed at an appropriate level of approximation. The previous section has developed the fundamental basis for such calculations.

Cylinder pressure is usually measured with piezoelectric pressure transducers. This type of transducer contains a quartz crystal. One end of the crystal is exposed through a diaphragm to the cylinder pressure; as the cylinder pressure increases, the crystal is compressed and generates an electric charge which is proportional to the pressure. A charge amplifier is then used to produce an output voltage proportional to this charge. Accurate cylinder pressure versus crank angle data can be obtained with these systems provided the following steps are carried out: (1) the correct reference pressure is used to convert the measured pressure signals to absolute pressures; (2) the pressure versus crank angle (or volume) phasing is accurate to within about 0.2°; (3) the clearance volume is estimated with sufficient accuracy; (4) transducer temperature variations (which can change the transducer calibration factor) due to the variation in wall heat flux during the engine cycle are held to a minimum. Log p versus log V plots can be used to check the quality of cylinder pressure data. The first three of the above requirements can be validated using log p–log V diagrams for a motored engine. If the effects of thermal cycling are significant, the expansion stroke on the log p–log V plot for a firing engine shows excessive curvature.[12]

Figure 9-10 shows pressure-volume data from a firing spark-ignition engine on both a linear p–V and a log p–log V diagram. On the log p–log V diagram the compression process is a straight line of slope 1.3. The start of combustion can be identified by the departure of the curve from the straight line. The end of com-

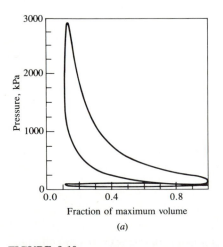

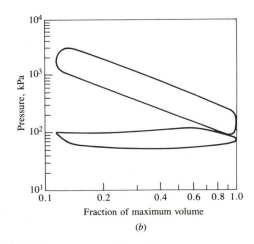

Fraction of maximum volume

(a)

Fraction of maximum volume

(b)

FIGURE 9-10
(a) Pressure-volume diagram; (b) log p–log (V/V_{max}) plot. 1500 rev/min, MBT timing, imep = 513 kPa, $\phi = 0.8$, $r_c = 8.72$, propane fuel.[12]

bustion can be located approximately in similar fashion; the expansion stroke following combustion is essentially linear with slope 1.33. Since both the compression of the unburned mixture prior to combustion and the expansion of the burned gases following the end of combustion are close to adiabatic isentropic processes (for which $pV^\gamma = $ constant; $\gamma = c_p/c_v$), the observed behavior is as expected. More extensive studies[12, 13] show that the compression and expansion processes are well fitted by a polytropic relation:

$$pV^n = \text{constant} \tag{9.20}$$

The exponent n for the compression and expansion processes is 1.3 (± 0.05) for conventional fuels. It is comparable to the average value of γ_u for the unburned mixture over the compression process, but is larger than γ_b for the burned gas mixture during expansion due to heat loss to the combustion chamber walls (see Figs. 4-13 and 4-16).

Log p–log V plots such as Fig. 9-10 approximately define the start and end of combustion, but do not provide a mass fraction burned profile. One well-established technique for estimating the mass fraction burned profile from the pressure and volume data is that developed by Rassweiler and Withrow.[2] They correlated cylinder pressure data with flame photographs, and showed how Eq. (9.20) could be used to account for the effect of cylinder volume change on the pressure during combustion. Assuming that the unburned gas filling the volume V_u ahead of the flame at any crank angle during combustion has been compressed polytropically by the advancing flame front, then the volume $V_{u,0}$ it occupied at time of spark is

$$V_{u,0} = V_u\left(\frac{p}{p_0}\right)^{1/n} \tag{9.21}$$

Similarly, the burned gas behind the flame filling the volume V_b would, at the end of combustion, fill a volume $V_{b,f}$ given by

$$V_{b,f} = V_b\left(\frac{p}{p_f}\right)^{1/n} \tag{9.22}$$

The mass fraction burned x_b is equal to $1 - (V_{u,0}/V_0)$ and to $V_{b,f}/V_f$, where V_0 amd V_f are the total cylinder volumes at time of spark and at the end of combustion, respectively. Since $V = V_u + V_b$, Eqs. (9.21) and (9.22) then give:

$$x_b = \frac{p^{1/n}V - p_0^{1/n}V_0}{p_f^{1/n}V_f - p_0^{1/n}V_0} \tag{9.23}$$

This method is widely used, though it contains several approximations. Heat-transfer effects are included only to the extent that the polytropic exponent n in Eq. (9.22) differs appropriately from γ. The pressure rise due to combustion is proportional to the amount of fuel chemical energy released rather than the mass of mixture burned. Also, the polytropic exponent n is not constant during combustion. Selecting an appropriate value for n (whether n is assumed to be constant or to vary through the combustion process) is the major difficulty in applying this pressure data analysis procedure.

The effects of heat transfer, crevices, and leakage can be explicitly incorporated into cylinder pressure data analysis by using a "heat release" approach based on the first law of thermodynamics. A major advantage of such an approach is that the pressure changes can be related directly to the amount of fuel chemical energy released by combustion, while retaining the simplicity of treating the combustion chamber contents as a single zone. Figure 9-11 shows the appropriate open-system boundary for the combustion chamber.[14] The first law for this open system is

$$\delta Q_{ch} = dU_s + \delta Q_{ht} + \delta W + \sum h_i \, dm_i \tag{9.24}$$

The change in sensible energy of the charge dU_s is separated from that due to change in composition: the term δQ_{ch} represents the "chemical energy" released by combustion. The work is piston work and equal to $p \, dV$. δQ_{ht} is heat transfer to the chamber walls. The mass flux term represents flow across the system boundary. In the absence of fuel injection, it represents flow into and out of crevice regions (see Sec. 8.6).

The accuracy with which this energy balance can be made depends on how adequately each term in the above equation can be quantified. Assuming that U_s is given by $mu(T)$, where T is the mean charge temperature and m is the mass within the system boundary, then

$$dU_s = mc_v(T) \, dT + u(T) \, dm$$

Note that this mean temperature determined from the ideal gas law is close to the mass-averaged cylinder temperature during combustion since the molecular weights of the burned and unburned gases are essentially identical. Crevice effects can usually be modeled adequately by flow into and out of a single volume at the cylinder pressure, with the gas in the crevice at a substantially lower temperature. Leakage to the crankcase can usually be neglected. Then Eq. (9.24), on substituting for dU_s and $dm_i \, (= dm_{cr} = -dm)$, becomes

$$\delta Q_{ch} = mc_v \, dT + (h' - u)dm_{cr} + p \, dV + \delta Q_{ht} \tag{9.25}$$

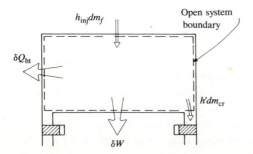

FIGURE 9-11
Open system boundary for combustion chamber for heat-release analysis.

where

dm_cr > 0 when flow is out of the cylinder into the crevice
dm_cr < 0 when flow is from the crevice to the cylinder
h' is evaluated at cylinder conditions when $dm_{cr} > 0$ and
  at crevice conditions when $dm_{cr} < 0$

Use of the ideal gas law (neglecting the change in gas constant R) with Eq. (9.25) then gives

$$\delta Q_{ch} = \left(\frac{c_v}{R}\right)V\,dp + \left(\frac{c_v}{R} + 1\right)p\,dV + (h' - u + c_v T)\,dm_{cr} + \delta Q_{ht} \quad (9.26)$$

This equation can be used in several ways. When the heat or energy release term, δQ_{ch}, is combined with the heat-transfer and crevice terms, the combination is termed *net heat release*—the combustion energy release less heat lost to the walls. It is equal to the first two terms on the right-hand side of Eq. (9.26) which, together, represent the sensible energy change and work transfer to the piston. While heat losses during combustion are a small fraction of the fuel energy (10 to 15 percent), the distributions of heat release and heat transfer with crank angle are different; heat transfer becomes more important as the combustion process ends and average gas temperatures peak. The net heat-release profile obtained from integrating the first two terms on the right-hand side of Eq. (9.26), normalized to give unity at its maximum value, is often interpreted as the burned mass fraction (or, more correctly, the energy-release fraction) versus crank angle profile.

Use of Eq. (9.26) requires a value for c_v/R $[=1/(\gamma - 1)]$. The ratio of specific heats γ for both unburned and burned gases decreases with increasing temperature and varies with composition (see Figs. 4-13, 4-16, and 4-18). As the mean charge temperature increases during compression and combustion and then decreases during expansion, γ should vary. An approximate approach, modeling $\gamma(T)$ with a linear function of temperature fitted to the appropriate curves in Figs. 4-13, 4-16, and 4-18 and with γ constant during combustion, has been shown to give adequate results.[15]

The convective heat-transfer rate to the combustion chamber walls can be calculated from the relation

$$\frac{dQ_{ht}}{dt} = Ah_c(T - T_w)$$

where A is the chamber surface area, T is the mean gas temperature, T_w is the mean wall temperature, and h_c is the heat-transfer coefficient (averaged over the chamber surface area). h_c can be estimated from engine heat-transfer correlations (see Sec. 12.4.3). Since crevice effects are usually small, a sufficiently accurate model for their overall effect is to consider a single aggregate crevice volume where the gas is at the same pressure as the combustion chamber, but at a different temperature. Since these crevice regions are narrow, an appropriate assumption is that the crevice gas is at the wall temperature. Inserting this crevice model into Eq. (9.26), with $\gamma(T) = a + bT$, gives the *chemical energy-* or *gross heat-release rate*:

$$\frac{dQ_{ch}}{d\theta} = \frac{\gamma}{\gamma - 1} p \frac{dV}{d\theta} + \frac{1}{\gamma - 1} V \frac{dp}{d\theta}$$

$$+ V_{cr}\left[\frac{T'}{T_w} + \frac{T}{T_w(\gamma - 1)} + \frac{1}{bT_w} \ln\left(\frac{\gamma - 1}{\gamma' - 1}\right) \right] \frac{dp}{d\theta} + \frac{dQ_{ht}}{d\theta} \quad (9.27)$$

An example of the use of Eq. (9.27) to analyze an experimental pressure versus crank angle curve for a conventional spark-ignition engine is shown in Fig. 9-12. The integrated heat release is plotted against crank angle. The lowest curve shown is the net heat release. The addition of heat transfer and crevice models gives the chemical energy release. The curve at the top of the figure is the mass of fuel within the combustion chamber times its lower heating value. It decreases slightly as p_{max} is approached due to flow into crevices. The difference between the final value of Q_{ch} and $(m_f Q_{LHV})$ should equal the combustion inefficiency (which is a few percent of $m_f Q_{LHV}$). The combustion inefficiency can be determined from the exhaust gas composition (see Sec. 4.9.4). Inaccuracies in the cylinder pressure data and the heat-release calculation will also contribute to this difference. An important advantage of a heat-release analysis that relates the pressure changes to the amount of fuel chemical energy within the cylinder is that this error can be determined. In the example in Fig. 9-12, the measured combustion inefficiency was close to the amount shown.

Two-zone models (one zone representing the unburned mixture ahead of the flame and one the burned mixture behind the flame) are used to calculate the mass fraction burned profile from measured cylinder pressure data.[10] Figure 9-8 shows results from such an analysis, using the methodology described in Sec. 9.2.1. The advantage of a two-zone analysis is that the thermodynamic properties of the cylinder contents can be quantified more accurately. The disadvantages are that the unburned and the burned zone heat-transfer areas must both now be estimated, and a model for the composition of the gas flowing into the crevice

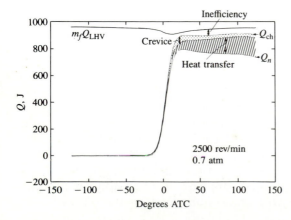

FIGURE 9-12
Results of heat-release analysis showing the effects of heat transfer, crevices, and combustion inefficiency.[14]

region must be developed. Due to this complexity, crevice models are usually omitted despite the fact that their impact can be significant.

9.2.3 Combustion Process Characterization

The mass fraction burned profiles as a function of crank angle in each individual cycle shown in Fig. 9-2 and the chemical energy- or gross heat-release curve in Fig. 9-12 have a characteristic S-shape. The rate at which fuel-air mixture burns increases from a low value immediately following the spark discharge to a maximum about halfway through the burning process and then decreases to close to zero as the combustion process ends. It proves convenient to use these mass fraction burned or energy-release fraction curves to characterize different stages of the spark-ignition engine combustion process by their duration in crank angles, thereby defining the fraction of the engine cycle that they occupy. The flame development process, from the spark discharge which initiates the combustion process to the point where a small but measurable fraction of the charge has burned, is one such stage. It is influenced primarily by the mixture state, composition, and motion in the vicinity of the spark plug (see Sec. 9.3). The stage during which the major portion of the charge burns as the flame propagates to the chamber walls is next. This stage is obviously influenced by conditions throughout the combustion chamber. The final stage, where the remainder of the charge burns to completion, cannot as easily be quantified because energy-release rates are comparable to other energy-transfer processes that are occurring.

The following definitions are most commonly used to characterize the energy-release aspects of combustion:

Flame-development angle $\Delta\theta_d$. The crank angle interval between the spark discharge and the time when a small but significant fraction of the cylinder mass has burned or fuel chemical energy has been released. Usually this fraction is 10 percent, though other fractions such as 1 and 5 percent have been used.†

Rapid-burning angle $\Delta\theta_b$. The crank angle interval required to burn the bulk of the charge. It is defined as the interval between the end of the flame-development stage (usually mass fraction burned or energy-release fraction of 10 percent) and the end of the flame-propagation process (usually mass fraction burned or energy-release fraction of 90 percent).‡

Overall burning angle $\Delta\theta_o$. The duration of the overall burning process. It is the sum of $\Delta\theta_d$ and $\Delta\theta_b$.

---

† This angle is sometimes called the *ignition delay*. Since the flame starts to propagate outward immediately following the spark discharge there is no delay, and the terminology used here is preferred (see Sec. 9.3).

‡ An alternative definition for $\Delta\theta_b$ uses the maximum burning rate to define an angle or time characteristic of the bulk charge burning process[4] (see Fig. 9-13). $\Delta\theta_b$ and $\Delta\theta_b^*$ are usually closely comparable.

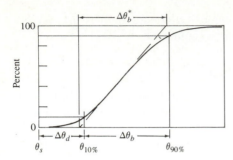

FIGURE 9-13
Definition of flame-development angle, $\Delta\theta_d$, and rapid-burning angle, $\Delta\theta_b$, on mass fraction burned versus crank angle curve.

Figure 9-13 illustrates these definitions on a mass fraction burned, or fraction of fuel energy released, versus crank angle plot. While the selection of the 10 and 90 percent points is arbitrary, such a choice avoids the difficulties involved in determining accurately the shape of the curve at the start and end of combustion. These angles can be converted to times (in seconds) by dividing by $6N$ (with N in revolutions per minute).

A functional form often used to represent the mass fraction burned versus crank angle curve is the Wiebe function:

$$x_b = 1 - \exp\left[-a\left(\frac{\theta - \theta_0}{\Delta\theta}\right)^{m+1}\right] \tag{9.28}$$

where θ is the crank angle, θ_0 is the start of combustion, $\Delta\theta$ is the total combustion duration ($x_b = 0$ to $x_b = 1$), and a and m are adjustable parameters. Varying a and m changes the shape of the curve significantly. Actual mass fraction burned curves have been fitted with $a = 5$ and $m = 2$.[16]

9.3   FLAME STRUCTURE AND SPEED

9.3.1   Experimental Observations

The combustion process in the spark-ignition engine takes place in a turbulent flow field. This flow field is produced by the high shear flows set up during the intake process and modified during compression, as described in Chap. 8. The importance of the turbulence to the engine combustion process was recognized long ago through experiments where the intake event, and the turbulence it generates, was eliminated and the rate of flame propagation decreased substantially. Understanding the structure of this engine flame as it develops from the spark discharge and the speed at which it propagates across the combustion chamber, and how that structure and speed depend on charge motion, charge composition, and chamber geometry, are critical to engine optimization. This section reviews experimental evidence that describes the essential features of the flame development and propagation processes.

Direct flame photographs such as those in Fig. 9-1 indicate the location and shape of the actual reaction zone which radiates in the blue region of the visible spectrum. An irregular front is apparent. Further insight into the structure of the flame can be obtained from photographs taken with techniques that are sensitive to density changes in the flow field, such as schlieren and shadowgraph. With these techniques, a parallel light beam is passed through the combustion chamber. Portions of the beam which pass through regions where density gradients normal to the beam exist are deflected, due to the refractive index gradients that result from the density gradients. In the schlieren technique, the beam is focused on a knife edge; the deflected parts of the beam are displaced relative to the knife edge and produce lighter or darker regions when subsequently refocused onto film. With the shadowgraph technique, the parallel beam emerging from the combustion chamber is photographed directly; deflected parts of the beam produce lighter and darker regions on the film. With these techniques, details of flame structure can be discerned.

Figure 9-14 shows a set of photographs from one engine cycle, from a high-speed schlieren movie taken in a special visualization spark-ignition engine operating at 1400 rev/min and 0.5 atm inlet pressure. Also shown are the cylinder pressure versus crank angle data, and the mass fraction burned profile calculated from the pressure data using the method of Rassweiler and Withrow[2] (see Sec. 9.2.2). This engine had a square-cross-section cylinder with two quartz walls to permit easy optical access, but otherwise operated normally.[17] Visualization of

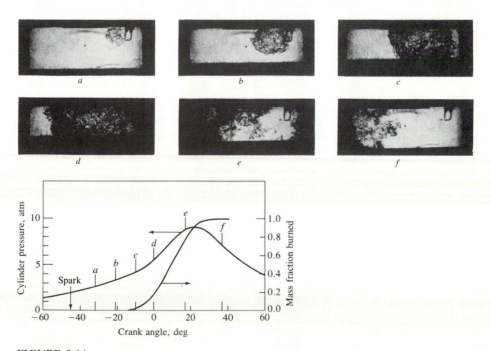

FIGURE 9-14
Sequence of movie frames from one engine cycle in a square-cross-section cylinder, single-cylinder, engine with two glass walls, and corresponding pressure and mass fraction burned curves. 1400 rev/min, 0.5 atm inlet pressure.

the flame is especially important during the early stages of flame development when the pressure rise due to combustion is too small to be detected.

These photographs show how the flame "ball," roughly spherical in shape, grows steadily from the time of spark discharge. The effect of turbulence is already visible in the convoluted flame surface in Fig. 9-14a. The volume enflamed behind the front continues to grow in a roughly spherical manner, except where intersected by the chamber walls, as seen in Fig. 9-14b and c. The mass fraction burned and the associated pressure rise due to combustion become significant by the time the flame front has traversed two-thirds to three-quarters of the field of view. Note that the fraction of the cylinder filled with enflamed charge is less than is suggested by the photos because the front of the flame is approximately spherical and the cylinder has a square cross section. Maximum cylinder pressure occurs close to the time the flame makes contact with the far wall, as seen in Fig. 9-14e. Finally, the unburned mixture ahead of and within the front burns out and the density gradients associated with the flame reaction zone disappear, clearing the field of view as shown in Fig. 9-14e and f.

A useful relationship between the mass fraction burned, $x_b(= m_b/m)$, and the volume fraction occupied by the burned gas, $y_b(= V_b/V)$, can be obtained from the identities

$$m = m_u + m_b \qquad V = V_u + V_b$$

and the ideal gas law:

$$x_b = \left[1 + \frac{\rho_u}{\rho_b} \left(\frac{1}{y_b} - 1 \right) \right]^{-1} \tag{9.29}$$

While the density ratio (ρ_u/ρ_b) does depend on the equivalence ratio, burned gas fraction in the unburned mixture, gas temperature, and pressure, its value is close to 4 for most spark-ignition engine operating conditions. Thus, the plot of x_b against y_b has a universal form,[10] as shown in Fig. 9-15. This curve is an important aid in interpreting flame geometry information.

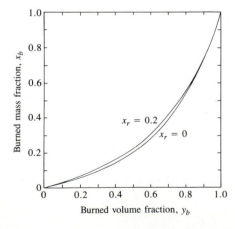

FIGURE 9-15
Relation between mass fraction burned x_b and volume fraction burned y_b. x_r is residual mass fraction.[4]

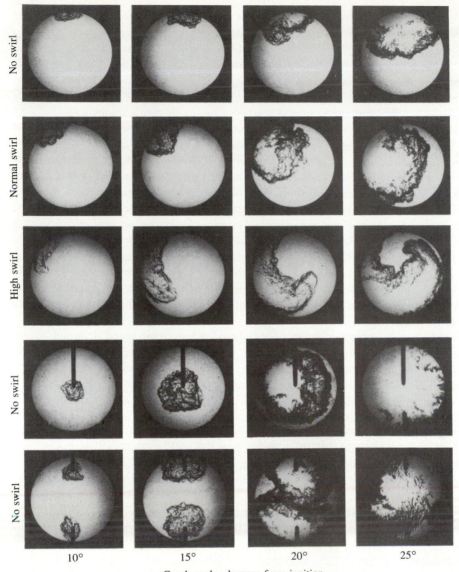

Crank angle, degrees from ignition

FIGURE 9-16
Laser shadowgraph photographs of engine combustion process taken in single-cylinder engine with transparent cylinder head. From top to bottom: side plug without swirl; side plug with normal swirl; side plug with high swirl; central plug without swirl; two plugs without swirl.[18]

The above-described features of the developing and propagating flame are common to almost all engine geometries and operating conditions. Figure 9-16 shows shadowgraph photographs of the flame at fixed crank angle intervals after ignition, taken through a transparent cylinder head with different geometric and flow configurations.[18] The approximately spherical development of the flame from the vicinity of the spark plug, except where it intercepts the chamber walls, is evident for side and center ignition with one plug, and for ignition with two plugs in the absence of any intake generated swirl. With normal levels of swirl, the flame center is convected with the swirling flow, but the flame front as it grows is still approximately spherical in shape. Only with unusually high levels of swirl and aerodynamic stabilization of the flame at the spark plug location does the flame become stretched out and distorted by the flow in a major way.[18]

At any given flame radius, the geometry of the combustion chamber and the spark plug location govern the flame front surface area—the area of the approximately spherical surface corresponding to the leading edge of the flame contained by the piston, cylinder head, and cylinder wall. The larger this surface area, the greater the mass of fresh charge that can cross this surface and enter the flame zone. The photos in Fig. 9-16 illustrate the importance of flame area. The center plug location gives approximately twice the flame area of the side plug geometry at a given flame radius, and burns about twice as fast (the fraction of the cylinder volume enflamed is about twice the size, at a fixed crank angle interval after spark). The arrangement with two spark plugs at opposite sides of the chamber is not significantly different in enflamed volume from the single center plug because, once the flame fronts are intersected by the cylinder wall, the flame front areas are comparable.

Mixture burning rate is strongly influenced by engine speed. It is well established that the duration of combustion in crank angle degrees only increases slowly with increasing engine speed.[19] Figure 9-17 shows how the interval between the spark discharge and 10 percent mass fraction burned, the flame development angle $\Delta\theta_d$, and the interval between the spark and 90 percent mass fraction burned, the overall burning angle $\Delta\theta_d + \Delta\theta_b$ (see Sec. 9.2.3), vary with engine speed.[20] Both intervals increase by a factor of about 1.6 for a factor of 4 increase in engine speed; i.e., the burning rate throughout the combustion process increases almost, though not quite, as rapidly as engine speed. Additionally, at a given engine speed, increasing in-cylinder gas velocities (e.g., with intake generated swirl) increases the burning rate: the flame size for the swirling flows in Fig. 9-16 is larger than for the quiescent case with the same plug location at the crank angle intervals after spark shown. Increasing engine speed and introducing swirl both increase the levels of turbulence in the engine cylinder at the time of combustion (see Sec. 8.2.2). Increased turbulence increases the rate of development and propagation of the turbulent premixed engine flame.

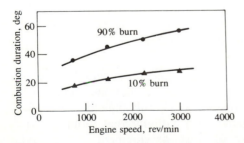

FIGURE 9-17
Effect of engine speed on flame-development angle (0 to 10 percent burned) and overall burning angle (0 to 90 percent burned). $\phi = 1.0$, intake pressure 0.54 atm, spark 30° BTC.[20]

It is also well established that unburned mixture composition and state affect the burning rate. Reducing the inlet pressure (and maintaining the ratio of exhaust to inlet pressure fixed to hold the residual gas fraction constant) increases both the flame development and rapid burning angles.[19] The fuel/air equivalence ratio affects the burning rate. Both flame development and burning angles show a minimum for slightly rich mixtures ($\phi \approx 1.2$) and increase significantly as the mixture becomes substantially leaner than stoichiometric.[19, 20] The burned gas fraction in the unburned mixture, due to the residual gas fraction *and* any recycled exhaust gases, affects the burning rate: increasing the burned gas fraction slows down both flame development and propagation.[20] Fuel composition changes can be significant also. While mixtures of isooctane or conventional gasolines with air and burned gases (at identical conditions) have closely comparable burning rates, propane, methane, methanol, and ethanol mixtures exhibit modest differences in burning rate and hydrogen-air mixtures substantial differences. The basic combustion chemistry of the fuel, air, burned gas mixture influences the combustion process. However, the relative importance of combustion chemistry effects depends on combustion chamber design and burn rate. Faster burning engines (which have higher turbulence) are less sensitive to changes in mixture composition, pressure, and temperature than are slower burning engines (which have lower turbulence). The effects of chamber geometry, gas motion, and gas composition and state are interrelated.[21]

9.3.2 Flame Structure

Laminar flames in premixed fuel, air, residual gas mixtures are characterized by a laminar flame speed S_L and a laminar flame thickness δ_L (see Sec. 9.3.3). The laminar flame speed is the velocity at which the flame propagates into quiescent premixed unburned mixture ahead of the flame. There are several ways to define the thickness of a laminar flame.[22] Given the molecular diffusivity D_L (see Sec. 4.8), dimensional arguments give the most commonly used definition: $\delta_L = D_L/S_L$. Turbulent flames are also characterized by the root mean square velocity fluctuation, the turbulence intensity u' [Eq. (8.3)], and the various length scales of the turbulent flow ahead of the flame. The integral length scale l_I [Eq. (8.8)] is a measure of the size of the large energy-containing structures of the flow. The Kolmogorov scale l_K [Eq. (8.11)] defines the smallest structures of the flow where small-scale kinetic energy is dissipated via molecular viscosity.

Several dimensionless parameters are used to characterize turbulent premixed flames. The dimensionless parameter used to define the turbulence is the turbulent Reynolds number, $\mathrm{Re}_T = u'l_I/\nu$. For homogeneous and isotropic (no preferred direction) turbulence, the integral and Kolmogorov scales are related by Eq. (8.14): $l_K/l_I = \mathrm{Re}_T^{-3/4}$. A characteristic turbulent eddy turnover time τ_T can be defined as

$$\tau_T = \frac{l_I}{u'}$$

A characteristic chemical reaction time is the residence time in a laminar flame:

$$\tau_L = \frac{\delta_L}{S_L}$$

The ratio of the characteristic eddy turnover time to the laminar burning time is called the *Damköhler number*:

$$Da = \frac{\tau_T}{\tau_L} = \left(\frac{l_I}{\delta_L}\right)\left(\frac{S_L}{u'}\right) \tag{9.30}$$

It is an inverse measure of the influence of the turbulent flow on the chemical processes occurring in the flame. Other ratios are of interest. The ratio δ_L/l_K is a measure of the stretch or local distortion to which a laminar flame is subjected by the turbulent flow. Unless $l_I/\delta_L \gg 1$ the concept of a localized flame region has little significance. The ratio u'/S_L is a measure of the relative strength of the turbulence.[22]

Different regimes of turbulent flames are apparent in the plot of Damköhler number versus turbulent Reynolds number in Fig. 9-18.[22] It has been assumed that $D_L \approx v$ and that the relationships for homogeneous isotropic turbulence are valid. Two regimes—distributed reactions and reaction sheets—are normally identified. In the distributed reaction regime, chemical reactions proceed in distributed reaction zones and thin-sheet flames do not occur. A sufficient condition for this regime is $l_I \ll \delta_L$. In the reaction sheet regime, propagating reaction fronts are wrinkled and convoluted by the turbulence. A sufficient condition for the existence of reaction sheets is $l_K \gg \delta_L$. For $Re_T > 1$, there is a region in Fig. 9-18 where $l_I > \delta_L > l_K$: the characteristics of flames in this regime are unclear.

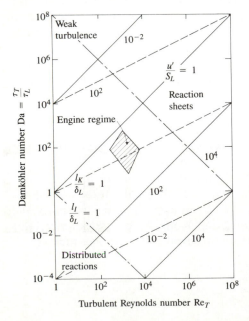

FIGURE 9-18
Different turbulent flame regimes on plot of Damköhler number versus turbulence Reynolds number, u' is turbulence intensity; S_L is laminar flame speed; l_I, l_K, and δ_L are integral scale, Kolmogorov scale, and laminar flame thickness. (*From Abraham et al.*[22])

Values of Da and Re_T for a typical spark-ignition engine (the cross-hatched region in Fig. 9-18) lie predominantly in the reaction sheet flame regime. Engine operation at high speed (the lower right boundary) and low load (the lower left boundary) gives values of Da and Re_T which fall below the $l_K/\delta_L = 1$ line. This is largely due to the low values of laminar flame speed that result from the high amounts of residual gas and EGR under these conditions (see Sec. 9.3.4 and Fig. 6-19). Whether the flame structure under these conditions is significantly different is not known. Observations of engine flames to date, described below, lie above the $l_K/\delta_L = 1$ line, within the reaction sheet regime. One would expect, then, the structure of the flame in a spark-ignition engine, once developed, to be that of a thin reaction sheet wrinkled and convoluted by the turbulent flow.

Detailed observations have been made of flame structure from ignition to flame extinguishing at the far cylinder wall. A flame develops from the spark discharge which causes ignition as follows. In the initial breakdown phase of ignition, a cylindrical discharge between the spark plug electrodes is established.[23] As electrical energy is fed into the discharge, the arc expands and exothermic chemical reactions capable of sustaining a propagating flame develop. Figure 9-19 shows how this development of a flame kernel occurs, with a set of shadowgraph photographs taken at 40-μs intervals of the spark plug electrode gap in one cylinder of a 2-liter conventional engine. The first photograph is between 20 and 50 μs after the spark breakdown occurred. The complete sequence ($\sim 200\ \mu$s) corresponds to 1.3 crank angle degrees. The outer boundary of this developing flame kernel is approximately spherical and is smooth with modest irregularities, corresponding to a thin reaction zone with high-temperature gases inside. As this developing sheetlike flame grows it interacts with the turbulent flow field in the vicinity of the spark plug: the flame outer surface becomes increasingly convoluted and the flame center can be convected away from the plug in a direction and with a velocity that can vary substantially cycle-by-cycle, as seen in Fig. 9-20.[4, 17]

The structure of the flame continues to develop as it propagates across the chamber. Evidence, largely from schlieren photographs and studies of flame structure with laser diagnostics, shows that early in the burning process the flame

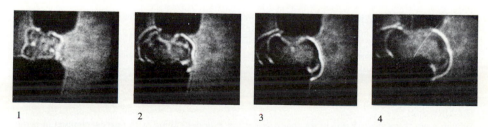

1   2   3   4

FIGURE 9-19
Shadowgraph photographs of spark-generated kernel between the spark plug electrodes. First photograph on left, 20 to 50 μs after breakdown; 40 μs between photos. Stoichiometric mixture, 1100 rev/min. (*Courtesy A. Douad, Institut Francais du Petrole.*)

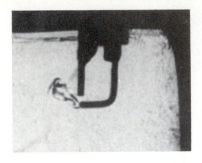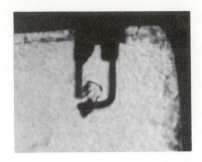

FIGURE 9-20
Schlieren photographs of developing flame, 5° after spark discharge, showing different convection processes in two different cycles. Spark plug wires 0.8 mm diameter. 1400 rev/min; 0.5 atm inlet pressure; propane fuel; $\phi = 0.9$; spark advance 45° BTC.[17]

is a thin, moderately wrinkled but simply connected, front or reaction sheet between unburned and burned gas. The thickness of the front is about 0.1 mm which is comparable to the thickness of a laminar flame under the prevailing conditions. The scale of the wrinkles is typically about 2 mm at engine speeds of 1000 to 2000 rev/min. As the flame propagates across the chamber, the thickness of the reaction sheet front remains roughly constant, the flame front becomes more convoluted, and the scale of the wrinkles tends to decrease with time.[24]

Further evidence that the thin reaction sheet front becomes highly wrinkled and convoluted by the turbulent flow field into a thick turbulent flame "brush" is provided by the schlieren photographs in Fig. 9-21. These show a flame propagating across the combustion chamber of a square-cross-section single-cylinder engine with a special transparent piston crown containing a 13-mm wide channel to isolate a small section of the flame.[25] The flame on the left is for a propane-air mixture; that on the right is hydrogen-air. The energy density per unit volume of mixture and the flow field are comparable for each fuel. The effective thickness of this turbulent flame—the average distance between the region ahead of the flame, where only unburned mixture exists, and the region behind the flame, where only

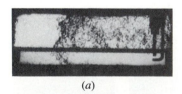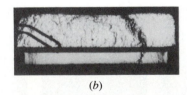

(a) (b)

FIGURE 9-21
Schlieren photographs of flame in square-cross-section cylinder engine, with narrow channel in piston crown (at bottom of pictures) which permits observation of 13-mm wide section of flame. (a) Propane fuel, spark timing 36° BTC, photograph at 14° ATC, flame thickness ≈ 4.6 mm. (b) Hydrogen fuel, spark timing 3° BTC, photograph at 10° ATC, flame thickness ≈ 1.5 mm. Stoichiometric mixtures, 1380 rev/min, 0.5 atm inlet pressure.[25]

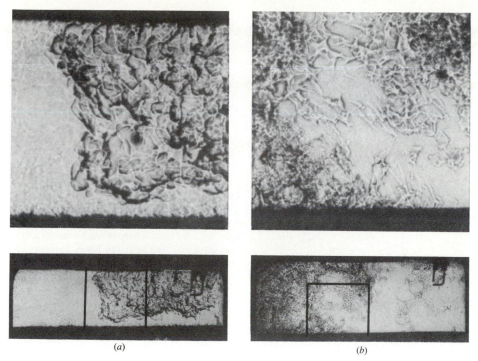

(a)   (b)

FIGURE 9-22
Enlarged schlieren photographs of (a) flame front and (b) flame back in square-cross-section cylinder
engine with two glass side walls. 1400 rev/min, 0.5 atm inlet pressure, propane fuel, $\phi = 0.9$.[17]

burned mixture exists—is apparent.† It was 4 to 5 mm for propane and 1.5 mm
for hydrogen. The difference is due to the substantially higher laminar flame
speed for the hydrogen, air, burned gas mixture (see Sec. 9.3.4) which increases
the Damköhler number and shifts the flame toward the weak-turbulence (i.e., less
wrinkled) flame regime in Fig. 9-18.

Additional insight into the structure of the developed engine flame can be
obtained by enlarging photographs of the leading and trailing edges of the flame,
obtained with the schlieren or shadowgraph technique. Figure 9-22a shows the
front of the flame 40° after the spark, when it has propagated about halfway
across the chamber. It shows the irregular but smoothly curved surfaces which
comprise the leading edge of the flame. Figure 9-22b shows the back of the flame
70° after the spark, when the front of the flame has just reached the wall of the
combustion chamber farthest from the spark plug. It shows large clear regions of
burned gas behind the flame and smaller clear regions connected by a lacelike

---

† These photographs were selected from a large number to give the minimum flame thickness corresponding to the flame front perpendicular to the channel length.

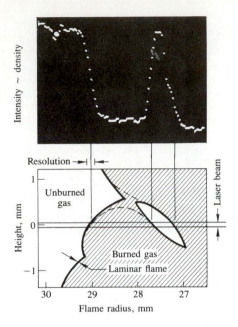

FIGURE 9-23
Upper picture: oscillogram of output of optical multichannel analyzer showing the intensity of light scattered from a narrow laser beam as a function of distance through the flame. Intensity is a measure of gas density. Lower picture: schematic of flame structure corresponding to this signal. 600 rev/min, 1 atm inlet pressure, propane fuel, $\phi = 1.0$.[24]

structure within which the remaining regions of unburned mixture are being consumed.[17] The analogy with a crumpled sheet of paper is appropriate.

Laser scattering experiments, where "snapshots" of the density profile along a laser beam passed through the flame were obtained using Rayleigh scattered light from the gas molecules, provide explicit evidence of this structure.[26] Figure 9-23 shows an oscillogram of the output of the optical analyzer. Each dot represents the intensity of the scattered light which is a measure of the gas density. The signal on the left corresponds to unburned gas. The flame is propagating from right to left. The oscillogram shows a thin transition zone of width 0.25 mm between unburned and burned gas, followed at a distance of 1.5 mm by an "island" of unburned gas. The fraction of oscillograms showing such "islands" varied from 0 at 300 rev/min to 20 percent at 1800 rev/min. An interpretation of this signal consistent with the available photographic evidence is shown underneath.[24]

The above results suggest that increasing engine speed, which increases turbulence levels in the unburned charge, increasingly convolutes and probably multiply-connects the thin reaction sheet flame front. Enlarged schlieren photographs of a 9-mm diameter section of the developed engine flame, viewed normal to the flame surface, indicate that increasing turbulence intensity and decreasing turbulence scales result in increasingly finely wrinkled flame structures. Figure 9-24 shows a set of such photographs, arranged in order of increasing turbulent Reynolds number. Increases in turbulence intensity, which was measured, were achieved by increases in engine speed and by modifying the inlet valve. Relevant parameters for each photograph are given in Table 9.1.

The above theoretical discussion and experimental evidence indicates that

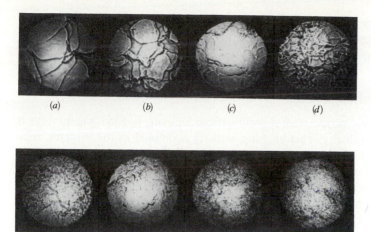

FIGURE 9-24
Shadowgraph photographs, with light beam normal to flame front, of 10-mm diameter section of flame. Photographs, arranged in order of increasing turbulent Reynolds number, suggest this is an appropriate scaling parameter (see Table 9.1).[27]

developed turbulent flames in spark-ignition engines, under normal operating conditions, are highly wrinkled and probably multiply-connected thin reaction sheets. The overall thickness of the turbulent flame "brush," front to back, is of order 1 cm. The thickness of the thin reaction sheet is comparable to estimates of the laminar flame thickness under the prevailing unburned mixture conditions which are of order 0.1 mm. The scale of the wrinkles is of order 1 mm.[24] Direct evidence to date is limited to the low to mid engine speed and low to high engine

TABLE 9.1

Parameters for shadowgraph photographs in Fig. 9-24

Photograph	Engine speed, rev/min	Turbulence intensity, m/s	Valve	Re_M	Re_T
a	300	0.44	S	106	229
b	600	0.88	S	157	503
c	300	1.07	US	173	611
d	900	1.33	S	193	760
e	1200	1.80	S	224	1024
f	600	1.95	US	234	1117
g	900	2.90	US	285	1658
h	1200	4.0	US	333	2263

Note: Valve: S, shrouded; US, unshrouded (produced higher turbulence due to less ordered flow). $Re_M = l_M u'/\nu$; $Re_T = l_I u'/\nu$ (see Sec. 8.2.1).[27]

load ranges. Whether the structure becomes significantly different at high engine speed is not known. Models of this turbulent flame development and propagation process are reviewed in Chap. 14.

9.3.3 Laminar Burning Speeds

An important intrinsic property of a combustible fuel, air, burned gas mixture is its laminar burning velocity. This burning velocity is defined as the velocity, relative to and normal to the flame front, with which unburned gas moves into the front and is transformed to products under laminar flow conditions. Some details of flame structure help explain the significance of this quantity. A flame is the result of a self-sustaining chemical reaction occurring within a region of space called the flame front where unburned mixture is heated and converted into products. The flame front consists of two regions: a preheat zone and a reaction zone. In the preheat zone, the temperature of the unburned mixture is raised mainly by heat conduction from the reaction zone: no significant reaction or energy release occurs and the temperature gradient is concave upward ($\partial^2 T/\partial x^2 > 0$). Upon reaching a critical temperature, exothermic chemical reaction begins. The release of chemical energy as heat results in a zone where the temperature gradient is concave downward ($\partial^2 T/\partial x^2 < 0$). The region between the temperature where exothermic chemical reaction begins and the hot boundary at the downstream equilibrium burned gas temperature is called the reaction zone. The thicknesses of the preheat and reaction zones can be calculated for one-dimensional flames from conservation equations of mass and energy. The thickness of the preheat zone $\delta_{L,\,ph}$ is

$$\delta_{L,\,ph} = \frac{4.6\bar{k}}{\bar{c}_p \rho_u S_L} \tag{9.31}$$

where $\bar{k}$ and $\bar{c}_p$ are the mean thermal conductivity and specific heat at constant pressure in the preheat zone and S_L is the laminar burning velocity.[28] Thus, the factors which govern the laminar burning velocity of a specific unburned mixture—the velocity at which this flame structure propagates relative to the unburned gas ahead of it—are the temperature and species concentration gradients within the flame and the mixture transport and thermodynamic properties.

Laminar burning velocities at pressures and temperatures typical of unburned mixture in engines are usually measured in spherical closed vessels by propagating a laminar flame radially outward from the vessel center. The laminar burning velocity is then given by

$$S_L = \frac{dm_b/dt}{A_f \rho_u} \tag{9.32}$$

where the mass burning rate is determined from the rate of pressure rise in the vessel and A_f is the flame area. Because the laminar flame thickness [e.g., given by Eq. (9.31)] under engine conditions is of order 0.2 mm[26] and is therefore much less than characteristic vessel dimensions, in applying Eq. (9.32) the flame can be treated as negligibly thin. Laminar burning velocities for methane, propane, iso-octane, methanol, gasoline, and hydrogen—premixed with air—at pressures, temperatures, and equivalence ratios which occur in engines have been measured using this technique.[29-33] Also, the effect of a burned gas diluent on laminar burning velocity with gasoline-air mixtures has been determined.[32] Correlations derived from these data are the most accurate means available for estimating laminar burning velocities for mixtures and conditions relevant to spark-ignition engines.

The effect of the mixture fuel/air equivalence ratio on laminar burning velocity for several hydrocarbon fuels and methanol is shown in Fig. 9-25. The burning velocity peaks slightly rich of stoichiometric for all the fuels shown. The values for isooctane and gasoline are closely comparable. Data at higher pressures and temperatures have been fitted to a power law of the form:

$$S_L = S_{L,0} \left(\frac{T_u}{T_0} \right)^\alpha \left(\frac{p}{p_0} \right)^\beta \tag{9.33}$$

where $T_0 = 298$ K and $p_0 = 1$ atm are the reference temperature and pressure, and $S_{L,0}$, α, and β are constants for a given fuel, equivalence ratio, and burned gas diluent fraction. For propane, isooctane, and methanol, these constants can be represented by

$$\alpha = 2.18 - 0.8(\phi - 1) \tag{9.34a}$$

$$\beta = -0.16 + 0.22(\phi - 1) \tag{9.34b}$$

and

$$S_{L,0} = B_m + B_\phi(\phi - \phi_m)^2 \tag{9.35}$$

where ϕ_m is the equivalence ratio at which $S_{L,0}$ is a maximum with value B_m.

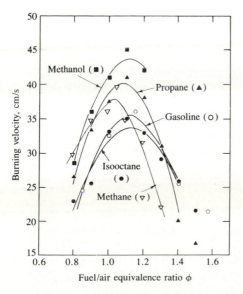

FIGURE 9-25
Laminar burning velocity for several fuels as function of equivalence ratio, at 1 atm and 300 K. Lines are least-squares polynomial fits to data.[29, 30]

TABLE 9.2
Parameters ϕ_m, B_m, and B_ϕ for Eq. (9.35)

Fuel	ϕ_m	B_m, cm/s	B_ϕ, cm/s	Ref.
Methanol	1.11	36.9	−140.5	30
Propane	1.08	34.2	−138.7	30
Isooctane	1.13	26.3	−84.7	30
Gasoline	1.21	30.5	−54.9	32

Note: Values of $S_{L,0}$ given by Eq. (9.35) are obtained from least-squares fits of Eq. (9.33) to data over the range $p = 1$–8 atm, $T_u = 300$–700 K. They do not correspond exactly to the laminar flame speed data at 1 atm and 298 K in Fig. 9-25.

Values of ϕ_m, B_m, and B_ϕ are given in Table 9.2.[30] For gasoline (a reference gasoline with average molecular weight of 107 and an H/C ratio of 1.69) additional data were available and were correlated by[32]

$$\alpha_g = 2.4 - 0.271\phi^{3.51} \tag{9.36a}$$

$$\beta_g = -0.357 + 0.14\phi^{2.77} \tag{9.36b}$$

For methane, simple equations such as (9.34a, b) do not adequately correlate the data over the range of p and T_u relevant to engines. However, laminar burning velocity data from a spherical constant-volume bomb experiment have been obtained along an unburned gas isentropic path, as the pressure in the bomb rises during combustion. Variation in laminar burning velocity along such unburned gas isentropes does correlate with a power law:

$$S_{L,s} = S_{L,0}\left(\frac{\rho_u}{\rho_{u0}}\right)_s^\varepsilon \tag{9.37}$$

Values for $S_{L,0}$ and ε from the literature are summarized in Table 9.3.

TABLE 9.3
Parameters for methane-air laminar burning velocity correlation [Eq. (9.37)]

ϕ	p_i, atm	$S_{L,0}$,† cm/s	ε	Ref.
1.0	0.5	49	0.51	31
1.0	1.0	35	0.2	31
0.8–1.2	1–8	‡	0.17–0.19	33

† At 298 K initial temperature.
‡ See Fig. 9-25.

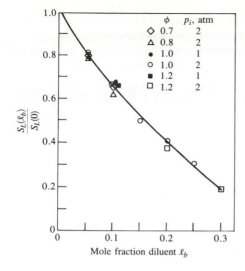

FIGURE 9-26
Effect of burned gas mole fraction $\tilde{x}_b$ in unburned mixture on laminar burning velocity. Fuel: gasoline.[32]

The presence of burned gas in the unburned cylinder charge due to residual gases and any recycled exhaust gases causes a substantial reduction in the laminar burning velocity. Any burned gas in the unburned mixture reduces the heating value per unit mass of mixture and, thus, reduces the adiabatic flame temperature. It acts as a diluent. The effect of increasing burned gas or diluent fraction on laminar flame speed is shown in Fig. 9-26. The diluent used was a mixture of CO_2 and N_2, chosen to match the heat capacity of actual gasoline-air combustion products.† The proportional reduction in laminar burning velocity is essentially independent of the unburned mixture equivalence ratio, pressure, and temperature over the range of interest in engines. The data in Fig. 9-26 are correlated by the relation:

$$S_L(\tilde{x}_b) = S_L(\tilde{x}_b = 0)(1 - 2.06\tilde{x}_b^{0.77}) \tag{9.38}$$

where $\tilde{x}_b$ is the mole fraction of burned gas diluent. Other studies corroborate the magnitude of this burned gas effect.[32]

Note that for equal heat capacity added to the unburned mixture, burned gases have a much larger effect on laminar burning velocity than does excess air. For example, the laminar burning velocity of a stoichiometric mixture as it is leaned to $\phi = 0.8$ is reduced by 23 percent. The excess air required has a heat capacity of about 0.2 times that of the combustion products of the undiluted mixture. Adding the same heat capacity by adding stoichiometric burned gases (which requires a burned gas mole fraction of 0.175) reduces the laminar burning

---

† The water in actual residual and exhaust gas was omitted. A mixture of 80 percent N_2 and 20 percent CO_2, by volume, was used.

velocity by 55 percent.[32] Proper allowance for the burned gas fraction in estimating laminar burning velocities for spark-ignition engines is most important.

The above correlations define the laminar burning velocity as a function of unburned mixture thermodynamic properties and composition, only. It has been assumed that flame thickness and curvature effects are negligible.[28] Our interest in laminar burning velocity is twofold: first, it is used to define the characteristic chemical reaction time of the mixture in Eq. (9.30); second, a presumed consequence of the wrinkled thin-reaction-sheet turbulent-flame structure is that, locally, the sheet propagates at the laminar burning velocity. The above correlations adequately characterize a quiescent burning process. However, laminar flame propagation can be influenced by the local flow field in the unburned gas. If the flame thickness is less than the Kolmogorov scale, the primary effect is one of straining which affects both the flame area (usually referred to as flame stretching for an area increase) and the local (laminar) burning velocity. While this problem is not yet well understood, it is known that straining can affect the laminar burning velocity and can cause flame extinction. The laminar burning velocity decreases with increasing strain rate, and the Lewis number of the unburned mixture has a significant influence on this rate of decrease. The Lewis number is the ratio of diffusivities of heat and mass. For stoichiometric mixtures it is close to one; it increases above about unity as the unburned fuel-air mixture is leaned out. Thus the local flow field may have a discernable effect on the local burning velocity of the thin laminarlike reaction-sheet flame, especially for lean or dilute mixtures.[34]

9.3.4   Flame Propagation Relations

If the heat-release or mass burning rate analysis of Sec. 9.2.2 is coupled with an analysis of flame geometry data, substantial additional insight into the behavior of spark-ignition engine flames is obtained. Flame photographs (such as those in Figs. 9-1, 9-14, and 9-16 and Refs. 4 and 35) effectively define the position of the front or leading edge of the turbulent engine flame. The "shadow" of the enflamed zone, under normal engine conditions, is close to circle: only in the presence of very high swirl does substantial distortion of the flame shape occur.[18] Thus, to a good approximation, the surface which defines the leading edge of the turbulent flame (ahead of which only unburned mixture exists) is a portion of the surface of a sphere. Figure 9-27 indicates the geometrical parameters which define this flame surface: r_c, α_c, z_c, the coordinates of the flame center; r_f, the radius of the best-fit circle to the flame front silhouette; and the geometry of the combustion chamber walls. The flame is initiated at the spark plug; however, it may move away from the plug during the early stages of its development as shown. We define the *flame front area* A_f as the spherical surface of radius r_f coinciding with the leading edge of the flame contained within the combustion chamber, and the *enflamed volume* V_f as the volume within the chamber behind this flame front.

The thermodynamic analysis of cylinder pressure data allows us to define additional geometrical parameters. The *burned gas radius* r_b is the radius of the

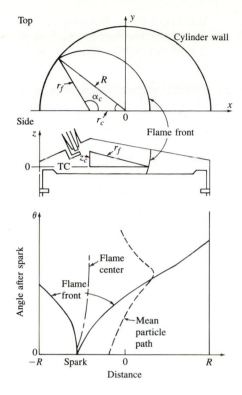

FIGURE 9-27
Schematic of spherical flame front in engine combustion chamber identifying parameters which define flame geometry. (*From Beretta et al.*[4])

spherical surface within the combustion chamber which would contain all the burned gas behind it; i.e.,

$$V_b(r_b, r_c, \alpha_c, z_c) = V_b(p, \theta) \tag{9.39}$$

The *spherical burning area* A_b is the area of this spherical surface; i.e.,

$$A_b = \frac{\partial V_b(r_b, r_c, \alpha_c, z_c)}{\partial r_b} \tag{9.40}$$

The *laminar burning area* A_L is the surface area the flame would have if it burned at the laminar flame speed, i.e.,

$$A_L = \frac{dm_b/dt}{\rho_u S_L} \tag{9.41}$$

where S_L is the laminar flame speed in the unburned mixture ahead of the flame (see Sec. 9.3.3).

Several velocities can be defined. The *mean expansion speed* of the front u_f is given by

$$u_f = \frac{dA_S/dt}{L_S} \tag{9.42}$$

where A_S is the "shadow" area enclosed by the "best-fit" circle through the leading edge of the flame and

$$L_S = \frac{\partial A_S}{\partial r_f}$$

is the arc length within the chamber of this "best-fit" circle. The *mean expansion speed* of the *burned gas* u_b is

$$u_b = \frac{\partial V_b/\partial t}{A_b} \qquad (9.43)$$

This derivative is taken with the piston position fixed since only burned volume changes due to combustion are of interest. The *burning speed* S_b is defined by

$$S_b = \frac{dm_b/dt}{\rho_u A_b} \qquad (9.44)$$

The *mean gas speed* just ahead of the flame front u_g is

$$u_g = u_b - S_b \qquad (9.45)$$

Note that combining Eqs. (9.41) and (9.44) gives the relation

$$S_b A_b = S_L A_L \qquad (9.46)$$

Also, it follows from Eqs. (9.29), (9.43), and (9.44) that

$$\frac{u_b}{S_b} = \frac{\rho_u}{\rho_b}(1 - y_b) + y_b = \frac{\rho_u/\rho_b}{[(\rho_u/\rho_b) - 1]x_b + 1} \qquad (9.47)$$

As x_b and $y_b \rightarrow 0$, u_b/S_b approaches the expansion ratio ρ_u/ρ_b. As x_b and $y_b \rightarrow 1$, u_b/S_b approaches unity.

The variation of the above quantities during the engine combustion process, coupled with the photographs and discussion in Sec. 9.3.2, provide substantial insight into the flame development and propagation process. Figure 9-28 shows results from an analysis of cylinder pressure data and the corresponding flame front location information (determined from high-speed movies through a window in the piston) of several individual engine operating cycles. The combustion chamber was a typical wedge design with a bore of 102 mm and a compression ratio of 7.86. The flame radius initially grows at a rate that increases with time and exhibits substantial cycle-by-cycle variation in its early development (Fig. 9-28a). Later ($r_f \gtrsim 30$ mm) the growth rate, which approximates the expansion speed u_b, reaches an essentially constant value. The flame radius r_f is initially equal to the burned gas radius r_b; it increases above r_b as the flame grows and becomes increasingly distorted by the turbulent flow field (Fig. 9-28b). Eventually $r_f - r_b$ goes to an essentially constant value of about 6 mm for $r_b \gtrsim 30$ mm. This difference, $r_f - r_b$, is approximately half the thickness of the turbulent flame brush.

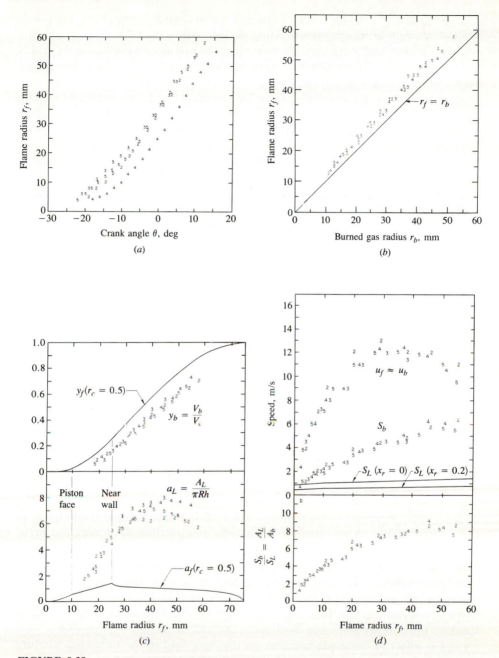

FIGURE 9-28
Variation of flame geometry and velocity parameters during four individual combustion cycles at 1044 rev/min, $\phi = 0.98$, 1 atm inlet pressure: (a) flame radius r_f versus crank angle; (b) flame radius r_f versus burned gas radius r_b; (c) normalized enflamed volume y_f, burned volume y_b, normalized flame front area a_f, and laminar area a_L versus flame radius; (d) front expansion speed u_b, burning speed S_b, and laminar flame speed S_L versus flame radius. (*From Beretta et al.[4]*)

Normalized enflamed and burned volumes, and flame front area and laminar burning area, are shown in Fig. 9-28c. Volumes are normalized by the cylinder volume, and areas by $\pi R h$, where h is the average clearance height and R the cylinder radius. Discontinuities occur in the flame area a_f at the points where the flame front contacts first the piston face and then the near cylinder wall. The laminar area A_L is initially close to the flame area A_f and then increases rapidly as the flame grows beyond 10 mm in radius. During the rapid burning combustion phase ($y_f \gtrsim 0.2$) the value of y_f is significantly greater than y_b. During this phase, the laminar area exceeds the flame area by almost an order of magnitude. These observations indicate the existence of substantial pockets of unburned mixture behind the leading edge of the flame.[4]

The ratio of the volume of the unburned mixture within the turbulent flame zone ($V_f - V_b$) to the reaction-sheet area within the flame zone ($A_L - A_f$) defines a characteristic length

$$l_T = \frac{V_f - V_b}{A_L - A_f} \tag{9.48}$$

which can be thought of as the scale of the pockets of unburned mixture within the flame. For the data set of Fig. 9-28, l_T is approximately constant and of order 1 mm.[24]

These flame geometry results would be expected from the previous photographic observations of how the flame grows from a small approximately spherical smooth-surfaced kernal shortly after ignition to a highly wrinkled reaction-sheet turbulent flame of substantial overall thickness. Initially, the amount of unburned gas within the enflamed volume is small. During the rapid burning phase of the combustion process, however, a significant fraction (some 25 percent; see Fig. 9-2) of the gas entrained into the flame zone is unburned.

The front expansion speed u_f, burning speed S_b, and laminar flame speed S_L are shown in Fig. 9-28d. The expansion speed increases as the flame develops to a maximum value that is several times the mean piston speed of 3.1 m/s and is comparable to the mean flow velocity through the inlet valve of 18 m/s.[24] The burning speed increases steadily from a value close to the laminar flame speed at early times to almost an order of magnitude greater than S_L during the rapid burning phase. During this rapid burning phase, since ($r_f - r_b$) is approximately constant, the flame front expansion speed and the mean burned gas expansion speed are essentially equal. The difference between $u_b \approx u_f$ and S_b is the unburned gas speed u_g just ahead of the flame front. Note that the ratio u_f/S_b ($\approx u_b/S_b$) decreases monotonically from a value equal to the expansion ratio (ρ_u/ρ_b) at spark to unity as the flame approaches the far wall, as required by Eq. (9.47).

The effect of flame propagation on the flow field in the unburned mixture ahead of the flame is important because it is the turbulence just ahead of the flame that determines the local burning velocity. Measurements of mean velocities, rms fluctuation velocities, and turbulence intensities have been made using laser doppler anemometry (see Sec. 8.2.2) at different locations within engine combustion chambers (e.g., Refs. 36 and 37). Such data are difficult to interpret

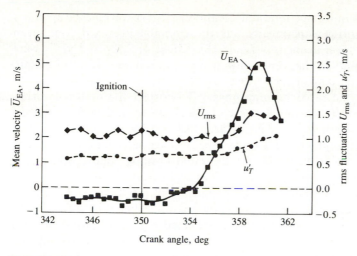

FIGURE 9-29
Laser doppler anemometer measurements of ensemble-averaged mean velocity $\bar{U}_{EA}$ [Eq. (8.20)], rms
fluctuation in individual-cycle mean velocity U_{rms} [Eq. (8.21)] and turbulence intensity u'_T [Eq. (8.22)],
close to the cylinder axis, from before ignition to after flame arrival. Disc-shaped chamber, spark plug
in cylinder wall, measurement at $x/B = 0.57$, $B = 76$ mm, 300 rev/min. $\bar{S}_p = 0.83$ m/s.[36]

because the mean flow varies cycle-by-cycle, the turbulence is not homogeneous,
and the flame motion and shape show substantial cyclic variations. The results in
Fig. 9-29 were taken in a special single-cylinder engine with a disc-shaped combustion chamber where the spark plug was located in the cylinder liner. Shown
are the ensemble-averaged mean velocity, cyclic variation in mean velocity, and
the turbulence intensity, normal to the front, during the major portion of the
combustion process, close to the chamber center.

The mean velocity normal to the front increases steadily from shortly after
ignition, as the combustion-produced gas expansion displaces unburned mixture
toward the wall. It peaks as the flame arrives at the measurement location. The
cyclic variation in mean velocity and the turbulence intensity normal to the front
remain essentially constant until a few degrees before the flame arrival. These two
quantities are comparable in magnitude; thus the turbulence intensity is lower
than the rms fluctuation velocity (in this case by about a factor of 2) (see Sec.
8.2.2). Whether the increase in turbulence as the flame approaches is due to rapid
distortion resulting from the compression of the unburned mixture which occurs
during combustion or is the result of inadequate resolution of cycle-to-cycle flow
variations is unclear. Rapidly imposed distortions of a turbulent flow field, such
as those imposed by combustion-produced gas expansion, would lead to an
increase in vorticity and turbulence intensity. Other studies, e.g., Ref. 37, indicate
there is little or no increase in turbulence intensity ahead of the flame.

The variation of burning speed with engine speed has also been carefully
examined in a study where flame position was determined from high-speed
movies, mass burning rates from cylinder pressure, and turbulence information

from experiments at equivalent motored conditions.[35] The turbulence quantity obtained during motoring experiments was the ensemble-averaged root mean square velocity fluctuation defined by Eq. (8.18). Values of S_b/S_L and u_F'/S_L were determined at two points in the combustion process: at a flame radius of 30 mm (the end of the flame development process) and at mass fraction burned equal to 0.5 (halfway through the rapid burning phase). To correct the motored turbulence data for the higher pressure levels corresponding to engine firing conditions, a simple rapid distortion model (see Sec. 14.4.2) based on conservation of angular momentum in turbulent eddies was used. A linear correlation between S_b and u_F' results, as shown in Fig. 9-30, for the rapid burning combustion phase. Note that as u_F'/S_L goes to zero, S_b/S_L approaches a value close to unity.

Once the flame front reaches the far cylinder wall (see Fig. 9-27) the front can no longer propagate: however, combustion continues behind the front until all the unburned mixture entrained into the enflamed region is consumed. This final burning or termination phase of the combustion process can be approximated by an exponential decay in the mass burning rate with a characteristic time constant τ_b of order 1 ms. Since these "islands" or "pockets" of unburned mixture behind the leading edge of the flame have a characteristic scale l_T based on the laminar flame area [Eqs. (9.41) and (9.48)], it follows that

$$l_T = \tau_b S_L \tag{9.49}$$

In summary, the above flame data analysis procedures show that the relationships between r_f and r_b, V_f and V_b, A_f and A_L, u_f, S_b and S_L are distinctly different in the three phases of combustion: (1) the development phase, where a highly wrinkled reaction-sheet "thick"-overall turbulent flame evolves from the essentially spherical flame kernal established by the spark discharge; (2) the rapid-burning phase, where this thick "developed" turbulent flame propagates across the combustion chamber to the far wall, during which most of the mass is burned; and (3) the termination phase after the flame front has reached the far wall and propagation of the front is no longer possible, when the remaining unburned mixture within the flame burns up. The burning velocity, in the rapid-burning phase of the combustion process, scales with turbulence intensity, which in turn scales with engine speed.

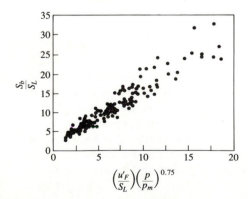

FIGURE 9-30
Variation of burning speed with turbulence intensity. The ensemble-averaged rms velocity fluctuation was measured during motoring engine operation. The ratio p/p_m (firing pressure/motoring pressure) corrects for the effect of additional compression on the turbulence intensity. Range of engine speeds and spark timings.[35]

9.4 CYCLIC VARIATIONS IN COMBUSTION, PARTIAL BURNING, AND MISFIRE

9.4.1 Observations and Definitions

Observation of cylinder pressure versus time measurements from a spark-ignition engine, for successive operating cycles, shows that substantial variations on a cycle-by-cycle basis exist. Since the pressure development is uniquely related to the combustion process, substantial variations in the combustion process on a cycle-by-cycle basis are occurring. In addition to these variations in each individual cylinder, there can be significant differences in the combustion process and pressure development between the cylinders in a multicylinder engine. Cyclic variations in the combustion process are caused by variations in mixture motion within the cylinder at the time of spark cycle-by-cycle, variations in the amounts of air and fuel fed to the cylinder each cycle, and variations in the mixing of fresh mixture and residual gases within the cylinder each cycle, especially in the vicinity of the spark plug. Variations between cylinders are caused by differences in these same phenomena, cylinder-to-cylinder.

Cycle-by-cycle variations in the combustion process are important for two reasons. First, since the optimum spark timing is set for the "average" cycle, faster-than-average cycles have effectively overadvanced spark-timing and slower-than-average cycles have retarded timing, so losses in power and efficiency result. Second, it is the extremes of the cyclic variations that limit engine operation. The fastest burning cycles with their overadvanced spark timing are most likely to knock. Thus, the fastest burning cycles determine the engine's fuel octane requirement and limit its compression ratio (see Sec. 9.6.3). The slowest burning cycles, which are retarded relative to optimum timing, are most likely to burn incompletely. Thus these cycles set the practical lean operating limit of the engine or limit the amount of exhaust gas recycle (used for NO emissions control) which the engine will tolerate. Due to cycle-by-cycle variations, the spark timing and average air/fuel ratio must always be compromises, which are not necessarily the optimum for the average cylinder combustion process. Variations in cylinder pressure have been shown to correlate with variations in brake torque which directly relate to vehicle driveability.

An example of the cycle-by-cycle variations in cylinder pressure and the variations in mixture burning rate that cause them are shown in Fig. 9-31. Pressure and gross heat-release rate [calculated from the cylinder pressure using Eq. (9.27)] for several successive cycles at a mid-load, mid-speed point are shown as a function of crank angle. The maximum heat-release rate and the duration of the heat release or burning process vary by a factor of two from the slowest to the fastest burning cycle shown. The peak cylinder pressure varies accordingly. The faster burning cycles have substantially higher values of maximum pressure than do the slower burning cycles; with the faster burning cycles peak pressure occurs closer to TC.

The heat-release rate data in Fig. 9-31 show that there are cycle-by-cycle variations in the early stages of flame development (from zero to a few percent of

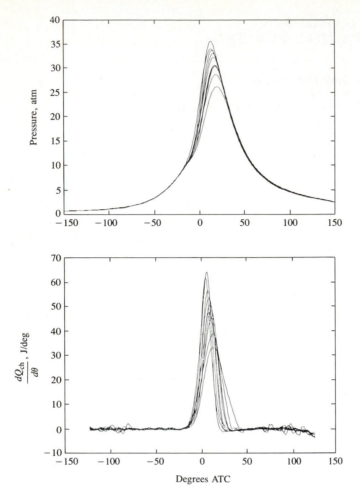

FIGURE 9-31
Measured cylinder pressure and calculated gross heat-release rate for ten cycles in single-cylinder spark-ignition engine operating at 1500 rev/min, $\phi = 1.0$, $p_{inlet} = 0.7$ atm, MBT timing 25° BTC.[14]

the total heat release) and in the major portion of the combustion process—the rapid-burning phase—indicated by the variations in the maximum burning rate.

As the mixture becomes leaner with excess air or more dilute with a higher burned gas fraction from residual gases or exhaust gas recycle, the magnitude of cycle-by-cycle combustion variations increases. Eventually, some cycles become sufficiently slow burning that combustion is not completed by the time the exhaust opens: a regime where *partial burning* occurs in a fraction of the cycles is encountered. For even leaner or more dilute mixtures, the *misfire* limit is reached. At this point, the mixture in a fraction of the cycles fails to ignite. While spark-ignition engines will continue to operate with a small percentage of the cycles in the partial-burn or misfire regimes, such operation is obviously undesirable from the point of efficiency, hydrocarbon emissions, torque variations, and roughness. The partial-burn and misfire regimes are discussed in Sec. 9.4.3.

Various measures of cycle-by-cycle combustion variability are used. It can be defined in terms of variations in the cylinder pressure between different cycles, or in terms of variations in the details of the burning process which cause the differences in pressure. The following quantities have been used:

1. *Pressure-related parameters.* The maximum cylinder pressure p_{max}; the crank angle at which this maximum pressure occurs θ_{pmax}; the maximum rate of pressure rise $(dp/d\theta)_{max}$; the crank angle at which $(dp/d\theta)_{max}$ occurs; the indicated mean effective pressure [which equals $\int p \, dV/V_d$, see Eqs. (2.14), (2.15), and (2.19)].

2. *Burn-rate-related parameters.* The maximum heat-release rate (net or gross, see Sec. 9.2.2); the maximum mass burning rate; the flame development angle $\Delta\theta_d$ and the rapid burning angle, $\Delta\theta_b$ (see Sec. 9.2.3).

3. *Flame front position parameters.* Flame radius, flame front area, enflamed or burned volume, all at given times; flame arrival time at given locations.

Pressure-related quantities are easiest to determine; however, the relation between variations in combustion rate and variations in cylinder pressure is complex.[38] Equation (9.26) defines the factors that govern this relationship. Because the rate of change of pressure is substantially affected by the rate of change of cylinder volume as well as rate of burning, changes in the phasing of the combustion process relative to TC (e.g., which result from changes in flame development angle) as well as changes in the shape and magnitude of the heat-release rate profile affect the pressure. Figure 9-32 illustrates how the magnitude of the maximum cylinder pressure p_{max} and the crank angle at which it occurs θ_{pmax} vary as the crank angle at which combustion effectively starts (e.g., θ at which 1 percent of the cylinder mass has burned) and the burning rate are varied. Curve *CABE* shows how p_{max} and θ_{pmax} vary for a fixed fast-burning heat-release profile (the duration of the heat-release process and its maximum value are held

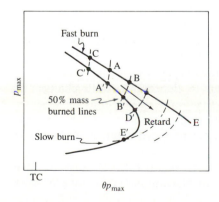

FIGURE 9-32
Schematic of variation in maximum cylinder pressure and crank angle at which it occurs, in individual cycles. *CABE* typical of fast heat-release process; *C'A'B'D'E'* typical of slow heat-release process. (*From Matekunas.*[38])

constant), as the phasing of this combustion process relative to TC is varied. A corresponds to MBT timing where the start of combustion is phased to give maximum brake torque, B corresponds to retarded timing, and C to over-advanced timing. $C'A'B'D'E'$ is a similar curve for a slow-burning heat-release profile. A' corresponds to MBT timing, and B' and D' to increasingly retarded timing. Note that with a sufficiently slow-burning heat-release profile, beyond D', θ_{pmax} decreases as the burn process is increasingly retarded. This occurs when the rate of increase of pressure due to combustion becomes so low that it is more than offset by the pressure decrease due to volume increase: eventually for extremely slow and late burning, the maximum pressure approaches the motored pressure at TC. The dashed lines show the constant start-of-combustion timing *relative* to MBT for each burn rate curve. Note that θ_{pmax} for constant relative timing varies little as the heat-release profile or burn rate varies.[38]

We can now explain the effects of variations in the heat-release profile (both in the development stage of the burning process, which effectively changes the location of the start of combustion, and in the rate of burning throughout the process) on p_{max} and θ_{pmax}, when the spark timing occurs at a fixed crank angle. For a fixed burning rate profile (duration of burn and maximum burning rate) as the start of combustion is delayed to be closer to TC, p_{max} decreases and θ_{pmax} initially increases (A to B or A' to B'). This is the effect of a change in relative timing or phase of the burning process due to a slower initial rate of flame development with fixed spark timing. If, in addition to the flame development being slower, the heat-release rate throughout the burning process is lower, then that combustion process is even more retarded from the optimum and p_{max} decreases and θ_{pmax} increases further, to their values at D'. The effect of a faster initial flame development and faster burning rate, with fixed spark time, is the opposite. The magnitudes of the changes in p_{max} and θ_{pmax} depend, obviously, on the extent of the cyclic variations; they also depend on whether the average burn process is fast or slow. For fast-burning engines, a larger fraction of the heat release occurs near TC when the chamber volume is changing relatively slowly. Thus pressure variations are mainly due to combustion variations. With slow-burning engines, where a significant fraction of the energy release occurs well after TC, the effect of volume change also becomes significant and augments the effect of combustion variations. For large variations and a slow average burning process, p_{max} can fall below E', and θ_{pmax} then decreases. A fast-burning combustion process significantly reduces the impact of cyclic combustion variations on engine performance.[39]

We can now evaluate the various measures of combustion variability. The maximum pressure variation has been shown to depend on both changes in phasing and burning rate. The magnitude of this variation depends on whether the combustion chamber is faster or slower burning, on average. It also depends on whether the burning process is substantially retarded relative to MBT. It depends, too, on cyclic cylinder fuel and air charging variations. Thus the interpretation of variations in p_{max} [or in the maximum rate of pressure rise $(dp/d\theta)_{max}$] in terms of variations in the rate and phasing of the burning process must be done with care. The location of maximum pressure θ_{pmax} also depends on relative phasing of combustion and on the burn rate profile. In addition, for slow-burning chambers and retarded timing (around E') variations produce little change in θ_{pmax}. However, for fast-burning chambers, with MBT or only slightly retarded timing, the location of peak pressure depends essentially on the phasing of each combustion process relative to its MBT phasing, and is independent of charging variations. For these reasons, θ_{pmax} is a useful measure of variability in combustion event phasing.[38]

One important measure of cyclic variability, derived from pressure data, is the *coefficient of variation* in *indicated mean effective pressure*. It is the standard deviation in imep divided by the mean imep, and is usually expressed in percent:

$$\text{COV}_{\text{imep}} = \frac{\sigma_{\text{imep}}}{\text{imep}} \times 100 \tag{9.50}$$

It defines the cyclic variability in indicated work per cycle, and it has been found that vehicle driveability problems usually result when COV_{imep} exceeds about 10 percent.

Figure 9-33 illustrates the relationships between p_{max}, θ_{pmax}, and imep for 120 cycles of an engine cylinder at fixed operating conditions and three different spark timings.[38] The MBT timing data show a spread in imep at a fixed value of θ_{pmax}. This imep data band is relatively flat and is centered around $\theta_{pmax} \approx 16°$; only at later values of θ_{pmax} does imep fall off significantly. The vertical spread in imep around $\theta_{pmax} = 16°$ is due to variations in the amount of fuel entering the cylinder each cycle; normal variations in the burn profile under these conditions, which effectively change the phasing of the combustion process, produce only modest reductions in imep. For early θ_{pmax} (the extreme upper left of Fig. 9-33a), the variations in p_{max} are also due mainly to these fuel-charging variations, cycle-by-cycle; these are the fastest burning cycles with the most advanced phasing. As

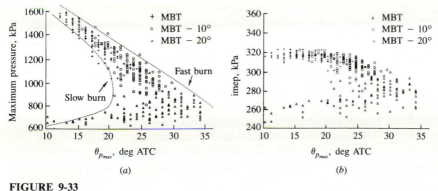

FIGURE 9-33
(a) Individual-cycle maximum pressure versus crank angle at which p_{max} occurs. (b) Individual-cycle indicated mean effective pressure versus θ_{pmax}.[38]

θ_{pmax} increases, the dispersion increases as cyclic variations in phasing and burning rate have increasing impact.

An important issue is whether variations in the early stages of flame development and variations in subsequent portions of the burning process are independent of each other or are correlated. Plots of early flame development angle (spark to 1 percent mass burned) against the burning angle (1 to 90 percent burned) from individual cycles for several different combustion chambers indicate the following. There is a trend with increasing flame development angle for the burning angle to increase (or the burning rate to decrease); however, there is much scatter about this trend (for a given value of flame development angle different cycles show a substantial range in burning angle), and the quantitative aspects of the trend depend on operating conditions and on combustion chamber design. In addition, as the mean rapid burning angle increases (due to changing operating conditions or a slower-burning chamber design) the mean flame development angle, the cyclic variation in the flame development angle, and the cyclic variation in the rapid burning angle all increase.[40] This topic is discussed more fully in the following section.

The shapes of the frequency distributions in individual-cycle pressure data (e.g., in p_{max}, $(dp/d\theta)_{max}$, θ_{pmax}) and in burn rate data such as $\Delta\theta_d$, $\Delta\theta_b$, $(dQ/d\theta)_{max}$ depend on whether the combustion process is fast and "robust" (e.g., with close-to-stoichiometric mixtures at higher loads at optimum timing—well away from the lean operating limit of the engine) or slower and less repeatable, closer to the lean or dilute-mixture operating limit. Under robust combustion conditions these distributions are close to normal distributions.[41-43] When the combustion process is much slower, the cyclic variability becomes large and the distribution becomes skewed toward the slower burning cycles which have low imep (due to the substantial retard of these slower cycles). When partial burning and then misfire occur, the low-pressure tail of the distribution approaches the motored pressure value at TC.[44] Examples of the frequency distributions of imep in these two combustion variability regimes are shown later in Fig. 9-36.

Cylinder pressure data are often averaged over many cycles to obtain the mean cylinder pressure at each crank angle. The primary use of this average pressure versus crank angle data is in calculating the average indicated mean effective pressure (which is a linear function of p). Since combustion parameters are not linearly related to the cylinder pressure [see Eq. (9-27)], analysis of the *average* pressure data will not necessarily yield accurate values of average combustion parameters. The error will be most significant when the combustion variability is largest. It is best to determine mean combustion parameters by averaging their values obtained from a substantial number of individual cycle analysis results. The number of cycles which must be averaged to obtain the desired accuracy depends on the extent of the combustion variability. For example, while 40 to 100 cycles may define imep to within a few percent when combustion is highly repeatable, several hundred cycles of data may be required when cyclic combustion variations are large.[12]

9.4.2 Causes of Cycle-by-Cycle and Cylinder-to-Cylinder Variations

Cycle-by-cycle combustion variations are evident from the beginning of the combustion process. Analysis of flame photographs from many engine cycles taken in special research engines with windows in the combustion chamber has shown that dispersion in the fraction of the combustion chamber volume inflamed is present from the start of combustion (e.g., see Refs. 3 and 23). Dispersion in burning rate is also evident throughout the combustion process (see Figs. 9-2 and 9-31). Three factors have been found to influence this dispersion:[45]

1. The variation in gas motion in the cylinder during combustion, cycle-by-cycle
2. The variation in the amounts of fuel, air, and recycled exhaust gas supplied to a given cylinder each cycle
3. Variations in mixture composition within the cylinder each cycle—especially near the spark plug—due to variations in mixing between air, fuel, recycled exhaust gas, and residual gas

The relative importance of these factors is not yet fully defined, and depends on engine design and operating variables. The variation in the velocity field within the engine cylinder throughout the cycle, and from one cycle to the next, has been reviewed in Sec. 8.2.2. Toward the end of the compression stroke, the ensemble-averaged rms velocity fluctuation is of comparable magnitude to the mean piston speed, and may be larger than the mean flow velocity if there is no strongly directed local mean flow pattern (see Figs. 8-8 and 8-9). This ensemble-averaged velocity fluctuation combines both cycle-by-cycle variation in the mean flow and the turbulent velocity fluctuations. During compression, these two components are of comparable magnitude (see Figs. 8-9 and 9-29). While this data base is limited, it indicates that substantial variations in the mean flow exist, cycle-by-cycle, both in the vicinity of the spark plug *and* throughout the combustion chamber. Velocity variations contribute in a major way to variations in the initial motion of the flame center as it grows from the kernel established by the spark, and in the initial rate of growth of the flame; they can also affect the burning rate once the flame has developed to fill a substantial fraction of the combustion chamber. Variations in gas motion near the spark plug convect the flame in its early stages in different directions and at different velocities, cycle-by-cycle. This affects the flame's interaction with the cylinder walls, changing the flame area development with time. Variations in the turbulent velocity fluctuations near the spark plug will result in variations in the rate at which the small initially laminarlike flame kernel develops into a turbulent flame. Variations in the mean flow throughout the chamber will produce differences in flame front shape; also, they may produce differences in turbulence which affect the propagation velocity of the front (see Fig. 9-30 for the relation between mean burning speed and turbulence intensity).

It is well known that, on a time-averaged basis, the fuel, air, and recycled exhaust gas flows into each cylinder of a multicylinder engine are not identical. These flow rate differences are typically a few percent (see Secs. 7.6.2 and 7.6.3). It is also known that the flow patterns within the different cylinders are not necessarily identical due to differences between the individual intake manifold runner and port geometries in many production engines. All these factors contribute to cylinder-to-cylinder variations in the combustion process: there can be significant differences in the mean burn rate parameters as well as in the cyclic variations in these parameters.[41] Also, the limited data available on the variation in mixture composition within each cylinder for each cycle indicates that cyclic charging variations in individual cylinders are comparable in magnitude to cylinder-to-cylinder differences (i.e., of order ± 5 percent[46]). Whether the amount of residual gas left in the cylinder varies significantly, cycle-by-cycle, is not known. At higher loads, where the combustion process is more repeatable (and always completed relatively early in the expansion stroke) and the residual gas fraction smaller (see Sec. 6.4), variations in the total amount of residual are not expected to be significant. At light loads (particularly at idle), where combustion variability is much higher and partial-burning cycles may occur, and especially with high valve overlap engine designs, variations in the residual gas mass and its composition may become important.

In addition, mixing of fuel, air, recycled exhaust, and residual is not complete: nonuniformities in composition exist within the cylinder at the start of combustion. Composition variations, cycle-by-cycle, in the vicinity of the spark plug electrode gap will affect the early stages of flame development, especially as the flame grows through the laminarlike burning phase following the creation of a small flame kernel by the spark discharge (see Sec. 9.5.1). Figure 9-34 indicates the extent of these composition nonuniformities. The available data comes from experiments where a small rapid-acting sampling valve located in the spark plug center electrode was used to extract gas from the vicinity of the electrode gap, close to the start of combustion, for individual cycle composition analysis. Figure 9-34a shows the cycle-by-cycle air/fuel ratio fluctuations in the burned gases sampled from one cylinder of a four-cylinder gasoline-fueled carbureted engine just after combustion has started. The standard deviation was typically 2 to 6 percent of the mean (A/F).[46–48] Figure 9-34b shows the relationship between total hydrocarbon and CO_2 concentrations in unburned mixture, sampled just before spark discharge. The CO_2 concentration is a measure of the burned gas fraction in the sampled unburned mixture; hence on average it correlates inversely with the total hydrocarbon concentration. However, there is substantial fluctuation in CO_2 concentration about the mean value, at a given fuel fraction, indicating significant fluctuations, cycle-by-cycle, in the mixing of fresh mixture with residual gas. Nonuniformities in EGR distribution between cylinders and EGR mixing within the cylinder would also increase the variations in burned gas fraction locally at the spark gap, cycle-by-cycle.[48]

Experiments in a multicylinder production SI engine, where the fuel/air ratio nonuniformities and the nonuniform mixing of fresh mixture with residual

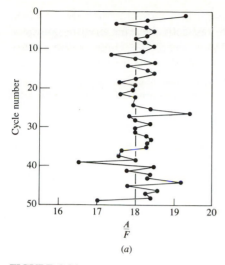

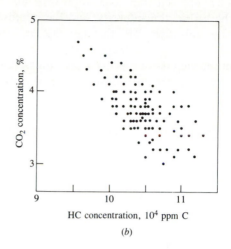

FIGURE 9-34

(a) Air/fuel ratio in 50 consecutive cycles, in vicinity of spark plug, measured just after ignition with a rapid-acting sampling valve located in the plug center electrode. Engine operated at 1400 rev/min, MBT timing, imep = 314 kPa.[47] (b) CO_2 and unburned HC concentrations in gas sampled in individual cycles from the vicinity of the spark plug just prior to ignition. Engine operated at 1200 rev/min, $\phi = 0.98$, $p_{inlet} = 0.5$ atm, gasoline fuel.[48]

gas were removed in turn as contributors to cycle-by-cycle variations (by comparing premixed propane operation with conventional carbureted operation with gasoline, and by removing residual gas by purging with nonfiring cycles), showed that the three contributing factors to cyclic combustion variations—velocity variations, fuel/air ratio variations, and residual gas mixing variations—are of comparable importance at road-load conditions.[45]

An explanation for cycle-by-cycle variations can be developed from the description of the turbulent flame propagation process in Sec. 9.3. Conditions in the *vicinity* of the spark plug will influence the initial stages of the flame propagation process—establishing a stable kernel and its development into a turbulent flame. During the developed flame propagation phase, the *average* conditions in the bulk gas within the combustion chamber will be the determining factors since the flame front spans the chamber, effectively averaging out local non-uniformities. By conditions are meant the turbulent velocity fluctuations and length scales in the flow, proportions of fuel, air, and burned gas in the mixture, and the mixture state.

Using the turbulent combustion model described in Sec. 14.4.2 (which is based on the description of the flame development and propagation process in Sec. 9.3), the flame development angle $\Delta\theta_d$ (the time to burn a few large eddies and establish a developed turbulent flame) can be expressed as[20]

$$\Delta\theta_d = C\left(\frac{l_I}{u'}\right)^{1/3}\left(\frac{l_M}{S_L}\right)^{2/3} \qquad (9.51)$$

The turbulent flow field influences $\Delta\theta_d$ through l_I, u', and l_M, the integral scale, the turbulence intensity, and the microscale, respectively. The mixture composition influences $\Delta\theta_d$ through the laminar flame speed S_L. There is therefore a variability in the flame development period, since all of these quantities can vary in the vicinity of the spark plug on a cycle-by-cycle basis.

The pressure development during the rapid-burning developed turbulent flame propagation phase, when the flame spans the combustion chamber, depends on the average rate of burning in the flame. Thus, variations in the turbulent flow field and mixture composition across the gas entering the flame front are averaged out and are not important. However, variation in chamber-average quantities are significant. The combustion model in Sec. 14.4.2 leads to the following expression for the maximum burning rate:

$$\left(\frac{dm_b}{dt}\right)_{max} = \frac{Cm_f(h^*/B)(\rho_u^*/\rho_i)^{10/9}[(\bar{u}'^*S_L^*)/h_i]^{2/3}}{\nu^{*1/3}} \tag{9.52}$$

Here, m_f is the mass of fuel in the chamber, h is the instantaneous (mean) clearance height, B is the bore, ρ the density, $\bar{u}'$ the average turbulence intensity across the flame front, and ν the kinematic viscosity; the * denotes the value at the time of the maximum burning rate and the subscript i the value at spark; C is a constant depending on engine design. It can be seen from Eq. (9.52) that cycle-by-cycle variations in the maximum burning rate can result from variations in the overall flow pattern within the combustion chamber (which vary $\bar{u}'^*$) and from variations in the amount of fuel (m_f) that enters the cylinder each cycle.† Also, it can be seen that variations in the flame development process will result in variations in the maximum burning rate because the crank angle at which the maximum burning rate occurs is shifted, and all the starred parameters in Eq. (9.52) will have different values.

From the discussion of flame development and structure in Sec. 9.3, and Eqs. (9.51) and (9.52) above, we would expect that mixture conditions and motion leading to slower flame development rates (longer flame development angles, $\Delta\theta_d$)—lower turbulence intensities and more dilute mixtures—would also give lower burning rates (longer rapid burning angles, $\Delta\theta_b$). Data from many different engines and a wide range of operating cases show that this is the case, on average, though there is substantial variation about the mean trend. Figure 9-35 shows these trends; it also shows that the standard deviation of $\Delta\theta_d$ and the standard deviation of $\Delta\theta_b$ for a given chamber and operating condition generally increase as the average burning process becomes slower.[40]

One final factor of importance is how variations in flame development and burning rate affect engine torque. With fixed spark timing, such variations in the combustion process cycle-by-cycle result in slower developing and/or burning

---

† Variations in the total amount of air, recycled exhaust gas, and residual in the chamber could also, for some operating regimes, be significant.

cycles being retarded and faster developing and/or burning cycles being over-advanced. The curve of torque versus combustion timing (relative to optimum timing), Fig. 9-3b, is almost independent of the burning rate; i.e., a given magnitude retard (of say 10°) relative to optimum timing gives almost the same reduction in torque for a very fast burn as it does for a very slow burn. This is because the burning process, for optimum timing, is centered at about 10° ATC independent of the burn rate, and retard or advance shifts this "center" by equal amounts for all burn rates.[16] One of the major advantages of fast-burn engines is now apparent. The magnitude of the variations in the flame development process and subsequent flame propagation rate are decreased as the burning rate is increased (see Fig. 9-35): the ratio of standard deviation in $\Delta\theta_d$ and $\Delta\theta_b$ to the mean values remains approximately constant. Thus, these smaller combustion variations in fast-burn engines, which correspond to modest retard and over-advance in nonaverage burn rate cycles, have little effect on torque. In contrast, the larger combustion variations of slow-burning engines result in significant cyclic torque variations.

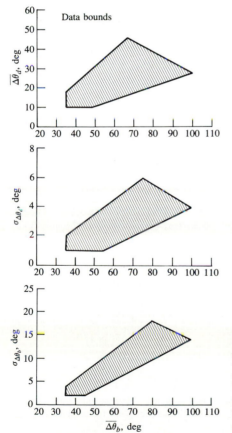

FIGURE 9-35
Variation in mean value of flame development angle $\overline{\Delta\theta_d}$ (spark to 1 percent mass burned) and standard deviations of flame development angle and rapid burning angle $\Delta\theta_b$ (1 to 90 percent mass burned) with mean rapid burning angle $\overline{\Delta\theta_b}$. Range of combustion chamber geometries and engine operating conditions.[40]

9.4.3   Partial Burning, Misfire, and Engine Stability

As the unburned mixture in a spark-ignition engine is leaned out with excess air or is diluted with increasing amounts of burned residual gas and exhaust gas recycle, the flame development period, the duration of the rapid burning phase, and the cycle-by-cycle fluctuations in the combustion process all increase. Eventually a point is reached where engine operation becomes rough and unstable, and hydrocarbon emissions increase rapidly. The point at which these phenomena occur effectively defines the engine's *stable operating limit*.† These phenomena result from the lengthening of all stages of the combustion process as the unburned mixture is diluted. With increasing dilution, first a fraction of the cycles burns so slowly that combustion is only just completed prior to exhaust valve opening. Then as burning lengthens further, in some cycles there is insufficient time to complete combustion within the cylinder; also, flame extinguishment before the exhaust valve opens and before the flame has propagated across the chamber may start to occur in some cycles. Finally, misfiring cycles where the mixture never ignites may start to occur. The proportion of partial burning or nonburning cycles increases rapidly if the mixture is made even more lean or dilute, and the point is soon reached where the engine will not run at all.

The impact on engine stability of increasing combustion variability, due to increased exhaust gas recycle at part-load, is shown in Fig. 9-36. Figure 9-36a shows the distributions of individual-cycle indicated mean effective pressure values for 0, 20, and 28 percent EGR. Without EGR at these conditions, the spread in imep is narrow. Increasing EGR widens the distribution significantly and cycles with low imep, and eventually zero imep, occur. Figure 9-36b shows how the coefficient of variation of imep and hydrocarbon emissions increase as EGR is increased. Slow burn, then partial burn, and then misfire cycles occur with increasing frequency. In the slow-burn cycles, combustion is complete but ends after 80° ATC and the indicated mean effective pressure is low (between 85 and 46 percent of the mean value). Imep in partial-burn cycles was less than 46 percent of the mean. In misfiring cycles, imep < 0. Empirically, it has been found that COV_{imep} [see Eq. (9.50)] is about 10 percent at the engine's stable operating limit, which here occurs just before the onset of partial-burning cycles[49].

An explanation of combustion phenomenon at the engine stable operating limit has been developed by Quader.[50] It involves the following terms:

> **Ignition-limited spark timing** or the **ignition limit.** The spark timing [advanced from maximum brake torque (MBT) timing] at which *misfire* (i.e., failure of flame initiation) first occurs at a given mixture composition, in a given small but arbitrary fraction of cycles (e.g., 0.5 to 1 percent).

---

† This limit has often been called the *lean operating limit*. Since what limits engine operation in practice is excessive torque fluctuations, cycle-by-cycle, and high hydrocarbon emissions, resulting from the use of mixtures made overly dilute with either air or burned gases (or with both), *stable operating limit* is a more appropriate term.

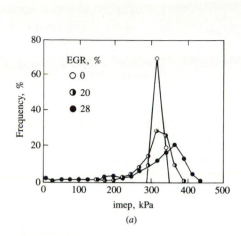

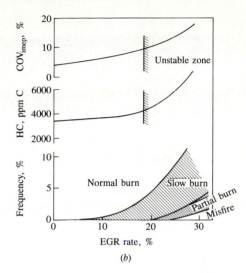

FIGURE 9-36
(a) Frequency distributions in indicated mean effective pressure at different EGR rates; 0 percent gave excellent engine stability, 20 percent acceptable stability, and 28 percent poor stability. (b) Coefficient of variation in imep, HC emissions, and percentage of normal, slow, partial-burn, and misfire cycles. Engine conditions: 1400 rev/min, $\phi = 1.0$, MBT timing, imep = 324 kPa.[49]

Partial-burn-limited spark timing or the **partial-burn limit.** The spark timing (retarded from MBT) at which incomplete flame propagation occurs at a given mixture composition in a given small percent of the cycles (again, this frequency is selected arbitrarily for experimental convenience).

Lean misfire limit at MBT spark. The leanest mixture stoichiometry at which the engine could be stabilized to operate at MBT spark timing with a misfire frequency below a specified value (again, this frequency, usually a percent or less of the cycles, is selected arbitrarily for convenience). A *dilute misfire limit,* the maximum amount of exhaust gas recycle that can be absorbed at a given stoichiometry for stable engine operation, can be similarly defined.

Engine experiments have defined the locations of the ignition limit line and the partial-burn limit line; they are shown qualitatively in Fig. 9-37. At a given spark timing, on this spark timing versus equivalence ratio plot, progressive leaning of the mixture fed to the engine will lead to the onset of misfire or to the onset of partial burning, depending on the location of the lines and the spark timing selected. The individual figures show the possible interactions of the maximum brake torque (MBT) timing line—leaner mixtures require greater advance—with the ignition limit and partial-burn limit lines. At MBT timing, the partial-burn limit may or may not be reached prior to misfire or the ignition limit. It will depend on the engine and ignition system design and operating conditions. For spark timings retarded relative to MBT, partial burning and not failure of flame initiation is the primary cause of unstable engine operation.

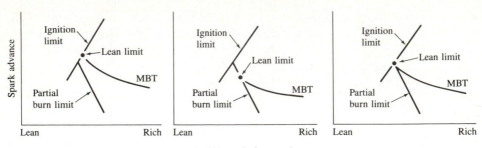

FIGURE 9-37
Schematics of three possible combinations of ignition limit, partial-burn limit, and MBT timing curves as function of fuel/air equivalence ratio. (*From Quader.*[50])

An example of these limiting combustion regimes for lean engine operation is shown in Fig. 9-38. With MBT timing, as the mixture is leaned out (at constant air flow rate), complete combustion in all cycles changes to partial burning in some cycles which changes to no ignition in some cycles at the ignition limit. In the partial-burning regime, the most common type of incomplete-combustion cycle was a slow-burning cycle which required more time to complete burning than was available: flame extinguishment during expansion was much less common. Engine performance measurements showed that the engine stability limit—evidenced by minimum fuel consumption and onset of rapid increase in HC emissions—occurred at $\phi = 0.65$, just before the partial-burn limit line where some slow-burning cycles occur but combustion is still complete in all cycles.[44]

From the above it is clear that flame initiation is a necessary but not sufficient condition for complete combustion. Too-slow flame development and prop-

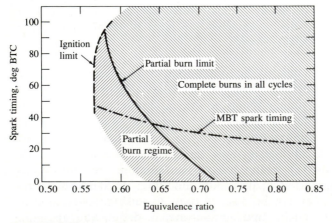

FIGURE 9-38
Actual limiting combustion regimes for lean-operating engine. 1200 rev/min, volumetric efficiency $\approx$ 60 percent, methane fuel, 40 mJ spark energy, 2.5 ms spark duration.[44]

agation following successful ignition is usually the factor which limits engine operation with dilute mixtures. Experiments have shown that a limited interval, of order 80 crank angle degrees (depending on engine geometry and spark plug location), is available during the engine cycle when conditions are favorable for complete flame propagation.[49, 51] Outside of this interval, the mixture pressure and temperature are too low, and the turbulence intensity is too low to sustain a sufficiently rapid rate of combustion. Thus, it is factors which increase the flame development and propagation rates which primarily extend the partial-burn limit.

9.5 SPARK IGNITION

In spark-ignition engines, the electrical discharge produced between the spark plug electrodes by the ignition system starts the combustion process close to the end of the compression stroke. The high-temperature plasma kernel created by the spark develops into a self-sustaining and propagating flame front—a thin reaction sheet where the exothermic combustion chemical reactions occur. The function of the ignition system is to initiate this flame propagation process, in a repeatable manner cycle-by-cycle, over the full load and speed range of the engine at the appropriate point in the engine cycle. Shadowgraph and schlieren photographs of the kernel created by the discharge between the plug electrodes, the growth of that kernal, and its transition to a propagating flame have already been presented in Figs. 9-19 and 9-20, and described in the accompanying text. A spark can arc from one electrode to another when a sufficiently high voltage is applied. Ignition systems commonly used to provide this spark are: battery ignition systems where the high voltage is obtained with an ignition coil (coil ignition systems); battery systems where the spark energy is stored in a capacitor and transferred as a high-voltage pulse to the spark plug by means of a special transformer (capacitive-discharge ignition systems); and magneto ignition systems where the magneto—a rotating magnet or armature—generates the current used to produce a high-voltage pulse.

This section reviews our basic understanding of electrical discharges in inflammable gas mixtures relevant to engine ignition (Sec. 9.5.1), the major design and operating characteristics of conventional engine ignition systems (Sec. 9.5.2), and, briefly, some alternative approaches to generating a propagating flame (Sec. 9.5.3).

9.5.1 Ignition Fundamentals

A spark can arc from one plug electrode to the other only if a sufficiently high voltage is applied. In a typical spark discharge, the electrical potential across the electrode gap is increased until breakdown of the intervening mixture occurs. Ionizing streamers then propagate from one electrode to the other. The impedance of the gap decreases drastically when a streamer reaches the opposite electrode, and the current through the gap increases rapidly. This stage of the discharge is called the *breakdown phase*. It is followed by the *arc phase*, where the thin cylindrical plasma expands largely due to heat conduction and diffusion and, with inflammable mixtures, the exothermic reactions which lead to a propagating flame develop. This may be followed by a *glow discharge phase* where, depending on the details of the ignition system, the energy storage device (e.g., the ignition coil) will dump its energy into the discharge circuit.[52, 53]

Figure 9-39 shows the behavior of the discharge voltage and current as a function of time for a conventional coil ignition system. Typical values are shown; actual values depend on the details of the electrical components. The breakdown phase is characterized by a high-voltage (~ 10 kV), high-peak current (~ 200 A), and an extremely short duration (~ 10 ns). A narrow (~ 40 μm diameter) cylindrical ionized gas channel is established very early. The energy supplied is transferred almost without loss to this plasma column. The temperature and pressure in the column rise very rapidly to values up to about 60,000 K and a few hundred atmospheres, respectively. A strong shock or blast wave propagates outward, the channel expands, and, as a result, the plasma temperature and pressure fall. Some 30 percent of the plasma energy is carried away by the shock wave; however, most of this is regained since spherical blast waves transfer most of their energy to the gas within a small (~ 2 mm diameter) sphere into which the breakdown plasma soon expands.[52, 53]

A breakdown phase always precedes arc and glow discharges: it creates the electrically conductive path between the electrodes. The arc phase voltage is low

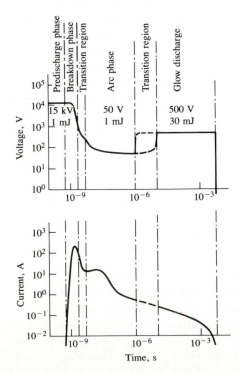

FIGURE 9-39
Schematic of voltage and current variation with time for conventional coil spark-ignition system. Typical values for energy and voltage in the three phases—breakdown, arc, and glow discharge—are given.[52]

(<100 V), though the current can be as high as the external circuit permits. In contrast to the breakdown phase where the gas in the channel is fully dissociated and ionized, in the arc phase the degree of dissociation may still be high at the center of the discharge, but the degree of ionization is much lower (about 1 percent). Voltage drops at the cathode and anode electrodes are a significant fraction of the arc voltage, and the energy deposited in these electrode sheath regions, which is conducted away by the metal electrodes, is a substantial fraction of the total arc energy (see Table 9.4 below). The arc requires a hot cathode spot, so evaporation of the cathode material occurs. The arc increases in size due primarily to heat conduction and mass diffusion. Due to these energy transfers the gas temperature in the arc is limited to about 6000 K: the temperature and degree of dissociation decrease rapidly with increasing distance from the arc axis. Currents less than 200 mA, a large electrode voltage drop at the cathode (300 to 500 V), a cold cathode, and less than 0.01 percent ionization are typical for the glow discharge. Energy losses are higher than in the arc phase, and peak equilibrium gas temperatures are about 3000 K.[52]

About 0.2 mJ of energy is required to ignite a quiescent stoichiometric fuel-air mixture at normal engine conditions by means of a spark. For substantially leaner and richer mixtures, and where the mixture flows past the electrodes, an order of magnitude greater energy (~3 mJ) may be required.[54] Conventional ignition systems deliver 30 to 50 mJ of electrical energy to the spark. Due to the physical characteristics of the discharge modes discussed above, only a fraction of the energy supplied to the spark gap is transmitted to the gas mixture. The energy balance for the breakdown, arc, and glow phases of the discharge is given in Table 9.4. Radiation losses are small throughout. The end of the breakdown phase occurs when a hot cathode spot develops, turning the discharge into an arc; heat losses to the electrodes then become substantial. The breakdown phase reaches the highest power level (~1 MW), but the energy supplied is small (0.3 to 1 mJ). The glow discharge has the lowest power level (~10 W) but the highest energy (30 to 100 mJ), due to its long discharge time. The arc phase lies between.

The proportions of the electrical energy supplied, which can be transferred to the plasma in these three phases of the discharge, are shown in Fig. 9-40.[53, 55] The different transfer capabilities for breakdown arc and glow discharges arise

TABLE 9.4

Energy distribution for breakdown, arc, and glow discharges†

	Breakdown, %	Arc, %	Glow, %
Radiation loss	<1	5	<1
Heat loss to electrodes	5	45	70
Total losses	6	50	70
Plasma energy	94	50	30

† Typical values, under idealized conditions with small electrodes.
Source: Maly and Vogel.[52]

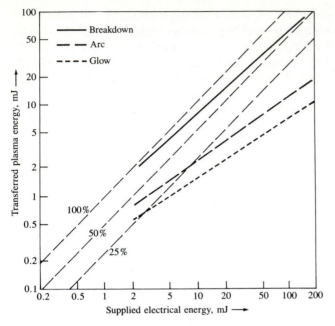

FIGURE 9-40
Energy transferred to the spark kernal as a function of supplied electrical energy for breakdown, arc, and glow discharges.[55]

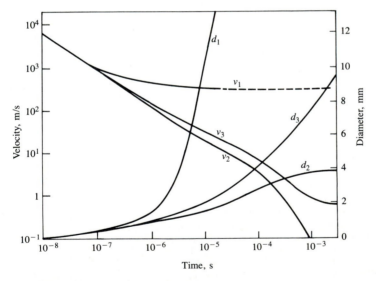

FIGURE 9-41
Diameters (d) and expansion velocities (v) of volumes activated by a capacitive-discharge ignition system: 3 mJ electrical energy, 100 μs duration. Subscripts denote: 1, shock in air at 1 atm; 2, plasma in air at 1 atm; 3, electrical and chemical plasma in stoichiometric methane-air mixture at 1 atm.[52]

primarily from the differences in heat losses to the electrodes, as explained above. These losses increase with increasing supplied energy. In arc and glow discharges, increases in either discharge time or discharge current (or in both) always lead to substantial decreases in energy-transfer efficiency. If the glow discharge current is increased above about 100 mA, the discharge changes to the arc mode and heavy electrode erosion will result. Thus there are practical limits to the arc and glow discharge currents; also, the time available for ignition in the engine limits increases in discharge time.

The initial expansion velocities in the discharge are much higher than those in self-propagating flames. Figure 9-41 shows the expansion velocities and diameters of the volumes activated by a 3-mJ, 100-μs discharge from a capacitor-discharge ignition system as a function of time after spark onset.[52] The curves shown are (1) for the shock wave following a discharge in air at 1 atm; (2) the plasma in air at 1 atm; (3) the electrical and chemical plasma following a discharge in a stoichiometric methane-air mixture at 1 atm. The initially strong shock wave attenuates rapidly to the local sound speed. Up to times of order microseconds, the effects of fuel combustion chemistry are small. The change in slope of the velocity curves for the plasmas at about 100 μs indicates the transition from an expansion caused by the initial high pressure in the breakdown discharge to expansion resulting from heat conduction and diffusion.

The temperature distributions within the three different types of discharge provide additional insight. During the breakdown discharge, on a time scale of nanoseconds, the temperature rises to 60,000 K. Increasing the breakdown energy does not produce *higher* kernel temperatures; instead the channel diameter increases, producing a larger plasma volume. The kernel temperatures then decrease to the order of 10,000 K on a microsecond time scale as the plasma expands behind the shock wave. Arc and glow discharges, because their power inputs are much lower, do not increase the kernel temperatures; rather, they extend the cooling period on a microsecond and millisecond time scale, respectively.[52]

The characters of the temperature profiles that each of these three types of discharge create in air, with essentially the same total electrical energy input (30 to 33 mJ), are indicated in Fig. 9-42. The radial profiles in the undisturbed midplane of the arc are shown. The expansion-wave-induced expansion of the plasma behind the shock with the breakdown discharge produces a larger plasma, earlier, with a steep temperature front. Thus it creates favorable conditions for transferring heat and radicals to the surrounding unburned mixture. In addition cold gas flows into the central region of the plasma, due to the boundary layers which the rapidly expanding flow sets up on the electrodes, effectively insulating the hottest plasma region from the cold electrode surfaces.[53] The arc and glow discharges, each preceded by a much lower energy initial breakdown process, show a much slower expansion rate and the more gradual temperature profile produced by heat conduction and diffusion.

Chemical reactions can be observed spectroscopically a few nanoseconds after spark onset. They are initiated by the very high radical density in the break-

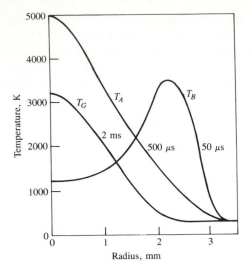

FIGURE 9-42
Radial temperature profiles at selected times after spark onset for ignition systems with different electrical energies and discharge times in air at 1 atm. T_B, breakdown discharge, 30 mJ energy, 60 ns duration; T_A, capacitive discharge, 3 mJ energy, 100 μs duration with superimposed current arc of 2 A, 30 mJ energy, for 230 μs; T_G, capacitive-discharge system, 3 mJ energy, 100 μs duration with superimposed constant-current glow discharge of 60 mA, 30 mJ energy, 770 μs duration.[52]

down plasma where all the heavy particles are present as highly excited atoms and ions. Since the kernel temperatures are much too high to allow the species present in normal combustion products to exist, combustion reactions take place at the outer plasma surface where the conditions are ideal for rapid chemical activity (temperatures of one thousand to a few thousand kelvins). The chemical energy released is added to the plasma energy and becomes evident when the plasma velocity falls below about 100 m/s (see Fig. 9-41). At this point, the inte-

TABLE 9.5
Equilibrium composition of stoichiometric isooctane-air combustion products

Species	Temperature, K†			
	2000	3000	4000	5000
CO	2.4($-$3)	6.1($-$2)	9.4($-$2)	9.0($-$2)
CO_2	1.2($-$1)	5.7($-$2)	5.0($-$3)	3.5($-$4)
H	2($-$5)	9.7($-$3)	1.2($-$1)	1.9($-$1)
H_2	6.1($-$4)	1.5($-$2)	2.4($-$2)	3.5($-$3)
H_2O	1.4($-$1)	1.0($-$1)	1.1($-$2)	1.2($-$4)
N	n	1($-$5)	6.6($-$4)	1.2($-$2)
NH	n	n	2($-$5)	6($-$5)
NO	5.7($-$4)	1.5($-$2)	2.9($-$2)	1.7($-$2)
N_2	7.3($-$1)	6.9($-$1)	5.7($-$1)	5.1($-$1)
O	1($-$5)	8.6($-$3)	9.8($-$2)	1.6($-$1)
OH	4.5($-$4)	2.2($-$2)	3.3($-$2)	5.6($-$3)
O_2	1.1($-$3)	2.3($-$2)	1.7($-$2)	2.2($-$3)

† At 4 atm pressure: mole fractions, $9.0(-2) = 9.0 \times 10^{-2}$.
$n = <5 \times 10^{-6}$

rior of the plasma still consists of a fully dissociated reacted gas mixture with most of its energy stored in radicals. An indication of the gas composition across the steep temperature profile at the plasma interface can be obtained from Table 9.5, which shows the equilibrium composition of C_8H_{18}-air combustion products over the relevant temperature range. Note, however, that the gas will not be in equilibrium. The different radicals have different diffusivities, with the hydrogen atom some five times that of other species. Thus the H radical will diffuse furthest into the as-yet unreacted mixture. On the high-temperature side of the inflamma-tion zone, the large number of radical particles transfer their energy to the mixture molecules within a few collisions. On the low-temperature side of the zone, above-equilibrium concentrations of combustion-initiating radicals (O and H) build up. In addition, a high heat flux into the region occurs, from the plasma core, by conduction down the steep temperature gradient. As a consequence of these conditions, reactions will occur and energy will be released more rapidly than in a normal flame.[53]

Figure 9-43a shows the size of the activated volume as a function of time for several types of discharge, both in a stoichiometric mixture and in air. It can be seen from the air curves for the capacitive discharge (CDI) and breakdown discharge that, after about 10 μs, the plasma ceases to be the energy source for continued growth of the activated volume. At this time, the plasma temperature at the interface has fallen to a value comparable to flame temperatures. With combustible mixtures, molecules such as OH, CH, C_2, CO, etc., appear, indicat-ing that combustion reactions are now occurring. Figure 9-43a shows that for $t \geq 20$ μs, the volume activated by the discharge with the combustible mixture grows much faster than the volume activated in air. Continuing the supply of electrical energy in the arc and glow discharge does produce a higher expansion rate in both the combustible mixture and in air, due to additional heat conduc-tion and diffusion to the interface, but the onset of inflammation is not signifi-cantly affected. The radial temperature profiles across the discharge at selected times, shown in Fig. 9-43b, illustrate these points. This, therefore, is the critical point in the inflammation process: at some 20 μs after onset of the discharge the flame reactions must be proceeding sufficiently rapidly to be self-sustaining; i.e., chemical energy release must more than offset heat losses across the front to the surrounding unburned mixture via diffusion and conduction.[53]

Thus the characteristics of the breakdown phase of the discharge have the greatest impact on inflammation. The size of the activated volume a given time interval after spark initiation, the temperature difference across the kernel inter-face, and the velocity of the interface are all substantially increased by increasing the breakdown phase energy (see Fig. 9-43). Additional energy input during the arc and glow discharge phases has a more modest effect on these critical kernel properties. This is graphically illustrated by Fig. 9-44 where the same energy input into breakdown, arc, and glow discharge modes produces substantially dif-ferent ignition limits.

Several models of the plasma–unburned mixture interface have been devel-oped in attempts to quantify the complex phenomena described above (e.g., Refs.

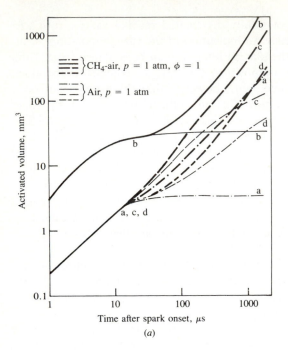

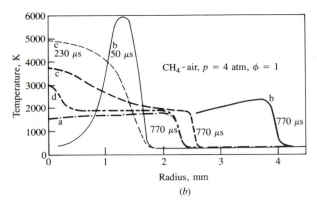

FIGURE 9-43
(*a*) Size of discharge-activated volume as function of time for several types of discharge in air and in stoichiometric methane-air mixture at 1 atm. a: CDI, 3 mJ energy, 100 μs duration; b: breakdown discharge, 30 mJ energy, 20 ns duration; c: CDI plus arc discharge, 1.5 A, 40 V, 500 μs duration; d: CDI + glow discharge, 30 mA, 500 V, 2 ms duration. (*b*) Temperature profiles for different discharge modes at different times in stoichiometric methane-air mixture at 300 K, 4 atm. a: CDI, 3 mJ energy, 100 μs duration; b: breakdown discharge, 20 mJ energy, 80 ns duration; c: CDI, 3 mJ energy, 100 μs duration, plus 2 A, 30 mJ energy, 230 μs duration arc; d: CDI, 3 mJ energy, 100 μs duration, plus 60 mA, 30 mJ energy, 770 μs duration glow discharge.[53]

56 and 57). They are based on the requirement that the energy release due to chemical reaction in the plasma front (and production of radical species) exceed the losses due to conduction and diffusion to the unburned gas ahead of the front. While these models are incomplete, they do provide a theoretical basis for the well-known fact that the minimum energy required to ignite a premixed fuel-air mixture depends strongly on mixture composition. Figure 9-45 shows a typical set of results on the minimum ignition energy as a function of the equivalence ratio under quiescent conditions.[58] The curve shows a minimum for slightly rich-of-stoichiometric mixtures; the minimum energy required for successful ignition increases rapidly as the mixture is leaned out. While the initial plasma kernel

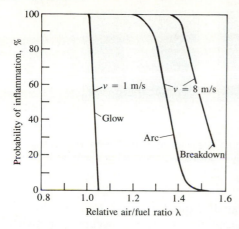

FIGURE 9-44
Probability of inflammation of stoichiometric methane-air mixture, at 300 K, 4 atm, as function of relative air/fuel ratio λ ($= 1/\phi$) for different ignition discharges with equal total electrical energy (30–33 mJ). Breakdown: 30 mJ, 60 ns duration. Arc: CDI, 3 mJ, 100 μs duration; plus 2 A, 30 mJ, 230 μs duration arc. Glow: CDI, 3 mJ, 100 μs duration; plus 60 mA, 30 mJ, 770 μs duration glow discharge. v = mixture velocity.[52]

growth (up to 10 to 100 μs) is not greatly affected by the mixture equivalence ratio, the inflammation process and the thickness and rate of propagation of the resulting flame are strongly affected. The lean side of the minimum is of more practical interest than the rich side. Because the chemical energy density of the mixture and flame temperature decrease as the mixture is leaned out, the flame speed decreases and the flame becomes thicker. Thus more time is available for heat losses from the inflammation zone, less energy is available to offset these losses, and the rate of energy transfer into the zone decreases. The consequence is that, as the mixture is leaned out, the approximately spherical discharge-created plasma must grow to a larger size before inflammation will occur. Substantially, more energy must therefore be supplied to the discharge.[53]

In engines, the mixture is not quiescent: mean and fluctuating velocities in the range 1 to 10 m/s exist in the clearance volume at TC (see Sec. 8.2.2). On the time scale of the breakdown discharge phase (10 ns), this fluid motion is not important. In the arc and glow discharge phases, however, the arc is convected

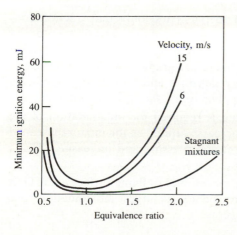

FIGURE 9-45
Effect of mixture equivalence ratio and flow velocity on minimum ignition energy for propane-air mixtures at 0.17 atm.[58]

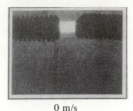 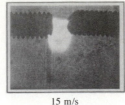 

0 m/s   15 m/s   35 m/s

FIGURE 9-46
Photographs of single glow discharge (30 mJ, 0.77–1.5 ms) in air at 2 atm flowing perpendicular to axis of electrodes. Below 25 m/s almost no multiple discharges; above 25 m/s only multiple discharges. 1.2 mm electrode gap.[56]

by the flow and lengthened accordingly, as illustrated in Fig. 9-46. For velocities below 15 m/s a steady increase in discharge channel length occurs. For higher velocities, an increasing number of reignitions occur, so the discharge energy is distributed into many separate channels. As the channel lengthens, the ratio of total discharge voltage to anode plus cathode voltage drop increases substantially, and the relative importance of heat losses to the electrodes decreases. Thus more energy is transferred to the gas—the energy-transfer efficiency increases. However, as the channel is lengthened, the energy transferred is spread over a larger volume. Depending on flow velocity, and mixture and discharge conditions, increasing velocity may increase or decrease the minimum ignition energy or the lean ignition limit for a specific ignition system. Both the mean flow velocity and turbulence levels are important.[53–58] With a conventional coil ignition system, increasing mean flow velocity up to the point where reignitions start to occur extends the lean limit. With breakdown discharge systems, the lean ignition limit decreases as flow velocity increases.[53] With capacitive-discharge systems at low flow velocities, the flow has little impact: at high flow velocities the minimum ignition energy increases.[59]

Maly[53] has summarized these fundamental aspects of spark-discharge ignited flames as follows:

1. Of the total electrical energy supplied to the spark, only that fraction contained within the outer surface layer of the plasma (of thickness of the order of the inflammation zone) is available for initiating the flame propagation process. The energy density and the temperature gradient in this layer depend on the discharge mode. Highest energy densities and temperature gradients are achieved if the ignition energy is supplied in the shortest time interval.

2. A minimum radius of the spark plasma is required for inflammation of the fuel-air mixture to occur. This radius increases rapidly as the mixture is leaned out (or diluted); it decreases with increasing pressure and increasing plasma expansion velocity.

3. After inflammation, burning rates are proportional to flame surface area. Thus discharges and plasma geometries that produce the largest inflammation zone surface area, most rapidly, are advantageous.

4. The time over which the ignition energy can be used effectively for inflammation decreases as the initial flame velocity increases. Ignition energy supplied after inflammation has occurred will have only a modest impact on flame propagation.

9.5.2 Conventional Ignition Systems

The ignition system must provide sufficient voltage across the spark plug electrodes to set up the discharge and supply sufficient energy to the discharge to ignite the combustible mixture adjacent to the plug electrodes under all operating conditions. It must create this spark at the appropriate time during the compression stroke. Usually spark timing is set to give maximum brake torque for the specific operating condition, though this maximum torque may be constrained by emission control or knock control requirements. For a given engine design, this optimum spark timing varies as engine speed, inlet manifold pressure, and mixture composition vary. Thus, in most applications, and especially the automotive applications, the system must have means for automatically changing the spark timing as engine speed and load vary.

With an equivalence ratio best suited for ignition and with homogeneous mixture distribution, spark energies of order 1 mJ and durations of a few microseconds would suffice to initiate the combustion process. In practice, circumstances are less ideal. The air, fuel, and recycled exhaust are not uniformly distributed between cylinders; the mixture of air, fuel, recycled exhaust gas, and residual gas within each cylinder is not homogeneous. Also, the pressure, temperature, and density of the mixture between the spark plug electrodes at the time the spark is needed affect the voltage required to produce a spark. These vary significantly over the load and speed range of an engine. The spark energy and duration, therefore, has to be sufficient to initiate combustion under the most unfavorable conditions expected in the vicinity of the spark plug over the complete engine operating range. Usually if the spark energy exceeds 50 mJ and the duration is longer than 0.5 ms reliable ignition is obtained.

In addition to the spark requirements determined by mixture quality, pressure, temperature, and density, there are others determined by the state of the plugs. The erosion of the plug electrodes over extended mileage increases the gap width and requires a higher breakdown voltage. Also, spark plug fouling due to deposit buildup on the spark plug insulator can result in side-tracking of the spark. When compounds formed by the burning of fuel, lubricating oil, and their additives are deposited on the spark plug insulator, these deposits provide an alternative path for the spark current. If the resistance of the spark plug deposits is sufficiently low, the loss of electrical energy through the deposits may prevent the voltage from rising to that required to break down the gas. The influence of side-tracking on spark generation decreases with lower source impedance of the high-voltage supply, and therefore with a higher available energy.

The fundamental requirements of the high-voltage ignition source can be summarized as: (1) a high ignition voltage to break down the gap between the plug electrodes; (2) a low source impedance or steep voltage rise; (3) a high energy storage capacity to create a spark kernel of sufficient size; (4) sufficient duration of the voltage pulse to ensure ignition. There are several commonly used concepts that partly or fully satisfy these requirements.[54, 60]

COIL IGNITION SYSTEMS. Breaker-operated inductive ignition systems have been used in automotive engines for many years. While they are being replaced with more sophisticated systems (such as transistorized coil ignition systems), they provide a useful introduction to ignition system design and operation. Figure 9-47 shows the circuit of a typical breaker ignition system. The system includes a battery, switch, resistor, coil, distributor, spark plugs, and the necessary wiring. The circuit functions as follows. If the breaker point is closed when the ignition is switched on, current flows from the battery, through the resistor, primary winding of the ignition coil, contacts, and back to the battery through ground. This current sets up a magnetic field within the iron core of the coil. When ignition is required, the breaker points are opened by the action of the distributor cam, interrupting the primary current flow. The resulting decay of magnetic flux in the coil induces a voltage in both the primary and secondary windings. The voltage induced in the secondary winding is routed by the distributor to the correct spark plug to produce the ignition spark.

The current and voltage waveforms are shown in Fig. 9-48. The primary current for any given time of contact closure t is given by

$$I_p = \frac{V_0}{R} (1 - e^{-Rt/L_p})$$

(9.53)

where I_p is the primary current, V_0 is the supply voltage, R is the total primary circuit resistance, and L_p is the primary circuit inductance. The primary current requires time to build up. At low speeds the time of contact closure is sufficient for the primary current to reach the maximum permitted by the circuit resistance; at high speeds the primary current may not reach its maximum. Thus, only at higher engine speeds does the term e^{-Rt/L_p} become significant. When the points open the primary current falls to zero and a voltage of order 15 kV is induced in the secondary winding. If the coil is not connected to a spark plug,

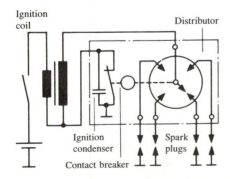

FIGURE 9-47
Schematic of conventional coil ignition system. (*Courtesy Robert Bosch GmbH and SAE.*[54])

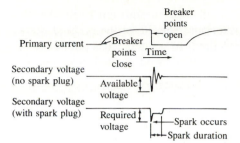

Primary current

Secondary voltage
(no spark plug)

Secondary voltage
(with spark plug)

FIGURE 9-48
Current and voltage waveforms for breaker ignition system.[61]

this induced voltage will have a damped sinusoidal waveform, as shown in the
center trace. The peak value of this voltage is the maximum voltage that can be
produced by the system and is called the *available voltage* V_a of the system. The
maximum energy transferred to the secondary system is given by

$$E_{s,\,max} = \frac{1}{2}\, C_s V_a^2$$

where C_s is the total capacitance of the secondary circuit. Hence, the available
voltage of the system is given by

$$V_a = \left(\frac{2E_{s,\,max}}{C_s}\right)^{1/2} \tag{9.54}$$

If all the energy stored in the primary circuit of the coil, $\frac{1}{2}L_p I_p^2$, is transferred to
the secondary,

$$V_a = I_p\left(\frac{L_p}{C_s}\right)^{1/2} \tag{9.55}$$

When the coil is connected to a spark plug, the secondary voltage will rise
to the breakdown potential of the spark plug, and a discharge between the plug
electrodes will occur. This alters the waveform as shown in the bottom trace of
Fig. 9-48. After the spark occurs, the voltage is reduced to a lower value until all
the energy is dissipated and the arc goes out. The value of this voltage which
caused breakdown to occur is called the *required voltage* of the spark plug. The
interval during which the spark occurs is called the *spark duration*. The available
voltage of the ignition system must always exceed the required voltage of the
spark plug to ensure breakdown. The spark must then possess sufficient energy
and duration to initiate combustion under all conditions of operation.

The major limitations of the breaker-operated induction-coil system are the
decrease in available voltage as engine speed increases due to limitations in the
current switching capability of the breaker system, and the decreasing time available to build up the primary coil stored energy. Also, because of the high source
impedance (about 500 kΩ) the system is sensitive to side-tracking across the
spark plug insulator. A further disadvantage is that due to their high current

load, the breaker points are subject to electrical wear in addition to mechanical wear, which results in short maintenance intervals. The life of the breaker points is dependent on the current they are required to switch. Acceptable life is obtained with $I_p \approx 4$ A; increased currents cause a rapid reduction in breaker point life and system reliability.

TRANSISTORIZED COIL IGNITION (TCI) SYSTEMS. In automotive applications, the need for much reduced ignition system maintenance, extended spark plug life, improved ignition of lean and dilute mixtures, and increased reliability and life has led to the use of coil ignition systems which provide a higher output voltage and which use electronic triggering to maintain the required timing without wear or adjustment (see Refs. 54 and 61). These are called transistorized coil ignition (TCI) or high-energy electronic-ignition systems. The higher output voltage is required because spark plugs are now set to wider gaps (e.g., about 1 mm) to extend the ability to ignite the fuel mixture over a wider range of engine operation, and because during the extended mileage between spark plug replacement electrode erosion further increases the gap. In automotive applications an available ignition voltage of 35 kV is now usually provided. In addition to higher voltage, longer spark duration (about 2 ms) has been found to extend the engine operating conditions over which satisfactory ignition is achieved.

Most of the solid-state ignition systems now in use operate on the same basic principle. Figure 9-49 shows the block circuit diagram of a transistorized coil ignition system. The distributor points and cam assembly of the conventional ignition system are replaced by a magnetic pulse generating system which detects the distributor shaft position and sends electrical pulses to an electronic control module. The module switches off the flow of current to the coil primary windings, inducing the high voltage in the secondary windings which is distributed to the

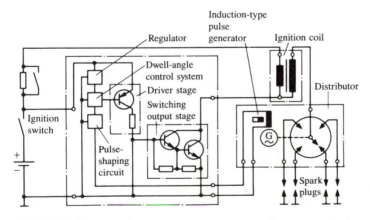

FIGURE 9-49
Schematic of transistorized coil ignition system with induction pulse generator. (*Courtesy Robert Bosch GmbH and SAE.*[54])

spark plugs as in the conventional breaker system. The control module contains timing circuits which then close the primary circuit so that buildup of primary circuit current can occur. There are many types of pulse generators that could trigger the electronic circuit of the ignition system.[60] A magnetic pulse generator, where a gear-shaped iron rotor driven by the distributor shaft rotates past the stationary pole piece of the pickup, is usually used. The number of teeth on the rotor is the same as the number of cylinders. A magnetic field is provided by a permanent magnet. As each rotor tooth passes the pole piece it first increases and then decreases the magnetic field strength ψ linked with the pickup coil, producing a voltage signal proportional to $d\psi/dt$. The electronic module switches off the coil current to produce the spark as the rotor tooth passes through alignment and the pickup coil voltage abruptly reverses and passes through zero. The increasing portion of the voltage waveform, after this voltage reversal, is used by the electronic module to establish the point at which the primary coil current is switched on for the next ignition pulse.

CAPACITIVE-DISCHARGE IGNITION (CDI) SYSTEMS. With this type of system (shown schematically in Fig. 9-50) a capacitor, rather than an induction coil, is used to store the ignition energy. The capacitance and charging voltage of the capacitor determine the amount of stored energy. The ignition transformer steps up the primary voltage, generated at the time of spark by the discharge of the capacitor through the thyristor, to the high voltage required at the spark plug. The CDI trigger box contains the capacitor, thyristor power switch, charging device (to convert battery voltage to the charging voltage of 300 to 500 V by means of pulses via a voltage transformer), pulse shaping unit, and control unit.

The principal advantage of CDI is its insensitivity to electrical shunts in the high-voltage ignition circuit that result from spark plug fouling. Because of the fast capacitive discharge, the spark is strong but short (0.1 to 0.3 ms). This can lead to ignition failure at operating conditions where the mixture is very lean or dilute.[54]

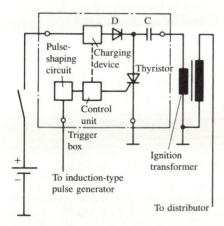

FIGURE 9-50
Schematic of capacitive-discharge system. (*Courtesy Robert Bosch GmbH and SAE.*[54])

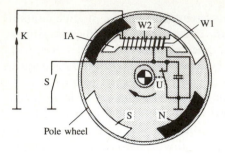

FIGURE 9-51
Schematic of breaker-triggered magneto system with ignition armature. (*Courtesy Robert Bosch GmbH and S.A.E.*[54])

MAGNETO IGNITION. With this type of ignition system, a magneto supplies the ignition voltage for the spark discharge independent of a battery or generator. Magneto ignition is commonly used in small four-stroke and two-stroke engines. Figure 9-51 illustrates the system and its operation. A time-varying magnetic flux Φ_0 is set up in the ignition armature (IA) as the rotating permanent magnets on the pole wheel generate a current in the closed primary winding W1. This primary current generates an additional flux Φ_I, giving a resultant flux $\Phi_R = \Phi_0 + \Phi_I$. To generate the ignition voltage, the primary current flow is interrupted and the flux collapses rapidly from Φ_R to Φ_0, producing a high-voltage pulse in the winding which is connected to the spark plug electrode. The current can be interrupted with contact breakers (breaker-triggered magneto) or with a transistor (semiconductor magneto). Since the flux generated by the rotating pole wheel depends on engine speed, the magnitude of the ignition voltage varies with speed.[54]

SPARK PLUG DESIGN. The function of the spark plug is to provide an electrode gap across which the high-voltage discharge occurs which ignites the compressed mixture of fuel vapor and air in the combustion chamber. In addition, it must provide a gas-tight conducting path from the high-voltage wire to the electrode gap. Figure 9-52 shows a typical spark plug design. There are three principal

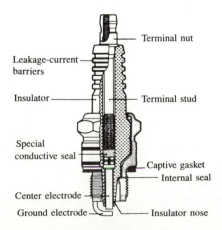

FIGURE 9-52
Cutaway drawing of conventional spark plug. (*Courtesy Robert Bosch GmbH and S.A.E.*[54])

components: an insulator, electrodes, and a shell. The insulated material must have adequate thermal shock resistance, tensile and compressive strength, and impact strength. It must also have low porosity to limit absorption of combustion gases and high resistivity to prevent leakage of high-voltage charge at both ambient temperatures and normal operating temperatures. Alumina is usually used as the insulator material.

The electrodes are normally made of high-nickel alloys to withstand the high ignition voltage, high temperatures, and corrosive gases with minimum erosion. The center-electrode surface temperatures can average 650–700°C under normal operating conditions. A wide range of electrode geometries are available.

The location of the special conductive seal within the shell affects the *heat rating* of the spark plug. For a "hot" plug, an insulator with a long conical nose is used; for a "cold" plug a short-nosed insulator is used. The length of the heat conduction path from the insulator nose to the shell is changed in this way to vary and control the temperature of the exposed part of the insulator. The spark plug insulator tip temperature increases with increasing speed. It is desirable to have the tip hotter than about 350°C to prevent fouling at low speed. High-speed high-load tip temperatures must be kept below about 950°C to prevent preignition. Normally, the gap between the center and ground electrodes is 0.7 to 0.9 mm. For extremely dilute mixtures this is usually increased to 1.2 mm. Magneto ignition systems use smaller gaps ($\sim$0.5 mm). High-compression-ratio racing engines use smaller gaps (0.3 to 0.4 mm).[54]

9.5.3 Alternative Ignition Approaches

A large number of methods for initiating combustion in spark-ignition engines with electrical discharges, in addition to those described in the previous section, have been proposed and examined. These include different designs of spark plug, use of more than one plug, use of higher power, higher energy, or longer-duration discharges, and ignition systems that initiate the main combustion process with a high-temperature reacting jet—plasma-jet and flame-jet ignition systems.[62] Conventional ignition systems normally ignite the unburned fuel, air, burned gas mixture within the cylinder and perform satisfactorily under conditions away from the lean or dilute engine stable operating limit. Thus, these alternative ignition approaches have the goal of extending the engine stability limit (and/or of reducing the cyclic combustion variability near the stability limit), usually by achieving a faster initial burning rate than can be obtained with conventional systems. This section describes the more interesting of these alternative ignition approaches.

ALTERNATIVE SPARK-DISCHARGE APPROACHES. There are many different designs of spark plugs. These use different geometry electrodes, gap widths, and gap arrangements. The effects of the major plug electrode design features on the engine's stable lean operating limit are illustrated in Fig. 9-53.[63] Ignition system effects are important when misfire due to the quenching effect of the spark plug

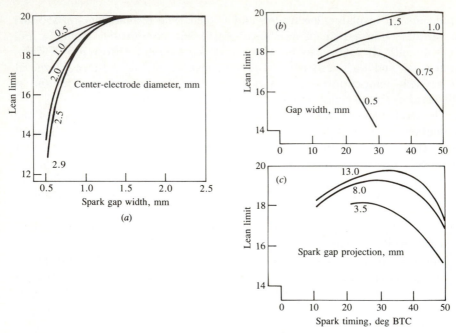

FIGURE 9-53
Effect of spark plug electrode diameter, plug gap width, and projection of gap into chamber on air/fuel ratio at engine's lean stable operating limit. Baseline conditions: 30 mJ spark energy, 3.5 mm projection, center electrode diameter 2.5 mm, gap width 0.75 mm, 40° BTC spark timing. 1600 rev/min, intake pressure 300 mmHg.[63]

electrodes determines the stable operating limit.† Thus, smaller spark plug center-electrode diameters (Fig. 9-53a), larger electrode gap widths (Fig. 9-53b), and higher electrode temperatures (obtained by projecting the gap further into the combustion chamber; Fig. 9-53c), all extend the lean stability limit to leaner (or more dilute) mixtures for the more advanced spark timings and smaller gap widths. For spark timing closer to TC and for larger gap widths, these data show the lean stability limit to be much less sensitive (or not sensitive at all) to plug geometry or spark energy. Multigap plugs, designed to produce a series of discharges which together form a long arc, have also been used to generate a larger initial flame kernel and thereby extend the lean limit.

Use of more than one plug, at separate locations in the combustion chamber and fired simultaneously, is also common.[49] The advantages are twofold. First, the effective flame area in the early stages of flame development is increased substantially (e.g., by almost a factor of two for two widely spaced plugs). Second, the variations in flow velocity and mixture composition in the vicinity of the (multiple) plugs produce less variability in the initial mixture burning rate than occurs with a single plug. Studies of heat-release rates, flame

---

† See Sec. 9.4.3 for a detailed discussion of the stable operating limit.

development angle, and rapid-burning angle, and imep and torque fluctuations have defined the effects of both increasing the number of ignition sites from 1 to 12 and of changing their geometric location.[64] These results confirm that increasing the number of simultaneously developing flame kernels increases the initial mixture burning rate, as anticipated. It also extends the lean stable operating limit and reduces cyclic combustion variability under conditions where slow and occasional partial-burning cycles would occur with fewer spark plug gaps.

Many studies have examined the effects of higher-energy discharges on engine operation near the lean operating limit. It is useful to differentiate between higher current discharges and longer duration discharges: most high-energy (conventional-type) ignition systems have both these features. The results of these studies show that away from the lean or dilute stable operating limit, increasing the discharge current or duration has no significant effect on engine operating characteristics. The higher current does, as would be expected, result in a larger flame kernel during the inflammation process and thereby modestly reduces the spark advance required for maximum brake torque with a given combustion chamber and set of operating conditions. Figure 9-54 shows these trends for rich, stoichiometric, and slightly lean mixtures. The figure also shows that both higher

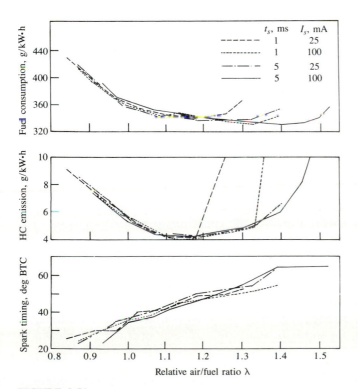

FIGURE 9-54
Effect of higher spark currents I_s and longer spark durations t_s on fuel consumption, HC emissions, and MBT spark timing, as a function of relative air/fuel ratio λ ($=1/\phi$). 2000 rev/min, bmep $=3$ atm, 2.8-dm³ six-cylinder engine.[65]

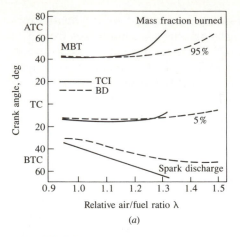

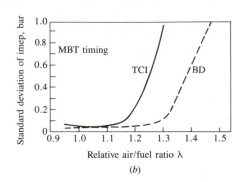

FIGURE 9-55
(a) MBT spark timing and location of 5 and 95 percent mass fraction burned points for conventional transistorized coil ignition (TCI) system (43 mJ energy, 2 ms duration) and breakdown system (BD) (43 mJ energy, ~10 ns duration) as function of relative air/fuel ratio λ ($=1/\phi$). (b) Standard deviation in indicated mean effective pressure as a function of relative air/fuel ratio λ for TCI and BD systems.[66]

currents and longer duration discharges do extend the lean engine stability limit [and also the dilute (with EGR) stability limit]. Note the HC emissions data, which indicate that longer discharges have a greater impact than higher currents on extending the *misfire* limit. The increase in fuel consumption as the lean operating limit is approached is due to the rapidly increasing cycle-by-cycle combustion variability.

The discussion of discharge fundamentals in Sec. 9.5.1 showed that depositing energy into the discharge during the initial short breakdown phase resulted in faster flame kernel growth than did depositing the same energy at slower rates. Ignition systems of this type have been used. In such systems, a capacitor is connected in parallel with the spark plug electrodes, and a low-impedance discharge path allows the energy stored in the capacitor to be discharged into the gap very rapidly. The anticipated effects on the engine's combustion process are observed. Away from the lean engine stability limit, the primary impact is a reduction in the flame development period due to the more rapid initial flame kernel growth. Thus MBT timing is less advanced with these breakdown systems than with conventional systems (see Fig. 9-55a). The lean limit can also be extended, and acceptable engine stability obtained (i.e., tolerable cycle-by-cycle combustion variations) for leaner or more dilute engine operation, as shown in Fig. 9-55b.[66, 67]

PLASMA-JET IGNITION. In a plasma-jet ignitor, the spark discharge is confined within a cavity that surrounds the plug electrodes, which connects with the combustion chamber via an orifice. The electrical energy supplied to the plug electrodes is substantially increased above values used in normal ignition systems by allowing a capacitor to discharge at a relatively low voltage and high current through the spark generated in a conventional manner with a high-voltage low-current ignition system. Stored energies of about 1 J are typically used, and this energy is discharged in some 20 μs. This high-power discharge creates a high-temperature plasma so rapidly that the pressure in the cavity increases substantially, causing a supersonic jet of plasma to flow from the cavity into the main combustion chamber. The plasma enters the combustion chamber as a turbulent jet, preceded by a hemispherical blast wave. The gas dynamic effects of the blast wave are dissipated by the time combustion starts, which is typically of order 1 ms after the discharge commences. Ignition in the main combustion chamber takes place in the turbulent jet; the flame starts out as a turbulent flame in contrast to the flame with conventional ignition systems, which is initially laminarlike. The penetration of the jet depends on its initial momentum; it thus depends on the amount of energy deposited, cavity size, and orifice area. If the cavity is filled prior to ignition with a hydrocarbon (or a mixture of hydrocarbons) the ignition capabilities are enhanced due to the large increase in hydrogen atoms created in the plasma.[62]

The effects of plasma-jet ignitors on engine combustion are similar to those of breakdown ignition systems: the flame development period is significantly shortened, and the engine's lean stable operating limit is extended. In addition, the phenomenon of misfire, which is the failure to initiate combustion in a fraction of the engine's operating cycles, no longer occurs.[68] The lean operating limit of the engine is, however, normally controlled by flame extinguishment.

FLAME-JET IGNITION. With this type of system, ignition occurs in a prechamber cavity which is physically separated from the main chamber above the piston and is connected to it via one or more orifices or nozzles. As the flame develops in this cavity the pressure of the gases in the prechamber rises, forcing gas out into the main chamber through the orifice (or orifices) as one or more turbulent burning jets. The jet or jets penetrate into the main chamber, igniting the unburned mixture in the main chamber, thereby initiating the primary combustion process. Ignition within the cavity is usually achieved with a conventional spark discharge. The function of the prechamber or cavity is to transform the initial flame around the spark plug electrodes into one or more flame jets in the main chamber, which have a substantial surface area that can ignite extremely lean or dilute mixtures in a repeatable manner. Many different systems for achieving this goal have been developed; some of these have been used in production spark-ignition engines. Examples of the three major types of flame-jet ignition systems are shown in Fig. 9-56.

Figure 9-56a shows an example of the simplest type of flame-jet ignition concept (often called a *torch cell*). The cavity has no separate valve so is unscavenged; nor is there any prechamber fuel metering system. The function of the prechamber cavity is to increase the initial growth rate of the flame immediately following spark discharge by having this flame growth take place in a

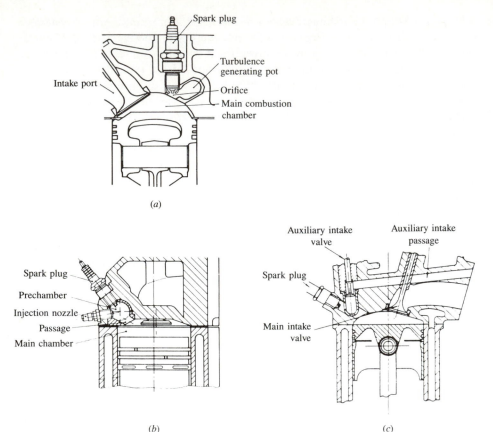

FIGURE 9-56
Flame-jet ignition concepts: (a) turbulence-generating torch cell;[69] (b) prechamber stratified-charge engine with auxiliary fuel injector with no prechamber scavenging;[70] (c) prechamber stratified-charge engine with prechamber inlet valve and auxiliary carburetor.[71]

more turbulent region than the main combustion chamber: the flame jet or jets which then emerge from the cavity produce a large initial flame surface area in the main chamber to start the bulk charge combustion process. The prechamber system shown was called a turbulent generating pot.[69] Another approach is to incorporate the cavity into the spark plug. Systems with prechamber volumes varying from 20 percent of the clearance volume to less than 1 percent have been developed. The flow pattern produced within the prechamber by the flow into the cavity during compression and the location of the spark plug electrodes within the cavity and of the nozzle or orifice are critical design issues. A major problem with these systems is that the prechamber is never completely scavenged by fresh mixture between cycles, so the burned gas fraction in the unburned mixture within the prechamber is always substantially higher than the burned gas fraction in the unburned main chamber mixture.[72]

Figure 9-56*b* and *c* shows two *prechamber stratified-charge* engine flame-jet ignition concepts. Here the mixture in the prechamber cavity is enriched by addition of fuel so that it ends up being slightly rich-of-stoichiometric at the time of spark discharge. The initial inflammation process in the cavity then occurs more rapidly and more repeatably. The operating principle of these stratified-charge systems is described briefly in Sec. 1.9. Figure 9-56*b* shows a system where the prechamber is unscavenged and fuel is injected directly into the prechamber cavity (in addition to the main fuel-injection process which occurs into the fresh charge in the intake system) to richen the mixture (which is lean overall) at the time of spark to an easily ignitable rich-of-stoichiometric composition. With this approach, the prechamber volume is usually 20 to 25 percent of the clearance volume. Figure 9-56*c* shows a prechamber stratified-charge flame-jet ignition system where the prechamber is scavenged between each combustion event. With this approach, a separate small intake valve feeds very rich mixture into the prechamber during the intake process, while the main fuel metering system feeds lean mixture to the main intake valve. During intake the prechamber is completely scavenged by the rich intake stream. During compression the lean mixture flowing from the main chamber to the prechamber brings the prechamber mixture equivalence ratio to slightly rich-of-stoichiometric at the time of spark discharge.[71]

The number and size of the orifices connecting the prechamber and the main chamber have a significant effect on the development of the main chamber burning process. Two different approaches are shown in the flame development stage in Fig. 9-57. Figure 9-57*a* shows the jets produced when the prechamber has more than one small nozzle which direct the burning prechamber mixture deep into the main chamber charge. A fast burning of the lean main-chamber charge results. With this approach, prechamber volumes of 2 to 3 percent of the clearance volume and nozzle area/prechamber volume ratios of 0.03 to 0.04 cm^{-1} are used. Figure 9-57*b* shows the approach used by Honda in their CVCC engine.

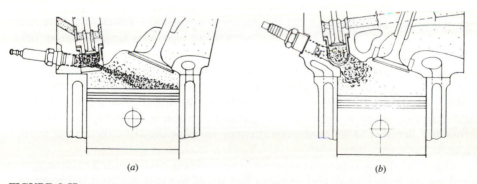

(*a*)   (*b*)

FIGURE 9-57
Two different approaches to prechamber orifice design with the prechamber stratified-charge carbureted and scavenged engine: (*a*) one or more small orifice(s) for deep jet penetration and faster burning process; (*b*) large orifice for lower-velocity jet and slower burn.[73]

A larger prechamber (5 to 12 percent) and larger orifice (orifice area/prechamber volume ratio of 0.04 to 0.2 cm^{-1}) gives a lower velocity jet which penetrates the main chamber charge more slowly, resulting in a slower burn.

All these concepts extend the engine's lean stable operating limit, relative to equivalent conventional engines, by several air/fuel ratios. For example, the unscavenged cavity without auxiliary fueling can operate satisfactorily at part-load with air/fuel ratios of 18 (equivalence ratio $\phi \approx 0.8$, relative air/fuel ratio $\lambda \approx 1.25$). The prechamber stratified-charge flame-jet ignition concepts can operate much leaner than this; however, the best combination of fuel consumption and emissions characteristics is obtained with $\phi \approx 0.9 - 0.75$, $\lambda \approx 1.1 - 1.3$.[70, 71] One performance penalty associated with all these flame-jet ignition concepts is the additional heat losses to the prechamber walls due to increased surface area and flow velocities. The stratified-charge prechamber concepts also suffer an efficiency penalty, relative to equivalent operation with uniform air/fuel ratios, due to the presence of fuel-rich regions during the combustion process.

9.6 ABNORMAL COMBUSTION: KNOCK AND SURFACE IGNITION

9.6.1 Description of Phenomena

Abnormal combustion reveals itself it many ways. Of the various abnormal combustion processes which are important in practice, the two major phenomena are knock and surface ignition. These abnormal combustion phenomena are of concern because: (1) when severe, they can cause major engine damage; and (2) even if not severe, they are regarded as an objectionable source of noise by the engine or vehicle operator. *Knock* is the name given to the noise which is transmitted through the engine structure when essentially spontaneous ignition of a portion of the end-gas—the fuel, air, residual gas, mixture ahead of the propagating flame—occurs. When this abnormal combustion process takes place, there is an extremely rapid release of much of the chemical energy in the end-gas, causing very high local pressures and the propagation of pressure waves of substantial amplitude across the combustion chamber. *Surface ignition* is ignition of the fuel-air mixture by a hot spot on the combustion chamber walls such as an overheated valve or spark plug, or glowing combustion chamber deposit: i.e., by any means other than the normal spark discharge. It can occur before the occurrence of the spark (*preignition*) or after (*postignition*). Following surface ignition, a turbulent flame develops at each surface-ignition location and starts to propagate across the chamber in an analogous manner to what occurs with normal spark ignition.

Because the spontaneous ignition phenomenon that causes knock is governed by the temperature and pressure history of the end gas, and therefore by the phasing and rate of development of the flame, various combinations of these two phenomena—surface ignition and knock—can occur. These have been categorized as indicated in Fig. 9-58. When autoignition occurs repeatedly, during

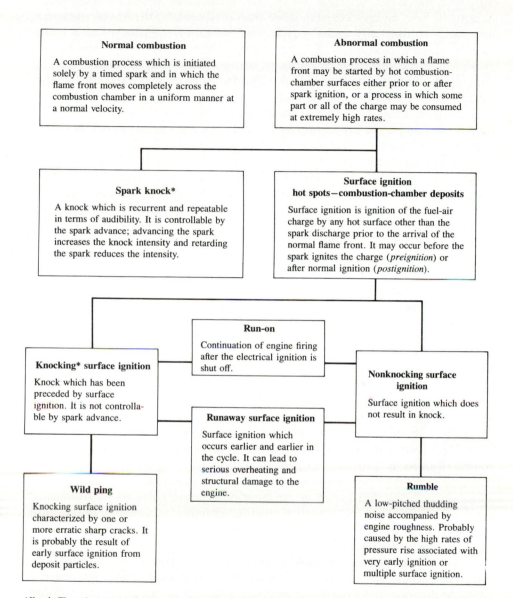

Normal combustion

A combustion process which is initiated solely by a timed spark and in which the flame front moves completely across the combustion chamber in a uniform manner at a normal velocity.

Abnormal combustion

A combustion process in which a flame front may be started by hot combustion-chamber surfaces either prior to or after spark ignition, or a process in which some part or all of the charge may be consumed at extremely high rates.

Spark knock*

A knock which is recurrent and repeatable in terms of audibility. It is controllable by the spark advance; advancing the spark increases the knock intensity and retarding the spark reduces the intensity.

Surface ignition
hot spots—combustion-chamber deposits

Surface ignition is ignition of the fuel-air charge by any hot surface other than the spark discharge prior to the arrival of the normal flame front. It may occur before the spark ignites the charge (*preignition*) or after normal ignition (*postignition*).

Run-on

Continuation of engine firing after the electrical ignition is shut off.

Knocking* surface ignition

Knock which has been preceded by surface ignition. It is not controllable by spark advance.

Nonknocking surface ignition

Surface ignition which does not result in knock.

Runaway surface ignition

Surface ignition which occurs earlier and earlier in the cycle. It can lead to serious overheating and structural damage to the engine.

Wild ping

Knocking surface ignition characterized by one or more erratic sharp cracks. It is probably the result of early surface ignition from deposit particles.

Rumble

A low-pitched thudding noise accompanied by engine roughness. Probably caused by the high rates of pressure rise associated with very early ignition or multiple surface ignition.

Knock: The noise associated with autoignition of a portion of the fuel-air mixture ahead of the advancing flame front. Autoignition is the spontaneous ignition and the resulting very rapid reaction of a portion or all of the fuel-air mixture.

FIGURE 9-58
Definition of combustion phenomena—normal and abnormal (knock and surface ignition)—in a spark-ignition engine. (*Courtesy Coordinating Research Council.*)

otherwise normal combustion events, the phenomena is called *spark-knock*. Repeatedly here means occurring more than occasionally: the knock phenomenon varies substantially cycle-by-cycle, and between the cylinders of a multi-cylinder engine, and does not necessarily occur every cycle (see below). Spark-knock is controllable by the spark advance: advancing the spark increases the knock severity or intensity and retarding the spark decreases the knock. Since surface ignition usually causes a more rapid rise in end-gas pressure and temperature than occurs with normal spark ignition (because the flame either starts propagating sooner, or propagates from more than one source), knock is a likely outcome following the occurrence of surface ignition. To identify whether or not surface ignition causes knock, the terms knocking surface ignition and non-knocking surface ignition are used. Knocking surface ignition usually originates from preignition caused by glowing combustion chamber deposits: the severity of knock generally increases the earlier that preignition occurs. Knocking surface ignition cannot normally be controlled by retarding the spark timing, since the spark-ignited flame is not the cause of knock. Nonknocking surface ignition is usually associated with surface ignition that occurs late in the operating cycle.

The other abnormal combustion phenomena in Fig. 9-58, while less common, have the following identifying names. Wild ping is a variation of knocking surface ignition which produces sharp cracking sounds in bursts. It is thought to result from early ignition of the fuel-air mixture in the combustion chamber by glowing loose deposit particles. It disappears when the particles are exhausted and reappears when fresh particles break loose from the chamber surfaces. Rumble is a relatively stable low-frequency noise (600 to 1200 Hz) phenomenon associated with deposit-caused surface ignition in high-compression-ratio engines. This type of surface ignition produces very high rates of pressure rise following ignition. Rumble and knock can occur together. Run-on occurs when the fuel-air mixture within the cylinder continues to ignite after the ignition system has been switched off. During run-on, the engine usually emits knocklike noises. Run-on is probably caused by compression ignition of the fuel-air mixture, rather than surface ignition. Runaway surface ignition is surface ignition that occurs earlier and earlier in the cycle. It is usually caused by overheated spark plugs or valves or other combustion chamber surfaces. It is the most destructive type of surface ignition and can lead to serious overheating and structural damage to the engine.[74]

After some additional description of surface-ignition phenomena, the remainder of Sec. 9.6 will focus on knock. This is because surface ignition is a problem that can be solved by appropriate attention to engine design, and fuel and lubricant quality. In contrast, knock is an inherent constraint on engine performance and efficiency since it limits the maximum compression ratio that can be used with any given fuel.

Of all the engine surface-ignition phenomena in Fig. 9-58, preignition is potentially the most damaging. Any process that advances the start of combustion from the timing that gives maximum torque will cause higher heat rejection because of the increasing burned gas pressures and temperatures that result.

Higher heat rejection causes higher temperature components which, in turn, can advance the preignition point even further until critical components can fail. The parts which can cause preignition are those least well cooled and where deposits build up and provide additional thermal insulation: primary examples are spark plugs, exhaust valves, metal asperities such as edges of head cavities or piston bowls. Under normal conditions, using suitable heat-range spark plugs, preignition is usually initiated by an exhaust valve covered with deposits coming from the fuel and from the lubricant which penetrates into the combustion chamber. Colder running exhaust valves and reduced oil consumption usually alleviate this problem: locating the exhaust valve between the spark plug and the end-gas region avoids contact with both the hottest burned gas near the spark plug and the end-gas. Engine design features that minimize the likelihood of preignition are: appropriate heat-range spark plug, removal of asperities, radiused metal edges, well-cooled exhaust valves with sodium-cooled valves as an extreme option.[75, 76]

Knock primarily occurs under wide-open-throttle operating conditions. It is thus a direct constraint on engine performance. It also constrains engine efficiency, since by effectively limiting the temperature and pressure of the end-gas, it limits the engine compression ratio. The occurrence and severity of knock depend on the knock resistance of the fuel and on the antiknock characteristics of the engine. The ability of a fuel to resist knock is measured by its octane number: higher octane numbers indicate greater resistance to knock (see Sec. 9.6.3). Gasoline octane ratings can be improved by refining processes, such as catalytic cracking and reforming, which convert low-octane hydrocarbons to high-octane hydrocarbons. Also, antiknock additives such as alcohols, lead alkyls, or an organomanganese compound can be used. The octane number requirement of an engine depends on how its design and the conditions under which it is operated affect the temperature and pressure of the end-gas ahead of the flame and the time required to burn the cylinder charge. An engine's tendency to knock, as defined by its *octane requirement*—the octane rating of the fuel required to avoid knock—is increased by factors that produce higher temperatures and pressures or lengthen the burning time. Thus knock is a constraint that depends on both the quality of available fuels and on the ability of the engine designer to achieve the desired normal combustion behavior while holding the engine's propensity to knock at a minimum.[74]

The pressure variation in the cylinder during knocking combustion indicates in more detail what actually occurs. Figure 9-59 shows the cylinder pressure variation in three individual engine cycles, for normal combustion, light knock, and heavy knock, respectively.[77] When knock occurs, high-frequency pressure fluctuations are observed whose amplitude decays with time. Figures 9-59a and b have the same operating conditions and spark advance. About one-third of the cycles in this engine at these conditions had no trace of knock and had normal, smoothly varying, cylinder pressure records as in Fig. 9-59a. Knock of varying severity occurred in the remaining cycles. With light or trace knock, knock occurs late in the burning process and the amplitude of the pressure fluctuations

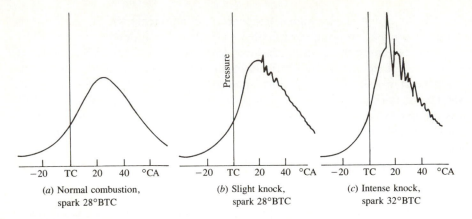

FIGURE 9-59

Cylinder pressure versus crank angle traces of cycles with (a) normal combustion, (b) light knock, and (c) heavy knock. 4000 rev/min, wide-open throttle, 381-cm^3 displacement single-cylinder engine.[77]

is small (Fig. 9-59b). With heavy knock, illustrated here with more advanced spark timing and by selecting an especially high intensity knocking cycle, knock occurs closer to top-center earlier in the combustion process and the initial amplitude of the pressure fluctuation is much larger. These pressure fluctuations produce the sharp metallic noise called "knock." They are the result of the essentially spontaneous release of much of the end-gas fuel's chemical energy. This produces a substantial *local* increase in gas pressure and temperature, thereby causing a shock wave to propagate away from the end-gas region across the combustion chamber. This shock wave, the expansion wave that accompanies it, and the reflection of these waves by the chamber walls create the oscillatory pressure versus time records shown in Fig. 9-59b and c. Note that once knock occurs, the pressure distribution across the combustion chamber is no longer uniform: transducers located at different points in the chamber will record different pressure levels at a given time until the wave propagation phenomena described above have been damped out.[78]

Many methods of knock detection and characterization have been used. The human ear is a surprisingly sensitive knock detector and is routinely used in determining the octane requirement of an engine—the required fuel quality the engine must have to avoid knock. Knock detectors used for knock control systems normally respond to the vibration-driven acceleration of parts of the engine block caused by knocking combustion pressure waves. A high-intensity flash is observed when knock occurs; this is accompanied by a sharp increase in ionization. Optical probes and ionization detectors have therefore been used. The spark plug can serve as an ionization detector. For more detailed studies of knock in engines, the piezoelectric pressure transducer is the most useful monitoring device. Often the transducer signal is filtered so that the pressure fluctuations caused by knock are isolated.[75]

The amplitude of the pressure fluctuation is a useful measure of the inten-

sity of knock because it depends on the amount of end-gas which ignites spontaneously and rapidly, and because engine damage due to knock results from the
high gas pressures (and temperatures) in the end-gas region. Use of this measure
of knock severity or intensity shows there is substantial variation in the extent of
knock, cycle-by-cycle. Figure 9-60 shows the knock intensity in one hundred consecutive cycles in a given cylinder of a multicylinder engine operating at fixed
conditions for knocking operation. The intensity varies randomly from essentially
no knock to heavy knock.[79] Cylinder-to-cylinder variations are also substantial
due to variations in compression ratio, mixture composition and conditions, burn
rate, and combustion chamber cooling. One or more cylinders may not knock at
all while others may be knocking heavily.[80]

Since the knock phenomenon produces a nonuniform state in the cylinder,
and since the details of the knock process in each cycle and in each cylinder are
different, a fundamental definition of knock intensity or severity is extremely difficult. The ASTM-CFR method for rating fuel octane quality (see Sec. 9.6.3) by the
severity of knocking combustion uses the time derivative of pressure during the
cycle. Cylinder pressure is determined with a pressure transducer. The low-
frequency component of pressure change due to normal combustion is filtered
out and the rate of pressure rise is averaged over many cycles during the pressure
fluctuations following knock. This approach obviously provides only an average
relative measure of knock intensity. The maximum rate of pressure rise has been
used to quantify knock severity. An accelerometer mounted on the engine can
give indications of relative knock severity provided that it is mounted in the same
location for all tests. The most precise measure of knock severity is the maximum
amplitude of the pressure oscillations that occur with knocking combustion. The
cylinder pressure signal (from a high-frequency response pressure transducer) is
filtered with a band-pass filter so that only the component of the pressure signal
that corresponds to the fluctuations occurring after knock remains. The filter is
set for the first transverse mode of gas vibration in the cylinder (in the 3 to
10 kHz range, depending on bore and chamber geometry). The maximum amplitude of pressure oscillation gives a good indication of the severity of knock.[81]
The knock intensities in individual cycles shown in Fig. 9-60 were determined in

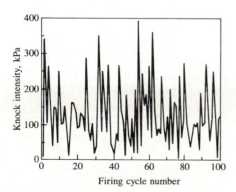

FIGURE 9-60
Knock intensity (maximum amplitude of band-
pass-filtered pressure signal) in one hundred individual consecutive cycles. One cylinder of V-8
engine, 2400 rev/min, wide-open throttle.[79]

this manner.[79] Note that because the pressure fluctuations are the consequence of a wave propagation phenomenon, the location of the pressure transducer in relation to the location of the knocking end-gas and the shape of the combustion chamber will affect the magnitude of the maximum recorded pressure-fluctuation amplitude.

The impact of knock depends on its intensity and duration. Trace knock has no significant effect on engine performance or durability. Heavy knock can lead to extensive engine damage. In automobile applications, a distinction is usually made between "acceleration knock" and "constant-speed knock." Acceleration knock is primarily an annoyance, and due to its short duration is unlikely to cause damage. Constant-speed knock, however, can lead to two types of engine damage. It is especially a problem at high engine speeds where it is masked by other engine noises and is not easily detected. Heavy knock at constant speed can easily lead to:

1. Preignition, if significant deposits are present on critical combustion chamber components. This could lead to runaway preignition.†
2. Runaway knock—spark-knock occurring earlier and earlier, and therefore more and more intensely. This soon leads to severe engine damage.
3. Gradual erosion of regions of the combustion chamber, even if runaway knock does not occur.

The engine can be damaged by knock in different ways: piston ring sticking; breakage of the piston rings and lands; failure of the cylinder head gasket; cylinder head erosion; piston crown and top land erosion; piston melting and holing. Examples of component damage due to preignition and knock are shown in Fig. 9-61.[82-84]

The mechanisms that cause this damage are thought to be the following. Preignition damage is largely thermal as evidenced by fusion of spark plugs or pistons. When knock is very heavy, substantial additional heat is transferred to the combustion chamber walls and rapid overheating of the cylinder head and piston results. Under these conditions, knock is not stable: the overheating increases the engine's octane requirement which in turn increases the intensity of knock. It becomes heavier and heavier, and the uncontrolled running away of this phenomena can lead to engine failure in minutes. This damage is due to overheating of the engine: the piston and rings seize in the bore. The damage due to heavy knock over extended periods—erosion of piston crowns and (aluminum) cylinder heads in the end-gas region—is due primarily to the high gas pressures in this region. Extremely high pressure pulses of up to 180 atm due to heavy knock can occur locally in the end-gas region, in the 5 to 10 kHz frequency

---

† Note that heavy knock can also remove deposits from the combustion chamber walls, thereby decreasing the octane requirement of the engine.

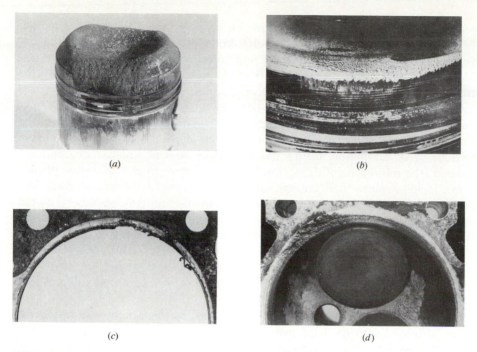

FIGURE 9-61
Examples of component damage from abnormal engine combustion. (a) Piston holing by preigni-tion;[83] (b) piston crown erosion after 10 hours of high-speed knocking;[82] (c) cylinder head gasket splitting failure due to heavy knock;[83] (d) erosion of aluminum cylinder head along the top of the cylinder liner due to heavy knock.[83]

range. These high local pressures are combined with the higher-than-normal local surface temperatures which occur with the higher knocking heat fluxes and weaken the material. Pitting and erosion due to fatigue with these excessive mechanical stresses, and breakage of rings and lands, can then occur.[78, 82–84]

9.6.2 Knock Fundamentals

As yet, there is no complete fundamental explanation of the knock phenomenon over the full range of engine conditions at which it occurs. It is generally agreed that knock originates in the extremely rapid release of much of the energy con-tained in the end-gas ahead of the propagating turbulent flame, resulting in high local pressures. The nonuniform nature of this pressure distribution causes pres-sure waves or shock waves to propagate across the chamber, which may cause the chamber to resonate at its natural frequency. Two theories have been advanced to explain the origin of knock: the autoignition theory and the detona-tion theory. The former holds that when the fuel-air mixture in the end-gas region is compressed to sufficiently high pressures and temperatures, the fuel oxi-dation process—starting with the preflame chemistry and ending with rapid energy release—can occur spontaneously in parts or all of the end-gas region. The latter theory postulates that, under knocking conditions, the advancing flame front accelerates to sonic velocity and consumes the end-gas at a rate much faster than would occur with normal flame speeds.[75] These theories attempt to describe what causes the rapid release of chemical energy in the end-gas which creates very high pressures, locally, in the end-gas region. The engine phenomenon "knock" includes also the propagation of strong pressure waves across the chamber, chamber resonance, and transmission of sound through the engine structure. The detonation theory has led many to call knock "detonation." However, the more general term "knock" is preferred, since this engine phenomenon includes more than the end-gas energy release, and there is much less evidence to support the detonation theory than the autoignition theory as the initiating process. Most recent evidence indicates that knock originates with the spontaneous or autoignition of one or more local regions within the end-gas. Additional regions (some adjacent to already ignited regions and some separated from these regions) then ignite until the end-gas is essentially fully reacted. This sequence of processes can occur extremely rapidly. Thus, the autoignition theory is most widely accepted.

Photographic studies of knocking combustion have been an important source of insight into the fundamentals of the phenomenon over the past fifty years.† Figure 9-62 shows two sets of schlieren photographs, one from a cycle with normal combustion and the other from a cycle with knock.[88] These photographs were taken in an overhead valve engine with a disc-shaped combustion chamber, with a window which permits observation of the chamber opposite to the spark plug. A reflecting mirror on the piston crown permits use of the schlieren technique which identifies regions where changes in gas density exist. Operating conditions, except for spark advance, were the same for both cycles. In the normal combustion sequence, the turbulent flame front moves steadily through the end-gas as combustion goes to completion. The cylinder pressure varies smoothly throughout this process. When the spark is advanced by 15°, the end-gas temperature and pressure are increased significantly and knock occurs. In this sequence of photographs (b), the initial flame propagation process (photographs 1 to 3) is like that of the normal combustion sequence: then almost

---

† See Refs. 85 to 87 for early high-speed photographic studies. See Refs. 88 to 91 for some recent studies.

FIGURE 9-62
Schlieren photographs from high-speed movies of (a) normal flame propagation through the end gas and (b) knocking combustion (autoignition occurs in photograph 4), with corresponding cylinder pressure versus crank angle traces. Disc-shaped combustion chamber with details of window shown in insert. 1200 rev/min, 80 percent volumetric efficiency, $(A/F) = 12.5$; spark timing: (a) 10° BTC, (b) 25° BTC.[88]

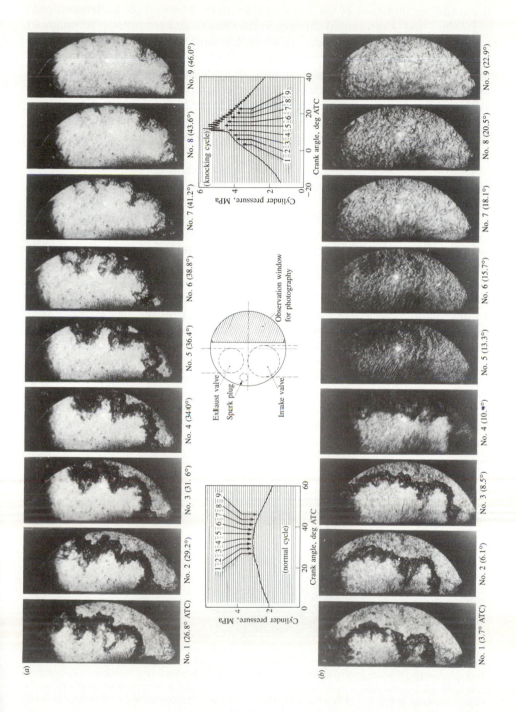

No. 9 (46.0°) No. 8 (43.6°) No. 7 (41.2°) No. 6 (38.8°) No. 5 (36.4°) No. 4 (34.0°) No. 3 (31.6°) No. 2 (29.2°) No. 1 (26.8° ATC)

(knocking cycle)

Cylinder pressure, MPa

1 2 3 4 5 6 7 8 9

Crank angle, deg ATC

Exhaust valve

Spark plug

Intake valve

Observation window for photography

(normal cycle)

Cylinder pressure, MPa

1 2 3 4 5 6 7 8 9

Crank angle, deg ATC

(a)

No. 9 (22.9°) No. 8 (20.5°) No. 7 (18.1°) No. 6 (15.7°) No. 5 (13.3°) No. 4 (10.9°) No. 3 (8.5°) No. 2 (6.1°) No. 1 (3.7° ATC)

(b)

the entire region ahead of the flame appears dark (photograph 4). Between photographs 3 and 4 substantial changes in the density and temperature throughout most of the end-gas region have occurred. Examination of the cylinder pressure trace shows that this corresponds closely to the time when the pressure recorded by the pressure transducer rises rapidly. Immediately after this, pressure oscillations (at 6 to 8 kHz) are detected. In photograph 5 the flame is no longer visible: the end-gas expansion has pushed the flame back out of the field of view. Subsequent photographs are alternatively lighter and darker, indicative of changing local density fields as pressure waves propagate back and forth across the chamber.

More extensive studies of this type, which relate photographs from high-speed movies of the combustion process to the cylinder pressure development, indicate that the location or locations where ignition of one or more portions of the end-gas first occur and the subsequent rate with which the ignition process develops throughout the rest of the end-gas vary substantially cycle-by-cycle and with the intensity of the knocking process. Figure 9-63 shows five shadowgraph photographs from a knocking engine cycle in a research engine similar to that shown in Fig. 9-62.[89] The photographs are 33 μs apart; the total sequence shown lasts 1°. The first photograph shows the flame front prior to onset of knock. The second photograph shows the onset of autoignition with the appearance of dark regions near the wall (two identified by arrows) where substantial density gradients resulting from local energy release exist. The third, fourth, and fifth photographs show the spread of these ignited regions with time through the remaining end-gas. The exact location where autoignition occurred was identified with a photodigitizing system which ranked regions of the photograph by their brightness or darkness. The digitized version of the second photograph is also shown in Fig. 9-63. Additional smaller regions of autoignition are evident adjacent to the wall and in the vicinity of the flame front. While the location of autoig-

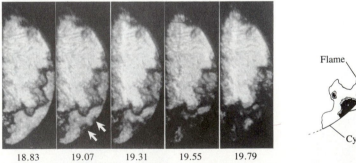
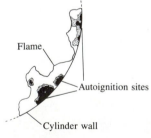

18.83   19.07   19.31   19.55   19.79

FIGURE 9-63
Five shadowgraph photographs of knocking combustion cycle identifying location of autoignition sites (arrows). Crank angle of each photo indicated: 33 μs between frames. Operating conditions and engine details as in Fig. 9-62. Photodigitized picture of second photograph showing additional details of the autoignition sites in the end-gas region on right.[89]

nition sites varied with engine operating conditions, the majority of occurrences in this study were in the vicinity of the cylinder wall. The rate of spread of the autoignited end-gas region also varied significantly. Under heavy knocking conditions the entire end-gas region became ignited very rapidly and high-amplitude pressure oscillations occurred. Under trace knock conditions, autoignition could occur; yet the spread of the autoignited region could be sufficiently slow for no pressure oscillations to be detected.

When the above-described end-gas ignition process occurs rapidly, the gas pressure in the end-gas region rises substantially due to the rapid release of the end-gas fuel's chemical energy. The erosion damage that knock can produce, due to stress-induced material fatigue as described in the previous section, indicates the location of this high-pressure region. With the chamber geometry typical of most engines where the flame propagates toward the cylinder wall, the damage is confined to the thin crescent-shaped region on the opposite side of the chamber to the spark plug, where one expects the end-gas to be located. A shock wave propagates from the outer edge of this high-pressure end-gas region across the chamber at supersonic velocity, and an expansion wave propagates into the high-pressure region toward the near wall. The presence of such a shock wave has been observed photographically.[90] The shock wave and expansion wave reflect off the walls of the chamber, eventually producing standing waves. Usually these standing waves are due to transverse gas vibration and are of substantial amplitude. The amplitude of the pressure oscillations builds up as the standing waves are established, and then decays as the gas motion is damped out. The frequency of the pressure oscillations (normally in the 5 to 10 kHz range) decreases with time as the initially finite-amplitude supersonic pressure waves decay to small-amplitude sound waves.[80] Thus the pressure signal detected with a transducer during a knocking combustion cycle will depend on the details of the end-gas ignition process, the combustion chamber geometry, and the location of the transducer in relation to the end-gas region.

The pressure variation across the cylinder bore, due to knock, can be illustrated by the following example. Consider the disc-shaped combustion chamber with the spark plug located in the cylinder wall shown in Fig. 9-64a. Figure 9-64b shows the pressure and temperature distribution across the combustion chamber due to the normal flame propagation process, at the time rapid ignition of the end-gas occurs. If the end-gas ignites completely and instantaneously, its pressure and temperature will suddenly rise, as shown. A shock wave will now propagate to the right and an expansion wave to the left, as shown in the distance-time diagram in Fig. 9-64c. These waves reflect off the walls and interact. The pressures at the cylinder wall, in the end-gas region and on the opposite side of the chamber at the spark plug, develop as shown in Fig. 9-64d. Figure 9-65 shows two simultaneously recorded pressure traces with heavy knock from two transducers at the top of the cylinder liner, one located near the spark plug and one in the end-gas region. In the end-gas region the pressure rises extremely rapidly when knock occurs, to a value considerably higher than that recorded on the opposite side of the chamber where the pressure rises more gradually. Standing

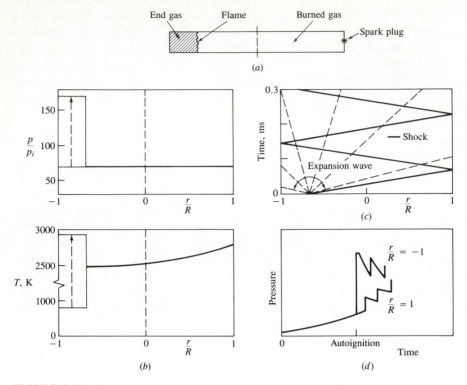

FIGURE 9-64
Illustration of how cylinder pressure distribution develops following knock. (a) Schematic of disc-shaped combustion chamber at time of knock. (b) Pressure and temperature across the diameter of a disc-shaped engine combustion chamber, before and after end-gas autoignition (assumed to occur very rapidly, at constant end-gas density). (c) Schematic of shock and expansion wave pattern following end-gas autoignition on distance-time plot. (d) Pressure variation with time at cylinder wall in the end-gas region, and at the opposite side of the cylinder.

waves are then set up and the amplitudes of the oscillations decay as the waves are damped out.[78]

The fundamental theories of knock are based on models for the autoignition of the fuel-air mixture in the end-gas. *Autoignition* is the term used for a rapid combustion reaction which is not initiated by any external ignition source. Often in the basic combustion literature this phenomenon is called an explosion. Before discussing the relevant theories of hydrocarbon oxidation, the necessary terminology will be defined and illustrated with the autoignition behavior of the much simpler hydrogen-oxygen system. For the hydrocarbons commonly found in practical fuels, the chemical reaction schemes by which the fuel molecules are broken down and react to form products are extremely complicated, and are as yet imperfectly understood.

The autoignition of a gaseous fuel-air mixture occurs when the energy re-

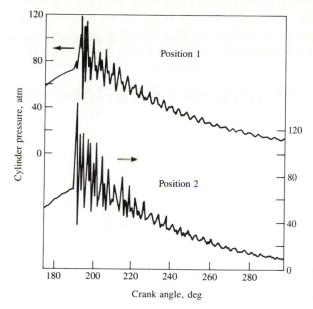

FIGURE 9-65
Simultaneously recorded pressure traces on opposite sides of spark-ignition engine combustion chamber in heavily knocking cycle. Position 1: at top of cylinder liner closest to spark plug. Position 2: at top of cylinder liner farthest from spark plug. 2000 rev/min, wide-open throttle, $(A/F) = 12.3$, spark advance 20° BTC.[78]

leased by the reaction as heat is larger than the heat lost to the surroundings; as a result the temperature of the mixture increases, thereby rapidly accelerating, due to their exponential temperature dependence, the rates of the reactions involved. The state at which such spontaneous ignition occurs is called the *self-ignition temperature* and the resulting self-accelerating event where the pressure and temperature increases rapidly is termed a *thermal explosion.*

In complex reacting systems such as exist in combustion, the "reaction" is not a single- or even a few-step process; the actual chemical mechanism consists of a large number of simultaneous, interdependent reactions or *chain reactions*. In such chains there is an *initiating* reaction where highly reactive intermediate species or *radicals* are produced from stable molecules (fuel and oxygen). This step is followed by *propagation* reactions where radicals react with the reactant molecules to form products and other radicals to continue the chain. The process ends with *termination* reactions where the chain propagating radicals are removed. Some propagating reactions produce two reactive radical molecules for each radical consumed. These are called *chain-branching* reactions. When, due to chain-branching, the number of radicals increases sufficiently rapidly, the reaction rate becomes extremely fast and a *chain-branching explosion* occurs. While the terms thermal and chain-branching explosions have been introduced separately, in many situations the self-accelerations in temperature and radicals occur simultaneously and the two phenomena must be combined.[92, 93]

The oxidation of hydrogen at high pressures and temperatures provides a good illustration of these phenomena. For stoichiometric hydrogen-oxygen mixtures, no reaction occurs below 400°C unless the mixture is ignited by an external source such as a spark; above 600°C explosion occurs spontaneously at all pressures.†

The initiating steps proceed primarily through hydrogen peroxide (H_2O_2) to form the hydroxyl radical (OH) at lower temperatures, or through dissociation of H_2 at higher temperatures to form the hydrogen atom radical H. The basic radical-producing chain sequence is composed of three reactions:

$$\text{(R1)} \qquad \qquad H + O_2 = O + OH$$
$$\text{(R2)} \qquad \qquad O + H_2 = H + OH \qquad \qquad (9.56)$$
$$\text{(R3)} \qquad \qquad H_2 + OH = H_2O + H$$

The first two reactions are chain-branching; two radicals are produced for each one consumed. The third reaction is necessary to complete the chain sequence: starting with one hydrogen atom, the sequence R1 then R2 and R3 produces two H; starting with OH, the sequence R3, then R1, then R2 produces two OH. Since all three reactions are required, the multiplication factor is less than 2 but greater than 1. Repeating this sequence over and over again rapidly builds up high concentrations of radicals from low initial levels.

However, these three reactions do not correspond to the overall stoichiometry

$$H_2 + \tfrac{1}{2}O_2 = H_2O$$

and other reactions must become important. In flames, the chain-branching process ceases when the reverse of reactions R1–R3 become significant. A quasi equilibrium is established; while the overall process has proceeded a considerable way toward completion, a substantial amount of the available energy is still contained in the high radical concentrations. Over a longer time scale, this energy is released through three-body recombination reactions (the principal chain terminating reactions):

$$\text{(R4)} \qquad \qquad H + H + M = H_2 + M$$
$$\text{(R5)} \qquad \qquad O + O + M = O_2 + M \qquad \qquad (9.57)$$
$$\text{(R6)} \qquad \qquad H + OH + M = H_2O + M$$

M refers to any available third-body species, required in these recombination reactions to remove the excess energy.[92]

With this introductory background let us now turn to the autoignition of hydrocarbon-air mixtures. The process by which a hydrocarbon is oxidized can exhibit four different types of behavior, or a sequential combination of them, depending on the pressure and temperature of the mixture: slow reactions; single

---

† In between, three separate explosion limits—pressure-temperature boundaries for specific mixture ratios of fuel and oxidizer that separate regions of slow and fast reaction—exist; these are not, however, of interest to us here.

or multiple cool flames (slightly exothermic reactions); two-stage ignition (cool flame followed by a hot flame); single-stage ignition (hot flame). Slow reactions are a low-pressure, low-temperature ($<200°C$) phenomenon not normally occurring in engines. At 300 to 400°C one or more combustion waves often appear, accompanied by faint blue light emission; the reaction is quenched, however, when only a small fraction of the reactants have reacted and the temperature rise is only tens of degrees. These are called *cool flames*. Depending on conditions and the fuel, a cool flame may be followed by a "hot flame" or high-temperature explosion where the reaction accelerates rapidly after ignition. This is termed *two-stage ignition*. As the temperature of the mixture increases, a transition from two-stage to single-stage ignition occurs. While all hydrocarbons exhibit induction intervals which are followed by a very rapid reaction rate, some hydrocarbon compounds do not exhibit the cool flame or two-stage ignition behavior.

Figure 9-66 shows these ignition limits for isooctane, methane, and benzene. For isooctane, ignition in the low-temperature regions is by a two-stage process: there is a first time interval before the cool flame appears and then a second time interval from the appearance of the cool flame to the hot flame combustion process. Ignition in the high-temperature region is by a continuous one-stage process. The cool flame phenomena vary enormously with hydrocarbon structure. Normal paraffins give strong cool flames, branched-chain paraffins are more resistant. Olefins give even lower luminosity cool flames with longer induction periods. Methane shows only the high-temperature ignition limit, as indicated in Fig. 9-66. Benzene, also, does not exhibit the cool flame phenomenon and other aromatics give hardly detectable luminosity.[75] It is thought that some com-

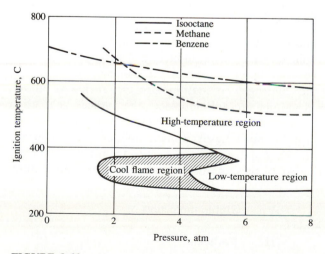

FIGURE 9-66
Ignition diagrams for isooctane, methane, and benzene. Two-stage ignition occurs in the low-temperature region; the first stage may be a cool flame. Single-stage ignition occurs in the high-temperature region.[94]

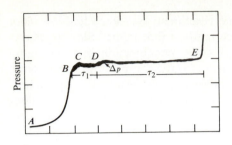

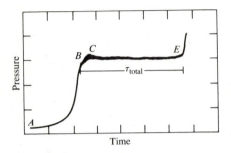

FIGURE 9-67
Pressure records of the autoignition of isooctane-air mixture ($\phi = 0.9$) in a rapid compression machine. $A \rightarrow B \rightarrow C$ is the compression process. Top: two-stage ignition (at D, then at E) at postcompression pressure of 1.86 MPa and temperature of 686 K. Bottom: single-stage ignition at E at post-compression conditions of 2.12 MPa and 787 K. Vertical scale: 690 kPa/division. Horizontal scale: 1 ms/division.[95]

pounds knock by a low-temperature two-stage ignition mechanism, some via a high-temperature single-stage ignition mechanism, and for some fuels both mechanisms may play a role.

Examples of these two mechanisms in rapid-compression machine experiments, where a homogeneous isooctane-air mixture was compressed to different final conditions in a piston-cylinder apparatus and allowed to autoignite, are shown in Fig. 9-67. $A \rightarrow B \rightarrow C$ is the piston-motion-produced compression. The top trace shows a well-defined cool flame at D, preceding hot ignition at E. The lower trace, at a higher temperature, shows a single-stage ignition process.[95]

Many of the above phenomena—long induction periods, initial slow increase in reaction rate, two-stage ignition process—cannot be explained with simple mechanisms like the hydrogen oxidation process reviewed above. Although chain-branching reactions are taking place, the radical generation process must be more complex. Explanations of the long induction periods are based on the formation of unstable but long-lived intermediates. These intermediates can then either react to form stable molecules or to form active radicals, the dominance and rate of either of these paths depending on the temperature: i.e.,[93]

$$A \rightarrow M^* \begin{cases} \text{I} & \text{Stable molecules} \\ \text{II} & \text{Radicals} \end{cases}$$

(9.58)

This is called a *degenerate-branching* mechanism. Hydroperoxides are an important metastable intermediate produced in the chain propagation process in the low-temperature ignition process. They have the form ROOH, where R is an organic radical (formed by abstracting a hydrogen atom from a fuel hydrocarbon molecule). However, at higher temperatures, ROOH is no longer the major product of the chain propagation process: instead, it is hydrogen peroxide, H_2O_2. While H_2O_2 is relatively stable at lower temperatures, above 500°C it decomposes into two OH radicals.

An outline of the basic hydrocarbon (RH) oxidation process due to Semenov is as follows:[93]

$$
\begin{array}{lll}
\text{(R1)} & RH + O_2 \dashrightarrow \dot{R} + H\dot{O}_2 & \left.\right\} \quad \text{Chain initiation} \\[6pt]
\text{(R2)} & \dot{R} + O_2 \dashrightarrow R\dot{O}_2 & \\[6pt]
\text{(R3)} & \dot{R} + O_2 \dashrightarrow \text{olefin} + H\dot{O}_2 & \\[6pt]
\text{(R4)} & R\dot{O}_2 + RH \dashrightarrow ROOH + \dot{R} & \left.\right\} \quad \text{Chain propagation} \\[6pt]
\text{(R5)} & R\dot{O}_2 \dashrightarrow R'CHO + R''O & \qquad\qquad\qquad (9.59) \\[6pt]
\text{(R6)} & H\dot{O}_2 + RH \dashrightarrow H_2O_2 + \dot{R} & \\[6pt]
\text{(R7)} & ROOH \dashrightarrow R\dot{O} + \dot{O}H & \left.\right\} \\[6pt]
\text{(R8)} & R'CHO + O_2 \dashrightarrow R'\dot{C}O + H\dot{O}_2 & \quad \text{Degenerate branching} \\[6pt]
\text{(R9)} & R\dot{O}_2 \dashrightarrow \text{destruction} & \left.\right\} \quad \text{Chain termination}
\end{array}
$$

The dot denotes an active radical; each dash denotes the number of free bonds on the organic radical R.

Reaction R1 is slow and explains the induction period in hydrocarbon combustion. R2 is fast and of near-zero activation energy. R3 leads to olefins known to occur in the oxidation of saturated hydrocarbons. R4 and R5 yield the main intermediates. The degenerate branching comes about from the delay in decomposition of the reactive species in R7 and R8. As one radical is used up to form the reactants in R7 and R8, the multiple radicals do not appear until these reactants decompose.[93] A more extensive discussion of hydrocarbon oxidation mechanisms can be found in Benson.[96]

The following evidence indicates the relevance of the above mechanism to knock in engines. End-gas sampling studies have identified products of slow combustion reactions of isooctane; these principally include olefins, cyclic ethers, aldehydes (R'CHO), and ketones (R''CO).[97] Such studies have shown increasing concentrations of peroxides (predominantly H_2O_2 with traces of organic peroxides) with isoparaffinic fuels which show two-stage ignition behavior. Higher temperature, single-stage ignition fuels such as benzene and toluene gave no detectable peroxide. Aldehydes and ketones have been measured in significant and increasing concentrations in motored engines where the peak cycle temperature was steadily increased. In motored engines, the occurrence of cool flames, the two-stage nature of the autoignition ignition process at intermediate compression temperatures, and the transition to a single-stage ignition process at very high compression ratios (i.e., motored-engine gas temperatures with a peak much higher than normal to simulate end-gas conditions in a firing engine) have also been demonstrated.[75, 94]

Two types of models of this autoignition process have been developed and used: (1) empirical induction-time correlations; (2) chemical mechanisms which embody many or all of the features of the "full" hydrocarbon oxidation process given in Eq. (9.59).

Induction-time correlations are derived by matching an Arrhenius function to measured data on induction or autoignition times, for given fuel-air mixtures, over the relevant mixture pressure and temperature ranges. It is then assumed that autoignition occurs when

$$\int_{t=0}^{t_i} \frac{dt}{\tau} = 1 \tag{9.60}$$

where τ is the induction time at the instantaneous temperature and pressure for the mixture, t is the elapsed time from the start of the end-gas compression process ($t = 0$), and t_i is the time of autoignition. This equation can be derived by assuming that the overall rate of production of the critical species in the induction period chemistry, for a given mixture, depends only on the gas state and that the concentration of the critical species required to initiate autoignition is fixed (i.e., independent of the gas state).[98]

A number of empirical relations for induction time for individual hydrocarbons and blended fuels have been developed from fundamental or engine studies of autoignition (see Ref. 99). These relations have the form

$$\tau = Ap^{-n} \exp\left(\frac{B}{T}\right) \tag{9.61}$$

where A, n, and B are fitted parameters that depend on the fuel. The ability of these types of equations to predict the onset of knock with sufficient accuracy is unclear. The most extensively tested correlation is that proposed by Douaud and Eyzat:[100]

$$\tau = 17.68\left(\frac{ON}{100}\right)^{3.402} p^{-1.7} \exp\left(\frac{3800}{T}\right) \tag{9.62}$$

where τ is in milliseconds, p is absolute pressure in atmospheres, and T is in kelvin. ON is the appropriate octane number of the fuel (see Sec. 9.6.3). If the temperature and pressure time history of the end-gas during an individual cycle are known, Eqs. (9.60) and (9.62) together can be used to determine whether autoignition occurs before the normally propagating flame consumes the end-gas.

An important question with any model of the end-gas autoignition process is characterizing the end-gas temperature. During intake, the combustion chamber walls are hotter than the entering gases: thus, heat is transferred from the walls to the fresh mixture. During compression, the mixture temperature rises to levels substantially above the wall temperature. A thermal boundary layer will build up adjacent to the wall, as heat is now transferred to the wall. Unburned mixture away from the wall will be compressed essentially adiabatically (see Sec. 12.6.5). In addition, any unburned mixture which for some portion of time during intake and compression has been in close proximity to the exhaust valve and piston (which run at higher temperatures than the water-cooled wall regions) could be at higher temperatures than adiabatically compressed mixture, due to substantial heat transfer to the unburned mixture. Thus, the end-gas temperature is not uniform and the distribution of temperature is extremely complex. Often, the *mean* unburned mixture temperature is used to characterize its state. Alternatively, the *core* temperature corresponding to adiabatic compression of mixture from conditions at the start of compression is used. In the absence of substantial heating by the exhaust valve and piston, the core temperature is a better representation of the maximum unburned mixture at any point in the cycle.

While more complex and complete chemical models of the autoignition process are being developed for simple paraffinic hydrocarbon fuel compounds (e.g., Refs. 91 and 101), no detailed models are yet available for use with real blended fuels in engines. However, a generalized kinetic model for hydrocarbon oxidation based on a degenerate branched-chain mechanism, known as the Shell model, has been developed and tested with some success. The model uses generic chemical entities representative of a variety of individual species which undergo a set of generalized reactions. This is justified by the broadly similar (though complex) ignition behavior of a variety of different fuel molecules and the similar kinetics exhibited by the organic radicals of the same type in the hydrocarbon oxidation process [Eq. (9.59)].[95, 102]

The Shell model is based on a generic eight-step degenerate chain-branching reaction scheme. The scheme involves the fuel (RH), oxygen, radicals formed from the fuel (R), products (P), intermediate product (Q), and degenerate-branching agent (B). The rate constants are either fixed at values consistent with the literature or fitted so that measured induction times (such as those illustrated in Fig. 9-67) are adequately predicted.[102] An example of results obtained with this scheme is shown in Fig. 9-68. It shows the calculated pressure, temperature, and species concentrations in the end-gas region of an operating spark-ignition

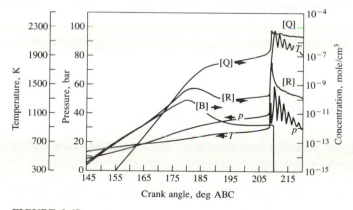

FIGURE 9-68

Pressure, temperature, and composition in the end-gas, before, during, and after autoignition, predicted by the Shell model in spark-ignition engine combustion process. B, Q, R defined in text.[103]

engine, leading up to knock at 209° ABC. In this example, the autoignition model has been incorporated in a multidimensional model of the flow and flame propagation processes within the combustion chamber (see Sec. 14.5).[103]

9.6.3   Fuel Factors

The tendency to knock depends on engine design and operating variables which influence end-gas temperature, pressure, and the time spent at high values of these two properties before flame arrival. Thus, for example, the tendency to knock is decreased through reductions in the end-gas temperature that follow from decreasing the inlet air temperature and retarding the spark from MBT timing. However, knock is a phenomenon that is governed by both engine and fuel factors; its presence or absence in an engine depends primarily on the antiknock quality of the fuel.

Individual hydrocarbon compounds vary enormously in their ability to resist knock, depending on their molecular size and structure. Their tendency to knock has been measured by the *critical compression ratio* of an engine: i.e., the compression ratio at which, under specified operating conditions, the specific fuel compound will exhibit incipient knock. Knocking tendency is related to molecular structure† as follows:[105, 106]

Paraffins
1. Increasing the length of the carbon chain increases the knocking tendency.
2. Compacting the carbon atoms by incorporating side chains (thereby shortening the length of the basic chain) decreases the tendency to knock.
3. Adding methyl groups (CH_3) to the side of the basic carbon chain, in the second from the end or center position, decreases the knocking tendency.

Olefins
4. The introduction of one double bond has little antiknock effect; two or three double bonds generally result in appreciably less knocking tendency.
5. Exceptions to this rule are acetylene (C_2H_2), ethylene (C_2H_4), and propylene (C_3H_6), which knock much more readily than the corresponding saturated hydrocarbons.

Napthenes and aromatics
6. Napthenes have significantly greater knocking tendency than have the corresponding size aromatics.
7. Introducing one double bond has little antiknock effect; two and three double bonds generally reduce knocking tendency appreciably.

---

† See Sec. 3.3 for a review of hydrocarbon structure and its nomenclature. A more extensive discussion is given by Goodger.[104]

8. Lengthening the side chain attached to the basic ring structure increases the knocking tendency in both groups of fuels, whereas branching of the side chain decreases the knocking tendency.

Figure 9-69 identifies the magnitude of these trends on a plot of the critical compression ratio against the number of carbon atoms in the molecule. The strong dependence of knocking tendency on fuel molecular size and structure is apparent.

Practical fuels are blends of a large number of individual hydrocarbon compounds from all the hydrocarbon series of classes: alkanes (paraffins), cyclanes (napthenes), alkenes (olefins), and aromatics (see Sec. 3.3). A practical measure of a fuel's resistance to knock is obviously required.[107, 108] This property is defined by the fuel's *octane number*. It determines whether or not a fuel will knock in a given engine under given operating conditions: the higher the octane number, the higher the resistance to knock. Octane number is not a single-valued quantity, and may vary considerably depending on engine design, operating conditions during test, ambient weather conditions during test, mechanical condition of engine, and type of oil and fuel used in past operation. The octane number (ON) scale is based on two hydrocarbons which define the ends of the scale. By definition, normal heptane (n-C_7H_{16}) has a value of zero and isooctane (C_8H_{18}: 2,2,4-trimethylpentane) has an octane number of 100. These hydrocarbons were chosen because of the great difference in their ability to resist knock and the fact that isooctane had a higher resistance to knock than any of the gasolines available at the time the scale was established. Blends of these two hydrocarbons define the knock resistance of intermediate octane numbers: e.g., a blend of 10 percent n heptane and 90 percent isooctane has an octane number of 90. A fuel's octane number is determined by measuring what blend of these two hydrocarbons matches the fuel's knock resistance.

Several octane rating methods for fuels have been developed. Two of these—the research method (ASTM D-2699)† and the motor method (ASTM D-2700)—are carried out in a standardized single-cylinder engine. In the motor method, the engine operating conditions are more severe; i.e., the conditions are more likely to produce knock. In addition, road octane rating methods have been developed to define the antiknock quality of fuels in cars operated on the road or on chassis dynamometers. The engine used in the ASTM research and motor methods is the single-cylinder engine developed under the auspices of the Cooperative Fuel Research Committee in 1931—the CFR engine.‡ This test engine is a robust four-stroke overhead-valve engine with an 82.6-mm (3.25-in) bore and 114.3-mm (4.5-in) stroke. The compression ratio can be varied from 3 to 30 while

---

† ASTM denotes American Society for Testing and Materials; the letter and number defines the specific testing code.
‡ The Cooperative Fuel Research Committee is now the Coordinating Research Council, Inc.

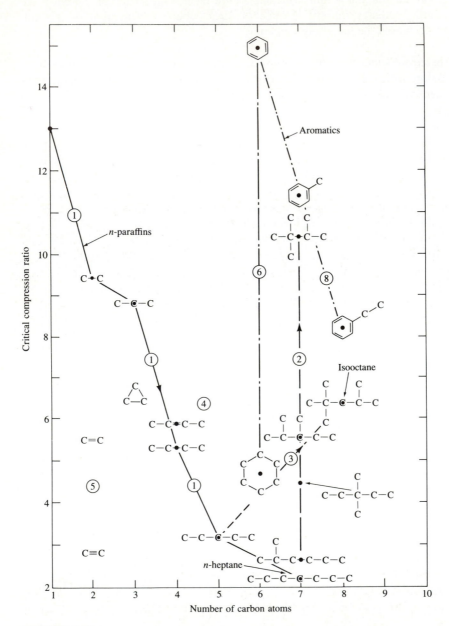

FIGURE 9-69
Critical compression ratio (for incipient knock at 600 rev/min and 450 K coolant temperature) as a function of number of carbon atoms in hydrocarbon molecule, illustrating the effects of changes in molecular structure. (*Developed from Lovell.*[105])

TABLE 9.6
Operating conditions for research and motor methods

	Research method	Motor method
Inlet temperature	52°C (125°F)	149°C (300°F)
Inlet pressure	Atmospheric	
Humidity	0.0036–0.0072 kg/kg dry air	
Coolant temperature	100°C (212°F)	
Engine speed	600 rev/min	900 rev/min
Spark advance	13° BTC (constant)	19–26° BTC (varies with compression ratio)
Air/fuel ratio	Adjusted for maximum knock	

the engine is operating, with a mechanism which raises or lowers the cylinder and cylinder head assembly relative to the crankcase. A special valve mechanism maintains a constant tappet clearance with vertical adjustment of the head. The engine is equipped with multiple-bowl carburetors so two reference fuels (usually blends of *n*-heptane and isooctane) and the fuel being rated can be placed in separate bowls. By means of a selector valve, the engine can be operated on any of the three fuels. The engine operating conditions of the research and motor methods are summarized in Table 9.6. The test conditions are chosen to represent the engine operating range where knock is most severe. With the fuel under test, the fuel/air ratio is adjusted for maximum knock. The compression ratio is then adjusted to produce knock of a standardized intensity, as measured with a magnetostriction knock detector. The level of knock obtained with the test fuel is bracketed by two blends of the reference fuels not more than two octane numbers apart (with one knocking more and one less than the test fuel). The octane number of the gasoline is then obtained by interpolation between the knock-meter scale readings for the two reference fuels and their octane numbers. For fuels below 100 ON, the *primary reference fuels* are blends of isooctane and *n*-heptane; the percent by volume of isooctane in the blend is the octane number. For fuels above 100 ON, the antiknock quality of the fuel is determined in terms of isooctane plus milliliters of the antiknock additive, tetraethyl lead, per U.S. gallon.†

The octane ratings of several individual hydrocarbon compounds and common blended fuels are summarized in App. D, Table D.4. Practically all fuels exhibit a difference between their research and motor octane numbers. The motor method of determining ON uses more severe operating conditions than the

---

† The octane number of the fuel is calculated from $ON = 100 + 28.28T/[1.0 + 0.736T + (1.0 + 1.472T - 0.035216T^2)^{1/2}]$, where T is milliliters of tetraethyl lead per U.S. gallon. Tetraethyl lead, $(C_2H_5)_4Pb$, contains 64.06 weight percent lead; 1 ml of TEL contains 1.06 grams of lead.

research method (higher inlet mixture temperature, more advanced timing). Thus, the motor octane number (MON) is usually lower than the research octane number (RON). The numerical difference between these octane numbers is called the *fuel sensitivity*:

$$\text{Fuel sensitivity} = \text{RON} - \text{MON} \qquad (9.63)$$

Fuel sensitivity varies with the source of crude petroleum and refining processes used. The primary reference fuels themselves (mixtures of isooctane and *n*-heptane), by definition, have the same octane numbers by both the research and motor methods. Since the primary reference fuels are paraffins, we would expect other paraffins to have little or no sensitivity. In contrast, olefins and aromatics have high sensitivity. In general, therefore, straight-run gasolines containing high percentages of saturated hydrocarbons have low sensitivity, while cracked or reformed gasolines containing large percentages of unsaturated hydrocarbons have high sensitivity. Fuels having high sensitivity generally, but not always, have lower road octane ratings (i.e., octane ratings determined in cars in on-the-road use) than do low-sensitivity fuels of the same research octane number. Regular grade unleaded gasoline typically has a RON of at least 91 and a MON of about 83, giving a sensitivity of 8.

Research and motor octane number fuel ratings are made in a single-cylinder engine run at constant speed, wide-open throttle and fixed spark timing. These methods do not always predict how a fuel will behave in an automobile engine operated under a variety of speed, load, and weather conditions. Several methods of rating a gasoline in actual vehicles, either on the road or on chassis dynamometers which duplicate outdoor road conditions, have, therefore, been developed. These methods determine the fuel's *road octane number*.[107] The road ratings of current fuels usually lie between the motor and research ratings. Road octane number can be related to motor and research ratings with equations of the form

$$\text{Road ON} = a\,(\text{RON}) + b\,(\text{MON}) + c$$

where *a*, *b*, and *c* are experimentally derived constants. Recent studies show $a \approx b \approx 0.5$ gives good agreement. An *antiknock index* which is the mean of the research and motor octane numbers is now used in the United States to characterize antiknock quality:

$$\text{Antiknock index} = \frac{\text{RON} + \text{MON}}{2} \qquad (9.64)$$

Refiners and automobile manufacturers are interested in the octane number requirement of engines or vehicles on the road. The *octane number requirement* (usually abbreviated to OR) of an engine or vehicle-engine combination is defined as the minimum fuel octane number that will resist knock throughout the engine's operating speed and load range. The octane number requirement of a single engine or vehicle does not usually provide adequate information for that particular model; every model has a range of requirements due to production tolerances and variations in engine and vehicle condition.

Modern gasolines contain a number of chemical additives designed to improve fuel quality. These additives are used to raise the octane number of the fuel, control surface ignition, reduce spark plug fouling, resist gum formation, prevent rust, reduce carburetor icing, remove carburetor or injector deposits, minimize deposits in the intake system, and prevent valve sticking. The octane number of hydrocarbon fuels can be increased by antiknock agents. Their use generally allows an increase in antiknock quality to be achieved at less expense than modifying the fuel's hydrocarbon composition by refinery processing. The most effective antiknock agents are lead alkyls. Tetraethyl lead (TEL), $(C_2H_5)_4$ Pb, was first introduced in 1923. Tetramethyl lead (TML), $(CH_3)_4Pb$, was introduced in 1960. Since TML boils in the mid-range of a gasoline (110°C), whereas TEL boils at the high end (200°C), the introduction of TML permitted better distribution of octane amongst the cylinders of an engine. In 1959 a manganese antiknock compound (methylcyclopentadienyl manganese tricarbonyl), now known as MMT, was introduced as a supplementary antiknock agent for TEL. It is also an antiknock agent in its own right.

About 1970, in the United States, low-lead and unleaded gasolines were introduced. Two factors influenced the reduction in the use of lead alkyls: concern about the toxicological aspects of lead in the urban environment and the use of catalytic devices for emission control that are poisoned by lead. Unleaded and reduced-lead-content gasolines are now required in the United States. Japan has almost completely converted to unleaded fuel. In Europe, requirements which reduce the lead content of gasolines and introduce unleaded fuel were implemented in the late 1980s. MMT is sometimes used as an antiknock additive in unleaded gasoline. However, its role as a deposit in plugging exhaust catalytic converters limits its use to low concentrations. The expanding use of unleaded fuels has increased interest in other methods of boosting the octane rating of gasolines. The use of oxygenates—alcohols and ethers—as gasoline extenders, which due to their excellent antiknock quality increase the fuel's octane rating, is becoming more common. A brief review of the mechanism by which these additives and compounds improve the knock resistance of gasolines follows.

Lead is the most effective antiknock element known. In the form of lead alkyls, it is stable and fuel-soluble. The precise mechanism by which lead alkyls control knock is not fully known. It is generally agreed that the alkyls decompose before they exert their antiknock action. The decomposed material—lead oxide, PbO—either as a vapor or as a fog (a dispersion of fine particles), inhibits the preflame chain-branching reaction which leads to autoignition of the fuel-air charge, thus slowing the reaction rate. However, lead has little effect on two-stage ignition until after the cool flame.[109] Commercially available tetraalkyl lead antiknock fluids are based on TEL, TML, physical mixtures of TEL and TML, and mixed ethyl-methyl compounds produced by reacting TML and TEL. This range of compounds offers volatilities between the extremes of TEL and TML. The individual alkyls vary in antiknock behavior as a function of fuel composition and combustion conditions. The average effect of various amounts of TEL in a large number of regular gasolines is summarized in Fig. 9-70: the effectiveness of

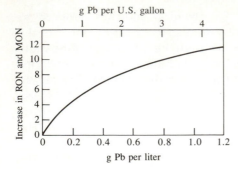

FIGURE 9-70
Gasoline octane number increase resulting from use of antiknock additive tetraethyl lead. Varies with fuel composition: average values shown.

each successive increment added steadily decreases. The addition of about 0.8 g lead per liter (3 g Pb per U.S. gallon, the maximum economic limit to lead concentration) provides an average gain of about 10 octane numbers in modern gasolines, though effectiveness varies with chemical composition of the fuel. TML offers a greater octane number gain than TEL in many gasolines, particularly in highly aromatic fuels with a low sulfur content. One of its major values, however, is in overcoming engine knock that results from fuel component segregation in the intake manifold; gasoline fractions of different volatility separate in the intake manifold of a multicylinder engine and the heavier fractions lag behind (see Sec. 7.6.3). When a gasoline containing a lead alkyl is burned in a spark-ignition engine, it produces nonvolatile combustion products. These deposit on the walls of the combustion chamber and on the spark plug, causing lead-fouling of the spark plug electrodes and tracking across the plug insulator, and hot corrosion of the exhaust valve. Commercial antiknock fluids, therefore, contain scavenging agents—combinations of ethylene dibromide and ethylene dichloride—which transfer the lead oxides which would otherwise deposit into volatile lead-bromine compounds which are largely exhausted with the combustion gases.

Low concentrations of methylcyclopentadienyl manganese tricarbonyl (MMT) act as an octane promoter; it is most effective in highly paraffinic gasolines. It is sometimes used as a supplement to TEL. MMT-TEL antiknock fluids are tailored to optimize octane cost effectiveness and the average fluid contains about 0.03 g Mn/g Pb. The octane gains vary significantly with fuel composition. In unleaded fuels MMT is sometimes used in low concentrations to provide from 0.5 to 1 octane number gain. On a weight of metal basis, MMT is about twice as effective as TEL in terms of research octane number gain and about equally effective in terms of motor octane number gain.[109]

The use of oxygenates (oxygen containing organic compounds) as extenders or substitutes for gasoline is increasing.[110] In some cases, this is because the oxygenate can be produced from nonpetroleum sources (e.g., biomass, coal) and thus may offer strategic or economic benefits. In other cases, the good antiknock blending characteristics of oxygenates can aid in meeting octane quality demands where increasingly stringent regulations limit lead alkyl use. Several oxygenates have been used as automotive fuels; those of major interest are methanol

(CH$_3$OH), ethanol (C$_2$H$_5$OH), tertiary butyl alcohol (TBA) (C$_4$H$_9$OH), and methyl tertiary butyl ether (MTBE).

Table 9.7 lists the antiknock characteristics of these compounds and their physical and chemical characteristics relative to gasoline. The blending value of antiknock index [Eq. (9.64)] given in the table is not necessarily the same as the compound's antiknock index [$(R + M)/2$ from App. D, Table D.4] when used alone as a fuel, and depends on the gasoline composition with which the compound is blended. MTBE-gasoline blends have good water stability, and MTBE has little effect on vapor pressure and material compatibility. TBA is moderately susceptible to water extraction and loss. Ethanol is technically feasible as a high octane supplement or substitute for gasoline; economically it is less attractive than methanol.

Methanol, because it can be made from natural gas, coal, or cellulose materials, has near- and long-term potential. Its high octane quality (130 RON, 95 MON), when used in low-concentration (~ 5 percent) methanol-cosolvent-gasoline blends, can help offset the octane loss from lead alkyl phase-out. Problems with these blends include poor solubility in gasoline in the presence of water; toxicity; an energy content about half that of gasoline; high latent heat of vaporization and oxygen content which contribute to poor driveability; incompatibility with many commonly used metals and elastomers; blending effects on gasoline volatility which may force the displacement of large volumes of butane. Some of these problems can be partially reduced by using cosolvents such as TBA or isobutanol. Use of methanol as a neat fuel in specially designed engines permits advantage to be taken of its high octane rating via high compression ratios. Problems include its energy content of one-half that of gasoline; engine starting problems which require starting aids such as 5 to 10 percent isopentane at temperatures below 10°C or intake system heaters; toxicity; extensive engine modifications.[110]

The octane requirement of an engine-vehicle combination usually increases during use, primarily due to the buildup of combustion chamber deposits within the engine cylinder. While these deposits increase the engine's compression ratio modestly, their largest effect is to increase the temperature of the outer surface of

TABLE 9.7
Oxygenate properties[110]

	Methanol	Ethanol	TBA	MTBE	Gasoline
Typical $(R + M)/2$ blending value	112	110	98	105	87–93
Weight percent oxygen	50	35	22	18	0
Stoichiometric (A/F)	6.5	9.0	11.2	11.7	14.5
Specific gravity	0.796	0.794	0.791	0.746	0.74
Lower heating value, MJ/kg	20.0	26.8	32.5	35.2	44.0
Latent heat of vaporization, MJ/kg	1.16	0.84	0.57	0.34	0.35
Boiling temperature, °C	65	78	83	55	27–227

TBA, tertiary butyl alcohol; MTBE, methyl teriary butyl ether.

the combustion chamber. This *increases* heat transfer *to* the fresh mixture during induction and *decreases* heat transfer from the unburned charge during compression. End-gas temperatures are therefore higher, thus increasing the likelihood of knock. As the combustion chamber deposits stabilize (over 15,000 to 25,000 km of driving), the engine's octane requirement typically increases by about 5 octane numbers; the increase can vary from between 1 to over 13 octane numbers.

Essentially all of the octane requirement increase results from the buildup of deposits on the combustion chamber walls: when the deposits are completely removed from the engine the octane requirement returns to close to its original value. The volume of deposits for leaded and unleaded fuels which build up inside the engine cylinder are similar in magnitude, and are in the range 0.3 to 1 cm^3 per cylinder. The compression ratio increase associated with this volume of deposits is small (0.1 compression ratio) and contributes therefore on the order of 10 percent of the octane requirement increase. The primary effect of the deposits is thought to be changes in heat transfer between the end-gas and the combustion chamber walls, as explained above. In an experiment where the deposits were removed from various regions of the combustion chamber in sequence and the octane requirement after each removal was determined, it was found that the deposits on the cylinder head around the end-gas region were the cause of about two-thirds of the total ORI observed. Though the volume of the deposits for leaded and unleaded fuels are comparable, the composition and density are substantially different. For unleaded fuels the major element is carbon (one-third to one-half by mass). The deposit density with leaded fuels is 2 to 5 g/cm^3; with unleaded fuels it is a factor of 5 lower.[111]

PROBLEMS

9.1. The table gives relevant properties of three different spark-ignition engine fuels. The design and operating parameters of a four-cylinder 1.6-liter displaced volume engine are to be optimized for each fuel over the engine's operating load and speed range. You may assume that for each engine-fuel combination, the gross indicated fuel conversion efficiency and imep at any operating condition are given by 0.8 times the fuel-air cycle efficiency at those conditions. Also, assume that for every five research octane number increase above 95 (the gasoline value) in fuel antiknock rating, the compression ratio can be increased by one unit. For gasoline, the engine compression ratio is 9, so if the octane number of the fuel is 100, a compression ratio of 10 can be used.

   (*a*) At part-throttle operation—at an intake pressure of 0.5 atm and a speed of 2500 rev/min—estimate the gross indicated fuel conversion efficiency and specific fuel consumption for each engine-fuel combination. Each engine-fuel combination operates at the lean limit given.

   (*b*) At the appropriate equivalence ratio for maximum power, with 1 atm inlet pressure at the same speed (2500 rev/min), the volumetric efficiencies are as shown. Explain these volumetric efficiency values. Each fuel is fully vaporized in the inlet manifold and the inlet mixture temperature is held constant for all fuels. Manifold and valve design remains the same.

(c) Estimate the ratios of the maximum indicated mean effective pressure, at the same conditions as in (b), of the natural gas and methanol fueled engines to the maximum imep of the gasoline engine.

(d) If the methane engine burns 33 percent faster than the gasoline engine (i.e., its combustion event takes three-quarters of the time), sketch a *carefully drawn qualitative* cylinder pressure versus crank angle curve for the two engines, from halfway through the compression stroke to halfway through the expansion stroke. Put both curves on the same graph. Conditions are as in (b). Show the motored *and* firing pressure curves for each engine. The spark timing should be set for maximum brake torque for each engine. Show the location of spark timing, the location of maximum cylinder pressure, and the approximate location of the end of combustion.

Fuel properties

Fuel	Formula	Research octane number	ϕ at lean operating limit	Stoichiometric air/fuel ratio	Heating value, MJ/kg	η_v
Natural gas	CH_4	120	0.7	17.2	50	0.78
Methanol	CH_3OH	105	0.8	6.4	20	0.75
Gasoline	$(CH_2)_n$	95	0.9	14.9	44	0.85

ϕ = fuel/air equivalence ratio.

9.2. The figure shows the flame propagating radially outward from the center of a disc-shaped combustion chamber. Combustion in such a device has many features in common with spark-ignition engine combustion. The chamber diameter is 10 cm and the height is 1.5 cm. For this problem, the flame can be thought of as a thin cylindrical sheet. Its radius increases approximately linearly with time. The volume of the chamber is constant. The fuel is a typical hydrocarbon fuel; the mixture is stoichiometric; the initial temperature is room temperature.

(a) Sketch qualitative but *carefully proportioned* graphs of the following quantities versus time from the start of combustion to the end of combustion:
  (1) The ratio of actual pressure in the chamber to the initial pressure
  (2) The ratio of average density of the gas *ahead* of the flame to the initial density
  (3) The ratio of the average density of the gas *behind* the flame to the initial density

(b) On a qualitative but *carefully proportioned* graph of r/R_0 versus time show how the radial positions of gas elements, initially at $r/R_0 = 0, 0.5$, and 1.0 before combustion, change during the combustion process as the flame propagates radially outward from the center ($r = 0$) to the outer wall ($r = R_0$).

Note: Accurate numerical calculations are not required to answer this question. You will be graded on the amount of physical insight your diagrams and the supporting brief explanations of your logic communicate. You should write down any equations or approximate numerical values for relevant quantities that help explain your reasoning.

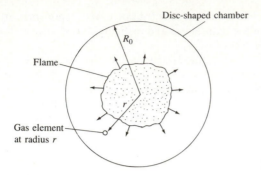

FIGURE P9-2

9.3. Highly compact bowl-in-piston or bowl-in-head combustion chambers permit SI
engine operation at higher-than-normal compression ratios and with leaner mixtures. A vigorous swirling flow in the combustion chamber just prior to combustion
is achieved by these chamber designs. The table compares the characteristics of a
compact-chamber spark-ignition engine (Fig. P9-3) with a more conventional spark-
ignition engine which has an open chamber (see Fig. 9-4).

	Compact chamber 1.5 liter	Conventional 1.5 liter
Compression ratio	14 : 1	9 : 1
Fuel octane requirement	97 RON	97 RON
Maximum bmep	980 kPa	965 kPa
Air/fuel ratio at WOT	17 : 1	13 : 1
MBT timing at 2000 rev/min and WOT	5° BTC	25° BTC
Air/fuel ratio at cruise conditions	22 : 1	16 : 1

For gasoline, the stoichiometric air/fuel ratio is 14.6. RON = research octane
number, WOT = wide-open throttle, MBT = maximum brake torque.

Use the data provided, and any additional quantitative information you have
or can generate easily, to answer the following questions:

(*a*) Explain whether the combustion process rate in the compact-chamber engine
will be faster, slower, or about the same rate as the conventional engine combustion process.

(*b*) The compact-chamber engine has a higher compression ratio than the conventional engine, yet it has about the same maximum bmep at WOT in the mid-
speed range. Explain quantitatively how the differences in compression ratio,
air/fuel ratio, and MBT spark timing at WOT between the two engines influence
maximum bmep and result in negligible total change.

(*c*) Both engines have been designed to have the same fuel octane requirement, yet
the compact-chamber engine has a much higher compression ratio. Explain
which features of the engine allow it to operate without knock at a 14 : 1 compression ratio while the conventional engine can only operate without knock at
9 : 1.

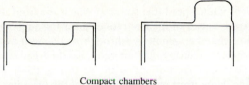

Compact chambers   **FIGURE P9-3**

(d) The efficiency of the compact-chamber engine at part-throttle conditions is higher than that of the conventional engine. Briefly explain why this is the case and estimate approximately the ratio of the two engine efficiencies.

9.4. In a spark-ignition engine, a turbulent flame propagates through the uniform premixed fuel-air mixture within the cylinder and extinguishes at the combustion chamber walls.

(a) Draw a carefully proportioned qualitative graph of cylinder pressure p and mass fraction burned x_b as a function of crank angle θ for $-90° < \theta < 90°$ for a typical SI engine at wide-open throttle with the spark timing adjusted for maximum brake torque. Mark in the crank angles of spark discharge, and of the flame development period (0 to 10 percent burned) and end of combustion, on both p and x_b versus θ curves relative to the top-center crank position.

(b) Estimate approximately the fraction of the cylinder volume occupied by burned gases when the mass fraction burned is 0.5 (i.e., halfway through the burning process).

(c) A simple model for this turbulent flame is shown on the left in Fig. P9-4. The rate of burning of the charge dm_b/dt is given by

$$\frac{dm_b}{dt} = A_f \rho_u S_T$$

where A_f is the area of the flame front (which can be approximated by a portion of a cylinder whose axis is at the spark plug position), ρ_u is the unburned mixture

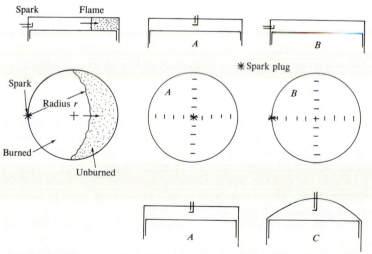

FIGURE P9-4

density, and S_T is the turbulent flame speed (the speed at which the front moves relative to the unburned mixture ahead of it). The rate of mass burning is influenced therefore by combustion chamber geometry (through A_f) as well as those factors that influence S_T (turbulent intensity, fuel/air ratio, residual gas fraction, and EGR). Compare combustion chambers A and B shown in Fig. P9-4. Sketch the *approximate* location of the flame front when 50 percent of the mass has been burned. (A careful qualitative sketch is sufficient; however, provide a quantitative justification for your sketch.) Sketch the mass fraction burned versus crank angle curves for these two combustion chambers on the same graph, each with its spark timing set for maximum brake torque. You may assume the value of S_T is the same for A and B.

(d) Compare combustion chambers A and C in Fig. P9-4 which have the same flame travel distance but have different chamber shapes. Which chamber has

  (1) the faster rate of mass burning during the first half of the combustion process;

  (2) the faster rate of mass burning during the second half of the combustion process;

  (3) the more advanced timing for maximum brake torque?

  Explain your answers.

9.5. A new synthetic fuel with chemical formula $(CH_2O)_n$ is being developed from coal for automotive use. You are making an evaluation of what changes in the spark-ignition engines you produce might be required if gasoline is replaced by this fuel "X." First make the following calculations:

(a) What is the stoichiometric air/fuel ratio for "X"? How does this compare with the stoichiometric air/fuel ratio for gasoline?

(b) Given that in a constant-volume calorimeter experiment to determine the heating value of "X" the combustion of 50 g of fuel with excess air at standard conditions resulted in a temperature rise of 1.25°C of the water and calorimeter (of combined heat capacity 650 kJ/K), what is the heating value of a stoichiometric mixture of "X" and air, *per kilogram of mixture*? How does this value compare with a stoichiometric gasoline-air mixture?

Then,

(c) Compare *approximately* the specific fuel consumption of a spark-ignition engine operated on stoichiometric gasoline and "X" mixtures. What do you conclude about the relative size of the fuel systems required to provide equal power?

(d) "X"-air mixtures take twice as long to burn as gasoline-air mixtures (the crank angle between the spark and end of combustion is twice as large). Sketch carefully drawn pressure-time curves over the entire engine four-stroke cycle for the two mixtures, for the same displacement and compression ratio engines, for the same *imep* (for throttled engine operation) for stoichiometric mixtures, with optimum spark timing, indicating the relative crank angle location of spark, peak cylinder pressure, and end of combustion, and the relative values of intake pressure, peak cylinder pressure, and pressure at the exhaust valve opening, for the two mixtures at the same equivalence ratio. (You do not need to make calculations to sketch these graphs.)

(e) Though "X"-air mixtures are slower burning than gasoline-air mixtures, the engine compression ratio can be increased to 14 : 1 from the 8 : 1 values typical for gasoline. Using typical fuel-air cycle efficiencies and the relation between the fuel-air cycle and real cycle efficiency, determine whether the slower-burning higher-compression-ratio engine using "X" will be more efficient, have about the same efficiency, or be less efficient than the faster-burning lower-compression-ratio gasoline engine.

9.6. Knock in spark-ignition engines is an abnormal combustion condition. Almost everyone who rides a motorcycle or drives a car experiences this phenomenon at some time and usually changing into a lower gear will take the engine away from this condition. Use your experience of what changes in other variables do, and consult this and other textbooks to complete a table with the dependent variables shown at the top of the columns.

The independent variables are: speed, compression ratio, chamber surface/volume ratio, spark plug distance from cylinder axis, percent EGR, inlet mixture temperature, inlet mixture pressure, (F/A), wall temperature, air swirl, squish motion, fuel octane number. In these columns show the corresponding influence on the dependent variables by a " + " for an increase and a " − " for a decrease. Show the effect of increase in engine system independent variables on: cylinder pressure and temperature, flame speed, total burn time, autoignition induction period, tendency to knock. Provide in the extreme right-hand column brief comments to explain your answers.

Effect of increase in engine independent variable	Dependent variables						
	Cylinder		Flame speed	Total burn time	Induction period	Tendency to knock	Explanation
	Pressure	Temperature					
Speed, rev/min							
Etc.							

9.7. The attached graph gives the pressure–crank angle curve for a spark-ignition engine running at $\phi = 1.0$. The mass fraction burned x_b is also shown. Estimate the tem-

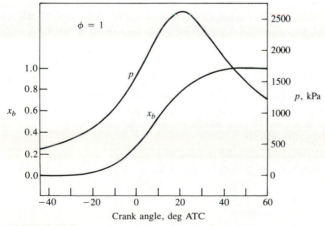

FIGURE P9-7

perature of the reactants at a number of crank angles and plot a graph of T_u versus θ. Assume the reactants in the cylinder are at 333 K and 1 atm pressure at the start of compression. It is necessary to make simplifications in order to do this and you should explain clearly what other assumptions you make; they should be compatible with Prob. 9.8 below.

9.8. If the combustion takes place progressively through a large number of very small zones of gas and there is no mixing between the zones, determine:

(a) the temperature at $-30°$, just after combustion, of the zone which burns at $-30°$;

(b) the temperature at $0°$, just after combustion, of the zone which burns at $0°$;

(c) the temperature at $+30°$, just after combustion of the zone which burns at $+30°$;

(d) the temperature of the products in these three zones at $+30°$.

Plot your results versus crank angle to show whether there is a spatial distribution of temperature in the cylinder after combustion.

Note: Each small unburned gas element burns at essentially constant pressure and is subsequently compressed and/or expanded. Use charts (Chap. 4) or an equilibrium computer code. The unburned gas state is given by Prob. 9.7.

9.9. An approximate way to calculate the pressure in the end-gas just after knock occurs is to assume that all the end-gas (the unburned gas ahead of the flame) burns instantaneously at constant volume. We assume that the inertia of the burned gases prevents significant gas motion while the end-gas autoignites.

For the pressure data in Fig. P9-7, assume autoignition occurs at 10 crank angle degrees after the top-center position. Determine the maximum pressure reached in the end-gas after knock occurs. From the mass fraction burned and approximate average burned gas conditions at this time, estimate the volume occupied by the end-gas as a fraction of the cylinder volume just before autoignition occurs. Use the charts (Chap. 4) or an equilibrium computer code.

9.10. At spark timing ($30°$ BTC) in an automobile spark-ignition engine (with bore = stroke = 85 mm and $r_c = 9$) at 2000 rev/min, operating on gasoline, the cylinder pressure is 7.5 atm and the mixture temperature is 650 K. The fuel-air mixture is stoichiometric with a residual gas fraction of 8 percent. The rapid burning angle $\Delta\theta_b$ (10 to 90 percent mass burned) is $35°$. Estimate (a) the mean piston speed; (b) the average flame travel speed based on $\Delta\theta_b$ (the spark plug is located 15 mm from the cylinder axis); (c) the turbulence intensity at TC [see Eq. (8.23)]; (d) the laminar flame speed at spark; (e) the turbulent burning speed S_b at TC using Fig. 9-30 (appropriate firing and motored pressures are given in Fig. 9-2a); (f) the mean expansion speed of the burned gases u_b at TC. Discuss briefly the relative magnitudes of these velocities.

9.11. The following combustion chamber design changes increase the mass burning rate in a spark-ignition engine at fixed compression ratio, bore, speed, and inlet mixture conditions. Explain how each change affects the burning rate.

(a) Reducing the amount of EGR.

(b) Using two spark plugs per cylinder instead of one.

(c) Generating swirl within the cylinder using a mask on the cylinder head (see Fig. 8-14).

(d) Using a combustion chamber with higher clearance height near the spark plug and a more central plug location.

9.12. Explain why SI engine torque varies, at fixed speed and inlet mixture conditions, as the spark timing is varied from very advanced (e.g., 60° before TC) to close to TC. What is the "best" spark timing? Explain how it varies with engine speed and load.

9.13. (a) Explain the causes of the observed variation in cylinder pressure versus crank angle and imep in spark-ignition engines, cycle-by-cycle.

(b) What impacts do these cylinder pressure variations have on engine operation?

9.14. (a) Describe briefly what occurs when a spark-ignition engine "knocks."

(b) SI engine knock is primarily a problem at wide-open throttle and lower engine speeds. Explain why this is the case.

(c) With a knock sensor, the normal knock control strategy is to retard the spark timing when knock is detected, until knock no longer occurs. Explain why this strategy is effective and is preferred over other possible approaches (e.g., throttling the inlet, adding EGR).

(d) In a knocking engine, the crank angle at which autoignition occurs and the magnitude of the pressure oscillations which result vary substantially, cycle-by-cycle. Suggest reasons why this happens.

9.15. (a) The electrical energy stored in a typical ignition system coil is 50 mJ. Almost all this energy is transferred from the coil during the glow discharge phase. If the glow discharge lasts for 2 ms, use the data in Fig. 9-39 to estimate the glow discharge voltage and current.

(b) Only a fraction of this energy is transferred to the fuel-air mixture between the spark plug electrodes. Estimate the energy transferred during the breakdown and glow phases of the discharge, using the data in Fig. 9-40.

(c) Overall about one-tenth of the coil energy is transferred to the fuel-air mixture. What fraction of the cylinder contents' chemical energy ($m_f Q_{LHV}$) does this correspond to at a typical part-load condition ($p_i = 0.5$ atm, $\phi = 1.0$)? Assume 500 cm^3 per cylinder displaced volume. If the average burned gas temperature within the flame kernal just after spark is 3500 K and the cylinder pressure is 6 atm, what radius of kernal has fuel chemical energy equal to the electrical energy transferred to the kernal?

REFERENCES

1. Nakamura, H., Ohinouye, T., Hori, K. Kiyota, Y., Nakagami, T., Akishino, K., and Tsukamoto, Y.: "Development of a New Combustion System (MCA-JET) in Gasoline Engine," SAE paper 780007, *SAE Trans.*, vol. 87, 1978.
2. Rassweiler, G. M., and Withrow, L.: "Motion Pictures of Engine Flames Correlated with Pressure Cards," *SAE Trans.*, vol. 83, pp. 185–204, 1938. Reissued as SAE paper 800131, 1980.
3. Nakanishi, K., Hirano, T., Inoue, T., and Ohigashi, S.: "The Effects of Charge Dilution on Combustion and Its Improvement—Flame Photograph Study," SAE paper 750054, *SAE Trans.*, vol. 84, 1975.
4. Beretta, G. P., Rashidi, M., and Keck, J. C.: "Turbulent Flame Propagation and Combustion in Spark Ignition Engines," *Combust. Flame*, vol. 52, pp. 217–245, 1983.
5. Amann, C. A.: "Cylinder-Pressure Measurement and Its Use in Engine Research," SAE paper 852067, 1985.
6. Lavoie, G. A., Heywood, J. B., and Keck, J. C.: "Experimental and Theoretical Investigation of Nitric Oxide Formation in Internal Combustion Engines," *Combust. Sci. Technol.*, vol. 1, pp. 313–326, 1970.
7. Rassweiler, G. M., and Withrow, L.: "Flame Temperatures Vary with Knock and Combustion-Chamber Position," *SAE Trans.*, vol. 36, pp. 125–133, 1935.

8. Lavoie, G. A.: "Spectroscopic Measurement of Nitric Oxide in Spark-Ignition Engines," *Combust. Flame*, vol. 15, pp. 97–108, 1970.

9. Heywood, J. B., and Keck, J. C.: "Formation of Hydrocarbons and Oxides of Nitrogen in Automobile Engines," *Env. Sci. Technol.*, vol. 7, pp. 216–223, 1973.

10. Kreiger, R. B., and Borman, G. L.: "The Computation of Apparent Heat Release for Internal Combustion Engines," ASME paper 66-WA/DGP-4, 1966.

11. LoRusso, J. A., and Tabaczynski, R. J.: "Combustion and Emissions Characteristics of Methanol, Methanol-Water, and Gasoline-Methanol Blends in a Spark-Ignition Engine," SAE paper 769019, *Proceedings of Eleventh Intersociety Energy Conversion Engineering Conference*, Lake Tahoe, Nev., pp. 122–132, Sept. 12–17, 1976.

12. Lancaster, D. R., Kreiger, R. B., and Lienesch, J. H.: "Measurement and Analysis of Engine Pressure Data," SAE paper 750026, *SAE Trans.*, vol. 84, 1975.

13. Caris, D. F., and Nelson, E. E.: "A New Look at High Compression Engines," *SAE Trans.*, vol. 67, pp. 112–124, 1959.

14. Gatowski, J. A., Balles, E. N., Chun, K. M., Nelson, F. E., Ekchian, J. A., and Heywood, J. B.: "Heat Release Analysis of Engine Pressure Data," SAE paper 841359, *SAE Trans.*, vol. 93, 1984.

15. Chun, K. M., and Heywood, J. B.: "Estimating Heat-Release and Mass-of-Mixture Burned from Spark-Ignition Engine Pressure Data," *Combust. Sci. Technol.*, vol. 54, pp. 133–144, 1987.

16. Heywood, J. B., Higgins, J. M., Watts, P. A., and Tabaczynski, R. J.: "Development and Use of a Cycle Simulation to Predict SI Engine Efficiency and NO_x Emissions," SAE paper 790291, 1979.

17. Gatowski, J. A., Heywood, J. B., and Deleplace, C.: "Flame Photographs in a Spark-Ignition Engine," *Combust. Flame*, vol. 56, pp. 71–81, 1984.

18. Witze, P. O.: "The Effect of Spark Location on Combustion in a Variable-Swirl Engine," SAE paper 820044, *SAE Trans.*, vol. 91, 1982.

19. Taylor, C. F.: *The Internal Combustion Engine in Theory and Practice*, vol. II, *Combustion, Fuels, Materials, Design*, MIT Press, 1968.

20. Hires, S. D., Tabaczynski, R. J., and Novak, J. M.: "The Prediction of Ignition Delay and Combustion Intervals for a Homogeneous Charge, Spark Ignition Engine," SAE paper 780232, *SAE Trans.*, vol. 87, 1978.

21. Heywood, J. B.: "Combustion Chamber Design for Optimum Spark-Ignition Engine Performance," *Int. J. Vehicle Design*, vol. 5, no. 3, pp. 336–357, 1984.

22. Abraham, J., Williams, F. A., and Bracco, F. V.: "A Discussion of Turbulent Flame Structure in Premixed Charges," SAE paper 850345, in *Engine Combustion Analysis: New Approaches*, P-156, 1985.

23. Namazian, M., Hansen, S. P., Lyford-Pike, E. J., Sanchez-Barsse, J., Heywood, J. B., and Rife, J.: "Schlieren Visualization of the Flow and Density Fields in the Cylinder of a Spark-Ignition Engine," SAE paper 800044, *SAE Trans.*, vol. 89, 1980.

24. Keck, J. C.: "Turbulent Flame Structure and Speed in Spark-Ignition Engines," *Proceedings of the Nineteenth International Symposium on Combustion*, The Combustion Institute, pp. 1451–1466, 1982.

25. Heywood, J. B., and Vilchis, F. R.: "Comparison of Flame Development in a Spark-Ignition Engine Fueled with Propane and Hydrogen," *Combust. Sci. Technol.*, vol. 38, pp. 313–324, 1984.

26. Smith, J. R.: "Turbulent Flame Structure in a Homogeneous-Charge Engine," SAE paper 820043, *SAE Trans.*, vol. 91, 1982.

27. Smith, J. R.: "The Influence of Turbulence on Flame Structure in an Engine," in T. Uzkan (ed.), *Flows in Internal Combustion Engines*, pp. 67–72, ASME, New York, 1982.

28. Rallis, C. J., and Garforth, A. M.: "The Determination of Laminar Burning Velocity," *Prog. Energy Combust. Sci.*, vol. 6, pp. 303–330, 1980.

29. Metghalchi, M., and Keck, J. C.: "Laminar Burning Velocity of Propane-Air Mixtures at High Temperature and Pressure," *Combust. Flame*, vol. 38, pp. 143–154, 1980.

30. Metghalchi, M., and Keck, J. C.: "Burning Velocities of Mixtures of Air with Methanol, Isooctane, and Indolene at High Pressure and Temperature," *Combust. Flame*, vol. 48, pp. 191–210, 1982.

31. Milton, B. E., and Keck, J. C.: "Laminar Burning Velocities in Stoichiometric Hydrogen and

Hydrogen-Hydrocarbon Gas Mixtures," *Combust. Flame*, vol. 58, pp. 13–22, 1984.

32. Rhodes, D. B., and Keck, J. C.: "Laminar Burning Speed Measurements of Indolene-Air-Diluent Mixtures at High Pressures and Temperature," SAE paper 850047, 1985.

33. Metghalchi, M.: "Laminar Burning Velocity of Isooctane-Air, Methane-Air, and Methanol-Air Mixtures at High Temperature and Pressure," M.S. Thesis, Department of Mechanical Engineering, MIT, 1976.

34. Daneshyar, H., Mendes-Lopes, J. M. C., Ludford, G. S. S., and Tromans, P. S.: "The Influence of Straining on a Premixed Flame and Its Relevance to Combustion in SI Engines," paper C50/83, in *Combustion in Engineering*, vol. 1, pp. 191–199, Proceedings of an International Conference of the Institution of Mechanical Engineers, Mechanical Engineering Publications Ltd., London, 1983.

35. Groff, E. G., and Matekunas, F. A.: "The Nature of Turbulent Flame Propagation in a Homogeneous Spark-Ignited Engine," SAE paper 800133, *SAE Trans.*, vol. 89, 1980.

36. Witze, P. O., and Mendes-Lopes, J. M. C.: "Direct Measurement of the Turbulent Burning Velocity in a Homogeneous-Charge Engine," SAE paper 851531, 1986.

37. Hall, M. J., Bracco, F. V., and Santavicca, D. A.: "Cycle-Resolved Velocity and Turbulence Measurements in an IC Engine with Combustion," SAE paper 860320, 1986.

38. Matekunas, F. A.: "Modes and Measures of Cyclic Combustion Variability," SAE paper 830337, *SAE Trans.*, vol. 92, 1983.

39. Young, M. B.: "Cyclic Dispersion in the Homogeneous-Charge Spark-Ignition Engine—A Literature Survey," SAE paper 810020, *SAE Trans.*, vol. 90, 1981.

40. Young, M. B.: "Cyclic Dispersion—Some Quantitative Cause and Effect Relationships," SAE paper 800459, 1980.

41. Patterson, D. J.: "Cylinder Pressure Variations Present a Fundamental Combustion Problem," SAE paper 660129, *SAE Trans.*, vol. 75, 1966.

42. Fisher, R. V., and Macey, J. P.: "Digital Data Acquisition with Emphasis on Measuring Pressure Synchronously with Crank Angle," SAE paper 750028, 1975.

43. Douaud, A., and Eyzat, P.: "DIGITAP—An On-Line Acquisition and Processing System for Instantaneous Engine Data—Applications," SAE paper 770218, 1977.

44. Peters, B. D.: "Mass Burning Rates in a Spark Ignition Engine Operating in the Partial-Burn Regime," paper C92/79, Conference Proceedings on *Fuel Economy and Emissions of Lean Burn Engines*, Institution of Mechanical Engineers, London, June 12–14, 1979.

45. Hansel, J. G.: "Lean Automotive Engine Operation—Hydrocarbon Exhaust Emissions and Combustion Characteristics," SAE 710164, *SAE Trans.*, vol. 80, 1971.

46. Yu, H. T. C.: "Fuel Distribution Studies—A New Look at an Old Problem," *SAE Trans.*, vol. 71, pp. 596–613, 1963.

47. Nakagawa, Y., Nakai, M., and Hamai, K.: "A Study of the Relationship between Cycle-to-Cycle Variations of Combustion and Heat Release Delay in a Spark-Ignited Engine," *Bull. JSME*, vol. 25, no. 199, pp. 54–60, 1982.

48. Matsui, K., Tanaka, T., and Ohigashi, S.: "Measurement of Local Mixture Strength at Spark Gap of S.I. Engines," SAE paper 790483, *SAE Trans.*, vol. 88, 1979.

49. Kuroda, H., Nakajima, Y., Sugihara, K., Takagi, Y., and Muranaka, S.: "The Fast Burn with Heavy EGR, New Approach for Low NO_x and Improved Fuel Economy," SAE 780006, *SAE Trans.*, vol. 87, 1978.

50. Quader, A. A.: "What Limits Lean Operation in Spark Ignition Engines—Flame Initiation or Propagation?," SAE paper 760760, *SAE Trans.*, vol. 85, 1976.

51. Quader, A. A.: "Lean Combustion and the Misfire Limit in Spark Ignition Engines," SAE paper 741055, *SAE Trans.*, vol. 83, 1974.

52. Maly, R., and Vogel, M.: "Ignition and Propagation of Flame Fronts in Lean CH_4-Air Mixtures by the Three Modes of the Ignition Spark," *Proceedings of Seventeenth International Symposium on Combustion*, pp. 821–831, The Combustion Institute, 1976.

53. Maly, R.: "Spark Ignition: Its Physics and Effect on the Internal Combustion Process," in J. C. Hilliard and G. S. Springer (eds.), *Fuel Economy in Road Vehicles Powered by Spark Ignition Engines*, chap. 3, Plenum Press, New York, 1984.

54. Bosch: *Automotive Handbook*, Robert Bosch GmbH, Stuttgart, West Germany, 1976.
55. Maly, R., Saggau, B., Wagner, E., and Ziegler, G.: "Prospects of Ignition Enhancement," SAE paper 830478, *SAE Trans.*, vol. 92, 1983.
56. Maly, R.: "Ignition Model for Spark Discharges and the Early Phase of Flame Front Growth," *Proceedings of Eighteenth International Symposium on Combustion*, pp. 1747–1754, The Combustion Institute, 1981.
57. Refael, S., and Sher, E.: "A Theoretical Study of the Ignition of a Reactive Medium by Means of an Electrical Discharge," *Combust. Flame*, vol. 59, pp. 17–30, 1985.
58. Ballal, D. R., and Lefebvre, A. H.: "Influence of Flow Parameters on Minimum Ignition Energy and Quenching Distance," *Proceedings of Fifteenth International Symposium on Combustion*, pp. 1473–1481, The Combustion Institute, 1974.
59. Kono, M., Linuma, K., Kumagai, S., and Sakai, T.: "Spark Discharge Characteristics and Igniting Ability of Capacitor Discharge Ignition Systems," *Combust. Sci. Technol.*, vol. 19, pp. 13–18, 1978.
60. Rittmannsberger, N.: "Developments in High-Voltage Generation and Ignition Control," paper C106/72, in the Automobile Division Conference on *Automotive Electric Equipment*, Institution of Mechanical Engineers, London, 1972.
61. Huntzinger, G. O., and Rigsby, G. E.: "HEI—A New Ignition System Through New Technology," SAE paper 750346, *SAE Trans.*, vol. 84, 1975.
62. Dale, J. D., and Oppenheim, A. K.: "Enhanced Ignition for I.C. Engines with Premixed Gases," SAE paper 810146, *SAE Trans.*, vol. 90, 1981.
63. Tanuma, T., Sasaki, K., Kaneko, T., and Kawasaki, H.: "Ignition, Combustion, and Exhaust Emissions of Lean Mixtures in Automotive Spark Ignition Engines," SAE paper 710159, *SAE Trans.*, vol. 80, 1971.
64. Nakamura, N., Baika, T., and Shibata, Y.: "Multipoint Spark Ignition for Lean Combustion," SAE paper 852092, 1985.
65. Schwartz, H.: "Ignition Systems for Lean Burn Engines," paper C95/79, *Proceedings of Conference on Fuel Economy and Emissions of Lean Burn Engines*, pp. 87–96, Institution of Mechanical Engineers, London, June 1979.
66. Ziegler, G. F. W., Wagner, E. P., Saggau, B., and Maly, R. R.: "Influence of a Breakdown Ignition System on Performance and Emission Characteristics," SAE paper 840992, *SAE Trans.*, vol. 93, 1984.
67. Anderson, R. W., and Asik, J. R.: "Lean Air-Fuel Ignition System Comparison in a Fast-Burn Engine," SAE paper 850076, 1985.
68. Edwards, C. F., Oppenheim, A. K., and Dale, J. D.: "A Comparative Study of Plasma Ignition Systems," SAE paper 830479, 1983.
69. Noguchi, M., Sanda, S., and Nakamura, N.: "Development of Toyota Lean Burn Engine," SAE paper 760757, *SAE Trans.*, vol. 85, 1976.
70. Brandstetter, W. R., Decker, G., and Reichel, K.: "The Water-Cooled Volkswagen PCI-Stratified Charge Engine," SAE paper 750869, *SAE Trans.*, vol. 84, 1975.
71. Date, T., and Yagi, S.: "Research and Development of the Honda CVCC Engine," SAE paper 740605, 1974.
72. Konishi, M., Nakamura, N., Oono, E., Baika, T., and Sanda, S.: "Effects of a Prechamber on NO_x Formation Process in the SI Engine," SAE paper 790389, 1979.
73. Turkish, M. C.: "Prechamber and Valve Gear Design for 3-Valve Stratified Charge Engines," SAE paper 751004, *SAE Trans.*, vol. 84, 1975.
74. Ethyl Corporation: "Engine Combustion Noises," Ethyl technical note PCDTN-MS 117768 Rev. 774.
75. Wheeler, R. W.: "Abnormal Combustion Effects on Economy," in J. C. Hilliard and G. S. Springer (eds.), *Fuel Economy in Road Vehicles Powered by Spark-Ignition Engines*, chap. 6, pp. 225–276, Plenum Press, 1984.
76. Guibet, J. C., and Duval, A.: "New Aspects of Preignition in European Automotive Engines," SAE paper 720114, *SAE Trans.*, vol. 81, 1972.
77. Douaud, A., and Eyzat, P.: "DIGITAP—An On-Line Acquisition and Processing System for Instantaneous Engine Data—Applications," SAE paper 770218, 1977.

78. Lee, W., and Schaefer, H. J.: "Analysis of Local Pressures, Surface Temperatures and Engine Damages under Knock Conditions," SAE paper 830508, *SAE Trans.*, vol. 92, 1983.
79. Leppard, W. R.: "Individual-Cylinder Knock Occurrence and Intensity in Multicylinder Engines," SAE paper 820074, 1982.
80. Nakajima, Y., Onoda, M., Nagai, T., and Yoneda, K.: "Consideration for Evaluating Knock Intensity," *JSAE Rev.*, vol. 9, pp. 27–35, 1982.
81. Benson, G., Fletcher, E. A., Murphy, T. E., and Scherrer, H. C.: "Knock (Detonation) Control by Engine Combustion Chamber Shape," SAE paper 830509, 1983.
82. Cornetti, G. M., DeCristofaro, F., and Gozzelino, R.: "Engine Failure and High Speed Knock," SAE paper 770147.
83. Renault, F.: "A New Technique to Detect and Control Knock Damage," SAE paper 820073, 1982.
84. *The Relationship between Knock and Engine Damage*, Tentative Code of Practice, CEC Report M-07-T-83, Co-ordinating European Council (CEC), London, England, 1984.
85. Withrow, L., and Rassweiler, G. M.: "Slow-Motion Shows Knocking and Non-Knocking Explosions," *SAE J.*, vol. 39, no. 2, pp. 297–303, 312, 1936.
86. Miller, C. D., Olsen, H. L., Logan, W. O., and Osterstrom, G. E.: NACA report 857, 1946.
87. Male, T.: "Photography at 500,000 Frames per Second of Combustion and Detonation in a Reciprocating Engine," *Proceedings of Third Symposium on Combustion, Flame and Explosion Phenomena*, p. 721, Williams and Wilkins, Combustion Institute, Pittsburgh, 1949.
88. Nakajima, Y., Nagai, T., Iijima, T., Yokoyama, J., and Nakamura, K.: "Analysis of Combustion Patterns Effective in Improving Anti-Knock Performance of a Spark-Ignition Engine," *JSAE Rev.*, vol. 13, pp. 9–17, 1984.
89. Nakagawa, Y., Takagi, Y., Itoh, T., and Iijima, T.: "Laser Shadowgraphic Analysis of Knocking in S.I. Engine," SAE paper 845001; also in *XX FISITA Congress Proceedings*, vol. P-143, 1984.
90. Hayashi, T., Taki, M., Kojima, S., and Kondo, T.: "Photographic Observation of Knock with a Rapid Compression and Expansion Machine," SAE paper 841336, *SAE Trans.*, vol. 93, 1984.
91. Smith, J. R., Green, R. M., Westbrook, C. K., and Pitz, W. J.: "An Experimental and Modeling Study of Engine Knock," *Proceedings of Twentieth International Symposium on Combustion*, pp. 91–100, The Combustion Institute, 1984.
92. Barnard, J. A.: *Flame and Combustion*, 2d ed., Chapman and Hall, 1985.
93. Glassman, I.: *Combustion*, Academic Press, 1977.
94. Downes, D.: "Chemical and Physical Studies of Engine Knock," in *Six Lectures on the Basic Combustion Processes*, pp. 127–155, Ethyl Corporation, Detroit, Mich., 1954.
95. Halstead, M. P., Kirsch, L. J., Prothero, A., and Quinn, C. P.: "A Mathematical Model for Hydrocarbon Autoignition at High Pressures," *Proc. R. Soc.*, ser. A, vol. 346, pp. 515–538, 1975.
96. Benson, S. W.: "The Kinetics and Thermochemistry of Chemical Oxidation with Application to Combustion and Flames," *Prog. Energy Combust. Sci.*, vol. 7, pp. 125–134, 1981.
97. Alperstein, M., and Bradow, R. L.: "Investigations into the Composition of End Gases from Otto Cycle Engines," SAE paper 660410, *SAE Trans.*, vol. 75, 1966.
98. Livengood, J. C., and Wu, P. C.: "Correlation of Autoignition Phenomenon in Internal Combustion Engines and Rapid Compression Machines," *Proceedings of Fifth International Symposium on Combustion*, p. 347, Reinhold, 1955.
99. By, A., Kempinski, B., and Rife, J. M.: "Knock in Spark Ignition Engines," SAE paper 810147, 1981.
100. Douaud, A. M., and Eyzat, P.: "Four-Octane-Number Method for Predicting the Anti-Knock Behavior of Fuels and Engines," SAE paper 780080, *SAE Trans.*, vol. 87, 1978.
101. Cox, R. A., and Cole, J. A.: "Chemical Aspects of the Autoignition of Hydrocarbon-Air Mixtures," *Combust. Flame*, vol. 60, pp. 109–123, 1985.
102. Halstead, M. P., Kirsch, L. J., and Quinn, C. P.: "Autoignition of Hydrocarbon Fuels at High Temperatures and Pressures—Fitting of a Mathematical Model," *Combust. Flame*, vol. 30, pp. 45–60, 1977.
103. Schäpertöns, H., and Lee, W.: "Multidimensional Modelling of Knocking Combustion in SI Engines," SAE paper 850502, 1985.
104. Goodger, E. M.: *Hydrocarbon Fuels*, Macmillan, London, 1975.

105. Lovell, W. G.: "Knocking Characteristics of Hydrocarbons," *Ind. Engng. Chem.*, vol. 40, pp. 2388–2438, 1948.
106. Lichty, L. C.: *Combustion Engine Processes*, McGraw-Hill, 1967.
107. Ethyl Corporation: "Determining Road Octane Numbers," Ethyl technical note PCDTN SP-347 (113) Rev. 573.
108. Benson, J.: "What Good are Octanes," in *Chemtech*, pp. 16–22, American Chemical Society, January 1976.
109. Barusch, M. R., Macpherson, J. B., and Amberg, G. H.: "Additives, Engine Fuel," in J. J. McKetta and W. A. Cunningham (eds.), *Encyclopedia of Chemical Processing and Design*, vol. 2, pp. 1–77, Marcel Dekker, New York and Basel, 1977.
110. McCabe, L. J., Fitch, F. B., and Lowther, H. V.: "Future Trends in Automotive Fuels and Engine Oils," SAE paper 830935, in *Proceedings of Second International Pacific Conference on Automotive Engineering*, pp. 678–697, Tokyo, Japan, Nov. 7–10, 1983.
111. Benson, J. D.: "Some Factors Which Affect Octane Requirement Increase," SAE paper 750933, *SAE Trans.*, vol. 84, 1975.

CHAPTER 10

COMBUSTION IN COMPRESSION-IGNITION ENGINES

10.1 ESSENTIAL FEATURES OF PROCESS

The essential features of the compression-ignition or diesel engine combustion process can be summarized as follows. Fuel is injected by the fuel-injection system into the engine cylinder toward the end of the compression stroke, just before the desired start of combustion. Figures 1-17, 1-18, and 1-19 illustrate the major components of common diesel fuel-injection systems. The liquid fuel, usually injected at high velocity as one or more jets through small orifices or nozzles in the injector tip, atomizes into small drops and penetrates into the combustion chamber. The fuel vaporizes and mixes with the high-temperature high-pressure cylinder air. Since the air temperature and pressure are above the fuel's ignition point, spontaneous ignition of portions of the already-mixed fuel and air occurs after a delay period of a few crank angle degrees. The cylinder pressure increases as combustion of the fuel-air mixture occurs. The consequent compression of the unburned portion of the charge shortens the delay before ignition for the fuel and air which has mixed to within combustible limits, which then burns rapidly. It also reduces the evaporation time of the remaining liquid fuel. Injection continues until the desired amount of fuel has entered the cylinder. Atomization, vaporization, fuel-air mixing, and combustion continue until essentially all the fuel has passed through each process. In addition, mixing of the air remaining in the cylinder with burning and already burned gases continues throughout the combustion and expansion processes.

It will be clear from this summary that the compression-ignition combustion process is extremely complex. The details of the process depend on the characteristics of the fuel, on the design of the engine's combustion chamber and fuel-injection system, and on the engine's operating conditions. It is an unsteady, heterogeneous, three-dimensional combustion process. While an adequate conceptual understanding of diesel engine combustion has been developed, to date an ability to describe many of the critical individual processes in a quantitative manner is lacking.

Some important consequences of this combustion process on engine operation are the following:

1. Since injection commences just before combustion starts, there is no knock limit as in the spark-ignition engine resulting from spontaneous ignition of the premixed fuel and air in the end-gas. Hence a higher engine compression ratio can be used in the compression-ignition engine, improving its fuel conversion efficiency relative to the SI engine.
2. Since injection timing is used to control combustion timing, the delay period between the start of injection and start of combustion must be kept short (and reproducible). A short delay is also needed to hold the maximum cylinder gas pressure below the maximum the engine can tolerate. Thus, the spontaneous ignition characteristics of the fuel-air mixture must be held within a specified range. This is done by requiring that diesel fuel have a cetane number (a measure of the ease of ignition of that fuel in a typical diesel environment; see Sec. 10.6.2) above a certain value.
3. Since engine torque is varied by varying the amount of fuel injected per cycle with the engine air flow essentially unchanged, the engine can be operated unthrottled. Thus, pumping work requirements are low, improving part-load mechanical efficiency relative to the SI engine.
4. As the amount of fuel injected per cycle is increased, problems with air utilization during combustion lead to the formation of excessive amounts of soot which cannot be burned up prior to exhaust. This excessive soot or black smoke in the exhaust constrains the fuel/air ratio at maximum engine power to values 20 percent (or more) lean of stoichiometric. Hence, the maximum indicated mean effective pressure (in a naturally aspirated engine) is lower than values for an equivalent spark-ignition engine.
5. Because the diesel always operates with lean fuel/air ratios (and at part load with very lean fuel/air ratios), the effective value of γ ($=c_p/c_v$) over the expansion process is higher than in a spark-ignition engine. This gives a higher fuel conversion efficiency than the spark-ignition engine, for a given expansion ratio (see Sec. 5.5.3).

The major problem in diesel combustion chamber design is achieving sufficiently rapid mixing between the injected fuel and the air in the cylinder to complete combustion in the appropriate crank angle interval close to top-center. The foregoing discussion indicates (and more detailed analysis will confirm) that mixing rates control the fuel burning rate. Commercial diesel engines are made with a very large range of cylinder sizes, with cylinder bores varying from about 70 to 900 mm. The mean piston speed at maximum rated power is approximately constant over this size range (see Sec. 2.2), so the maximum rated *engine* speed will be inversely proportional to the stroke [see Eq. (2.9)]. For a fixed crank angle interval for combustion (of order 40 to 50° to maintain high fuel conversion efficiency), the time available for combustion will, therefore, scale with the stroke. Thus, at the small end of the diesel engine size range, the mixing between the injected fuel and the air must take place on a time scale some 10 times shorter than in engines at the large end of this range. It would be expected, therefore, that the design of the engine combustion chamber (including the inlet port and valve) and the fuel-injection system would have to change substantially over this size range to provide the fuel and air motion inside the cylinder required to achieve the desired fuel-air mixing rate. As engine size decreases, more vigorous air motion is required while less fuel jet penetration is necessary. It is this logic, primarily, that leads to the different diesel combustion chamber designs and fuel injection systems found in practice over the large size range of commercial diesel engines.

10.2 TYPES OF DIESEL COMBUSTION SYSTEMS

Diesel engines are divided into two basic categories according to their combustion chamber design: (1) *direct-injection (DI) engines*, which have a single open combustion chamber into which fuel is injected directly; (2) *indirect-injection (IDI) engines*, where the chamber is divided into two regions and the fuel is injected into the "prechamber" which is connected to the main chamber (situated above the piston crown) via a nozzle, or one or more orifices. IDI engine designs are only used in the smallest engine sizes. Within each category there are several different chamber geometry, air-flow, and fuel-injection arrangements.

10.2.1 Direct-Injection Systems

In the largest-size engines, where mixing rate requirements are least stringent, quiescent direct-injection systems of the type shown in Fig. 10-1a are used. The momentum and energy of the injected fuel jets are sufficient to achieve adequate fuel distribution and rates of mixing with the air. Additional organized air motion is not required. The combustion chamber shape is usually a shallow bowl in the crown of the piston, and a central multihole injector is used.

As engine size decreases, increasing amounts of air swirl are used to achieve faster fuel-air mixing rates. Air swirl is generated by suitable design of the inlet port (see Sec. 8.3); the swirl rate can be increased as the piston approaches TC by forcing the air toward the cylinder axis, into a bowl-in-piston type of combustion

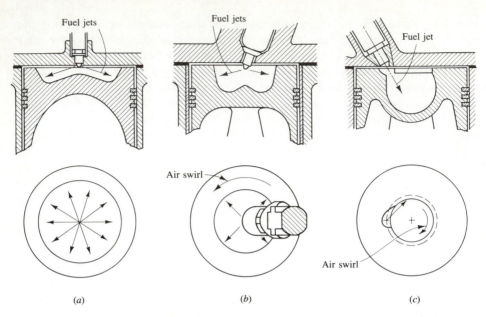

FIGURE 10-1
Common types of direct-injection compression-ignition or diesel engine combustion systems: (a) quiescent chamber with multihole nozzle typical of larger engines; (b) bowl-in-piston chamber with swirl and multihole nozzle; (c) bowl-in-piston chamber with swirl and single-hole nozzle. (b) and (c) used in medium to small DI engine size range.

chamber. Figure 10-1b and c shows the two types of DI engine with swirl in common use. Figure 10-1b shows a DI engine with swirl, with a centrally located multihole injector nozzle. Here the design goal is to hold the amount of liquid fuel which impinges on the piston cup walls to a minimum. Figure 10-1c shows the M.A.N. "M system" with its single-hole fuel-injection nozzle, oriented so that most of the fuel is deposited on the piston bowl walls. These two types of designs are used in medium-size (10- to 15-cm bore) diesels and, with increased swirl, in small-size (8- to 10-cm bore) diesels.

10.2.2   Indirect-Injection Systems

Inlet generated air swirl, despite amplification in the piston cup, has not provided sufficiently high fuel-air mixing rates for small high-speed diesels such as those used in automobiles. Indirect-injection or divided-chamber engine systems have been used instead, where the vigorous charge motion required during fuel injection is generated during the compression stroke. Two broad classes of IDI systems can be defined: (1) swirl chamber systems and (2) prechamber systems, as illustrated in Fig. 10-2a and b, respectively. During compression, air is forced from the main chamber above the piston into the auxiliary chamber, through the nozzle or orifice (or set of orifices). Thus, toward the end of compression, a vigorous flow in the auxiliary chamber is set up; in swirl chamber systems the connect-

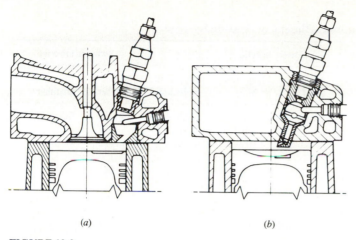

(a)   (b)

FIGURE 10-2
Two common types of small indirect-injection diesel engine combustion system: (a) swirl prechamber; (b) turbulent prechamber.

ing passage and chamber are shaped so that this flow within the auxiliary chamber rotates rapidly.

Fuel is usually injected into the auxiliary chamber at lower injection-system pressure than is typical of DI systems through a pintle nozzle as a single spray, as shown in Fig. 1-18. Combustion starts in the auxiliary chamber; the pressure rise associated with combustion forces fluid back into the main chamber where the jet issuing from the nozzle entrains and mixes with the main chamber air. The glow plug shown on the right of the prechamber in Fig. 10-2 is a cold-starting aid. The plug is heated prior to starting the engine to ensure ignition of fuel early in the engine cranking process.

10.2.3   Comparison of Different Combustion Systems

The number of different combustion chamber types proposed and tried since the beginnings of diesel engine development is substantial. Over the years, however, through the process of evolution and the increased understanding of the physical and chemical processes involved, only a few designs based on a sound principle have survived. The important characteristics of those chambers now most commonly used are summarized in Table 10.1. The numbers for dimensions and operating characteristics are typical ranges for each different type of diesel engine and combustion system.

The largest, slowest speed, engines for power generation or marine applications use open quiescent chambers which are essentially disc shaped; the motion of the fuel jets is responsible for distributing and mixing the fuel. These are usually two-stroke cycle engines. In the next size range, in large truck and locomotive engines, a quiescent chamber consisting of a shallow dish or bowl in the

TABLE 10.1
Characteristics of Common Diesel Combustion Systems

System	Direct injection				Indirect injection	
	Quiescent	Medium swirl	High swirl "M"	High swirl multispray	Swirl chamber	Pre-chamber
Size	Largest	Medium	Medium—smaller	Medium—small	Smallest	Smallest
Cycle	2-/4-stroke	4-stroke	4-stroke	4-stroke	4-stroke	4-stroke
Turbocharged/supercharged/naturally aspirated	TC/S	TC/NA	TC/NA	NA/TC	NA/TC	NA/TC
Maximum speed, rev/min	120–2100	1800–3500	2500–5000	3500–4300	3600–4800	4500
Bore, mm	900–150	150–100	130–80	100–80	95–70	95–70
Stroke/bore	3.5–1.2	1.3–1.0	1.2–0.9	1.1–0.9	1.1–0.9	1.1–0.9
Compression ratio	12–15	15–16	16–18	16–22	20–24	22–24
Chamber	Open or shallow dish	Bowl-in-piston	Deep bowl-in-piston	Deep bowl-in-piston	Swirl pre-chamber	Single/multi-orifice pre-chamber
Air-flow pattern	Quiescent	Medium swirl	High swirl	Highest swirl	Very high swirl in pre-chamber	Very turbulent in pre-chamber
Number of nozzle holes	Multi	Multi	Single	Multi	Single	Single
Injection pressure	Very high	High	Medium	High	Lowest	Lowest

piston crown is often used. The air utilization in these engines is low, but they are invariably supercharged or turbocharged to obtain high power density.

In the DI category, as engine size decreases and maximum speed rises, swirl is used increasingly to obtain high-enough fuel-air mixing rates. The swirl is generated by suitably shaped inlet ports, and is amplified during compression by forcing most of the air toward the cylinder axis into the deep bowl-in-piston combustion chamber. In about the same size range, an alternative system to the multihole nozzle swirl system is the M.A.N. "M" system (or wall-wetting system), where most of the fuel from the single-hole pintle nozzle is placed on the wall of the spherical bowl in the piston crown.

In the smallest engine sizes, the IDI engine has traditionally been used to obtain the vigorous air motion required for high fuel-air mixing rates. There are several different geometries in use. These either generate substantial swirl in the auxiliary chamber during the latter part of the compression stroke, using a nozzle or connecting passage that enters the auxiliary chamber tangentially, or they generate intense turbulence in the prechamber through use of several small orifices and obstructions to the flow within the prechamber. The most common design of swirl chamber is the Ricardo Comet design shown in Fig. 10-2a. An alternative IDI engine to the two types listed in Table 10-1 is the air cell system. In that system the fuel is injected into the main chamber and not the auxiliary "air cell." The auxiliary chamber acts as a turbulence generator as gas flows into and out of the cell.

10.3 PHENOMENOLOGICAL MODEL OF COMPRESSION-IGNITION ENGINE COMBUSTION

Studies of photographs of diesel engine combustion, combined with analyses of engine cylinder pressure data, have led to a widely accepted descriptive model of the compression-ignition engine combustion process. The concept of *heat-release rate* is important to understanding this model. It is defined as the rate at which the chemical energy of the fuel is released by the combustion process. It can be calculated from cylinder pressure versus crank angle data, as the energy release required to create the measured pressure, using the techniques described in Sec. 10.4. The combustion model defines four separate phases of diesel combustion, each phase being controlled by different physical or chemical processes. Although the relative importance of each phase does depend on the combustion system used, and engine operating conditions, these four phases are common to all diesel engines.

10.3.1 Photographic Studies of Engine Combustion

High-speed photography at several thousand frames per second has been used extensively to study diesel engine combustion. Some of these studies have been carried out in combustion chambers very close to those used in practice, under normal engine operating conditions (e.g., Refs. 1 and 2). Sequences of individual frames from movies provide valuable information on the nature of the combustion process in the different types of diesel engines. Figure 10-3 shows four combustion chamber geometries that have been studied photographically. These are: (a) a quiescent chamber typical of diesel engines in the 3 to 20 dm^3/cylinder displacement used for industrial, marine, and rail traction applications (only the burning of a single fuel spray of the multispray combustion system could be studied[2]); (b) a smaller high-speed DI engine with swirl and four fuel jets centrally injected; (c) an M.A.N. "M" DI system; and (d) a Ricardo Comet V swirl chamber IDI system.[1]

The combustion sequences were recorded on color film and show the following features:

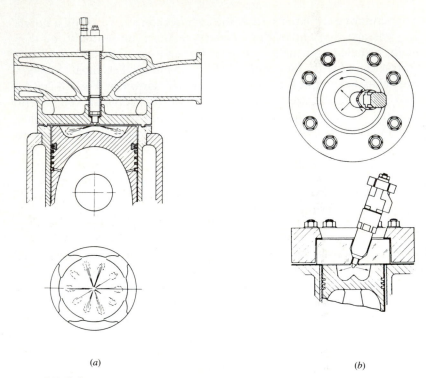

(a)   (b)

FIGURE 10-3
Four diesel combustion chambers used to obtain photographs of the compression-ignition combustion process shown in Fig. 10-4 on color plate: (a) quiescent DI chamber; (b) multihole nozzle DI chamber with swirl: on p. 499; (c) M.A.N. "M" DI chamber; (d) Ricardo Comet IDI swirl chamber.[1,2]

Fuel spray(s). The fuel droplets reflect light from spot lamps and define the extent of the liquid fuel spray prior to complete vaporization.

Premixed flame. These regions are of too low a luminosity to be recorded with the exposure level used. The addition of a copper additive dope to the fuel gives these normally blue flames a green color bright enough to render them visible.

Diffusion flame. The burning high-temperature carbon particles in this flame provide more than adequate luminosity and appear as yellow-white. As the flame cools, the radiation from the particles changes color through orange to red.

Over-rich mixture. The appearance of a brown region, usually surrounded by a white diffusion flame, indicates an excessively rich mixture region where substantial soot particle production has occurred. Where this fuel-rich soot-laden cloud contacts unburned air there is a hot white diffusion flame.

Table 10.2 summarizes the characteristics of these different regions, discernable in the photographs shown in Fig. 10-4 on the color plate.

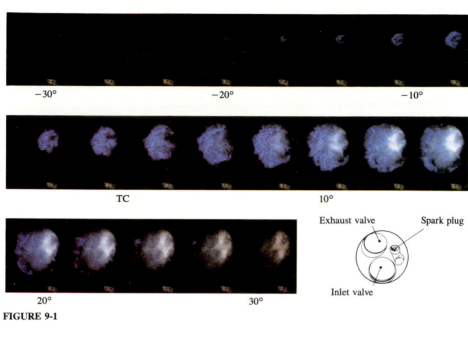

−30°   −20°   −10°

TC   10°

Exhaust valve   Spark plug

Inlet valve

20°   30°

FIGURE 9-1

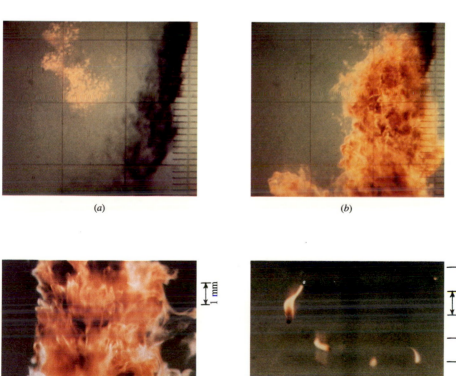

(a)   (b)

(c)   (d)

FIGURE 10-5

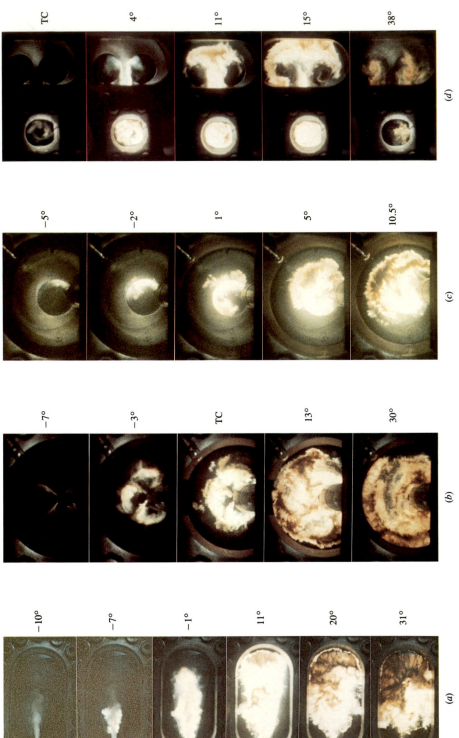

FIGURE 10-4

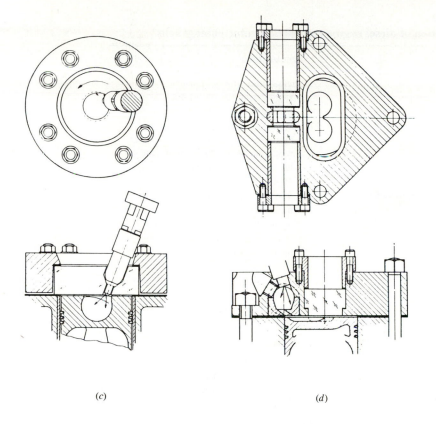

(c)   (d)

Figure 10-4a shows a sequence of photographs from one combustion event of the single spray, burning under conditions typical of a large quiescent DI engine. The fuel spray is shown penetrating into the chamber. Ignition occurs at $-8°$ in the fuel-air mixture left behind on the edge of the spray not far from the injector. The flame then spreads rapidly ($-7°$) along the outside of the spray to the spray tip. Here some of the fuel, which has had a long residence time in the chamber, burns with a blue-green low-luminosity flame (colored green by the copper fuel additive). The flame engulfing the remainder of the spray is brilliant white-yellow from the burning of the soot particles which have been formed in

---

FIGURE 10-4   **(On Color Plate, facing this page)**
Sequence of photographs from high-speed movies taken in special visualization diesel engines shown in Fig. 10-3: (a) combustion of single spray burning under large DI engine conditions; (b) combustion of four sprays in DI engine with counterclockwise swirl; (c) combustion of single spray in M.A.N. "M" DI diesel; (d) combustion in prechamber (on left) and main chamber (on right) in Ricardo Comet IDI swirl chamber diesel. 1250 rev/min, imep = 827 kPa (120 lb/in²)[1,2] (*Courtesy Ricardo Consulting Engineers.*)

TABLE 10.2
Interpretation of diesel engine combustion color photographs[1]

Color	Interpretation
Grey	Background; the gas (air in early stages, combustion products later) is transparent and not glowing
Green	Early in combustion process; low luminosity "premixed"-type flame, rendered visible by copper added to fuel. Later; burned gas above about 1800°C
White, and yellow-white	Carbon particle burnup in diffusion flame, 2000–2500°C
Yellow, orange-red	Carbon burnup in diffusion flame at lower temperatures; last visible in film at 1000°C
Brown	Soot clouds from very fuel-rich mixture regions. Where these meet air (grey) there is always a white fringe of hot flame

the fuel-rich spray core. At this stage ($-1°$), about 60 percent of the fuel has been injected. The remainder is injected into this enflamed region, producing a very fuel-rich zone apparent as the dark brown cloud ($11°$). This soot cloud moves to the outer region of the chamber ($11°$ to $20°$); white-yellow flame activity continues near the injector, probably due to combustion of ligaments of fuel which issued from the injector nozzle as the injector needle was seating. Combustion continues well into the expansion stroke ($31°C$).

This sequence shows that fuel distribution is always highly nonuniform during the combustion process in this type of DI engine. Also the air which is between the individual fuel sprays of the quiescent open-chamber diesel mixes with each burning spray relatively slowly, contributing to the poor air utilization with this type of combustion chamber.

Figure 10-4b shows a combustion sequence from the DI engine with swirl (the chamber shown in Fig. 10-3b). The inner circle corresponds to the deep bowl in the piston crown, the outer circle to the cylinder liner. The fuel sprays (of which two are visible without obstruction from the injector) first appear at $-13°$. At $-7°$ they have reached the wall of the bowl; the tips of the sprays have been deflected slightly by the anticlockwise swirl. The frame at $-3°$ shows the first ignition. Bright luminous flame zones are visible, one on each spray. Out by the bowl walls, where fuel vapor has been blown around by the swirl, larger greenish burning regions indicating the presence of premixed flame can be seen. The fuel downstream of each spray is next to ignite, burning yellow-white due to the soot formed by the richer mixture. Flame propagation back to the injector follows extremely rapidly and at TC the bowl is filled with flame. At 5° ATC the flame spreads out over the piston crown toward the cylinder wall due to combustion-produced gas expansion and the reverse squish flow (see Sec. 8.4). The brown regions (13°) are soot-laden fuel-rich mixture originating from the fuel which impinges on the wall. The last frame (30° ATC) shows the gradual diminution of the soot-particle-laden regions as they mix with the excess air and burn up. The last dull-red flame visible on the film is at about 75° ATC, well into the expansion stroke.

Figure 10-4c shows the combustion sequence for the M.A.N. "M"-type DI engine. In the version of the system used for these experiments, the fuel was injected through a two-hole nozzle which produces a main jet directed tangentially onto the walls of the spherical cup in the piston crown, and an auxiliary spray which mixes a small fraction of the fuel directly with the swirling air flow. More recent "M" systems use a pintle nozzle with a single variable orifice.[3] At −5° the fuel spray is about halfway round the bowl. Ignition has just occurred of fuel adjacent to the wall which has mixed sufficiently with air to burn. The flame spreads rapidly (−2°, 1°) to envelop the fuel spray, and is convected round the cup by the high swirl air flow. By shortly after TC the flame has filled the bowl and is spreading out over the piston crown. A soot cloud is seen near the top right of the picture at 5° ATC which spreads out around the circumference of the enflamed region. There is always a rim of flame between the soot cloud and the cylinder liner as excess air is mixed into the flame zone (10.5°). The flame is of the carbon-burning type throughout; little premixed green flame is seen even at the beginning of the combustion process.

Figure 10-4d shows the combustion sequence in a swirl chamber IDI engine of the Ricardo Comet V design. The swirl chamber (on the left) is seen in the view of the lower drawing of Fig. 10-3d (with the connecting passageway entering the swirl chamber tangentially at the bottom left to produce clockwise swirl). The main chamber is seen in the plan view of the upper drawing of Fig. 10-3d. Two sprays emerge from the Pintaux nozzle after the start of injection at −11°. The smaller auxiliary spray which is radial is sharply deflected by the high swirl. Frame 1 shows how the main spray follows the contour of the chamber; the auxiliary spray has evaporated and can no longer be seen. The first flame occurs at −1° in the vaporized fuel from the auxiliary spray and is a green premixed flame. The flame then spreads to the main spray (TC), becoming a yellow-white carbon-particle-burning flame with a green fringe. At 4° ATC the swirl chamber appears full of carbon-burning flame, which is being blown down the throat and into the recesses in the piston crown by the combustion generated pressure rise in the prechamber. The flame jet impinges on the piston recesses entraining the air in the main chamber, leaving green patches where all carbon is burned out (4°, 11°, 15°). A brown soot cloud is emerging from the throat. By 15° ATC this soot cloud has spread around the cylinder, with a bright yellow-white flame at its periphery. This soot then finds excess air and burns up, while the yellow-white flame becomes yellow and then orange-red as the gases cool on expansion. By 38° ATC most of the flame is burnt out.

Magnified color photographs of the flame around a single fuel spray under conditions typical of a direct-injection diesel engine, shown in Fig. 10-5 on the color plate, provide additional insight into the compression-ignition and flame-development processes.[4] These photographs were obtained in a rapid compression machine: this device is a cylinder-piston apparatus in which air is rapidly compressed by moving the piston to temperatures and pressures similar to those in the diesel engine combustion chamber at the time of injection. A single fuel spray was then injected into the disc-shaped combustion chamber. The air flow prior to compression was forced to swirl around the cylinder axis and much of that swirl remains after compression.

Figure 10-5a shows a portion of the liquid fuel spray (which appears black due to back lighting) and the rapidly developing flame 0.4 ms after ignition occurs. Ignition commences in the *fuel vapor*–air mixture region, set up by the jet motion and swirling air flow, away from the liquid core of the spray. In this region the smaller fuel droplets have evaporated in the hot air atmosphere that surrounds them and mixed with sufficient air for combustion to occur. Notice that the fuel vapor concentration must be nonuniform; combustion apparently occurs around small "lumps" of mixture of the appropriate composition and temperature. Figure 10-5b shows the same flame at a later time, 3.2 ms after ignition. The flame now surrounds most of the liquid spray core. Its irregular boundary reflects the turbulent character of the fuel spray and its color variation indicates that the temperature and composition in the flame region are not uniform.

Figure 10-5c shows a portion of this main flame region enlarged to show its internal structure. A highly convoluted flame region is evident, which has a similar appearance to a gaseous turbulent diffusion flame. The major portion of the diesel engine flame has this character, indicative of the burning of fuel vapor–air pockets or lumps or eddies of the appropriate composition. Only at the end of the combustion process is there visible evidence of individual fuel droplets burning with an envelope flame. Figure 10-5d shows the same region of the combustion chamber as Fig. 10-5c, but at the end of the burning process well after injection has been completed. A few large droplets are seen burning with individual droplet flames. It is presumed that such large drops were formed at the end of the injection process as the injector nozzle was closing.

FIGURE 10-5   (On Color Plate, facing page 498)
Photographs from high-speed movie of single fuel spray injected into a swirling air flow in a rapid-compression machine. (a) Spray and flame 0.4 ms after ignition; scale on right in millimeters. (b) Flame surrounding spray 3.2 ms after ignition. (c) Magnified photograph of main portion of flame. (d) Individual droplet burning late in combustion process after injection completed. Air temperature ~500°C. 50 mg fuel injected.[4] (*Courtesy Professor M. Ogasawara, Osaka University.*)

10.3.2 Combustion in Direct-Injection, Multispray Systems

Figure 10-6 shows typical data for cylinder pressure (p), fuel-injector needle-lift, and fuel pressure in the nozzle gallery through the compression and expansion strokes of a direct-injection diesel. The engine had central fuel injection through a four-hole nozzle into a disc-shaped bowl-in-piston combustion chamber. The rate of fuel injection can be obtained from the fuel-line pressure, cylinder pressure, nozzle geometry, and needle-lift profiles by considering the injector as one or more flow restrictions;[5] it is similar in phasing and comparable in shape to the needle-lift profile. There is a delay of 9° between the start of injection and start of combustion [identified by the change in slope of the $p(\theta)$ curve]. The pressure rises rapidly for a few crank angle degrees, then more slowly to a peak value about 5° after TC. Injection continues after the start of combustion. A rate-of-heat-release diagram† from the same study, corresponding to this rate of fuel injection and cylinder pressure data, is shown in Fig. 10-7. The general shape of the rate-of-heat-release curve is typical of this type of DI engine over its load and speed range. The heat-release-rate diagram shows negligible heat release until toward the end of compression when a slight loss of heat during the delay period (which is due to heat transfer to the walls and to fuel vaporization and heating) is

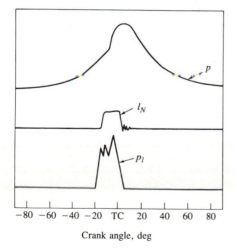

Crank angle, deg

FIGURE 10-6
Cylinder pressure p, injector needle lift l_N, and injection-system fuel-line pressure p_l, as functions of crank angle for small DI diesel engine.[5]

---

† The heat-release rate plotted here is the net heat-release rate (see Sec. 10.4). It is the sum of the change of sensible internal energy of the cylinder gases and the work done on the piston. It differs from the rate of fuel energy released by combustion by the heat transferred to the combustion chamber walls. The heat loss to the walls is 10 to 25 percent of the fuel heating value in smaller engines; it is less in larger engine sizes. This net heat release can be used as an indicator of actual heat release when the heat loss is small.

evident. During the combustion process the burning proceeds in three distinguishable stages. In the first stage, the rate of burning is generally very high and lasts for only a few crank angle degrees. It corresponds to the period of rapid cylinder pressure rise. The second stage corresponds to a period of gradually decreasing heat-release rate (though it initially may rise to a second, lower, peak as in Fig. 10-7). This is the main heat-release period and lasts about 40°. Normally about 80 percent of the total fuel energy is released in the first two periods. The third stage corresponds to the "tail" of the heat-release diagram in which a small but distinguishable rate of heat release persists throughout much of the expansion stroke. The heat release during this period usually amounts to about 20 percent of the total fuel energy.

From studies of rate-of-injection and heat-release diagrams such as those in Fig. 10-7, over a range of engine loads, speeds, and injection timings, Lyn[6] developed the following summary observations. First, the total burning period is much longer than the injection period. Second, the absolute burning rate increases proportionally with increasing engine speed; thus on a crank angle basis, the burning interval remains essentially constant. Third, the magnitude of the initial peak of the burning-rate diagram depends on the ignition delay period, being higher for longer delays. These considerations, coupled with engine combustion photographic studies, lead to the following model for diesel combustion.

Figure 10-8 shows schematically the rate-of-injection and rate-of-burning diagrams, where the injected fuel as it enters the combustion chamber has been divided into a number of elements. The first element which enters mixes with air and becomes "ready for burning" (i.e., mixes to within combustible limits), as shown conceptually by the lowest triangle along the abscissa in the rate-of-burning figure. While some of this fuel mixes rapidly with air, part of it will mix much more slowly. The second and subsequent elements will mix with air in a similar manner, and the total "ready-for-burning" diagram, enclosed by the dashed line, is obtained. The total area of this diagram is equal to that of the rate-of-injection diagram. Ignition does not occur until after the delay period is over, however. At the ignition point, some of the fuel already injected has mixed with enough air to be within the combustible limits. That "premixed" fuel-air

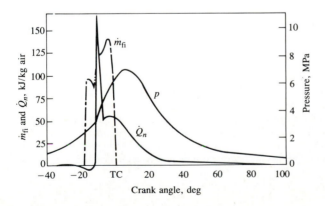

FIGURE 10-7
Cylinder pressure p, rate of fuel injection $\dot{m}_{fi}$, and net heat-release rate $\dot{Q}_n$ calculated from p for small DI diesel engine, 1000 rev/min, normal injection timing, bmep = 620 kPa.[5]

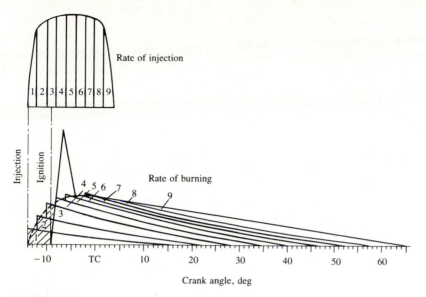

FIGURE 10-8
Schematic of relationship between rate of fuel injection and rate of fuel burning or energy release.[6]

mixture (the shaded region in Fig. 10-8) is then added to the mixture which becomes ready for burning after the end of the delay period, producing the high initial rate of burning as shown. Such a heat-release profile is generally observed with this type of naturally aspirated DI diesel engine. Photographs (such as those in Fig. 10-4a and b) show that up to the heat-release-rate peak, flame regions of low green luminosity are apparent because the burning is predominantly of the premixed part of the spray. After the peak, as the amount of premixed mixture available for burning decreases and the amount of fresh mixture mixed to be "ready for burning" increases, the spray burns essentially as a turbulent diffusion flame with high yellow-white or orange luminosity due to the presence of carbon particles.

To summarize, the following stages of the overall compression-ignition diesel combustion process can be defined. They are identified on the typical heat-release-rate diagram for a DI engine in Fig. 10-9.

Ignition delay (ab). The period between the start of fuel injection into the combustion chamber and the start of combustion [determined from the change in slope on the p-θ diagram, or from a heat-release analysis of the $p(\theta)$ data, or from a luminosity detector].

Premixed or rapid combustion phase (bc). In this phase, combustion of the fuel which has mixed with air to within the flammability limits during the ignition delay period occurs rapidly in a few crank angle degrees. When this burning mixture is added to the fuel which becomes ready for burning and burns during this phase, the high heat-release rates characteristic of this phase result.

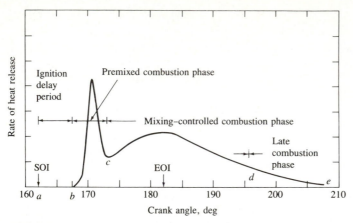

FIGURE 10-9
Typical DI engine heat-release-rate diagram identifying different diesel combustion phases.

Mixing-controlled combustion phase (cd). Once the fuel and air which pre-mixed during the ignition delay have been consumed, the burning rate (or heat-release rate) is controlled by the rate at which mixture becomes available for burning. While several processes are involved—liquid fuel atomization, vapor-ization, mixing of fuel vapor with air, preflame chemical reactions—the rate of burning is controlled in this phase primarily by the fuel vapor–air mixing process. The heat-release rate may or may not reach a second (usually lower) peak in this phase; it decreases as this phase progresses.

Late combustion phase (de). Heat release continues at a lower rate well into the expansion stroke. There are several reasons for this. A small fraction of the fuel may not yet have burned. A fraction of the fuel energy is present in soot and fuel-rich combustion products and can still be released. The cylinder charge is nonuniform and mixing during this period promotes more complete combustion and less-dissociated product gases. The kinetics of the final burnout processes become slower as the temperature of the cylinder gases fall during expansion.

10.3.3 Application of Model to Other Combustion Systems

In the M.A.N. "M" DI engine system, and in IDI systems, the shapes of the heat-release-rate curve are different from those of the quiescent or moderate swirl DI shown in Figs. 10-7 and 10-9. With the "M" system, the initial heat-release "spike" is much less pronounced (in spite of the fact that a large fraction of the fuel is injected during the delay period) though the total burning period is about the same. Lyn[6] has suggested that the lower initial burning rate is due to the fact that the smaller number of nozzle holes (one or two instead of about four or more) and the directing of the main spray tangentially to the wall substantially reduce the free mixing surface area of the fuel jets. However, since the burning rates after ignition are relatively high, mixing must speed up. This occurs due to the centrifugal forces set up in the swirling flow. Initially, the fuel is placed near the wall, and mixing is inhibited by the effect of the high centrifugal forces on the fuel vapor which is of higher density than the air and so tends to remain near the wall. Once ignition occurs, the hot burning mixture expands, decreases in density, and is then moved rapidly toward the center of the chamber. This strong radial mixing is the rate-determining process. An additional delaying mechanism exists if significant fuel is deposited on the wall. At compression air temperatures, the heat transferred to the fuel film on the wall from the gases in the cylinder is too small to account for the observed burning rates. Only after combustion starts will the gas temperature and heat-transfer rates be high enough to evaporate the fuel off the wall at an adequate rate.

In the swirl chamber IDI engine, where the air in the main chamber is not immediately available for mixing, again the rate-determining processes are different.[6] There is no initial spike on the rate-of-heat-release curve as was the case with DI engines. The small size of the chamber, together with the high swirl rate generated just before injection, results in considerable fuel impingement on the walls. This and the fact that the ignition delay is usually shorter with the IDI engine due to the higher compression ratio used account for the low initial burning rate.

Based on the above discussion Lyn[6] proposed three basic injection, mixing, and burning patterns important in diesel engines:

A. Fuel injection across the chamber with substantial momentum. Mixing proceeds immediately as fuel enters the chamber and is little affected by combustion.

B. Fuel deposition on the combustion chamber walls. Negligible mixing during the delay period due to limited evaporation. After ignition, evaporation becomes rapid and its rate is controlled by access of hot gases to the surface, radial mixing being induced by differential centrifugal forces. Burning is therefore delayed by the ignition lag.

C. Fuel distributed near the wall: mixing proceeds during the delay, but at a rate smaller than in mechanism A. After ignition, mixing is accelerated by the same mechanism as in mechanism B.

Figure 10-10 shows, schematically, the construction of the burning-rate or heat-release-rate diagrams (from the same injection-rate diagram) for the DI diesel combustion system with a central multihole nozzle, for the "M"-type DI diesel, and for the swirl chamber IDI. For the DI engine with multihole nozzle, mechanism A is predominant. For the DI engine with fuel sprayed tangentially to the wall, mechanisms B and C prevail; the delayed mixing prevents excessively high initial burning rates. For the IDI swirl chamber engine, the shorter ignition delay together with mixing process C during the delay period produces a gradual increase in burning rate, as shown in Fig. 10-10c.

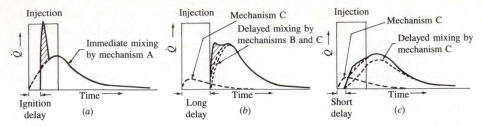

FIGURE 10-10

Schematic injection-rate and burning-rate diagrams in three different types of naturally aspirated diesel combustion system: (a) DI engine with central multihole nozzle; (b) DI "M"-type engine with fuel injected on wall; (c) IDI swirl chamber engine. Mechanisms A, B, and C defined in text.[6]

10.4 ANALYSIS OF CYLINDER PRESSURE DATA

Cylinder pressure versus crank angle data over the compression and expansion strokes of the engine operating cycle can be used to obtain quantitative information on the progress of combustion. Suitable methods of analysis which yield the rate of release of the fuel's chemical energy (often called heat release), or rate of fuel burning, through the diesel engine combustion process will now be described. The methods of analysis are similar to those described in Sec. 9.2.2 for spark-ignition engines and start with the first law of thermodynamics for an open system which is quasi static (i.e., uniform in pressure and temperature). The first law for such a system (see Fig. 9-11) is

$$\frac{dQ}{dt} - p\frac{dV}{dt} + \sum_i \dot{m}_i h_i = \frac{dU}{dt} \tag{10.1}$$

where dQ/dt is the heat-transfer rate across the system boundary into the system, $p(dV/dt)$ is the rate of work transfer done by the system due to system boundary displacement, $\dot{m}_i$ is the mass flow rate into the system across the system boundary at location i (flow out of the system would be negative), h_i is the enthalpy of flux i entering or leaving the system, and U is the energy of the material contained inside the system boundary.

The following problems make the application of this equation to diesel combustion difficult:

1. Fuel is injected into the cylinder. Liquid fuel is added to the cylinder which vaporizes and mixes with air to produce a fuel/air ratio distribution which is nonuniform and varies with time. The process is not quasi static.
2. The composition of the burned gases (also nonuniform) is not known.
3. The accuracy of available correlations for predicting heat transfer in diesel engines is not well defined (see Chap. 12).
4. Crevice regions (such as the volumes between the piston, rings, and cylinder wall) constitute a few percent of the clearance volume. The gas in these regions is cooled to close to the wall temperature, increasing its density and, therefore,

the relative importance of these crevices. Thus crevices increase heat transfer and contain a nonnegligible fraction of the cylinder charge at conditions that are different from the rest of the combustion chamber.

Due to difficulties in dealing with these problems, both sophisticated methods of analysis and more simple methods give only approximate answers.

10.4.1 Combustion Efficiency

In both heat-release and fuel mass burned estimations, an important factor is the completeness of combustion. Air utilization in diesels is limited by the onset of black smoke in the exhaust. The smoke is soot particles which are mainly carbon. While smoke and other incomplete combustion products such as unburned hydrocarbons and carbon monoxide represent a combustion inefficiency, the magnitude of that inefficiency is small. At full load conditions, if only 0.5 percent of the fuel supplied is present in the exhaust as black smoke, the result would be unacceptable. Hydrocarbon emissions are the order of or less than 1 percent of the fuel. The fuel energy corresponding to the exhausted carbon monoxide is about 0.5 percent. Thus, the combustion inefficiency [Eq. (4.69)] is usually less than 2 percent; the combustion efficiency is usually greater than about 98 percent (see Fig. 3-9). While these emissions are important in terms of their air-pollution impact (see Chap. 11), from the point of view of energy conversion it is a good approximation to regard combustion and heat release as essentially complete.

10.4.2 Direct-Injection Engines

For this type of engine, the cylinder contents are a single open system. The only mass flows across the system boundary (while the intake and exhaust valves are closed) are the fuel and the crevice flow. An approach which incorporates the crevice flow has been described in Sec. 9.2.2; crevice flow effects will be omitted here. Equation (10.1) therefore becomes

$$\frac{dQ}{dt} - p\frac{dV}{dt} + \dot{m}_f h_f = \frac{dU}{dt} \tag{10.2}$$

Two common methods are used to obtain combustion information from pressure data using Eq. (10.2). In both approaches, the cylinder contents are assumed to be at a uniform temperature at each instant in time during the combustion process. One method yields fuel energy- or heat-release rate; the other method yields a fuel mass burning rate. The term *apparent* is often used to describe these quantities since both are approximations to the real quantities which cannot be determined exactly.

HEAT-RELEASE ANALYSIS. If U and h_f in Eq. (10.2) are taken to be the sensible internal energy of the cylinder contents and the sensible enthalpy of the injected fuel, respectively,† then dQ/dt becomes the difference between the chemical energy or heat released by combustion of the fuel (a positive quantity) and the heat transfer from the system (in engines, the heat transfer is from the system and by thermodynamic convention is a negative quantity). Since $h_{s,f} \approx 0$, Eq. (10.2) becomes

$$\frac{dQ_n}{dt} = \frac{dQ_{ch}}{dt} - \frac{dQ_{ht}}{dt} = p \frac{dV}{dt} + \frac{dU_s}{dt} \tag{10.3}$$

The apparent *net heat-release* rate, dQ_n/dt, which is the difference between the apparent *gross heat-release* rate dQ_{ch}/dt and the heat-transfer rate to the walls dQ_{ht}/dt, equals the rate at which work is done on the piston plus the rate of change of sensible internal energy of the cylinder contents.

If we further assume that the contents of the cylinder can be modeled as an ideal gas, then Eq. (10.3) becomes

$$\frac{dQ_n}{dt} = p \frac{dV}{dt} + mc_v \frac{dT}{dt} \tag{10.4}$$

From the ideal gas law, $pV = mRT$, with R assumed constant, it follows that

$$\frac{dp}{p} + \frac{dV}{V} = \frac{dT}{T} \tag{10.5}$$

Equation (10.5) can be used to eliminate T from Eq. (10.4) to give

$$\frac{dQ_n}{dt} = \left(1 + \frac{c_v}{R}\right) p \frac{dV}{dt} + \frac{c_v}{R} V \frac{dp}{dt}$$

or

$$\frac{dQ_n}{dt} = \frac{\gamma}{\gamma - 1} p \frac{dV}{dt} + \frac{1}{\gamma - 1} V \frac{dp}{dt} \tag{10.6}$$

Here γ is the ratio of specific heats, c_p/c_v. An appropriate range for γ for diesel heat-release analysis is 1.3 to 1.35; Eq. (10.6) is often used with a constant value of γ within this range. More specifically, we would expect γ for diesel engine heat-release analysis to have values appropriate to air at end-of-compression-stroke temperatures prior to combustion (≈ 1.35) and to burned gases at the overall equivalence ratio following combustion (≈ 1.26–1.3). The appropriate values for γ during combustion which will give most accurate heat-release information are not well defined.[7, 8]

More complete methods of heat-release analysis based on Eq. (10.2) have been proposed and used. These use more sophisticated models for the gas properties before, during, and after combustion, and for heat transfer and crevice effects.[8] However, it is also necessary to deal with the additional issues of: (1) mixture nonuniformity (fuel/air ratio nonuniformity *and* burned and unburned gas nonuniformities); (2) accuracy of any heat-transfer model used (see Chap. 12);

---

† That is, $U = U_s = U(T) - U(298 \text{ K})$ and $h_f = h_{s,f} = h_f(T) - h_f(298 \text{ K})$; see Sec. 5.5 for definition.

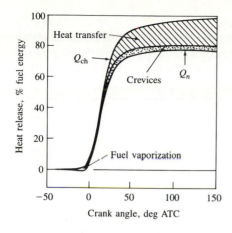

FIGURE 10-11
Gross and net heat-release profile during combustion, for a turbocharged DI diesel engine in mid-load, mid-speed range, showing relative magnitude of heat transfer, crevice, and fuel vaporization and heatup effects.

and (3) the effects of the crevice regions. These additional phenomena must be dealt with at an equivalent level of accuracy for more complex heat-release models to be worth while. For many engineering applications, Eq. (10.6) is adequate for diesel engine combustion analysis.

Additional insight can be obtained by incorporating a model for the largest of the effects omitted from Eq. (10.6), the heat transfer dQ_{ht}/dt (see Chap. 12); we thereby obtain a close approximation to the *gross* heat-release rate. The integral of the gross heat-release rate over the complete combustion process should then equal (to within a few percent only, since the analysis is not exact) the mass of fuel injected m_f times the fuel lower heating value Q_{LHV}: i.e.,

$$Q_{ch} = \int_{t_{start}}^{t_{end}} \frac{dQ_{ch}}{dt} \, dt = m_f Q_{LHV} \tag{10.7}$$

Of course, Eqs. (10.1) to (10.4), (10.6), and (10.7) also hold with crank angle θ as the independent variable instead of time t.

Figure 10-11 illustrates the relative magnitude of gross and net heat release, heat transfer, crevice effects, and heat of vaporization and heating up of the fuel for a turbocharged DI diesel engine operating in the mid-load, mid-speed range. The net heat release is the gross heat release due to combustion, less the heat transfer to the walls, crevice effects, and the effect of fuel vaporization and heatup (which was omitted above by neglecting the mass addition term in dU/dt). This last term is sufficiently small to be neglected. The enthalpy of vaporization of diesel fuel is less than 1 percent of its heating value; the energy change associated with heating up fuel vapor from injection temperature to typical compression air temperatures is about 3 percent of the fuel heating value. The heat transfer integrated over the duration of the combustion period is 10 to 25 percent of the total heat released.

FUEL MASS BURNING RATE ANALYSIS. If the internal energies of the fuel, air, and burned gases in Eq. (10.1) are evaluated relative to a consistent datum (such as that described in Sec. 4.5.2), then this equation can be used to obtain an apparent *fuel mass burning rate* from cylinder pressure versus crank angle data. (With such a species energy datum the "heat release" is properly accounted for in the internal energy and enthalpy terms.) Following Krieger and Borman,[9] Eq. (10.2) can be written as

$$\frac{d}{dt}(mu) = -p\frac{dV}{dt} + \frac{dQ}{dt} + h_f\frac{dm}{dt} \tag{10.8}$$

Here Q is the heat transfer to the gas within the combustion chamber (that is, $Q = -Q_{ht}$), m is the mass within the combustion chamber, and dm/dt has been substituted for $\dot{m}_f$.

Since the properties of the gases in the cylinder during combustion (assumed to be uniform and in chemical equilibrium at the pressure p and average temperature T) are in general a function of p, T, and the equivalence ratio ϕ,

$$u = u(T, p, \phi) \quad \text{and} \quad R = R(T, p, \phi)$$

Therefore

$$\frac{du}{dt} = \frac{\partial u}{\partial T}\frac{dT}{dt} + \frac{\partial u}{\partial p}\frac{dp}{dt} + \frac{\partial u}{\partial \phi}\frac{d\phi}{dt} \tag{10.9a}$$

$$\frac{dR}{dt} = \frac{\partial R}{\partial T}\frac{dT}{dt} + \frac{\partial R}{\partial p}\frac{dp}{dt} + \frac{\partial R}{\partial \phi}\frac{d\phi}{dt} \tag{10.9b}$$

Also,

$$\phi = \phi_0 + \left(\frac{m}{m_0} - 1\right)\frac{1 + (F/A)_0}{(F/A)_s} \tag{10.10}$$

and

$$\frac{d\phi}{dt} = \frac{1 + (F/A)_0}{(F/A)_s m_0}\frac{dm}{dt} \tag{10.11}$$

(F/A) is the fuel/air ratio; the subscript 0 denotes the initial value prior to fuel injection and the subscript s denotes the stoichiometric value. It then follows that

$$\frac{1}{m}\frac{dm}{dt} = \frac{-(RT/V)(dV/dt) - (\partial u/\partial p)(dp/dt) + (1/m)(dQ/dt) - CB}{u - h_f + D(\partial u/\partial \phi) - C[1 + (D/R)(\partial R/\partial \phi)]} \tag{10.12}$$

where

$$B = \frac{1}{p}\frac{dp}{dt} - \frac{1}{R}\frac{\partial R}{\partial p}\frac{dp}{dt} + \frac{1}{V}\frac{dV}{dt}$$

$$C = \frac{T(\partial u/\partial T)}{1 + (T/R)(\partial R/\partial T)}$$

$$D = \frac{[1 + (F/A)_0]m}{(F/A)_s m_0}$$

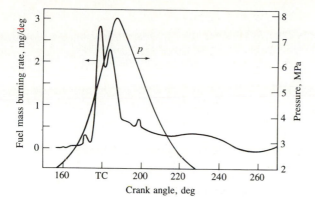

FIGURE 10-12
Cylinder pressure p and fuel mass burning rate calculated from p, as a function of crank angle, using the Krieger and Borman method[9] for DI diesel engine at 3200 rev/min and full load.

Equation (10.12) can be solved numerically for $m(t)$ given m_0, ϕ_0, $p(t)$, and appropriate models for the working fluid properties (see Sec. 4.7) and for the heat-transfer term dQ/dt (see Chap. 12).

Figure 10-12 shows cylinder pressure data for an open chamber DI diesel and fuel mass burning rate $dm/d\theta$ calculated from that data using the above method. The heat-transfer model of Annand was used (see Sec. 12.4.3). The result obtained is an *apparent* fuel mass burning rate. It is best interpreted, after multiplying by the heating value of the fuel, as the fuel chemical-energy or heat-release rate. The *actual* fuel burning rate is unknown because not all the fuel "burns" with sufficient air available locally to produce products of *complete* combustion. About 60 percent of the fuel has burned in the first one-third of the total combustion period. The integral of the fuel mass burning rate over the combustion process should equal the total fuel mass burned; in this case it is 3 percent less than the total fuel mass injected. Note that chemical energy continues to be released well into the expansion process. The accuracy of this type of calculation then decreases, however, since errors in estimating heat transfer significantly affect the apparent fuel burning rate.

Krieger and Borman also carried out sensitivity analyses for the critical assumptions and variables. They found that the effect of dissociation of the product gases was negligible. This permits a substantial simplification of Eq. (10.12). With no dissociation, $u = u(T, \phi)$, and $R = \tilde{R}/M$ can be taken as constant, since the molecular weight M changes little. Then

$$\frac{dm}{dt} = \frac{[1 + (c_v/R)]p(dV/dt) + (c_v/R)V(dp/dt) - (dQ/dt)}{h_f + (c_v/R)(pV/m) - u - D(\partial u/\partial \phi)} \qquad (10.13)$$

where D, as before, is $[1 + (F/A)_0]m/[(F/A)_s m_0]$. Given the uncertainties inherent in the heat-transfer model and the neglect of nonuniformities and crevices, Eq. (10.13) represents an adequate level of sophistication.

The other sensitivity variations studied by Krieger and Borman were: shifting of the phasing of the pressure data 2° forward and 2° backward; translating the pressure data ± 34 kPa (5 lb/in²); changing the heat transfer ± 50 percent;

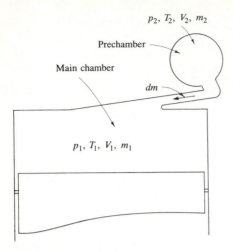

FIGURE 10-13
Schematic defining variables in main chamber (subscript 1) and prechamber (subscript 2) for IDI engine heat-release analysis.

increasing the initial mass 5 percent. The initial mass change had a negligible effect on the fuel burning rate calculations. The heat-transfer changes of ± 50 percent changed the mass of fuel burned by about ± 5 percent. The change in phasing of the pressure data was more significant. It needs to be stressed that *accurate* (in magnitude and phasing) pressure data are a most important requirement for useful heat-release or fuel mass burning rate analysis.

10.4.3 Indirect-Injection Engines

In IDI diesel engines, the pressures in each of the two chambers, main and auxiliary, are not the same during the combustion process. Since combustion starts in the auxiliary or prechamber, the fuel energy release in the prechamber causes the pressure there to rise above the main chamber pressure. Depending on combustion chamber design and operating conditions, the prechamber pressure rises to be 0.5 to 5 atm above that in the main chamber. This pressure difference causes a flow of fuel, air, and burning and burned gases into the main chamber, where additional energy release now occurs. The analysis of the DI diesel in the previous section was based on uniform pressure throughout the combustion chamber. For IDI engines the effect of the pressure difference between the chambers must usually be included.

Figure 10-13 shows an IDI combustion chamber divided at the nozzle into two open systems. Applying the first law [Eq. (10.1)] to the *main chamber* yields

$$\frac{dQ_1}{dt} - p_1 \frac{dV_1}{dt} + h_{2,1} \frac{dm}{dt} = \frac{dU_1}{dt} \tag{10.14}$$

and to the *auxiliary chamber* yields

$$\frac{dQ_2}{dt} - h_{2,1} \frac{dm}{dt} + h_f \frac{dm_f}{dt} = \frac{dU_2}{dt} \tag{10.15}$$

Here dm/dt is the mass flow rate between the chambers with positive flow from the prechamber to the main chamber. If $dm/dt > 0$, $h_{2,1} = h_2$; if $dm/dt < 0$, $h_{2,1} = h_1$. If we define U_1 and U_2 as sensible internal energies and h_f as the sensible enthalpy of the fuel, then dQ_1/dt and dQ_2/dt represent the net heat-release rates—the difference between the combustion energy-release rates and the rates of heat transfer to the walls.

If we use an ideal gas model for the working fluid in each chamber, with c_v, c_p, and M constant, the relation $p_1 V_1 = m_1 R T_1$ and $p_2 V_2 = m_2 R T_2$ can be used to eliminate m and T from the dU/dt terms and, with the fact that $h_{s,f} = 0$, can be used to write Eqs. (10.14) and (10.15) as

$$\frac{dQ_1}{dt} = \frac{\gamma}{\gamma - 1} p_1 \frac{dV_1}{dt} + \frac{1}{\gamma - 1} V_1 \frac{dp_1}{dt} - c_p T_{2,1} \frac{dm}{dt} \qquad (10.16)$$

$$\frac{dQ_2}{dt} = \frac{1}{\gamma - 1} V_2 \frac{dp_2}{dt} + c_p T_{2,1} \frac{dm}{dt} \qquad (10.17)$$

When Eqs. (10.16) and (10.17) are added together, the term representing the enthalpy flux between the two chambers cancels out, and the following equation for *total* net heat-release results:

$$\frac{dQ}{dt} = \frac{dQ_1}{dt} + \frac{dQ_2}{dt} = \frac{\gamma}{\gamma - 1} p_1 \frac{dV_1}{dt} + \frac{1}{\gamma - 1} \left(V_1 \frac{dp_1}{dt} + V_2 \frac{dp_2}{dt} \right) \qquad (10.18)$$

The comments made in the previous section regarding the interpretation of the net heat release (it is the gross heat release due to combustion less the heat transfer to the walls, and other smaller energy transfers due to crevices, fuel vaporization, and heatup) also hold here.

In practice, Eq. (10.18) is difficult to use since it requires experimental data for both the main and auxiliary chamber pressures throughout the combustion process. Access for two pressure transducers through the cylinder head is not often available; even when access can be achieved, the task of obtaining pressure data from two different transducers under the demanding thermal loading conditions found in IDI diesels, of sufficient accuracy such that the difference between the pressures (of order 0.5 to 5 atm) at pressure levels of 60 to 80 atm can be interpreted, requires extreme diligence in technique.[10, 11] Figure 10-14a and b shows apparent net heat-release rate profiles for an IDI diesel obtained using Eq. (10.18) with $\gamma = 1.35$.[11] Curves of dQ/dt and $dQ/d\theta$ are shown at three different speeds and essentially constant fuel mass injected per cycle. While the absolute heat-release rates increase with increasing speed, the relative rates are essentially independent of speed, indicating that combustion rates, which depend on fuel-air mixing rates, scale approximately with engine speed.

Equation (10.18) (or its equivalent) has been used assuming $p_2 = p_1$ and using either main chamber or auxiliary chamber pressure data alone. The error associated with this approximation can be estimated as follows. If we write $p_2 =$

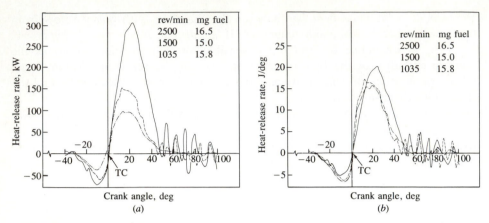

FIGURE 10-14
Calculated net heat-release-rate profiles for IDI diesel engine at constant load ($0.29 \le \phi \le 0.32$). (a) Heat-release rate in kilowatts. (b) Heat-release rate in joules per degree.[11]

$p_1 + \Delta p$ then Eq. (10.18) becomes

$$\frac{dQ}{dt} = \frac{\gamma}{\gamma - 1} p_1 \frac{dV_1}{dt} + \frac{V_1 + V_2}{\gamma - 1} \frac{dp_1}{dt} + \frac{V_2}{\gamma - 1} \frac{d(\Delta p)}{dt} \tag{10.19}$$

If the last term is omitted, Eq. (10.19) is identical to Eq. (10.6) derived for the DI diesel. Since the term $V(dp_1/dt)/(\gamma - 1)$ is much larger than the first term on the right-hand side of Eq. (10.19) during the early stages of the combustion process, the error involved in omitting the last term is given to a good approximation by $[V_2/(V_1 + V_2)]d(\Delta p)/dp_1$. In the initial stages of combustion this error can be quite large (of order 0.25 based on data in Ref. 10 close to TC). Later in the combustion process it becomes negligible (of order a few percent after 20° ATC).[10] Thus, neglecting Δp will lead to errors in predicting the initial heat-release *rate*. The magnitude of the error will depend on the design of the combustion chamber and on engine speed and load (with more restricted passageways, higher loads and speeds, giving higher values of Δp and, therefore, greater error). Later in the combustion process the error is much less, so *integrated* heat-release data derived ignoring Δp will show a smaller error.

A model analogous to the above, but using the approach of Krieger and Borman[9] (see Sec. 10.4.2), for the IDI diesel has been developed and used by Watson and Kamel.[10] The energy conservation equation for an *open* system developed in Sec. 14.2.2, with energy and enthalpy modeled using a consistent datum (see Sec. 4.5.2), with appropriate models for convective and radiation heat transfer and for gas properties, was applied to the main chamber and also to the prechamber. These equations were solved using accurately measured main chamber and prechamber pressure data to determine the apparent rate of heat release (here, the rate of fuel burning multiplied by the fuel heating value) in the main chamber and prechambers through the combustion process. The engine was a Ricardo Comet swirl chamber IDI design. Some results are shown in Fig.

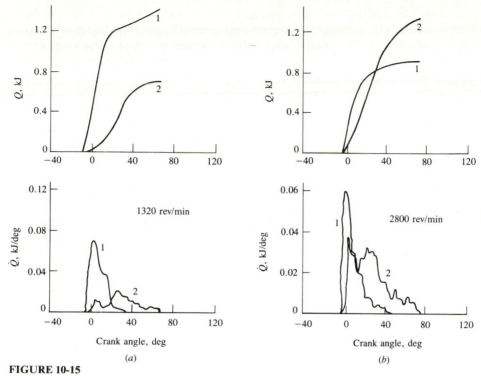

FIGURE 10-15
Calculated gross heat-release rates in IDI swirl-chamber diesel engine at full load. 1 Prechamber heat release. 2 Main chamber heat release. Top figures: integrated heat release. Bottom figures: heat-release rate. (a) 1320 rev/min; (b) 2800 rev/min.[10]

10-15. For this particular engine, at low engine speeds two-thirds of the heat release occurs in the prechamber; at higher engine speeds about two-thirds of the heat release occurs in the main chamber.

10.5 FUEL SPRAY BEHAVIOR

10.5.1 Fuel Injection

The fuel is introduced into the cylinder of a diesel engine through a nozzle with a large pressure differential across the nozzle orifice. The cylinder pressure at injection is typically in the range 50 to 100 atm. Fuel injection pressures in the range 200 to 1700 atm are used depending on the engine size and type of combustion system employed. These large pressure differences across the injector nozzle are required so that the injected liquid fuel jet will enter the chamber at sufficiently high velocity to (1) atomize into small-sized droplets to enable rapid evaporation and (2) traverse the combustion chamber in the time available and fully utilize the air charge.

Examples of common diesel fuel-injection systems were described briefly in Sec. 1.7 and illustrated in Figs. 1-17 to 1-19. (See also Refs. 12 and 13 for more extensive descriptions of diesel fuel-injection systems.) The task of the fuel-injection system is to meter the appropriate quantity of fuel for the given engine speed and load to each cylinder, each cycle, and inject that fuel at the appropriate time in the cycle at the desired rate with the spray configuration required for the particular combustion chamber employed. It is important that injection begin and end cleanly, and avoid any secondary injections.

To accomplish this task, fuel is usually drawn from the fuel tank by a supply pump, and forced through a filter to the injection pump. The injection pump sends fuel under pressure to the nozzle pipes which carry fuel to the injector nozzles located in each cylinder head. Excess fuel goes back to the fuel tank. Figures 1-17 and 1-19 show two common versions of fuel systems used with multicylinder engines in the 20 to 100 kW per cylinder brake power range which operate with injection pressures between about 300 and 1200 atm.

In-line injection pumps (Fig. 1-17) are used in engines in the 40 to 100 kW per cylinder maximum power range. They contain a plunger and barrel assembly for each engine cylinder. Each plunger is raised by a cam on the pump camshaft and is forced back by the plunger return spring. The plunger stroke is fixed. The plunger fits sufficiently accurately within the barrel to seal without additional sealing elements, even at high pressures and low speeds. The amount of fuel delivered is altered by varying the *effective* plunger stroke. This is achieved by means of a control rod or rack, which moves in the pump housing and rotates the plunger via a ring gear or linkage lever on the control sleeve. The plunger chamber above the plunger is always connected with the chamber below the plunger helix by a vertical groove or bore in the plunger. Delivery ceases when the plunger helix exposes the intake port (port opening), thus connecting the plunger chamber with the suction gallery. When this takes place depends on the rotational position of the plunger. In the case of a lower helix, delivery always starts (port closing) at the same time, but ends sooner or later depending on the rotational position of the plunger. With a plunger with an upper helix, port closing (start of delivery) not port opening is controlled by the helix and is varied by rotating the plunger. Figure 1-18 illustrates how the plunger helix controls fuel delivery.[14]

Distributor-type fuel-injection pumps (such as that illustrated in Fig. 1-19) are normally used in multicylinder engines with less than 30 kW per cylinder maximum power with injection pressures up to 750 atm. These pumps have only one plunger and barrel. The pump plunger is made to describe a combined rotary and stroke movement by the rotating cam plate. The fuel is accurately metered to each injection nozzle in turn by this plunger which simultaneously acts as the distributor. Such units are more compact and cheaper than in-line pumps but cannot achieve such high injection pressures. The distributor-type fuel-injection pump is combined with the automatic timing device, governor, and supply pump to form a single unit.

Single-barrel injection pumps are used on small one- and two-cylinder diesel engines, as well as large engines with outputs of more than 100 kW per cylinder. Figure 10-16 shows the layout of the injection system and a section

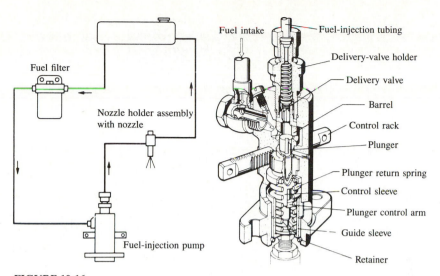

FIGURE 10-16
Fuel-injection system with single-barrel pump. Left: system layout. Right: section through fuel-injection pump. (*Courtesy Robert Bosch GmbH and SAE.*[14])

through the fuel-injection pump.[14] Such pumps are driven by an auxiliary cam on the engine camshaft. Also used extensively on larger engines are unit injectors where the pump and injector nozzle are combined into a single unit. An example of a unit injector and its driving mechanism used on a large two-stroke cycle diesel engine is shown in Fig. 10-17. Fuel, supplied to the injector through a fuel-distributing manifold, enters the cavity or plunger chamber ahead of the plunger through a metering orifice. When fuel is to be injected, the cam via the rocker arm pushes down the plunger, closing the metering orifice and compressing the fuel, causing it to flow through check valves and discharge into the cylinder through the injector nozzles or orifices. The amount of fuel injected is controlled by the rack, which controls the spill of fuel into the fuel drain manifold by rotating the plunger with its helical relief section via the gear.

The most important part of the injection system is the nozzle. Examples of different nozzle types and a nozzle holder assembly are shown in Fig. 1-18. The nozzles shown are fluid-controlled needle valves where the needle is forced against the valve seat by a spring. The pressure of the fuel in the pressure chamber above the nozzle aperture opens the nozzle by the axial force it exerts on the conical surface of the nozzle needle. Needle valves are used to prevent dribble from the nozzles when injection is not occurring. It is important to keep the volume of fuel left between the needle and nozzle orifices (the sac volume) as small as possible to prevent any fuel flowing into the cylinder after injection is over, to control hydrocarbon emissions (see Sec. 11.4.4). Multihole nozzles are used with most direct-injection systems; the M.A.N. "M" system uses a single-hole nozzle. Pintle nozzles, where the needle projects into and through the nozzle hole, are used in indirect-injection engine systems. The shape of the pin on the

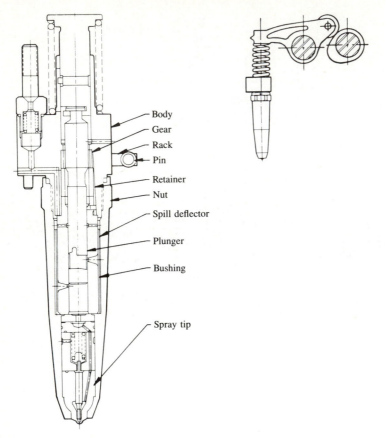

Body
Gear
Rack
Pin
Retainer
Nut
Spill deflector
Plunger
Bushing
Spray tip

FIGURE 10-17
Unit fuel injector and its driving mechanism, typically used in large diesel engines.[15]

end of the nozzle needle controls the spray pattern and fuel-delivery characteristics. Auxiliary nozzle holes are sometimes used to produce an auxiliary smaller spray to aid ignition and starting. Open nozzle orifices, without a needle, are also used.

The technology for electronic control of injection is now available. In an electronic injector, such as that shown in Fig. 10-18, a solenoid operated control valve performs the injection timing and metering functions in a fashion analogous to the ports and helices of the mechanical injector. Solenoid valve closure initiates pressurization and injection, and opening causes injection pressure decay and end of injection. Duration of valve closure determines the quantity of fuel injected. The unit shown uses camshaft/rocker arm driven plungers to generate the injection pressure, and employs needle-valve nozzles of conventional design. Increased flexibility in fuel metering and timing and simpler injector mechanical design are important advantages.[16]

Accurate predictions of fuel behavior within the injection system require

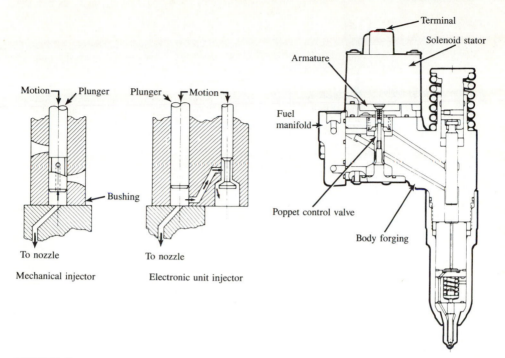

FIGURE 10-18
Electronically controlled unit fuel-injection system.[16]

sophisticated hydraulic models: Hiroyasu[17] provides an extensive reference list of such models. However, approximate estimates of the injection rate through the injector nozzles can be made as follows. If the pressure upstream of the injector nozzle(s) can be estimated or measured, and assuming the flow through each nozzle is quasi steady, incompressible, and one dimensional, the mass flow rate of fuel injected through the nozzle $\dot{m}_f$ is given by

$$\dot{m}_f = C_D A_n \sqrt{2\rho_f \Delta p} \tag{10.20}$$

where A_n is the nozzle minimum area, C_D the discharge coefficient, ρ_f the fuel density, and Δp the pressure drop across the nozzle. If the pressure drop across the nozzle and the nozzle open area are essentially constant during the injection period, the mass of fuel injected is then

$$m_f = C_D A_n \sqrt{2\rho_f \Delta p} \, \frac{\Delta\theta}{360N} \tag{10.21}$$

where $\Delta\theta$ is the nozzle open period in crank angle degrees and N is engine speed. Equations (10.20) and (10.21) illustrate the dependence of injected amounts of fuel on injection system and engine parameters.

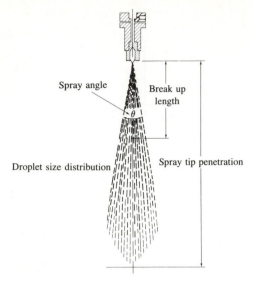

FIGURE 10-19
Schematic of diesel fuel spray defining its major parameters.[18]

10.5.2   Overall Spray Structure

The fuel is introduced into the combustion chamber of a diesel engine through one or more nozzles or orifices with a large pressure differential between the fuel supply line and the cylinder. Different designs of nozzle are used (e.g., single-orifice, multiorifice, throttle, or pintle; see Fig. 1-18), depending on the needs of the combustion system employed. Standard diesel injectors usually operate with fuel-injection pressures between 200 and 1700 atm. At time of injection, the air in the cylinder has a pressure of 50 to 100 atm, temperature about 1000 K, and density between 15 and 25 kg/m^3. Nozzle diameters cover the range 0.2 to 1 mm diameter, with length/diameter ratios from 2 to 8. Typical distillate diesel fuel properties are: relative specific gravity of 0.8, viscosity between 3 and 10 $kg/m \cdot s$ and surface tension about 3×10^{-2} N/m (at 300 K). Figure 10-19 illustrates the structure of a typical DI engine fuel spray. As the liquid jet leaves the nozzle it becomes turbulent and spreads out as it entrains and mixes with the surrounding air. The initial jet velocity is greater than 10^2 m/s. The outer surface of the jet breaks up into drops of order 10 μm diameter, close to the nozzle exit. The liquid column leaving the nozzle disintegrates within the cylinder over a finite length called the *breakup length* into drops of different sizes. As one moves away from the nozzle, the mass of air within the spray increases, the spray diverges, its width increases, and the velocity decreases. The fuel drops evaporate as this air-entrainment process proceeds. The tip of the spray penetrates further into the combustion chamber as injection proceeds, but at a decreasing rate. Figure 10-20 shows photographs of a diesel fuel spray injected into quiescent air in a rapid-compression machine which simulates diesel conditions.[19] Two different pho-

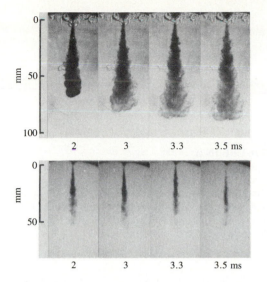

FIGURE 10-20
Shadowgraph and back-illuminated photographs of evaporating spray injected into nitrogen at 3.4 MPa and 670 K in rapid-compression machine. Times in milliseconds are after start of injection: injection duration 3.3 ms. Top (shadowgraph) photographs show full vapor and liquid region. Bottom (back-illuminated) photographs only show liquid-containing core.[19]

tographic techniques, back lighting and shadowgraph,† have been used to distinguish the liquid-containing core of the jet and the extent of the fuel vapor region of the spray which surrounds the liquid core. The region of the jet closest to the nozzle (until injection ceases at 3.3 ms) contains liquid drops and ligaments; the major region of the spray is a substantial vapor cloud around this narrow core which contains liquid fuel.

Different spray configurations are used in the different diesel combustion systems described earlier in this chapter. The simplest configuration involves multiple sprays injected into quiescent air in the largest-size diesels (Fig. 10-1a). Figures 10-19 and 10-20 illustrate the essential features of each spray under these circumstances until interactions with the wall occur. Each liquid fuel jet atomizes into drops and ligaments at the exit from the nozzle orifice (or shortly thereafter). The spray entrains air, spreads out, and slows down as the mass flow in the spray increases. The droplets on the outer edge of the spray evaporate first, creating a fuel vapor–air mixture sheath around the liquid-containing core. The highest velocities are on the jet axis. The equivalence ratio is highest on the centerline (and fuel-rich along most of the jet), decreasing to zero (unmixed air) at the spray boundary. Once the sprays have penetrated to the outer regions of the combustion chamber, they interact with the chamber walls. The spray is then forced to flow tangentially along the wall. Eventually the sprays from multihole nozzles

---

† The back lighting identifies regions where sufficient liquid fuel (as ligaments or drops) is present to attenuate the light. The shadowgraph technique responds to density gradients in the test section, so it identifies regions where fuel vapor exists.

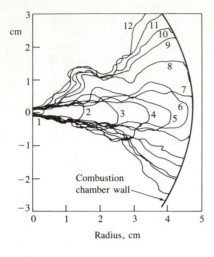

FIGURE 10-21
Sketches of outer vapor boundary of diesel fuel spray from 12 successive frames of rapid-compression-machine high-speed shadowgraph movie showing interaction of vaporizing spray with cylindrical wall of combustion chamber. Injection pressure 60 MPa. Time between frames 0.14 ms.[20]

interact with one another. Figure 10-21 shows diesel fuel sprays interacting with the cylindrical outer wall of a disc-shaped combustion chamber in a rapid-compression machine, under typical diesel-injection conditions. The cylinder wall causes the spray to split with about half flowing circumferentially in either direction. Adjacent sprays then interact forcing the flow radially inward toward the chamber axis.[20]

Most of the other combustion systems in Figs. 10-1 and 10-2 use air swirl to increase fuel-air mixing rates. A schematic of the spray pattern which results when a fuel jet is injected radially outward into a swirling flow is shown in Fig. 10-22. Because there is now relative motion in both radial and tangential directions between the initial jet and the air, the structure of the jet is more complex. As the spray entrains air and slows down it becomes increasingly bent toward the swirl direction; for the same injection conditions it will penetrate less with swirl than without swirl. An important feature of the spray is the large vapor-containing region downstream of the liquid-containing core. Figure 10-23 shows schlieren photographs of four fuel jets injected on the axis of an IDI diesel engine

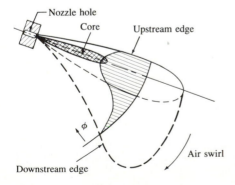

FIGURE 10-22
Schematic of fuel spray injected radially outward from the chamber axis into swirling air flow. Shape of equivalence ratio (ϕ) distribution within jet is indicated.

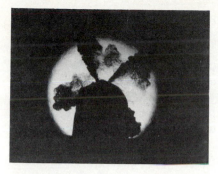

FIGURE 10-23
Schlieren photographs of vaporizing sprays injected into swirling air flow in transparent prechamber of special IDI diesel.[21] Left: high sensitivity, showing boundaries of the vapor regions of spray. Right: low sensitivity, showing liquid-containing core (dark) in relation to vapor regions (mottled).

prechamber with high clockwise swirl. The photograph on the left, with high sensitivity, shows the outer boundary of the fuel vapor region of the spray; the low-sensitivity photograph on the right locates the liquid phase regions of the spray.[21] The interaction between the swirl and both liquid and vapor spray regions is evident, as is the spray interaction with the chamber wall.

Other spray flow patterns are used. The spray may enter the swirling air flow tangentially as in the M.A.N. " M " system shown in Fig. 10-1c. The spray then interacts immediately with the combustion chamber walls.

To couple the spray-development process with the ignition phase of the combustion, it is important to know which regions of the spray contain the fuel injected at the beginning of the injection process. These regions of the sprays are likely to autoignite first. Each spray develops as follows. At the start of injection the liquid fuel enters the quiescent air charge, atomizes, moves outward from the nozzle, and slows down rapidly as air is entrained into the spray and accelerated. This start-up process forms a vortex or "puff" at the head of the spray. The injected fuel which follows encounters less resistance; thus drops from that fuel overtake the drops from first-injected fuel, forcing them outward toward the periphery of the spray. At the tip of the unsteady spray the drops meet the highest aerodynamic resistance and slow down, but the spray continues to penetrate the air charge because droplets retarded at the tip are continually replaced by new higher-momentum later-injected drops.[22] Accordingly, droplets in the periphery of the spray and behind the tip of the spray come from the earliest injected fuel.[23] As Figs. 10-20 and 10-23 indicate, these drops evaporate quickly.

10.5.3 Atomization

Under diesel engine injection conditions, the fuel jet usually forms a cone-shaped spray at the nozzle exit. This type of behavior is classified as the atomization breakup regime, and it produces droplets with sizes very much less than the nozzle exit diameter. This behavior is different from other modes of liquid jet breakup. At low jet velocity, in the Rayleigh regime, breakup is due to the unstable growth of surface waves caused by surface tension and results in drops larger than the jet diameter. As jet velocity is increased, forces due to the relative motion of the jet and the surrounding air augment the surface tension force, and lead to drop sizes of the order of the jet diameter. This is called the first wind-induced breakup regime. A further increase in jet velocity results in breakup characterized by divergence of the jet spray after an intact or undisturbed length downstream of the nozzle. In this second wind-induced breakup regime, the unstable growth of short-wavelength waves induced by the relative motion between the liquid and surrounding air produces droplets whose average size is much less than the jet diameter. Further increases in jet velocity lead to breakup in the atomization regime, where the breakup of the outer surface of the jet occurs at, or before, the nozzle exit plane and results in droplets whose average diameter is much smaller than the nozzle diameter. Aerodynamic interactions at the liquid/gas interface appear to be one major component of the atomization mechanism in this regime.[22, 24]

A sequence of very short time exposure photographs of the emergence of a liquid jet from a nozzle of 0.34 mm diameter and $L_n/d_n = 4$ into high-pressure nitrogen at ambient temperature is shown in Fig. 10-24. The figure shows how the spray tip penetrates and the spray spreads during the early part of its travel.[25] Data such as these were used to examine the dependence of the spray development on gas and liquid density, liquid viscosity, and nozzle geometry.[24-26] The effects of the most significant variables, gas/liquid density ratio and nozzle geometry, on initial jet spreading angle are shown in Fig. 10-25. For a given geometry (cylindrical hole and length/diameter = 4), the initial jet spreading or spray angle increases with increasing gas/liquid density ratio as shown in Fig. 10-25a. Typical density ratios for diesel injection conditions are between 15×10^{-3} and 30×10^{-3}. Of several different nozzle geometry parameters examined, the length/diameter ratio proved to be the most significant (see Fig. 10-25b).

For jets in the atomization regime, the spray angle θ was found to follow the relationship

$$\tan \frac{\theta}{2} = \frac{1}{A} \, 4\pi \left(\frac{\rho_g}{\rho_l} \right)^{1/2} \frac{\sqrt{3}}{6} \tag{10.22}$$

where ρ_g and ρ_l are gas and liquid densities and A is a constant for a given nozzle geometry.† The data in Fig. 10-25a are fitted by $A = 4.9$. This behavior is in accord with the theory that aerodynamic interactions are largely responsible for jet breakup. Note that the data in Fig. 10-25b show a continuous trend as the jet breakup regime makes a transition from second wind-induced breakup (solid

---

† An empirical equation for A is $A = 3.0 + 0.28 \, (L_n/d_n)$, where L_n/d_n is the length/diameter ratio of the nozzle.[25]

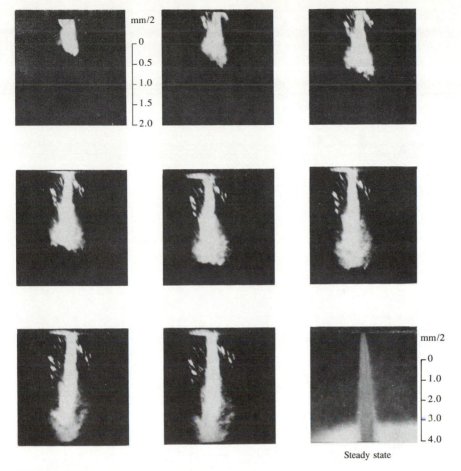

FIGURE 10-24
Photographs showing initial emergence and steady state (bottom right) of high-pressure liquid spray.
Time between frames 2.1 μs. Liquid: water. Gas: nitrogen at 1380 kPa. Δp across nozzle 11 MPa.
Nozzle diameter 0.34 mm.[25]

symbols) to atomization regime breakup (open symbols). The growth of aerodynamic surface waves is known to be responsible for jet breakup in the second
wind-induced breakup regime. Such a mechanism can explain the observed data
trends in the atomization regime, if an additional mechanism is invoked to
explain nozzle geometry effects. One possible additional mechanism is liquid
cavitation. A criterion for the onset of jet atomization at the nozzle exit plane was
developed. For $(\rho_l/\rho_g)(\text{Re}_l/\text{We}_l)^2 > 1$ (which is true for distillate fuel injection
applications) the design criterion is

$$\left(\frac{\rho_l}{\rho_g}\right)^{1/2} < k \tag{10.23}$$

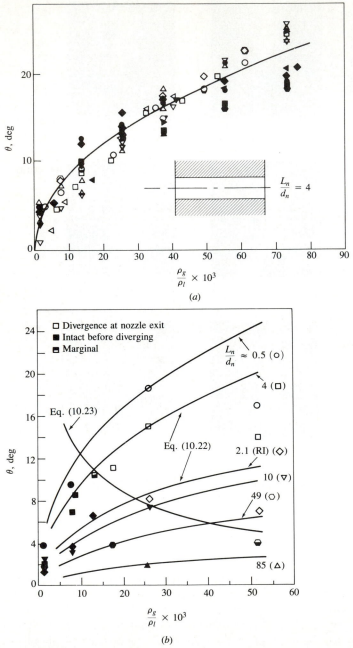

FIGURE 10-25
(a) Initial spray angle of atomizing jets versus density ratio (ρ_g/ρ_l = gas density/liquid density) for fixed nozzle geometry shown. Various fluids and gases at liquid pressures of 3.4–92 MPa. Nozzle diameters d_n = 0.254, 0.343, and 0.61 mm.[22] (b) Initial spray angle versus density ratio for a wide range of nozzle geometries. L_n/d_n = nozzle length/diameter ratio (RI = rounded inlet geometry). Solid symbols indicate jets which break up and diverge downstream of nozzle exit. Open symbols indicate jet breakup at nozzle exit.[25]

where k is an empirical constant depending on nozzle geometry in the range 6 to 12 ($k = 18.3/\sqrt{A}$).

Jet breakup trends can be summarized as follows. The initial jet divergence angles increase with increasing gas density. Divergence begins progressively closer to the nozzle as gas density increases until it reaches the nozzle exit. Jet divergence angles increase with decreasing fuel viscosity; divergence begins at the nozzle exit once the liquid viscosity is below a certain level. Nozzle design affects the onset of the jet atomization regime. Jet divergence angles decrease with increasing nozzle length. For the same length, rounded inlet nozzles produce less divergent jets than sharp-edged inlet nozzles. The initial jet divergence angle and intact spray length are quasi steady with respect to changes in operating conditions which occur on time scales longer than about 20 μs.[25] Note that while all these results were obtained under conditions where evaporation was not occurring, the initial spray-development processes are not significantly affected by evaporation (see Sec. 10.5.6).

10.5.4 Spray Penetration

The speed and extent to which the fuel spray penetrates across the combustion chamber has an important influence on air utilization and fuel-air mixing rates. In some engine designs, where the walls are hot and high air swirl is present, fuel impingement on the walls is desired. However, in multispray DI diesel combustion systems, overpenetration gives impingement of liquid fuel on cool surfaces which, especially with little or no air swirl, lowers mixing rates and increases emissions of unburned and partially burned species. Yet underpenetration results in poor air utilization since the air on the periphery of the chamber does not then contact the fuel. Thus, the penetration of liquid fuel sprays under conditions typical of those found in diesel engines has been extensively studied.

Many correlations based on experimental data and turbulent gas jet theory have been proposed for fuel spray penetration.[17] These predict the penetration S of the fuel spray tip across the combustion chamber for injection into quiescent air, as occurs in larger DI engines, as a function of time. An evaluation of these correlations[27] indicated that the formula developed by Dent,[28] based on a gas jet mixing model for the spray, best predicts the data:†

$$S = 3.07\left(\frac{\Delta p}{\rho_g}\right)^{1/4}(td_n)^{1/2}\left(\frac{294}{T_g}\right)^{1/4} \tag{10.24}$$

where Δp is the pressure drop across the nozzle, t is time after the start of injection, and d_n is the nozzle diameter. All quantities are expressed in SI units: t in

---

† For nozzles where $2 \leq L_n/d_n \leq 4$, and for $t > 0.5$ ms. At exceptionally high chamber densities ($p > 100$ atm) Eq. (10.24) overpredicts penetration.

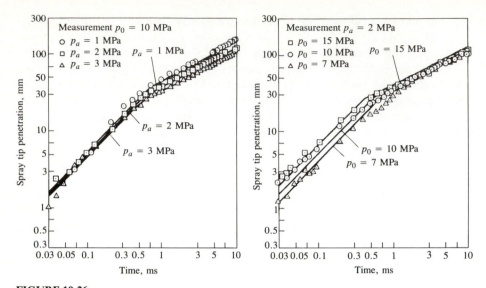

FIGURE 10-26
Spray tip penetration as function of time at various ambient pressures (p_a) and injection pressures (p_0). Fuel jets injected into quiescent air at room temperature.[29]

seconds, S and d_n in meters, Δp in pascals, ρ_g in kilograms per cubic meter, and T_g in kelvins.

More detailed studies have examined the spray tip location as a function of time, following start of a diesel injection process in high-pressure bombs. Data taken by Hiroyasu et al.,[29] shown in Fig. 10-26, illustrate the sensitivity of the spray tip position as a function of time to ambient gas state and injection pressure for fuel jets injected into quiescent air at room temperature. These data show that the initial spray tip penetration increases linearly with time t (i.e., the spray tip velocity is constant) and, following jet breakup, then increase as $\sqrt{t}$. Injection pressure has a more significant effect on the initial motion before breakup; ambient gas density has its major impact on the motion after breakup. Hiroyasu et al. correlated their data for spray tip penetration S(m) versus time as

$$t < t_{\text{break}}: \qquad S = 0.39 \left(\frac{2\Delta p}{\rho_l} \right)^{1/2} t$$

$$t > t_{\text{break}}: \qquad S = 2.95 \left(\frac{\Delta p}{\rho_g} \right)^{1/4} (d_n t)^{1/2}$$

(10.25)

where

$$t_{\text{break}} = \frac{29 \rho_l d_n}{(\rho_g \Delta p)^{1/2}}$$

(10.26)

and Δp is the pressure drop across the nozzle (pascals), ρ_l and ρ_g are the liquid and gas densities, respectively (in kilograms per cubic meter), d_n is the nozzle

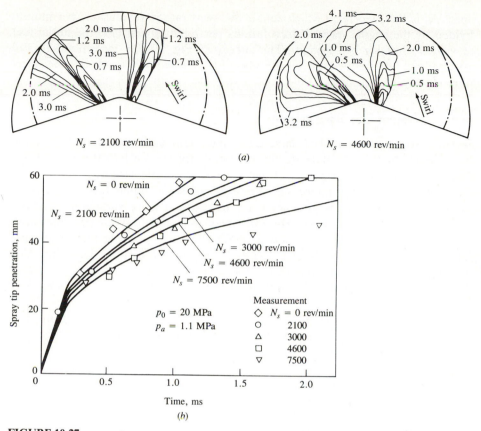

FIGURE 10-27
(a) Measured outer boundary of sprays injected into swirling air flow. (b) Spray tip penetration as a function of time for different swirl rates. Solid lines show Eq. (10.27).[29]

diameter (meters), and t is time (seconds). The results of Reitz and Bracco[25] indicate that the breakup or intact length depends on nozzle geometry details in addition to the diameter (see Fig. 10-25b). Note that under high injection pressures and nozzle geometries with short length/diameter ratios, the intact or breakup length becomes very short; breakup can occur at the nozzle exit plane.

The effect of combustion air swirl on spray penetration is shown in Fig. 10-27. Figure 10-27a shows how the spray shape and location changes as swirl is increased; Fig. 10-27b shows how spray tip penetration varies with time and swirl rate.[29] These authors related their data on spray tip penetration with swirl, S_s, through a correlation factor to the equivalent penetration, S, without swirl given by Eq. (10.25):

$$\frac{S_s}{S} = \left(1 + \frac{\pi R_s N S}{30 v_j}\right)^{-1} \tag{10.27}$$

where R_s is the swirl ratio which equals the swirl rate in revolutions per minute divided by the engine speed N (revolutions per minute), and v_j is the initial fuel jet velocity (meters per second). The curves in Fig. 10-27b correspond to Eq. (10.27). Swirl both reduces the penetration of the spray and spreads out the spray more rapidly.

10.5.5   Droplet Size Distribution

Previous sections in Sec. 10.5 have discussed the overall characteristics of the diesel engine fuel spray—its spreading rate and penetration into the combustion chamber. While the distribution of fuel via the spray trajectory throughout the combustion chamber is important, atomization of the liquid fuel into a large number of small drops is also necessary to create a large surface area across which liquid fuel can evaporate. Here we review how the drop size distribution in the fuel spray depends on injection parameters and the air and fuel properties. Since the measurement of droplet characteristics in an operating diesel engine is extremely difficult, most results have come from studies of fuel injection into constant-volume chambers filled with high-pressure quiescent air at room temperature.

During the injection period, the injection conditions such as injection pressure, nozzle orifice area, and injection rate may vary. Consequently, the droplet size distribution at a given location in the spray may also change with time during the injection period. In addition, since the details of the atomization process are different in the spray core and at the spray edge, and the trajectories of individual drops depend on their size, initial velocity, and location within the spray, the drop size distribution will vary with position within the spray.[29] None of these variations has yet been adequately quantified.

The aerodynamic theory of jet breakup in the atomization regime summarized in Sec. 10.5.3 (which is based on work by G. I. Taylor) leads to the prediction that the initial average drop diameter D_d is proportional to the length of the most unstable surface waves:[22]

$$\bar{D}_d = C \frac{2\pi\sigma}{\rho_g v_r^2} \lambda^* \tag{10.28}$$

where σ is the liquid-fuel surface tension, ρ_g is the gas density, v_r is the relative velocity between the liquid and gas (taken as the mean injection velocity v_j), C is a constant of order unity, and λ^* is the dimensionless wavelength of the fastest growing wave. λ^* is a function of the dimensionless number $(\rho_l/\rho_g)(\text{Re}_j/\text{We}_j)^2$, where the jet Reynolds and Weber numbers are given by $\text{Re}_j = \rho_l v_j d_n/\mu_l$ and $\text{We}_j = \rho_l v_j^2 d_n/\sigma$ and d_n is the nozzle orifice diameter. λ^* goes to 3/2 as this number increases above unity. Near the edge of the spray close to the nozzle, this equation predicts observed drop size trends with respect to injection velocity, fuel properties, nozzle L/d, and nozzle diameter, though measured mean drop sizes are larger by factors of 2 to 3.[30] However, within the dense early region of the spray, secondary atomization phenomena—coalescence and breakup—occur

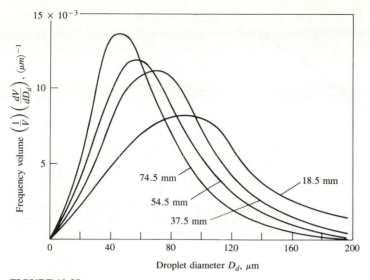

FIGURE 10-28
Droplet size distribution in diesel fuel spray injected through throttling pintle nozzle into quiescent room-temperature air at 11 atm. Nozzle opening pressure 9.9 MPa. Pump speed 500 rev/min. Droplets are sampled well downstream of injector at given radial distances from spray axis.[32]

which will change the droplet size distribution and mean diameter. The downstream drop size in the solid-cone sprays used in diesel-injection systems is markedly influenced by both drop coalescence and breakup. Eventually a balance is reached as coalescence decreases (due to the expansion of the spray) and breakup ceases (due to the reduced relative velocity between the drops and the entrained gas).[31]

Measurements of droplet size distributions within a simulated diesel spray indicate how size varies with location. Figure 10-28 shows the variation in drop size distribution with radial distance from the spray axis, at a fixed axial location. The drop sizes were measured with a liquid immersion technique where a sample of drops is collected in a small cell filled with an immiscible liquid. Size distributions can be expressed in terms of:

1. The incremental number of drops Δn within the size range $D_d - \Delta D_d/2 < D_d < D_d + \Delta D_d/2$
2. The incremental volume ΔV of drops in this size range
3. The cumulative number of drops n less than a given size D_d
4. The cumulative volume V of drops less than a given size D_d

Since the drops are spherical:

$$\frac{dn}{dD_d} = \frac{6}{\pi D_d^3}\frac{dV}{dD_d}$$

(10.29)

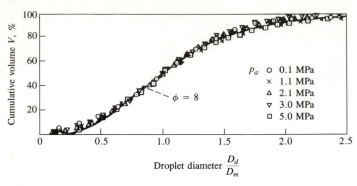

FIGURE 10-29
Normalized drop-size cumulative frequency distribution in spray injected into ambient-temperature air for air pressures from 0.1 to 5 MPa. Throttling pintle nozzle with nozzle opening pressure of 9.9 MPa. Median drop diameter $D_m = 1.224 D_{SM}$.[32]

The distributions shown in the figure are frequency distributions of drop volume.[32] The peak in the distribution shifts to larger drop diameters as the radial position decreases: on average, the drops are smaller at the periphery of the spray.

To characterize the spray, expressions for drop size distribution and mean diameter are desirable. An appropriate and commonly used mean diameter is the *Sauter mean diameter*:

$$D_{SM} = \left(\int D_d^3 \, dn \right) \Big/ \left(\int D_d^2 \, dn \right) \tag{10.30}$$

where dn is the number of drops with diameter D_d in the range $D_d - dD_d/2 < D_d < D_d + dD_d/2$. The integration is usually carried out by summing over an appropriate number of drop size groups. The Sauter mean diameter is the diameter of the droplet that has the same surface/volume ratio as that of the total spray.

Various expressions for the distribution of drop sizes in liquid sprays have been proposed. One proposed by Hiroyasu and Kadota[32] based on the chi-square statistical distribution fits the available experimental data. Figure 10-29 shows how the chi-square distribution with a degree of freedom equal to 8 fits well to experimental measurements of the type shown in Fig. 10-28. Here D_m is the median drop diameter which for this chi-square curve is $1.224 D_{SM}$. The non-dimensional expression for drop size distribution in sprays injected through hole nozzles, pintle nozzles, and throttling pintle nozzles given by the chi-square distribution is

$$\frac{dV}{V} = 13.5 \left(\frac{D_d}{D_{SM}} \right)^3 \exp\left[-3\left(\frac{D_d}{D_{SM}} \right) \right] d\left(\frac{D_d}{D_{SM}} \right) \tag{10.31}$$

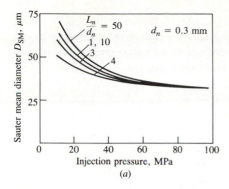
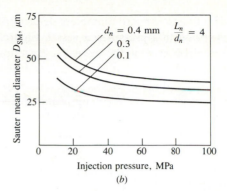

FIGURE 10-30
Effect of fuel-injection pressure and nozzle geometry and size on Sauter mean drop diameter. (a) Effect of nozzle length/diameter ratio L_n/d_n and injection pressure. (b) Effect of nozzle diameter d_n and injection pressure.[18]

An empirical expression for the Sauter mean diameter D_{SM} (in micrometers) for typical diesel fuel properties given by Hiroyasu and Kadota[32] is

$$D_{SM} = A(\Delta p)^{-0.135} \rho_a^{0.121} V_f^{0.131} \qquad (10.32)$$

where Δp is the mean pressure drop across the nozzle in megapascals, ρ_a is the air density in kilograms per cubic meter, and V_f is the amount of fuel delivered per cycle per cylinder in cubic millimeters per stroke. A is a constant which equals 25.1 for pintle nozzles, 23.9 for hole nozzles, and 22.4 for throttling pintle nozzles. Other expressions for predicting D_{SM} can be found in Ref. 17.

The effects of injection pressure, nozzle geometry and size, air conditions, and fuel properties on Sauter mean diameter in sprays obtained with diesel fuel-injection nozzles have been extensively studied. Various immersion, photographic, and optical techniques for making such measurements have been used.[17] Some of the major effects are illustrated in Figs. 10-30 and 10-31 which show

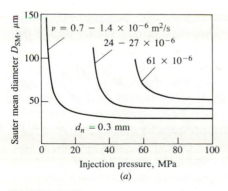
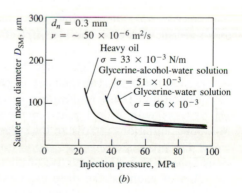

FIGURE 10-31
Effect of (a) liquid viscosity v and (b) liquid surface tension σ on Sauter mean drop diameter as a function of injection pressure. Air conditions: 3 MPa and ambient temperature.[17]

average Sauter mean diameters determined optically from studies of steady diesel fuel sprays in a pressurized vessel. Figure 10-30 shows that nozzle size affects the mean drop size in the expected direction. Nozzle length/diameter ratio is also shown to be important: an $L_n/d_n = 4$ gives the minimum mean drop size at low and intermediate injection pressures. This L_n/d_n also corresponds to the minimum value of spray breakup length and to the maximum spray cone angle. Fuel viscosity and surface tension also affect mean drop size as shown in Fig. 10-31, with the effects being most significant at lower injection pressures.

10.5.6   Spray Evaporation

The injected liquid fuel, atomized into small drops near the nozzle exit to form a spray, must evaporate before it can mix with air and burn. Figure 10-20 showed the basic structure of an evaporating diesel spray under conditions typical of a large direct-injection engine. Back illumination showed that a core exists along the axis of the spray where the liquid fuel ligaments or drops are sufficiently dense to attenuate the light beam. Once the start-up phase of the injection process is over, the length of this core remains essentially constant until injection ends. This core is surrounded by a much larger vapor-containing spray region which continues to penetrate deeper into the combustion chamber: the core extends only partway to the spray tip. Additional insight into the physical structure of evaporating sprays can be obtained from the schlieren photographs taken just after the end of injection in a prechamber engine with air swirl, shown in Fig. 10-32. The lowest magnification picture (A) shows the overall structure of the spray. The only liquid-containing region evident is that part of the core nearest the nozzle which shows black on the left of the photograph. The spreading vapor region of the spray, carried around the chamber by the swirling air flow, appears mottled due to local turbulent vapor concentration and temperature fluctuations. The dark region within the spray vapor region is due to soot formed where the fuel vapor concentration is sufficiently high. It is probable that, after the breakup length, the dense black liquid core of the spray is composed of individual droplets but the concentration is so high along the optical path that the light beam is fully extinguished. However, the last part of the core close to the nozzle tip (B) disperses sufficiently for individual features to be resolved. The small black dots are liquid fuel drops in the size range 20 to 100 μm. Fuel drop vapor trails can be observed in the highest magnification photo (C) corresponding to various stages of evaporation. These range from drops showing little surrounding vapor to vapor trails with little liquid remaining at the head. The vapor trails show random orientations relative to the spray axis, presumably due to local air turbulence. The process of droplet evaporation under normal engine operating conditions appears to be rapid relative to the total combustion period.[21]

Let us examine the drop evaporation process in more detail. Consider a liquid drop at close to ambient temperature injected into air at typical end-of-compression engine conditions. Three phenomena will determine the history of the drop under these conditions:

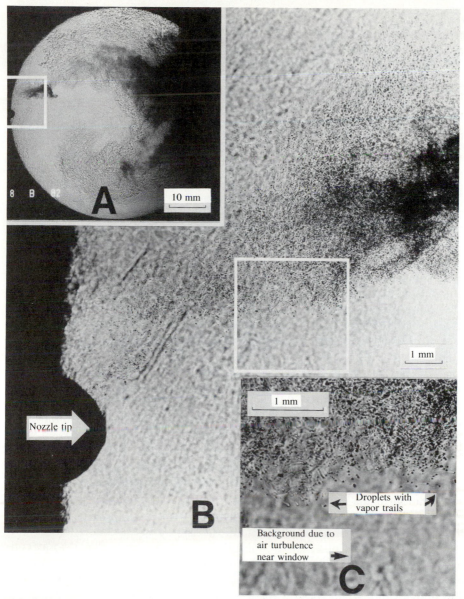

FIGURE 10-32
Shadowgraph photographs at three magnifications taken just after the end of injection of diesel-fuel spray into swirling air flow in prechamber of special diesel engine. Nozzle hole diameter = 0.25 mm.[21]

1. Deceleration of the drop due to aerodynamic drag
2. Heat transfer to the drop from the air
3. Mass transfer of vaporized fuel away from the drop

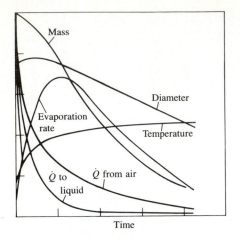

FIGURE 10-33
Schematic of variation of mass, diameter, temperature, evaporation rate, heat-transfer rate from air, and heat-transfer rate to liquid drop core as function of time during evaporation process of individual drop in diesel environment at the time of injection.

As the droplet temperature increases due to heat transfer, the fuel vapor pressure increases and the evaporation rate increases. As the mass transfer rate of vapor away from the drop increases, so the fraction of the heat transferred to the drop surface which is available to increase further the drop temperature decreases. As the drop velocity decreases, the convective heat-transfer coefficient between the air and the drop decreases. The combination of these factors gives the behavior shown in Fig. 10-33 where drop mass, temperature, velocity, vaporization rate, and heat-transfer rate from the air are shown schematically as a function of time following injection.[33] Analysis of individual fuel drops 25 μm in diameter, injected into air at typical diesel conditions, indicates that evaporation times are usually less than 1 ms.[34]

Such an analysis is relevant to drops that are widely separated (e.g., at the edge of the spray). In the spray core, where drop number densities are high, the evaporation process has a significant effect on the temperature and fuel-vapor concentration in the air within the spray. As fuel vaporizes, the local air temperature will decrease and the local fuel vapor pressure will increase. Eventually thermodynamic equilibrium would pertain: this is usually called adiabatic saturation.[33] Calculated thermodynamic equilibrium temperatures for diesel spray conditions are plotted in Fig. 10-34 as a function of the fuel/air mass ratio for *n*-dodecane and *n*-heptane. The initial liquid fuel temperature was 300 K. The ratio of fuel vapor to air mass at these equilibrium conditions is also shown. To the left of the peaks in the m_{fv}/m_a curves, only fuel vapor is present. To the right of these peaks, liquid fuel is also present because the vapor phase is saturated.[35] Liquid fuel vaporization causes substantial reductions in gas temperature. While this equilibrium situation may not be reached within the spray, these results are useful for understanding the temperature distribution within an evaporating spray.

To quantify accurately the fuel vaporization rate within a diesel fuel spray requires the solution of the coupled conservation equations for the liquid drop-

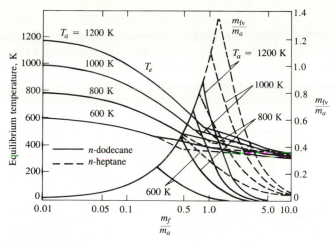

FIGURE 10-34
Adiabatic-saturation conditions for equilibrium mixtures formed by injecting n-dodecane and n-heptane, initially liquid at 300 K, into air at initial temperature T_a between 600 and 1200 K and initial density 6.5 kg/m^3. Equilibrium mixture temperature (T_e) and ratio of fuel vapor mass (m_{fv}) to air mass (m_a) shown as function of ratio of total fuel mass m_f to m_a. Fuel vapor only present to left of peaks in m_{fv}/m_a curves: liquid fuel also present to right of peaks.[35]

lets and the air within the combustion chamber. Various phenomenological models and computational fluid dynamic models have been developed for this purpose (see Secs. 14.4.3 and 14.5.5). In the most sophisticated of these, the spray is assumed to be composed of discrete computational particles each of which represents a group of droplets of similar size, temperature, etc. The distribution functions in droplet size, velocity, temperature, etc., produced by the fuel injector are statistically sampled and the resulting discrete particles are followed along lagrangian trajectories as they interact and exchange mass, momentum, and energy with the surrounding gas. Drops interact directly with each other via collisions and indirectly via evaporation by modifying the ambient vapor concentration and gas temperature. Studies with such models indicate that, under normal diesel engine conditions, 70 to 95 percent of the injected fuel is in the vapor phase at the start of combustion. Evaporation is more than 90 percent complete after 1 ms. However, only 10 to 35 percent of the vaporized fuel has mixed to within flammability limits in a typical medium-speed DI diesel engine. Thus combustion is largely mixing-limited, rather than evaporation-limited.[36] Of course, under cold-starting conditions, evaporation becomes a major constraint.

10.6   IGNITION DELAY

10.6.1   Definition and Discussion

The ignition delay in a diesel engine was defined as the time (or crank angle) interval between the start of injection and the start of combustion. The start of injection is usually taken as the time when the injector needle lifts off its seat (determined by a needle-lift indicator). The start of combustion is more difficult to determine precisely. It is best identified from the change in slope of the heat-release rate, determined from cylinder pressure data using the techniques described in Sec. 10.4, which occurs at ignition. Depending on the character of the combustion process, the pressure data alone may indicate when pressure change due to combustion first occurs; in DI engines under normal conditions ignition is well defined, but in IDI engines the ignition point is harder to identify. Flame luminosity detectors are also used to determine the first appearance of the flame. Experience has shown that under normal conditions, the point of appearance of the flame is later than the point of pressure rise and results in greater uncertainty or error in determining the ignition point.

Both physical and chemical processes must take place before a significant fraction of the chemical energy of the injected liquid fuel is released. The physical processes are: the atomization of the liquid fuel jet; the vaporization of the fuel droplets; the mixing of fuel vapor with air. The chemical processes are the pre-combustion reactions of the fuel, air, residual gas mixture which lead to autoignition. These processes are affected by engine design and operating variables, and fuel characteristics, as follows.

Good atomization requires high fuel-injection pressure, small injector hole diameter, optimum fuel viscosity, and high cylinder air pressure at the time of injection (see Sec. 10.5.3). The rate of vaporization of the fuel droplets depends on the size of the droplets, their distribution, and their velocity, the pressure and temperature inside the chamber, and the volatility of the fuel. The rate of fuel-air mixing is controlled largely by injector and combustion chamber design. Some combustion chamber and piston head shapes are designed to amplify swirl and create turbulence in the air charge during compression. Some engine designs use a prechamber or swirl chamber to create the vigorous air motion necessary for rapid fuel-air mixing (see Sec. 10.2). Also, injector design features such as the number and spatial arrangement of the injector holes determine the fuel spray pattern. The details of each nozzle hole affect the spray cone angle. The penetration of the spray depends on the size of the fuel droplets, the injection pressure, the air density, and the air-flow characteristics. The arrangement of the sprays, the spray cone angle, the extent of spray penetration, and the air flow all affect the rate of air entrainment into the spray. These physical aspects of fuel-injection and fuel-spray behavior are reviewed in Sec. 10.5.

The chemical component of the ignition delay is controlled by the precombustion reactions of the fuel. A fundamental discussion of autoignition or spontaneous hydrocarbon oxidation in premixed fuel-air mixtures is given in Sec. 9.6.2. Since the diesel engine combustion process is heterogeneous, its spontaneous ignition process is even more complex. Though ignition occurs in vapor phase regions, oxidation reactions can proceed in the liquid phase as well between the fuel molecules and the oxygen dissolved in the fuel droplets. Also, cracking of large hydrocarbon molecules to smaller molecules is occurring. These chemical processes depend on the composition of the fuel and the cylinder charge temperature and pressure, as well as the physical processes described above which govern the distribution of fuel throughout the air charge.

Since the ignition characteristics of the fuel affect the ignition delay, this property of a fuel is very important in determining diesel engine operating characteristics such as fuel conversion efficiency, smoothness of operation, misfire, smoke emissions, noise, and ease of starting. The ignition quality of a fuel is defined by its cetane number. Cetane number is determined by comparing the ignition delay of the fuel with that of primary reference fuel mixtures in a standardized engine test (see below). For low cetane fuels with too long an ignition delay, most of the fuel is injected before ignition occurs, which results in very rapid burning rates once combustion starts with high rates of pressure rise and high peak pressures. Under extreme conditions, when autoignition of most of the injected fuel occurs, this produces an audible knocking sound, often referred to as "diesel knock." For fuels with very low cetane numbers, with an exceptionally long delay, ignition may occur sufficiently late in the expansion process for the burning process to be quenched, resulting in incomplete combustion, reduced power output, and poor fuel conversion efficiency. For higher cetane number fuels, with shorter ignition delays, ignition occurs before most of the fuel is injected. The rates of heat release and pressure rise are then controlled primarily by the rate of injection and fuel-air mixing, and smoother engine operation results.

10.6.2 Fuel Ignition Quality

The ignition quality of a diesel fuel is defined by its *cetane number*. The method used to determine the ignition quality in terms of cetane number is analogous to that used for determining the antiknock quality of gasoline in terms of octane number. The cetane number scale is defined by blends of two pure hydrocarbon reference fuels. Cetane (*n*-hexadecane, $C_{16}H_{34}$), a hydrocarbon with high ignition quality, represents the top of the scale with a cetane number of 100. An isocetane, heptamethylnonane (HMN), which has a very low ignition quality, represents the bottom of the scale with a cetane number of 15.† Thus, cetane number (CN) is given by

$$CN = \text{percent } n\text{-cetane} + 0.15 \times \text{percent HMN} \qquad (10.33)$$

The engine used in cetane number determination is a standardized single-cylinder, variable compression ratio engine with special loading and accessory equipment and instrumentation. The engine, the operating conditions, and the test procedure are specified by ASTM Method D613.[37] The operating requirements include: engine speed—900 rev/min; coolant temperature—100°C; intake air temperature—65.6°C (150°F); injection timing—13° BTC; injection

---

† In the original procedure α-methylnapthalene ($C_{11}H_{10}$) with a cetane number of zero represented the bottom of the scale. Heptamethylnonane, a more stable compound, has replaced it.

pressure—10.3 MPa (1500 lb/in^2). With the engine operating under these conditions, on the fuel whose cetane number is to be determined, the compression ratio is varied until combustion starts at TC: i.e., an ignition delay period of 13° (2.4 ms at 900 rev/min) is produced. The above procedure is then repeated using reference fuel blends. Each time a reference fuel is tried, the compression ratio is adjusted to give the same 13° ignition delay. When the compression ratio required by the actual fuel is bracketed by the values required by two reference blends differing by less than five cetane numbers, the cetane number of the fuel is determined by interpolation between the compression ratios required by the two reference blends.

Because of the expense of the cetane number test, many correlations which predict ignition quality based on the physical properties of diesel fuels have been developed.[38, 39] A calculated *cetane index* (CCI) is often used to estimate ignition quality of diesel fuels (ASTM D976[40]). It is based on API gravity and the mid-boiling point (temperature 50 percent evaporated). It is applicable to straight-run fuels, catalytically cracked stocks, and blends of the two. Its use is suitable for most diesel fuels and gives numbers that correspond quite closely to cetane number. A *diesel index* is also used. It is based on the fact that ignition quality is linked to hydrocarbon composition: *n*-paraffins have high ignition quality, and aromatic and napthenic compounds have low ignition quality. The aniline point (ASTM D611[41]—the lowest temperature at which equal volumes of the fuel and aniline become just miscible) is used, together with the API gravity, to give the diesel index:

$$\text{Diesel index} = \text{aniline point (°F)} \times \frac{\text{API gravity†}}{100} \qquad (10.34)$$

The diesel index depends on the fact that aromatic hydrocarbons mix completely with aniline at comparatively low temperatures, whereas paraffins require considerably higher temperatures before they are completely miscible. Similarly, a high API gravity denotes low specific gravity and high paraffinicity, and, again, good ignition quality. The diesel index usually gives values slightly above the cetane number. It provides a reasonable indication of ignition quality in many (but not all) cases.

10.6.3 Autoignition Fundamentals

Basic studies in constant-volume bombs, in steady-flow reactors, and in rapid-compression machines have been used to study the autoignition characteristics of fuel-air mixtures under controlled conditions. In some of these studies the fuel and air were premixed; in some, fuel injection was used. Studies with fuel injec-

---

† API gravity is based on specific gravity and is calculated from: API gravity, deg = (141.5/specific gravity at 60°F) − 131.5.

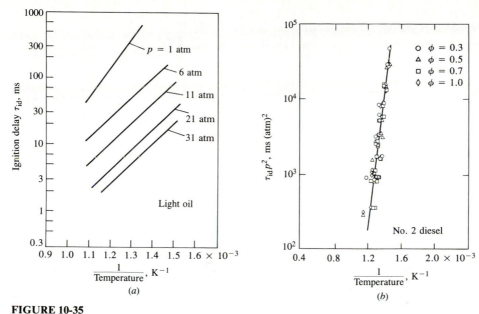

FIGURE 10-35
(a) Ignition delay as function of reciprocal air temperature for light oil spray injected into constant-volume combustion bomb. Injection pressure 9.8 MPa (100 atm). Air pressures indicated.[42] (b) Ignition delay × (pressure)[2] measured in steady-flow reactor for No. 2 diesel fuel as function of reciprocal temperature. Fuel/air equivalence ratio ϕ varied from 0.3 to 1.0.[43]

tion into constant-temperature and pressure environments have shown that the temperature and pressure of the air are the most important variables for a given fuel composition. Ignition delay data from these experiments have usually been correlated by equations of the form:

$$\tau_{id} = Ap^{-n} \exp\left(\frac{E_A}{\tilde{R}T}\right) \tag{10.35}$$

where τ_{id} is the ignition delay (the time between the start of injection and the start of detectable heat release), E_A is an apparent activation energy for the fuel autoignition process, $\tilde{R}$ is the universal gas constant, and A and n are constants dependent on the fuel (and, to some extent, the injection and air-flow characteristics).

Figure 10-35a shows ignition delay data obtained by injecting liquid fuel sprays into a high-pressure heated constant-volume bomb.[42] Figure 10-35b shows ignition delay data from a steady-flow high-pressure reactor where vaporized fuel was mixed rapidly with the heated air stream.[43] The match between the form of Eq. (10.35) and the data is clear. Figure 10-35b also shows an equivalence ratio dependence of the ignition delay. Representative values for A, n, and E_A for Eq. (10.35), taken from these and other studies, are given in Table 10.3. Ignition delay times calculated with these formulas for various diesel engines are given in

TABLE 10.3

Constants for Arrhenius equation for ignition delay:[44]

$$\tau_{id}(ms) = A p^{-n} \exp\left[E_A/(\tilde{R}T)\right]$$

Investigator	Apparatus	Fuel	p, atm	T, K	n	A	$E_A/\tilde{R}$, K
		Conditions				**Parameters**	
Spadaccini and TeVelde[43] No. 1	Steady flow	No. 2 diesel	10–30	650–900	2	2.43×10^{-9}	20,926
Spadaccini and TeVelde[43] No. 2	Steady flow	No. 2 diesel	10–30	650–900	1	4.00×10^{-10}	20,080
Stringer et al.[45]	Steady flow	Diesel 45–50 cetane number	30–60	770–980	0.757	0.0405	5,473
Wolfer[46]	Constant-volume bomb	Fuel with cetane number > 50	8–48	590–782	1.19	0.44	4,650
Hiroyasu et al.[29]	Constant-volume bomb	Kerosene	1–30	673–973	1.23	0.0276	7,280

Table 10.4. Air pressures and temperatures at TC piston position were estimated from measured cylinder pressure data. Measured ignition delay times in these types of engines are: 0.6 to 3 ms for low-compression-ratio DI diesel engines over a wide range of operating conditions; 0.4 to 1 ms for high-compression-ratio and turbocharged DI diesel engines; 0.6 to 1.5 ms for IDI diesel engines.[44]

The variation in the calculated delay times can be attributed to several factors:

1. In some cases the correlations are being extrapolated outside their original range of operating conditions.
2. The methods used to detect the start of combustion, and hence the duration of the delay, are not identical.
3. The experimental apparatus and the method of fuel-air mixture preparation are different.

The third factor is probably the most significant. As has been explained, the phenomenon of autoignition of a fuel spray consists of sequences of physical and chemical processes of substantial complexity. The relative importance of each process depends on the ambient conditions, on fuel properties, and on how the fuel-air mixture is prepared. For example, fuel evaporation times are significant in cold engines, but not under fully warmed-up conditions. Thus, an equation of the simple form of Eq. (10.35) can only fit the data over a limited range of conditions. The correlations of Spadaccini and TeVelde[43] have much higher activation energies. Normally, lower values of $E_A/\tilde{R}$ imply that physical processes such as vaporization and mixing are important and relevant to chemical processes. Thus, fuel preparation, mixture inhomogeneity, heat loss, and nonuniform flow patterns

TABLE 10.4
Calculated ignition delay times[44]

	Conditions			Spadaccini and TeVelde[43]		τ_{id}, ms		
Engine	Speed, rev/min	p, atm	T, K	No. 1	No. 2	Stringer et al.[45]	Wolfer[46]	Hiroyasu et al.[29]
IDI Diesel								
1. Low swirl	600	45.6	690	17.3	38.2	6.26	3.94	9.60
	1200	49.3	747	1.47	3.83	3.22	2.15	3.90
	1800	52.5	758	0.86	2.44	2.76	1.82	3.13
2. High swirl	600	45.2	674	36.3	76.9	7.60	4.68	12.5
	1200	48.4	721	4.18	10.3	4.25	2.75	5.67
	1800	51.8	744	1.47	4.07	3.19	2.08	3.82
DI Diesel								
1. Low compression ratio		42.8	781	0.57	1.37	2.60	1.92	2.39
2. High compression ratio	1500	58.8	975	0.0015	0.0060	0.508	0.407	0.322

affect the ignition delay. While the work of Spadaccini and TeVelde probably describes the chemical ignition delay more accurately, since great care was taken to obtain a uniform mixture and flow, the experiments in constant-volume bombs with diesel-type fuel injectors are more relevant to the diesel engine compression-ignition combustion process because they include the appropriate physical and chemical processes. The available engine ignition delay data suggest that for delays less than about 1 ms, the rate of decrease in delay with increasing temperature becomes much less than that indicated by the data in Fig. 10-35. This is due to the increasing relative importance of physical processes relative to chemical processes during the delay period. Thus relations of the form of Eq. (10.35) should be used with caution.

In general, τ_{id} is a function of mixture temperature, pressure, equivalence ratio, and fuel properties (though no accepted form for the variation with equivalence ratio is yet established). In the above referenced studies, the fuel was injected into a uniform air environment where the pressure and temperature only changed due to the cooling effect of the fuel-vaporization and fuel-heating processes. In an engine, pressure and temperature change during the delay period due to the compression resulting from piston motion. To account for the effect of changing conditions on the delay the following empirical integral relation is usually used:

$$\int_{t_{si}}^{t_{si} + \tau_{id}} \left(\frac{1}{\tau}\right) dt = 1 \tag{10.36}$$

where t_{si} is the time of start of injection, τ_{id} is the ignition delay period, and τ is the ignition delay at the conditions pertaining at time t. Whether the variation in conditions is significant depends on the amount of injection advance before TC that is used and the length of the delay.

10.6.4   Physical Factors Affecting Delay

The physical factors that affect the development of the fuel spray and the air charge state (its pressure, temperature, and velocity) will influence the ignition delay. These quantities depend on the design of the fuel-injection system and combustion chamber, and the engine operating conditions. The injection system variables affecting the fuel-spray development are injection timing, quantity, velocity, rate, drop size, and spray form or type. The relevant charge conditions depend on the combustion system employed, the details of the combustion chamber design, inlet air pressure and temperature, compression ratio, injection timing, the residual gas conditions, coolant and oil temperature, and engine speed. Data on these interactions are available for different types of diesel engines. The trends observed with the different diesel combustion systems are generally similar, though some of the details are different. In this section the ignition delay trends during normal (fully warmed-up) engine operation are considered. The dependence of the ignition delay on engine design and operating variables during engine starting and warm-up is also very important, and may be different from fully warmed-up behavior due to lower air temperature and pressure.

INJECTION TIMING. At normal engine conditions (low to medium speed, fully warmed engine) the minimum delay occurs with the start of injection at about 10 to 15° BTC.[47] The increase in the delay with earlier or later injection timing occurs because the air temperature and pressure change significantly close to TC. If injection starts earlier, the initial air temperature and pressure are lower so the delay will increase. If injection starts later (closer to TC) the temperature and pressure are initially slightly higher but then decrease as the delay proceeds. The most favorable conditions for ignition lie in between.

INJECTION QUANTITY OR LOAD. Figure 10-36 shows the effect of increasing injection quantity or engine load on ignition delay. The delay decreases approximately linearly with increasing load for this DI engine. As load is increased, the residual gas temperature and the wall temperature increase. This results in higher charge temperature (and also, to a lesser extent, charge pressure) at injection, thus shortening the ignition delay. When adjustment is made for this increasing temperature, it is found that increasing the quantity of fuel injected has no significant effect on the delay period under normal operating conditions. Under engine starting conditions, however, the delay increases due to the larger drop in mixture temperature associated with evaporating and heating the increased amount of fuel.[47] This latter result should be expected since it is the first part of the injected fuel which ignites first; subsequent injected fuel (above the minimum required to maintain the fuel-air mixture within the flammability limits during the delay) does not influence the delay.

DROP SIZE, INJECTION VELOCITY, AND RATE. These quantities are determined by injection pressure, injector nozzle hole size, nozzle type, and geometry.

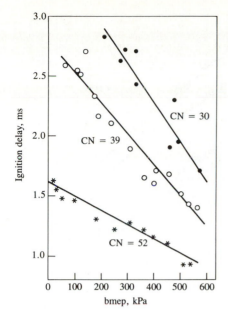

FIGURE 10-36
Ignition delays measured in a small four-stroke cycle DI diesel engine with $r_c = 16.5$ as a function of load at 1980 rev/min. Fuel cetane numbers 30, 39, and 52.[48]

Experiments by Lyn and Valdmanis[47] have shown that none of these factors has a significant effect on the delay. At normal operating conditions, increasing injection pressure produces only modest decreases in the delay. Doubling the nozzle hole diameter at constant injection pressure to increase the fuel flow rate (by a factor of about 4) and increase the drop size (by about 30 percent) had no significant effect on the delay. Studies of different nozzle hole geometries showed that the length/diameter ratio of the nozzle was not significant; nor did changes in nozzle type (multihole, pintle, pintaux) cause any substantial variation in delay at normal engine conditions.

INTAKE AIR TEMPERATURE AND PRESSURE. Figure 10-35 showed values of ignition delay for diesel fuels plotted against the reciprocal of charge temperature for several charge pressures at the time of injection. The intake air temperature and pressure will affect the delay via their effect on charge conditions during the delay period. Figure 10-37 shows the effects of inlet air pressure and temperature as a function of engine load. The fundamental ignition data available show a strong dependence of ignition delay on charge temperatures below about 1000 K at the time of injection. Above about 1000 K, the data suggest that the charge temperature is no longer so significant. Through this temperature range there is an effect of pressure at the time of injection on delay: the higher the pressure the shorter the delay, with the effect decreasing as charge temperatures increase and delay decreases. Since air temperature and pressure during the delay period are such important variables, other engine variables that affect the relation between the inlet air state and the charge state at the time of injection will influence the delay. Thus, an increase in the compression ratio will decrease the ignition delay,

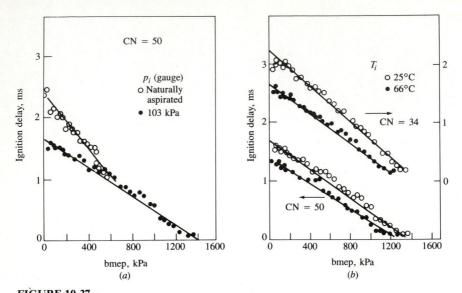

FIGURE 10-37
Effect of inlet air pressure and temperature on ignition delay over load range of small DI diesel at 1980 rev/min. (*a*) Engine naturally aspirated and with 1 atm boost; inlet air temperature $T_i = 25°C$; 50 cetane number fuel. (*b*) Engine naturally aspirated; $T_i = 25$ and $66°C$; 34 and 50 cetane number fuel.[48]

and injection timing will affect the delay (as was discussed earlier), largely due to the changes in charge temperature and pressure at the time of injection.

ENGINE SPEED. Increases in engine speed at constant load result in a slight decrease in ignition delay when measured in milliseconds; in terms of crank angle degrees, the delay increases almost linearly.[48] A change in engine speed changes the temperature/time and pressure/time relationships. Also, as speed increases, injection pressure increases. The peak compression temperature increases with increasing speed due to smaller heat loss during the compression stroke.[47]

COMBUSTION CHAMBER WALL EFFECTS. The impingement of the spray on the combustion chamber wall obviously affects the fuel evaporation and mixing processes. Impingement of the fuel jet on the wall occurs, to some extent, in almost all of the smaller, higher speed engines. With the "M" system, this impingement is desired to obtain a smooth pressure rise. The ignition delay with the "M" system is longer than in conventional DI engine designs.[47] Engine and combustion bomb experiments have been carried out to examine the effect of wall impingement on the ignition delay. Figure 10-38 shows the effect of jet wall impingement on ignition delay measured in a constant-volume combustion bomb, for a range of air pressures and temperatures, and wall temperatures.[29] The wall was perpendicular to the spray and was placed 100 mm from the nozzle tip. The data shows that the presence of the wall reduces the delay at the lower

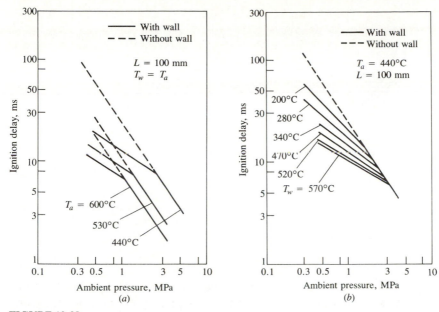

FIGURE 10-38
Effect of spray impingement on wall 100 mm from nozzle on ignition delay from combustion bomb studies. (a) Effect of air temperature as a function of air pressure; $T_w = T_{air}$. (b) Effect of wall temperature at 440°C air temperature.[29]

pressures and temperatures studied, but has no significant effect at the high pressures and temperatures more typical of normal diesel operation. Engine experiments where the delay was measured while the jet impingement process was varied showed analogous trends. The jet impingement angle (the angle between the fuel jet axis and the wall) was varied from almost zero (jet and wall close to parallel) to perpendicular. The delay showed a tendency to become longer as the impingement angle decreased. The most important result is not so much the modest change in delay but the difference in the initial rate of burning that results from the differences in fuel evaporation and fuel-air mixing rates.

SWIRL RATE. Changes in swirl rate change the fuel evaporation and fuel-air mixing processes. They also affect wall heat transfer during compression and, hence, the charge temperature at injection. Only limited engine studies of the effect of swirl rate variations on ignition delay have been made. At normal operating engine speeds, the effect of swirl rate changes on the delay are small. Under engine starting conditions (low engine speeds and compression temperatures) the effect is much more important,[47] due presumably to the higher rates of evaporation and mixing obtained with swirl.

OXYGEN CONCENTRATION. The oxygen concentration in the charge into which the fuel is injected would be expected to influence the delay. The oxygen concentration is changed, for example, when exhaust gas is recycled to the intake

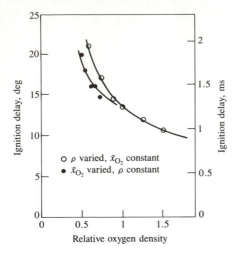

FIGURE 10-39

Effect of oxygen density in gas on ignition delay in single-cylinder DI engine of 1.3-dm^3 displacement with $r_c = 15$ at 1800 rev/min. Oxygen density changed by recycling exhaust gas at constant inlet density and by varying inlet pressure from 0.5 to 3 atm with air.[49]

for the control of oxides of nitrogen emissions (see Chap. 11). Results of a study carried out in a single-cylinder DI engine operated at a constant air/fuel ratio (30 : 1), manifold temperature, injection timing, and speed (1800 rev/min), where the oxygen concentration was varied by recirculating known amounts of cooled exhaust, are shown in Fig. 10-39.[49] Oxygen density is normalized by the naturally aspirated no-recirculation test value. As oxygen concentration is decreased, ignition delay increases.

10.6.5   Effect of Fuel Properties

Since both physical and chemical processes take place during the ignition delay, the effects of changes in the physical and chemical properties of fuels on the delay period have been studied. The chemical characteristics of the fuel are much the more important. The ignition quality of the fuel, defined by its cetane number, will obviously affect the delay. The dependence of cetane number on fuel molecular structure is as follows. Straight-chain paraffinic compounds (normal alkanes) have the highest ignition quality, which improves as the chain length increases. Aromatic compounds have poor ignition quality as do the alcohols (hence, the difficulties associated with using methanol and ethanol, possible alternative fuels, in compression-ignition engines). Figure 10-40 illustrates these effects. A base fuel was blended with pure paraffinic (normal, iso-, and cycloalkanes), aromatic, and olefinic hydrocarbons of various carbon numbers, by up to 20 percent by volume. The base fuel, a blend of 25 percent n-hexadecane and 75 percent isooctane, had a cetane number of 38.3. The figure shows that the resulting ignition delays correlate well as a function of cetane number at constant compression ratio and engine operating conditions. Addition of normal alkanes (excluding n-pentane and lower carbon number alkanes) improve the ignition quality. As the chain length of the added paraffin gets longer (higher carbon number) the cetane number improve-

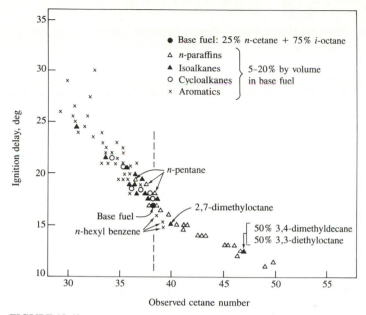

FIGURE 10-40
Effect of type of hydrocarbon structure on ignition quality of fuels in DI diesel combustion process at
constant compression ratio and engine operating conditions.[50]

ment increases. Isoalkanes, depending on the degree of branching, degrade ignition quality (unless the branching is concentrated at one end of the molecule,
when these types of isoalkanes improve ignition quality). Cycloalkanes and aromatics generally reduce the cetane number, unless they have a long n-alkane
chain attached to the ring. The cetane number of a fuel (a measure of its ability to
autoignite) generally varies inversely with its octane number (a measure of its
ability to resist autoignition; see Fig. 9-69 for the effect of hydrocarbon structure
on knock). The cetane number of commercial diesel fuel is normally in the range
40 to 55.

Engine ignition delay data with diesel fuels of different cetane number, at
various constant loads and speeds, shown in Fig. 10-41, demonstrate similar
trends. Within the normal diesel fuel cetane number range of 40 to 55, an approximately linear variation is evident. However, decreasing fuel cetane number below
about 38 may result in a more rapid increase in ignition delay.

Cetane number is controlled by the source of crude oil, by the refining
process, and by additives or ignition accelerators. Just as it is possible to reduce
the tendency to knock or autoignite in spark-ignition engine fuels by adding
antiknock agents, so there are additives that improve the ignition quality of
compression-ignition engine fuels. Generally, substances that increase the tendency to knock enhance ignition, and vice versa. Ignition-accelerating additives
include organic peroxides, nitrates, nitrites, and various sulfur compounds. The
most important of these commercially are the alkyl nitrates (isopropyl nitrate,

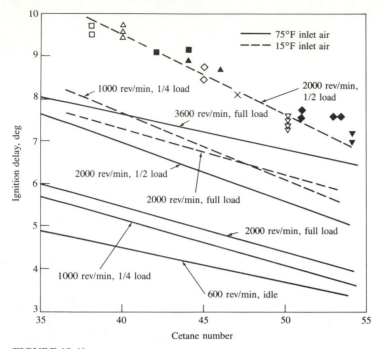

FIGURE 10-41
Effect of fuel cetane number on ignition delay over the load and speed range of 6.2-dm^3 eight-cylinder IDI swirl-chamber diesel engine. Top curve indicates typical fit between data and least-squares straight line over this cetane number range.[51]

primary amyl nitrates, primary hexyl nitrates, octyl nitrate).[52] Typically, about 0.5 percent of these additives by volume in a distillate fuel gives about a 10 cetane number increase in a fuel's ignition quality, though their effectiveness may depend on the composition of the base fuel. The incremental effect of increasing amounts of ignition-accelerating additives on cetane number decreases.[48] Usually, the ignition delay obtained with cetane improved blends are found to be equivalent to those obtained with natural diesel fuels of the same cetane number. Two potential practical uses for ignition accelerators are in upgrading the ignition characteristics of poorer quality diesel fuel and (in much larger amounts) making possible the use of alcohols in compression-ignition engines.

The physical characteristics of diesel fuel do not significantly affect the ignition delay in fully or partially warmed-up engines. Tests with fuels of different front-end volatility (over the cetane number range 38 to 53), and with substantially different front-end ignition quality for the same average cetane number, showed no discernible differences. Fuel viscosity variations over a factor of 2.5 were also tested and showed no significant effect.[48] Thus, in a warmed-up engine, variations in fuel atomization, spray penetration, and vaporization rate over reasonable ranges do not appear to influence the duration of the delay period significantly (see also Sec. 10.5.6 on fuel vaporization).

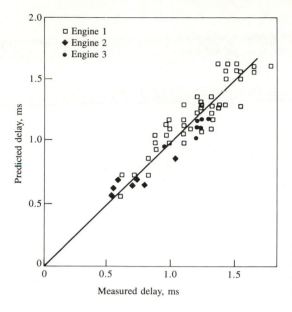

FIGURE 10-42
Comparison of engine ignition delays predicted with Eq. (10.37) with corresponding measured values.[54]

10.6.6   Correlations for Ignition Delay in Engines

Many correlations have been proposed for predicting ignition delay as a function of engine and air charge variables. These usually have the form of Eq. (10.35) and have been based on data from more fundamental experiments in combustion bombs and flow reactors. An important factor in assessing the appropriateness of any correlation is how it is to be used to predict the magnitude of the delay. If an equation for predicting the complete delay process (including all the physical and chemical processes from injection to combustion) is required, then the data show that such a simple form for the equation is unlikely to be adequate for the full range of engine conditions (see Table 10.4). However, if a model for the autoignition process of a premixed fuel-air mixture during the delay period is required, for use in conjunction with models for the physical processes of fuel evaporation and fuel-air mixing, then correlations such as those listed in Table 10.3 may be sufficiently accurate.

An empirical formula, developed by Hardenberg and Hase[53] for predicting the duration of the ignition delay period in DI engines, has been shown to give good agreement with experimental data over a wide range of engine conditions (see Fig. 10-42).[54] This formula gives the ignition delay (in crank angle degrees) in terms of charge temperature T (kelvins) and pressure p (bars) during the delay (taken as TC conditions) as

$$\tau_{id}(CA) = (0.36 + 0.22\bar{S}_p) \exp\left[E_A\left(\frac{1}{\tilde{R}T} - \frac{1}{17,190}\right)\left(\frac{21.2}{p - 12.4}\right)^{0.63}\right] \quad (10.37)$$

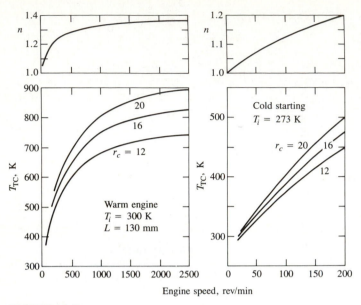

FIGURE 10-43
Exponent n for polytropic model of compression process in Eq. (10.39) and corresponding end-of-compression air temperature at TC. Warm and cold DI diesel engine with 130 mm stroke.[53]

where $\bar{S}_p$ is the mean piston speed (meters per second) and $\tilde{R}$ is the universal gas constant (8.3143 J/mol·K). E_A (joules per mole) is the apparent activation energy, and is given by

$$E_A = \frac{618,840}{CN + 25} \tag{10.38}$$

where CN is the fuel cetane number. The apparent activation energy decreases with increasing fuel cetane number. The delay in milliseconds is given by

$$\tau_{id}(ms) = \frac{\tau_{id}(CA)}{0.006N}$$

where N, engine speed, is in revolutions per minute. Values for T and p can be estimated using a polytropic model for the compression process:

$$T_{TC} = T_i r_c^{n-1} \qquad p_{TC} = p_i r_c^n \tag{10.39a,b}$$

where n is the polytropic exponent (see Sec. 9.2.2), r_c is the compression ratio, and the subscript i denotes intake manifold conditions. Values of the polytropic exponent are given in Fig. 10-43 for a direct-injection diesel under warm and cold engine operating conditions.[49,53]

10.7 MIXING-CONTROLLED COMBUSTION

10.7.1 Background

Earlier sections of this chapter have developed our current understanding of the individual processes which together make up the total injection-mixing-burning sequence—atomization, vaporization, fuel spray development, air entrainment, ignition, and combustion. While the phenomenological model developed by Lyn[6] provides satisfactory logical links between these processes, quantitative links are still lacking. Especially difficult to quantify are the relations between fuel spray behavior, flame structure, and fuel burning rate—the area of focus of this section. The color photographs of the compression-ignition combustion process in different types of diesel engines in Figs. 10-4 and 10-5 (see color plate between 498 and 499) show how the flame immediately following ignition spreads rapidly and envelops the spray. Depending on the spray configuration, the visible flame may then fill almost the entire combustion chamber. The flame and spray geometries are closely related. Mixing processes are also critical during the ignition delay period: while the duration of the delay period is not influenced in a major way by the rates of spray processes which together control "mixing," the amount of fuel mixed with air to within combustible limits during the delay (which affects the rate of pressure rise once ignition has occurred) obviously is directly related to mixing rates. Thus substantial observational evidence supports the mixing-controlled character of diesel engine combustion.

However, while it is well accepted that diesel combustion is normally controlled by the fuel-air mixing rate, fundamentally based quantitatively accurate models for this coupled mixing and combustion process are not yet available. The difficulties are twofold. First, the spray geometry in real diesel combustion systems is extremely complex. Second, the phenomena which must be described (and especially the unsteady turbulent diffusion diesel engine flame) are inadequately understood. Current capabilities for modeling these phenomena are reviewed in Chap. 14. Thermodynamic-based models of the diesel combustion process with atomization, vaporization, and spray development described by empirical or turbulent-jet-based submodels have been developed and used to predict burning rates. These are described in Secs. 14.4.3 and 14.4.4, and show reasonable but not precise agreement with data. Fluid-mechanic-based models of air flow, fuel spray behavior, and combustion are under active development (see Sec. 14.5). While realistic air-flow predictions are now feasible, predictions of spray behavior are less well developed and combustion-rate predictions are still exploratory.

In the following sections, the evidence linking spray characteristics to flame structure and burning rates is summarized.

10.7.2 Spray and Flame Structure

The structure of each fuel spray is that of a narrow liquid-containing core (densely filled with drops of order 20 μm in diameter) surrounded by a much

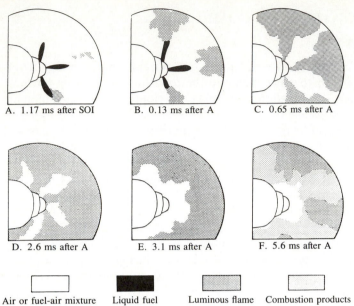

A. 1.17 ms after SOI   B. 0.13 ms after A   C. 0.65 ms after A

D. 2.6 ms after A   E. 3.1 ms after A   F. 5.6 ms after A

Air or fuel-air mixture   Liquid fuel   Luminous flame   Combustion products

FIGURE 10-44
Tracings of outer boundary of liquid fuel spray and flame from high-speed movies of diesel combustion taken in a rapid-compression machine, looking down on piston through transparent head. First occurrence of luminous flame at A, 1.17 ms after start of injection. End of injection at D.[57]

larger gaseous-jet region containing fuel vapor (see Fig. 10-20). The fuel concentration in the core is extremely high: local fuel/air equivalence ratios near the nozzle of order 10 have been measured during the injection period. Fuel concentrations within the spray decrease with increasing radial and axial position at any given time, and with time at a fixed location once injection has ended.[55] The fuel distribution within the spray is controlled largely by turbulent-jet mixing processes. Fuel vapor concentration contours determined from interferometric studies of unsteady vaporizing diesel-like sprays, presented by Lakshminarayan and Dent,[56] confirm this gaseous turbulent-jet-like structure of the spray, with its central liquid-containing core which evaporates relatively quickly once fuel injection ends.

The location of the autoignition sites and subsequent spreading of the enflamed region in relation to the fuel distribution in the spray provides further evidence of the mixing-controlled character of combustion. Figure 10-44 shows how this process occurs under conditions typical of direct-injection quiescent-chamber diesel engines. It shows tracings of the liquid fuel spray and flame boundaries taken from high-speed movies of the injection and combustion processes with central injection of five fuel jets into a disc-shaped chamber in a rapid-compression machine.[57] These and other similar studies show that autoignition first occurs toward the edge of the spray, between the spray tip (which may by then have interacted with the combustion chamber walls, and which contains

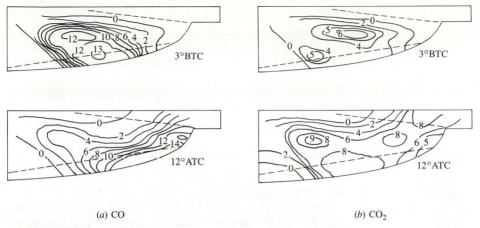

(a) CO   (b) CO_2

FIGURE 10-45
Contours of constant CO and CO_2 concentration in a plane along the spray axis calculated from gas composition data obtained with a rapid-acting sampling valve in a large 30.5-cm bore DI quiescent-chamber diesel engine with $r_c = 12.85$ operated at 500 rev/min. Injection starts 17° BTC, ignition occurs 8° BTC, injection ends 5° ATC.[60]

later-injected fuel as explained in Sec. 10.5.2) and the injector nozzle. Experiments where air/fuel ratio contours for a gaseous fuel jet injected into a swirling air flow in a rig simulated the fuel-air mixing process in open-chamber diesels,[58] under conditions chosen to match a set of diesel combustion rapid-compression-machine experiments where the autoignition sites and subsequent flame develop-ment were recorded on movies,[59] showed that autoignition occurred in a concentration band between the equivalence ratios of 1 and 1.5. Subsequent flame development, along mixture contours close to stoichiometric, occurs rapidly, as indicated in Fig. 10-44. Initially, this is thought to be due to sponta-neous ignition of regions close to the first ignition site due to the temperature rise associated with the strong pressure wave which emanates from each ignition site due to local rapid chemical energy release. Also, spontaneous ignition at addi-tional sites on the same spray, well separated from the original ignition location, can occur. Turbulent mixing provides another flame-spreading mechanism. From this point flame development is rapid, and the gas expansion which occurs on burning deforms the original spray form. These processes take place in each fuel spray in a closely comparable though not necessarily identical manner. Com-bustion movies such as those in Figs. 10-4 and 10-44 show that flame rapidly envelops each spray once spontaneous ignition occurs.

Gas-sampling data indicate that the burned gases within the flame-enveloped spray are only partially reacted and may be fuel-rich. Figure 10-45 shows CO and CO_2 concentration contours determined from rapid-acting sample valve measurements from the combustion chamber of a large quiescent-chamber two-stroke cycle diesel engine.[60] The contour maps shown correspond to the centerline of one of the five injected fuel sprays. Injection commenced at

17° BTC and ended about 5° ATC; ignition occurred at 8° BTC. The contours at 3° BTC show high CO concentrations in the burned gases which now occupy most of the spray region, indicating locally very fuel-rich conditions. Later, at 12° ATC, fuel injection has ceased, this rich core has moved outward to the piston-bowl wall, and combustion within the expanded spray region is much more complete. This oxidation of CO, as air is entrained into the spray region, mixes, and burns, releases substantial additional chemical energy.

The role of air swirl in promoting much more rapid fuel-air mixing in medium-size and smaller diesel engines is evident from similar gas-sampling studies in engines with these different combustion systems. The variation of gas species and unburned hydrocarbon concentrations within critical regions of the combustion chamber have been mapped out by a number of investigators.[61-63] These data show that during the early stages of injection and combustion, the boundaries of the individual sprays can be identified as they are convected around the combustion chamber bowl by the swirl. The fuel distribution within the combustion chamber is highly nonuniform. However, within each spray, sufficient air has mixed into the spray to bring the peak fuel/air equivalence ratios within the spray, in the outer regions of the chamber, to close to stoichiometric values.[63] This substantially different character of the spray with swirl is clear from the data in Fig. 10-46. Figure 10-46a shows equivalence ratio values determined from gas sampling, versus crank angle, from several studies with different designs of combustion chamber. While the local values obviously depend on the relation of the sample valve location to spray position at any given crank angle, the much lower values of equivalence ratio with swirl relative to quiescent chambers, during injection and the early stages of combustion, clearly indicate that swirl enhances mixing rates substantially. As combustion ends, these data indicate relatively uniform fuel distribution within the combustion chamber, at least on a gross geometric scale. However, early in the combustion process the high CO levels, found in all these combustion systems as shown in Fig. 10-46b, indicate that the burned gases are only partially reacted. With quiescent chambers this is largely due to lack of oxygen. With swirl, however, substantial oxygen is present. Whether the high CO with swirl is due to kinetic limitations or to smaller-scale mixture nonhomogeneities is unclear.

10.7.3 Fuel-Air Mixing and Burning Rates

The model of diesel combustion obtained from heat-release analyses of cylinder pressure data identifies two main stages of combustion (see Fig. 10-9). The first is the premixed-combustion phase, when the fuel which has mixed sufficiently with air to form an ignitable mixture during the delay period is consumed. The second is the mixing-controlled combustion phase, where rates of burning (at least in naturally aspirated engines) are lower. Experimental evidence from heat-release analysis indicates that the majority of the fuel (usually more than 75 percent) burns during the second mixing-controlled phase. Such evidence forms the basis for the heat-release models used in diesel engine cycle simulations. For example,

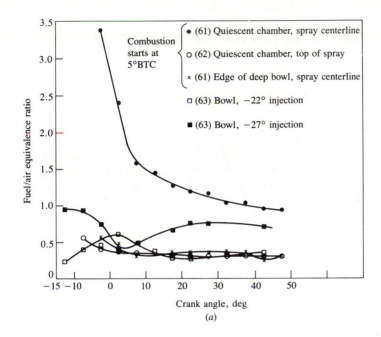

(a)

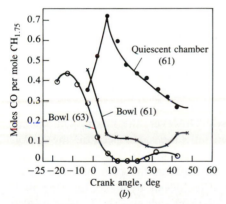

(b)

FIGURE 10-46

Time and space-resolved gas-composition data obtained from rapid-acting sampling valves from within the combustion chambers of quiescent and high-swirl bowl-in-piston DI diesel engines. (a) Local fuel/air equivalence ratios on spray centerline and periphery with quiescent chamber, edge of deep bowl with swirl, and within a shallow bowl with swirl, three-quarters of the way out to the bowl wall, for two injection timings ($-22°$ and $-27°$). (b) CO concentration on spray centerline with quiescent chamber, edge of deep bowl, and within shallow bowl with swirl.[61-63]

the fraction of the fuel β which burns in the premixed phase has been correlated by Watson *et al.* (see Sec. 14.4.3) by the relation

$$\beta = 1 - \frac{a\phi^b}{\tau_{id}^c} \tag{10.40}$$

where ϕ is the fuel/air equivalence ratio, τ_{id} the ignition delay (in milliseconds), and $a \approx 0.9$, $b \approx 0.35$, and $c \approx 0.4$ are constants depending on engine design. Equation (10.40) shows the expected trends for the premixed fraction, with changes in the overall equivalence ratio ϕ (increasing injection duration as load is increased) and changes in the ignition delay.

That the fuel-burning or heat-release rate is predominantly mixing controlled is supported by the following types of evidence. Estimates of the rate at which fuel-air mixture with composition within the combustible limits is produced in diesel sprays under typical engine conditions, based on a variety of turbulent-jet models of the spray (e.g., see Refs. 29, 36, and 59 and also Sec. 14.4.3), show that mixing rates and burning rates are comparable in magnitude. Estimates of characteristic times for the turbulent-jet mixing processes in diesel combustion chambers show these to be comparable to the duration of the heat-release process, and much longer than characteristic times for evaporation and the combustion chemical kinetics.[36, 44]

Then, measured diesel-combustion heat-release profiles show trends with engine design and operating parameter changes that correspond to fuel-air mixing being the primary controlling factor. Examples of heat-release profiles measured in rapid-compression-machine studies of diesel combustion, shown in Fig. 10-47, illustrate this clearly. The rapid-compression machine had a disc-shaped chamber of 10 cm diameter with a 3.1 cm clearance height at the end of a compression process through a volume ratio of 15.4; a five-hole centrally located fuel-injector nozzle was used. Figure 10-47a shows the heat-release profiles for different initial air temperatures which produce different ignition delays. Longer delays allow more fuel to mix to within combustible limits during the delay, so the peak premixed heat-release rate increases. However, the mixing-controlled-phase heat-release-rate magnitudes are essentially the same because the spray-mixing processes are little affected by these changes in air temperature. Figure 10-47b and c shows that heat-release rates throughout the combustion process are increased by increased fuel-injection rate (achieved by increasing the fuel-injection pressure) and by swirl. Both these changes increase the fuel-air mixing rates within the fuel spray and therefore increase the heat-release rate during the mixing-controlled combustion phase.

Diesel engine heat-release rate trends, as design and operating variables are changed, can be related to mixing rates in analogous fashion. Table 10.5 summarizes the trends that have been investigated. The directional effects of changes in engine parameters on the ignition delay period and the fuel-air mixing rate are all consistent with the measured changes in premixed and mixing-controlled heat-release rates. The controlling role of fuel-air mixing in the diesel engine fuel spray on combustion is clear.

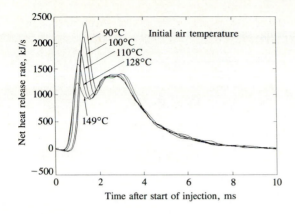

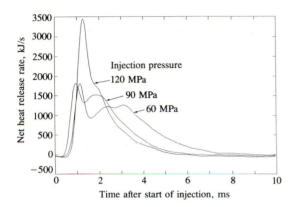

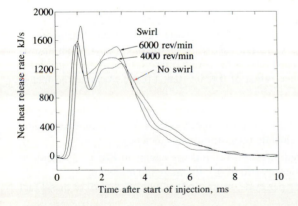

FIGURE 10-47
Net heat-release rates, as a function of time after start of injection, calculated from cylinder pressure data from rapid-compression-machine studies of DI diesel combustion. (a) Effect of varying initial air temperature: 4000 rev/min swirl, injection pressure 60 MPa. (b) Effect of varying injection pressure: no swirl. (c) Effect of varying swirl: injection pressure 60 MPa.[20]

TABLE 10.5

Effects of engine design and operating variables on heat-release rates

Reference	Parameter varied	τ_{id}	$\dot{m}_m$	Q_p	Q_m
			Effect on		
5, 64	Injection rate ↑	*	↑	↑	↑
65	Turbocharger boost ↑	↓	*	↓	*
66	Compression ratio ↓	↑	*	↑	*
66	Number of injector holes ↑	*	↑	↑	↑
67, 68	Injection advance ↑	↑	*	↑	*
67, 68	Swirl ↑	*	↑	↑	↑
67	Intake-air temperature ↓	↑	*	↑	*
68, 69	Injection pressure ↑	*	↑	↑	↑
11, 69	Speed ↑	*	↑	↑	↑

τ_{id}, ignition delay; $\dot{m}_m = (dm/dt)_m$, fuel-air mixing rate; $Q_p = (dQ/dt)_p$, heat-release rate during premixed-combustion phase; $Q_m = (dQ/dt)_m$, heat-release rate during mixing-controlled-combustion phase. ↑ increase; ↓ decrease; * minor effect.
Source: From Plee and Ahmad.[44]

PROBLEMS

10.1. Describe the sequence of processes which must occur before the liquid fuel in the injection system in a direct-injection compression-ignition engine is fully burned.

10.2. Small high-swirl direct-injection CI engines have fuel conversion efficiencies which are about 10 percent higher than values typical of equivalent indirect-injection engines. (IDI engines are used because they achieve higher bmep.) What combustion-system-related differences contribute to this higher efficiency?

10.3. In a diesel engine, because the fuel distribution is nonuniform the burned gas temperature is nonuniform. Consider small fuel-air mixture elements initially at 1000 K and 6.5 MPa at top-center with a range of equivalence ratios. Each element burns at essentially constant pressure. Calculate (using the charts in Chap. 4, or an appropriate chemical equilibrium thermodynamic computer code) the burned gas temperature for mixture equivalence ratios of 0.4, 0.6, 0.8, 1.0, 1.2. Assume the fuel is isooctane.

10.4. The levels of combustible species in the exhaust of a direct-injection diesel engine are: HC, 0.8 g/kW·h; CO, 3 g/kW·h; particulates, 0.7 g/kW·h. If the specific fuel consumption is 210 g/kW·h calculate the combustion efficiency.

10.5. Consider the naturally aspirated direct-injection diesel engine in Fig. 1-23 operating at 2300 rev/min and an equivalence ratio of 0.7. Estimate the following:

(a) Mass of air in each cylinder per cycle, mass, and volume (as liquid) of diesel fuel injected per cylinder per cycle.

(b) Estimate the average drop size [e.g., use Eq. (10.32)]. The cylinder pressure at time of injection (close to TC) is 50 atm; the fuel injection pressure is 500 atm.

(c) Assuming all fuel droplets are the same size as the average drop, how many drops are produced per injection? If these drops are uniformly distributed throughout the air charge at TC, what is the approximate distance between drops? (Neither of these assumptions is correct; however, the calculations illustrate the difficulty of the fuel-air mixing process.)

10.6. Estimate the following quantities for a typical direct-injection diesel fuel spray. The injection pressure is 500 atm; the cylinder pressure during injection is 50 atm.
  (a) Assuming that the flow through the nozzle orifice is incompressible and quasi steady, estimate the liquid fuel velocity at the orifice exit. At this velocity, how long would the fuel take to reach the cylinder wall? The bore is 125 mm.
  (b) Each nozzle orifice diameter d_n is 0.34 mm and $L_n/d_n = 4$. Determine the spray angle and plot spray tip penetration versus time.
  (c) Use Eq. (10.32) to estimate the initial average drop size assuming that the injection process in (a) above continues for 1 millisecond and the injector nozzle has four orifices.

10.7. Diesel fuel is injected as a liquid at room temperature into air at 50 atm and 800 K, close to TC at the end of compression. If the overall equivalence ratio is 0.7, estimate the reduction in average air temperature which would occur when the fuel is fully vaporized and uniformly mixed. Assume such mixing takes place at constant volume prior to any combustion.

10.8. Using Eq. (10.37) estimate the ignition delay in milliseconds and crank angle degrees for these operating conditions in Table 10.4: low swirl IDI diesel 600 and 1800 rev/min; high swirl IDI diesel 1800 rev/min; DI diesel low and high compression ratio. The fuel cetane number is 45; stroke = 0.1 m. Discuss whether the predicted trends with speed, swirl, and compression ratio are consistent with Sec. 10.6.4.

10.9. The compression ratio of truck diesel engines must be set at about 18 so that the engine will start when cold. Using Eqs. (10.37) to (10.39) develop a graph of τ_{id} (in degrees) as a function of compression ratio for $r_c = 12$ to 20. Assume $p_i = 1$ atm, $T_i = 255$ K, $n = 1.13$, speed = 100 rev/min, bore = stroke = 120 mm, fuel cetane number = 45. If the ignition delay must be less than 20° CA for satisfactory starting, what compression ratio is required?

10.10. Equation (10.40) predicts the fraction β of the fuel injected into a direct-injection diesel engine which burns in the premixed phase. Plot β as a function of τ_{id} for $\phi = 0.4$. Show that for turbocharged DI diesel engines where τ_{id} is 0.4 to 1 ms, the premixed combustion phase is much less important than it normally is for naturally aspirated engines where τ_{id} is between 0.7 and 3 ms.

REFERENCES

1. Alcock, J. F., and Scott, W. M.: "Some More Light on Diesel Combustion," *Proc. Auto. Div., Instn Mech. Engrs*, No. 5, pp. 179–191, 1962–1963.
2. Scott, W. M.: "Understanding Diesel Combustion through the Use of High Speed Moving Pictures in Color," SAE paper 690002, *SAE Trans.*, vol. 78, 1969.
3. Neitz, A., and D'Alfonso, N.: "The M.A.N. Combustion System with Controlled Direct Injection for Passenger Car Diesel Engines," SAE paper 810479, 1979.
4. Ogasawara, M., Tokunaga, Y., Horio, K., Uryu, M., and Hirofumi, N.: "Photographic Study on the Intermittent Spray Combustion by a Rapid Compression Machine" (in Japanese), *Internal Combustion Engine*, vol. 15, 1976.
5. Austen, A. E. W., and Lyn, W.-T.: "Relation between Fuel Injection and Heat Release in a Direct-Injection Engine and the Nature of the Combustion Processes," *Proc. Instn Mech. Engrs*, No. 1, pp. 47–62, 1960–1961.
6. Lyn, W.-T.: "Study of Burning Rate and Nature of Combustion in Diesel Engines," in *Proceedings of Ninth International Symposium on Combustion*, pp. 1069–1082, The Combustion Institute, 1962.
7. Cheng, W., and Gentry, R.: "Effects on Charge Non-Uniformity on Diesel Heat Release Analysis," SAE paper 861568, 1986.

8. Gatowski, J. A., Balles, E. N., Chun, K. M., Nelson, F. E., Ekchian, J. A., and Heywood, J. B.: "Heat Release Analysis of Engine Pressure Data," SAE paper 841359, *SAE Trans.*, vol. 93, 1984.

9. Krieger, R. B., and Borman, G. L.: "The Computation of Apparent Heat Release for Internal Combustion Engines," ASME paper 66-WA/DGP-4, in *Proc. Diesel Gas Power*, ASME, 1966.

10. Watson, N., and Kamel, M.: "Thermodynamic Efficiency Evaluation of an Indirect Injection Diesel Engine," SAE paper 790039, *SAE Trans.*, vol. 88, 1979.

11. Kort, R. T., Mansouri, S. H., Heywood, J. B., and Ekchian, J. A.: "Divided-Chamber Diesel Engine, Part II: Experimental Validation of a Predictive Cycle-Simulation and Heat Release Analysis," SAE paper 820274, *SAE Trans.*, vol. 91, 1982.

12. Obert, E. F.: *Internal Combustion Engines and Air Pollution*, Intext Educational Publishers, 1973 edition.

13. Weathers, Jr., T., and Hunter, C.: *Diesel Engines for Automobiles and Small Trucks*, Reston Publishing Company, a Prentice-Hall Company, Reston, Va. 1981.

14. Bosch: *Automotive Handbook*, 1st English ed., Robert Bosch GmbH, 1976.

15. Williams, Jr., H. A.: "The GM/EMD Model 710G Series Engine," in *Marine Engine Development*, SP-625, SAE, 1985. Also ASME paper 85-GGP-24, 1985.

16. Hames, R. J., Straub, R. D., and Amann, R. W.: "DDEC Detroit Diesel Electronic Control," SAE paper 850542, 1985.

17. Hiroyasu, H.: "Diesel Engine Combustion and Its Modeling," in *Diagnostics and Modeling of Combustion in Reciprocating Engines*, pp. 53–75, COMODIA 85, Proceedings of Symposium, Tokyo, Sept. 4–6, 1985.

18. Arai, M., Tabata, M., and Hiroyasu, H.: "Disintegrating Process and Spray Characterization of Fuel Jet Injected by a Diesel Nozzle," SAE paper 840275, *SAE Trans.*, vol. 93, 1984.

19. Kamimoto, T., Kobayashi, H., and Matsuoka, S.: "A Big Size Rapid Compression Machine for Fundamental Studies of Diesel Combustion," SAE paper 811004, *SAE Trans.*, vol. 90, 1981.

20. Balles, E.: "Fuel-Air Mixing and Diesel Combustion in a Rapid Compression Machine," Ph.D. Thesis, Department of Mechanical Engineering, MIT, June 1987.

21. Browne, K. R., Partridge, I. M., and Greeves, G.: "Fuel Property Effects on Fuel/Air Mixing in an Experimental Diesel Engine," SAE paper 860223, 1986.

22. Bracco, F. V.: "Modeling of Engine Sprays," SAE paper 850394, 1985.

23. Kuo, T., and Bracco, F. V.: "Computations of Drop Sizes in Pulsating Sprays and of Liquid-Core Length in Vaporizing Sprays," SAE paper 820133, *SAE Trans.*, vol. 91, 1982.

24. Reitz, R. D., and Bracco, F. V.: "Mechanism of Atomization of a Liquid Jet," *Phys. Fluid*, vol. 25, no. 10, pp. 1730–1742, 1982.

25. Reitz, R. D., and Bracco, F. V.: "On the Dependence of Spray Angle and Other Spray Parameters on Nozzle Design and Operating Conditions," SAE paper 790494, 1979.

26. Wu, K.-J., Su, C.-C., Steinberger, R. L., Santavicca, D. A., and Bracco, F. V.: "Measurements of the Spray Angle of Atomizing Jets," *J. Fluids Engng*, vol. 105, pp. 406–413, 1983.

27. Hay, N., and Jones, P. L.: "Comparison of the Various Correlations for Spray Penetration," SAE paper 720776, 1972.

28. Dent, J. C.: "Basis for the Comparison of Various Experimental Methods for Studying Spray Penetration," SAE paper 710571, *SAE Trans.*, vol. 80, 1971.

29. Hiroyasu, H., Kadota, T., and Arai, M.: "Supplementary Comments: Fuel Spray Characterization in Diesel Engines," in James N. Mattavi and Charles A. Amann (eds.), *Combustion Modeling in Reciprocating Engines*, pp. 369–408, Plenum Press, 1980.

30. Wu, K.-J., Reitz, R. D., and Bracco, F. V.: "Measurements of Drop Size at the Spray Edge near the Nozzle in Atomizing Liquid Jets," *Phys. Fluids*, vol. 29, no. 4, pp. 941–951, 1986.

31. Reitz, R. D., and Diwakar, R.: "Effect of Drop Breakup on Fuel Sprays," SAE paper 860469, 1986.

32. Hiroyasu, H., and Kadota, T.: "Fuel Droplet Size Distribution in Diesel Combustion Chamber," SAE paper 740715, *SAE Trans.*, vol. 83, 1974.

33. El Wakil, M. M., Myers, P. S., and Uyehara, O. A.: "Fuel Vaporization and Ignition Lag in Diesel Combustion," in *Burning a Wide Range of Fuels in Diesel Engines*, *SAE Progress in Technol.*, vol. 11, pp. 30–44, SAE, 1967.

34. Borman, G. L., and Johnson, J. H.: "Unsteady Vaporization Histories and Trajectories of Fuel Drops Injected into Swirling Air," in *Burning a Wide Range of Fuels in Diesel Engines, SAE Progress in Technology*, vol. 11, pp. 13–29, SAE, 1967; also SAE paper 598C, 1962.

35. Kamimoto, T., and Matsuoka, S.: "Prediction of Spray Evaporation in Reciprocating Engines," SAE paper 770413, *SAE Trans.*, vol. 86, 1977.

36. Kuo, T., Yu, R. C., and Shahed, S. M.: "A Numerical Study of the Transient Evaporating Spray Mixing Process in the Diesel Environment," SAE paper 831735, *SAE Trans.*, vol. 92, 1983.

37. ASTM Method D613, Cetane Test Procedure.

38. Henein, N. A., and Fragoulis, A. N.: "Correlation between Physical Properties and Autoignition Parameters of Alternate Fuels," SAE paper 850266, 1985.

39. Gulder, O. L., Glavincevski, B., and Burton, G. F.: "Ignition Quality Rating Methods for Diesel Fuels—A Critical Appraisal," SAE paper 852080, 1985.

40. ASTM D976, Calculated Cetane Index.

41. ASTM D611.

42. Igura, S., Kadota, T., and Hiroyasu, H.: "Spontaneous Ignition Delay of Fuel Sprays in High Pressure Gaseous Environments," *Trans. Japan Soc. Mech. Engrs*, vol. 41, no. 345, pp. 24–31, 1975.

43. Spadaccini, L. J., and TeVelde, J. A.: "Autoignition Characteristics of Aircraft-Type Fuels," *Combust. Flame*, vol. 46, pp. 283–300, 1982.

44. Plee, S. L., and Ahmad, T.: "Relative Roles of Premixed and Diffusion Burning in Diesel Combustion," SAE paper 831733, *SAE Trans.*, vol. 92, 1983.

45. Stringer, F. W., Clarke, A. E., and Clarke, J. S.: "The Spontaneous Ignition of Hydrocarbon Fuels in a Flowing System," *Proc. Instn Mech. Engrs*, vol. 184, pt. 3J, 1969–1970.

46. Wolfer, H. H.: "Ignition Lag in Diesel Engines," VDI-Forschungsheft 392, 1938; Translated by Royal Aircraft Establishment, Farnborough Library No. 358, UDC 621–436.047, August 1959.

47. Lyn, W.-T., and Valdmanis, E.: "Effects of Physical Factors on Ignition Delay," SAE paper 680102, 1968.

48. Wong, C. L., and Steere, D. E.: "The Effects of Diesel Fuel Properties and Engine Operating Conditions on Ignition Delay," SAE paper 821231, *SAE Trans.*, vol. 91, 1982.

49. Andree, A., and Pachernegg, S. J.: "Ignition Conditions in Diesel Engines," SAE paper 690253, *SAE Trans.*, vol. 78, 1969.

50. Glavincevski, B., Gülder, O. L., and Gardner, L.: "Cetane Number Estimation of Diesel Fuels from Carbon Type Structural Composition," SAE paper 841341, 1984.

51. Olree, R., and Lenane, D.: "Diesel Combustion Cetane Number Effects," SAE paper 840108, *SAE Trans.*, vol. 93, 1984.

52. Schaefer, A. J., and Hardenberg, H. O.: "Ignition Improvers for Ethanol Fuels," SAE paper 810249, *SAE Trans.*, vol. 90, 1981.

53. Hardenberg, H. O., and Hase, F. W.: "An Empirical Formula for Computing the Pressure Rise Delay of a Fuel from its Cetane Number and from the Relevant Parameters of Direct-Injection Diesel Engines," SAE paper 790493, *SAE Trans.*, vol. 88, 1979.

54. Dent, J. C., and Mehta, P. S.: "Phenomenological Combustion Model for a Quiescent Chamber Diesel Engine," SAE paper 811235, *SAE Trans.*, vol. 90, 1981.

55. Chang, Y. J., Kobayashi, H., Matsuzawa, K., and Kamimoto, T.: "A Photographic Study of Soot Formation and Combustion in a Diesel Flame with a Rapid Compression Machine," in *Diagnostics and Modeling of Combustion in Reciprocating Engines*, pp. 149–157, COMODIA 85, Proceedings of Symposium, Tokyo, Sept. 4–6, 1985.

56. Lakshminarayan, P. A., and Dent, J. C.: "Interferometric Studies of Vaporising and Combustion Sprays," SAE paper 830244, *SAE Trans.*, vol. 92, 1983.

57. Colella, K. J., Balles, E. N., Ekchian, J. A., Cheng, W. K., and Heywood, J. B.: "A Rapid Compression Machine Study of the Influence of Charge Temperature on Diesel Combustion," SAE paper 870587, 1987.

58. Morris, C. J., and Dent, J. C.: "The Simulation of Air Fuel Mixing in High Swirl Open Chamber Diesel Engines," *Proc. Instn Mech. Engrs*, vol. 190, no. 47/76, pp. 503–513, 1976.

59. Rife, J., and Heywood, J. B.: "Photographic and Performance Studies of Diesel Combustion with

a Rapid Compression Machine," SAE paper 740948, *SAE Trans.*, vol. 83, 1974.

60. Whitehouse, N. D., Cough, E., and Jeje, A. B.: "The Study of Combustion in a Quiescent Combustion Chamber Diesel Engine," ASME paper 82-HT-35, 1982.

61. Nightingale, D. R.: "A Fundamental Investigation into the Problem of NO Formation in Diesel Engines," SAE paper 750848, *SAE Trans.*, vol. 84, 1975.

62. Bennethum, J. E., Mattavi, J. N., and Toepel, R. R.: "Diesel Combustion Chamber Sampling-Hardware, Procedures, and Data Interpretation," SAE paper 750849, *SAE Trans.*, vol. 84, 1975.

63. Rhee, K. T., Myers, P. S., and Uyehara, O. A.: "Time- and Space-Resolved Species Determination in Diesel Combustion Using Continuous Flow Gas Sampling," SAE paper 780226, *SAE Trans.*, vol. 87, 1978.

64. Shipinsky, J., Uyehara, O. A., and Myers, P. S.: "Experimental Correlation between Rate-of-Injection and Rate-of-Heat-Release in a Diesel Engine," ASME paper 68–DGP-11, 1968.

65. Grigg, H. C., and Syed, M. H.: "The Problem of Predicting Rate of Heat Release in Diesel Engine," *Proc. Instn Mech. Engrs*, vol. 184, pt. 3J, 1969–1970.

66. Whitehouse, N. D., Clough, E., and Way, J. B.: "The Effect of Changes in Design and Operating Conditions on Heat Release in Direct-Injection Diesel Engines," SAE paper 740085, 1974.

67. Meguerdichian, M., and Watson, N.: "Prediction of Mixture Formation and Heat Release in Diesel Engines," SAE paper 780225, 1978.

68. Kamimoto, T., Aoyagi, Y., Matsui, Y., and Matsuoka, S.: "The Effects of Some Engine Variables on Measured Rates of Air Entrainment and Heat Release in a DI Diesel Engine," SAE paper 8000253, *SAE Trans.*, vol. 89, 1980.

69. Dent, J. C., Mehta, P. S., and Swan, J.: "A Predictive Model for Automotive DI Diesel Engine Performance and Smoke Emissions," paper C126/82, presented at the International Conference on *Diesel Engines for Passenger Cars and Light Duty Vehicles*, Institution of Mechanical Engineers, London, England, Oct. 5–7, 1982.

POLLUTANT FORMATION AND CONTROL

11.1 NATURE AND EXTENT OF PROBLEM

Spark-ignition and diesel engines are a major source of urban air pollution. The spark-ignition engine exhaust gases contain oxides of nitrogen (nitric oxide, NO, and small amounts of nitrogen dioxide, NO_2—collectively known as NO_x), carbon monoxide (CO), and organic compounds which are unburned or partially burned hydrocarbons (HC). The relative amounts depend on engine design and operating conditions but are of order: NO_x, 500 to 1000 ppm or 20 g/kg fuel; CO, 1 to 2 percent or 200 g/kg fuel; and HC, 3000 ppm (as C_1) or 25 g/kg fuel. Piston blowby gases, and fuel evaporation and release to the atmosphere through vents in the fuel tank and carburetor after engine shut-down, are also sources of unburned hydrocarbons. However, in most modern engines these nonexhaust sources are effectively controlled by returning the blowby gases from the crankcase to the engine intake system and by venting the fuel tank and carburetor float bowl through a vapor-absorbing carbon cannister which is purged by some of the engine intake air during normal engine operation. In diesel engine exhaust, concentrations of NO_x are comparable to those from SI engines. Diesel hydrocarbon emissions are significant though exhaust concentrations are lower by about a factor of 5 than typical SI engine levels. The hydrocarbons in the exhaust may also condense to form white smoke during engine starting and warm-up.

Specific hydrocarbon compounds in the exhaust gases are the source of diesel odor. Diesel engines are an important source of particulate emissions; between about 0.2 and 0.5 percent of the fuel mass is emitted as small (~ 0.1 μm diameter) particles which consist primarily of soot with some additional absorbed hydrocarbon material. Diesel engines are not a significant source of carbon monoxide.

Use of alcohol fuels in either of these engines substantially increases aldehyde emissions. While these are not yet subject to regulation, aldehydes would be a significant pollutant if these fuels were to be used in quantities comparable to gasoline and diesel. Currently used fuels, gasoline and diesel, contain sulfur: gasoline in small amounts (≤ 600 ppm by weight S), diesel fuel in larger amounts (≤ 0.5 percent). The sulphur is oxidized (or burned) to produce sulfur dioxide, SO_2, of which a fraction can be oxidized to sulfur trioxide, SO_3, which combines with water to form a sulfuric acid aerosol.

In general, the concentrations of these pollutants in internal combustion engine exhaust differ from values calculated assuming chemical equilibrium. Thus the detailed chemical mechanisms by which these pollutants form and the kinetics of these processes are important in determining emission levels. For some pollutant species, e.g., carbon monoxide, organic compounds, and particulates, the formation and destruction reactions are intimately coupled with the primary fuel combustion process. Thus an understanding of the formation of these species requires knowledge of the combustion chemistry. For nitrogen oxides and sulfur oxides, the formation and destruction processes are not part of the fuel combustion process. However, the reactions which produce these species take place in an environment created by the combustion reactions, so the two processes are still intimately linked. A summary of the mechanisms by which these pollutants form in internal combustion engines provides an introduction to this chapter. In subsequent sections, the details of the basic formation mechanisms of each pollutant and the application of these mechanisms to the combustion process in both spark-ignition and compression-ignition engines will be developed.

The processes by which pollutants form within the cylinder of a conventional spark-ignition engine are illustrated qualitatively in Fig. 11-1. The schematic shows the combustion chamber during four different phases of the engine operating cycle: compression, combustion, expansion, and exhaust. Nitric oxide (NO) forms throughout the high-temperature burned gases behind the flame through chemical reactions involving nitrogen and oxygen atoms and molecules, which do not attain chemical equilibrium. The higher the burned gas temperature, the higher the rate of formation of NO. As the burned gases cool during the expansion stroke the reactions involving NO freeze, and leave NO concentrations far in excess of levels corresponding to equilibrium at exhaust conditions. Carbon monoxide also forms during the combustion process. With rich fuel-air mixtures, there is insufficient oxygen to burn fully all the carbon in the fuel to CO_2; also, in the high-temperature products, even with lean mixtures, dissociation ensures there are significant CO levels. Later, in the expansion stroke, the CO oxidation process also freezes as the burned gas temperature falls.

The unburned hydrocarbon emissions have several different sources.

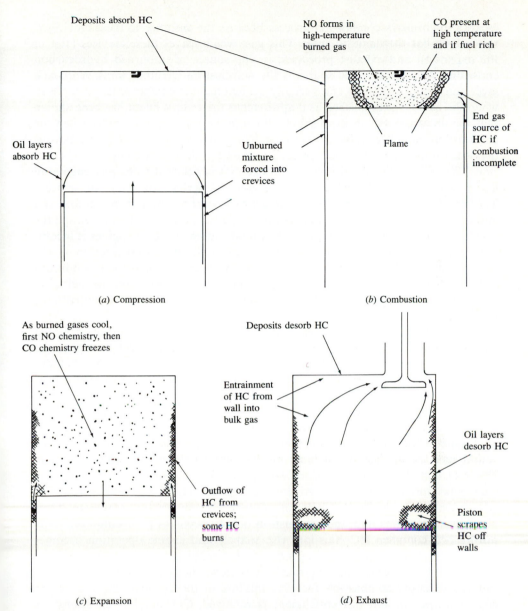

Deposits absorb HC

NO forms in
high-temperature
burned gas

CO present at
high temperature
and if fuel rich

Oil layers
absorb HC

Unburned
mixture
forced into
crevices

Flame

End gas
source of
HC if
combustion
incomplete

(a) Compression

(b) Combustion

As burned gases cool,
first NO chemistry, then
CO chemistry freezes

Deposits desorb HC

Entrainment
of HC from
wall into
bulk gas

Oil layers
desorb HC

Outflow of
HC from
crevices;
some HC
burns

Piston
scrapes
HC off
walls

(c) Expansion

(d) Exhaust

FIGURE 11-1
Summary of HC, CO, and NO pollutant formation mechanisms in a spark-ignition engine.

During compression and combustion, the increasing cylinder pressure forces
some of the gas in the cylinder into crevices, or narrow volumes, connected to the
combustion chamber: the volumes between the piston, rings, and cylinder wall
are the largest of these. Most of this gas is unburned fuel-air mixture; much of it

escapes the primary combustion process because the entrance to these crevices is too narrow for the flame to enter. This gas, which leaves these crevices later in the expansion and exhaust processes, is one source of unburned hydrocarbon emissions. Another possible source is the combustion chamber walls. A quench layer containing unburned and partially burned fuel-air mixture is left at the wall when the flame is extinguished as it approaches the wall. While it has been shown that the unburned HC in these thin (≤ 0.1 mm) layers burn up rapidly when the combustion chamber walls are clean, it has also been shown that the porous deposits on the walls of engines in actual operation do increase engine HC emissions. A third source of unburned hydrocarbons is believed to be any engine oil left in a thin film on the cylinder wall, piston and perhaps on the cylinder head. These oil layers can absorb and desorb fuel hydrocarbon components, before and after combustion, respectively, thus permitting a fraction of the fuel to escape the primary combustion process unburned. A final source of HC in engines is incomplete combustion due to bulk quenching of the flame in that fraction of the engine cycles where combustion is especially slow (see Sec. 9.4.3). Such conditions are most likely to occur during transient engine operation when the air/fuel ratio, spark timing, and the fraction of the exhaust recycled for emission control may not be properly matched.

The unburned hydrocarbons exit the cylinder by being entrained in the bulk-gas flow during blowdown and at the end of the exhaust stroke as the piston pushes gas scraped off the wall out of the exhaust valve. Substantial oxidation of the hydrocarbons which escape the primary combustion process by any of the above processes can occur during expansion and exhaust. The amount of oxidation depends on the temperature and oxygen concentration time histories of these HC as they mix with the bulk gases.

One of the most important variables in determining spark-ignition engine emissions is the fuel/air equivalence ratio, ϕ. Figure 11-2 shows qualitatively how NO, CO, and HC exhaust emissions vary with this parameter. The spark-ignition engine has normally been operated close to stoichiometric, or slightly fuel-rich, to ensure smooth and reliable operation. Figure 11-2 shows that leaner mixtures give lower emissions until the combustion quality becomes poor (and eventually misfire occurs), when HC emissions rise sharply and engine operation becomes erratic. The shapes of these curves indicate the complexities of emission control. In a cold engine, when fuel vaporization is slow, the fuel flow is increased to provide an easily combustible fuel-rich mixture in the cylinder. Thus, until the engine warms up and this enrichment is removed, CO and HC emissions are high. At part-load conditions, lean mixtures could be used which would produce lower HC and CO emissions (at least until the combustion quality deteriorates) and moderate NO emissions. Use of recycled exhaust to dilute the engine intake mixture lowers the NO levels, but also deteriorates combustion quality. Exhaust gas recirculation (EGR) is used with stoichiometric mixtures in many engine control systems. Note that the highest power levels are obtained from the engine with slightly rich-of-stoichiometric mixtures and no recycled exhaust to dilute the incoming charge. As we will see, several emission control techniques are required

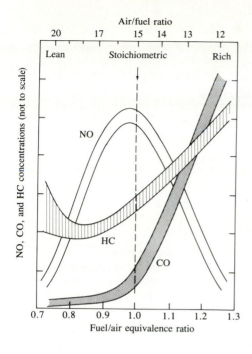

FIGURE 11-2
Variation of HC, CO, and NO concentration in the exhaust of a conventional spark-ignition engine with fuel/air equivalence ratio.

to reduce emissions of all three pollutants, over all engine operating modes, and achieve acceptable average levels.

In the diesel engine, the fuel is injected into the cylinder just before combustion starts, so throughout most of the critical parts of the cycle the fuel distribution is nonuniform. The pollutant formation processes are strongly dependent on the fuel distribution and how that distribution changes with time due to mixing. Figure 11-3 illustrates how various parts of the fuel jet and the flame affect the formation of NO, unburned HC, and soot (or particulates) during the "premixed" and "mixing-controlled" phases of diesel combustion in a direct-injection engine with swirl. Nitric oxide forms in the high-temperature burned gas regions as before, but temperature and fuel/air ratio distributions within the burned gases are now nonuniform and formation rates are highest in the close-to-stoichiometric regions. Soot forms in the rich unburned-fuel-containing core of the fuel sprays, within the flame region, where the fuel vapor is heated by mixing with hot burned gases. Soot then oxidizes in the flame zone when it contacts unburned oxygen, giving rise to the yellow luminous character of the flame. Hydrocarbons and aldehydes originate in regions where the flame quenches both on the walls and where excessive dilution with air prevents the combustion process from either starting or going to completion. Fuel that vaporizes from the nozzle sac volume during the later stages of combustion is also a source of HC. Combustion generated noise is controlled by the early part of the combustion process, the initial rapid heat release immediately following the ignition-delay period.

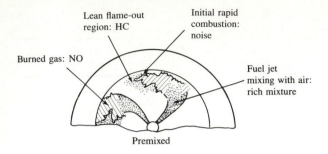

Lean flame-out region: HC

Initial rapid combustion: noise

Burned gas: NO

Fuel jet mixing with air: rich mixture

Premixed

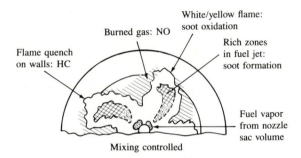

Burned gas: NO

White/yellow flame: soot oxidation

Flame quench on walls: HC

Rich zones in fuel jet: soot formation

Fuel vapor from nozzle sac volume

Mixing controlled

FIGURE 11-3
Summary of pollutant formation mechanisms in a direct-injection compression-ignition engine during "premixed" and "mixing-controlled" combustion phases.

11.2 NITROGEN OXIDES

11.2.1 Kinetics of NO Formation

While nitric oxide (NO) and nitrogen dioxide (NO_2) are usually grouped together as NO_x emissions, nitric oxide is the predominant oxide of nitrogen produced inside the engine cylinder. The principal source of NO is the oxidation of atmospheric (molecular) nitrogen. However, if the fuel contains significant nitrogen, the oxidation of the fuel nitrogen-containing compounds is an additional source of NO. Gasolines contain negligible amounts of nitrogen; although diesel fuels contain more nitrogen, current levels are not significant.

The mechanism of NO formation from atmospheric nitrogen has been studied extensively.[1] It is generally accepted that in combustion of near-stoichiometric fuel-air mixtures the principal reactions governing the formation of NO from molecular nitrogen (and its destruction) are†

$$O + N_2 = NO + N \qquad (11.1)$$

$$N + O_2 = NO + O \qquad (11.2)$$

$$N + OH = NO + H \qquad (11.3)$$

---

† This is often called the extended Zeldovich mechanism. Zeldovich[1] was the first to suggest the importance of reactions (11.1) and (11.2). Lavoie et al.[2] added reaction (11.3) to the mechanism; it does contribute significantly.

TABLE 11.1
Rate constants for NO formation mechanism[1]

Reaction	Rate constant, $cm^3/mol \cdot s$	Temperature range, K	Uncertainty, factor of or %
(1) $O + N_2 \rightarrow NO + N$	$7.6 \times 10^{13} \exp[-38,000/T]$	2000–5000	2
(-1) $N + NO \rightarrow N_2 + O$	1.6×10^{13}	300–5000	$\pm 20\%$ at 300 K
			2 at 2000–5000 K
(2) $N + O_2 \rightarrow NO + O$	$6.4 \times 10^9 \, T \exp[-3150/T]$	300–3000	$\pm 30\%$ 300–1500 K
			2 at 3000 K
(-2) $O + NO \rightarrow O_2 + N$	$1.5 \times 10^9 \, T \exp[-19,500/T]$	1000–3000	$\pm 30\%$ at 1000 K
			2 at 3000 K
(3) $N + OH \rightarrow NO + H$	4.1×10^{13}	300–2500	$\pm 80\%$
(-3) $H + NO \rightarrow OH + N$	$2.0 \times 10^{14} \exp[-23,650/T]$	2200–4500	2

The forward and reverse rate constants (k_i^+ and k_i^-, respectively) for these reactions have been measured in numerous experimental studies. Recommended values for these rate constants taken from a critical review of this published data are given in Table 11.1. Note that the equilibrium constant for each reaction, $K_{c,i}$ (see Sec. 3.7.2), is related to the forward and reverse rate constants by $K_{c,i} = k_i^+/k_i^-$. The rate of formation of NO via reactions (11.1) to (11.3) is given by [see Eqs. (3.55) and (3.58)]

$$\frac{d[NO]}{dt} = k_1^+[O][N_2] + k_2^+[N][O_2] + k_3^+[N][OH]$$

$$- k_1^-[NO][N] - k_2^-[NO][O] - k_3^-[NO][H] \quad (11.4)$$

where [ ] denote species concentrations in moles per cubic centimeter when k_i have the values given in Table 11.1. The forward rate constant for reaction (11.1) and the reverse rate constants for reactions (11.2) and (11.3) have large activation energies which results in a strong temperature dependence of NO formation rates.

A similiar relation to (11.4) can be written for $d[N]/dt$:

$$\frac{d[N]}{dt} = k_1^+[O][N_2] - k_2^+[N][O_2] - k_3^+[N][OH]$$

$$- k_1^-[NO][N] + k_2^-[NO][O] + k_3^-[NO][H] \quad (11.5)$$

Since [N] is much less than the concentrations of other species of interest ($\sim 10^{-8}$ mole fraction), the steady-state approximation is appropriate: $d[N]/dt$ is set equal to zero and Eq. (11.5) used to eliminate [N]. The NO formation rate then becomes

$$\frac{d[NO]}{dt} = 2k_1^+[O][N_2] \frac{1 - [NO]^2/(K[O_2][N_2])}{1 + k_1^-[NO]/(k_2^+[O_2] + k_3^+[OH])} \quad (11.6)$$

where $K = (k_1^+/k_1^-)(k_2^+/k_2^-)$.

TABLE 11.2
Typical values of R_1, R_1/R_2, and $R_1/(R_2 + R_3)$†

Equivalence ratio	R_1‡	R_1/R_2	$R_1/(R_2 + R_3)$
0.8	5.8×10^{-5}	1.2	0.33
1.0	2.8×10^{-5}	2.5	0.26
1.2	7.6×10^{-6}	9.1	0.14

† At 10 atm pressure and 2600 K.
‡ Units gmol/cm$^3 \cdot$ s.

NO forms in both the flame front and the postflame gases. In engines, however, combustion occurs at high pressure so the flame reaction zone is extremely thin (~ 0.1 mm) and residence time within this zone is short. Also, the cylinder pressure rises during most of the combustion process, so burned gases produced early in the combustion process are compressed to a higher temperature than they reached immediately after combustion. Thus, NO formation in the postflame gases almost always dominates any flame-front–produced NO. It is, therefore, appropriate to assume that the combustion and NO formation processes are decoupled and to approximate the concentrations of O, O_2, OH, H, and N_2 by their equilibrium values at the local pressure and equilibrium temperature.

To introduce this equilibrium assumption it is convenient to use the notation $R_1 = k_1^+[O]_e[N_2]_e = k_1^-[NO]_e[N]_e$, where $[\ ]_e$ denotes equilibrium concentration, for the one-way equilibrium rate for reaction (11.1), with similiar definitions for $R_2 = k_2^+[N]_e[O_2]_e = k_2^-[NO]_e[O]_e$ and $R_3 = k_3^+[N]_e[OH]_e = k_3^-[NO]_e[H]_e$. Substituting $[O]_e$, $[O_2]_e$, $[OH]_e$, $[H]_e$, and $[N_2]_e$ for $[O]$, $[O_2]$, $[OH]$, $[H]$, and $[N_2]$ in Eq. (11.6) yields

$$\frac{d[NO]}{dt} = \frac{2R_1\{1 - ([NO]/[NO]_e)^2\}}{1 + ([NO]/[NO]_e)R_1/(R_2 + R_3)} \tag{11.7}$$

Typical values of R_1, R_1/R_2 and $R_1/(R_2 + R_3)$ are given in Table 11.2. The difference between R_1/R_2 and $R_1/(R_2 + R_3)$ indicates the relative importance of adding reaction (11.3) to the mechanism.

The strong temperature dependence of the NO formation rate can be demonstrated by considering the initial value of $d[NO]/dt$ when $[NO]/[NO]_e \ll 1$. Then, from Eq. (11.7),

$$\frac{d[NO]}{dt} = 2R_1 = 2k_1^+[O]_e[N_2]_e \tag{11.8}$$

The equilibrium oxygen atom concentration is given by

$$[O]_e = \frac{K_{p(O)}[O_2]_e^{1/2}}{(\tilde{R}T)^{1/2}} \tag{11.9}$$

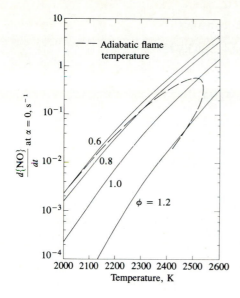

FIGURE 11-4
Initial NO formation rate, mass fraction per second (for $[NO]/[NO]_e \ll 1$), as a function of temperature for different equivalence ratios (ϕ) and 15 atm pressure. Dashed line shows adiabatic flame temperature for kerosene combustion with 700 K, 15 atm air.[3]

where $K_{p(O)}$ is the equilibrium constant for the reaction

$$\tfrac{1}{2}O_2 = O$$

and is given by

$$K_{p(O)} = 3.6 \times 10^3 \exp\left(\frac{-31,090}{T}\right) \qquad \text{atm}^{1/2} \qquad (11.10)$$

The initial NO formation rate may then be written [combining Eqs. (11.8), (11.9), and (11.10) with k_1^+ from Table 11.1] as

$$\frac{d[NO]}{dt} = \frac{6 \times 10^{16}}{T^{1/2}} \exp\left(\frac{-69,090}{T}\right)[O_2]_e^{1/2}[N_2]_e \qquad \text{mol/cm}^3 \cdot \text{s} \quad (11.11)$$

The strong dependence of $d[NO]/dt$ on temperature in the exponential term is evident. High temperatures and high oxygen concentrations result in high NO formation rates. Figure 11-4 shows the NO formation rate as a function of gas temperature and fuel/air equivalence ratio in postflame gases. Also shown is the adiabatic flame temperature attained by a fuel-air mixture initially at 700 K at a constant pressure of 15 atm. For adiabatic constant-pressure combustion (an appropriate model for each element of fuel that burns in an engine), this initial NO formation rate peaks at the stoichiometric composition, and decreases rapidly as the mixture becomes leaner or richer.

A characteristic time for the NO formation process, τ_{NO}, can be defined by

$$\tau_{NO}^{-1} = \frac{1}{[NO]_e}\frac{d[NO]}{dt} \qquad (11.12)$$

$[NO]_e$ can be obtained from the equilibrium constant

$$K_{NO} = 20.3 \times \exp(-21,650/T)$$

for the reaction

$$O_2 + N_2 = 2NO$$

as $[NO]_e = (K_{NO}[O_2]_e[N_2]_e)^{1/2}$. Equations (11.11) and (11.12) can be combined to give

$$\tau_{NO} = \frac{8 \times 10^{-16} T \exp(58,300/T)}{p^{1/2}} \tag{11.13}$$

where τ_{NO} is in seconds, T in kelvins, and p in atmospheres. Use has been made of the fact that $\tilde{x}_{N_2} \approx 0.71$. For engine combustion conditions, τ_{NO} is usually comparable to or longer than the times characteristic of changes in engine conditions so the formation process is kinetically controlled. However, for close-to-stoichiometric conditions at the maximum pressures and burned gas temperatures, τ_{NO} is of the same order as typical combustion times (1 ms) and equilibrium NO concentrations may be attained.

Evidence that this formation model is valid under conditions typical of those found in engines is provided by high-pressure combustion bomb studies. Newhall and Shahed[4] have measured the NO production, using the q-band absorption technique, behind hydrogen-air and propane-air planar flames propagating axially in a cylindrical bomb. Some results are compared with predictions made with this kinetic scheme (coupled with an "unmixed" combustion calculation to determine local pressure and temperature; see Sec. 9.2.1) in Fig. 11-5. The agreement is excellent. Note that the NO concentration rises smoothly from

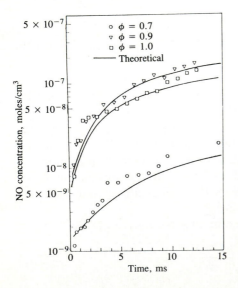

FIGURE 11-5
Measured and calculated rate-limited NO concentrations behind flame in high-pressure cylindrical bomb experiments with H_2-air mixture. ϕ = equivalence ratio.[4]

TABLE 11.3
Typical nitrogen content of distillate fuels[1]

Fraction	Average nitrogen, wt %	Range, wt %
Crude	0.65	—
Heavy distillates	1.40	0.60–2.15
Light distillates	0.07	0–0.60

close to zero, indicating that at these high pressures there is negligible NO production within the flame front itself.

Fuel nitrogen is also a source of NO via a different and yet to be fully explained mechanism. Table 11.3 shows the typical nitrogen content of petroleum-derived fuels. During distillation, the fuel nitrogen is concentrated in the higher boiling fractions. In distillate fuels, the nitrogen can exist as amines and ring compounds (e.g., pyridine, quinoline, and carbazoles). During combustion these compounds are likely to undergo some thermal decomposition prior to entering the combustion zone. The precursors to NO formation will therefore be low molecular weight nitrogen-containing compounds such as NH_3, HCN, and CN. The detailed information on the kinetics of NO formation from these compounds is limited. Oxidation to NO is usually rapid, occurring on a time scale comparable to that of the combustion reactions. The NO yield (amount of fuel nitrogen converted to NO) is sensitive to the fuel/air equivalence ratio. Relatively high NO yields (approaching 100 percent) are obtained for lean and stoichiometric mixtures; relatively low yields are found for rich mixtures. NO yields are only weakly dependent on temperature, in contrast to the strong temperature dependence of NO formed from atmospheric nitrogen.[1]

11.2.2 Formation of NO_2

Chemical equilibrium considerations indicate that for burned gases at typical flame temperatures, NO_2/NO ratios should be negligibly small. While experimental data show this is true for spark-ignition engines, in diesels NO_2 can be 10 to 30 percent of the total exhaust oxides of nitrogen emissions.[5] A plausible mechanism for the persistence of NO_2 is the following.[6] NO formed in the flame zone can be rapidly converted to NO_2 via reactions such as

$$NO + HO_2 \rightarrow NO_2 + OH \tag{11.14}$$

Subsequently, conversion of this NO_2 to NO occurs via

$$NO_2 + O \rightarrow NO + O_2 \tag{11.15}$$

unless the NO_2 formed in the flame is quenched by mixing with cooler fluid. This explanation is consistent with the highest NO_2/NO ratio occurring at light load in diesels, when cooler regions which could quench the conversion back to NO are widespread.[5]

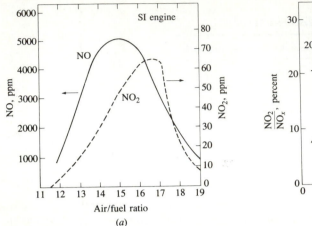
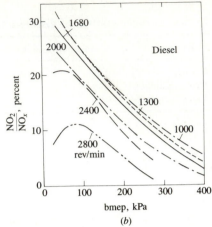

FIGURE 11-6
(a) NO and NO_2 concentrations in SI engine exhaust as function of air/fuel ratio, 1500 rev/min, wide-open throttle; (b) NO_2 as percent of total NO_x in diesel exhaust as function of load and speed.[5]

Figure 11-6 shows examples of NO and NO_2 emissions data from a spark-ignition and a diesel engine. The maximum value for the ratio (NO_2/NO) for the SI engine is 2 percent, at an equivalence ratio of about 0.85. For the diesel this ratio is higher, and is highest at light load and depends on engine speed.

It is customary to measure total oxides of nitrogen emissions, NO plus NO_2, with a chemiluminescence analyzer and call the combination NO_x. It is always important to check carefully whether specific emissions data for NO_x are given in terms of mass of NO or mass of NO_2, which have molecular weights of 30 and 46, respectively.

11.2.3   NO Formation in Spark-Ignition Engines

In conventional spark-ignition engines the fuel and air (and any recycled exhaust) are mixed together in the engine intake system, and vigorous mixing with the residual gas within the cylinder occurs during the intake process. Thus the fuel/air ratio and the amount of diluent (residual gas plus any recycled exhaust) is approximately uniform throughout the charge within the cylinder during combustion.† Since the composition is essentially uniform, the nature of the NO formation process within the cylinder can be understood by coupling the kinetic mechanism developed in Sec. 11.2.1 with the burned gas temperature distribution and pressure in the cylinder during the combustion and expansion processes. The

---

† It is well known that the mixture composition within the cylinder is not completely uniform and varies from one cycle to the next. Both these factors contribute to cycle-by-cycle combustion variations. For the present discussion, the assumption of mixture uniformity is adequate.

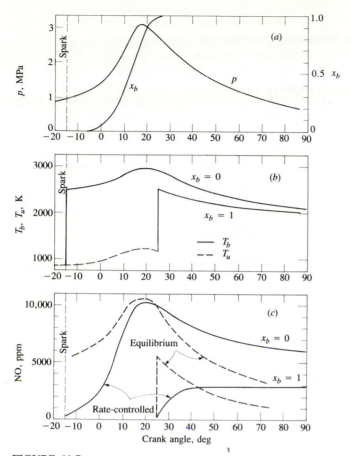

FIGURE 11-7
Illustration of SI engine NO formation model: (*a*) measured cylinder pressure *p* and calculated mass fraction burned x_b; (*b*) calculated temperature of unburned gas T_u and burned gas T_b in early- and late-burning elements; (*c*) calculated NO concentrations in early- and late-burning elements for rate-controlled model and at equilibrium.[7]

temperature distribution which develops in the burned gases due to the passage of the flame across the combustion chamber has been discussed in Sec. 9.2.1. Mixture which burns early is compressed to higher temperatures after combustion, as the cylinder pressure continues to rise; mixture which burns later is compressed primarily as unburned mixture and ends up after combustion at a lower burned gas temperature. Figure 11-7*a* and *b* shows measured cylinder pressure data from an operating engine, with estimates of the mass fraction burned (x_b) and the temperatures of a gas element which burned just after spark discharge and a gas element which burned at the end of the burning process. The model used to estimate these temperatures assumed no mixing between mixture elements which burn at different times. This assumption is more realistic than the alternative idealization that the burned gases mix rapidly and are thus uniform (see Sec. 9.2.1). If the NO formation kinetic model [Eq. (11.7)] is used to calculate NO concentrations in these burned gas elements, using the equilibrium concentrations of the species O, O_2, N_2, OH, and H corresponding to the average fuel/air equivalence ratio and burned gas fraction of the mixture and these pressure and temperature profiles, the rate-limited concentration profiles in Fig. 11-7c are obtained. Also shown are the NO concentrations that would correspond to chemical equilibrium at these conditions. The rate-controlled concentrations rise from the residual gas NO concentration, lagging the equilibrium levels, then cross the equilibrium levels and "freeze" well above the equilibrium values corresponding to exhaust conditions. Depending on details of engine operating conditions, the rate-limited concentrations may or may not come close to equilibrium levels at peak cylinder pressure and gas temperature. Also, the amount of decomposition from peak NO levels which occurs during expansion depends on engine conditions as well as whether the mixture element burned early or late.[7]

Once the NO chemistry has frozen during the early part of the expansion stroke, integration over all elements will give the final average NO concentration in the cylinder which equals the exhaust concentration. Thus, if {NO} is the local mass fraction of NO, then the average exhaust NO mass fraction is given by

$$\overline{\{NO\}} = \int_0^1 \{NO\}_f \, dx_b \tag{11.16}$$

where $\{NO\}_f$ is the final frozen NO mass fraction in the element of charge which burned when the mass fraction burned was x_b. Note that $\{NO\} = [NO]M_{NO}/\rho$, where $M_{NO} = 30$, the molecular weight of NO. The average exhaust concentration of NO as a mole fraction is given by

$$\tilde{x}_{NOav} = \overline{\{NO\}} \frac{M_{exh}}{M_{NO}} \tag{11.17}$$

and the exhaust concentration in ppm is $\tilde{x}_{NOav} \times 10^6$. The earlier burning fractions of the charge contribute much more to the exhausted NO than do later burning fractions of the charge: frozen NO concentrations in these early-burning elements can be an order of magnitude higher than concentrations in late-burning elements. In the absence of vigorous bulk gas motion, the highest NO concentrations occur nearest the spark plug.

Substantial experimental evidence supports this description of NO formation in spark-ignition engines. The NO concentration gradient across the burned gas in the engine cylinder, due to the temperature gradient, has been demonstrated using gas sampling techniques[8, 9] and using measurements of the chemiluminescent radiation from the reaction $NO + O \rightarrow NO_2 + h\nu$ to determine the local NO concentration. Figure 11-8 shows NO concentration data as a function of crank angle, taken by Lavoie[10] through two different windows in the cylinder head of a specially constructed L-head engine where each window was a different distance from the spark plug. The stars indicate the estimated initial NO concentration that results from mixing of the residual gas with the fresh charge, at the

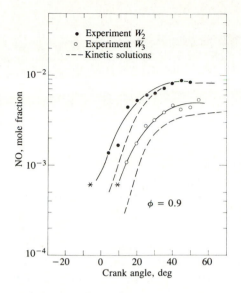

- • Experiment W_2
- ○ Experiment W_3
- --- Kinetic solutions

NO, mole fraction

10^{-2}

10^{-3}

10^{-4}

$\phi = 0.9$

−20　0　20　40　60

Crank angle, deg

FIGURE 11-8
Spectroscopically measured NO concentrations through two windows W_3 and W_2 in special L-head SI engine (W_2 is closer to spark than W_3). The asterisks mark estimated initial conditions and flame arrival times. The dashed lines are calculated rate-limited concentrations for parts of charge burning at these flame arrival times with zero initial NO concentration.[10]

time of arrival of the flame at each window. The observed NO mole fractions rise smoothly from these initial values and then freeze about one-third of the way through the expansion process. NO levels observed at window W_2, closest to the spark plug, were substantially higher than those observed at window W_3. The dashed lines show calculated NO concentrations obtained using the NO formation kinetic model with an "unmixed" thermodynamic analysis for elements that burned at the time of flame arrival at each window. Since the calculated values started from zero NO concentration at the flame front (and not the diluted residual gas NO level indicated by the star), the calculations initially fall below the data. However, the difference between the two measurement locations and the frozen levels are predicted with reasonable accuracy. Thus, the rate-limited formation process, freezing of NO chemistry during expansion, and the existence of NO concentration gradients across the combustion chamber have all been observed.

The most important engine variables that affect NO emissions are the fuel/air equivalence ratio, the burned gas fraction of the *in-cylinder* unburned mixture, and spark timing. The burned gas fraction depends on the amount of diluent such as recycled exhaust gas (EGR) used for NO_x emissions control, as well as the residual gas fraction. Fuel properties will affect burned gas conditions; the effect of normal variations in gasoline properties is modest, however. The effect of variations in these parameters can be explained with the NO formation mechanism described above: changes in the time history of temperature and oxygen concentration in the burned gases during the combustion process and early part of the expansion stroke are the important factors.[11]

EQUIVALENCE RATIO. Figure 11-9 shows the effect of variations in the fuel/air equivalence ratio on NO emissions. Maximum burned gas temperatures occur at

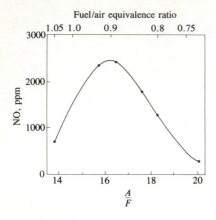

FIGURE 11-9
Variation of exhaust NO concentration with A/F and fuel/air equivalence ratio. Spark-ignition engine, 1600 rev/min, $\eta_v = 50$ percent, MBT timing.[12]

$\phi \approx 1.1$; however, at this equivalence ratio oxygen concentrations are low. As the mixture is enriched, burned gas temperatures fall. As the mixture is leaned out, increasing oxygen concentration initially offsets the falling gas temperatures and NO emissions peak at $\phi \approx 0.9$. Detailed predictions of NO concentrations in the burned gases suggest that the concentration versus time histories under fuel-lean conditions are different in character from those for fuel-rich conditions. In lean mixtures NO concentrations freeze early in the expansion process and little NO decomposition occurs. In rich mixtures, substantial NO decomposition occurs from the peak concentrations present when the cylinder pressure is a maximum. Thus in lean mixtures, gas conditions at the time of peak pressure are especially significant.[7]

BURNED GAS FRACTION. The unburned mixture in the cylinder contains fuel vapor, air, and burned gases. The burned gases are residual gas from the previous cycle and any exhaust gas recycled to the intake for NO_x emissions control. The residual gas fraction is influenced by load, valve timing (especially the extent of valve overlap), and, to a lesser degree, by speed, air/fuel ratio, and compression ratio as described in Sec. 6.4. The burned gases act as a diluent in the unburned mixture; the absolute temperature reached after combustion varies inversely with the burned gas mass fraction. Hence increasing the burned gas fraction reduces NO emissions levels. However, it also reduces the combustion rate and, therefore, makes stable combustion more difficult to achieve (see Secs. 9.3 and 9.4).

Figure 11-10 shows the effect of increasing the burned gas fraction by recycling exhaust gases to the intake system just below the throttle plate. Substantial reductions in NO concentrations are achieved with 15 to 25 percent EGR, which is about the maximum amount of EGR the engine will tolerate under normal part-throttle conditions. Of course, increasing the EGR at fixed engine load and speed increases the inlet manifold pressure, while fuel flow and air flow remain approximately constant.

The primary effect of the burned gas diluent in the unburned mixture on the NO formation process is that it reduces flame temperatures by increasing the

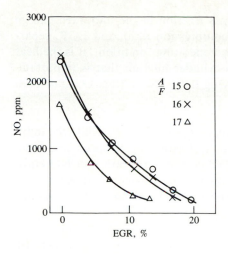

FIGURE 11-10
Variation of exhaust NO concentration with percent recycled exhaust gas (EGR). Spark-ignition engine, 1600 rev/min, $\eta_v = 50$ percent, MBT timing.[12]

heat capacity of the cylinder charge, per unit mass of fuel. Figure 11-11 shows the effect of different diluent gases added to the engine intake flow, in a single-cylinder engine operated at constant speed, fuel flow, and air flow.[13] The data in Fig. 11-11a show that equal volume percentages of the different diluents produce different reductions in NO emissions. The same data when plotted against diluent heat capacity (diluent mass flow rate × specific heat, c_p) collapse to a single

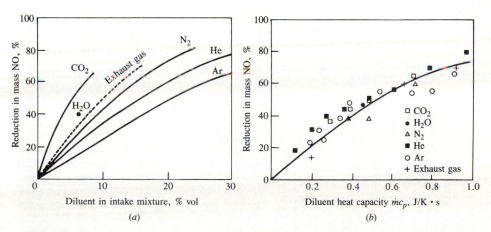

FIGURE 11-11
(a) Percentage reduction in mass NO emissions with various diluents. (b) Correlation of NO reduction with diluent heat capacity. Spark-ignition engine operated at 1600 rev/min, constant brake load (intake pressure ~ 0.5 atm), with MBT spark timing.[13]

curve.† A similiar study where the burned gas fraction in the unburned charge was varied by changing the valve overlap, compression ratio, and EGR, separately, showed that, under more realistic engine operating conditions, it is the heat capacity of the total diluent mass in the in-cylinder mixture that is important. Whether the diluent mass is changed by varying the valve overlap, EGR, or even the compression ratio is not important.[14]

EXCESS AIR AND EGR. Because of the above, it is possible to correlate the influence of engine operating variables (such as air/fuel ratio, engine speed, and load) and design variables (such as valve timing and compression ratio) on NO emissions with two parameters which define the in-cylinder mixture composition: the fuel/air equivalence ratio (often the air/fuel ratio is used instead) and the gas/fuel ratio. The gas/fuel ratio (G/F) is given by

$$\frac{G}{F} = \frac{\text{total mass in cylinder}}{\text{fuel mass in cylinder}} = \frac{A}{F}\left(1 + \frac{x_b}{1 - x_b}\right) \tag{11.18}$$

where x_b is the burned gas fraction [Eq. (4.3)]. These together define the relative proportions of fuel, air, and burned gases in the in-cylinder mixture, and hence

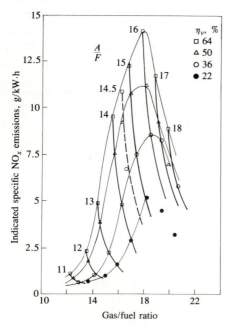

FIGURE 11-12

Correlation between gas/fuel ratio (G/F) and indicated specific NO_x emissions at various air/fuel ratios (A/F) and volumetric efficiencies (η_v). Spark-ignition engine operated at 1400 rev/min with spark timing retarded to give 0.95 of maximum brake torque.[15]

† Some of the scatter in Fig. 11-11 is due to the fact that the residual gas fraction is slightly different for each diluent.

will correlate NO emissions.† Figure 11-12 shows the correlation of specific NO emissions, from a four-cylinder engine, over a wide range of engine operating conditions with the air/fuel ratio and gas/fuel ratio. Lines of constant air/fuel ratio and volumetric efficiency are shown; the direction of increasing dilution with residual gas and EGR at constant air/fuel ratio is to the right. Excessive dilution results in poor combustion quality, partial burning, and, eventually, misfire (see Sec. 9.4.3). Lowest NO emissions consistent with good fuel consumption (avoiding the use of rich mixtures) are obtained with a stoichiometric mixture, with as much dilution as the engine will tolerate without excessive deterioration in combustion quality.[15]

Comparisons between predictions made with the NO formation model (described at the beginning of this section) and experimental data show good agreement with normal amounts of dilution.[16] With extreme dilution, at NO levels of about 100 ppm or less, the NO formed within the flame reaction zone cannot, apparently, be neglected. Within the flame, the concentrations of radicals such as O, OH, and H can be substantially in excess of equilibrium levels, resulting in much higher formation rates within the flame than in the postflame gases. It is believed that the mechanism [reactions (11.1) to (11.3)] and the formation rate equation (11.6) are valid. However, neglecting flame-front-formed NO is no longer an appropriate assumption.[17]

SPARK TIMING. Spark timing significantly affects NO emission levels. Advancing the timing so that combustion occurs earlier in the cycle increases the peak cylinder pressure (because more fuel is burned before TC and the peak pressure moves closer to TC where the cylinder volume is smaller); retarding the timing decreases the peak cylinder pressure (because more of the fuel burns after TC). Higher peak cylinder pressures result in higher peak burned gas temperatures, and hence higher NO formation rates. For lower peak cylinder pressures, lower NO formation rates result. Figure 11-13 shows typical NO emission data for a spark-ignition engine as a function of spark timing. NO emission levels steadily decrease as spark timing is retarded from MBT timing and moved closer to TC. Since exact determination of MBT timing is difficult (and not critical for fuel consumption and power where the variation with timing around MBT is modest), there is always considerable uncertainty in NO emissions at MBT timing. Often, therefore, an alternative reference timing is used, where spark is retarded from MBT timing to the point where torque is decreased by 1 or 2 percent from the maximum value. Great care with spark timing is necessary to obtain accurate NO emissions measurements under MBT-timing operating conditions.

---

† Spark timing also affects NO emissions, as discussed next. The above discussion relates to engines run with timing at MBT or with torque at a fixed percentage of (and close to) the maximum.

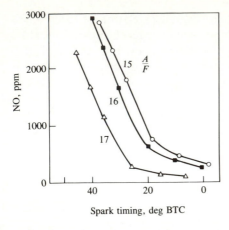

FIGURE 11-13

Variation of exhaust NO concentration with spark retard. 1600 rev/min, $\eta_v = 50$ percent; left-hand end of curve corresponds to MBT timing for each A/F.[12]

11.2.4   NO$_x$ Formation in Compression-Ignition Engines

The kinetic mechanisms for NO and NO$_2$ formation described in Secs. 11.2.1 and 11.2.2 and the assumptions made regarding equilibration of species in the C—O—H system apply to diesels as well as to spark-ignition engines. The critical difference, of course, is that injection of fuel into the cylinder occurs just before combustion starts, and that nonuniform burned gas temperature and composition result from this nonuniform fuel distribution during combustion. The fuel-air mixing and combustion processes are extremely complex. During the "premixed" or uncontrolled diesel combustion phase immediately following the ignition delay, fuel-air mixture with a spread in composition about stoichiometric burns due to spontaneous ignition and flame propagation. During the mixing controlled combustion phase, the burning mixture is likely to be closer to stoichiometric (the flame structure is that of a turbulent, though unsteady, diffusion flame). However, throughout the combustion process mixing between already burned gases, air, and lean and rich unburned fuel vapor–air mixture occurs, changing the composition of any gas elements that burned at a particular equivalence ratio. In addition to these composition (and hence temperature) changes due to mixing, temperature changes due to compression and expansion occur as the cylinder pressure rises and falls.

The discussion in Sec. 11.2.1 showed that the critical equivalence ratio for NO formation in high-temperature high-pressure burned gases typical of engines is close to stoichiometric. Figure 11-4 is relevant here: it shows the initial NO formation rate in combustion products formed by burning a mixture of a typical hydrocarbon fuel with air (initially at 700 K, at a constant pressure of 15 atm). NO formation rates are within a factor of 2 of the maximum value for $0.85 \lesssim \phi \lesssim 1.1$.

The critical time period is when burned gas temperatures are at a maximum: i.e., between the start of combustion and shortly after the occurrence of peak cylinder pressure. Mixture which burns early in the combustion process

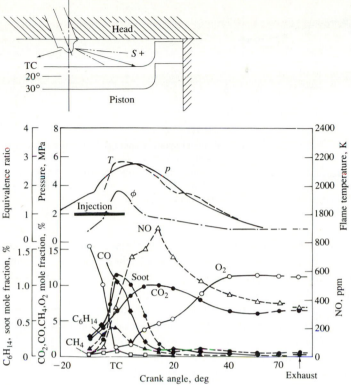

FIGURE 11-14
Concentrations of soot, NO, and other combustion product species measured at outer edge of bowl-in-piston combustion chamber (location S) of quiescent DI diesel with rapid sampling valve. Cylinder gas pressure p, mean temperature T, and local equivalence ratio ϕ shown. Bore = 95 mm, stroke = 110 mm, r_c = 14.6. Four-hole nozzle with hole diameter = 0.2 mm.[18]

is especially important since it is compressed to a higher temperature, increasing the NO formation rate, as combustion proceeds and cylinder pressure increases. After the time of peak pressure, burned gas temperatures decrease as the cylinder gases expand. The decreasing temperature due to expansion *and* due to mixing of high-temperature gas with air or cooler burned gas freezes the NO chemistry. This second effect (which occurs only in the diesel) means that freezing occurs more rapidly in the diesel than in the spark-ignition engine, and much less decomposition of the NO occurs.

The above description is supported by the NO concentration data obtained from experiments where gas was sampled from within the cylinder of normally operating diesel engines with special gas-sampling valves and analyzed. Figure 11-14 shows time histories of major species concentrations, through the combustion process, determined with a rapid-acting sampling valve (1 ms open time) in a quiescent *direct-injection diesel* engine. Concentrations at different positions in the combustion chamber were obtained; the sample valve location for the Fig.

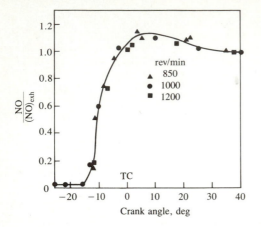

FIGURE 11-15
Ratio of cylinder-average NO concentration at given crank angle (determined from cylinder-dumping experiments) to exhaust NO concentration. DI diesel, equivalence ratio = 0.6, injection timing at 27° BTC.[19]

11-14 data is shown. Local NO concentrations rise from the residual gas value following the start of combustion, to a peak at the point where the local burned gas equivalence ratio changes from rich to lean (where the CO_2 concentration has its maximum value). As the local burned gas equivalence ratio becomes leaner due to mixing with excess air, NO concentrations decrease since formation becomes much slower as dilution occurs. At the time of peak NO concentrations within the bowl (15° ATC), most of the bowl region was filled with flame. The total amount of NO within the cylinder of this type of direct-injection diesel during the NO formation process has also been measured.[19] At a predetermined time in one cycle, once steady-state warmed-up engine operation had been achieved, the contents of the cylinder were dumped into an evacuated tank by rapidly cutting open a diaphragm which had previously sealed off the tank system. Figure 11-15 shows how the ratio of the average cylinder NO concentration divided by the exhaust concentration varies during the combustion process. NO concentrations reach a maximum shortly after the time of peak pressure. There is a modest amount of NO decomposition. Variations in engine speed have little effect on the shape of this curve. The 20 crank angle degrees after the start of combustion is the critical time period.

Results from similar cylinder-dumping experiments where injection timing and load (defined by the overall equivalence ratio) were varied also showed that almost all of the NO forms within the 20° following the start of combustion. As injection timing is retarded, so the combustion process is retarded; NO formation occurs later, and concentrations are lower since peak temperatures are lower. The effect of the overall equivalence ratio on NO_x concentrations is shown in Fig. 11-16. At high load, with higher peak pressures, and hence temperatures, and larger regions of close-to-stoichiometric burned gas, NO levels increase. Both NO and NO_x concentrations were measured; NO_2 is 10 to 20 percent of total NO_x. Though NO levels decrease with a decreasing overall equivalence ratio, they do so much less rapidly than do spark-ignition engine NO emissions (see Fig. 11-9) due to the nonuniform fuel distribution in the diesel. Though the

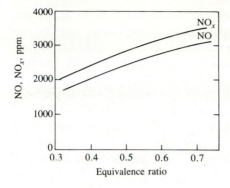

FIGURE 11-16
Exhaust NO_x and NO concentrations as a function of overall equivalence ratio or engine load. DI diesel, 1000 rev/min, injection timing at 27° BTC.[19]

amount of fuel injected decreases proportionally as the overall equivalence ratio is decreased, much of the fuel still burns close to stoichiometric. Thus NO emissions should be roughly proportional to the mass of fuel injected (provided burned gas pressures and temperatures do not change greatly).

Similar gas-sampling studies have been done with *indirect-injection* diesel engines. Modeling studies suggest that most of the NO forms within the prechamber and is then transported into the main chamber where the reactions freeze as rapid mixing with air occurs. However, the prechamber, except at light load, operates rich overall so some additional NO can form as the rich combustion products are diluted through the stoichiometric composition.[20] Figure 11-17 shows local NO concentrations and equivalence ratios as a function of crank angle determined with a rapid-acting sampling valve at different locations

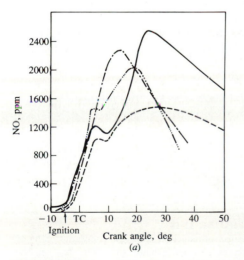

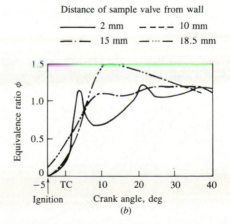

FIGURE 11-17
(a) NO concentrations measured with rapid sampling valve and (b) calculated equivalence ratios at different distances from the wall in swirl chamber of IDI diesel engine, as function of crank angle. Engine speed = 1000 rev/min, injection at 13° BTC, ignition at 5° BTC.[21]

within the prechamber of a Comet swirl chamber IDI engine.[21] The gas mixture rapidly becomes stoichiometric or fuel-rich. Composition nonuniformities across the prechamber are substantial. Peak NO concentrations, as expected, correspond approximately to locally stoichiometric regions. Because the mixture remains fuel-rich in the prechamber as the burned gases expand (after the time of peak pressure which occurs between 6 and 10° ATC), substantial NO decomposition within the prechamber can occur. However, by this time much of the gas (and NO) in the prechamber has been transferred to the main chamber where freezing of the NO chemistry will occur. Cylinder-gas dumping experiments, where both main chamber and prechamber gases were dumped and quenched, confirm this description. Cylinder average NO concentrations, determined by rapidly opening a diaphram which separated the engine cylinder from an evacuated tank at predetermined points in the cycle of an otherwise normally operated IDI engine, rise rapidly once combustion starts, until the NO chemistry is effectively frozen at about 15° ATC. Little net NO decomposition occurs.[22] Heat-release-rate diagrams obtained from pressure data analysis for the same IDI engine indicate that combustion is only about one-half complete at the time the NO formation process ceases.

Diluents added to the intake air (such as recycled exhaust) are effective at reducing the NO formation rate, and therefore NO_x exhaust emissions. As with spark-ignition engines, the effect is primarily one of reducing the burned gas temperature for a given mass of fuel and oxygen burned. Figure 11-18 shows the effect of dilution of the intake air with N_2, CO_2, and exhaust gas on NO_x exhaust levels.[23] The heat capacity of CO_2 (per mole) at the temperatures relevant to diesel combustion is about twice that of N_2. That of exhaust gas is slightly higher than that of N_2. Therefore these data show that the effect is primarily one of reduced burned gas temperatures. Note that the composition of the exhaust gas of a diesel varies with load. At idle, there is little CO_2 and H_2O, and the composition does not differ much from that of air. At high load the heat

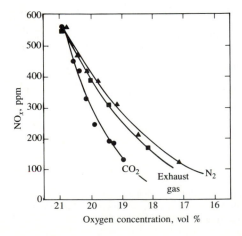

FIGURE 11-18
Effect of reduction in oxygen concentration by different diluents (exhaust gas, CO_2, N_2) on NO_x emissions in DI diesel. Bore = 140 mm, stroke = 152 mm, r_c = 14.3. Speed = 1300 rev/min, fuel rate = 142 mm^3/stroke, injection timing at 4° BTC.[23]

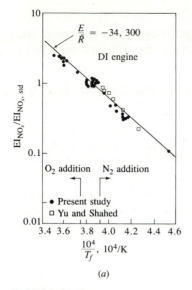

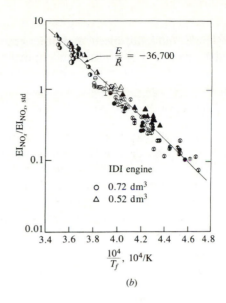

(a)   (b)

FIGURE 11-19
Correlation of NO_x emissions index EI_{NO_x} for a wide range of operating conditions with reciprocal of stoichiometric mixture flame temperature for: (a) DI engines; (b) IDI engines. Flame temperatures varied by addition of different diluents and oxygen.[25, 26] Values of EI_{NO_x} normalized with value at standard conditions.

capacity increases as the concentrations of CO_2 and H_2O are substantially higher. Similar studies in an *indirect-injection* engine show comparable trends. Addition of diluents [exhaust gas (EGR) and nitrogen] reduce peak flame temperatures and NO_x emissions; also, addition of oxygen (which corresponds to a *reduction* in diluent fraction) increases flame temperatures and therefore increases NO_x emissions.[24]

Confirmation that NO forms in the close-to-stoichiometric burned gas regions and the magnitude of the stoichiometric burned gas temperature controls NO_x emissions is given by the following. Plee et al.[25, 26] have shown that the effects of changes in intake gas composition (with EGR, nitrogen, argon, and oxygen addition) and temperature on NO_x emissions can be correlated by

$$EI_{NO_x} = \text{constant} \times \exp\left(\frac{E}{\tilde{R}T_f}\right) \qquad (11.19)$$

T_f(kelvin) is the stoichiometric adiabatic flame temperature (evaluated at a suitable reference point: fuel-air mixture at top-center pressure and air temperature) and E is an overall activation energy. Figure 11-19 shows EI_{NO_x} for a range of intake air compositions and temperatures, and two DI and two IDI engines for several loads and speeds, normalized by the engine NO_x level obtained for standard air, plotted on a log scale against the reciprocal of the stoichiometric adiabatic flame at TC conditions. A single value of $E/\tilde{R}$ correlates the data over two orders of magnitude. There is, of course, some scatter since the model used is overly simple, and load, speed, and other engine design and operating parameters affect the process. The overriding importance of the burned gas temperature of close-to-stoichiometric mixture is clear, however.

11.3 CARBON MONOXIDE

Carbon monoxide (CO) emissions from internal combustion engines are controlled primarily by the fuel/air equivalence ratio. Figure 11-20 shows CO levels in the exhaust of a conventional spark-ignition engine for several different fuel compositions.[27] When the data are plotted against the relative air/fuel ratio or the equivalence ratio, they are correlated by a single curve. For fuel-rich mixtures CO concentrations in the exhaust increase steadily with increasing equivalence ratio, as the amount of excess fuel increases. For fuel-lean mixtures, CO concentrations in the exhaust vary little with equivalence ratio and are of order 10^{-3} mole fraction.

Since spark-ignition engines often operate close to stoichiometric at part load and fuel rich at full load (see Sec. 7.1), CO emissions are significant and must be controlled. Diesels, however, always operate well on the lean side of stoichiometric; CO emissions from diesels are low enough to be unimportant, therefore, and will not be discussed further.

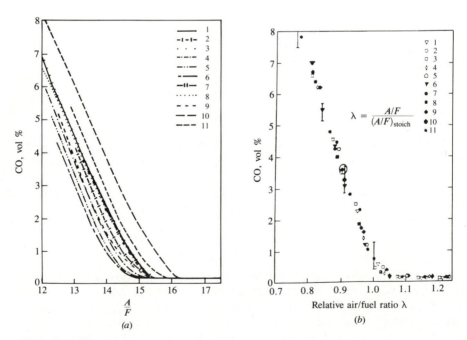

FIGURE 11-20
Variation of SI engine CO emissions with eleven fuels of different H/C ratio: (a) with air/fuel ratio; (b) with relative air/fuel ratio λ.[27]

The levels of CO observed in spark-ignition engine exhaust gases are lower than the maximum values measured within the combustion chamber, but are significantly higher than equilibrium values for the exhaust conditions. Thus the processes which govern CO exhaust levels are kinetically controlled. In premixed hydrocarbon-air flames, the CO concentration increases rapidly in the flame zone to a maximum value, which is larger than the equilibrium value for adiabatic combustion of the fuel-air mixture. CO formation is one of the principal reaction steps in the hydrocarbon combustion mechanism, which may be summarized by[1]

$$RH \rightarrow R \rightarrow RO_2 \rightarrow RCHO \rightarrow RCO \rightarrow CO \tag{11.20}$$

where R stands for the hydrocarbon radical. The CO formed in the combustion process via this path is then oxidized to CO_2 at a slower rate. The principal CO oxidation reaction in hydrocarbon-air flames is

$$CO + OH = CO_2 + H \tag{11.21}$$

The rate constant for this reaction is[1]

$$k_{CO}^+ = 6.76 \times 10^{10} \exp\left(\frac{T}{1102}\right) \qquad cm^3/gmol \tag{11.22}$$

It is generally assumed that in the postflame combustion products in a spark-ignition engine, at conditions close to peak cycle temperatures (2800 K) and pressures (15 to 40 atm), the carbon-oxygen-hydrogen system is equilibrated. Thus CO concentrations in the immediate postflame burned gases are close to equilibrium. However, as the burned gases cool during the expansion and exhaust strokes, depending on the temperature and cooling rate, the CO oxidation process [reaction (11.21)] may not remain locally equilibrated.

Newhall carried out a series of kinetic calculations for an engine expansion stroke assuming the burned gas at the time of peak cylinder pressure was uniform and in equilibrium.[28] Of the reactions important to CO chemistry, only three-body radical-recombination reactions such as

$$H + H + M = H_2 + M \tag{11.23}$$

$$H + OH + M = H_2O + M \tag{11.24}$$

$$H + O_2 + M = HO_2 + M \tag{11.25}$$

were found to be rate controlling. The bimolecular exchange reactions and the CO oxidation reaction (11.21) were sufficiently fast to be continuously equilibrated. Only during the later stages of the expansion stroke was the CO concentration predicted to depart from equilibrium, as shown in Fig. 11-21. Using this technique to predict average CO levels at the end of expansion over a range of equivalence ratios (rich to lean), Newhall obtained a good match to experimental data (see Fig. 11-22). The kinetically controlled aspects of the CO emissions mechanism have thus been confirmed.

These calculations showed that a partial equilibrium amongst the bimolecular exchange reactions occurred *a posteriori*. Analyses based explicitly on this partial equilibrium assumption (which are considerably simpler) have been

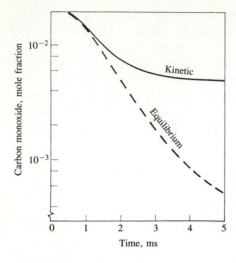

FIGURE 11-21
Results of kinetic calculations of CO concentrations during expansion stroke following TC combustion in SI engine; stoichiometric mixture.[28]

carried out.[29, 30] The appropriate three-body atom and radical recombination reactions [e.g., (11.23) to (11.25)] were treated as the rate-limiting constraint on the total number of particles or moles per unit volume of burnt gases, i.e.,

$$\left(\frac{1}{V}\right)\frac{dn}{dt} = \sum_{i=1}^{k}(R_i^- - R_i^+) \tag{11.26}$$

V is the volume of the elemental system considered, n is the total number of moles, R_i^+ and R_i^- are the forward and backward rates for reaction i, and k represents the number of three-body recombination reactions included. All other

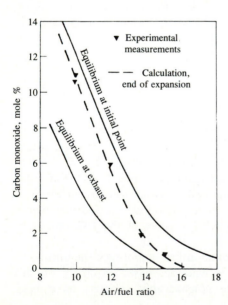

FIGURE 11-22
Predicted CO concentration at end of expansion stroke, compared with measured exhaust concentrations, as function of air/fuel ratio. Equilibrium levels at TC combustion and exhaust conditions also shown.[28]

reactions were assumed to be equilibrated. The studies using this simplified kinetic model have confirmed that at peak cylinder pressures and temperatures, equilibration times for CO are faster than times characteristic of changes in burnt gas conditions due to compression or expansion. Thus the CO concentration rapidly equilibrates in the burnt gases just downstream of the reaction zone following combustion of the hydrocarbon fuel. However, it has already been pointed out in Sec. 9.2.1 that the burnt gases are not uniform in temperature. Also, the blowdown of cylinder pressure to the exhaust manifold level during the exhaust process and the decrease in gas temperature that accompanies it occupies a substantial portion of the cycle—about 60 crank angle degrees. Thus, the temperature- and pressure-time profiles of parts of the charge at different locations throughout the cylinder differ, depending on when these parts of the charge burn and when they exit the cylinder through the exhaust valve and enter the exhaust manifold.

The results of an idealized calculation which illustrate these effects are shown in Fig. 11-23. The CO mole fractions in different elements or parts of the burnt gas mixture are plotted versus crank angle; x_b is the fraction of the total charge which had burned when each element shown burned; z is the mass fraction which had left the cylinder at the time each element left the cylinder. The partial equilibrium calculations show the burned gases are close to equilibrium until about 60 crank angle degrees after top-center. During the exhaust blowdown process after the exhaust valve opens, gas which leaves the cylinder early

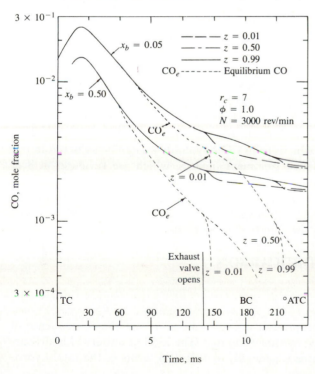

FIGURE 11-23
CO concentrations in selected elements of SI engine cylinder charge, which burn at different times and which exit the cylinder at different times. x_b is mass fraction burned when element burned. z is fraction of gas which has already left cylinder during exhaust process prior to element leaving cylinder. Speed = 3000 rev/min, $r_c = 7$, equivalence ratio = 1.0.[30]

($z \ll 1$) cools more rapidly than gas which leaves late ($z \approx 1$). In these calculations, mixing between gas elements which burn at different times was neglected. It can be seen that a CO gradient exists across the burned gases and that the CO concentration in the exhaust gases is unlikely to be uniform. Experiments with single-cylinder engines support these conclusions that CO is in equilibrium during the combustion process but deviates from equilibrium late in the expansion stroke (e.g., see Refs. 10 and 31).

Conclusions from these detailed studies are as follows. The measured average exhaust CO concentrations for fuel-rich mixtures are close to equilibrium concentrations in the burned gases during the expansion process. For close-to-stoichiometric mixtures, the partial equilibrium CO predictions are in agreement with measurements and are orders of magnitude above CO equilibrium concentrations corresponding to exhaust conditions. For fuel-lean mixtures, measured CO emissions are substantially higher than predictions from any of the models based on kinetically controlled bulk gas phenomena. One possible explanation of this lean-mixture discrepancy is that only partial oxidation to CO may occur of some of the unburned hydrocarbons emerging during expansion and exhaust from crevices in the combustion chamber and from any oil layers or deposits on the chamber walls.

While many questions about details of the CO oxidation mechanisms remain, as a practical matter exhaust emissions are determined by the fuel/air equivalence ratio. The degree of control achieved within the engine to date has come from improving mixture uniformity and leaning-out the intake mixture. In multicylinder engines, because CO increases rapidly as the inlet mixture becomes richer than stoichiometric, cylinder-to-cylinder variations in equivalence ratio about the mean value are important; nonuniform distribution can significantly increase average emissions. Thus improved cylinder-to-cylinder fuel/air ratio distribution has become essential. Also, because it is necessary to enrich the fuel-air mixture when the engine is cold, CO emissions during engine warm-up are much higher than emissions in the fully warmed-up state. Further, in transient engine operation during acceleration and deceleration, control of fuel metering has had to be improved. Additional reductions in CO beyond what can be achieved in the engine are possible with exhaust treatment devices, which are reviewed in Sec. 11.6. Oxidation of CO in the exhaust system without use of special exhaust treatment devices does not occur to any significant degree because the exhaust gas temperature is too low (Fig. 11-23 shows that the CO oxidation reactions effectively freeze as the gas passes through the exhaust valve).

11.4   UNBURNED HYDROCARBON EMISSIONS

11.4.1   Background

Hydrocarbons, or more appropriately organic emissions, are the consequence of incomplete combustion of the hydrocarbon fuel. The level of unburned hydrocarbons (HC) in the exhaust gases is generally specified in terms of the total hydro-

TABLE 11.4
Hydrocarbon composition of spark-ignition engine exhaust (by class)[33]

	Carbon, percent of total HC			
	Paraffins	**Olefins**	**Acetylene**	**Aromatics**
Without catalyst	33	27	8	32
With catalyst	57	15	2	26

carbon concentration expressed in parts per million carbon atoms (C_1).† While total hydrocarbon emission is a useful measure of combustion inefficiency, it is not necessarily a significant index of pollutant emissions. Engine exhaust gases contain a wide variety of hydrocarbon compounds. Table 11.4 shows the average breakdown by class of the hydrocarbons in spark-ignition engine exhaust gases, both with and without a catalytic converter, with gasoline fuel. Some of these hydrocarbons are nearly inert physiologically and are virtually unreactive from the standpoint of photochemical smog. Others are highly reactive in the smog-producing chemistry. (Some hydrocarbons are known carcinogens; see Sec. 11.5.2). Based on their potential for oxidant formation in the photochemical smog chemistry, hydrocarbon compounds are divided into nonreactive and reactive categories. Table 11.5 shows one reactivity scale which has been used to estimate the overall reactivity of exhaust gas hydrocarbon mixtures. Other scales are used for the same purpose.[34] Scales that assign high values for reactivity to the olefins (like Table 11.5), which react most rapidly in the photochemical smog reaction, probably best approximate smog-formation potential near the sources of hydrocarbon pollution. The simplest scale, which divides the HC into two classes—methane and nonmethane hydrocarbons—probably best approximates the end result for all HC emissions. All hydrocarbons except methane react, given enough time. More detailed breakdowns of the composition of spark-ignition and diesel engine exhaust HC are available in the literature.[33, 35]

Fuel composition can significantly influence the composition and magnitude of the organic emissions. Fuels containing high proportions of aromatics and olefins produce relatively higher concentrations of reactive hydrocarbons. However, many of the organic compounds found in the exhaust are not present

---

† This is because the standard detection instrument, a flame ionization detector (FID), is effectively a carbon atom counter: e.g., one propane molecule generates three times the response generated by one methane molecule. Some data in the literature are presented as ppm propane (C_3), or ppm hexane (C_6); to convert to ppm C_1 multiply by 3 or by 6, respectively. Older measurements of hydrocarbon emissions were made with nondispersive infrared (NDIR) detectors which had different sensitivities for the different hydrocarbon compounds. For gasoline-fueled engines, HC emissions determined by FID analyzers are about twice the levels determined by NDIR analyzers,[32] though this scaling is not exact.

TABLE 11.5
Reactivity of classes of hydrocarbons

Hydrocarbons	Relative reactivity†
C_1–C_4 paraffins Acetylene Benzene	0
C_4 and higher molecular weight paraffins Monoalkyl benzenes *Ortho-* and *para*-dialkyl benzenes Cyclic paraffins	2
Ethylene *Meta*-dialkyl benzenes Aldehydes	5
1-Olefins (except ethylene) Diolefins Tri- and tetraalkyl benzenes	10
Internally bonded olefins	30
Internally bonded olefins with substitution at the double bond Cycloolefins	100

† General Motors Reactivity Scale (0–100). Based on the NO_2 formation rate for the hydrocarbon relative to the NO_2 formation rate
for 2,3-dimethyl-2-benzene.[34]

in the fuel, indicating that significant pyrolysis and synthesis occur during the combustion process.

Oxygenates are present in engine exhaust, and are known to participate in the formation of photochemical smog. Some oxygenates are also irritants and odorants. The oxygenates are generally categorized as carbonyls, phenols, and other noncarbonyls. The carbonyls of interest are low molecular weight alde-hydes and aliphatic ketones. The volatile aldehydes are eye and respiratory tract irritants. Formaldehyde is a major component ($\lesssim 20$ percent of total carbonyls). Carbonyls account for about 10 percent of the HC emissions from diesel pas-senger car engines, but only a few percent of spark-ignition engine HC emissions. Phenols are odorants and irritants: levels are much lower than aldehyde levels. Other noncarbonyls include methanol, ethanol, nitromethane, methyl formate. Whether these are significant with conventional hydrocarbon fuels is unclear.[35] Use of alcohol fuels increases oxygenate emissions. Both methanol and aldehyde emissions increase substantially above gasoline-fueled levels with methanol-fueled spark-ignition engines.

11.4.2 Flame Quenching and Oxidation Fundamentals

Flame quenching or extinction occurs at the walls of engine combustion chambers. The cool walls of the chamber act as a sink for heat and the active radical species generated within the flame. Quenching of the flame occurs under several different geometrical configurations: the flame may be propagating normal to, parallel to, or at an angle to the wall; the flame may quench at the entrance to a crevice—a thin volume with a narrow entrance to the combustion chamber such as the region between the piston crown and the cylinder wall. When the flame quenches, it leaves a layer or volume of unburned mixture ahead of the flame. (Whether this results in unburned hydrocarbon emissions depends on the extent to which these quench region hydrocarbons can subsequently be oxidized.)

Flame-quenching processes are analyzed by relating the heat release within the flame to the heat loss to the walls under conditions where quenching just occurs. This ratio, a Peclet number (Pe), is approximately constant for any given geometrical configuration. The simplest configuration for study is the two-plate quench process, where the minimum spacing between two parallel plates through which a flame will propagate is determined. The Peclet number for this two-plate configuration is given by:

$$Pe_2 = \frac{\rho_u S_L c_{p,f}(T_f - T_u)}{k_f(T_f - T_u)/d_{q2}} = \frac{\rho_u S_L c_{p,f} d_{q2}}{k_f} \tag{11.27}$$

which is approximately constant over a wide range of conditions. ρ, S_L, c_p, T, and k are the density, laminar flame speed, specific heat at constant pressure, gas temperature, and thermal conductivity, respectively, with the subscripts u and f referring to unburned and flame conditions. d_{q2} is the two-plate quench distance. The wall temperature and unburned gas temperature are assumed to be equal; this assumption is also appropriate in the engine context since there is ample time during the compression stroke for a thermal boundary layer to build up to a thickness of at least the quench distance.

Lavoie[36] has developed empirical correlations for two-plate quench-distance data for propane-air mixtures: only limited data for liquid hydrocarbon fuels such as isooctane are available. The data in the pressure range 3 to 40 atm are well fitted by

$$Pe_2 = \frac{9.5}{\phi} \left(\frac{p}{3}\right)^{0.26 \min(1,\, 1/\phi^2)} \tag{11.28}$$

where p is the pressure in atmospheres and ϕ is the fuel/air equivalence ratio. The two-plate quench distance d_{q2} is then obtained from Eq. (11.27) and Prandtl number and viscosity relations for the flame conditions (see Sec. 4.8 or Ref. 36). Thus the minimum size crevice region into which a flame will propagate can be determined.

For the process of a flame front quenching on a single wall, there are many possible geometries. The simplest is where the flame front is parallel to the wall and approaches it head on. The one-wall quench distance d_{q1}, defined as the position of closest approach of the reaction zone to the wall, scales with flame properties in a similar way to the two-plate quench distance. Thus, a one-wall Peclet number relation can be formed:

$$\text{Pe}_1 = \frac{\rho_u S_L c_{p,u} d_{q1}}{k_u} \approx 8 \tag{11.29}$$

where the subscript u denotes properties evaluated at unburned gas conditions.

Using the wall temperature as representative of the unburned gas temperature (because the thermal boundary-layer thickness is greater than typical quench distances), Lavoie showed that

$$\frac{d_{q1}}{d_{q2}} = \frac{\text{Pe}_1}{\text{Pe}_2} = 0.2 \tag{11.30}$$

is a reasonable fit to the single-wall quench data. Typical two-wall quench distances for spark-ignition engine conditions are 0.2 to 1 mm; these distances represent the minimum crevice size the flame will enter. Single-wall quench distances are, therefore, in the range 0.04 to 0.2 mm.

While a fraction of the fuel hydrocarbons can escape the primary combustion process unburned or only partially reacted, oxidation of some of these hydrocarbons can occur during the expansion and exhaust processes. Hydrocarbon oxidation rates have been determined in a number of different studies and several different empirical correlations of the data in the form of overall reaction rate equations have been proposed. A reasonable fit to the experimental data on unburned HC burnup is the rate expression:[36]

$$\frac{d[\text{HC}]}{dt} = -6.7 \times 10^{15} \exp\left(\frac{-18{,}735}{T}\right)\tilde{x}_{\text{HC}}\,\tilde{x}_{\text{O}_2}\left(\frac{p}{RT}\right)^2 \tag{11.31}$$

where [ ] denotes concentration in moles per cubic centimeter, $\tilde{x}_{\text{HC}}$ and $\tilde{x}_{\text{O}_2}$ are the mole fractions of HC and O_2, respectively, t is in seconds, T in kelvins, and the density term (p/RT) has units of moles per cubic centimeter. The spread in the data about this equation is substantial, however.

Studies of combustion of premixed fuel-air mixtures at high pressure in closed vessels or bombs have been useful in identifying the mechanisms by which hydrocarbons escape complete combustion. The residual unburned hydrocarbons left in the bomb following a combustion experiment have been shown to come primarily from crevices in the bomb walls. Unburned HC levels were proportional to total crevice volume, and decreased to very low values (~ 10 ppm C) as all the crevices were filled with solid material. Thus wall quench hydrocarbons apparently diffuse into the burned gases and oxidize following the quenching event.[37] Analytical studies of the flame quenching process, and postquench diffusion and oxidation with kinetic models of the hydrocarbon oxidation process, are in agreement with these bomb data.[38] Flame quenching can be thought of as a two-stage process. The first step is the extinction of the flame at a short distance from the cold wall, determined by a balance between thermal conduction of heat from the hot reaction zone to the wall and heat released in the reaction zone by the flame reactions. The second step is the postquench diffusion and oxidation occurring on a time scale of one or a few milliseconds after quenching. The diffusion and oxidation process ultimately reduces the mass of wall quench hydrocarbons to several orders of magnitude below its value at the time of quenching.

Closed-vessel combustion experiments have also been used to show that oil layers on the walls of the bomb cause an increase in residual unburned HC levels after combustion is complete. The additional HC that result in experiments with oil films present are primarily (>95 percent) fuel molecules, and are directly proportional to the amount of oil placed on the walls of the reactor and the solubility of the specific fuel in the oil. These results show that absorption of fuel in the oil occurs prior to ignition. This dissolved fuel is then desorbed into the burned gases well after combustion is complete. Thus fuel absorption into and desorption from any oil layers is a potentially important engine HC mechanism.[39]

11.4.3 HC Emissions from Spark-Ignition Engines

Unburned hydrocarbon levels in the exhaust of a spark-ignition engine under normal operating conditions are typically in the range 1000 to 3000 ppm C_1. This corresponds to between about 1 and $2\frac{1}{2}$ percent of the fuel flow into the engine; the engine combustion efficiency is high. As indicated in Fig. 11-2, HC emissions rise rapidly as the mixture becomes substantially richer than stoichiometric. When combustion quality deteriorates, e.g., with very lean mixtures, HC emissions can rise rapidly due to incomplete combustion or misfire in a fraction of the engine's operating cycles. As outlined in Sec. 11.1, there are several mechanisms that contribute to total HC emissions. Also, any HC escaping the primary combustion process may oxidize in the expansion and exhaust processes. While a complete description of the HC emissions process cannot yet be given, there are sufficient fundamental data available to indicate which mechanisms are likely to be most important, and thus how major engine variables influence HC emission levels.

Four possible HC emissions formation mechanisms for spark-ignition engines (where the fuel-air mixture is essentially premixed) have been proposed: (1) flame quenching at the combustion chamber walls, leaving a layer of unburned fuel-air mixture adjacent to the wall; (2) the filling of crevice volumes with unburned mixture which, since the flame quenches at the crevice entrance, escapes the primary combustion process; (3) absorption of fuel vapor into oil layers on the cylinder wall during intake and compression, followed by desorption of fuel vapor into the cylinder during expansion and exhaust; (4) incomplete combustion in a fraction of the engine's operating cycles (either partial burning or complete misfire), occurring when combustion quality is poor (e.g., during engine transients when A/F, EGR, and spark timing may not be adequately controlled). In addition, as deposits build up on the combustion chamber walls,

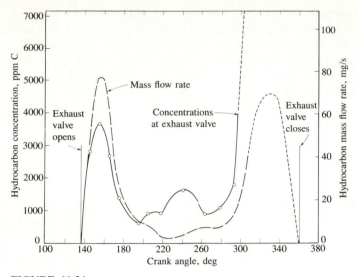

FIGURE 11-24
Variation in HC concentration and HC mass flow rate at the exhaust valve during the exhaust process. SI engine operating at 1200 rev/min and $\phi = 1.2$, unthrottled.[40]

HC emissions increase. Whether the deposits constitute an additional mechanism or merely modify one of the above mechanisms is unclear.

All these processes (except misfire) result in unburned hydrocarbons close to the combustion chamber walls, and not in the bulk of the cylinder gases. Thus, the distribution of HC in the exhaust gases would not be expected to be uniform. Experiments have been done to determine the unburned HC concentration distribution in the exhaust port during the exhaust process to provide insight into the details of the formation mechanisms. Gas concentrations were measured with a rapid-acting sampling valve placed at the exhaust port exit. Figure 11-24 shows results from these time-resolved HC concentration measurements. HC concentrations vary significantly during the exhaust process. Gas which remains in the exhaust port between exhaust pulses has a high HC concentration, so purging techniques where air or nitrogen was bled into the exhaust port were used to displace this high HC gas while the exhaust valve was closed. The high HC concentration in the blowdown exhaust gases is clearly discernible, as is the rapid rise in HC concentration toward the end of the exhaust stroke. The cylinder-exit HC concentrations were then multiplied by the instantaneous exhaust gas mass flow rate to obtain the instantaneous HC mass emission rate from the cylinder throughout the exhaust process, also shown in Fig. 11-24. The unburned HC are exhausted in two peaks of approximately equal mass: the first of these coincides with the exhaust blowdown mass flow pulse (which removes the majority of the *mass* from the cylinder); the second occurs toward the end of the exhaust stroke where HC concentrations are very high and the mass flow rate is relatively low.[40] Other experiments have confirmed these observations.[41] Clearly, mixing of unburned HC with the bulk cylinder gases occurs during expansion and/or the exhaust blowdown process. Then, the final stages of piston motion during the exhaust stroke push most of the remaining fraction of the cylinder mass with its high HC concentration into the exhaust. This would be expected to leave a high concentration of HC in the residual gas in the cylinder. Experiments conducted in which the valve mechanism of a single-cylinder engine was arranged to disengage during operation and trap residual gases in the cylinder confirm this. For one set of typical engine operating conditions, approximately one-third of the hydrocarbons left unburned in an engine combustion event was retained in the cylinder and recycled.[42]

FLAME QUENCHING AT THE WALLS. The existence of quench layers on the cold combustion chamber walls of a spark-ignition engine was shown photographically by Daniel.[43] Photographs of the flame region immediately after flame arrival at the wall through a window in the cylinder head showed a thin non-radiating layer adjacent to the wall. The quench layer thicknesses measured were in the range 0.05 to 0.4 mm (thinnest at high load), in rough agreement with predictions based on experiments in combustion bombs. However, more recent work in bombs and engines indicates that diffusion of hydrocarbons from the quench layer into the burned gases and its subsequent oxidation occur on a time scale of a few milliseconds, at least with smooth clean combustion chamber walls. The constant-volume combustion bomb data which suggested this conclusion and the kinetic calculations which support this explanation of why quench layers are not significant with smooth clean walls have already been described in Sec. 11.4.2. The following evidence shows these conclusions are also valid in an engine.

A special rapid-acting poppet valve was used in a single-cylinder engine to sample the gases from a torus-shaped region, of height of order 0.25 mm and diameter about 6 mm, adjacent to the wall over a 1-ms period. Sampling was repeated every cycle to provide a steady stream of sampled gases for analysis. Figure 11-25 shows the variation in concentrations of HC species through the combustion, expansion, and exhaust processes. The fuel was propane (C_3H_8). The fuel concentration drops rapidly to a low value when the flame arrives at the valve; at the same time, intermediate hydrocarbon product concentrations rise and then fall sharply to values below 1 ppm. Beginning at 60° ATC, all HC concentrations rise and vary somewhat during the remainder of the cycle in a way that depends strongly on engine operating conditions. The observed rapid rise in partial oxidation products immediately after flame arrival is consistent with the flame quenching short of the wall. The presence of CH_2O and CH_3CHO in significant quantities indicates that low-temperature oxidation processes are occurring. However, since all HC product concentrations fall rapidly within 2 ms of flame arrival to very low values, the unburned HC in the quench layer diffuse into the bulk burned gases and oxidize. The increase in HC concentrations later in the cycle results from the sampling of hydrocarbons from sources other than quench layers.[44]

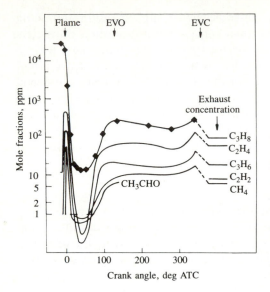

FIGURE 11-25

Concentrations (mole fractions) of selected hydrocarbons adjacent to combustion chamber wall, as a function of crank angle during combustion, expansion, and exhaust processes. Mass sampled with rapid-acting valve held constant at 7.6×10^{-6} g per pulse. Total exhaust HC = 400 ppm C. Engine speed = 1250 rev/min, imep = 380 kPa, equivalence ratio = 0.9, MBT spark timing, no EGR.[44]

Though quench layers on clean smooth combustion chamber walls are not a significant source of unburned hydrocarbons, it has been shown that wall surface finish does affect exhaust HC levels. Comparisons have been made between the standard "rough" as-cast cylinder head surfaces and the same cylinder heads when smoothed. The average exhaust HC concentration decreased by 103 ppm C, or 14 percent; the smoothed surface area was 32 percent of the total combustion chamber surface area.[45] Buildup of deposits on the combustion chamber surfaces also affect HC emission levels, as will be discussed later.

CREVICE HC MECHANISM. The crevices in the combustion chamber walls—small volumes with narrow entrances—into which the flame is unable to penetrate have been shown to be a major source of unburned HC. The largest of these crevice regions is the volumes between the piston, piston rings, and cylinder wall. Other crevice volumes in production engines are the threads around the spark plug, the space around the plug center electrode, crevices around the intake and exhaust valve heads, and the head gasket crevice. Table 8.1 shows the size and relative importance of these crevice regions in one cylinder of a production V-6 engine determined from measurements of cold-engine components. Total crevice volume is a few percent of the clearance volume and the piston and ring pack crevices are the dominant contributor.

The important crevice processes occurring during the engine cycle are the following. As the cylinder pressure rises during compression, unburned mixture is forced into the crevice regions. Since these volumes are thin they have a large surface/volume ratio; the gas flowing into each crevice cools by heat transfer to close to the wall temperature. During combustion, while the pressure continues to rise, unburned mixture continues to flow into the crevice volumes. When the flame arrives at each crevice, it can either propagate into the crevice and fully or partially burn the fuel and air within the crevice or it can quench at the crevice entrance. Whether the flame quenches depends on crevice entrance geometry, the composition of the unburned mixture, and its thermodynamic state as described in Sec. 11.4.2. After flame arrival and quenching, burned gases will flow into each crevice until the cylinder pressure starts to decrease. Once the crevice gas pressure is higher than the cylinder pressure, gas flows back from each crevice into the cylinder.

The most important of these crevices, the volumes between the piston, piston rings, and cylinder wall, is shown schematically in Fig. 8-27. This crevice consists of a series of volumes, connected by flow restrictions such as the ring side clearance and ring gap whose geometry changes as the ring moves up and down in the ring groove sealing either at the top or bottom ring surface. The gas flow, pressure distribution, and ring motion are therefore coupled, and their behavior during the compression and expansion strokes has already been discussed in Sec. 8.6. During compression and combustion, mass flows *into* the volumes in this total crevice region. Once the cylinder pressure starts to decrease (after about 15° ATC) gas flows out of the top of these crevice regions in Fig. 8-27 into the cylinder at low velocity adjacent to the cylinder wall. The important result is that the fraction of the total cylinder charge (5 to 10 percent) trapped in these regions at the time of peak cylinder pressure escapes the primary combustion process. Most of this gas flows back into the cylinder during the expansion process. Depending on spark plug location in relation to the position of the top ring gap, well above 50 percent of this gas can be unburned fuel-air mixture. Its potential contribution to unburned HC emissions is obvious.

There is substantial evidence to support the above description of crevice HC phenomena and the piston ring crevice region in particular. Visualization studies in a special engine have identified the spark plug crevice outflow, low-velocity gas expansion out of the volume above the first ring after the time of peak pressure, and the jet-like flows through the top ring gap later in the expansion process when the pressure difference across the ring changes sign.[46] Gas sampling from the volume above the top ring, using a rapid-acting sample valve mounted in the piston crown, has shown that the gas composition in this region corresponds to unburned fuel-air mixture until flame arrival at the crevice entrance closest to the sampling valve location. Next, product gases enter the crevice as the cylinder pressure continues to rise. Then, during expansion as gas flows out of this region, the composition of the gas sampled reverts back toward that of the unburned mixture which enters the crevice region earlier.[47]

Direct evidence that the piston and ring crevice regions are a major contributor to exhaust HC emissions comes from experiments where the volume of this crevice region was substantially changed. Wentworth[48] almost completely eliminated this crevice by moving the top piston ring as close to the crown of the piston as possible, and sealing this ring at top and bottom in its groove with O rings. Tests of this sealed ring-orifice design in a production engine showed reductions of between 47 and 74 percent from baseline HC levels over a range of

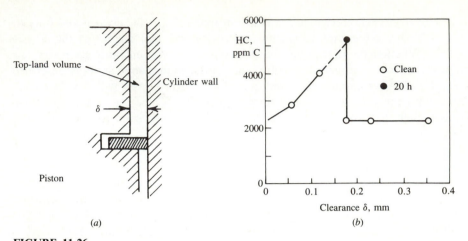

FIGURE 11-26
(a) Piston top-land crevice volume. (b) Effect of increasing top-land clearance on exhaust hydrocarbon emissions. Unthrottled spark-ignition engine, $r_c = 6, 885$ rev/min, $A/F = 13$, MBT timing.[49]

speeds and loads. Haskell and Legate,[49] in experiments in a single-cylinder CFR engine, steadily increased the piston top-land clearance (see Fig. 11-26a) and measured the effect on exhaust HC emissions. Figure 11-26b shows the results: HC emissions increase as the top-land clearance increases until the clearance equals about 0.18 mm, when emissions drop to the zero clearance level. This clearance (0.18 mm) is close to the two-plate quench distance estimated from Eq. (11.27). For piston top-land clearances above this value, the flame can enter the crevice and burn up much of the crevice HC.

The relative importance of the different crevices in the combustion chamber walls has been examined by using the cylinder head and piston of a four-cylinder production engine to form two constant-volume reactors or combustion bombs.[50] The cylinder head was sealed with a steel plate across the head gasket plane to make one reactor; the piston and ring pack and cylinder wall, again sealed with a plate at the head gasket plane, formed the second reactor. Each reactor was filled with a propane-air mixture and combustion initiated with a spark discharge across a spark plug; following combustion the burned gases were exhausted, sampled, and analyzed. The crevices were sequentially filled with epoxy or viton rubber, and after filling each crevice, the exhaust HC emission level determined. It was found that the ring pack crevices produced approximately 80 percent of the total scaled HC emissions, the head gasket crevice about 13 percent, and the spark plug threads 5 percent. All other HC sources in these reactors produced less than 2 percent of the total HC. While these numbers cannot be applied directly to an operating engine (the crevice filling and emptying rates in the bomb experiments are substantially different from these rates in an engine), they do underline the importance of the ring pack crevice region.

Blowby is the gas that flows from the combustion chamber, past the piston and into the crankcase. It is forced through any leakage paths afforded by the

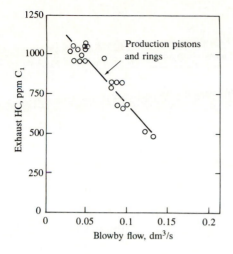

FIGURE 11-27
Effect of increasing crankcase blowby on exhaust hydrocarbon emissions. Production pistons and rings. SI engine at 1200 rev/min, intake manifold pressure 0.6 atm, $A/F = 14.2$.[51]

piston-bore-ring assembly in response to combustion chamber pressure. Blowby of gases from the cylinder to the crankcase removes gas from this crevice region and thereby prevents some of the crevice gases from returning to the cylinder. Crankcase blowby gases used to be vented directly to the atmosphere and constituted a significant source of HC emissions. The crankcase is now vented to the engine intake system, the blowby gases are recycled, and this source of HC emissions is now fully controlled. Blowby at a given speed and load is controlled primarily by the greatest flow resistance in the flow path between the cylinder and the crankcase. This is the smallest of the compression ring ring-gap areas. Figure 8-30 shows how blowby increases linearly with the smallest gap area. Figure 11-27 shows how exhaust HC levels decrease as blowby increases and more crevice HC flows to the crankcase. Crankcase blowby gases represent a direct performance loss. They are a smaller efficiency loss because crankcase gases are now recycled to the engine intake system.

The location of the ring gap in relation to the spark plug also affects HC emission levels. Experiments have shown that HC emissions are highest when the top ring gap is farthest from the spark plug; the gas flowing into the crevice directly above the gap is then unburned mixture for the longest possible time. With the top ring gap closest to the spark plug, HC exhaust levels are lowest because burned gas reaches the gap location at the earliest time in the combustion process. The difference, highest to lowest, was between 9 and 42 percent of the average level for any set of operating conditions, and in most cases was above 20 percent.[51]

The fate of these crevice HC when they flow back into the cylinder during expansion and exhaust is not well understood. Both jet-like flows (e.g., that from the ring gap) and low-velocity creeping flows (e.g., that from the piston top-land crevice) have been observed (see Fig. 8-29). While the former could mix rapidly with the high-temperature bulk burned gases, the latter will enter the cool gases

TABLE 11.6
Amount of gas flowing into and out of crevice regions†

	% mass	ppm C
Total gas in all crevice regions	8.2	
Total gas back to combustion chamber	7.0	
Unburned back to combustion chamber	3.7–7.0‡	5000–9400
Unburned to blowby	0.5–1.2‡	
Total unburned escaping primary combustion	4.2–8.2‡	

† For V-6 engine operating at 2000 rev/min and wide-open throttle.
‡ Depends on spark plug and ring gap location.

in the cylinder wall boundary layer and mix and (probably) burn much more slowly. Hydrocarbon transport and oxidation processes are discussed more fully below.

Table 11.6 presents a summary of estimates of the total mass of gas and mass of unburned mixture in the piston, ring, and cylinder wall crevice region for a typical spark-ignition engine.[46] When compared to exhaust HC levels, it is clear that these crevices are a major source of unburned hydrocarbons.

ABSORPTION AND DESORPTION IN ENGINE OIL. The presence of lubricating oil in the fuel or on the walls of the combustion chamber is known to result in an increase in exhaust hydrocarbon levels. In experiments where exhaust HC concentrations rose irregularly with time, with engine operating conditions nominally constant, it was shown that oil was present on the piston top during these high emission periods. When engine oil was added to the fuel, HC emissions increased, the amount of additional HC in the exhaust increasing with the increasing amount of oil added. The increase in exhaust HC was primarily unreacted fuel (isooctane) and not oil or oil-derived compounds.[51] The increase in HC can be substantial: exhaust HC levels from a clean engine can double or triple when operated on a fuel containing 5 percent lubricating oil over a period of order 10 minutes. (With deposits from leaded-fuel operation present on the combustion chamber walls, however, a much smaller increase in exhaust HC was observed.) It has been proposed that fuel vapor absorption into and desorption from oil layers on the walls of the combustion chamber could explain these phenomena.[49]

The absorption and desorption mechanism would work as follows. The fuel vapor concentration within the cylinder is close to the inlet manifold concentration during intake and compression. Thus, for about one crankshaft revolution, any oil film on the walls will absorb fuel vapor. During the latter part of compression, the fuel vapor pressure is increasing so, by Henry's law, absorption will continue even if the oil was saturated during intake. During combustion the fuel vapor concentration in the bulk gases goes essentially to zero so the absorbed fuel vapor will desorb from the liquid oil film into the gaseous combustion products. Desorption could continue throughout the expansion and exhaust strokes.

Some of the desorbed fuel vapor will mix with the high-temperature combustion products and oxidize. However, desorbed vapor that remains in the cool boundary layer or mixes with the cooler bulk gases late in the cycle may escape full oxidation and contribute to unburned HC emissions.

Experiments, where measured amounts of oil were placed on the piston crown, confirm that oil layers on the combustion chamber surface increase exhaust HC emissions. The exhaust HC levels increased in proportion to the amount of oil added when the engine was fueled with isooctane. Addition of 0.6 cm^3 of oil produced an increase of 1000 ppm C in exhaust HC concentration. Fuel and fuel oxidation species, not oil oxidation products, were responsible for most of this increase. Similar experiments performed with propane fuel showed no increase in exhaust HC emissions when oil was added to the cylinder. The increase in exhaust HC is proportional to the solubility of the fuel in the oil. The exhaust HC levels decreased steadily back to the normal engine HC level before oil addition, over a period of several minutes. At higher coolant temperatures, the increase in HC on oil addition is less, and HC concentrations decreased back to the normal level more quickly. Increasing oil temperature would decrease viscosity, increasing the rate of drainage into the sump. It also changes the solubility and diffusion rate of the fuel in the oil.[52]

At the outer surface of the oil layer, the concentration of fuel vapor dissolved in the oil is given by Henry's law for dilute solutions in equilibrium:

$$\tilde{x}_f = \frac{p_f}{H} \tag{11.32}$$

where $\tilde{x}_f$ is the mole fraction of fuel vapor in the oil, p_f is the partial pressure of fuel vapor in the gas, and H is Henry's constant. If the oil layer is sufficiently thin, and hence diffusion sufficiently rapid, Eq. (11.32) can be used to estimate the mole fraction of the fuel dissolved in the oil. Since $p_f = n_{f,c}\tilde{R}T/V$ (where $n_{f,c}$ is the number of moles of fuel in the cylinder, T is the temperature, and V the cylinder volume) and $x_f = n_{f,o}/(n_{f,o} + n_o) = n_{f,o}/n_o$ for $n_o \gg n_{f,o}$ (where $n_{f,o}$ is the number of moles of fuel dissolved in the oil and n_o is the number of moles of oil),[53] then

$$\frac{n_{f,o}}{n_{f,c}} = \frac{n_o\,\tilde{R}T}{HV} \tag{11.33}$$

Diffusion is sufficiently rapid for Eq. (11.33) to be valid if the diffusion time constant τ_d is much less than characteristic engine times: i.e.,

$$\tau_d \approx \frac{\delta^2}{D} \ll N^{-1}$$

where δ is the oil layer thickness, D is the diffusion coefficient for fuel vapor in the oil, and N is engine speed. D for a hydrocarbon through a motor oil is of order 10^{-6} cm^2/s at 300 K and of order 10^{-5} cm^2/s at 400 K. Oil film thicknesses on the cylinder wall vary during the operating cycle between about 1 and 10 μm.[54, 55] Thus diffusion times for engine conditions are 10^{-1} to 10^{-3} s; for the thinnest oil layers approximate equilibration would be achieved. A theoretical

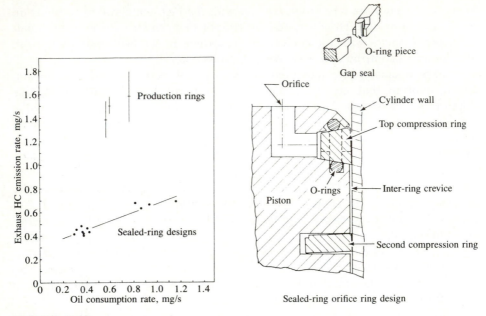

FIGURE 11-28
Correlation between exhaust hydrocarbon emissions and oil consumption rate. Production piston rings and sealed ring–orifice ring designs. SI engine at 1600 rev/min, imep = 422 kPa, equivalence ratio = 0.9, r_c = 8.0, intake pressure = 54 kPa, MBT spark timing.[57]

study of this problem—the one-dimensional cyclic absorption and desorption of a dilute amount of gas in a thin (constant thickness) isothermal liquid layer where diffusion effects are important—has been carried out. It suggests that oil layers on the cylinder wall could be a significant contributor to HC emission levels.[56]

Correlations between engine oil consumption and exhaust HC emissions provide a perspective on the relative importance of oil absorption/desorption and crevice mechanisms. Wentworth measured oil consumption and HC emissions in a spark-ignition engine for a range of piston ring designs.[57] Some of these designs were of the sealed ring–orifice type which effectively eliminates all the crevices between the piston, piston rings, and cylinder, and prevents any significant gas flow into or out of the ring region. HC emissions increase with increasing oil consumption for both production ring designs and the sealed ring–orifice designs, as shown in Fig. 11-28. Extrapolation to zero oil consumption from normal consumption levels shows a reduction in exhaust HC levels, but this decrease is significantly less than the difference in emission levels between the production and the sealed ring–orifice designs which effectively remove the major crevice region. The production piston used had a chamfered top land. The HC emissions for a normal piston top-land design would probably be higher.

POOR COMBUSTION QUALITY. Flame extinction in the bulk gas, before all of the flame front reaches the wall, is a source of HC emissions under certain engine operating conditions. As the cylinder pressure falls during the expansion stroke, the temperature of the unburned mixture ahead of the flame decreases. This slows the burning rate [the laminar flame speed decreases so the burning rate in Eq. (9.52) decreases]. If the pressure and temperature fall too rapidly, the flame can be extinguished. This type of bulk quench has been observed in spark-ignition engines; it results in very high HC concentrations for that particular cycle. Engine conditions where bulk quenching is most likely to occur are at idle and light load where engine speed is low and the residual gas fraction is high, with high dilution with excessive EGR or overly lean mixtures, and with substantially retarded combustion. Even if steady-state engine calibrations of A/F, EGR, and spark-timing are such that bulk quenching does not occur, under transient engine operation these variables may not be appropriately set to avoid bulk quenching in some engine cycles due to the different dynamic characteristics of the engine subsystems which control these variables.

The existence of zones of stable and unstable engine operation with lean or dilute mixtures has already been discussed (see Sec. 9.4.3). Detailed engine combustion studies have shown that, as mixture composition becomes more dilute (e.g., by increasing EGR) and unburned gas temperature and pressure during combustion become lower, combustion quality (or variability) and engine stability deteriorate. The standard deviation in a parameter such as indicated mean effective pressure (which depends for its magnitude on the proper timing of the start of combustion and on the duration of the combustion process) increases due first to an increase in the number of slower burning cycles, then as conditions worsen to the occurrence of partial burning cycles, and finally to some misfiring cycles. Figure 9-36 showed how unburned hydrocarbon emissions from a spark-ignition engine rise as the EGR rate is increased at constant load and speed, and combustion quality (defined by the ratio of standard deviation in imep to the average imep) deteriorates. Initially the increase in HC is modest and is caused by changes in the other HC emission mechanisms described above. However, when partial burning cycles are detected, HC emissions rise more rapidly due to incomplete combustion of the fuel in the cylinder in these cycles. When misfiring cycles—no combustion—occur the rise in HC becomes more rapid still.

The relative importance of bulk gas quenching in a fraction of the engine's operating cycles due to inadequate combustion quality as a source of HC, compared with the other sources described in this section, has yet to be established. However, one obvious technique for reducing its importance, burning the mixture faster so that combustion is completed before conditions conducive to slow and partial burning exist in the cylinder, does reduce engine exhaust HC emissions. Figure 11-29 shows a comparison of HC emissions from a moderate burn rate engine with HC emissions with a faster burn rate [i.e., with improved combustion quality—lower coefficient of variation in imep, COV_{imep}, Eq. (9.50)], achieved by the use of two spark plugs instead of one.[58] The exhaust measurements show lower HC emissions when significant amounts of EGR are used for NO_x control for the faster, and hence less variable, combustion process. Such evidence suggests that occasional partial burning cycles may occur, even under

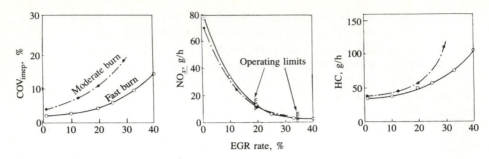

FIGURE 11-29
Effect of increasing burn rate on tolerance to recycled exhaust gas (EGR) and HC and NO_x emissions levels. COV_{imep} defined by Eq. (9.50). SI engine at 1400 rev/min, 324 kPa imep, equivalence ratio = 1.0, MBT timing.[58]

conditions where combustion appears to be "normal," and that this mechanism is important in practice.

EFFECT OF DEPOSITS. Deposit buildup on the combustion chamber walls (which occurs in vehicles over several thousands of miles) is known to increase unburned HC emissions. With leaded gasoline operation, the increase in HC emissions varies between about 7 and 20 percent. The removal of the deposits results typically in a reduction in HC emissions to close to clean engine levels. With unleaded gasoline, while the deposit composition is completely different (carbonaceous rather than lead oxide), the increase in HC emissions with accumulated mileage is comparable. Soft sooty deposits, such as those which accumulate after running the engine on a rich mixture, also cause an increase in HC emissions. Again, when the deposits were removed the emission rate fell about 25 percent to the original level.[59] Studies with simulated deposits (pieces of metal-foam sheet 0.6 mm thick) attached to the cylinder head and piston also showed increases in HC emissions. The increase varied between about 10 and 100 ppm C/cm^2 of simulated deposit area. The effect for a given area of deposit varied with deposit location. Locations close to the exhaust valve, where the flow direction during the exhaust process would be expected to be directly into the exhaust port, showed the highest increase in emissions.[45]

It is believed that absorption and desorption of hydrocarbons by these surface deposits is the mechanism that leads to an increase in emissions. Deposits can also build up in the piston ring crevice regions. A reduction in volume of these crevice regions would decrease HC emissions (and such a decrease has been observed). However, changes in piston–cylinder wall clearance due to deposits can affect the flame-quenching process and could increase emissions.[49]

HYDROCARBON TRANSPORT MECHANISMS. All of the above mechanisms (except misfire) result in high hydrocarbon concentrations adjacent to the combustion chamber walls. While any jet-type flows out of crevices during the expan-

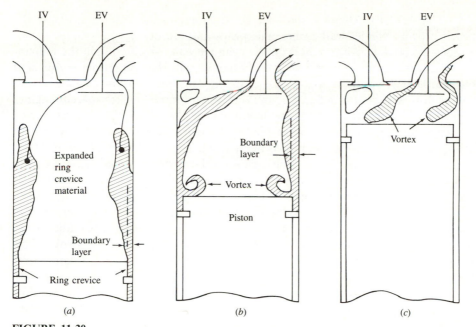

FIGURE 11-30
Schematic of flow processes by which ring crevice HC and HC desorbed from cylinder wall oil film
exit the cylinder: (a) exhaust blowdown process; (b) during exhaust stroke; (c) end of exhaust stroke.[60]

sion and exhaust strokes can transport unburned HC into the bulk gases, most of
the HC will remain near the wall. Two mechanisms by which gas near the cylinder wall exits the cylinder have been demonstrated. One is entrainment in the
vigorous gas flow out of the cylinder which occurs during the exhaust blowdown
process. The other is the vortex generated in the piston crown–cylinder wall
corner during the exhaust stroke.

Figure 11-30 illustrates these flow processes. In Fig. 11-30a the engine cylinder is shown as the exhaust valve opens during the blowdown process. At this
time the unburned HC from the ring crevice regions, laid along the wall during
expansion (and possibly HC from the oil film on the cylinder wall), is expanding
into the cylinder as the cylinder pressure falls. Some of this material will be
entrained by the bulk gases in the rapid motion which occurs during exhaust
blowdown (see Sec. 6.5). The rapid thinning of the thermal boundary-layer
regions on the combustion chamber walls during blowdown, which would result
from entrainment of the denser hydrocarbon-containing gas adjacent to the wall,
has been observed in schlieren movies taken in a transparent engine.[46] This
process, plus entrainment of any HC from the spark plug and head gasket crevices, would contribute to unburned HC in the blowdown gases which contain
about half the total HC emissions (see Fig. 11-24). During the exhaust stroke this
bulk gas entrainment process will continue, exhausting additional unburned HC,
as shown in Fig. 11-30b.

The second mechanism starts at the beginning of the exhaust stroke in the piston crown–cylinder wall corner. The piston motion during the exhaust stroke scrapes the boundary-layer gases off the cylinder wall (which contain the remainder of the piston and ring crevice hydrocarbons), rolls them up into a vortex, and pushes them toward the top of the cylinder. This piston crown–cylinder wall corner flow is discussed in Sec. 8.7, and has been observed in transparent engines as well as in water-flow engine analog studies. At the end of the exhaust stroke, the height of this vortex is comparable to the engine clearance height. As shown in Fig. 11-30c, a recirculation flow is likely to build up in the upper corner of the cylinder away from the exhaust valve, causing the vortex to detach from the wall and partly sweep out of the cylinder. In the corner nearest the valve, the flow is deflected around the valve, also tending to pull part of the vortex out of the chamber. In this way it is possible for a large part of the vortex, which now contains a substantial fraction of the unburned HC originally located adjacent to the cylinder wall, to leave the cylinder at the end of the exhaust stroke. This vortex flow is thought to be the mechanism that leads to the high HC concentrations measured at the end of the exhaust process, which contributes the other half of the exhausted HC mass (see Fig. 11-24), and to be responsible for the HC concentrations measured in the residual gases being much higher than average exhaust HC levels.[42] This study showed that at close to wide-open-throttle conditions, only about two-thirds of the HC which fail to oxidize inside the cylinder were exhausted, though 95 percent of the gas within the cylinder flows out through the exhaust valve. The residual gas HC concentration was about 11 times the average exhaust level. At part-throttle conditions, where the residual gas fraction is higher, it has been estimated that only about half of the unreacted HC in the cylinder will enter the exhaust.[61]

HYDROCARBON OXIDATION. Unburned hydrocarbons which escape the primary engine combustion process by the mechanisms described above must then survive the expansion and exhaust process without oxidizing if they are to appear in the exhaust. Since the formation mechanisms produce unburned HC at temperatures close to the wall temperature, mixing with bulk burned gas must take place first to raise the HC temperature to the point where reaction can occur. The sequence of processes which links the source mechanisms to hydrocarbons at the exhaust exit is illustrated in Fig. 11-31; it involves mixing and oxidation in both the cylinder and the exhaust system.

There is considerable evidence that substantial oxidation does occur. The oxidation of unburned HC in the quench layers formed on the combustion chamber walls on a time scale of order 1 ms after the flame is extinguished has already been discussed. Because the quench layers are thin, diffusion of HC into the bulk burned gas is rapid. Because the burned gases are still at a high temperature, oxidation then occurs quickly. Measurements of in-cylinder HC concentrations by gas sampling prior to exhaust valve opening show levels about 1.5 to 2 times the average exhaust level.[44, 63] The exhaust unburned HC are a mixture of fuel hydrocarbon compounds and pyrolysis and partial oxidation pro-

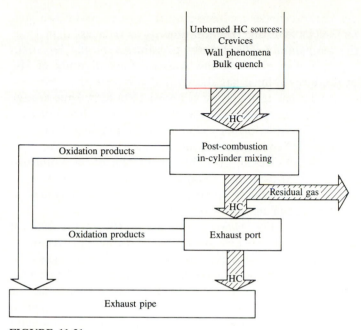

FIGURE 11-31
Schematic of complete SI engine HC formation and oxidation mechanism within the cylinder and exhaust system.[62]

ducts. While the relative proportion of fuel compared to reaction product hydrocarbon compounds varies substantially with engine operating conditions, an average value for passenger car vehicle exhausts is that fuel compounds comprise 40 percent of the total HC. Though partially reacted HC are produced in the flame-quenching process, these are closest to the high-temperature burned gases and are likely to mix and burn rapidly. That such a large fraction of the exhaust HC are reaction products indicates that substantial postformation HC reactions are occurring. There is direct evidence that HC oxidation in the exhaust system occurs.[64] Since in-cylinder gas temperatures are higher, it is likely that mixing of unburned HC with the bulk cylinder gases limits the amount of oxidation rather than the reaction kinetics directly.

Overall empirically based expressions for the rate of oxidation of hydrocarbons of the form of Eq. (11.31) have been developed and used to examine in-cylinder and exhaust burnup. A characteristic time τ_{HC} for this burnup process can be defined:

$$\frac{1}{\tau_{HC}} = \frac{1}{[HC]} \frac{d[HC]}{dt} \qquad (11.34)$$

Using an expression similar to Eq. (11.31) for $d[HC]/dt$, Weiss and Keck[63] have shown that any HC mixing with the burned gases in the cylinder prior to exhaust blowdown will oxidize. The in-cylinder gas temperature prior to blowdown generally exceeds 1250 K; the characteristic reaction time τ_{HC} is then less than 1 ms. During blowdown the temperature falls rapidly to values typically less than 1000 K; τ_{HC} is then greater than about 50 ms. An experimental study of HC exiting from a simulated crevice volume has shown that complete HC oxidation only occurs when the cylinder gas temperature is above 1400 K.[65] Thus a large fraction of the HC leaving crevice regions or oil layers during the exhaust process can be expected to survive with little further oxidation. Gas-sampling data show little decrease in in-cylinder HC concentrations during the exhaust stroke, thus supporting this conclusion.[44, 63] Overall, probably about half of the unburned HC formed by the source mechanisms described above will oxidize within the engine cylinder (the exact amount cannot yet be predicted with any accuracy; it is likely to depend on engine design and operating conditions[61]).

As shown schematically in Fig. 11-31, oxidation of HC in the exhaust system can occur. Often this is enhanced by air addition into the port region to ensure that adequate oxygen for burnup is available. However, since the gas temperature steadily decreases as the exhaust gases flow through the exhaust port and manifold, the potential for HC burnup rapidly diminishes. To oxidize the hydrocarbons in the gas phase, a residence time of order 50 ms or longer at temperatures in excess of 600°C are required. To oxidize carbon monoxide temperatures in excess of 700°C are required. Average exhaust gas temperatures at the cylinder exit (at the exhaust valve plane) are about 800°C; average gas temperatures at the exhaust port exit are about 600°C.† Figure 6-21 shows an example of the measured cylinder pressure, measured gas temperature at the exhaust port exit, and estimated mass flow rate into the port and gas temperature in the cylinder, during the exhaust process at a part-throttle operating condition. Port residence time and gas temperatures vary significantly through the process. Precise values of these variables obviously depend on engine operating conditions. It is apparent that only in the exhaust port and upstream end of the manifold can any significant gas-phase HC oxidation occur.

The importance of exhaust gas temperature to exhaust system emissions burnup is illustrated by the results shown in Fig. 11-32.[66] The exhaust system of a four-cylinder engine was modified by installing a section of heated and insulated pipe to maintain the exhaust gas temperature constant in the absence of any HC or CO burnup. The exhaust temperature entering this test section was varied by adjusting the engine operating conditions. The figure shows CO and HC concentrations as functions of residence time in the exhaust test section (or effectively as a function of distance from the engine). T_e is the entering gas temperature. The exhaust composition was fuel lean with 3 percent O_2 in the burned

---

† Note that there is a significant variation in the temperature of the exhaust gases throughout the exhaust process. The gas exhausted first is about 100 K hotter than the gas exhausted at the end of the process (see Sec. 6.5).

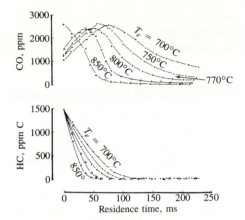

FIGURE 11-32
Effect of exhaust gas temperature on HC and CO burnup in the exhaust. SI engine at 1600 rev/min, engine air flow = 7.7 dm^3/s, lean mixture with 3% O_2 and 13% $(CO + CO_2)$ in exhaust.[66]

gas stream. HC oxidation starts immediately (for $T \gtrsim 600°C$), the rate of oxidation increasing rapidly with increasing temperature. Under fuel-lean conditions, incomplete HC oxidation can result in an increase in CO levels. CO oxidation commences later, when the gas temperature rises above the entering value due to heat released by the already occurring HC oxidation. The further heat released by CO oxidation accelerates the CO burnup process. These data underline the importance of the exhaust port heat-transfer and mixing processes. Both mixing between the hotter blowdown gases (with their lower HC concentration) and the cooler end-of-exhaust gases (with their higher HC concentration) and mixing between burned exhaust gas and secondary air are important.

Engine experiments where the exhaust gas reactions were quenched by timed injection of cold carbon dioxide at selected locations within the exhaust port have shown that significant reductions in HC concentration in the port can occur. Parallel modeling studies of the HC burnup process (based on instantaneous mass flow rate, estimated exhaust gas temperature, and an overall hydrocarbon reaction-rate expression), which predicted closely comparable magnitudes and trends, indicated that gas temperature and port residence time are the critical variables. The percent of unburned HC exiting the cylinder which reacted in the exhaust system (with most of the reaction occurring in the port) varied between a few and 40 percent. Engine operating conditions that gave highest exhaust temperatures (stoichiometric operation, higher speeds, retarded spark timing, lower compression ratio) and longest residence times (lighter load) gave relatively higher percent reductions. Air injection at the exhaust valve-stem base, phased to coincide with the exhaust process, showed that for stoichiometric and slightly rich conditions secondary air flow rates up to 30 percent of the exhaust flow substantially increased the degree of burnup. The timing of the secondary air flow relative to the exhaust flow and the location of the air injection point in the port are known to be critical.[64]

Reductions in exhaust port heat losses through the use of larger port cross-sectional areas (to reduce flow velocity and surface area per unit volume), insertion of port liners to provide higher port wall temperatures, and attention to port design details to minimize hot exhaust gas impingement on the walls are known to increase the degree of reaction occurring in the port.

SUMMARY. It will be apparent from the above that the HC emissions formation process in spark-ignition engines is extremely complex and that there are several paths by which a small but important amount of the fuel escapes combustion. It is appropriate here to summarize the overall structure of the spark-ignition engine hydrocarbon emission problem and identify the key factors and engine variables that influence the different parts of that problem. Table 11.7 provides such a summary. The total process is divided into four sequential steps: (1) the formation of unburned hydrocarbon emissions; (2) the oxidation of a fraction of these HC emissions within the cylinder, following mixing with the bulk gases; (3) the flow of a fraction of the unoxidized HC from the cylinder into the exhaust; (4) the oxidation in the exhaust system of a fraction of the HC that exit the cylinder. The detailed processes and the design and operating variables that influence each of these steps in a significant way are listed.

The four separate formation mechanisms identified in step 1 have substantial, though as yet incomplete, evidence behind them. They are listed in the most likely order of importance. Each has been extensively described in this section. It is through each of these mechanisms that fuel or fuel-air mixture escapes the primary combustion process. That fuel must then survive the expansion and exhaust processes and pass through the exhaust system without oxidation if it is to end up in the atmosphere as HC emissions. The rate of mixing of these unburned HC with the hot bulk cylinder gases, the temperature and composition of the gases with which these HC mix, and the subsequent temperature-time and composition-time histories of the mixture will govern the amount of in-cylinder oxidation that occurs. The distribution of these HC around the combustion chamber is nonuniform (and changes with time); they are concentrated close to the walls of the chamber. The fraction of these HC that will exit the chamber during the exhaust process will depend on the details of the in-cylinder flow patterns that take them through the exhaust valve. Overall, the magnitude of the residual fraction will be one major factor; the residual gas is known to be much richer in HC than the average exhaust. In particular, the flow patterns in the cylinder toward the end of the exhaust stroke as the gas scraped off the cylinder wall by the piston moves toward the exhaust valve will be important. Finally, a fraction of the unburned HC which leave the cylinder through the exhaust valve will burn up within the exhaust system. Gas-phase oxidation in the exhaust ports and hotter parts of the exhaust manifold is significant. The amount depends on the gas temperature, composition, and residence time. If catalysts or a thermal reactor are included in the exhaust system, very substantial additional reduction in HC emission levels can occur. These devices and their operating characteristics are described in Sec. 11.6.

TABLE 11.7
Critical factors and engine variables in HC emissions mechanisms

1. *Formation of HC*
   (*a*) Crevices
      (1) Crevice volume
      (2) Crevice location
          (relative to spark plug)
      (3) Load
      (4) Crevice wall temperature
      (5) Mixture composition†
   (*b*) Oil layers
      (1) Oil consumption
      (2) Wall temperature
      (3) Speed
   (*c*) Incomplete combustion
      (1) Burn rate and variability
      (2) Mixture composition†
      (3) Load
      (4) Spark timing‡
   (*d*) Combustion chamber walls
      (1) Deposits
      (2) Wall roughness

3. *Fraction HC flowing
   out of cylinder*
   (*a*) Residual fraction
      (1) Load
      (2) Exhaust pressure
      (3) Valve overlap
      (4) Compression ratio
      (5) Speed
   (*b*) In-cylinder flow during
       exhaust stroke
      (1) Valve overlap
      (2) Exhaust valve size and
          location
      (3) Combustion chamber shape
      (4) Compression ratio
      (5) Speed

2. *In-cylinder mixing and oxidation*
   (*a*) Mixing rate with bulk gas
      (1) Speed
      (2) Swirl ratio
      (3) Combustion chamber shape
   (*b*) Bulk gas temperature during
       expansion and exhaust
      (1) Speed
      (2) Spark timing‡
      (3) Mixture composition†
      (4) Compression ratio
      (5) Heat losses to walls
   (*c*) Bulk gas oxygen concentration
      (1) Equivalence ratio
   (*d*) Wall temperature
      (1) Important if HC source
          near wall
      (2) For crevices: importance
          depends on geometry

4. *Oxidation in exhaust system*
   (*a*) Exhaust gas temperature
      (1) Speed
      (2) Spark timing‡
      (3) Mixture composition†
      (4) Compression ratio
      (5) Secondary air flow
      (6) Heat losses in cylinder
          and exhaust
   (*b*) Oxygen concentration
      (1) Equivalence ratio
      (2) Secondary air flow
          and addition point
   (*c*) Residence time
      (1) Speed
      (2) Load
      (3) Volume of critical
          exhaust system component
   (*d*) Exhaust reactor§
      (1) Oxidation catalyst
      (2) Three-way catalyst
      (3) Thermal reactor

† Fuel/air equivalence ratio and burned gas fraction (residual plus recycled exhaust gas).
‡ Relative to MBT timing.
§ Of at least as great an importance as engine details if present in total emission control system. See Sec. 11.6.

11.4.4   Hydrocarbon Emission Mechanisms in Diesel Engines

BACKGROUND. Diesel fuel contains hydrocarbon compounds with higher boiling points, and hence higher molecular weights, than gasoline. Also, substantial pyrolysis of fuel compounds occurs within the fuel sprays during the diesel combustion process. Thus, the composition of the unburned and partially oxidized hydrocarbons in the diesel exhaust is much more complex than in the spark-ignition engine and extends over a larger molecular size range. Gaseous hydrocarbon emissions from diesels are measured using a hot particulate filter (at 190°C) and a heated flame ionization detector. Thus the HC constituents vary from methane to the heaviest hydrocarbons which remain in the vapor phase in the heated sampling line (which is also maintained at about 190°C). Any hydrocarbons heavier than this are therefore condensed and, with the solid-phase soot, are filtered from the exhaust gas stream upstream of the detector. The particulate emission measurement procedure measures a portion of total engine hydrocarbon emissions also. Particulates are collected by filtering from a diluted exhaust gas stream at a temperature of 52°C or less. Those hydrocarbons that condense at or below this temperature are absorbed onto the soot. They are the extractable fraction of the particulate: i.e., that fraction which can be removed by a powerful solvent, typically between about 15 and 45 percent of the total particulate mass. This section discusses gaseous hydrocarbon emissions; particulate emissions—soot and extractable material—are discussed in Sec. 11.5.

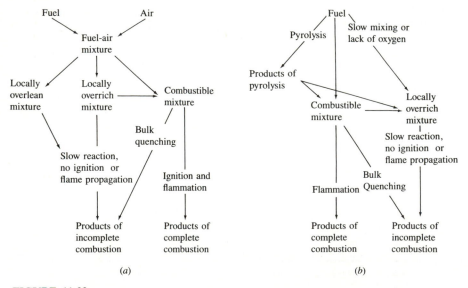

FIGURE 11-33
Schematic representation of diesel hydrocarbon formation mechanisms: (a) for fuel injected during delay period; (b) for fuel injected while combustion is occurring.[23]

The complex heterogeneous nature of diesel combustion, where fuel evapo-ration, fuel-air and burned-unburned gas mixing, and combustion can occur simultaneously, has been discussed extensively in Chap. 10. As a result of this complexity, there are many processes that could contribute to diesel engine hydrocarbon emissions. In Chap. 10 the diesel's compression-ignition combustion process was divided into four stages: (1) the ignition delay which is the time between the start of injection and ignition; (2) the premixed or rapid combustion phase, during which the fuel that has mixed to within combustible limits during the delay period burns; (3) the mixing controlled combustion phase, during which the rate of burning depends on the rate of fuel-air mixing to within the combusti-ble limits; (4) the late combustion phase where heat release continues at a low rate governed by the mixing of residual combustibles with excess oxygen and the kinetics of the oxidation process. There are two primary paths by which fuel can escape this normal combustion process unburned: the fuel-air mixture can become too lean to autoignite or to support a propagating flame at the condi-tions prevailing inside the combustion chamber, or, during the primary com-bustion process, the fuel-air mixture may be too rich to ignite or support a flame. This fuel can then be consumed only by slower thermal oxidation reactions later in the expansion process after mixing with additional air. Thus, hydrocarbons remain unconsumed due to incomplete mixing or to quenching of the oxidation process.†

Figure 11-33 shows schematically how these processes can produce incom-plete combustion products. Fuel injected during the ignition delay (Fig. 11-33a) will mix with air to produce a wide range of equivalence ratios. Some of this fuel will have mixed rapidly to equivalence ratios lower than the lean limit of com-bustion (locally overlean mixture), some will be within the combustible range, and some will have mixed more slowly and be too rich to burn (locally overrich mixture). The overlean mixture will not autoignite or support a propagating flame at conditions prevailing inside the combustion chamber (though some of this mixture may burn later if it mixes with high-temperature burned products early in the expansion stroke). In the "premixed" combustible mixture, ignition occurs where the local conditions are most favorable for autoignition. Unless quenched by thermal boundary layers or rapid mixing with air, subsequent autoignition or flame fronts propagating from the ignition sites consume the combustible mixture. Complete combustion of overrich mixture depends on further mixing with air or lean already-burned gases within the time available before rapid expansion and cooling occurs. Of all these possible mechanisms, the overlean mixture path is believed to be the most important.[23]

For the fuel injected after the ignition delay period is over (Fig. 11-33b), rapid oxidation of fuel or the products of fuel pyrolysis, as these mix with air,

---

† Note that under normal engine operating conditions, the combustion inefficiency is less than 2 percent; see Sec. 4.9.4 and Fig. 3-9.

results in complete combustion. Slow mixing of fuel and pyrolysis products with air, resulting in overrich mixture or quenching of the combustion reactions, can result in incomplete combustion products, pyrolysis products, and unburned fuel being present in the exhaust.[23]

Hydrocarbon emission levels from diesels vary widely with operating conditions, and different HC formation mechanisms are likely to be most important at different operating modes. Engine idling and light-load operation produce significantly higher hydrocarbon emissions than full-load operation. However, when the engine is overfueled, HC emissions increase very substantially. As will be explained more fully below, overmixing (overleaning) is an important source of HC, especially under light-load operation. Undermixing, resulting in overrich mixture during the combustion period, is the mechanism by which some of the fuel remaining in the injector nozzle sac volume escapes combustion, and is also the cause of very high HC emissions during overfueling. Wall temperatures affect HC emissions, suggesting that wall quenching is important, and under especially adverse conditions very high cyclic variability in the combustion process can cause an increase in HC due to partial burning and misfiring cycles.

OVERLEANING. As soon as fuel injection into the cylinder commences, a distribution in the fuel/air equivalence ratio across the fuel sprays develops. The amount of fuel that is mixed leaner than the lean combustion limit ($\phi_L \sim 0.3$) increases rapidly with time.[23] Figure 11-34 illustrates this equivalence ratio distribution in the fuel spray at the time of ignition. In a swirling flow, ignition occurs in the slightly lean-of-stoichiometric region downstream of the spray core

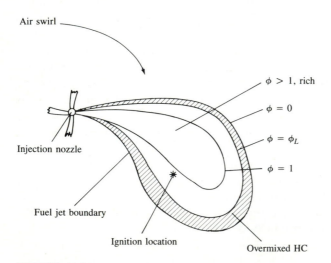

FIGURE 11-34
Schematic of diesel engine fuel spray showing equivalence ratio (ϕ) contours at time of ignition. ϕ_L = equivalence ratio at lean combustion limit (≈ 0.3). Shaded region contains fuel mixed leaner than ϕ_L.[67]

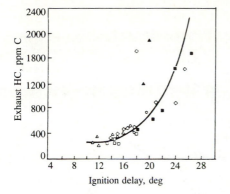

FIGURE 11-35
Correlation of exhaust HC concentration with duration of ignition delay for DI diesel engine. Various fuels, engine loads, injection timings, boost pressures, at 2800 rev/min.[67]

where the fuel which has spent most time within the combustible limits is located. However, the fuel close to the spray boundary has already mixed beyond the lean limit of combustion and will not autoignite or sustain a fast reaction front. This mixture can only oxidize by relatively slow thermal-oxidation reactions which will be incomplete. Within this region, unburned fuel, fuel decomposition products, and partial oxidation products (aldehydes and other oxygenates) will exist; some of these will escape the cylinder without being burned. The magnitude of the unburned HC from these overlean regions will depend on the amount of fuel injected during the ignition delay, the mixing rate with air during this period, and the extent to which prevailing cylinder conditions are conducive to autoignition. A correlation of unburned HC emissions with the length of the ignition delay would be expected. The data in Fig. 11-35 from a direct-injection naturally aspirated engine show that a good correlation between these variables exists. As the delay period increases beyond its minimum value (due to changes in engine operating conditions), HC emissions increase at an increasing rate.[67] Thus, overleaning of fuel injected during the ignition delay period is a significant source of hydrocarbon emissions, especially under conditions where the ignition delay is long.

UNDERMIXING. Two sources of fuel which enter the cylinder during combustion and which result in HC emissions due to slow or under mixing with air have been identified. One is fuel that leaves the injector nozzle at low velocity, often late in the combustion process. The most important source here is the nozzle sac volume, though secondary injections can increase HC emissions if the problem is severe. The second source is the excess fuel that enters the cylinder under overfueling conditions.

At the end of the fuel-injection process, the injector sac volume (the small volume left in the tip of the injector after the needle seats) is left filled with fuel. As the combustion and expansion processes proceed, this fuel is heated and vaporizes, and enters the cylinder at low velocity through the nozzle holes. This fuel vapor (and perhaps large drops of fuel also) will mix relatively slowly with air

Standard sac, volume $= 1.35$ mm^3

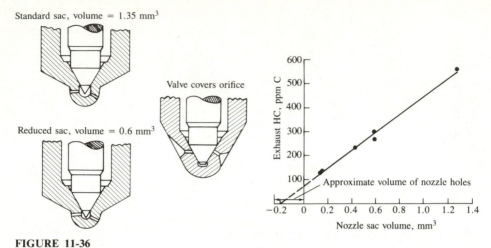

FIGURE 11-36
Effect of nozzle sac volume on exhaust HC concentration, DI diesel engine, at minimum ignition delay. $V_d = 1$ dm^3/cylinder, 1700–2800 rev/min.[67]

and may escape the primary combustion process. Figure 11-36 shows HC emissions at the minimum ignition delay for a direct-injection diesel engine as a function of sac volume, along with drawings of some of the injector nozzles used. The correlation between HC emissions (under conditions when the overleaning mechanism is least significant) and sac volume is striking. The extrapolation to zero HC emissions suggests that the fuel in the nozzle holes also contributes. Not all the fuel in the sac volume is exhausted as unburned hydrocarbons. For example, in Fig. 11-36 a volume of 1 mm^3 gives 350 ppm C_1 while 1 mm^3 of fuel would give 1660 ppm C_1. The sac volume may not be fully filled with fuel. Also, the higher-boiling-point fractions of the fuel may remain in the nozzle. Significant oxidation may also occur. In indirect-injection engines, similar trends have been observed, but the HC emission levels at short ignition delay conditions are substantially lower. The sac volume in current production nozzles helps to equalize the fuel pressures immediately upstream of the nozzle orifices. A small sac volume makes this equalization less complete and exhaust smoke deteriorates. The contribution of sac and hole volumes to exhaust HC can be reduced to below 0.75 g/ kW · h for a 1 dm^3 per cylinder displacement DI engine.[67]

In DI engines, exhaust smoke limits the full-load equivalence ratio to about 0.7. Under transient conditions as the engine goes through an acceleration process, overfueling can occur. Even though the overall equivalence ratio may remain lean, locally overrich conditions may exist through the expansion stroke and into the exhaust process. Figure 11-37 shows the effect of increasing the amount of fuel injected at constant speed, with the injection timing adjusted to keep the ignition delay at its minimum value (when HC emissions from overleaning are lowest). HC emissions are unaffected by an increasing equivalence ratio until a critical value of about 0.9 is reached when levels increase dramatically. A

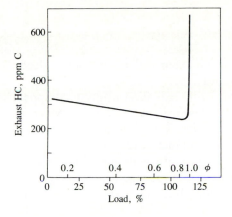

FIGURE 11-37
Effect of overfueling on exhaust HC concentration.
DI diesel engine, speed = 1700 rev/min, injection
timing at full-load 15° BTC.[67]

similar trend exists for IDI engines.[67] This mechanism is not significant under
normal operating conditions, but can contribute HC emissions under acceleration conditions if overfueling occurs. However, it produces less HC than does
overleaning at light load and idle.[23]

QUENCHING AND MISFIRE. Hydrocarbon emissions have been shown to be
sensitive to oil and coolant temperature: when these temperatures were increased
from 40 to 90°C in a DI diesel, HC emissions decreased by 30 percent. Since
ignition delay was maintained constant, overmixing phenomena should remain
approximately constant. Thus, wall quenching of the flame may also be a significant source of HC, depending on the degree of spray impingement on the combustion chamber walls.

While cycle-by-cycle variation in the combustion process in diesel engines is
generally much less than in spark-ignition engines, it can become significant
under adverse conditions such as low compression temperatures and pressures
and retarded injection timings. Substantial variations, cycle-by-cycle, in HC emissions are thought to result. In the limit, if misfire (no combustion) occurs in a
fraction of the operating cycles, then engine HC emissions rise as the percentage
of misfires increases. However, complete misfires in a well-designed and adequately controlled engine are unlikely to occur over the normal operating
range.[23]

SUMMARY. There are two major causes of HC emissions in diesel engines under
normal operating conditions: (1) fuel mixed to leaner than the lean combustion
limit during the delay period; (2) undermixing of fuel which leaves the fuel injector nozzle at low velocity, late in the combustion process. At light load and idle,
overmixing is especially important, particularly in engines of relatively small
cylinder size at high speed. In IDI engines, the contribution from fuel in the
nozzle sac volume is less important than with DI engines. However, other sources
of low velocity and late fuel injection such as secondary injection can be significant.

11.5 PARTICULATE EMISSIONS

11.5.1 Spark-Ignition Engine Particulates

There are three classes of spark-ignition engine particulate emissions: lead, organic particulates (including soot), and sulfates.

Significant sulfate emissions can occur with oxidation-catalyst equipped engines. Unleaded gasoline contains 150 to 600 ppm by weight sulfur, which is oxidized within the engine cylinder to sulfur dioxide, SO_2. This SO_2 can be oxidized by the exhaust catalyst to SO_3 which combines with water at ambient temperatures to form a sulfuric acid aerosol. Levels of sulfate emissions depend on the fuel sulfur content, the operating conditions of the engine, and the details of the catalyst system used. Typical average automobile sulfate emission rates are 20 mg/km or less.[68]

For automobile engines operated with regular and premium leaded gasolines (which contain about 0.15 g Pb/liter or dm³) the particulate emission rates are typically 100 to 150 mg/km. This particulate is dominated by lead compounds: 25 to 60 percent of the emitted mass is lead.[69] The particulate emission rates are considerably higher when the engine is cold, following start-up. The exhaust temperature has a significant effect on emission levels. The particle size distribution with leaded fuel is about 80 percent by mass below 2 μm diameter and about 40 percent below 0.2 μm diameter. Most of these particles are presumed to form and grow in the exhaust system due to vapor phase condensation enhanced by coagulation. Some of the particles are emitted directly, without settling. Some of the particles either form or are deposited on the walls where agglomeration may occur. Many of these are removed when the exhaust flow rate is suddenly increased, and these particles together with rust and scale account for the increase in mass and size of particles emitted during acceleration. Only a fraction (between 10 and 50 percent) of the lead consumed in the fuel is exhausted, the remainder being deposited within the engine and exhaust system.

Use of unleaded gasoline reduces particulate emissions to about 20 mg/km in automobiles without catalysts. This particulate is primarily soluble (condensed) organic material. Soot emissions (black smoke) can result from combustion of overly rich mixtures. In properly adjusted spark-ignition engines, soot in the exhaust is not a significant problem.

11.5.2 Characteristics of Diesel Particulates

MEASUREMENT TECHNIQUES. Diesel particulates consist principally of combustion generated carbonaceous material (soot) on which some organic compounds have become absorbed. Most particulate material results from incomplete combustion of fuel hydrocarbons; some is contributed by the lubricating oil. The emission rates are typically 0.2 to 0.6 g/km for light-duty diesels in an automobile. In larger direct-injection engines, particulate emission rates are 0.5 to 1.5 g/brake kW · h. The composition of the particulate material depends on the conditions in the engine exhaust and particulate collection system. At temperatures above 500°C, the individual particles are principally clusters of many small spheres or spherules of carbon (with a small amount of hydrogen) with individual spherule diameters of about 15 to 30 nm. As temperatures decrease below 500°C, the particles become coated with adsorbed and condensed high molecular weight organic compounds which include: unburned hydrocarbons, oxygenated hydrocarbons (ketones, esters, ethers, organic acids), and polynuclear aromatic hydrocarbons. The condensed material also includes inorganic species such as sulfur dioxide, nitrogen dioxide, and sulfuric acid (sulfates).

The objective of most particulate measurement techniques is to determine the amount of particulate being emitted to the atmosphere. Techniques for particulate measurement and characterization range from simple smoke meter opacity readings to analyses using dilution tunnels. Most techniques require lengthy sample-collection periods because the emission rate of individual species is usually low. The physical conditions under which particulate measurements are made are critical because the emitted species are unstable and may be altered through loss to surfaces, change in size distribution (through collisions), and chemical interactions among other species in the exhaust at any time during the measurement process (including sampling, storage, or examination). The most basic information is normally obtained on a mass basis: for example, grams per kilometer for a vehicle, grams per kilowatt-hour for an engine, grams per kilogram of fuel or milligrams per cubic meter of exhaust (at standard conditions). Smoke meters measure the relative quantity of light that passes through the exhaust or the relative reflectance of particulate collected on filter paper. They do not measure mass directly. They are used to determine visible smoke emissions and provide an approximate indication of mass emission levels. Visible smoke from heavy-duty diesels at high load is regulated. In the standard mass emission measurement procedure, dilution tunnels are used to simulate the physical and chemical processes the particulate emissions undergo in the atmosphere. In the dilution tunnel, the raw exhaust gases are diluted with ambient air to a temperature of 52°C or less, and a sample stream from the diluted exhaust is filtered to remove the particulate material.

PARTICULATE COMPOSITION AND STRUCTURE. The structure of diesel particulate material is apparent from the photomicrographs shown in Fig. 11-38 of particulates collected from the exhaust of an IDI diesel engine. The samples are seen to consist of collections of primary particles (spherules) agglomerated into aggregates (hereafter called particles). Individual particles range in appearance from clusters of spherules to chains of spherules. Clusters may contain as many as 4000 spherules. Occasional liquid hydrocarbon and sulfate droplets have been identified. The spherules are combustion generated soot particles which vary in diameter between 10 and 80 nm, although most are in the 15 to 30 nm range. Figure 11-39 shows a typical distribution of spherule size (solid line) determined by sizing and counting images in the photomicrographs. The number-mean diameter $(= \sum N_i d_i / N)$ is 28 nm. The volume contribution of these

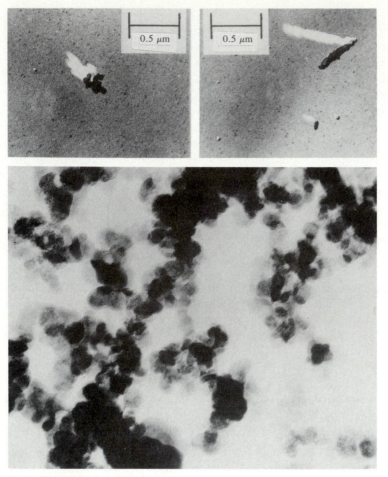

FIGURE 11-38
Photomicrographs of diesel particulates: cluster (upper left), chain (upper right), and collection from filter (bottom).[70]

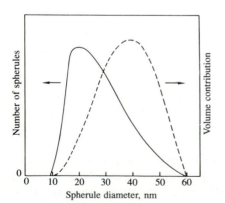

FIGURE 11-39
Typical distributions of spherule diameter and volume.[70]

TABLE 11.8
Chemical composition of particular matter[70]

	Idle	48 km/h
Extractable		
composition	$C_{23}H_{29}O_{4.7}N_{0.21}$	$C_{24}H_{30}O_{2.6}N_{0.18}$
H/C	1.26	1.63
Dry soot		
composition	$CH_{0.27}O_{0.22}N_{0.01}$	$CH_{0.21}O_{0.15}N_{0.01}$
H/C	0.27	0.21

spherules is shown as the dashed curve in Fig. 11-39. The volume-mean diameter, $(\sum N_i d_i^3 / N)^{1/3}$, is 31 nm.

Determination of the *particle* size distribution with a similar technique involves assigning a single dimension to a complex and irregular aggregate, and introduces uncertainties arising from only having two-dimensional images of particles available. Other approaches based on inertial impactors and electrical aerosol analysers have been used. Some of the data suggest that the particle size distribution is bimodal. The smaller-size range is thought to be liquid hydrocarbon drops and/or individual spherules characterized by number-mean diameters of 10 to 20 nm; the larger-size range is thought to be the particles of agglomerated spherules characterized by number-mean diameters of 100 to 150 nm. However, other particulate samples have not shown a bimodal distribution: volume-mean diameters ranged from 50 to 220 nm with no notable trend with either speed or load.[70]

The exhaust particulate is usually partitioned with an extraction solvent into a soluble fraction and a dry-soot fraction. Two commonly used solvents are dichloromethane and a benzene-ethanol mixture. Typically 15 to 30 mass percent is extractable, though the range of observations is much larger ($\sim$10 to 90 percent). Thermogravimetric analysis (weighing the sample as it is heated) produces comparable results. Typical average chemical compositions of the two particulate fractions are given in Table 11.8. Dry soot has a much lower H/C ratio than the extractable material. Although most of the particulate emissions are formed through incomplete combustion of fuel hydrocarbons, engine oil may also contribute significantly. The number-average molecular weight of the extractable material shown in Table 11.8 ranged from about 360 to 400 for a variety of engine conditions. This fell between the average molecular weight of the fuel (199) and that of the lubricating oil (443 when fresh and 489 when aged).[70] Radioactive tracer studies in a light-duty IDI diesel have shown that the oil was the origin of between 2 to 25 percent by mass of the total particulate and 16 to 80 percent of the extractable organic portion, the greatest percentages being measured at the highest engine speed studied (3000 rev/min). All of the oil contribution appeared in the extractable material. The contributions from the different individual compounds in the fuel have also been studied. All the compounds tested—paraffins,

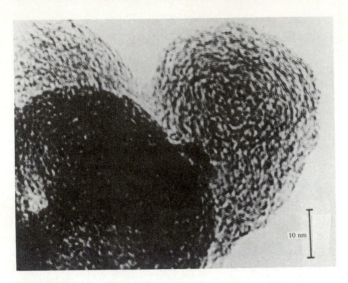

10 nm

FIGURE 11-40
Lattice-imaging micrograph of a diesel particulate.[72]

olefins, and aromatics—contributed to the particulate emissions; as a group, aromatics were the greatest contributors. Eighty percent of the carbon-14 used to tag individual fuel compounds was found in the *insoluble* fraction and 20 percent in the *soluble* particulate fraction.[71]

In addition to the elements listed in Table 11.8, trace amounts of sulfur, zinc, phosphorus, calcium, iron, silicon, and chromium have been found in particulates. Sulfur and traces of calcium, iron, silicon, and chromium are found in diesel fuel; zinc, phosphorus, and calcium compounds are frequently used in lubricating oil additives.[70]

A lattice image of a diesel particle is shown in Fig. 11-40; it suggests a concentric lamellate structure arranged around the center of each spherule. This arrangement of concentric lamellas is similar to the structure of carbon black. This is not surprising; the environment in which diesel soot is produced is similar to that in which oil furnace blacks are made. The carbon atoms are bonded together in hexagonal face-centered arrays in planes, commonly referred to as platelets. As illustrated in Fig. 11-41, the mean layer spacing is 0.355 nm (only slightly larger than graphite). Platelets are arranged in layers to form crystallites. Typically, there are 2 to 5 platelets per crystallite, and on the order of 10^3 crys-

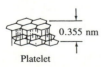

0.355 nm

Platelet

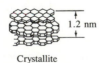

1.2 nm

Crystallite

Particle

FIGURE 11-41
Substructure of carbon particle.[73]

tallites per spherule. The crystallites are arranged with their planes more or less parallel to the particle surface. This structure of unordered layers is called turbostatic. The spherules, 10 to 50 nm in diameter, are fused together to form particles as shown in Fig. 11-40. A single spherule contains 10^5 to 10^6 carbon atoms.[70, 73]

A surface area of about 200 m^2/g has been measured for diesel soot. A smooth-surfaced 30-nm diameter sphere with a density of 2 g/cm^3 would have a surface/mass ratio of 100 m^2/g, so the measured value is about twice the superficial area. Approximating a particle of agglomerated spherules by a single sphere of 200 nm diameter gives a surface/mass ratio of 15 m^2/g.[70] These data and estimates of superficial area per unit mass indicate that diesel soot has low porosity.

SOLUBLE FRACTION COMPONENTS. The extractable organic fraction of diesel particulate emissions includes compounds that may cause health and environmental hazards. Thus chemical and biological characterization of the soluble organic fraction are important. Both soxhlet and sonification methods are used to extract the organic fraction from particulate samples. Because the particulates are mixtures of polar and nonpolar components, full extraction requires different solvents; any one solvent is a compromise. Methylene chloride is the most commonly used extractant, however. Since a complex mixture of organic compounds is associated with diesel particulates, a preliminary fractionation scheme is used to group similar types of compounds before final separation and identification. The scheme most frequently used results in seven fractions generally labeled as: basics, acidics, paraffins, aromatics, transitionals, oxygenates, and ether insolubles. Table 11.9 indicates the types of components in each fraction and the approximate proportions. The biological activity of the soluble organic fraction and its subfractions is most commonly assessed with the Ames *Salmonella*/microsomal test. With this test, a quantitative dose-response curve showing the mutagenicity of a sample compound is obtained. The Ames test uses a mutant strain of *Salmonella typhimurium* that is incapable of producing histidine. Mutagenicity is defined as the ability of a tested compound to revert—back-mutate—this bacterium to its wild state, where it regains its ability to produce histidine.[35]

11.5.3 Particulate Distribution within the Cylinder

Measurements have been made of the particulate distribution within the combustion chamber of operating diesel engines. The results provide valuable information on the particulate formation and oxidation processes and how these relate to the fuel distribution and heat-release development within the combustion chamber. Techniques used to obtain particulate concentration data include: use of rapid-acting poppet or needle valves which draw a small gas sample from the cylinder at a specific location and time for analysis (e.g., Refs. 21

TABLE 11-9
Components of the soluble organic fraction[35]

Fraction	Components of fraction	Percent of total
Acidic	Aromatic or aliphatic Acidic functional groups Phenolic and carboxylic acids	3–15
Basic	Aromatic or aliphatic Basic functional groups Amines	<1–2
Paraffin	Aliphatics, normal and branched Numerous isomers From unburned fuel and/or lubricant	34–65
Aromatic	From unburned fuel, partial combustion, and recombination of combustion products; from lubricants Single ring compounds Polynuclear aromatics	3–14
Oxygenated	Polar functional groups but not acidic or basic Aldehydes, ketones, or alcohols Aromatic phenols and quinones	7–15
Transitional	Aliphatic and aromatic Carbonyl functional groups Ketones, aldehydes, esters, ethers	1–6
Insoluble	Aliphatic and aromatic Hydroxyl and carbonyl groups High molecular weight organic species Inorganic compounds Glass fibers from filters	6–25

and 74), optical absorption techniques (e.g., Refs. 75 and 76), and cylinder dumping where the cylinder contents are rapidly emptied into an evacuated tank at a preset time in the cycle (e.g., Ref. 77). Both DI and IDI engines have been studied. Of course, concentration data taken at specific locations in the cylinder during the engine cycle are not necessarily representative of the cylinder contents in general; nor do they represent the time history of a given mass of gas. The fuel distribution, mixing, and heat-release patterns in the cylinder are highly nonuniform during the soot-formation process, and the details of gas motion in the vicinity of the sampling location as the piston changes position are usually unknown.

In direct-injection diesel engines, the highest particulate concentrations are found in the core region of each fuel spray where local average equivalence ratios are very rich (see Secs. 10.5.6 and 10.7.2). Soot concentrations rise rapidly soon after combustion starts. Figure 11-42 shows a set of sample-valve soot-concentration data from a large (30.5-cm bore, 38.1-cm stroke), quiescent, direct-

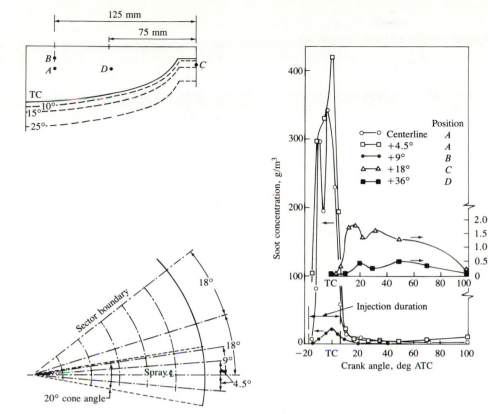

FIGURE 11-42
Particulate concentrations, in g/m³ at standard temperature and pressure, in various regions of the fuel spray as a function of crank angle in quiescent DI diesel engine, measured with rapid sampling valve. Different sample valve locations in combustion chamber and spray indicated on left. Cylinder bore = 30.5 cm, stroke = 38.1 cm, r_c = 12.9, engine speed = 500 rev/min, bmep = 827 kPa.[74]

injection diesel engine which illustrates these points.[74] The particulate concentrations on the fuel spray axis close to the injector orifice are remarkably high (~ 200 to 400 g/m³ at standard temperature and pressure). This corresponds to a large fraction of the fuel carbon in the extremely rich fuel vapor core being sampled as particulate (as soot and condensed HC species). Such high particulate fractions of the local fuel carbon (~ 50 percent) have also been found in the very fuel rich cores of high-pressure liquid-fueled turbulent diffusion flames. Pyrolysis of the fuel is therefore an important source of soot. These very high local soot concentrations decrease rapidly once fuel injection ceases and the rich core mixes to leaner equivalence ratios. Soot concentrations in the spray close to the piston bowl outer radius and at the cylinder wall rise later, are an order of magnitude less, and decay more slowly. Away from the fuel spray core, soot concentrations

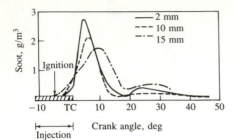

FIGURE 11-43
Particulate concentrations at various distances from wall of prechamber in swirl-chamber IDI diesel engine, measured with rapid sampling valve. Engine speed = 1000 rev/min, injection at 12° BTC, ignition at 5° BTC.[21]

decrease rapidly with increasing distance from the centerline. A useful comparison with these soot concentrations is the fuel concentration in a *stoichiometric* mixture, about 75 g fuel/m³. Approximate estimates of the mean soot concentration inside the cylinder through the combustion process suggests that almost all (over 90 percent) of the soot formed is oxidized prior to exhaust. Similar results have been obtained in a small direct-injection engine with swirl.[78, 79] Peak soot concentrations in the outer regions of the fuel spray were comparable ($\sim$10 g/m³). Measurements were not made in the spray core near the injector orifice; however, based on the equivalence ratio results in Fig. 10-46, soot concentrations would be expected to be lower due to the more rapid mixing with air that occurs with swirl.

Similar data are available from sampling in the prechamber of an IDI swirl chamber engine.[21] Figure 11-43 shows soot concentrations 2, 10, and 15 mm from the wall of the prechamber. Equivalence ratio distributions from this study have already been shown in Fig. 11-17. Concentrations peak 5 to 10° ATC at levels $\sim$2 g/m³; these are substantially lower than DI engine peak soot concentrations (presumably due to the more rapid mixing of fuel and air in the IDI engine). Concentrations in the prechamber at these locations then decrease substantially.

A better indication of average concentrations within the cylinder is given by total cylinder sampling experiments. Measurements of the total number of soot particles and soot volume fraction through the combustion process have been made in an IDI passenger car diesel. The contents of the engine cylinder, at a preselected point in the cycle, were rapidly expelled through a blowdown port, diluted, and collected in a sample bag. Figure 11-44 shows one set of results. Particles first appear shortly after the start of combustion (4 to 5° ATC). The number density rises to a maximum at 20° ATC and then falls rapidly as a result of particle coagulation and, possibly, oxidation. The exhaust particulate number density is less than one-tenth of the peak value. The volume fraction soot data (soot mass concentration is proportional to volume fraction) show a much flatter maximum earlier in the combustion process and a decrease (due to oxidation) from 20 to 40° ATC to about one-third of the peak value. Oxidation apparently ceases at about 40° ATC at these conditions.

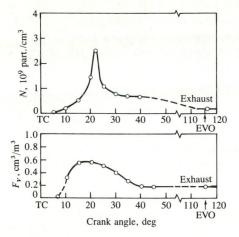

FIGURE 11-44
Cylinder-average particle-number density N and particle-volume fraction F_V, as a function of crank angle in IDI diesel engine determined from cylinder-dumping experiments. 1000 rev/min, $\phi = 0.32$, injection starts at 3.5° BTC. Gas volumes at standard temperature and pressure.[77]

11.5.4   Soot Formation Fundamentals

The soot particles, whose characteristics have been described in the above two sections, form primarily from the carbon in the diesel fuel. Thus, the formation process starts with a fuel molecule containing 12 to 22 carbon atoms and an H/C ratio of about 2, and ends up with particles typically a few hundred nanometers in diameter, composed of spherules 20 to 30 nm in diameter each containing some 10^5 carbon atoms and having an H/C ratio of about 0.1. Most of the information available on the fundamentals of soot formation in combustion comes from studies in simple premixed and diffusion flames, stirred reactors, shock tubes, and constant-volume combustion bombs. A recent review[80] summarizes the extensive literature available from such studies. Also, the production of carbon black requires a high yield of soot from pyrolysis of a hydrocarbon feedstock, and the literature from that field has much to contribute (see Ref. 81). However, the characteristics of diesel combustion which make it unsuitable for more fundamental studies—the high gas temperatures and pressures, complex fuel composition, dominance of turbulent mixing, the unsteady nature of the process, and the three-dimensional geometry—also make it difficult to interpret fundamental ideas regarding soot formation in the diesel context. There is much about the soot formation process in diesel engines, therefore, that is poorly and incompletely understood.

Soot formation takes place in the diesel combustion environment at temperatures between about 1000 and 2800 K, at pressures of 50 to 100 atm, and with sufficient air *overall* to burn fully all the fuel. The time available for the formation of solid soot particles from a fraction of the fuel is in the order of milliseconds. The resulting aerosol—dispersed solid-phase particles in a gas—can be characterized by the total amount of condensed phase (often expressed as the soot volume fraction, F_V, the volume of soot/total volume), the number of soot particles per unit volume (N), and the size of the particles (e.g., average diameter d). F_V, N, and d are mutually dependent [e.g., for spherical particles $F_V =$

$(\pi/6)Nd^3$], and any two of these variables characterize the system. It is most convenient to consider N and F_V as the independent variables since they each relate to the "almost-separate" stages of soot particle generation (the source of N) and soot particle growth (the source of F_V).

These stages can be summarized as follows:[80]

1. **Particle formation**, where the first condensed phase material arises from the fuel molecules via their oxidation and/or pyrolysis products. These products typically include various unsaturated hydrocarbons, particularly acetylene and its higher analogues ($C_{2n}H_2$), and polycyclic aromatic hydrocarbons (PAH). These two types of molecules are considered the most likely precursors of soot in flames. The condensation reactions of gas-phase species such as these lead to the appearance of the first recognizable soot particles (often called nuclei). These first particles are very small ($d < 2$ nm) and the formation of large numbers of them involve negligible soot loading in the region of their formation.

2. **Particle growth**, which includes both surface growth, coagulation, and aggregation. Surface growth, by which the bulk of the solid-phase material is generated, involves the attachment of gas-phase species to the surface of particles and their incorporation into the particulate phase. Figure 11-45, where the log of the molecular weight of a species is plotted against its hydrogen mole fraction $\tilde{x}_H$, illustrates some important points about this process. Starting with a fuel molecule of $\tilde{x}_H \gtrsim 0.5$ it is apparent that neither purely polyacetylene chain

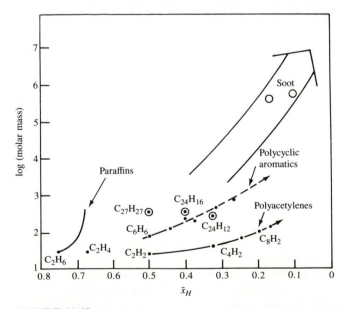

FIGURE 11-45
Paths to soot formation on plot of species molecular weight M versus hydrogen mole fraction $\tilde{x}_H$.[80]

growth nor purely PAH growth would lead to soot particles which have $\tilde{x}_H$ in the range 0.1 to 0.2. What is required is condensation of species with the right hydrogen content, or condensation of species with higher hydrogen content followed by dehydrogenation, or a combination of both these processes. Obviously some polyacetylenes and some PAH can satisfy these requirements, as can saturated platelets (e.g., $C_{27}H_{27}$; see Sec. 11.5.2). Surface growth reactions lead to an increase in the amount of soot (F_V) but the number of particles (N) remains unchanged. The opposite is true for growth by coagulation, where the particles collide and coalesce, which decreases N with F_V constant. Once surface growth stops, continued aggregation of particles into chains and clusters can occur.

These stages of particle generation and growth constitute the soot formation process. At each stage in the process oxidation can occur where soot or soot precursors are burned in the presence of oxidizing species to form gaseous products such as CO and CO_2. The eventual emission of soot from the engine will depend on the balance between these processes of formation and burnout. The emitted soot is then subject to a further mass addition process as the exhaust gases cool and are diluted with air. Adsorption into the soot particle surface and condensation to form new particles of hydrocarbon species in the exhaust gases occurs in the exhaust system and in the dilution tunnel which simulates what happens in the atmosphere. Figure 11-46 illustrates the relationship between these processes.[70] Although they are illustrated as discrete processes, there is some overlap, and they may occur concurrently in a given elemental mixture region within the diesel combustion chamber. Of course, due also to the non-homogeneous nature of the mixture and the duration of fuel injection and its overlap with combustion, at any given time different processes are in progress in different regions or packets of fluid. The fundamentals of each of these processes will now be reviewed.

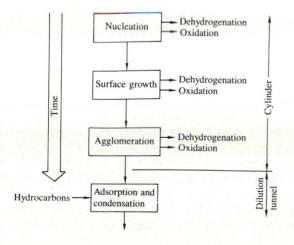

FIGURE 11-46
Processes leading to net production of diesel particulates.[70]

SOOT PARTICLE FORMATION. Empirically, it has been found useful to define the composition of the fuel-oxidizer mixture at the onset of soot formation in flames by the carbon/oxygen ratio. Equilibrium considerations indicate that soot formation should occur when, in

$$C_m H_n + y O_2 \rightarrow 2_y CO + \frac{n}{2} H_2 + (m - 2y) C_S \qquad (11.35)$$

m becomes larger than $2y$: i.e., the C/O ratio exceeds unity. The corresponding fuel/air equivalence ratio is given by

$$\phi = 2\left(\frac{C}{O}\right)(1 + \delta) \qquad (11.36)$$

where $\delta = n/(4m)$; ϕ is 3 for (C/O) = 1, with $n/m = 2$. The experimentally observed critical C/O ratios are less than unity, however, varying with fuel composition and details of the experimental setup from about 0.5 to 0.8. The critical C/O ratio for soot formation increases with increasing temperature but is only weakly dependent on pressure. Beyond the carbon formation limit, the yield of soot increases rapidly with increasing C/O ratio and is strongly enhanced by increasing pressure.[80]

It is obvious that soot formation is a nonequilibrium process. Yet despite decades of study, the precise details of the chemistry leading to the establishment of soot nuclei still elude investigators. Several different theories have been advanced to explain the pyrolysis process—the extensive decomposition and atomic rearrangement of the fuel molecules—that culminates in nucleation. Reviews of these theories can be found in Refs. 73, 80, and 81. Often-cited mechanisms are thermal cracking that results in fragmentation of fuel molecules into smaller ones, condensation reactions and polymerization that result in larger molecules, and dehydrogenation that lowers the H/C ratio of the hydrocarbons destined to become soot. Three different paths to the production of soot appear to exist, depending on the formation temperature. At the lowest temperatures ($\lesssim 1700$ K) only aromatics or highly unsaturated aliphatic compounds of high molecular weight are very effective in forming solid carbon through pyrolysis. At intermediate temperatures typical of diffusion flames ($\gtrsim 1800$ K), all normally used hydrocarbon fuels produce soot if burned at sufficiently rich stoichiometry, but appear to do so by following a different path. At very high temperatures, above the range of interest for diesel combustion, a third nucleation process seems likely that involves carbon vapor.[70]

A simple mechanistic model for nucleation in the low and intermediate temperature ranges which has considerable experimental support for its basic features has been advanced by Graham et al.[82] It is illustrated in Fig. 11-47. At low temperatures, an aromatic hydrocarbon can produce soot via a relatively fast direct route that involves condensation of the aromatic rings into a graphitelike structure. Above about 1800 K, however, a slower, less-direct route is favored that entails ring breakup into smaller hydrocarbon fragments. These fragments then polymerize to form larger unsaturated molecules that ultimately produce

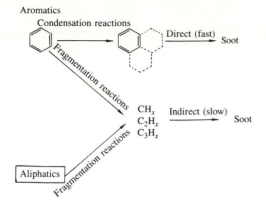

Aromatics

Condensation reactions

Direct (fast) Soot

Fragmentation reactions

CH_x
C_2H_x Indirect (slow) Soot
C_3H_x

Fragmentation reactions

Aliphatics

Fragmentation

FIGURE 11-47
Mechanistic model for formation of soot from aromatic and aliphatic hydrocarbon compounds.[70]

soot nuclei. Aliphatic molecules can only follow this latter less-direct route. Experimental measurements in flames suggest that polyunsaturated hydrocarbon compounds are involved in nucleation, and acetylenes and polyacetylenes have been detected that decrease in concentration as the mass of carbon formed increases. Such observations fit the indirect path in Fig. 11-47. Results of studies of pyrolysis of benzene between 1300 and 1700 K support a physical condensation mechanism for the low-temperature path. This mechanism begins with the transformation of the initial hydrocarbon into macromolecules by a gas-phase reaction. The partial pressure of these macromolecules grows until supersaturation becomes sufficient to force their condensation into liquid microdroplets. These become nuclei, and subsequently formed gaseous macromolecules then contribute to nuclei growth.[70]

SOOT PARTICLE GROWTH. Nucleation produces a large number of very small particles with an insignificant soot loading. The bulk of the solid-phase material is generated by surface growth, which involves the gas-phase deposition of hydrocarbon intermediates on the surfaces of the spherules that develop from the nuclei. A qualitative description of the changes that occur as a function of time in a premixed flame during nucleation and surface growth is shown in Fig. 11-48. The soot fraction F_V, in units of soot volume per unit volume of gas, is related to the number density N and the volume-mean diameter of the soot particles by

$$F_V = \frac{\pi}{6} N d^3 \tag{11.37}$$

d is the actual diameter of the spherules, or the diameter of a sphere of equivalent volume to an agglomerated particle. The rate of change of particle number density with time t can be written

$$\frac{dN}{dt} = \dot{N}_n - \dot{N}_a \tag{11.38}$$

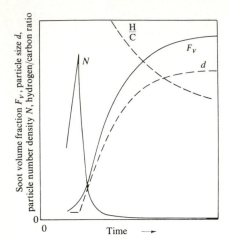

FIGURE 11-48
Variation in soot volume fraction F_V, particle number density N, particle size d, and soot hydrogen/carbon ratio with time in a flame.[70]

where $\dot{N}_n$ is the rate at which fresh nuclei appear and $\dot{N}_a$ is the rate of agglomeration of spherules or particles that collide and stick. At the peak of the N curve, $\dot{N}_n = \dot{N}_a$. To the left of the peak, $\dot{N}_n > \dot{N}_a$, the particle diameter remains essentially constant at the minimum detectable diameter and the (small) rise in soot volume is dominated by nucleation. To the right of the peak in the N curve, $\dot{N}_a > \dot{N}_n$. The number of agglomerating collisions is high because of the high number density; at the same time nucleation ends because there is enough dispersed surface area for gaseous deposition of hydrocarbon intermediates so the probability of generating new nuclei falls to zero. With nucleation halted slightly to the right of the N curve peak, all the subsequent increase in soot volume fraction (the majority) stems from surface growth. To the right of the N curve peak, the number density falls in the case illustrated by three orders of magnitude. This is the result of agglomeration, which is responsible for a portion of the increase in particle diameter. Agglomeration does not contribute to the rise in soot volume fraction, F_V. Surface growth that takes place on nuclei and on spherules is responsible for forming the concentric shells (somewhat distorted and warped) that constitute the outer portions of spherules and which are distinct from the less-organized spherule center (see Figs. 11-40 and 11-41). Surface growth on agglomerated particles may partly fill in the crevices at the junctures of adjoining spherules to provide the nodular structure evident in Fig. 11-40.[70]

Once particles have formed, interparticle collisions can lead to agglomeration, thereby decreasing the number of particles and increasing their size. Three types of agglomeration have been identified in soot formation. During the early stages of particle growth, collision of two spherical particles may result in their *coagulation* into a single spheroid. This is easy to visualize in hydrocarbon pyrolysis where the beginnings of a soot particle may have the viscosity of a tarry liquid.[70] Also, when the individual particles are small, rapid surface growth will quickly restore the original spherical shape.[73] This process occurs up to diameters of about 10 nm. On the other hand, if spherules have solidified before collision and surface growth rates have diminished, the resulting particles resemble a cluster in which the original spherules retain much of their individual identity. After surface growth essentially ceases, continued coalescence of the soot particles results in the formation of chainlike structures of discrete spherules. This suggests electrostatic forces are significant. Positive charge measured on these particle chains is claimed to be the cause of their chainlike structure.[70, 73] This latter coalescence once surface growth ceases is termed *aggregation*.

It has been shown experimentally that during coagulation the rate of decrease of particle number density was proportional to the product of a coagulation coefficient and the square of the number density:

$$-\frac{dN}{dt} = KN^2 \qquad (11.39)$$

This is the Smoluchowski equation for coagulation of a liquid colloid. Based on brownian motion, this equation is applicable when the Knudsen number (ratio of mean free path to particle diameter) exceeds 10. K depends on such factors as particle size and shape, size distribution, and the temperature, pressure, and density of the gas. Equation (11.39) has been used to predict coagulation rates in low-pressure sooting flames.[73, 80] It has also been modified so that it applies where the particle size and mean free path are comparable by using a more complex expression for K (see Ref. 83). These studies show that under conditions approximating those in engine flames, the fraction of the initial number density N_0 remaining at time t is given approximately by

$$\frac{N}{N_0} \approx (KN_0 t)^{-1} \qquad (11.40)$$

Thus as t increases, N/N_0 decreases rapidly. Although these coagulation calculations are simplistic (in that many of the assumptions made are not strictly valid since soot particles are not initially distributed homogeneously in the combustion space, they are not monodisperse, and surface growth and oxidation may be taking place during agglomeration), an overall conclusion is that the rate of coagulation of spherules and particles to larger particles is very sensitive to number density. Thus the number of particles decreases rapidly with advancing crank angle in the diesel engine during the early part of the expansion process (see Fig. 11-44) and agglomeration is essentially complete well before the exhaust valve opens.

Throughout the soot formation process in a flame, the H/C ratio of the hydrocarbons formed in the pyrolysis and nucleation process and of the soot particles continually decreases. The H/C ratio decreases from a value of about 2, typical of common fuels, to of order 1 in the youngest soot particles that can be sampled, and then to 0.2 to 0.3 once surface growth has ceased in the fully agglomerated soot.[80] The latter stages of this process are indicated in Fig. 11-48. The addition of mass to the soot particles occurs by reaction with gas-phase molecules. The reacting gas-phase hydrocarbons appear to be principally acetylenes, with larger polymers adding faster than the smaller. Small polyacetylenes undergo further polymerization in the gas phase, presumably by the same mechanism leading to nucleation. As a result of preferential addition of the larger polymers, the H/C ratio of the particles decreases toward its steady-state value. Thus most of the polyacetylenes added must be of very high molecular weight or dehydrogenation must also take place.[73, 80]

11.5.5 Soot Oxidation

In the overall soot formation process, shown schematically in Fig. 11-46, oxidation of soot at the precursor, nuclei, and particle stages can occur. The engine cylinder soot-concentration data reviewed in Sec. 11.5.3 indicate that a large fraction of the soot formed is oxidized within the cylinder before the exhaust process commences. In the discussion of diesel combustion movies in Sec. 10.3.1, dark brown regions were observed in the color photographs (see color plate, Fig. 10-4); these were interpreted as soot particle clouds, and were seen to be surrounded by a diffusion flame which appeared white from the luminosity of the high-temperature soot particles consumed in this flame. As air mixed with this soot-rich region, the white flame eradicated the dark soot clouds as the particles were burned up.

   In general, the rate of heterogeneous reactions such as the oxidation of soot depends on the diffusion of reactants to and products from the surface as well as the kinetics of the reaction. For particles less than about 1 μm diameter, diffusional resistance is minimal. The soot oxidation process in the diesel cylinder is kinetically controlled, therefore, since particle sizes are smaller than this limit. There are many species in or near the flame that could oxidize soot: examples are O_2, O, OH, CO_2, and H_2O. Recent reviews of soot formation[70, 73, 80] have concluded that at high oxygen partial pressures, soot oxidation can be correlated with a semiempirical formula based on pyrographite oxidation studies. For fuel-rich and close-to-stoichiometric combustion products, however, oxidation by OH has been shown to be more important than O_2 attack, at least at atmospheric pressure.

   It is argued on the basis of structural similarities that the rates of oxidation of soot and of pyrographites should be the same. This is a significant simplification. It has proved difficult to follow the oxidation of soot aerosols in flames, and if care is taken to avoid diffusional resistance, studies of bulk samples of pyrographite can then be used as a basis for understanding soot oxidation. The semiempirical formula of Nagle and Strickland-Constable has been shown[84] to correlate pyrographite oxidation for oxygen partial pressures $p_{O_2} < 1$ atm and temperatures between 1100 and 2500 K. This formula is based on the concept that there are two types of sites on the carbon surface available for O_2 attack. For the more reactive type A sites, the oxidation rate is controlled by the fraction of sites not covered by surface oxides (and therefore is of mixed order, between 0 and 1 in p_{O_2}). Type B sites are less reactive, and react at a rate which is first order in p_{O_2}. A thermal rearrangement of A sites into B sites is also allowed (with rate constant k_T). A steady-state analysis of this mechanism gives a surface mass oxi-

TABLE 11.10

Rate constants for Nagle and Strickland-Constable soot oxidation mechanism[84]

Rate constant	Units
$k_A = 20 \exp(-15{,}100/T)$	$g/cm^2 \cdot s \cdot atm$
$k_B = 4.46 \times 10^{-3} \exp(-7640/T)$	$g/cm^2 \cdot s \cdot atm$
$k_T = 1.51 \times 10^5 \exp(-48{,}800/T)$	$g/cm^2 \cdot s$
$k_Z = 21.3 \exp(2060/T)$	atm^{-1}

dation rate w (g $C/cm^2 \cdot s$):

$$\frac{w}{12} = \left(\frac{k_A p_{O_2}}{1 + k_Z p_{O_2}}\right)x + k_B p_{O_2}(1 - x) \tag{11.41}$$

where x is the fraction of the surface occupied by type A sites and is given by

$$x = \left(1 + \frac{k_T}{p_{O_2} k_B}\right)^{-1} \tag{11.42}$$

The empirical rate constants determined by Nagle and Strickland-Constable for this model are listed in Table 11.10. According to this mechanism, the reaction is first order at low oxygen partial pressures, but approaches zero order at higher pressures. At a given oxygen pressure, the rate initially increases exponentially with temperature (equivalent activation energy is k_A/k_Z or 34,100 cal/mol). Beyond a certain temperature the rate decreases as the thermal rearrangement favors formation of less reactive B sites. When, at sufficiently high temperature, the surface is completely covered with B sites, the rate is first order in oxygen partial pressure and increases again with temperature.[80]

Park and Appleton[84] have compared this formula with oxidation rate data obtained from pyrographite samples, carbon black particles, and with the available flame soot oxidation data. Figure 11-49 shows both the soot oxidation rate predicted by Eqs. (11.41) and (11.42) as a function of temperature and oxygen partial pressure, and the above-mentioned data. The formula correlates the data shown to within a factor of 2. Under diesel engine conditions, the O_2 partial pressure can be high ($\sim$several atmospheres), as can the temperatures of close-to-stoichiometric mixtures ($\lesssim$2800 K).

Equations (11.41) and (11.42) have been used to estimate the amount of soot that can be oxidized in a typical IDI diesel engine. It was assumed that soot was present in stoichiometric combustion products at selected times in the cycle and that mixing with air leaned out the burned gas mixture at different rates until the overall fuel/air equivalence ratio was reached. The surface recession rate during this process was computed. Figure 11-50 shows sample results at an engine speed of 1600 rev/min and an overall cylinder equivalence ratio of 0.58. Fast, intermediate, and slow mixing occurred in 30, 70, and 140°, respectively. The surface recession rate rises to a maximum as p_{O_2} increases and then decreases as the

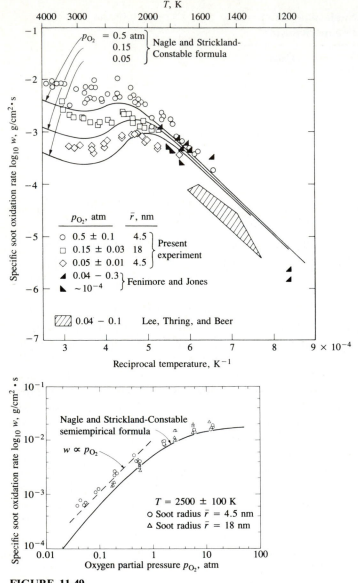

FIGURE 11-49
Specific soot oxidation rate measurements and predictions as a function of temperature and oxygen partial pressure.[84]

falling gas temperature more than offsets the increasing oxygen concentration. While the shape of the recession rate versus time curves depends on the mixing rate, the total amount of carbon burned (the area under each curve in Fig. 11-50*b*) is about the same (0.1 $\mu g/mm^2$). However, the point in the cycle at which the soot-containing burned gas mixture passes through stoichiometric is much

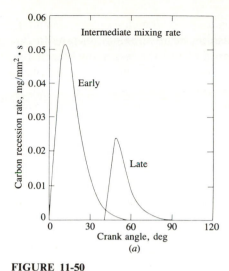
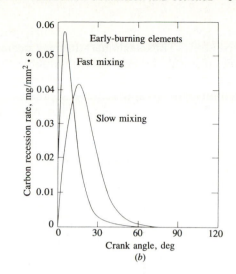

FIGURE 11-50
Soot particle burnup rate in diesel combustion environment: (a) in early- and late-burned fuel-air elements with intermediate mixing rate; (b) for fast and slow mixing for early-burning elements.[83]

more important. For the late mixing element shown (mixing lean of stoichiometric at 40° ATC), the total carbon mass oxidized is only 40 percent of that for the early mixing calculation. This is due primarily to the decreasing gas temperatures as the expansion stroke proceeds, and not the longer time available for burnup.[83]

For a spherical particle, the mass burning rate w (g/cm$^2 \cdot$ s) can be converted to a surface recession rate using

$$\frac{dr}{dt} = \frac{-w}{\rho}$$

where ρ is the density (≈ 2 g/cm^3). The integrated values of $w(t)$ when divided by ρ then give the maximum radius of a soot particle that can be burned up. Integrated values of 0.1 μg/mm^2 (estimated for TC start of burnup) correspond to a radius of about 50 nm or diameter of 100 nm. Individual spherule diameters are about 30 nm, so soot which mixes with air early in the expansion stroke is likely to be fully burned. Thus the soot present in the exhaust would be expected to come from regions which mix with air too late for the oxidation rate to be sufficient for particle burnup.

Agglomeration will have an indirect influence on the amount of soot oxidized through its effect on surface area. In the limiting case of a spherical cluster, n monodisperse spherules ($10 \lesssim n \lesssim 100$) can be imagined as compacted into a single solid sphere of equal volume. Alternatively, the same n spherules can be imagined compacted into a cylinder of diameter equal to that of the original spherules. Since oxidative attack is essentially an exterior surface phenomenon, the surface/volume ratio is the appropriate measure of the effect of particle shape on soot mass burnup rate. It can be shown that the surface/volume ratios for the single sphere, cylinder, and individual spherule are in the ratio $n^{-1/3}$, $\frac{2}{3}$, and 1, respectively. Thus agglomeration will decrease the relative oxidation rate. In the limit spherical clusters are less desirable than a chain; the larger the cluster the bigger the relative reduction in surface area. However, the densely packed spherule limit does not appear to be approached in practice. A specific surface area, of about 200 m^2/g for diesel soot, has been measured.[85] A smooth-surfaced 30-nm diameter spherule with a 2-g/cm^3 density has a surface/mass ratio of 100 m^2/g; the measured value is about twice this value, indicating low porosity and an agglomerate structure which is loosely rather than densely packed.[83]

Equation (11.41) shows a maximum recession rate in combustion products corresponding to a fuel/air equivalence ratio of about 0.9. Recent evidence shows that in an atmospheric pressure environment with rich and close-to-stoichiometric combustion products where O_2 mole fractions are low, oxidation by OH radical attack is much more significant than oxidation by O or O_2. The OH radical may be important in oxidizing soot in the flame zone under close-to-stoichiometric conditions.

11.5.6 Adsorption and Condensation

The final process in the particulate formation sequence illustrated in Fig. 11-46 is adsorption and condensation of hydrocarbons. This occurs primarily after the cylinder gases have been exhausted from the engine, as these exhaust gases are diluted with air. In the standard particulate mass emission measurement process this occurs in a dilution tunnel which simulates approximately the actual atmospheric dilution process. A diluted exhaust gas sample is filtered to remove the particulate. After equilibrating the collection filter at controlled conditions to remove water vapor, the particulate mass is obtained by weighing. In the prescribed EPA procedure, the filter temperature must not exceed 52°C. For a given exhaust gas temperature, the filter (and sample) temperature depends on the dilution ratio, as shown in Fig. 11-51.

The effect of the dilution ratio (and the dependent sample temperature) on collected particulate mass is shown in Fig. 11-52 for a standard dilution tunnel,

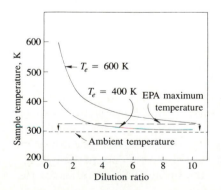

FIGURE 11-51
Effect of exhaust gas dilution ratio on the temperature of the collected particulate sample as a function of engine exhaust temperature T_e.[70]

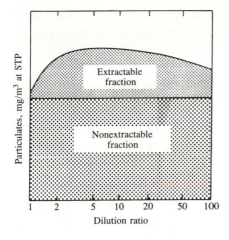

where the total sample is partitioned into extractable and nonextractable fractions. The nonextractable fraction is the carbonaceous soot generated during combustion and is not affected by the dilution process. With no dilution (dilution ratio of unity) the difference between the total and nonextractable mass is small; the bulk of the extractable fraction is acquired after the exhaust gas is mixed with dilution air. Extensive studies of this dilution process have shown that both adsorption and condensation occur. Adsorption involves the adherence of molecules of unburned hydrocarbons to the surfaces of the soot particles by chemical or physical (van der Waals) forces. This depends on the fraction of the available particle surface area occupied by hydrocarbons and on the partial pressure of the gaseous hydrocarbons that drives the adsorption process. As the dilution ratio increases from unity, the effect of decreasing temperature on the number of active sites dominates and, as shown in Fig. 11-52, the extractable fraction increases. At high dilution ratios, the sample temperature becomes insensitive to the dilution ratio (see Fig. 11-51) but the decreasing hydrocarbon partial pressure causes the extractable mass to fall again. Condensation will occur whenever the vapor pressure of the gaseous hydrocarbon exceeds its saturated vapor pressure. Increasing dilution decreases hydrocarbon concentrations and hence vapor pressure. However, the associated reduction in temperature does reduce the saturation pressure. High exhaust concentrations of hydrocarbons are the conditions where condensation is likely to be most significant, and the hydrocarbons most likely to condense are those of low volatility. Sources of low-volatility hydrocarbons are the high-boiling-point end of the fuel, unburned hydrocarbons that have been pyrolyzed but not consumed in the combustion process, and the lubricating oil.[70]

Experiments with a passenger car IDI diesel, where the oil was tagged with a radioactive tracer, have shown that the oil can contribute from 2 to 25 percent of the total particulate mass, with the greatest contribution occurring at high speed. On average, over half of the extractable mass was traceable to the oil. All the material traceable to the oil was found in the extractable fraction, indicating that the oil did not participate in the combustion process. However, the oil is not always a significant contributer: in another engine, fuel was the dominant source of extractable material.[70, 71]

11.6 EXHAUST GAS TREATMENT

11.6.1 Available Options

Our discussion so far has focused on *engine* emissions. Further reductions in emissions can be obtained by removing pollutants from the exhaust gases in the engine exhaust system. Devices developed to achieve this result include catalytic converters (oxidizing catalysts for HC and CO, reducing catalysts for NO_x, and three-way catalysts for all three pollutants), thermal reactors (for HC and CO), and traps or filters for particulates.

The temperature of exhaust gas in a spark-ignition engine can vary from 300 to 400°C during idle to about 900°C at high-power operation. The most common range is 400 to 600°C. Spark-ignition engines usually operate at fuel/air equivalence ratios between about 0.9 and 1.2 (see Sec. 7.1). The exhaust gas may therefore contain modest amounts of oxygen (when lean) or more substantial amounts of CO (when rich). In contrast, diesel engines, where load is controlled by the amount of fuel injected, always operate lean. The exhaust gas therefore contains substantial oxygen and is at a lower temperature (200 to 500°C). Removal of gaseous pollutants from the exhaust gases after they leave the engine cylinder can be either thermal or catalytic. In order to oxidize the hydrocarbons in the gas phase without a catalyst, a residence time of order or greater than 50 ms and temperatures in excess of 600°C are required. To oxidize CO, temperatures in excess of 700°C are required. Temperatures high enough for some homogeneous thermal oxidation can be obtained by spark retard (with some loss in efficiency) and insulation of the exhaust ports and manifold. The residence time can be increased by increasing the exhaust manifold volume to form a *thermal reactor* (see Sec. 11.6.3). However, this approach has limited application.

Catalytic oxidation of CO and hydrocarbons in the exhaust can be achieved at temperatures as low as 250°C. Thus effective removal of these pollutants occurs over a much wider range of exhaust temperatures than can be achieved with thermal oxidation. The only satisfactory method known for the removal of NO from exhaust gas involves catalytic processes. Removal of NO by catalytic oxidation to NO_2 requires temperatures <400°C (from equilibrium considerations) and subsequent removal of the NO_2 produced. Catalytic reaction of NO with added ammonia NH_3 is not practical because of the transient variations in NO produced in the engine. Reduction of NO by CO, hydrocarbons, or H_2 in the exhaust to produce N_2 is the preferred catalytic process. It is only feasible in spark-ignition engine exhausts. Use of catalysts in spark-ignition engines for CO, HC, and NO removal has become widespread. Catalysts are discussed in Sec. 11.6.2.

Particulates in the exhaust gas stream can be removed by a trap. Due to the small particle size involved, some type of filter is the most effective trapping method. The accumulation of mass within the trap and the increase in exhaust manifold pressure during trap operation are major development problems. Diesel particulates, once trapped, can be burned up either by initiating oxidation within the trap with an external heat source or by using a trap which contains catalytically active material. The operation of particulate traps is reviewed briefly in Sec. 11.6.4.

11.6.2 Catalytic Converters

The catalytic converters used in spark-ignition engines consist of an active catalytic material in a specially designed metal casing which directs the exhaust gas flow through the catalyst bed. The active material employed for CO and HC oxidation or NO reduction (normally noble metals, though base metals oxides can be used) must be distributed over a large surface area so that the mass-transfer characteristics between the gas phase and the active catalyst surface are sufficient to allow close to 100 percent conversion with high catalytic activity. The two configurations commonly used are shown in Fig. 11-53. One system employs a ceramic honeycomb structure or monolith held in a metal can in the exhaust stream. The active (noble metal) catalyst material is impregnated into a highly porous alumina washcoat about 20 μm thick that is applied to the passageway walls. The typical monolith has square-cross-section passageways with inside dimensions of ~ 1 mm separated by thin (0.15 to 0.3 mm) porous walls. The number of passageways per square centimeter varies between about 30 and 60. The washcoat, 5 to 15 percent of the weight of the monolith, has a surface area of 100 to 200 m^2/g. The other converter design uses a bed of spherical ceramic pellets to provide a large surface area in contact with the flow. With pellet catalysts, the noble metal catalyst is impregnated into the highly porous surface of the spherical alumina pellets (typically 3 mm diameter) to a depth of about 250 μm. The pellet material is chosen to have good crush and abrasion resistance after exposure to temperatures of order 1000°C. The gas flow is directed down through the bed as shown to provide a large flow area and low pressure drop. The gas flow is turbulent which results in high mass-transfer rates; in the monolith catalyst passageways, it is laminar.

OXIDATION CATALYSTS. The function of an oxidation catalyst is to oxidize CO and hydrocarbons to CO_2 and water in an exhaust gas stream which typically contains ~ 12 percent CO_2 and H_2O, 100 to 2000 ppm NO, ~ 20 ppm SO_2, 1 to 5 percent O_2, 0.2 to 5 percent CO, and 1000 to 6000 ppm C_1 HC, often with small amounts of lead and phosphorus. About half the hydrocarbons emitted by the SI engine are unburned fuel compounds. The saturated hydrocarbons (which comprise some 20 to 30 percent) are the most difficult to oxidize. The ease of oxidation increases with increasing molecular weight. Sufficient oxygen must be present to oxidize the CO and HC. This may be supplied by the engine itself

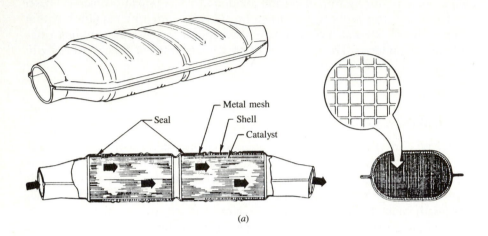

(a)

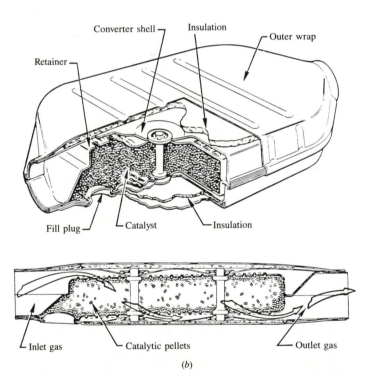

(b)

FIGURE 11-53
Catalytic converters for spark-ignition engine emission control: (a) monolith design; (b) pelletized design.[62]

running lean of stoichiometric or with a pump that introduces air into the exhaust ports just downstream of the valve. Venturi air addition into the exhaust port using the pressure pulsations generated by the exhaust process can also be used to add the required air.

Because of their high intrinsic activity, noble metals are most suitable as the catalytic material. They show higher specific activity for HC oxidation, are more thermally resistant to loss of low-temperature activity, and are much less deactivated by the sulfur in the fuel than base metal oxides. A mixture of platinum (Pt) and palladium (Pd) is most commonly used. For the oxidation of CO, olefins, and methane, the specific activity of Pd is higher than that of Pt. For the oxidation of aromatic compounds, Pt and Pd have similar activity. For oxidation of paraffin hydrocarbons (with molecular size greater than C_3), Pt is more active than Pd. Pure noble metals sinter rapidly in the 500 to 900°C temperature range experienced by exhaust catalysts. Since catalytic behavior is manifested exclusively by surface atoms, the noble metals are dispersed as finely as possible on an inert support such as γ-A_2O_3 which prevents particle-to-particle metal contact and suppresses sintering. The particle size of the noble metal particles in a fresh catalyst is less than 50 nm. This can increase to ~ 100 nm when the catalyst is exposed to the high temperatures of the exhaust in vehicle operation. Typical noble metal concentrations in a commercial honeycomb catalyst are between 1 and 2 g/dm^3 of honeycomb volume, with Pt/Pd = 2 on a weight basis. As a rough rule of thumb, the ceramic honeycomb volume required is about half the engine displaced volume. This gives a space velocity through the converter (volume flow rate of exhaust divided by converter volume) over the normal engine operating range of 5 to 30 per second.[68]

The *conversion efficiency* of a catalyst is the ratio of the rate of mass removal in the catalyst of the particular constituent of interest to the mass flow rate of that constituent into the catalyst: e.g., for HC,

$$\eta_{cat} = \frac{\dot{m}_{HC, in} - \dot{m}_{HC, out}}{\dot{m}_{HC, in}} = 1 - \frac{\dot{m}_{HC, out}}{\dot{m}_{HC, in}} \qquad (11.43)$$

The variation of conversion efficiency of a typical oxidizing catalytic converter with temperature is shown in Fig. 11-54. At high enough temperatures, the steady-state conversion efficiencies of a new oxidation catalyst are typically 98 to 99 percent for CO and 95 percent or above for HC. However, the catalyst is ineffective until its temperature has risen above 250 to 300°C. The term *light-off temperature* is often used to describe the temperature at which the catalyst becomes more than 50 percent effective.

The above numbers apply to fresh noble metal oxidation catalysts; as catalysts spend time in service their effectiveness deteriorates. Catalysis involves the adsorption of the reactants onto surface sites of high activity, followed by chemical reaction, then desorption of the products. Catalyst degradation involves both the deactivation of these sites by catalyst poisons and a reduction in the effective area of these sites through sintering. Poisoning affects both the warm-up and

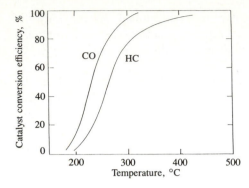

FIGURE 11-54
Conversion efficiency for CO and HC as a function of temperature for typical oxidizing catalytic converter.[62]

steady-state performance of the catalyst. When poisoning occurs, catalytic activity is impeded through prolonged contact with interfering elements that either physically block the active sites or interact chemically with the active material. The lead in fuel antiknock agents and the phosphorus in oil additives are the most important poisons. Though lead antiknock agents are not added to the gasoline used with catalyst-equipped vehicles, this "unleaded" fuel can be contaminated with small amounts (~ 10 mg Pb/dm^3) from the fuel distribution system. Between 10 and 30 percent of the lead in the fuel ends up on the catalyst. Its effect on catalyst conversion efficiency depends on the amount of lead on the catalyst, as shown in Fig. 11-55. Lead depresses the catalytic oxidation of HC to a greater extent than oxidation of CO. The oxidation activity of saturated hydrocarbons is particularly depressed. The extent of the poisoning that results from traces of critical elements in the fuel and oil depends on which elements are present and the amounts absorbed, as well as the composition of the catalyst and its operating conditions (especially its temperature).[68] Sintering is promoted by exposure of the catalyst to high operating temperatures. It involves the migration and agglomeration of sites, thus decreasing their active surface area. Sintering slows warm-up but has minimal effect on the steady-state conversion efficiency.

The oxidation kinetics of CO over Pt and Pd noble metal catalysts can be described by

$$\frac{d[CO]}{dt} = \frac{K_1 p_{CO} p_{O_2}}{(1 + K_2 p_{CO} + K_3 p_{HC})^2 (1 + K_4 p_{NO}^n)} \tag{11.44}$$

where K_1 to K_4 and n are constants at any given temperature, and p_{CO}, p_{O_2}, p_{HC}, and p_{NO} are the partial pressures of carbon monoxide, oxygen, hydrocarbons, and nitric oxide, respectively. A similar relationship can be written for the olefinic and aromatic HC oxidation rate (these being the most reactive hydrocarbons). These relationships incorporate the fact that the rates of CO and HC oxidation are inhibited by high CO and reactive HC concentrations, and that NO concentrations in the range 0 to 1000 ppm strongly inhibit oxidation also. The oxidation rate of paraffin hydrocarbons varies with the first power of the HC partial pres-

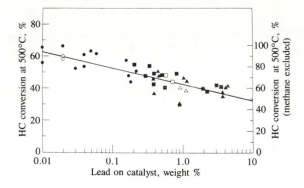

FIGURE 11-55
HC conversion efficiency as a function of lead concentration on catalyst. Total HC conversion on left; non-methane HC conversion on right. 0.001–0.013 g Pb/dm³ in fuel.[68]

sure, is inhibited by CO, olefins, and NO, and increases as the O_2 partial pressure is decreased to near-stoichiometric values.[68]

It will be apparent from the above that two extremely important considerations for successful use of catalysts for automotive applications are the test procedure that is used to measure emissions and the methods used to determine if the catalyst has the required durability. The U.S. Federal Test Procedure requires that the vehicle under test be at a temperature of 16 to 30°C for 12 hours prior to the test and that emissions are measured from the time the ignition key is turned on until the test has ended. In spark-ignition engines the mixture fed into the engine during start-up is enriched substantially (carburetors have a choke to accomplish this; additional fuel is injected with port or manifold fuel injection). The rationale is that if sufficient fuel is added to the inlet air, enough will evaporate to start the engine. However, until the rest of the fuel is consumed, the engine then runs rich and emits high concentrations of CO and HC. The catalyst is cold at this time, and until it warms up, these emissions will pass through without reaction. It is important that the catalyst be brought to its light-off temperature as quickly as possible (preferably in less than 60 s) and that mixture enrichment during start-up be held to a minimum. Thus catalysts should have low thermal inertia for rapid warm-up and low light-off temperatures for CO and HC, so they become effective quickly. The closer they are placed to the engine the faster they will reach light-off. However, they will then experience higher temperatures when fully warmed up and so be more susceptible to thermal degradation. While it is not too difficult to prepare catalysts that are highly effective when fresh, it is much more difficult to maintain effectiveness over extended mileage (50,000 miles) in which the catalyst is exposed to high temperatures and catalyst poisons. These can degrade both cold-start and warmed-up performance. Also, catalyst durability is affected by engine durability. Any engine malfunction that will expose the catalyst to excessive amounts of unburned fuel (such as ignition failure, misfire with too lean a mixture, or excessively rich operation) will severely overheat the catalyst.

Oxidation-catalyst-equipped vehicles may emit sulfuric acid aerosol. Unleaded gasoline contains 150 to 600 ppm by weight of S, which leaves the

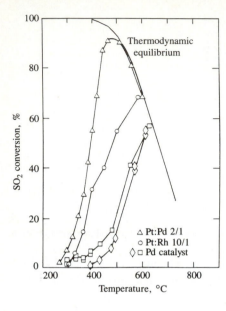

FIGURE 11-56
SO_2 conversion to SO_3 as a function of temperature with 5% O_2 concentration and no reducing gases present. Space velocity (volume flow per unit volume) $\sim 10 \ s^{-1}$. Results for Pt–Pd, Pt–Rh, and Pd catalysts.[68]

combustion chamber as SO_2. This SO_2 can be oxidized by the catalyst to SO_3 which combines with water at ambient conditions to form an H_2SO_4 aerosol. The SO_3 can be chemisorbed on the alumina catalyst surface; when large pellet beds are used, considerable storage of SO_3 at temperatures $<500°C$ can occur. At higher catalyst temperatures, this stored SO_3 is emitted as an SO_3–SO_2 mixture. SO_3 production can be controlled by lowering or raising the catalyst temperature. Figure 11-56 shows that at low temperatures SO_3 production is kinetically limited; at high temperatures SO_3 production is thermodynamically limited. Palladium and rhodium produce less SO_3 than Pt and have comparable HC and CO catalytic activity. By decreasing oxygen concentrations leaving the catalyst to ~ 1 percent, SO_3 production can be substantially reduced.[68]

NO CATALYSIS. NO is removed by reduction using the CO, hydrocarbons, and H_2 in the exhaust. The reactions are shown in Table 11.11. No catalyst is available for the decomposition of NO to O_2 and N_2 (thermodynamically favored at exhaust temperatures) which is sufficiently active for use in engine exhausts. NO reduction can be carried out under rich conditions where there is an excess of reducing species over oxidizing species. The catalyst used under these conditions is referred to as an *NO reduction catalyst*. Such a system requires a follow-up oxidation catalyst, together with addition of air from an air pump before the oxidation catalyst, to remove the remaining CO and hydrocarbons. Such a two-bed system can remove all three pollutants (NO, CO, and HC) from the exhaust. However, the rich operation necessary for NO reduction results in a fuel consumption penalty and constrains the performance of the NO catalyst since a fraction of the NO removed is converted to ammonia NH_3 rather than N_2. NH_3

TABLE 11.11

Possible NO reactions under reducing conditions[68]

1. $NO + CO \rightarrow \frac{1}{2}N_2 + CO_2$
2. $2NO + 5CO + 3H_2O \rightarrow 2NH_3 + 5CO_2$
3. $2NO + CO \rightarrow N_2O + CO_2$
4. $NO + H_2 \rightarrow \frac{1}{2}N_2 + H_2O$
5. $2NO + 5H_2 \rightarrow 2NH_3 + 2H_2O$
6. $2NO + H_2 \rightarrow N_2O + H_2O$

Reactions 3 and 6 occur at 200°C, which is below that usually found in auto exhausts.

formation under rich operation in the first bed must be small in this two-bed system because the second (oxidation) catalyst readily oxidizes NH_3 back to NO. Reduction of NO by CO or H_2 can be accomplished by base metal catalysts (e.g., CuO, NiO) in the temperature range 350 to 600°C. However, these catalyst materials are deactivated by sulfur and have shown limited thermal stability when used in vehicle exhausts. Alumina-supported noble metal catalysts reduce NO with CO–H_2 mixtures. Their NO-reduction activity is in the order $Ru > Rh > Pd > Pt$. Ruthenium (Ru) and rhodium (Rh) produce considerably less NH_3 than Pd or Pt under slightly rich conditions. While these properties make ruthenium a desirable NO catalyst, it forms volatile oxides under oxidizing conditions which results in loss of ruthenium from the alumina support.[68]

THREE-WAY CATALYSTS. If an engine is operated at all times with an air/fuel ratio at or close to stoichiometric, then both NO reduction and CO and HC oxidation can be done in a single catalyst bed. The catalyst effectively brings the exhaust gas composition to a near-equilibrium state at these exhaust conditions; i.e., a composition of CO_2, H_2O, and N_2. Enough reducing gases will be present to reduce NO and enough O_2 to oxidize the CO and hydrocarbons. Such a catalyst is called a *three-way catalyst* since it removes all three pollutants simultaneously. Figure 11-57 shows the conversion efficiency for NO, CO, and HC as a function of the air/fuel ratio. There is a narrow range of air/fuel ratios near stoichiometric in which high conversion efficiencies for all three pollutants are achieved. The width of this window is narrow, about 0.1 air/fuel ratios (7×10^{-3} in equivalence ratio units) for catalyst with high mileage use, and depends on catalyst formulation and engine operating conditions.

This window is sufficiently narrow to be beyond the control capabilities of an ordinary carburetor, though it can sometimes be achieved with sophisticated carburetors and fuel-injection systems. Thus closed-loop control of equivalence ratio has been introduced. An oxygen sensor in the exhaust is used to indicate whether the engine is operating on the rich or lean side of stoichiometric, and provide a signal for adjusting the fuel system to achieve the desired air-fuel mixture (see Sec. 7.4). Holding the equivalence ratio precisely on the chosen near-

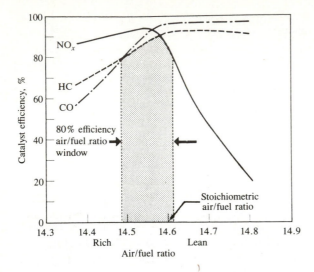

FIGURE 11-57

Conversion efficiency for NO, CO, and HC for a three-way catalyst as a function of exhaust gas air/fuel ratio.[68]

stoichiometric value is not a practical expectation of such a feedback system, and the equivalence ratio oscillates around the set point in an approximately periodic manner as the fuel flow is varied. Experimental data show that there is a considerable widening of the air/fuel ratio window where all three pollutants are effectively removed, with cyclic variation of the fuel flow. The maximum conversion in the middle of the window is reduced, however, from its value when there are no fluctuations. The effect of fluctuations depends on the frequency; frequencies of about 0.5 to 1 hertz are most effective and the usable window (at lower conversion efficiencies) can be broadened to about 1 air/fuel ratio. Some of the benefits of fluctuations in equivalence or air/fuel ratios are available even without any deliberate attempt to produce such variations with closed-loop feedback. Open-loop systems exhibit variations in the air/fuel ratio during normal vehicle operation.

Because of these cyclic variations in exhaust gas composition about a set point close to stoichiometric, it is desirable that the catalyst be able to reduce NO when a slight excess of oxygen is present (on the lean side) and remove CO and HC when there is a slight deficiency of oxygen (on the rich side). Rhodium is the principal ingredient used in commercial catalysts to remove NO. It is very active for NO reduction, is much less inhibited by CO and sulfur compounds, and produces less NH_3 than Pt. To remove NO under slightly lean-of-stoichiometric conditions, the catalyst must react the CO, H_2, or HC with NO rather than with O_2, as the exhaust gas passes through the catalyst bed. Rhodium shows some NO reduction activity slightly lean of stoichiometric. On the rich side, the three-way catalyst window is determined by hydrocarbon and CO removal. Platinum is most commonly used for HC and CO oxidation; it has good activity under stoichiometric and slightly lean conditions. When sufficient rhodium is present, the participation of Pt in NO removal is minimal. In the rich regime, the three-way catalyst consumes all the oxygen that is present in the exhaust, and as a consequence removes an equivalent amount of CO, H_2, and hydrocarbons; it is thought that the H_2 is removed first. In addition, the water-gas shift reaction

$$CO + H_2O = H_2 + CO_2$$

and the steam-reforming reaction

$$Hydrocarbon + H_2O \rightarrow CO, CO_2, H_2$$

can consume CO and HC. The exhaust contains an H_2/CO ratio of about $\frac{1}{3}$ (see Sec. 4.9.1), where the equilibrium ratio at 500°C is about 4. Considerable CO removal can be expected if the water-gas shift equilibrium is approached. Platinum is active in promoting this equilibrium. For large molecular weight paraffin hydrocarbons, and for olefins and aromatic hydrocarbons, the equilibrium for the steam-reforming reactions lies to the right. This reaction can therefore lead to considerable hydrocarbon removal. Rhodium is particularly active in the steam-reforming reaction; platinum is also active.[68]

The conversions of NO, CO, and hydrocarbons in a three-way catalyst operated with cyclical variations in equivalence ratio are larger than estimates based on summation of steady-state values during the cycle. At least part of the improved performance is thought to be due to the ability of the catalyst to undergo reduction-oxidation reactions. Such a catalyst component is usually referred to as an oxygen-storage component. In its oxidized state it can provide oxygen for CO and hydrocarbon oxidation in a rich exhaust gas environment, and in the process be reduced. When the exhaust cycles to lean conditions, this reduced component can react with O_2 or NO (which removes NO directly or indirectly by reducing the O_2 concentration). The oxidized component can then oxidize CO and HC in the next rich cycle, etc. Components such as ReO_2 or CeO_2 which exhibit this "redox" behavior can be included in three-way catalyst formulations. Commercial three-way catalysts contain platinum and rhodium (the ratio Pt/Rh varying substantially in the range 2 to 17) with some A_2O_3, NiO, and CeO_2. Alumina is the preferred support material.[68]

11.6.3 Thermal Reactors

In Secs. 11.3 and 11.4.2 it was explained that oxidation of CO and HC occurred during the expansion and exhaust processes in the cylinder of a conventional spark-ignition engine and, under certain circumstances, in the exhaust system. Oxidation after passage through the exhaust port can be enhanced with a *thermal reactor*—an enlarged exhaust manifold that bolts directly onto the cylinder head. Its function is to promote rapid mixing of the hot exhaust gases with any secondary air injected into the exhaust port (required with fuel-rich engine operation to produce a net oxidizing atmosphere), to remove nonuniformities in temperature and composition in the exhaust gases, and to retain the gases at a high enough

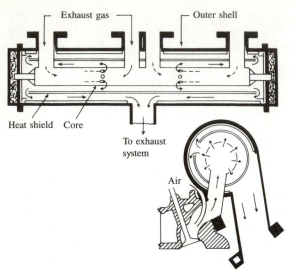

FIGURE 11-58
Schematic of exhaust thermal reactor for HC and CO oxidation.

temperature for sufficient time to oxidize much of the HC and CO which exits the cylinder. An example of a thermal reactor design is shown in Fig. 11-58.

The temperature levels typically required for bulk gas oxidation of HC and CO in a reactor are about 600 and 700°C, respectively. Note that they are considerably higher than those required for equivalent conversion in a catalytic converter and that higher temperatures are required for CO oxidation than for HC oxidation. The exhaust gas temperature in the manifold of a conventional engine is not sufficient to achieve any substantial reduction in engine exhaust port emissions. To achieve greater reductions, the reactor must be designed to reduce heat losses and increase residence time. In addition, to achieve rapid warm-up after engine start, a low thermal inertia reactor is desirable. Typically, a thin steel liner acts as the core of the reactor inside a cast-iron outer casing; with suitably arranged flow paths, this construction holds heat losses to a minimum by thermally isolating the core.

The effectiveness of the reactor depends on its operating temperature, the availability of excess oxygen mixed throughout the reacting gases, and the reactor volume. The operating temperature depends on the reactor inlet gas temperature, heat losses, and the amount of HC, CO, and H_2 burned up in the reactor. This latter factor is important: 1.5 percent CO removal results in a 220 K temperature rise. As a consequence, reactors with fuel-rich cylinder exhaust gas and secondary air give greater fractional reductions in HC and CO emissions than reactors with fuel-lean cylinder exhaust (which do not require any secondary air). As has already been explained, a higher core gas temperature is required to burn up the same fraction of CO which enters the reactor as of HC which enters. For lean engine exhaust gas, where the reactor core gas temperatures are a hundred degrees K lower than under fuel-rich operation, substantial reductions in CO emissions are difficult to achieve. For very lean operation, HC burnup becomes marginal.

A practical limitation to reactor effectiveness with fuel-rich engine operation is mixing of secondary air and engine exhaust gases in the exhaust port and the reactor core. The secondary air flow with a conventional air pump is effectively shut off by the exhaust blowdown process, and virtually no oxidation occurs in the exhaust port because the air and exhaust gases are segregated. Mixing in the reactor itself is promoted by suitably arranging the reactor inlet and exit ports and by using baffles. In systems with conventional secondary air pumps, maximum reductions in CO and HC occur with 10 to 20 percent excess air in the mixture. However, even with very high reactor core gas temperatures, 100 percent HC and CO oxidation is not achieved due to incomplete mixing. Improved control of secondary air flow has been shown to increase significantly CO emissions burnup.

11.6.4 Particulate Traps

An exhaust treatment technology that substantially reduces diesel engine particulate emissions is the trap oxidizer. A temperature-tolerant filter or trap removes the particulate material from the exhaust gas; the filter is then "cleaned off" by oxidizing the accumulated particulates. This technology is difficult to implement because: (1) the filter, even when clean, increases the pressure in the exhaust system; (2) this pressure increase steadily rises as the filter collects particulate matter; (3) under normal diesel engine operating conditions the collected particulate matter will not ignite and oxidize; (4) once ignition of the particulate occurs, the burnup process must be carefully controlled to prevent excessively high temperatures and trap damage or destruction. Trap oxidizers have been put into production for light-duty automobile diesel engines. Their use with heavy-duty diesel engines poses more difficult problems due to higher particulate loading and lower exhaust temperatures.

Types of particulate filters include: ceramic monoliths, alumina-coated wire mesh, ceramic foam, ceramic fiber mat, woven silica-fiber rope wound on a porous tube. Each of these has different inherent pressure loss and filtering efficiency. Regeneration of the trap by burning up the filtered particulate material can be accomplished by raising its temperature to the ignition point while providing oxygen-containing exhaust gas to support combustion and carry away the heat released. Diesel particulate matter ignites at about 500 to 600°C. This is above the normal temperature of diesel exhaust so either the exhaust gas flowing through the trap during regeneration must be heated (positive regeneration) or ignition must be made to occur at a lower temperature with catalytic materials on the trap or added to the fuel (catalytic regeneration). Catalytic coatings on the trap reduce the ignition temperature by up to 200°C.

Figure 11-59 shows a ceramic-coated trap oxidizer mounted on the exhaust system of a turbocharged IDI diesel engine. The trap is a ceramic honeycomb with half the cells closed at the inlet end and the other half of the cells closed at

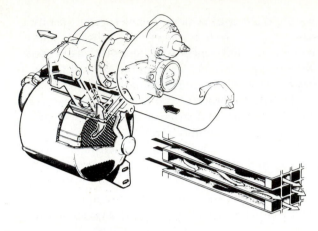

FIGURE 11-59
Catalytic ceramic-monolith particulate trap oxidizer mounted on exhaust of turbocharged automobile diesel engine.[86]

the exit end. Thus the particulate laden exhaust is forced to flow through the porous ceramic cell walls. The outside of the honeycomb is insulated and the trap is mounted close to the engine to maintain as high a trap temperature as possible. The pressure drop across the unloaded trap increases from 0.02 atm at 1000 rev/min to 0.15 atm at the maximum engine speed of 4500 rev/min. As the trap loads up, the pressure drop increases, requiring more fuel to be injected to compensate for the loss in power. This leads to higher exhaust temperature which eventually results in catalytic ignition of the particulate. The particulate oxidation rate depends on the trap temperature. With suitable trap location and design, the regeneration process is largely self-regulating. The particulate emissions from the engine are reduced by 70 percent or more.[86]

PROBLEMS

11.1. Figure 11-2 shows concentrations of NO, CO, and HC in a spark-ignition engine exhaust as a function of fuel/air equivalence ratio. Assume the concentration scale is parts per million. Explain the trends shown as the mixture is first made richer and then leaner than stoichiometric.

11.2. Figure 11-2 is for a spark-ignition engine. Construct a similar qualitative graph of NO, CO, and HC concentrations versus equivalence ratio for a direct-injection four-stroke cycle diesel engine.

11.3. A spark-ignition engine driving a car uses, on average, 120 grams of gasoline per mile traveled. The average emissions from the engine (upstream of the catalyst) are 1.5, 2, and 20 grams per mile of NO_x (as NO_2), HC, and CO, respectively. The engine operates with a stoichiometric gasoline-air mixture. Find the average concentrations in parts per million of NO_x, HC (as ppm C_1), and CO in the engine exhaust.

11.4. Calculate the average combustion inefficiency corresponding to the spark-ignition engine emissions levels given in Prob. 11.3. Include any hydrogen you estimate would be present in the exhaust stream.

11.5. A three-way catalytic converter is used with the spark-ignition engine in Prob. 11.3. For 10 percent of the driving time, the catalyst is cold and ineffective, and does not reduce the engine's emissions. For 90 percent of the time, the catalyst is hot and has conversion efficiencies as given in Fig. 11-57. Estimate the average *vehicle* emissions of NO_x, HC, and CO in grams per mile.

11.6. Figure 15-11 shows the variation in NO and HC emissions as concentrations (ppm) in the exhaust of a spark-ignition engine as a function of speed and load. Convert these data to graphs of indicated specific NO and HC emissions (g/kW · h) versus speed and imep. Assume η_v (based on atmospheric air density) = imep (kPa) × 10^{-3}.

11.7. Use the data in Fig. 11-44 to estimate:

  (a) The exhaust particulate emissions as a fraction of the maximum particulate loading during the cycle.

  (b) The maximum measured soot loading and the exhaust soot loading as fractions of the fuel carbon.

  (c) The equivalent sphere size of each soot particle at the number density peak (22° ATC) and in the exhaust.

  Assume a particulate density of 2 g/cm^3. Note that the gas volumes in Fig. 11-44 are determined at standard temperature and pressure.

11.8. Explain the following emissions trends. Highest marks will be given for succinct summaries of the *important* technical issues.

  (a) Nitric oxide (NO) emissions from diesels and spark-ignition engines as the equivalence ratio is varied show significantly different behavior (see Figs. 11-9 and 11-16). Redraw these graphs on the same plot and explain the different trends for these two types of engines as ϕ decreases on the lean side of stoichiometric.

  (b) Recirculation of a fraction of the exhaust gases to the intake is used to control engine nitric oxide emissions at part load. Exhaust gas recycle is usually more effective with spark-ignition engines than with diesels, as shown in Fig. P11-8. Explain why these trends are different.

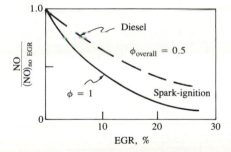

FIGURE P11-8

  (c) Brake specific particulate emissions from diesels are a major problem. Particulate emissions from conventional spark-ignition engines are negligible. Briefly explain why the particulate emission levels from these two types of engines are so different in magnitude.

(d) Diesels have low carbon monoxide (CO) emissions. Spark-ignition engine CO emissions when *averaged* over a typical urban automobile trip (cold engine start, warm-up, cruise, idle, acceleration, etc.) are substantial and require a catalyst for effective control. Explain this difference in average CO emissions (upstream of any catalyst) from these two types of engines.

11.9. The following questions refer to an engine with these geometric and operating characteristics (see Fig. 11-26a): $\phi = 1.0$; compression ratio $= 8:1$; bore $= 100$ mm; stroke $= 100$ mm; piston diameter above top ring $= 99.4$ mm; distance from piston crown top to top ring $= 9.52$ mm; volumetric efficiency $= 0.8$; temperature in cylinder at the start of compression $= 333$ K; pressure in cylinder at start of compression $= 1$ atm; mixture temperature before entering cylinder $= 30°C$; brake specific fuel consumption $= 300$ g/kW · h.

A substantial fraction of spark-ignition engine hydrocarbon emissions comes from the crevice between the piston crown and cylinder wall. Gas is forced into this crevice as the cylinder pressure increases and flows out of this crevice as the cylinder pressure decreases. The gas in the crevice can be assumed to be at the wall temperature, 400 K. The gas pushed into the crevice ahead of the flame is unburned mixture; the gas pushed in behind the flame is burned mixture. About two-thirds of the crevice gas is unburned. The maximum cylinder pressure is 3 MPa.

(a) Calculate the mass fraction of the cylinder gas which is in the crevice between the piston and cylinder wall and above the first piston ring, at the time of peak pressure.

(b) Assuming that half of the unburned fuel in this region is oxidized within the cylinder and a further one-third is oxidized in the exhaust port, calculate the engine HC emissions from this source in parts per million (ppm C_1) by volume.

(c) Calculate the ratio of brake specific hydrocarbon emissions to brake specific fuel consumption.

(d) Calculate the brake specific hydrocarbon emissions in grams of HC per kilowatt-hour.

11.10. Nitric oxide, NO, forms via reactions (11.1) to (11.3). Reaction (11.1) is "slow" and reactions (11.2) and (11.3) are "fast," so the initial rate of formation of NO is given by Eq. (11.8):

$$\frac{d[NO]}{dt} = 2k_1^+[N_2]_e[O]_e$$

where [ ] denote concentrations in gram-moles per cubic centimeter, k_1^+ is the rate constant for reaction (11.1), and the factor of 2 enters because the N atom formed in (11.1) immediately reacts via (11.2) or (11.3) to give an additional NO molecule:

$$k_1^+ = 7.6 \times 10^{13} \exp\left(\frac{-38,000}{T}\right) \quad cm^3/gmol \cdot s$$

where T is in kelvin.

Using the equilibrium composition data provided for mole fraction atomic oxygen (O), molecular nitrogen (N_2), and nitric oxide (NO):

(a) Plot the formation rate of NO as a function of the equivalence ratio at 3000 K and 5.5 MPa, and as a function of temperature for a stoichiometric mixture at 5.5 MPa.

(b) Estimate approximately the time taken to reach equilibrium NO levels at $\phi = 1$, 2750 K and 3000 K, 5.5 MPa.

(c) If the stoichiometric mixture inducted into the engine reaches 3000 K and 5.5 MPa after combustion, in the absence of any exhaust gas recirculation, calculate the percentage of the exhaust that must be recycled to the intake (at the initial intake temperature) to reduce the NO formation rate by a factor of 4 (assume the final pressure 5.5 MPa stays the same; of course, the final temperature decreases as the exhaust gas is recycled).

$p = 5.5$ MPa

ϕ	T(K)	Mole fraction O	Mole fraction N_2
0.9	3000	2.1×10^{-3}	0.73
1.0	3000	1.5×10^{-3}	0.73
1.1	3000	1×10^{-3}	0.73

$\phi = 1.0$, $p = 5.5$ MPa

T(K)	Mole fraction O	Mole fraction N_2	NO
2500	6×10^{-5}	0.73	—
2750	5×10^{-4}	0.73	4×10^{-3}
3000	1.5×10^{-3}	0.73	8×10^{-3}

REFERENCES

1. Bowman, C. T.: "Kinetics of Pollutant Formation and Destruction in Combustion," *Prog. Energy Combust. Sci.*, vol. 1, pp. 33–45, 1975.
2. Lavoie, G. A., Heywood, J. B., and Keck, J. C.: "Experimental and Theoretical Investigation of Nitric Oxide Formation in Internal Combustion Engines," *Combust. Sci. Technol.*, vol. 1, pp. 313–326, 1970.
3. Heywood, J. B., Fay, J. A., and Linden, L. H.: "Jet Aircraft Air Pollutant Production and Dispersion," *AIAA J.*, vol. 9, no. 5, pp. 841–850, 1971.
4. Newhall, H. K., and Shahed, S. M.: "Kinetics of Nitric Oxide Formation in High-Pressure Flames," in *Proceedings of Thirteenth International Symposium on Combustion*, pp. 381–390, The Combustion Institute, 1971.
5. Hilliard, J. C., and Wheeler, R. W.: "Nitrogen Dioxide in Engine Exhaust," SAE paper 790691, *SAE Trans.*, vol. 88, 1979.
6. Merryman, E. L., and Levy, A.: "Nitrogen Oxide Formation in Flames: The Roles of NO_2 and Fuel Nitrogen," in *Proceedings of Fifteenth International Symposium on Combustion*, p. 1073, The Combustion Institute, 1975.
7. Komiyama, K., and Heywood, J. B.: "Predicting NO_x Emissions and Effects of Exhaust Gas Recirculation in Spark-Ignition Engines," SAE paper 730475, *SAE Trans.*, vol. 82, 1973.
8. Alperstein, M., and Bradow, R. L.: "Exhaust Emissions Related to Engine Combustion Reactions," SAE paper 660781, *SAE Trans.*, vol. 75, 1966.
9. Starkman, E. S., Stewart, H. E., and Zvonow, V. A.: "Investigation into Formation and Modification of Exhaust Emission Precursors," SAE paper 690020, 1969.
10. Lavoie, G. A.: "Spectroscopic Measurement of Nitric Oxide in Spark-Ignition Engines," *Combust. Flame*, vol. 15, pp. 97–108, 1970.
11. Blumberg, P., and Kummer, J. K.: "Prediction of NO Formation in Spark-Ignition Engines—An Analysis of Methods of Control," *Combust. Sci. Technol.*, vol. 4, pp. 73–96, 1971.
12. Sakai, Y., Miyazaki, H., and Mukai, K.: "The Effect of Combustion Chamber Shape on Nitrogen Oxides," SAE paper 730154, 1973.
13. Quader, A. A.: "Why Intake Charge Dilution Decreases Nitric Oxide Emission from Spark Ignition Engines," SAE paper 710009, *SAE Trans.*, vol. 80, 1971.

14. Benson, J. D., and Stebar, R. F.: "Effects of Charge Dilution on Nitric Oxide Emission from a Single-Cylinder Engine," SAE paper 710008, *SAE Trans.*, vol. 80, 1971.

15. Toda, T., Nohira, H., and Kobashi, K.: "Evaluation of Burned Gas Ratio (BGR) as a Predominant Factor to NO_x," SAE paper 760765, *SAE Trans.*, vol. 85, 1976.

16. Lavoie, G. A., and Blumberg, P. N.: "A Fundamental Model for Predicting Fuel Consumption, NO_x and HC Emissions of the Conventional Spark-Ignited Engine," *Combust. Sci. Technol.*, vol. 21, pp. 225–258, 1980.

17. Lavoie, G. A., and Blumberg, P. N.: "Measurements of NO Emissions from a Stratified Charge Engine: Comparison of Theory and Experiment," *Combust. Sci. Technol.*, vol. 8, p. 25, 1973.

18. Aoyagi, Y., Kamimoto, T., Matsui, Y., and Matsuoka, S.: "A Gas Sampling Study on the Formation Processes of Soot and NO in a DI Diesel Engine," SAE paper 800254, SAE Trans., vol. 89, 1980.

19. Vioculescu, I. A., and Borman, G. L.: "An Experimental Study of Diesel Engine Cylinder-Averaged NO_x Histories," SAE paper 780228, *SAE Trans.*, vol. 87, 1978.

20. Mansouri, S. H., Heywood, J. B., and Radhakrishnan, K.: "Divided-Chamber Diesel Engine, Part I: Cycle-Simulation Which Predicts Performance and Emissions," SAE paper 820273, *SAE Trans.*, vol. 91, 1982.

21. Duggal, V. K., Priede, T., and Khan, I. M.: "A Study of Pollutant Formation within the Combustion Space of a Diesel Engine," SAE paper 780227, *SAE Trans.*, vol. 87, 1978.

22. Liu, X., and Kittelson, D. B.: "Total Cylinder Sampling from a Diesel Engine (Part II)," SAE paper 820360, 1982.

23. Yu, R. C., and Shahed, S. M.: "Effects of Injection Timing and Exhaust Gas Recirculation on Emissions from a D.I. Diesel Engine," SAE paper 811234, *SAE Trans.*, vol. 90, 1981.

24. Plee, S. L., Myers, J. P., and Ahmed, T.: "Flame Temperature Correlation for the Effects of Exhaust Gas Recirculation on Diesel Particulate and NO_x Emissions," SAE paper 811195, *SAE Trans.*, vol. 90, 1981.

25. Plee, S. L., Ahmad, T., and Myers, J. P.: "Diesel NO_x Emissions—A Simple Correlation Technique for Intake Air Effects," in *Proceedings of Nineteenth International Symposium on Combustion*, pp. 1495–1502, The Combustion Institute, Pittsburgh, 1983.

26. Ahmad, T., and Plee, S. L.: "Application of Flame Temperature Correlations to Emissions from a Direct-Injection Diesel Engine," SAE paper 831734, *SAE Trans.*, vol. 92, 1983.

27. Harrington, J. A., and Shishu, R. C.: "A Single-Cylinder Engine Study of the Effects of Fuel Type, Fuel Stoichiometry, and Hydrogen-to-Carbon Ratio and CO, NO, and HC Exhaust Emissions," SAE paper 730476, 1973.

28. Newhall, H. K.: "Kinetics of Engine-Generated Nitrogen Oxides and Carbon Monoxide," in *Proceedings of Twelfth International Symposium on Combustion*, pp. 603–613, Mono of Maryland, 1968.

29. Keck, J. C., and Gillespie, D.: "Rate-Controlled Partial-Equilibrium Method for Treating Reacting Gas Mixtures," *Combust. Flame*, vol. 17, pp. 237–241, 1971.

30. Delichatsios, M. M.: "The Kinetics of CO Emissions from an Internal Combustion Engine," S.M. Thesis, Department of Mechanical Engineering, MIT, June 1972.

31. Johnson, G. L., Caretto, L. S., and Starkman, E. S.: "The Kinetics of CO Oxidation in Reciprocating Engines," paper presented at the Western States Section, The Combustion Institute, Spring Meeting, April 1970.

32. Jackson, M. W.: "Analysis for Exhaust Gas Hydrocarbons—Nondispersive Infrared Versus Flame-Ionization," *J. Air Pollution Control Ass.*, vol. 16, p. 697–702, 1966.

33. Jackson, M. W.: "Effect of Catalytic Emission Control on Exhaust Hydrocarbon Composition and Reactivity," SAE paper 780624, *SAE Trans.*, vol. 87, 1978.

34. Patterson, D. J., and Henein, N. A.: *Emissions from Combustion Engines and Their Control*, Ann Arbor Science Publishers, Ann Arbor, Michigan, 1972.

35. "Diesel Technology, Impacts of Diesel-Powered Light-Duty Vehicles," report of the Technology Panel of the Diesel Impacts Study Committee, National Research Council, National Academy Press, Washington, D.C., 1982.

36. Lavoie, G. A.: "Correlations of Combustion Data for S.I. Engine Calculations—Laminar Flame

Speed, Quench Distance and Global Reaction Rates," SAE paper 780229, *SAE Trans.*, vol. 87, 1978.

37. Adamczyk, A. A., Kaiser, E. W., Cavolowsky, J. A., and Lavoie, G. A.: "An Experimental Study of Hydrocarbon Emissions from Closed Vessel Explosions," in *Proceedings of Eighteenth International Symposium on Combustion*, pp. 1695–1702, The Combustion Institute, 1981.

38. Westbrook, C. K., Adamczyk, A. A., and Lavoie, G. A.: "A Number Study of Laminar Wall Quenching," *Combust. Flame*, vol. 40, pp. 81–91, 1981.

39. Kaiser, E. W., Adamczyk, A. A., and Lavoie, G. A.: "The Effect of Oil Layers on the Hydrocarbon Emissions Generated During Closed Vessel Combustion," in *Proceedings of Eighteenth International Symposium on Combustion*, pp. 1881–1890, The Combustion Institute, 1981.

40. Tabaczynski, R. J., Heywood, J. B., and Keck, J. C.: "Time-Resolved Measurements of Hydrocarbon Mass Flow Rate in the Exhaust of a Spark-Ignition Engine," SAE paper 720112, *SAE Trans.*, vol. 81, 1972.

41. Ekchian, A., Heywood, J. B., and Rife, J. M.: "Time Resolved Measurements of the Exhaust from a Jet Ignition Prechamber Stratified Charge Engine," SAE paper 770043, *SAE Trans.*, vol. 86, 1977.

42. Daniel, W. A., and Wentworth, J. T.: "Exhaust Gas Hydrocarbons—Genesis and Exodus," SAE paper 486B, March 1962; also SAE Technical Progress Series, vol. 6, p. 192, 1964.

43. Daniel, W. A.: "Flame Quenching at the Walls of an Internal Combustion Engine," in *Proceedings of Sixth International Symposium on Combustion*, p. 886, Reinhold, New York, 1957.

44. LoRusso, J. A., Lavoie, G. A., and Kaiser, E. W.: "An Electrohydraulic Gas Sampling Valve with Application to Hydrocarbon Emissions Studies," SAE paper 800045, *SAE Trans.*, vol. 89, 1980.

45. Wentworth, J. T.: "More on Origins of Exhaust Hydrocarbons—Effects of Zero Oil Consumption, Deposit Location, and Surface Roughness," SAE paper 720939, *SAE Trans.*, vol. 81, 1972.

46. Namazian, M., and Heywood, J. B.: "Flow in the Piston-Cylinder-Ring Crevices of a Spark-Ignition Engine: Effect on Hydrocarbon Emissions, Efficiency and Power," SAE paper 820088, *SAE Trans.*, vol. 91, 1982.

47. Furuhama, S., and Tateishi, Y.: "Gases in Piston Top-Land Space of Gasoline Engine," *Trans. SAEJ*, no. 4, 1972.

48. Wentworth, J. T.: "The Piston Crevice Volume Effect on Exhaust Hydrocarbon Emission," *Combust. Sci. Technol.*, vol. 4, pp. 97–100, 1971.

49. Haskell, W. W., and Legate, C. E.: "Exhaust Hydrocarbon Emissions from Gasoline Engines—Surface Phenomena," SAE paper 720255, 1972.

50. Adamczyk, A. A., Kaiser, E. W., and Lavoie, G. A.: "A Combustion Bomb Study of the Hydrocarbon Emissions from Engine Crevices," *Combust. Sci. Technol.*, vol. 33, pp. 261–277, 1983.

51. Wentworth, J. T.: "Piston and Ring Variables Affect Exhaust Hydrocarbon Emissions," SAE paper 680109, *SAE Trans.*, vol. 77, 1968.

52. Kaiser, E. W., LoRusso, J. A., Lavoie, G. A., and Adamczyk, A. A.: "The Effect of Oil Layers on the Hydrocarbon Emissions from Spark-Ignited Engines," *Combust. Sci. Technol.*, vol. 28, pp. 69–73, 1982.

53. Kaiser, E. W., Adamczyk, A. A., and Lavoie, G. A.: "The Effect of Oil Layers on the Hydrocarbon Emissions Generated during Closed Vessel Combustion," in *Proceedings of Eighteenth International Symposium on Combustion*, pp. 1881–1890, The Combustion Institute, 1981.

54. McGeehan, J. A.: "A Literature Review of the Effects of Piston and Ring Friction and Lubricating Oil Viscosity on Fuel Economy," SAE paper 780673, *SAE Trans.*, vol. 87, 1978.

55. Shin, K., Tateishi, Y., and Furuhama, S.: "Measurement of Oil-Film-Thickness between Piston Ring and Cylinder," SAE paper 830068, *SAE Trans.*, vol. 92, 1983.

56. Carrier, G. F., Fendell, F. E., and Feldman, P. S.: "Cyclic Absorption/Desorption of Gas in a Liquid Wall Film," *Combust. Sci. Technol.*, vol. 25, pp. 9–19, 1981.

57. Wentworth, J. T.: "Effects of Top Compression Ring Profile on Oil Consumption and Blowby with the Sealed Ring–Orifice Design," SAE paper 820089, 1982.

58. Kuroda, H., Nakajima, Y., Sugihara, K., Takagi, Y., and Muranaka, S.: "The Fast Burn with Heavy EGR, New Approach for Low NO_x and Improved Fuel Economy," SAE paper 780006, *SAE Trans.*, vol. 87, 1978.

59. Jackson, M. W., Wiese, W. M., and Wentworth, J. T.: "The Influence of Air-Fuel Ratio, Spark Timing, and Combustion Chamber Deposits on Exhaust Hydrocarbon Emissions," SAE paper 486A, in *Vehicle Emissions*, vol. TP-6, SAE, 1962.

60. Lavoie, G. A., Lorusso, J. A., and Adamczyk, A. A.: "Hydrocarbon Emissions Modeling for Spark Ignition Engines," in J. N. Mattavi and C. A. Amann (eds.), *Combustion Modeling in Reciprocating Engines*, pp. 409–445, Plenum Press, 1978.

61. Daniel, W. A.: "Why Engine Variables Affect Exhaust Hydrocarbon Emission," SAE paper 700108, *SAE Trans.*, vol. 79, 1970.

62. Amann, C. A.: "Control of the Homogeneous-Charge Passenger-Car Engine—Defining the Problem," SAE paper 801440, 1980.

63. Weiss, P., and Keck, J. C.: "Fast Sampling Valve Measurements of Hydrocarbons in the Cylinder of a CFR Engine," SAE paper 810149, *SAE Trans.*, vol. 90, 1981.

64. Caton, J. A., Heywood, J. B., and Mendillo, J. V.: "Hydrocarbon Oxidation in a Spark Ignition Engine Exhaust Port," *Combust. Sci. Technol.*, vol. 37, nos. 3 and 4, pp. 153–169, 1984.

65. Green, R. M., Smith, J. R., and Medina, S. C.: "Optical Measurement of Hydrocarbons Emitted from a Simulated Crevice Volume in an Engine," SAE paper 840378, *SAE Trans.*, vol. 93, 1984.

66. Nakagawa, Y., Etoh, Y., and Maruyama, R.: "A Fundamental Analysis of HC and CO Oxidation Reaction in the Exhaust System," *JSAE Rev.*, no. 1, pp. 98–106, 1978.

67. Greeves, G., Khan, I. M., Wang, C. H. T., and Fenne, I.: "Origins of Hydrocarbon Emissions from Diesel Engines," SAE paper 770259, *SAE Trans.*, vol. 86, 1977.

68. Kummer, J. T.: "Catalysts for Automobile Emission Control," *Prog. Energy Combust. Sci.*, vol. 6, pp. 177–199, 1981.

69. Cadle, S. H., Nebel, G. J., and Williams, R. L.: "Measurements of Unregulated Emissions from General Motors' Light-Duty Vehicles," SAE paper 790694, *SAE Trans.*, vol. 88, 1979.

70. Amann, C. A., and Siegla, D. C.: "Diesel Particulates—What They Are and Why," *Aerosol Sci. Technol.*, vol. 1, pp. 73–101, 1982.

71. Mayer, W. J., Lechman, D. C., and Hilden, D. L.: "The Contribution of Engine Oil to Diesel Exhaust Particulate Emissions," SAE paper 800256, *SAE Trans.*, vol. 89, 1980.

72. Lahaye, J., and Prado, G.: "Morphology and Internal Structure of Soot and Carbon Blacks," in D. C. Siegla and G. W. Smith (eds.), *Particulate Carbon Formation during Combustion*, pp. 33–55, Plenum Press, 1981.

73. Smith, O. I.: "Fundamentals of Soot Formation in Flames with Application to Diesel Engine Particulate Emissions," *Prog. Energy Combust. Sci.*, vol. 7, pp. 275–291, 1981.

74. Whitehouse, N. D., Clough, E., and Uhunmwangho, S. O.: "The Development of Some Gaseous Products during Diesel Engine Combustion," SAE paper 800028, 1980.

75. Greeves, G., and Meehan, J. O.: "Measurement of Instantaneous Soot Concentration in a Diesel Combustion Chamber," paper C88/75, in *Combustion in Engines*, pp. 73–82, Institution of Mechanical Engineers, 1975.

76. Chang, Y. J., Kobayashi, H., Matsuzawa, K., and Kamimoto, T.: "A Photographic Study of Soot Formation and Combustion in a Diesel Flame with a Rapid Compression Machine," Proceedings of International Symposium on *Diagnostics and Modeling of Combustion in Reciprocating Engines*, COMODIA 85, pp. 149–157, Tokyo, Japan, September 4–6, 1985.

77. Du, C. J., and Kittelson, D. B.: "Total Cylinder Sampling from a Diesel Engine: Part III—Particle Measurements," SAE paper 830243, *SAE Trans.*, vol. 92, 1983.

78. Aoyagi, Y., Kamimoto, T., Matsui, Y., and Matsuoka, S.: "A Gas Sampling Study on the Formation Processes of Soot and NO in a DI Diesel Engine," SAE paper 800254, *SAE Trans.*, vol. 89, 1980.

79. Matsui, Y., Kamimoto, T., and Matsuoka, S.: "Formation and Oxidation Processes of Soot Particles in a D.I. Diesel Engine—An Experimental Study Via the Two-Color Method," SAE paper 820464, *SAE Trans.*, vol. 91, 1982.

80. Haynes, B. S., and Wagner, H. G.: "Soot Formation," *Prog. Energy Combust. Sci.*, vol. 7, pp. 229–273, 1981.

81. Lahaye, J., and Prado, G.: "Mechanisms of Carbon Black Formation," in P. L. Walker and P. A.

Thrower (eds.), *Chemistry and Physics of Carbon*, vol. 14, pp. 168–294, Marcel Dekker, New York, 1978.

82. Graham, S. C., Homer, J. B., and Rosenfeld, J. L. J.: "The Formation and Coagulation of Soot Aerosols Generated by the Pyrolysis of Aromatic Hydrocarbons," *Proc. R. Soc. Lond.*, vol. A344, pp. 259–285, 1975.

83. Amann, C. A., Stivender, D. L., Plee, S. L., and MacDonald, J. S.: "Some Rudiments of Diesel Particulate Emissions," SAE paper 800251, *SAE Trans.*, vol. 89, 1980.

84. Park, C., and Appleton, J. P.: "Shock-Tube Measurements of Soot Oxidation Rates," *Combust. Flame*, vol. 20, pp. 369–379, 1973.

85. Otto, K., Sieg, M. H., Zinbo, M., and Bartosiewicz, L.: "The Oxidation of Soot Deposits from Diesel Engines," SAE paper 800336, *SAE Trans.*, vol. 89, 1980.

86. Abthoff, J., Schuster, H., Langer, H., and Loose, G.: "The Regenerable Trap Oxidizer—An Emission Control Technique for Diesel Engines," SAE paper 850015, 1985.

ENGINE
HEAT
TRANSFER

12.1 IMPORTANCE OF HEAT TRANSFER

The peak burned gas temperature in the cylinder of an internal combustion engine is of order 2500 K. Maximum metal temperatures for the inside of the combustion chamber space are limited to much lower values by a number of considerations, and cooling for the cylinder head, cylinder, and piston must therefore be provided. These conditions lead to heat fluxes to the chamber walls that can reach as high as 10 MW/m² during the combustion period. However, during other parts of the operating cycle, the heat flux is essentially zero. The flux varies substantially with location: regions of the chamber that are contacted by rapidly moving high-temperature burned gases generally experience the highest fluxes. In regions of high heat flux, thermal stresses must be kept below levels that would cause fatigue cracking (so temperatures must be less than about 400°C for cast iron and 300°C for aluminum alloys). The gas-side surface of the cylinder wall must be kept below about 180°C to prevent deterioration of the lubricating oil film. Spark plug and valves must be kept cool to avoid knock and preignition problems which result from overheated spark plug electrodes or exhaust valves. Solving these engine heat-transfer problems is obviously a major design task.

Heat transfer affects engine performance, efficiency, and emissions. For a given mass of fuel within the cylinder, higher heat transfer to the combustion chamber walls will lower the average combustion gas temperature and pressure, and reduce the work per cycle transferred to the piston. Thus specific power and efficiency are affected by the magnitude of engine heat transfer. Heat transfer between the unburned charge and the chamber walls in spark-ignition engines affects the onset of knock which, by limiting the compression ratio, also influences power and efficiency. Most critical is heat transfer from the hot exhaust valve and piston to mixture in the end-gas region. Changes in gas temperature due to the heat-transfer impact on emission formation processes, both within the engine's cylinder and in the exhaust system where afterburning of CO and HC occurs. The exhaust temperature also governs the power that can be obtained from exhaust energy recovery devices such as a turbocharger turbine. Friction is both affected by engine heat transfer and contributes to the coolant load. The cylinder liner temperature governs the piston and ring lubricating oil film temperature, and hence its viscosity. Piston and liner distortion due to temperature nonuniformities have a significant impact on the piston component of engine friction. Some of the mechanical energy dissipated due to friction must be rejected to the atmosphere by the cooling system. The fan and water pump power requirements are determined by the magnitude of the heat rejected. The importance of engine heat transfer is clear.

To examine heat transfer more fully, it is helpful to divide the engine into its subsystems. The intake system consists of intake manifold and inlet ports and valves. Heat transfer to the inflowing charge reduces volumetric efficiency (see Sec. 6.2.1). However, in spark-ignition engines, the intake mixture is heated, with carbureted and single-point injected engines, to aid in vaporizing the fuel (see Sec. 7.6.3). Within the engine cylinder, the temperature of the charge relative to the wall temperature and the flow field vary enormously throughout the cycle. Both of these variables have a major influence on heat transfer. During the intake process, the incoming charge is usually cooler than the walls and the flow velocities are high. During compression the charge temperature rises above the wall temperature, and gas velocities decrease (see Sec. 8.2.2). Heat transfer is now from the cylinder gases to the chamber walls. During combustion gas temperatures increase substantially and the gas expansion which occurs on combustion produces increased gas motion. This is the period when heat-transfer rates to the walls are highest. Also, as the cylinder pressure rises, a small fraction of the cylinder charge is forced into crevice regions, resulting in additional heat transfer (see Sec. 8.6). During expansion, gas temperatures decrease so heat-transfer rates decrease. When the exhaust valve opens, however, the blowdown process (Sec. 6.5) produces high velocities within the cylinder, and past the exhaust valve and in the exhaust port. Substantial heat transfer from the exhausting gases to the valve, port, and (to a lesser extent) manifold occurs during the exhaust process. An example of how the heat-transfer rate to the total combustion chamber walls varies throughout the four-stroke operating cycle of a spark-ignition engine is shown in Fig. 14-9. The heat-transfer rate was estimated from the cylinder pressure, unburned and burned gas temperatures, combustion chamber surface area, and wall temperature, assuming gas velocities scaled with mean piston speed. An ability to predict the magnitude of the heat transfer between the working fluid, the walls of the intake system, combustion chamber, and exhaust system, and to the coolant is of obvious importance to the engine designer.

12.2 MODES OF HEAT TRANSFER

The following modes of heat transfer are important.

12.2.1 Conduction

Heat is transferred by molecular motion, through solids and through fluids at rest, due to a temperature gradient. The heat transfer by conduction, per unit area per unit time, $\dot{q}$, in a steady situation is given by Fourier's law:

$$\dot{q} = -k\nabla T \tag{12.1}$$

where k is the thermal conductivity. For a steady one-dimensional temperature variation

$$\dot{q}_x = \frac{\dot{Q}}{A} = -k\frac{dT}{dx}$$

Heat is transferred by conduction through the cylinder head, cylinder walls, and piston; through the piston rings to the cylinder wall; through the engine block and manifolds.

12.2.2 Convection

Heat is transferred through fluids in motion and between a fluid and solid surface in relative motion. When the motion is produced by forces other than gravity, the term *forced convection* is used. In engines the fluid motions are turbulent (see Chap. 8).

Heat is transferred by forced convection between the in-cylinder gases and the cylinder head, valves, cylinder walls, and piston during induction, compression, expansion, and exhaust processes. Heat is transferred by forced convection from the cylinder walls and head to the coolant (which may be liquid or gas), and from the piston to the lubricant or other piston coolant. Substantial convective heat transfer occurs to the exhaust valve, exhaust port, and exhaust manifold during the exhaust process. Heat transfer by convection in the inlet system is used to raise the temperature of the incoming charge. Heat is also transferred from the engine to the environment by convection.

In steady-flow forced-convection heat-transfer problems, the heat flux $\dot{q}$ transferred to a solid surface at temperature T_w from a flowing fluid stream at temperature T is determined from the relation

$$\dot{q} = h_c(T - T_w) \tag{12.2}$$

where h_c is called the heat-transfer coefficient. For many flow geometries (such as flow through pipes or over a plate), h_c is given by relations of the form

$$\left(\frac{h_c L}{k}\right) = \text{constant} \times \left(\frac{\rho v L}{\mu}\right)^m \left(\frac{c_p \mu}{k}\right)^n \tag{12.3}$$

where L and v are a characteristic length and velocity. The terms in brackets from left to right are the Nusselt, Reynolds, and Prandtl dimensionless numbers, respectively. For gases, the Prandtl number $(c_p \mu/k)$ varies little and is about 0.7 (see Sec. 4.8).

When boiling occurs at the surface (i.e., vapor is formed in the liquid), as may be the case in high heat flux areas on the coolant side in water-cooled engines, then different relationships for h_c must be used.

12.2.3   Radiation

Heat exchange by radiation occurs through the emission and absorption of electromagnetic waves. The wavelengths at which energy is transformed into thermal energy are the visible range (0.4 to 0.7 μm) and the infrared (0.7 to 40 μm). Heat transfer by radiation occurs from the high-temperature combustion gases and the flame region to the combustion chamber walls (although the magnitude of this radiation heat transfer relative to convective heat transfer is only significant in diesel engines). Heat transfer by radiation to the environment occurs from all the hot external surfaces of the engine.

The theory of radiant heat transfer starts from the concept of a "black body," i.e., a body that has a surface that emits or absorbs equally well radiation of all wavelengths and that reflects none of the radiation falling on it. The heat flux from one plane black body at temperature T_1 to another at temperature T_2 parallel to it across a space containing no absorbing material is given by

$$\dot{q} = \sigma(T_1^4 - T_2^4) \tag{12.4}$$

where σ is the Stefan-Boltzmann constant 5.67×10^{-8} W/m$^2 \cdot$K^4. Real surfaces are not "black" but reflect radiation to an extent which depends on wavelength. Gases are far from this black-body idealization. They absorb and emit radiation almost exclusively within certain wavelength bands characteristic of each species. These departures from black-body behavior are usually dealt with by applying a multiplying factor (an emissivity, ε) to Eq. (12.4). Similarly, a "shape factor" is applied to account for the fact that the angle of incidence of the radiation usually varies over any actual surface. These factors can be calculated for simple cases.

12.2.4   Overall Heat-Transfer Process

Figure 12-1 shows, schematically, the overall heat-transfer process from the gases within the cylinder through the combustion chamber wall to the coolant flow. The heat flux into the wall has in general both a convective and a radiation component. The heat flux is conducted through the wall and then convected from

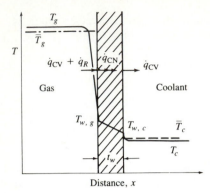

FIGURE 12-1
Schematic of temperature distribution and heat flow across the combustion chamber wall.

the wall to the coolant. A schematic temperature profile, and mean gas and coolant temperatures, $\bar{T}_g$ and $\bar{T}_c$, are shown.

In internal combustion engines, throughout each engine operating cycle, the heat transfer takes place under conditions of varying gas pressure and temperature, and with local velocities which vary more or less rapidly depending on intake port and combustion chamber configuration (see Chap. 8). In addition, the surface area of the combustion chamber varies through the cycle. The heat flux into the containing walls changes continuously from a small negative value during the intake process to a positive value of order several megawatts per square meter early in the expansion process. The flux variation lags behind the change in gas temperature. This lag between heat flux and driving temperature difference is clearly perceptible[1] but the precision of measurements to date suffice only for a rough estimate of its magnitude. Generally, investigators have concluded that the assumption that the heat-transfer process is *quasi steady* is sufficiently accurate for most calculation purposes. However, gas temperature and gas velocities vary significantly across the combustion chamber. The heat flux distribution over the combustion chamber walls is, therefore, nonuniform.

For a steady one-dimensional heat flow through a wall as indicated in Fig. 12-1, the following equations relate the heat flux $\dot{q} = \dot{Q}/A$ and the temperatures indicated:

Gas side:
$$\dot{q} = \dot{q}_{CV} + \dot{q}_R = h_{c,g}(\bar{T}_g - T_{w,g}) + \sigma\varepsilon(\bar{T}_g^4 - T_{w,g}^4) \tag{12.5}$$

where ε is the emissivity. The radiation term is generally negligible for SI engines.

Wall:
$$\dot{q} = \dot{q}_{CN} = \frac{k(T_{w,g} - T_{w,c})}{t_w} \tag{12.6}$$

Coolant side:
$$\dot{q} = \dot{q}_{CV} = h_{c,c}(T_{w,c} - \bar{T}_c) \tag{12.7}$$

If $h_{c,g}$ and $h_{c,c}$ are known, the temperatures $\bar{T}_g$, $T_{w,g}$, $T_{w,c}$, and $\bar{T}_c$ can be related to each other.

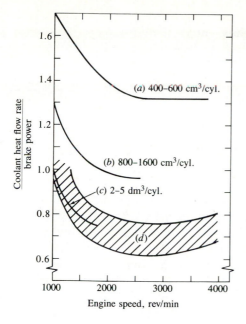

FIGURE 12-2
Ratio of coolant heat flow rate to brake power as a function of engine speed. Different size and types of engines: (a) small automotive diesels; (b) larger automotive diesels; (c) various diesels; (d) spark-ignition engines. (*Developed from Howarth.[2]*)

12.3 HEAT TRANSFER AND ENGINE ENERGY BALANCE

Figure 12-2 shows the magnitude of the heat-rejection rate to the coolant relative to the brake power for a range of engine types and sizes at maximum power. This ratio decreases with increasing engine speed and with increasing engine size. The smaller diesel engine designs use higher gas velocities to achieve the desired fuel-air mixing rates and have less favorable surface/volume ratios (see Sec. 10.2).

An overall first law energy balance for an engine provides useful information on the disposition of the initial fuel energy. For a control volume which surrounds the engine (see Fig. 3-8), the steady-flow energy-conservation equation is

$$\dot{m}_f h_f + \dot{m}_a h_a = P_b + \dot{Q}_{cool} + \dot{Q}_{misc} + (\dot{m}_f + \dot{m}_a)h_e \qquad (12.8)$$

where P_b is the brake power, $\dot{Q}_{cool}$ is the heat-transfer rate to the cooling medium, $\dot{Q}_{misc}$ is the heat rejected to the oil (if separately cooled) plus convection and radiation from the engine's external surface. It proves convenient to divide the exhaust enthalpy h_e into a sensible part $h_{e,s} = h_e(T) - h_e(298 \text{ K})$, plus the exhaust reference state enthalpy (see Sec. 4.5). Then Eq. (12.8) can be written:

$$P_b + \dot{Q}_{cool} + \dot{Q}_{misc} + \dot{H}_{e,ic} + \dot{m}h_{e,s} = \dot{m}_f Q_{LHV} \qquad (12.9)$$

where $\dot{H}_{e,ic}$ represents the exhaust enthalpy loss due to incomplete combustion. Typical values of each of these terms relative to the fuel flow × heating value are given in Table 12.1.

The energy balance within an engine is more complicated and is illustrated

TABLE 12.1

Energy balance for automotive engines at maximum power

	P_b	Q_{cool}	Q_{misc}	$H_{e,ic}$	$\dot{m}h_{e,s}$
	(percentage of fuel heating value)				
SI engine	25–28	17–26	3–10	2–5	34–45
Diesel	34–38	16–35	2–6	1–2	22–35

Sources: From Khovakh,[3] Sitkei,[4] and Burke *et al.*[5]

in the energy flow diagram in Fig. 12-3. The indicated power is the sum of the brake power and the friction power. A substantial part of the friction power (about half) is dissipated between the piston and piston rings and cylinder wall and is transferred as thermal energy to the cooling medium. The remainder of the friction power is dissipated in the bearings, valve mechanism, or drives auxiliary devices, and is transferred as thermal energy to the oil or surrounding environment (in Q_{misc}). The enthalpy initially in the exhaust gases can be subdivided into the following components: a sensible enthalpy (60 percent), an exhaust kinetic energy (7 percent), an incomplete combustion term (20 percent), and a heat transfer to the exhaust system (12 percent) (part of which is radiated

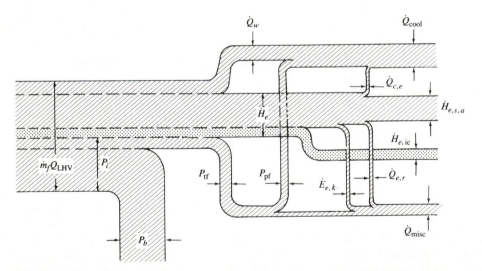

FIGURE 12-3
Energy flow diagram for IC engine. $(\dot{m}_f Q_{LHV})$ = fuel flow rate × lower heating value, $\dot{Q}_w$ = heat-transfer rate to combustion chamber wall, $\dot{H}_e$ = exhaust gas enthalpy flux, P_b = brake power, P_{tf} = total friction power, P_i = indicated power, P_{pf} = piston friction power, $\dot{Q}_{cool}$ = heat-rejection rate to coolant, $\dot{Q}_{c,e}$ = heat-transfer rate to coolant in exhaust ports, $\dot{H}_{e,s,a}$ = exhaust sensible enthalpy flux entering atmosphere, $\dot{H}_{e,ic}$ = exhaust chemical enthalpy flux due to incomplete combustion, $\dot{Q}_{e,r}$ = heat flux radiated from exhaust system, $\dot{E}_{e,k}$ = exhaust kinetic energy flux, $\dot{Q}_{misc}$ = sum of remaining energy fluxes and transfers.

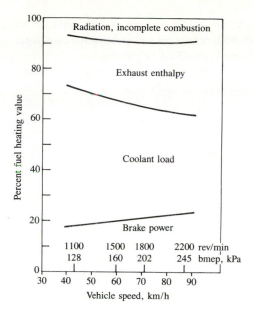

FIGURE 12-4
Brake power, coolant load, sensible exhaust enthalpy, and miscellaneous energy transfers as percent of fuel flow × heating value for spark-ignition engine at road-load operating conditions.[6]

to the environment and the remainder ends up in the cooling medium).† Thus the heat carried away by the coolant medium consists of heat transferred to the combustion chamber walls from the gases in the cylinder, heat transferred to the exhaust valve and port in the exhaust process, and a substantial fraction of the friction work.

At part-load, a greater fraction of the fuel heating value is absorbed into the coolant. Figure 12-4 shows data for a six-cylinder SI engine operated at road-load over a range of vehicle speeds. At low speeds and loads, the coolant heat-transfer rate is 2 to 3 times the brake power.

Although the heat losses are such a substantial part of the fuel energy input, elimination of heat losses would only allow a fraction of the heat transferred to the combustion chamber walls to be converted to useful work. The remainder would leave the engine as sensible exhaust enthalpy. Consider this example for an automotive high-speed naturally aspirated CI engine with a compression ratio of 15. The indicated efficiency is 45 percent, and 25 percent of the fuel energy is carried away by the cooling water. Of this 25 percent, about 2 percent is due to friction. Of the remaining 23 percent, about 8 percent is heat loss during combustion, 6 percent heat loss during expansion, and 9 percent heat loss during exhaust. Of the 8 percent lost during combustion about half (or 4 percent of the fuel energy) could be converted into useful work on the piston (see Fig. 5-9). Of the 6 percent heat loss during expansion, about one-third (or 2 percent) could

---

† The percentages are approximate.

have been utilized. Thus, of the 25 percent lost to the cooling system, only about 6 percent could have been converted to useful work on the piston, which would increase the indicated efficiency of the engine from 45 to 51 percent.

For a spark-ignition engine, the conversion to useful work will be lower, because the compression ratio is lower. However, as shown in Fig. 12-4, the heat losses at part-load (an important operating regime for automobile use) are a substantially larger fraction of the fuel heating value. Studies with computer simulations of the SI engine operating cycle indicate that at typical part-load conditions a proportional reduction in combustion-chamber-wall heat losses of 10 percent results in a proportional increase (improvement) in brake fuel conversion efficiency of about 3 percent.[7]

12.4 CONVECTIVE HEAT TRANSFER

12.4.1 Dimensional Analysis

While the overall time-averaged heat transfer to the coolant medium is adequate for some design purposes, the instantaneous heat flux during the engine cycle is a necessary input for realistic cycle calculations (see Sec. 14.4) and provides the fundamental input for obtaining the heat flux distribution to various parts of an operating engine. Equations (12.5), (12.6), and (12.7) provide the framework for calculating the heat flux $\dot{q}$, based on the assumption that at each point in the cycle the heat-transfer process is quasi steady. For example, neglecting radiation, if $\bar{T}_g$, $\bar{T}_{w,g}$, and $h_{c,g}$ can be calculated at each point in the cycle, $q(\theta)$ is obtained. Alternatively, if $\bar{T}_c$, $h_{c,c}$, $\bar{T}_g$, and $h_{c,g}$ are known, $\dot{q}$, $T_{w,g}$, and $T_{w,c}$ can be computed.

Dimensional analysis can be used to develop the functional form of relationships which govern the gas-side heat-transfer coefficient.[8] The engine convective heat-transfer process can be characterized geometrically by a length dimension—say the bore B—and a number of length ratios y_1, y_2, y_3, etc. (of which one will be the axial cylinder length z divided by the bore z/B), which define the cylinder and combustion chamber geometry. The flow pattern, similarly, may be characterized by one chosen velocity v and a set of velocity ratios u_1, u_2, u_3, etc. The gas properties of importance are the thermal conductivity k, the dynamic viscosity μ, the specific heat c_p, and the density ρ. If there is combustion, then the chemical energy release rate per unit volume $\dot{q}_{ch}$ may be important. The engine speed N and relative position in the cycle denoted by crank angle θ introduce the cyclical nature of the process. Thus

$$f(h_c, B, z, y_1, y_2, \ldots, y_m, v, u_1, u_2, \ldots, u_n, k, \mu, c_p, \rho, \dot{q}_{ch}, N, \theta) = 0$$

Applying dimensional analysis, with mass, length, time, and temperature as the independent dimensions, reduces the variables to four-fewer nondimensional groups:

$$F\left(\frac{h_c B}{k}, \frac{\rho v B}{\mu}, \frac{c_p \mu}{k}, \frac{c_p T}{v^2}, \frac{NB}{v}, \frac{\dot{q}_{ch}}{\rho c_p NT}, \frac{z}{B}, y_1, \ldots, y_m, u_1, \ldots, u_n, \theta\right) = 0$$

The first three groups are the familiar Nusselt, Reynolds, and Prandtl numbers, respectively. The next has the nature of a Mach number since $c_p T$ is proportional to the square of the sound speed. For Mach numbers much less than 1, the Mach number dependence is known to be small and can be omitted. It is usual to take for v the mean piston speed $\bar{S}_p = 2LN$. Then, by introducing the bore/stroke ratio B/L, the term NB/v is eliminated. z/B is a function of the compression ratio r_c, the ratio of connecting rod to crank radius $R = l/a$, and θ. Thus

$$F\left(\frac{h_c B}{k}, \frac{\rho \bar{S}_p B}{\mu}, \frac{c_p \mu}{k}, \frac{B}{L}, \frac{\dot{q}_{ch}}{\rho c_p N T}, r_c, R, y_1, \ldots, y_m, \theta, u_1, \ldots, u_n\right) = 0 \quad (12.10)$$

The dimensionless groups may be varied (but not reduced in number) by combination. While Eq. (12.10) reveals nothing about the functional form of the relationship between the groups, it does provide a basis for evaluating the correlations which have been proposed.

Many formulas for calculating instantaneous engine heat-transfer coefficients have been proposed (see Ref. 9 for a review). Only those with a functional form which fits Eq. (12.10) will be summarized here. The basis of these correlations is the assumption that the Nusselt, Reynolds, and Prandtl number relationship follows that found for turbulent flow in pipes or over flat plates:

$$\mathrm{Nu} = a \, \mathrm{Re}^m \, \mathrm{Pr}^n \qquad (12.11)$$

Distinctions should be made between correlations intended to predict the *time-averaged* heat flux to the combustion chamber walls, the *instantaneous spatially averaged* heat flux to the chamber walls (which is required for engine performance analysis), and the *instantaneous local* heat fluxes (which are not uniform over the combustion chamber and may be required for thermal stress calculations). In using these heat-transfer correlations, the critical choices to be made are (1) the velocity to be used in the Reynolds number; (2) the gas temperature at which the gas properties in Eq. (12.11) are evaluated; and (3) the gas temperature used in the convective heat-transfer equation (12.2).

The most widely used correlations and the basis of their derivation will now be summarized. Because the experimental data for evaluating these correlations in CI engines includes both convective and radiative heat fluxes, comparison of these correlations with data is deferred to Sec. 12.6.

12.4.2 Correlations for Time-Averaged Heat Flux

Taylor and Toong[10] have correlated overall heat-transfer data from 19 different engines. It was assumed that coolant and wall temperatures varied little between designs and that the effects of geometrical differences were small. Thus, at a given fuel/air ratio, the convective part of the heat flux should correlate with Reynolds number. To allow for variations in fuel/air ratio, Taylor and Toong defined an average effective gas temperature $T_{g,a}$ such that

$$\int Ah_c(T - T_{g,a})d\theta = 0$$

over the engine cycle. $T_{g,a}$ is the temperature at which the wall would stabilize if no heat was removed from outside. $T_{g,a}$ was obtained by extrapolating average heat-transfer data plotted versus gas-side combustion chamber surface temperature back to the zero heat-transfer axis. The Nusselt number, defined as

$$\text{Nu} = \frac{\bar{Q}B}{(\pi B^2/4)(T_{g,a} - T_c)k_g} = \frac{4\bar{Q}}{\pi Bk_g(T_{g,a} - T_c)} \tag{12.12}$$

plotted against Reynolds number, defined as

$$\text{Re} = \frac{\dot{m}B}{\mu_g(\pi B^2/4)} = \frac{4\dot{m}}{\pi \mu_g B} \tag{12.13}$$

where $\dot{m}$ is the charge mass flow rate, is shown in Fig. 12-5. Taylor and Toong proposed a power law of 0.75. Annand[8] suggests three separate lines for the three different types of engines covered, with slope 0.7. The diesel line is about 25 percent higher than the spark-ignition engine line (which corresponds in part to the radiative heat flux component present in diesels). The air-cooled engine line is lower than the liquid-cooled line, presumably because surface temperatures are higher. The average gas temperature values developed by Taylor and Toong are shown in the insert in Fig. 12-5.

12.4.3   Correlations for Instantaneous Spatial Average Coefficients

Annand[8] developed the following convective heat-transfer correlation to match previously published experimental data on instantaneous heat fluxes to selected cylinder head locations:

$$\left(\frac{h_c B}{k}\right) = a\left(\frac{\rho \bar{S}_p B}{\mu}\right)^b \tag{12.14}$$

The value of a varied with intensity of charge motion and engine design. With normal combustion, $0.35 \leq a \leq 0.8$ with $b = 0.7$, and a increases with increasing intensity of charge motion. Gas properties are evaluated at the cylinder-average charge temperature $\bar{T}_g$:

$$\bar{T}_g = \frac{pVM}{m\tilde{R}} \tag{12.15}$$

The same temperature is used in Eq. (12.2) to obtain the convective heat flux.

Note that in developing this correlation, the effects of differences in geometry and flow pattern between engines [the ratios $y_1, \ldots, y_m$ and $u_1, \ldots, u_n$ in Eq. (12.10)] have been incorporated in the proportionality constant a, and the

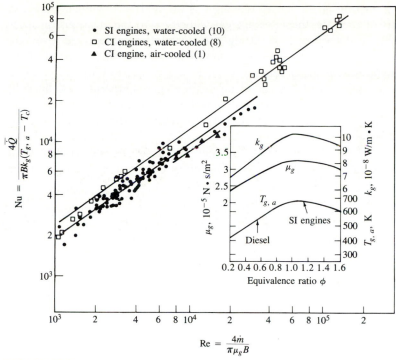

FIGURE 12-5
Overall engine heat-transfer correlation: gas-side Nusselt number versus Reynolds number for different types of IC engines. See text for definition of symbols. Insert gives effective gas temperature (wall
temperature for adiabatic operation), gas viscosity μ_g and thermal conductivity k_g. Lines have slope
0.7.[8, 10]

effect of chemical energy release is omitted. While only data from cylinder head
thermocouple locations were used as a basis for this correlation, it has often been
used to estimate instantaneous spatial average heat fluxes for the entire combustion chamber.

Woschni[11] assumed a correlation of the form

$$\text{Nu} = 0.035\,\text{Re}^m \qquad (12.16)$$

With the cylinder bore B taken as the characteristic length, with w as a local
average gas velocity in the cylinder, and assuming $k \propto T^{0.75}$, $\mu \propto T^{0.62}$, and
$p = \rho RT$, the above correlation can be written

$$h_c = CB^{m-1}p^m w^m T^{0.75-1.62m} \qquad (12.17)$$

During intake, compression, and exhaust, Woschni argued that the average
gas velocity should be proportional to the mean piston speed. During combustion and expansion, he attempted to account directly for the gas velocities
induced by the change in density that results from combustion (~ 10 m/s), which
are comparable to mean piston speeds. Thus a term proportional to the pressure

rise due to combustion $(p - p_m)$ was added (p_m is the motored cylinder pressure). The coefficients relating the local average gas velocity w to the mean piston speed and $(p - p_m)$ were determined by fitting the correlation, integrated over the engine cycle, to time-averaged measurements of heat transfer to the coolant for a wide range of engine operating conditions for a direct-injection four-valve diesel without swirl. T in Eq. (12.17) is the mean cylinder gas temperature defined by Eq. (12.15); the same temperature is used to obtain the heat flux from the heat-transfer coefficient h_c. Thus this correlation represents *spatially averaged* combustion chamber heat fluxes.

The average cylinder gas velocity w (meters per second) determined for a four-stroke, water-cooled, four-valve direct-injection CI engine without swirl was expressed as follows:

$$w = \left[C_1 \bar{S}_p + C_2 \frac{V_d T_r}{p_r V_r} (p - p_m) \right] \qquad (12.18)$$

where V_d is the displaced volume, p is the instantaneous cylinder pressure, p_r, V_r, T_r are the working-fluid pressure, volume, and temperature at some reference state (say inlet valve closing or start of combustion), and p_m is the motored cylinder pressure at the same crank angle as p.

For the gas exchange period: $C_1 = 6.18,$ $C_2 = 0$
For the compression period: $C_1 = 2.28,$ $C_2 = 0$
For the combustion and expansion period: $C_1 = 2.28,$ $C_2 = 3.24 \times 10^{-3}$

Subsequent studies in higher-speed engines with swirl indicated higher heat transfer than these velocities predicted. For engines with swirl, cylinder averaged gas velocities were given by Eq. (12.18) with:

For the gas exchange period: $C_1 = 6.18 + 0.417 \dfrac{v_s}{\bar{S}_p}$

For the rest of cycle: $C_1 = 2.28 + 0.308 \dfrac{v_s}{\bar{S}_p}$

where $v_s = B\omega_p/2$ and ω_p is the rotation speed of the paddle wheel used to measure the swirl velocity (see Sec. 8.3.1).[12] Spark-ignition engine tests showed that the above velocities gave acceptable predictions for this type of engine also.[13]

Woschni's correlation, with the exponent in Eq. (12.17) equal to 0.8, can be summarized as:

$$h_c(\text{W/m}^2 \cdot \text{K}) = 3.26B(\text{m})^{-0.2} p(\text{kPa})^{0.8} T(\text{K})^{-0.55} w(\text{m/s})^{0.8} \qquad (12.19)$$

with w defined above.

Hohenberg[14] examined Woschni's formula and made changes to give better predictions of time-averaged heat fluxes measured with probes in a direct-injection diesel engine with swirl. The modifications include use of a length based on instantaneous cylinder volume instead of bore, changes in the effective gas velocity, and in the exponent of the temperature term.

12.4.4   Correlations for Instantaneous Local Coefficients

LeFeuvre *et al.*[15] and Dent and Sulaiman[16] have proposed the use of the flat-plate forced convection heat-transfer correlation formula

$$\left(\frac{h_c l}{k}\right) = 0.036\left(\frac{\rho v l}{\mu}\right)^{0.8}\left(\frac{\mu c_p}{k}\right)^{0.333} \tag{12.20}$$

where l is the length of the plate and v the flow velocity over the plate. This formula has been applied to DI diesel engines with swirl, with l and v evaluated at a radius r as

$$l = 2\pi r \qquad v = r\omega$$

ω being the solid-body angular velocity of the charge. The heat flux at any radius r (with $Pr = 0.73$) is then given by

$$\dot{q}(r) = 0.023 \frac{k}{r}\left(\frac{\omega r^2}{v}\right)^{0.8}[\bar{T}_g(r) - T_w(r)] \tag{12.21}$$

This equation can be used if the swirl variation with crank angle is known (see Sec. 8.3) and an appropriate local gas temperature can be determined. It would not be consistent to use the cylinder average gas temperature given by Eq. (12.15) because during combustion substantial temperature nonuniformities exist between burned gases and air or mixture which has yet to burn or mix with already burned gas.

   An alternative approach is zonal modeling, where the combustion chamber is divided into a relatively small number of zones each with its own temperature, heat-transfer coefficient, and heat-transfer surface area history. This approach has been applied to spark-ignition engines (e.g., Ref. 17), where the division of the in-cylinder gases during combustion into a higher-temperature burned gas region behind the flame and lower-temperature unburned gas region ahead of the flame is clear (see Fig. 9-4). The heat transfer to the combustion chamber surfaces in contact with the unburned and burned gas zones [analagous to Eq. (12.2)] is given by

$$\dot{Q}_u = A_{u,w}\,h_{c,u}(T_u - T_w) \qquad \dot{Q}_b = A_{b,w}\,h_{c,b}(T_b - T_w) \tag{12.21a,b}$$

respectively. Since h_c depends on local gas properties and velocities, $h_{c,u}$ and $h_{c,b}$ are not necessarily the same. Examples of how the burned gas wetted areas on the piston, cylinder head, and liner vary during the combustion process are given in Fig. 14-8. Since the burned gas temperature T_b is much larger than the unburned gas temperature, the heat flux from the burned gas zone dominates.

   One useful development of this two-zone approach is the division of the burned gas zone into an adiabatic core and a thermal boundary layer. The advantages are: (1) this corresponds more closely to the actual temperature distribution (see Sec. 12.6.5); (2) a model for the boundary-layer flow provides a more fundamental basis for evaluating the heat-transfer coefficient. The local heat flux is then given by[18]

$$\dot{q} = \frac{k_e(T_{ac} - T_w)}{\delta} \tag{12.22}$$

where k_e is the effective thermal conductivity in the boundary layer, T_{ac} is the adiabatic core temperature, and δ the boundary-layer thickness. In the laminar regime δ would grow as $t^{1/2}$; in the turbulent regime δ would grow as $t^{0.8}$. Both growth regimes are observed.

Zonal models have also been used to describe DI diesel engine heat transfer.[19] A bowl-in-piston chamber was divided into three flow regions and two gas-temperature zones during combustion. An effective velocity

$$w = (U_x^2 + U_y^2 + 2k)^{1/2}$$

was used to obtain the heat-transfer coefficient, where U_x and U_y are the two velocity components parallel to the surface outside the boundary layer and k is the turbulent kinetic energy. Zonal models would be expected to be more accurate than global models. However, only limited validation has been carried out.

12.4.5   Intake and Exhaust System Heat Transfer

Convective heat transfer in the intake and exhaust systems is driven by much higher flow velocities than in-cylinder heat transfer. Intake system heat transfer is usually described by steady, turbulent pipe flow correlations.[9] With liquid fuel present in the intake, the heat-transfer phenomena become especially complicated (see Sec. 7.6). Exhaust flow heat-transfer rates are the largest in the entire cycle due to the very high gas velocities developed during the exhaust blowdown process and the high gas temperature (see Sec. 6.5). Exhaust system heat transfer is important since it affects emissions burnup in the exhaust system, catalyst, or particulate trap, it influences turbocharger performance, and it contributes significantly to the engine cooling requirements.

The highest heat-transfer rates occur during blowdown, to the exhaust valve and port. Detailed exhaust port convective heat-transfer correlations have been developed and tested. These are based on Nusselt-Reynolds number correlations. For the valve open period, relations of the form

$$\text{Nu} = K \, \text{Re}_j^n \tag{12.23}$$

have been proposed and evaluated.[20] For $L_v/D_v \lesssim 0.2$, the flow exits the valve as a jet, and $\text{Re}_j = v_j D_v/v$, where D_v is the valve diameter, v_j the velocity of the exhaust gases through the valve opening, and v the kinematic velocity. For $L_v/D_v \gtrsim 0.2$, the port is the limiting area and a pipe flow model with $\text{Re} = v_p D_p/v$ is more appropriate. v_p is the velocity in the port and D_p is the port diameter. For the valve closed period, the correlation

$$\text{Nu} = 0.022 \, \overline{\text{Re}}_D^{0.8} \tag{12.24}$$

was developed. Here $\overline{\mathrm{Re}}_D = \bar{v}_p D_p / v$, where $\bar{v}_p$ is the time-averaged exhaust port gas velocity and D_p is the port diameter. For straight sections of exhaust pipe downstream of the port, an empirical correlation based on measurements of average heat-transfer rates to the pipe has been derived:[21]

$$\mathrm{Nu} = 0.0483 \, \mathrm{Re}^{0.783} \qquad (12.25)$$

The Reynolds number is based on pipe diameter and velocity. Heat-transfer correlations for steady developing turbulent flow in pipes predict values about half that given by Eq. (12.25).

12.5 RADIATIVE HEAT TRANSFER

There are two sources of radiative heat transfer within the cylinder: the high-temperature burned gases and the soot particles in the diesel engine flame. In a spark-ignition engine, the flame propagates across the combustion chamber from the point of ignition through previously mixed fuel and air. Although the flame front is slightly luminous (see color plate, Fig. 9-1), all the chemical intermediaries in the reaction process are gaseous. Combustion is essentially complete early in the expansion stroke. In the compression-ignition engine (and in fuel-injected stratified-charge engines), most of the fuel burns in a turbulent diffusion flame as fuel and air mix together. There can be many ignition locations, and the flame conforms to the shape of the fuel spray until dispersed by air motion (see color plate, Figs. 10-4 and 10-5). The flame is highly luminous, and soot particles (which are mostly carbon) are formed at an intermediate step in the combustion process.

The radiation from soot particles in the diesel engine flame is about five times the radiation from the gaseous combustion products. Radiative heat transfer in conventional spark-ignition engines is small in comparison with convective heat transfer. However, radiative heat transfer in diesel engines is not negligible; it contributes 20 to 35 percent of the total heat transfer and a higher fraction of the maximum heat-transfer rate.

12.5.1 Radiation from Gases

Gases absorb and emit radiation in narrow wavelength bands rather than in a continuous spectrum as do solid surfaces. The simpler gas molecules such as H_2, O_2, and N_2 are essentially transparent to radiation. Of the gases important in combustion, CO, CO_2, and H_2O emit sufficient energy to warrant consideration. In gases, emission and absorption will occur throughout the gas volume. These processes will be governed by the number of molecules along the radiation path. For each species, this will be proportional to the product of the species partial pressure p_i and the path length l. In addition the radiative capacity depends on gas temperature T_g. Thus the emissivity of the gas ε_g can be expressed as

$$\varepsilon_g = f(T_g, p_1 l, \ldots, p_n l) \qquad (12.26)$$

The mean path length for a volume V with surface area A is given with sufficient accuracy by

$$l = 0.9 \times \frac{4V}{A} \tag{12.27}$$

$4V/A$ is the mean path length for a hemispherical enclosure.

Standard methods have been developed for estimating ε_g for mixtures of CO_2 and H_2O by Hottel and others (see Ref. 22). Charts based on experimental data give ε_{CO_2} and ε_{H_2O} as functions of $p_i l$ and T_g. Correction factors are applied for total pressures above one atmosphere and for the overlapping of spectral bands of CO_2 and H_2O. Estimates for engine combustion gases at peak conditions give $\varepsilon_g \approx 0.1$ and peak heat fluxes due to gas radiation of order 0.2 MW/ m^2. This amounts to ~ 5 percent of the peak convective heat transfer. Since gas radiation is proportional to T_g^4, this radiative flux falls off more rapidly from peak values than convective heat flux and, when integrated over the cycle, can be neglected.

12.5.2   Flame Radiation

Flame radiation is a much more complex process because the detailed geometry and chemical composition of the radiating region are not well known. Since the radiation from the optically transparent or nonluminous flames of spark-ignition engines is small, we will deal only with luminous nontransparent flames where the radiation comes from incandescent soot particles and has a continuous spectrum. Because the particle size distribution, number density and temperature, and flame geometry in a diesel engine are not well defined, flame emissivities cannot be calculated from first principles. Direct measurements of flame emissivities are required. A number of measurements of the magnitude and spectral distribution of radiation from a diesel engine combustion chamber have been made (see Ref. 9 for a summary). The most extensive of these by Flynn et al.[23] in a direct-injection engine used a monochromator to measure intensity of radiation at seven wavelengths. The viewing path cut through the piston crown into the central region of the bowl-in-piston combustion chamber. Fuel was injected through a five-hole nozzle and some air swirl was provided. At any given crank angle, the distribution of energy over the seven wavelengths was used to reconstruct the complete energy spectrum and to calculate the apparent radiation temperature and optical thickness.

The energy distribution was skewed from that of a grey-body model (for which emissivity is independent of wavelength), and the monochromatic emissivity was well fitted by the equation

$$\varepsilon_\lambda = 1 - \exp\left(\frac{-kl}{\lambda^{0.95}}\right) \tag{12.28}$$

used to describe the emissivity variation from clouds of small particles. Figure 12-6 shows sample results for the monochromatic emissive power and monochro-

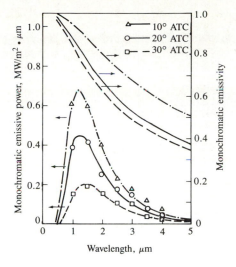

FIGURE 12-6

Variation in monochromatic emissive power and emissivity, with wavelength, at three different crank angles. DI diesel engine with 114 mm bore, 1995 rev/min, overall equivalence ratio 0.46. Radiation from piston bowl measured through cutout in piston crown.[23]

matic emissivity. Equation (12.28), combined with Planck's equation for black-body monochromatic emissive power

$$e_{b,\lambda} = \frac{2\pi K_1}{\lambda^5(e^{K_2/\lambda T_R} - 1)}$$ (12.29)

where $K_1 = 0.59548 \times 10^{-16}$ W·m² and $K_2 = 1.43879$ cm·K, defined an apparent radiation temperature T_R and optical thickness kl for the radiating medium. An apparent grey-body emissivity ε_a was also calculated from the standard equation

$$\varepsilon_a = \frac{\int \varepsilon_\lambda e_{b,\lambda} d\lambda}{\int e_{b,\lambda} d\lambda}$$

Figure 12-7 shows sample results for four equivalence ratios at an engine speed of 2000 rev/min. The radiation flux has approximately the same shape and time span as the net heat-release rate curve (which was determined from the cylinder pressure curve). During the period of maximum radiation, the apparent emissivity is 0.8 to 0.9; it then decreases as the expansion process proceeds.[23] In previous experiments on the same engine, instantaneous total heat fluxes had been measured at various locations on the cylinder head.[24] A comparison of radiant and total heat fluxes (both peak and average) showed that the radiation heat flux can be a substantial fraction of the peak heat flux. The average radiant flux is about 20 percent of the average total flux: the percentage varies significantly with load. These conclusions are supported by other experimental data summarized in the next section.

The radiation or apparent flame temperatures measured in diesels by several investigators show consistent results (see Fig. 12-8). Also included in the figure during the combustion and expansion process are typical values of: (1) the

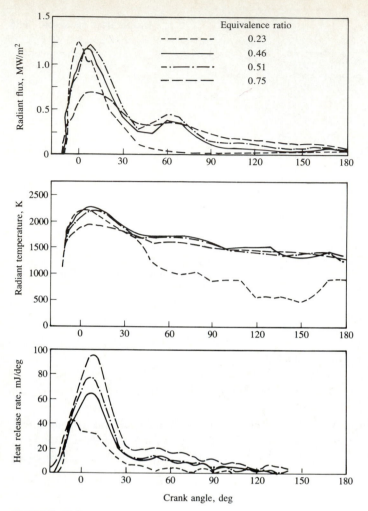

FIGURE 12-7
Radiant heat flux, apparent radiant temperature and net heat-release rate as function of crank angle for DI diesel engine at four different loads. Engine and measurement details as in Fig. 12-6; 2000 rev/min.[23]

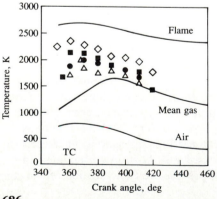

FIGURE 12-8
Apparent radiation temperatures measured in diesel engines, compared to calculated maximum adiabatic flame, cylinder-mean and air temperatures, during the combustion period. Adiabatic flame temperature is temperature attained by burning air at air temperature with fuel for equivalence ratio of 1.1. (*Data from Dent and Sulaiman,*[16] *Flynn et al.,*[23] *Lyn,*[25] *Kamimoto et al.;*[26] *calculated curves from Assanis and Heywood.*[27])

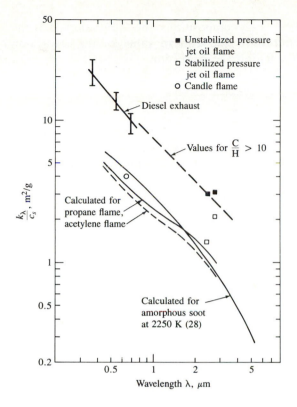

FIGURE 12-9

Measured and calculated values of k_λ/c_s as a function of wavelength λ. k_λ is monochromatic absorption coefficient, c_s is the soot concentration. (*From Field* et al.[28] *and Greeves and Meehan.*[29])

temperature of any air not yet mixed with fuel or burned gases; (2) the average temperature of the cylinder contents; and (3) the maximum possible flame temperature [corresponding to combustion of a slightly rich mixture ($\phi = 1.1$) with air at the temperature shown]. The measured radiation temperatures fall between the maximum flame temperatures and the bulk temperature. Such a model has been proposed, fits the available data, and has been used.[27] Zonal models have been proposed (e.g., the burned gas region is stoichiometric[16, 19]) to define an appropriate flame temperature.

The emissivity of an incandescent soot-burning flame can be calculated from a knowledge of the monochromatic absorption coefficient, which is given by

$$k_\lambda = 36\pi \frac{c_s}{\rho_s} \frac{n^2\kappa}{\lambda[(n^2 + n^2\kappa^2)^2 + 4(n^2 - n^2\kappa^2 + 1)]} = 36\pi \frac{c_s}{\rho_s} f(\lambda, T) \quad (12.30)$$

where n is the refractive index, κ is the absorption term in the complex refractive index, c_s is the soot concentration in kilograms per cubic meter, and ρ_s is the density of the soot particles (≈ 2 g/m³).[28] The absorption coefficient depends on wavelength and temperature, is independent of particle size, and depends only on the soot mass loading. Figure 12-9 shows several estimates of k_λ/c_s as a function of λ. The strong dependence on λ shows that clouds of soot particles are markedly not grey. There is a considerable spread in the different estimates shown.

This could be the result of different soot compositions (the C/H ratio affects the optical properties) and temperatures. To find a mean value for the absorption coefficient of a wide spectrum of radiation, the expression

$$k_a = \frac{\int k_\lambda e_{b,\lambda} \, d\lambda}{\int e_{b,\lambda} \, d\lambda}$$

must be evaluated. For example, for radiation from a black body at temperature 1800 K, the mean absorption coefficient is $\approx 1300c_s$ per meter, where c_s is in kilograms per cubic meter. At higher black-body temperatures, the value of k_a would be higher.

Annand[30] has applied this approach to a diesel engine. The apparent flame emissivity was related to the apparent mean absorptivity by

$$\varepsilon_a = 1 - \exp\left(-k_a l\right) \tag{12.31}$$

For Flynn's data, the peak emissivity is 0.8 which gives $k_a l = 1.6$. Since $l \approx 0.07$ m, this gives $k_a \approx 22$ m^{-1} and $c_s \approx 16$ g/m^3 (≈ 1 g/m^3 at NTP), which is a soot loading comparable with values measured in diesel engines during combustion (see Figs. 11-42 to 11-44).

12.5.3 Prediction Formulas

Well-accepted prediction formulas for radiant heat flux in an engine are not available. Annand has proposed a radiation term of the form

$$\dot{q}_R = \beta\sigma(\bar{T}_g^4 - T_w^4) \tag{12.32}$$

where σ is the Stefan-Boltzmann constant, $\bar{T}_g$ is the mean gas temperature, and T_w is the wall temperature. This term was coupled with a convective heat flux term to give a correlation for predicting total heat flux. In a first evaluation, when coupled with Eq. (12.14), $\beta = 0.6$ was proposed.[8] In a later study with a modified convective heat-transfer correlation,[31] $\beta \approx 1.6$ was proposed. (Note that since the temperature used is the average gas temperature and not the apparent flame temperature, β is not an emissivity.) The limited evaluation of this approach shows that $\beta = 0.6$ gave approximately correct magnitude for $\dot{q}_R$ for one engine and was too low for another.[32] $\beta = 1.6$ gave radiant heat fluxes higher than experimental data.[33]

Flynn et al.[23] developed an empirical expression for instantaneous radiant heat flux to fit their data, of the form

$$\dot{q}_R = 2\bar{\dot{q}}_R \, b(a + 1)\left(\frac{\theta - \theta_s}{360}\right)^a \exp\left[\frac{(\theta - \theta_s)^{a+1}}{360}\right] \tag{12.33}$$

where θ_s is the crank angle at the start of the radiation pulse. Correlations for $\bar{\dot{q}}_R$, a, b, and θ_s in terms of engine speed, manifold pressure, crank angle at the start of fuel injection, and the equivalence ratio were obtained and presented. This correlation gave a reasonable match to Flynn's data, an example of which was shown in Fig. 12-7.

It has been proposed that the apparent absorptivity should be proportional to pressure.[4, 33] The assumption is then made that the proportionality constant would be a unique function of the equivalence ratio and crank angle. Although k_a is dependent on p for gas radiation, it is not clear that the same proportionality should apply to soot radiation.

12.6 MEASUREMENTS OF INSTANTANEOUS HEAT-TRANSFER RATES

12.6.1 Measurement Methods

Values of instantaneous heat flux into the combustion chamber walls have been obtained from measurements of the instantaneous surface temperature. The temperature variation at the wall is a result of the time-varying boundary condition at the gas/wall interface. It is damped out within a small distance (~ 1 mm) from the wall surface, so measurements must be made at the surface. Various types of thermocouple or thermistor have been used.[9] One-dimensional unsteady heat conduction into the wall is then assumed:

$$\frac{\partial T}{\partial t} = \frac{1}{\rho c}\frac{\partial}{\partial x}\left(k\,\frac{\partial T}{\partial x}\right) \tag{12.34}$$

A sinusoidal variation with time of heat flux into a semi-infinite solid can be shown to produce a sinusoidal variation of surface temperature of the same frequency displaced in phase by 90°. The surface temperature T_w is expressed as a Fourier series:

$$T_w = T_m + \sum_{n=1}^{N}\left[A_n \cos{(n\omega t)} + B_n \sin{(n\omega t)}\right] \tag{12.35}$$

where T_m is the time-averaged value of T_w, A_n and B_n are Fourier coefficients, n is a harmonic number, and ω is the angular frequency (radians per second). The boundary conditions are $T = T_w(t)$ at $x = 0$ and $T = T_l$ (constant) at $x = l$. The solution of Eq. (12.35) is[1]

$$T(x, t) = T_m - (T_m - T_l)\frac{x}{l} + \sum_{n=1}^{N} \exp{(-\phi_n x)}F_n(x, t) \tag{12.36}$$

where

$$F_n = A_n \cos{(n\omega t - \phi_n x)} + B_n \sin{(n\omega t - \phi_n x)}$$

and $\phi_n = (n\omega/2\alpha)^{1/2}$, where α is the thermal diffusivity of the wall material $k/(\rho c)$.

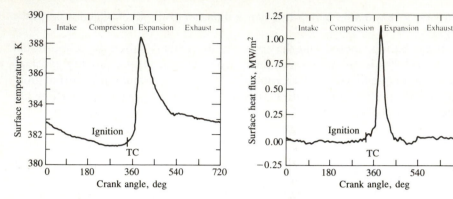

FIGURE 12-10
Surface temperature measured with thermocouple in cylinder head, and surface heat flux calculated from surface temperature, as a function of crank angle. Spark-ignition engine operated at part-load.[34]

The heat flux components at each frequency that caused that variation can be calculated via Fourier's law [Eq. (12.1)], and summed to give the total fluctuation of heat flux with time:

$$\dot{q}_w = \frac{k}{l}(T_m - T_i) + k \sum_{n=1}^{N} \phi_n[(A_n + B_n)\cos(n\omega t) - (A_n - B_n)\sin(n\omega t)] \quad (12.37)$$

Alternative approaches for solving Eq. (12.34) are through use of an electrical analogy to heat flow and by numerical methods. The latter become necessary if wall material properties depend significantly on temperature, as do combustion chamber deposits and some insulating ceramic materials. Several measurements of this type in spark-ignition and diesel engines have been made. A summary of these measurements can be found in Ref. 9.

Radiant heat fluxes are determined by a variety of techniques: e.g., photo-detector and infrared monochromator; thermocouple shielded by a sapphire window; pyroelectric thermal detector.

12.6.2   Spark-Ignition Engine Measurements

Figure 12-10 shows the surface temperature variation with crank angle, and the heat flux variation calculated from it, on the cylinder head of a spark-ignition engine at a part-load low-speed operating condition. The swing in surface temperature at this point (about halfway from the on-the-cylinder-axis spark plug to the cylinder wall) is 7 K. The heat flux rises rapidly when the flame arrives at the measurement location, has its maximum at about the time of peak cylinder pressure when gas temperatures peak (see Section 9.2.1), and then decays to relatively low levels by 60° ATC as expansion cools the burned gases. Peak heat fluxes on the cylinder head of 1.5 to 3 MW/m² were measured over the normal engine speed and load range.[34, 35]

The heat flux profile varies significantly with location and from one cycle to the next. Figure 12-11 illustrates both these effects. When the rapid rise in heat flux occurs depends on the flame arrival time at the measurement location. Thus the heat fluxes determined from surface temperature data averaged over many cycles show their rapid rise later, the further the distance from the spark plug location (Fig. 12-11a). The individual cycle data in Fig. 12-11b show how variations in the flame arrival time from one cycle to the next essentially shift the rising portion of the heat flux profile in time. Note that due to this cyclic variation, the average profile shows a less rapid rise in heat flux than do individual cycles.

Such measurements show that increasing engine speed and increasing engine load increase the surface heat flux. Retarding timing delays the rise in heat

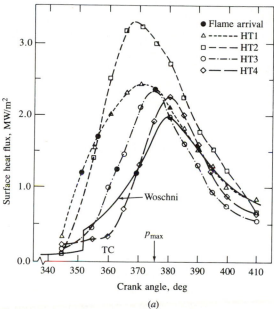

(a)

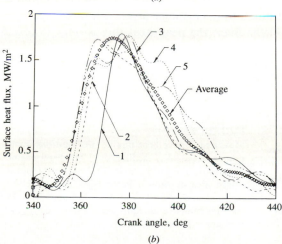

(b)

FIGURE 12-11

(a) Variation of surface heat flux with crank angle at four temperature measurement locations in the cylinder head of a spark-ignition engine. Each curve is an average over many cycles. Distances from on-axis spark plug are: HT1, 18.7 mm; HT2, 27.5 mm; HT3, 37.3 mm; HT4, 46.3 mm. Bore = 104.7 mm, 2000 rev/min, part-load η_v = 40 percent, A/F = 18, MBT timing. Solid curve shows heat flux predicted by Woschni's correlation.[34] (b) Heat flux histories for five consecutive individual cycles and 198-cycle average at location HT1. 1500 rev/min, A/F = 18, η_v = 40 percent, MBT timing.[35]

and reduces the peak value. Maximum heat fluxes occur with close-to-stoichiometric mixtures.[34, 35] All these trends would be expected from the variations in burned gas temperature that result from these changes in engine operation.

12.6.3   Diesel Engine Measurements

Measurements of instantaneous heat fluxes in diesel engines show similar features. The heat flux distribution is usually highly nonuniform. There may also be significant variations between the heat flux profiles from individual cycles. Figure 12-12 shows surface heat fluxes at two locations in a medium-swirl DI diesel, one over the piston bowl (higher heat flux) and the other over the piston squish area, in relation to the heat-release profile. The heat flux increases rapidly once combustion starts, reaches a maximum at close to the time of maximum cylinder pressure, and decreases to a low value by 40 to 60° ATC. Peak heat fluxes to the primary combustion chamber surfaces (the piston bowl and head directly above the bowl) are of order 10 MW/m².

In smaller diesel engines with swirl, the mean heat flux to the piston within the piston bowl is usually higher than the mean heat fluxes to the cylinder head and the annular squish portion of the piston crown (by about a factor of two).[16, 31] This would be expected since the piston bowl is the zone where most of the combustion takes place and gas velocities are highest. There are, in addition, substantial variations in heat flux at different locations within the piston bowl, on the head, and on the annular region of the piston crown.

In contrast to the above results obtained on smaller high-speed (~ 10-cm bore) diesels with swirl with deep bowl-in-piston combustion chambers of diameter about half the piston diameter, results from tests on a medium-speed 30-cm bore quiescent shallow-bowl piston direct-injection supercharged diesel showed a much more uniform heat flux distribution over the combustion chamber walls.[37]

Heat fluxes to the cylinder liner are much lower still (an order of magnitude less than the peak flux to the combustion chamber surface) and are also nonuniform. Figure 12-13 compares heat fluxes to the cylinder head with three locations along the liner. Even at the top of the liner, the peak heat flux is only 15 to 20

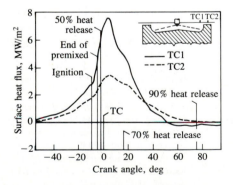

FIGURE 12-12
Measured surface heat flux at two locations in the cylinder head of a medium-swirl DI diesel engine. TC1 above the piston bowl, TC2 above the piston squish area as shown. Percentages of heat release are indicated. Bore = stroke = 114 mm, $r_c = 16$, 2000 rev/min, overall equivalence ratio = 0.5, intake pressure = 1.5 atm.[36]

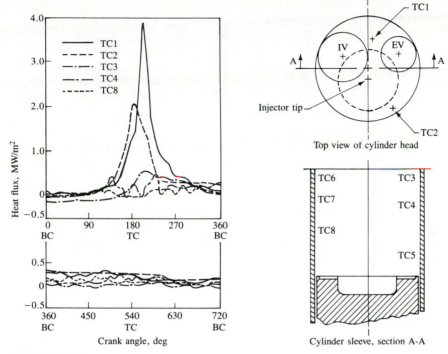

FIGURE 12-13
Measured surface heat fluxes at different locations in cylinder head and liner of naturally aspirated four-stroke cycle DI diesel engine. Bore = stroke = 114 mm, 2000 rev/min, overall equivalence ratio = 0.45.[15]

percent of the flux to the primary combustion chamber surfaces. Again this would be expected, since the combustion gases do not contact the lower parts of the cylinder wall until later in the expansion stroke when their temperature is much below the peak value.

Figure 12-14 shows examples of radiant heat flux measured above the piston bowl of a medium-swirl DI diesel engine as a function of engine speed and load.[16] The limited radiant heat flux data available exhibit the following trends. The rapid increase in radiant heat flux following combustion is delayed relative to the start of pressure rise due to combustion (by about 5°); this delay increases with increasing speed. The peak radiant flux remains approximately constant with increasing equivalence ratios up to $\phi \approx 0.5$. Further increases in the equivalence ratio produce a drop in level of radiant flux. The time-averaged radiant heat flux increased approximately linearly with increasing manifold pressure; however, peak radiative flux levels remained essentially unchanged. Peak and time-averaged values of the radiant heat flux decreased as injection timing was retarded.

In diesel engines, the relative importance of radiant heat transfer (as a percentage of the total heat flux) depends on the location on the combustion

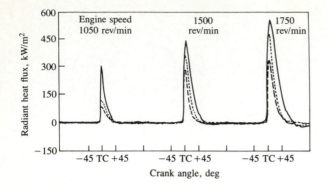

FIGURE 12-14
Measured radiant heat flux to cylinder head above the piston bowl in a high-swirl DI diesel engine when load and speed are varied. Solid curve: 80 percent load. Dotted curve: 40 percent load. Dashed curve: no load.[16]

chamber surface, crank angle, engine load, engine size, and engine design. The time-averaged radiant heat transfer increases as a proportion of the total heat transfer, with increasing load, as indicated in Fig. 12-15.[9] At high load, the total radiant heat flux is between 25 and 40 percent of the total time-averaged heat flux.

12.6.4 Evaluation of Heat-Transfer Correlations

The convective (or combined convective plus radiative) heat flux correlations have been compared with instantaneous engine heat-transfer measurements. One difficulty in this evaluation is the determination of spatially averaged combustion-chamber heat fluxes from the experimental data for comparison with correlations intended to predict the mean chamber heat flux as a function of crank angle. The area-averaged instantaneous heat flux prediction using Woschni's equation (12.19) for the spark-ignition engine conditions shown in Fig. 12-11a is comparable in magnitude to, though lower than, the measured heat fluxes to the cylinder head.

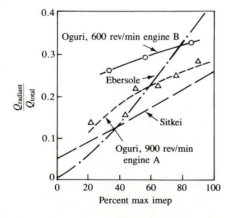

FIGURE 12-15
Radiant heat flux as fraction of total heat flux over the load range of several different diesel engines.[9]

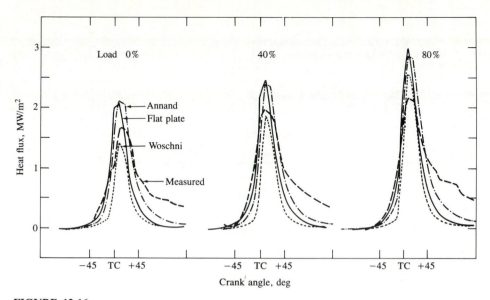

FIGURE 12-16
Comparison of measured mean heat fluxes on the cylinder head at 1050 rev/min in a fired high-swirl DI diesel engine with various prediction equations. Annand, Eq. (12.14), Woschni, Eq. (12.19), flat plate, Eq. (12.20) using measured gas motion.[16]

More extensive comparisons have been made of predictions with data for diesel engines. Annand's correlations, Eqs. (12.14) and (12.32) with $a = 0.06$, $b = 0.85$, and β (for the combustion phase only) $= 0.57$, gave reasonable agreement with instantaneous cylinder head heat fluxes, and overall time-averaged heat fluxes for a medium-speed quiescent DI engine design.[37] In a small high-speed diesel with swirl, values of $a = 0.13$, $b = 0.7$, and $\beta = 1.6$ gave an approximate fit to estimates of the instantaneous area-mean heat flux[31] and time-averaged heat flux to the piston.[38] Comparisons of the Annand and Woschni correlations generally show that the Annand correlation predicts higher heat fluxes at the same crank angle.[11, 24] The most careful comparison of all three correlations summarized in Secs. 12.4.3 and 12.4.4 with experimental data has been made by Dent and Sulaiman in a small high-speed diesel engine with swirl.[16] The mean experimental heat flux was estimated from a number of thermocouple measurements located at different points around the combustion chamber surface. Figure 12-16 shows the comparison. The Woschni correlation falls below the others at light load because the combustion-induced velocity term is smaller. Expansion stroke heat fluxes are underpredicted by all three correlations. Given the uncertainty in converting the measurements to an average heat flux value, the agreement is reasonably good.

Dent examined additional modifications to the flat plate formula [Eq. (12.20)], which was based on a cylinder-mean gas temperature. During combustion a two-zone model is more appropriate. Assuming an equivalence ratio

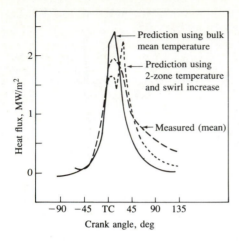

FIGURE 12-17
Measured heat fluxes on cylinder head in fired high-swirl DI diesel engine at 1050 rev/min and 40 percent load compared with predictions based on bulk mean gas temperature and using temperatures based on two-zone (air and burned gas) model.[16]

for the burned gas (Dent assumed stoichiometric), a combustion zone temperature can be determined from the relation

$$\bar{T}_b = \frac{m\bar{T}_g - m_a T_a}{m_b}$$

where m_a is the mass of air, T_a is the air temperature, and m_b is the burned gas mass, $m_b = m - m_a$. In addition, the observed swirl enhancement which occurs due to the combustion was included by multiplying the swirl velocity used in the heat-transfer correlation by the square root of the ratio (density of air in the motored case)/(density of burned gas in fired case). The combination of both effects (see Fig. 12-17) improves the shape of the predictions by broadening and lowering the peak.

Each of the convective-heat-transfer correlations described has limited experimental support. However, under engine design and operating conditions different from those under which they were derived, the predictions should be viewed with caution. Woschni's correlation is the correlation used most extensively for predicting spatially averaged instantaneous convective heat fluxes. However, the empirical constants which relate the mean piston speed and combustion-induced motion to the velocity used in the Reynolds number, determined by Woschni, will not necessarily apply to all the different types of engine. If local velocities are known, the flat-plate-based correlations provide the best available approach. Annand's correlation has the advantage of being the simplest correlation to use. Since the radiation component in diesel engine heat transfer is normally 20 to 40 percent of the total an approximate estimate of its value may suffice.

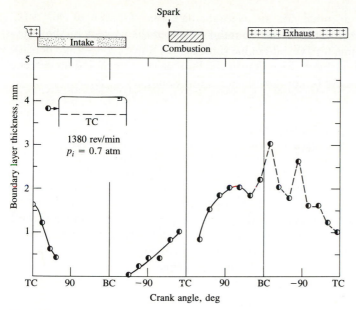

FIGURE 12-18
Thermal boundary-layer thickness, at the top of the cylinder wall in the clearance volume, determined from schlieren photographic measurements in a special visualization square-piston spark-ignition engine.[39]

12.6.5 Boundary-Layer Behavior

Measurements of thermal boundary-layer thickness in an operating spark-ignition engine have been made using schlieren photography in a special flow-visualization engine. Figure 12-18 shows one set of measurements on the cylinder wall in the clearance volume opposite the spark plug. The boundary-layer thickness decreases during intake and increases steadily during compression and expansion to about 2 mm. It stops growing and becomes unstable during the exhaust process, separating from the cylinder wall and becoming entrained into the bulk gas leaving the cylinder. The thickness of the thermal boundary layer varies substantially at different locations throughout the chamber. While the trends with crank angle were similar, the layers on the cylinder head and piston crown were substantially (up to 2 to 3 times) thicker during compression and expansion in the simple disc-shaped chamber studied.[39] This different behavior probably results because there is no bulk flow adjacent to the head and piston crown.

Estimates of thermal boundary-layer thickness in spark-ignition engines, based on convective-heat-transfer correlations and thermal energy conservation for the growing layer give thicknesses comparable to these measurements. Note that a substantial fraction of the cylinder mass is contained within the thermal boundary layer. For example, for an average thickness of 3 mm at 90° ATC during expansion, the volume of the boundary layer is 20 percent of the combustion chamber volume for typical engine dimensions. Since the average density in the boundary layer is about twice that in the bulk gases, some 30 to 40 percent of the cylinder mass would be contained within this boundary layer.

12.7 THERMAL LOADING AND COMPONENT TEMPERATURES

The heat flux to the combustion chamber walls varies with engine design and operating conditions. Also, the heat flux to the various parts of the combustion chamber is not the same. As a result of this nonuniform heat flux and the different thermal impedances between locations on the combustion chamber surface and the cooling fluid, the temperature distribution within engine components is nonuniform. This section reviews the variation in temperature and heat flux in the components that comprise the combustion chamber.

12.7.1 Component Temperature Distributions

Figures 12-19 to 12-22 show illustrative examples of measured temperature distributions in various engine components. Normally, the heat flux is highest in the center of the cylinder head, in the exhaust valve seat region, and to the center of the piston. It is lowest to the cylinder walls. Cast-iron pistons run about 40 to 80°C hotter than aluminum pistons. With flat-topped pistons (typical of spark-ignition engines) the center of the crown is hottest and the outer edge cooler by 20 to 50°C. Diesel engine piston crown surface temperatures are about 50°C higher than SI engine equivalent temperatures. As shown in Fig. 12-19, the maximum piston temperatures with DI diesel engine pistons are at the lip of the

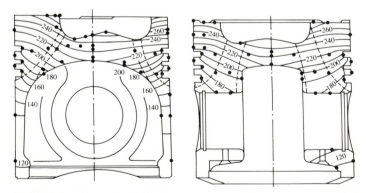

FIGURE 12-19
Isothermal contours (solid lines) and heat flow paths (dashed lines) determined from measured temperature distribution in piston of high-speed DI diesel engine. Bore 125 mm, stroke 110 mm, $r_c = 17$, 3000 rev/min, and full load.[40]

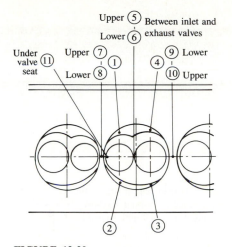

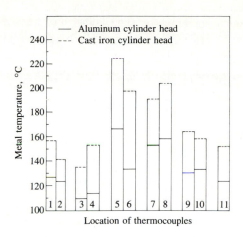

FIGURE 12-20
Variation of cylinder head temperature with measurement location in spark-ignition engine operating at 2000 rev/min, wide-open throttle with coolant water at 95°C and 2 atm.[41]

bowl. In IDI diesel engines, maximum piston temperatures occur where the pre-chamber jet impinges on the piston crown.

Figure 12-20 shows the temperatures at various locations on a four-cylinder SI engine cylinder head. The maximum temperatures occur where the heat flux is high and access for cooling is difficult. Such locations are the bridge between the valves and the region between the exhaust valves of adjacent cylinders. Figure 12-21 shows how the average heat flux and temperature vary along the length of a DI diesel engine liner. Because the lower regions of the liner are only exposed to combustion products for part of the cycle after significant gas expansion has occurred, the heat flux and temperature decrease significantly with distance from

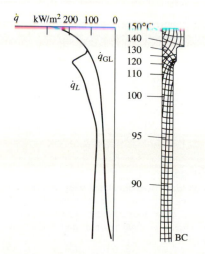

FIGURE 12-21
Temperature and heat flux distribution in the cylinder liner of a high-speed DI diesel engine at 1500 rev/min and bmep = 11 bar. $\dot{q}_L$ is heat flux into the liner; $\dot{q}_{GL}$ is heat flux from the gas to the liner. Difference is friction-generated heat flux.[42]

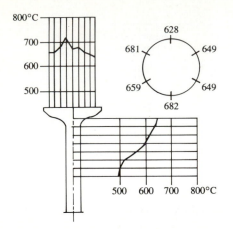

FIGURE 12-22
Temperature distribution in one of the four exhaust valves of a two-stroke-cycle uniflow DDA 4-53 DI diesel engine. Bore = 98 mm, stroke = 114 mm.[43]

the cylinder head. Note that the heat generated by friction between the piston and the liner, the difference between $\dot{q}_{GL}$ (the gas to liner heat flux) and $\dot{q}_L$ (the total heat flux into the liner), is a significant fraction of the liner thermal loading.

The exhaust valve is cooled through the stem and the guide, and the valve seat. In small-size valves the greater part of the heat transfer occurs through the stem; with large-size valves, the valve seat carries the higher thermal load.

Temperature distributions in engine components can be calculated from a knowledge of the heat fluxes across the component surface using finite element analysis techniques. For steady-state engine operation, the depth within a component to which the unsteady temperature fluctuations (caused by the variations in heat flux during the cycle) penetrate is small, so a quasi-steady solution is satisfactory. Results from such calculations for a spark-ignition engine piston illustrate the method.[44] A mean heat-transfer coefficient from the combustion chamber gases to the piston crown and a mean chamber gas temperature were defined (using the input from a cycle simulation of the type described in Sec.

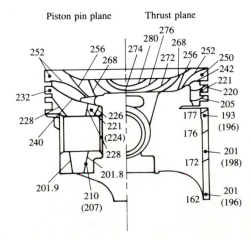

FIGURE 12-23
Measured (dots) and calculated temperature (°C) distributions in piston pin and thrust planes of the piston of a four-cylinder 2.5-dm³ spark-ignition engine at 4600 rev/min and wide-open throttle.[44]

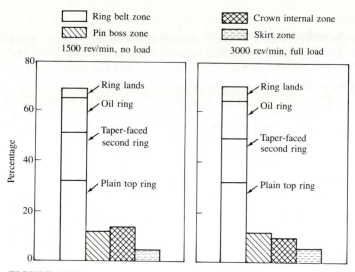

FIGURE 12-24
Heat outflow from various zones of piston as percentage of heat flow in from combustion chamber. High-speed DI diesel engine, 125 mm bore, 110 stroke, $r_c = 17$.[40]

14.4). These define the time-averaged heat flux into the piston. Heat-transfer coefficients for the different surfaces of the piston (underside of dome, ring-land areas, ring regions, skirt outer and inner surfaces, wrist pin bearings, etc.) were estimated. The actual piston shape was approximated with a three-dimensional grid for one quadrant of the piston. A standard finite element analysis of the heat flow through the piston yields the temperature distribution within the piston. The thermal stresses can therefore be calculated and added to the mechanical stress field to determine the total stress distribution. This can be used to define the potential fatigue regions in the actual piston design. Figure 12-23 shows the temperature distribution calculated with this approach, compared with measurements (indicated by dots). The agreement is acceptable, except in the piston skirt where the heat-transfer rate between the skirt and cylinder liner has been overestimated.

Detailed measurements of the temperature distribution in the piston allow the relative amounts of heat which flow out of the different piston surfaces to be estimated. Figure 12-24 shows examples of such estimates for a DI diesel engine at no-load and full-load. About 70 percent of the heat flows out through the ring zone, and much smaller amounts through the pin boss zone, underside of the crown and skirt. In larger diesel engines and highly loaded diesel engines, one or more cooling channels are incorporated into the piston crown. This reduces the heat flow out through the ring area significantly.[45]

12.7.2 Effect of Engine Variables

The following variables affect the magnitude of the heat flux to the different surfaces of the engine combustion chamber and the temperature distribution in the components that comprise the chamber: engine speed; engine load; overall equivalence ratio; compression ratio; spark or injection timing; swirl and squish motion; mixture inlet temperature; coolant temperature and composition; wall material; wall deposits. Of these variables, speed and load have the greatest effect. Equation (12.19), derived from the Nusselt-Reynolds number relation

$$h_c = \text{constant} \times B^{-0.2} p^{0.8} T^{-0.55} w^{0.8}$$

and the relation for heat-transfer rate per unit area [Eq. (12.2)]

$$\dot{q} = h_c(T - T_w)$$

are useful for predicting trends as engine operating and design variables change.

The effect of the above variables on engine and component heat flux will now be summarized. The comments which follow apply primarily to spark-ignition engines. In compression-ignition engines, the distribution of heat flux and temperature varies greatly with the size of cylinder and form of the combustion chamber.

SPEED, LOAD, AND EQUIVALENCE RATIO. Predictions of spark-ignition engine heat transfer as a function of speed and load are shown in Fig. 12-25. The cycle heat transfer is expressed as a percent of the fuel's chemical energy (mass of fuel $\times Q_{\text{LHV}}$). The heat transfer to the total combustion chamber surface (excluding the exhaust port) was calculated using a thermodynamic-based cycle simulation (see Sec. 14.4). The relative importance of heat losses *per cycle* decreases as speed and load increase: the *average* heat transfer *per unit time*, however, increases as speed and load increase.

Since speed and load affect p, T, and w in Eq. (12.19), simpler correlations have been developed to predict component heat fluxes from experimental data. Time-averaged heat fluxes at several combustion chamber locations, determined from measurements of the temperature gradient in the chamber walls, have been fitted with the empirical expression

$$\dot{q} = \text{constant} \times \left(\frac{\dot{m}_f}{A_p}\right)^n \tag{12.38}$$

with n between 0.5 and 0.75 (the value of n depending on engine type and location within the combustion chamber). Results for a modern four-cylinder SI

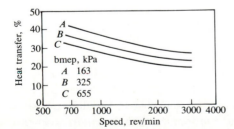

FIGURE 12-25
Predicted average heat-transfer rate (as percent of fuel flow rate $\times Q_{\text{LHV}}$) to combustion chamber walls of an eight-cylinder 5.7-dm³ spark-ignition engine as a function of speed and load. Stoichiometric operation; MBT timing.[46]

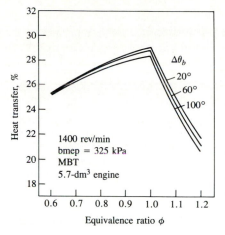

FIGURE 12-26
Predicted average heat-transfer rate (as percent of fuel flow rate $\times Q_{\text{LHV}}$) to combustion chamber walls of an eight-cylinder 5.7-dm^3 spark-ignition engine as a function of equivalence ratio and burn rate ($\Delta\theta_b$ = combustion duration).[46]

engine and several diesel engine designs, with appropriate values of n, can be found in Refs. 47 to 49. While this correlation is not dimensionless and does not satisfy Eq. (12.19), it provides a convenient method for reducing the experimental data. The heat flux to the cylinder head and liner for a spark-ignition engine were well correlated by Eq. (12.38) with $n = 0.6$. The flux distribution over the cylinder head at a fuel flow rate per unit piston area of 0.195 kg/s·m^2 for several different DI diesel engines were comparable in magnitude. The effect of speed at wide-open throttle on component temperatures for a spark-ignition engine can be found in Ref. 47. Exhaust valve, piston crown and top ring groove, and nozzle throat temperatures for a Comet prechamber diesel as a function of fuel flow rate can be found in Ref. 49.

The peak heat flux in an SI engine occurs at the mixture equivalence ratio for maximum power $\phi \approx 1.1$, and decreases as ϕ is leaned out or enriched from this value.[50] The major effect is through the gas temperature in Eqs. (12.2) and (12.19). However, as a fraction of the fuel's chemical energy, the heat transfer per cycle is a maximum at $\phi = 1.0$ and decreases for richer and leaner mixtures, as shown by the thermodynamic-based cycle-simulation predictions in Fig. 12-26. In CI engines, the air/fuel ratio variation is incorporated directly in the load variation effects.

COMPRESSION RATIO. Increasing the compression ratio in an SI engine decreases the total heat flux to the coolant until $r_c \approx 10$; thereafter heat flux increases slightly as r_c increases.[50] The magnitude of the change is modest; e.g., a 10 percent decrease in the maximum heat flux (at the valve bridge) occurs for an increase in r_c from 7.1 to 9.4.[47] Several gas properties change with increasing compression ratio (at fixed throttle setting): cylinder gas pressures and peak burned gas temperatures increase; gas motion increases; combustion is faster; the surface/volume ratio close to TC increases; the gas temperature late in the expansion stroke and during the exhaust stroke is reduced. Measured mean exhaust temperatures confirm the last point, which probably dominates the trend at lower

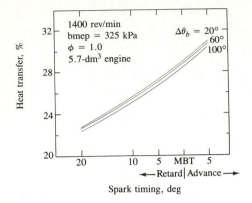

FIGURE 12-27
Predicted average heat-transfer rate (as percentage of fuel flow rate $\times Q_{LHV}$) to combustion chamber walls of an eight-cylinder 5.7-dm³ spark-ignition engine as a function of spark timing and burn rate ($\Delta\theta_b$ = combustion duration).[46]

compression ratios. As the compression ratio increases further, the other factors (which all increase heat transfer) become important.

The effect of changes in compression ratio on component temperatures depends on location. Generally, head and exhaust valve temperatures decrease with increasing compression ratio, due to lower expansion and exhaust stroke temperatures. The piston and spark plug electrode temperatures increase, at constant throttle setting, due to the higher peak combustion temperatures at higher compression ratios. If knock occurs (see Sec. 9.6), increases in heat flux and component temperatures result; see below.

SPARK TIMING. Retarding the spark timing in an SI engine decreases the heat flux as shown in Fig. 12-27. A similar trend in CI engines with retarding injection timing would be expected. The burned gas temperatures are decreased as timing is retarded because combustion occurs later when the cylinder volume is larger. Temperature trends vary with component. Piston and spark plug electrode temperatures change most with timing variations; exhaust valve temperature increases as timing is retarded due to higher exhausting gas temperatures.[47]

SWIRL AND SQUISH. Increased gas velocities, due to swirl or squish motion, will result in higher heat fluxes. Equation (12.19) indicates that the effect on local heat flux, relative to quiescent engine designs, should be proportional to (local gas velocity)$^{0.8}$. There is no direct evidence to support this correlation but there is evidence that use of a shrouded value to increase gas velocities within the cylinder increases the total heat transfer.[50]

INLET TEMPERATURE. The heat flux increases linearly with increasing inlet temperature; the gas temperatures throughout the cycle are increased. An increase of 100 K gives a 13 percent increase in heat flux.[51]

COOLANT TEMPERATURE AND COMPOSITION. Increasing liquid coolant temperature increases the temperature of components directly cooled by the liquid coolant. Figure 12-28 shows the result of a 50-K rise in coolant tem-

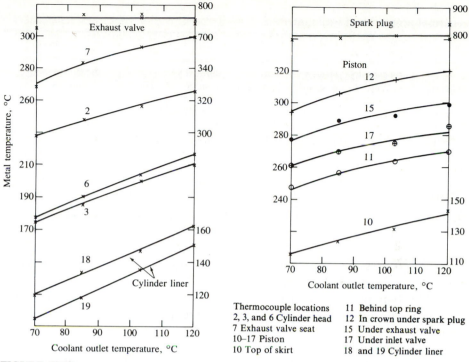

Thermocouple locations
2, 3, and 6 Cylinder head
7 Exhaust valve seat
10–17 Piston
10 Top of skirt

11 Behind top ring
12 In crown under spark plug
15 Under exhaust valve
17 Under inlet valve
18 and 19 Cylinder liner

FIGURE 12-28
Effect of coolant temperature on cylinder head, liner, exhaust valve, valve seat, piston, and spark plug metal temperatures. Spark-ignition engine at 5520 rev/min and wide-open throttle. $r_c = 8.5$.[47]

perature in a spark-ignition engine. The exhaust valve and spark plug temperatures are unchanged. The smaller response of the metal temperatures to coolant temperature change occurs at higher heat flux locations (such as the valve bridge), and indicates that heat transfer to the coolant has entered the nucleate-boiling regime in that region. The response is greater where heat fluxes are lower (e.g., the cylinder liner), indicating that there heat transfer to the coolant is largely by forced convection. When nucleate boiling occurs (i.e., when steam bubbles are formed in the liquid at the metal surface, although the bulk temperature of the coolant is below the saturation temperature), the metal temperature is almost independent of coolant temperature and velocity. Addition of antifreeze (ethylene glycol) to coolant water changes the thermodynamic properties of the coolant.

WALL MATERIAL. While the common metallic component materials of cast iron and aluminum have substantially different thermal properties, they both operate with combustion chamber surface temperatures (200 to 400°C) that are low relative to burned gas temperatures. There is substantial interest in using materials that could operate at much higher temperatures so that the heat losses from the working fluid would be reduced. Ceramic materials, such as silicon nitride and

TABLE 12.2
Thermal properties of wall materials

Material	Thermal conductivity k, W/m·K	Density ρ, kg/m³	Specific heat c, J/kg·K	Thermal diffusivity α, m²/s	$k\rho c$	Skin depth δ, mm	Peak temperature swing, K
Cast iron	54	7.2×10^3	480	1.57×10^{-5}	1.8×10^8	2.8	18
Aluminum	155	2.75×10^3	915	6.2×10^{-5}	3.9×10^8	5.4	12
Reaction-bonded silicon nitride	5–10	2.5×10^3	710	2.8×10^{-6}	1.3×10^7	1.2	70
Sprayed zirconia	1.2	5.2×10^3	732	3.2×10^{-7}	4.6×10^6	0.39	95

zirconia, have lower thermal conductivity than cast iron, would operate at higher temperatures, and thereby insulate the engine. The thermal properties of some of these materials are listed in Table 12.2. With these thermally insulating materials it is possible to reduce the heat transfer through the wall by a substantial amount.

This approach is most feasible for diesel engines where there is the possibility of eliminating the conventional engine coolant system and improving engine efficiency. Since the coolant-side heat transfer is essentially steady during each cycle, a high enough thermal resistance in the wall material can bring the net heat transfer close to zero. However, there is still substantial heat transfer between the working fluid in the cylinder and the combustion chamber walls. Figure 12-29 illustrates these heat-transfer processes by comparing the mean gas temperature to the piston surface temperature for metal and ceramic combustion chamber wall materials. The results come from a thermodynamic simulation of a turbocompounded diesel engine system operating cycle. From Eq. (12.2) the heat transfer is *from* the gas when $T_g > T_w$ and *to* the gas when $T_g < T_w$. With the ceramic material at about 800 K surface temperature, the *net* heat transfer is much reduced compared with the metal case. However, there is substantial heat transfer to the gas from the ceramic walls during intake (which reduces volumetric efficiency) and compression (which increases compression stroke work), and still substantial heat transfer from the gas during combustion and expansion.

The heat transfer from the hot walls to the incoming charge makes thermally insulating materials unattractive for spark-ignition engines. Such heat transfer would increase the unburned mixture temperature leading to earlier onset of knock (see Sec. 9.6).

The variation in ceramic surface temperature in Fig. 12-29 indicates the inherently unsteady nature of the heat-transfer interaction with the wall. During combustion and expansion, the thermal energy transferred from the gas to the wall is stored in a thin layer of wall material adjacent to the surface. While some

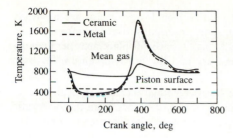

FIGURE 12-29
Mean gas temperature and piston surface temperature profiles predicted by turbocompounded DI diesel engine simulation for water-cooled metal combustion chamber walls and for partly insulated engine with ceramic walls.[27]

of this thermal energy diffuses through the wall, during intake and compression much of it is transferred back to the now low-temperature cylinder contents. The depth of penetration of the thermal wave into the material, the skin depth δ, is proportional to $\sqrt{\alpha/\omega}$, where $\alpha = k/(\rho c)$ is the thermal diffusivity and ω the frequency of the wave (proportional to engine speed). Values of α and δ are given in Table 12.2 for an engine speed of 1900 rev/min: $\sqrt{\alpha}/\delta = 1.4$, a constant. The magnitude of the temperature fluctuation (important because it is a source of fluctuating thermal stress) is proportional to $(\delta\rho c)^{-1}$: this varies as $(k\rho c)^{-1/2}$. Estimated peak temperature swings for the materials in Table 12.2 are tabulated.

KNOCK. Knock in an SI engine is the spontaneous ignition of the unburned "end-gas" ahead of the flame as the flame propagates across the combustion chamber. It results in an increase in gas pressure and temperature above the normal combustion levels (see Sec. 9.6). Knock results in increased local heat fluxes to regions of the piston, the cylinder head, and liner in contact with the end-gas. Increases to between twice and three times the normal heat flux in the end-gas region have been measured.[13, 52] It is thought that the primary knock damage to the piston crown in this region is due to the combination of extremely high local pressures and higher material temperatures.

PROBLEMS

12.1. If radiation in the combustion chamber is negligible, Eqs. (12.5), (12.6), and (12.7) can be combined into the following overall equation approximating the time-averaged heat transfer from the engine:

$$\dot{q} = h_{c,o}(\bar{T}_g - \bar{T}_c)$$

Derive the expression for $h_{c,o}$.

12.2. Given that the average heat flux through a particular zone in a cast iron liner 1 cm thick is 0.2 MW/m², the coolant temperature is 85°C, and the coolant side heat-transfer coefficient is 7500 W/m² · K, find the average surface temperature on the combustion chamber and coolant sides of the liner at that zone.

12.3. Figure 12-1 gives a schematic of the temperature profile from the gas inside the combustion chamber out to the coolant. Draw an equivalent figure showing schematically the temperature profiles at the following points in the engine cycle: (a) intake; (b) just prior to combustion; (c) just after combustion; (d) during the exhaust stroke. Your sketch should be carefully proportioned.

12.4. Using dimensional analysis, compare the relative heat losses of two geometrically similar SI engines (same bore/stroke ratio, same connecting rod/stroke ratio) operating at the same imep and power. Engine A has twice the displacement per cylinder of engine B. Assume that the wall temperature and the gas temperature for both engines are the same.

12.5. (a) Using Woschni's correlation, evaluate the percentage increase in heat transfer expected from an engine with a mean piston speed of 10 m/s when the swirl ratio is raised from 0 to 5. Do your comparison for the intake process only. The engine bore is 0.15 m and the engine speed is 2000 rev/min.

(b) Explain how both the generation of swirl and the change in heat transfer that results affect the volumetric efficiency of an engine.

12.6. (a) Explain how you would estimate the thermal boundary-layer thickness on the combustion chamber wall of an internal combustion engine.

(b) Using representative data, make a rough estimate of the thickness of the thermal boundary layer in the combustion chamber of an SI engine just after the completion of combustion and the fraction of the cylinder mass contained within the boundary layer. $B \approx 100$ mm. Transport properties given in Sec. 4.8.

12.7. (a) Using the analysis found in Sec. 12.6.1, calculate the depth below the surface where the amplitude of the temperature oscillations has attenuated to 1 percent of the amplitude at the surface. The wall material is aluminum and the four-stroke cycle engine is operating at 2500 rev/min. For this estimate, consider only the temperature oscillations which have a frequency equal to the engine firing frequency ($\omega = 2\pi N/n_R$).

(b) Repeat the calculation for the engine operating at 5000 rev/min.

(c) What is the dependence of the penetration depth on the amplitude of surface temperature fluctuations?

12.8. The instantaneous heat-transfer rate $\dot{Q}$ from the cylinder gases to the combustion chamber walls in a spark-ignition engine may be estimated approximately from the equation

$$\dot{Q} = h_c A(\bar{T}_g - \bar{T}_w)$$

where h_c is the heat-transfer coefficient, A is the surface area, $\bar{T}_g$ is the average temperature of the gas in the cylinder, and $\bar{T}_w$ is the average wall temperature. The heat-transfer coefficient can be obtained from the Nusselt, Reynolds, and Prandtl number relationship:

$$\text{Nu} = C(\text{Re})^m(\text{Pr})^n$$

where $C = 0.4$, $m = 0.75$, $n = 0.4$. The characteristic velocity and length scale used in this relation are the mean piston speed and the cylinder bore.

Assuming appropriate values for the engine geometry and operating conditions at wide-open throttle with the wall temperature at 400 K, at an engine speed of 2500 rev/min, and using the cylinder pressure versus crank angle curve of Fig. 14-9, calculate the following:

(a) The *average* temperature of the gas in the cylinder at $\theta = -180°, -90°, 0°, 20°, 40°, 90°, 150°$.

(b) The instantaneous heat-transfer coefficient h_c and heat-transfer rate $\dot{Q}$ from the gas to the combustion chamber walls of one cylinder at these crank angles. Plot these results versus θ.

(c) Estimate the fraction of the fuel energy that is transferred to the cylinder walls during compression and expansion.

Assume for the gas that the viscosity $\mu = 7 \times 10^{-5}$ kg/m·s, the thermal conductivity $k = 1.5 \times 10^{-1}$ J/m·s·K, the molecular weight = 28, and the Prandtl number is 0.8. Assume that the combustion chamber is disc-shaped with $B = 102$ mm, $L = 88$ mm, and $r_c = 9$. (The calculations required for this problem are straightforward; do not attempt anything elaborate.)

REFERENCES

1. Overbye, V. D., Bennethum, J. E., Uyehara, O. A., and Myers, P. S.: "Unsteady Heat Transfer in Engines," SAE paper 201C, *SAE Trans.*, vol. 69, pp. 461–494, 1961.
2. Howarth, M. H.: *The Design of High Speed Diesel Engines*, chap. 5, Constable, London, 1966.
3. Khovakh, M. (ed.): *Motor Vehicle Engines*, chap. 12, Mir Publishers, Moscow, 1971.
4. Sitkei, G.: *Heat Transfer and Thermal Loading in Internal Combustion Engines*, Akademiai Kiado, Budapest, 1974.
5. Burke, C. E., Nagler, L. H., Campbell, E. C., Zierer, W. E., Welch, H. L., Lundstrom, L. C., Kosier, T. D., and McConnell, W. A.: "Where Does All the Power Go," *SAE Trans.*, vol. 65, pp. 713–738, 1957.
6. Ament, F., Patterson, D. J., and Mueller, A.: "Heat Balance Provides Insight into Modern Engine Fuel Utilization," SAE paper 770221, 1977.
7. Novak, J. M., and Blumberg, P. N.: "Parametric Simulation of Significant Design and Operating Alternatives Affecting the Fuel Economy and Emissions of Spark-Ignited Engines," SAE paper 780943, *SAE Trans.*, vol. 87, 1978.
8. Annand, W. J. D.: "Heat Transfer in the Cylinders of Reciprocating Internal Combustion Engines," *Proc. Instn Mech. Engrs*, vol. 177, no. 36, pp. 973–990, 1963.
9. Borman, G., and Nishiwaki, K.: "A Review of Internal Combustion Engine Heat Transfer," *Prog. Energy Combust. Sci.*, vol. 13, pp. 1–46, 1987.
10. Taylor, C. F., and Toong, T. Y.: "Heat Transfer in Internal Combustion Engines," ASME paper 57-HT-17, 1957.
11. Woschni, G.: "Universally Applicable Equation for the Instantaneous Heat Transfer Coefficient in the Internal Combustion Engine," SAE paper 670931, *SAE Trans.*, vol. 76, 1967.
12. Sihling, K., and Woschni, G.: "Experimental Investigation of the Instantaneous Heat Transfer in the Cylinder of a High Speed Diesel Engine," SAE paper 790833, 1979.
13. Woschni, G., and Fieger, J.: "Experimental Investigation of the Heat Transfer at Normal and Knocking Combustion in Spark Ignition Engines," *MTZ*, vol. 43, pp. 63–67, 1982.
14. Hohenberg, G. F.: "Advanced Approaches for Heat Transfer Calculations," SAE paper 790825, *SAE Trans.*, vol. 88, 1979.
15. LeFeuvre, T., Myers, P. S., and Uyehara, O. A.: "Experimental Instantaneous Heat Fluxes in a Diesel Engine and Their Correlation," SAE paper 690464, *SAE Trans.*, vol. 78, 1969.
16. Dent, J. C., and Sulaiman, S. J.: "Convective and Radiative Heat Transfer in a High Swirl Direct Injection Diesel Engine," SAE paper 770407, *SAE Trans.*, vol. 86, 1977.
17. Krieger, R. B., and Borman, G. L.: "The Computation of Apparent Heat Release for Internal Combustion Engines," ASME paper 66-WA/DGP-4, in *Proceedings of Diesel Gas Power*, ASME, 1966.
18. Borgnakke, C., Arpaci, V. S., and Tabaczynski, R. J.: "A Model for the Instantaneous Heat Transfer and Turbulence in a Spark Ignition Engine," SAE paper 800287, 1980.
19. Morel, T., and Keribar, R.: "A Model for Predicting Spatially and Time Resolved Convective Heat Transfer in Bowl-in-Piston Combustion Chambers," SAE paper 850204, 1985.
20. Caton, J. A., and Heywood, J. B.: "An Experimental and Analytical Study of Heat Transfer in an Engine Exhaust Port," *Int. J. Heat Mass Transfer*, vol. 24, no. 4, pp. 581–595, 1981.
21. Malchow, G. L., Sorenson, S. C., and Buckius, R. O.: "Heat Transfer in the Straight Section of an Exhaust Port of a Spark Ignition Engine," SAE paper 790309, 1979.

22. Hottel, H. C., and Sarofim, A. F.: *Radiative Transfer*, McGraw-Hill, 1967.
23. Flynn, P., Mizuszwa, M., Uyehara, O. A., and Myers, P. S.: "Experimental Determination of Instantaneous Potential Radiant Heat Transfer within an Operating Diesel Engine," SAE paper 720022, *SAE Trans.*, vol. 81, 1972.
24. LeFeuvre, T., Myers, P. S., and Uyehara, O. A.: "Experimental Instantaneous Heat Fluxes in a Diesel Engine and Their Correlation," SAE paper 690464, *SAE Trans.*, vol. 78, 1969.
25. Lyn, W. T.: "Diesel Combustion Study by Infra Red Emission Spectroscopy," *J. Instn Petroleum*, vol. 43, 1957.
26. Kamimoto, T., Matsuoka, S., Matsui, Y., and Aoyagi, Y.: "The Measurement of Flame Temperature and Thermodynamic Analysis of the Combustion Process in a Direct Injection Diesel Engine," paper C96/75, in *Proceedings of Conference on Combustion in Engines*, Cranfield, Institution of Mechanical Engineers, 1975.
27. Assanis, D. N., and Heywood, J. B.: "Development and Use of Computer Simulation of the Turbocompounded Diesel System for Engine Performance and Components Heat Transfer Studies," SAE paper 860329, 1986.
28. Field, M. A., Gill, D. W., Morgan, B. B., and Hawksley, P. G. W.: "*Combustion of Pulverised Coal*, British Coal Utility Research Association, Leatherhead, U.K., 1967.
29. Greeves, G., and Meehan, J. O.: "Measurements of Instantaneous Soot Concentration in a Diesel Combustion Chamber," paper C88/75, in *Proceedings of Conference on Combustion in Engines*, Cranfield, Institution of Mechanical Engineers, 1975.
30. Annand, W. J. D.: "Heat Transfer from Flames in Internal Combustion Engines," in N. Afgan and J. M. Beer (eds.), *Heat Transfer from Flames*, chap. 24, John Wiley, 1974.
31. Annand, W. J. D., and Ma, T. H.: "Instantaneous Heat Transfer Rates to the Cylinder Head Surface of a Small Compression-Ignition Engine," *Proc. Instn Mech. Engrs*, vol. 185, no. 72/71, pp. 976–987, 1970–1971.
32. Oguri, T., and Inaba, S.: "Radiant Heat Transfer in Diesel Engines," SAE paper 720023, *SAE Trans.*, vol. 81, 1972.
33. Kunitomo, T., Matsuoka, K., and Oguri, T.: "Prediction of Radiative Heat Flux in a Diesel Engine," SAE paper 750786, *SAE Trans.*, vol. 84, 1975.
34. Alkidas, A. C.: "Heat Transfer Characteristics of a Spark-Ignition Engine," *Trans. ASME, J. Heat Transfer*, vol. 102, pp. 189–193, 1980.
35. Alkidas, A. C., and Myers, J. P.: "Transient Heat-Flux Measurements in the Combustion Chamber of a Spark-Ignition Engine," *Trans. ASME, J. Heat Transfer*, vol. 104, pp. 62–67, 1982.
36. Van Gerpen, J. H., Huang, C.-W., and Borman, G. L.: "The Effects of Swirl and Injection Parameters on Diesel Combustion and Heat Transfer," SAE paper 850265, 1965.
37. Whitehouse, N. D.: "Heat Transfer in a Quiescent Chamber Diesel Engine," *Proc. Instn Mech. Engrs*, vol. 185, no. 72/71, pp. 963–975, 1970–1971.
38. Ramchandani, M., and Whitehouse, N. D.: "Heat Transfer in a Piston of a Four Stroke Diesel Engine," SAE paper 760007, 1976.
39. Lyford-Pike, E. J., and Heywood, J. B.: "Thermal Boundary Layer Thickness in the Cylinder of a Spark-Ignition Engine," *Int. J. Heat Mass Transfer*, vol. 27, no. 10, pp. 1873–1878, 1984.
40. Furuhama, S., and Suzuki, H.: "Temperature Distribution of Piston Rings and Piston in High Speed Diesel Engine," *Bull. JSME*, vol. 22, no. 174, pp. 1788–1795, 1979.
41. Finlay, I. C., Harris, D., Boam, D. J., and Parks, B. I.: "Factors Influencing Combustion Chamber Wall Temperatures in a Liquid-Cooled, Automotive, Spark-Ignition Engine," *Proc. Instn Mech. Engrs*, vol. 199, no. D3, pp. 207–214, 1985.
42. Woschni, G.: "Prediction of Thermal Loading of Supercharged Diesel Engines," SAE paper 790821, 1979.
43. Worthen, R. P., and Raven, D. G.: "Measurements of Valve Temperatures and Strain in a Firing Engine," SAE paper 860356, 1986.
44. Li, C-H.: "Piston Thermal Deformation and Friction Considerations," SAE paper 820086, 1982.
45. Woschni, G., and Fieger, J.: "Determination of Local Heat Transfer Coefficients at the Piston of a High Speed Diesel Engine by Evaluation of Measured Temperature Distribution," SAE paper 790834, *SAE Trans.*, vol. 88, 1979.

46. Watts, P. A., and Heywood, J. B.: "Simulation Studies of the Effects of Turbocharging and Reduced Heat Transfer on Spark-Ignition Engine Operation," SAE paper 800289, 1980.

47. French, C. C. J., and Atkins, K. A.: "Thermal Loading of a Petrol Engine," *Proc. Instn Mech. Engrs*, vol. 187, 49/73, pp. 561–573, 1973.

48. Brock, E. K., and Glasspoole, A. J.: "The Thermal Loading of Cylinder Heads and Pistons on Medium-Speed Oil Engines," in *Proceedings of Symposium on Thermal Loading of Diesel Engines*, vol. 179, part 3C, pp. 3–18, Institution of Mechanical Engineers, 1964–1965.

49. French, C. C. J., and Hartles, E. R.: "Engine Temperatures and Heat Flows under High Load Conditions," in *Proceedings of Symposium on Thermal Loading of Diesel Engines*, vol. 179, part 3C, pp. 126–135, Institution of Mechanical Engineers, 1964–1965.

50. Kerley, R. V., and Thurston, K. W.: "The Indicated Performance of Otto-Cycle Engines," *SAE Trans.*, vol. 70, pp. 5–30, 1962.

51. Taylor, C. F.: *The Internal Combustion Engine in Theory and Practice*, vol. 1, chap. 8, MIT Press, 1966.

52. Lee, W., and Schaefer, H. J.: "Analysis of Local Pressures, Surface Temperatures and Engine Damages Under Knock Conditions," SAE paper 830508, *SAE Trans.*, vol. 92, 1983.

CHAPTER
13

ENGINE FRICTION AND LUBRICATION

13.1 BACKGROUND

Not all the work transferred to the piston from the gases contained inside the cylinder—the indicated work—is available at the drive shaft for actual use. That portion of the work transferred which is not available is usually termed *friction* work. It is *dissipated* in a variety of ways within the engine and engine auxiliaries. The friction work or power is a sufficiently large fraction of the indicated work or power—varying between about 10 percent at full load and 100 percent at idle or no-load—for the topic to be of great practical importance in engine design. Friction losses affect the maximum brake torque and minimum brake specific fuel consumption directly; often the difference between a good engine design and an average engine is the difference in their frictional losses. A large part of the friction losses appear as heat in the coolant and oil which must be removed in the radiator and oil cooler system. Thus, friction losses influence the size of the coolant systems. A knowledge of friction power is required to relate the combustion characteristics of an engine—which influence the indicated power—and the useful output—the brake power.

The friction work, defined as the difference between the work delivered to the piston while the working fluid is contained within the cylinder (i.e., during the

712

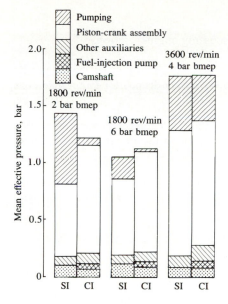

FIGURE 13-1
Comparison of major categories of friction losses: friction mean effective pressure at different loads and speeds for 1.6-liter four-cylinder overhead-cam automotive spark-ignition (SI) and compression-ignition (CI) engines.[1]

compression and expansion strokes) and the usable work delivered to the drive shaft, is expended as follows:

1. To draw the fresh mixture through the intake system and into the cylinder, and to expel the burned gases from the cylinder and out of the exhaust system. This is usually called the pumping work.
2. To overcome the resistance to relative motion of all the moving parts of the engine. This includes the friction between the piston rings, piston skirt, and cylinder wall; friction in the wrist pin, big end, crankshaft, and camshaft bearings; friction in the valve mechanism; friction in the gears, or pulleys and belts, which drive the camshaft and engine accessories.
3. To drive the engine accessories. These can include: the fan, the water pump, the oil pump, the fuel pump, the generator, a secondary air pump for emission control, a power-steering pump, and an air conditioner.

All this work is eventually dissipated as heat; the term friction work or power is therefore appropriate. Figure 13-1 indicates the relative importance of these components in typical four-cylinder automotive SI and diesel engines at different loads and speeds. The magnitudes of the friction from the major items in 1, 2, and 3 above are shown for an SI and a CI engine. The absolute value of the total friction work varies with load, and increases as speed increases. The pumping work for SI engines is larger than for equivalent CI engines and becomes comparable to rubbing friction at light loads as the engine is increasingly throttled. The piston and crank assembly contributes the largest friction component.

13.2   DEFINITIONS

The following terminology will be used in discussing engine friction.

Pumping work W_p. The net work per cycle done by the piston on the in-cylinder gases during the inlet and exhaust strokes. W_p is only defined for four-stroke cycle engines. It is the area $(B + C)$ in Fig. 2-4.†

Rubbing friction work W_{rf}. The work per cycle dissipated in overcoming the friction due to relative motion of adjacent components within the engine. This includes all the items listed in 2 above.

Accessory work W_a. The work per cycle required to drive the engine accessories; e.g., pumps, fan, generator, etc. Normally, only those accessories essential to engine operation are included.

Total friction work W_{tf}. The total friction work is the sum of these three components, i.e.,

$$W_{tf} = W_p + W_{rf} + W_a \qquad (13.1)$$

It is convenient to discuss the difference between indicated and brake output in terms of *mean effective pressure*, mep, the work per cycle per unit displaced volume:

$$\text{mep} = \frac{W_c}{V_d}$$

and power. Power and mep are related by

$$P = \text{mep} \times V_d \times \frac{N}{n_R}$$

where n_R (the number of revolutions per cycle) $= 1$ or 2 for the two-stroke or four-stroke cycle, respectively. Hence, from W_p, W_{rf}, W_a, and W_{tf} we can define pumping mean effective pressure and power (pmep and P_p), rubbing friction mean effective pressure and power (rfmep and P_{rf}), accessory mean effective pressure and power (amep and P_a), and total friction mean effective pressure and power (tfmep and P_{tf}), respectively.

Brake mean effective pressure and power (bmep and P_b), indicated mean effective pressure and power (imep and P_i), and mechanical efficiency have already been defined in Secs. 2.3, 2.4, 2.5, and 2.7. Note that for four-stroke cycle engines, two definitions of indicated output are in common use. These have been designated as:

---

† This definition gives $W_p > 0$ for naturally aspirated engines. For supercharged and turbocharged engines at high load, where p_i is usually greater than p_e, this definition gives $W_p < 0$. For such engines the sign convention for pumping work is often changed in order to maintain W_p as a positive quantity.

Gross indicated mean effective pressure, imep$_g$. The work delivered to the piston over the compression and expansion strokes, per cycle per unit displaced volume.

Net indicated mean effective pressure, imep$_n$. The work delivered to the piston over the entire four strokes of the cycle, per unit displaced volume.

From the above definitions it follows that

$$\text{imep}_g = \text{imep}_n + \text{pmep} \tag{13.2a}$$

$$\text{tfmep} = \text{pmep} + \text{rfmep} + \text{amep} \tag{13.2b}$$

$$\text{bmep} = \text{imep}_g - \text{tfmep} \tag{13.2c}$$

$$\text{bmep} = \text{imep}_n - \text{rfmep} - \text{amep} \tag{13.2d}$$

[Note that all the quantities in Eqs. (13.2a to d) are positive, except for pmep when $p_i > p_e$.]

That two different definitions of indicated output are in common use follows from two different approaches to determining friction work or power. In the standard engine test code procedures[2] friction power is measured in a hot motoring test: the engine is motored with water and oil temperatures held at the firing engine values, with the throttle setting unchanged from its firing engine position (in an SI engine). This measures (approximately) the sum of pumping, rubbing friction, and auxiliary power. The sum of brake power, and friction power determined in this way, is the gross indicated power. Alternatively, when an accurate record of cylinder pressure throughout the whole cycle is available, pumping power can be determined directly: the sum of rubbing friction and accessory power is then the difference between the net indicated power— determined from $\int p\, dV$ over the whole cycle—and the brake power.

For the reasons explained in Sec. 2.4, the gross indicated output is preferred and used in this text. The distinction is most important for SI engines at part load where the pumping power and rubbing friction power are comparable in magnitude. For unthrottled engines at low speeds, the distinction becomes less important (Fig. 13-1 shows the relative importance of pumping work under both these conditions).

13.3 FRICTION FUNDAMENTALS

The friction losses outlined in Sec. 13.1 can be classified into two groups, depending on the type of dissipation which occurs. One type is friction between two metal surfaces in relative motion, with a lubricant in between. The other type is turbulent dissipation.

13.3.1 Lubricated Friction

A primary problem in understanding friction between lubricated surfaces in engines is the wide variation in the magnitude of the forces involved. Thus

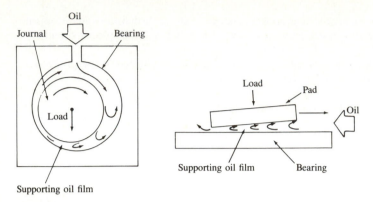

FIGURE 13-2
Schematic of a lubricated journal and a slider bearing.

various regimes of lubrication can occur. Figure 13-2 shows the operating conditions of two common geometries for lubricated parts: a journal and a slider bearing. The different regimes of lubricated friction can be illustrated by means of the Stribeck diagram shown in Fig. 13-3, where the coefficient of friction f (tangential force/normal force) for a journal bearing is plotted against a dimensionless duty parameter $\mu N/\sigma$, where μ is the dynamic viscosity of the lubricant, N is the rotational speed of the shaft, and σ is the loading force per unit area. For sliding surfaces the dimensionless duty parameter becomes $\mu U/(\sigma b)$, where U is the relative velocity of the two surfaces and b is the width of the sliding pad in the direction of motion.

The coefficient of friction can be expressed as

$$f = \alpha f_s + (1 - \alpha)f_L \qquad (13.3)$$

where f_s is the metal-to-metal coefficient of dry friction, f_L is the hydrodynamic coefficient of friction, and α is the metal-to-metal contact constant, varying between 0 and 1. As $\alpha \to 1, f \to f_s$ and the friction is called *boundary*, i.e., close to solid friction. The lubricating film is reduced to one or a few molecular layers and cannot prevent metal-to-metal contact between surface asperities. As $\alpha \to 0, f \to f_L$ and the friction is called *hydrodynamic* or *viscous* or *thick film*. The lubricant film is sufficiently thick to separate completely the surfaces in relative motion. In between these regimes, there is a *mixed* or *partial* lubrication regime where the transition from boundary to hydrodynamic lubrication occurs. While Fig. 13-3 applies to journal bearings, this discussion holds for any pair of engine parts in relative motion with lubricant in between.

Under boundary lubrication conditions, the friction between two surfaces in relative motion is determined by surface properties as well as by lubricant properties. The important surface properties are roughness, hardness, elasticity, plasticity, shearing strength, thermal conductivity, and wettability with respect to the lubricant. The important lubricant properties are mainly surface ones or

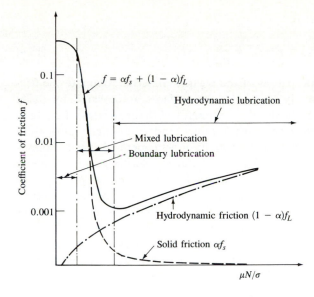

FIGURE diagram.

FIGURE 13-3
Stribeck diagram for journal bearing: coefficient of friction f versus dimensionless duty parameter $\mu N/\sigma$, where μ is the lubricant dynamic viscosity, N is rotational speed of shaft, σ is the loading force per unit area.

chemical ones, which govern the ability of lubricant (or additive) molecules to attach themselves to the solid surfaces. Figure 13-4 shows two surfaces under boundary lubrication conditions. Due to the surface asperities, the real contact area is much less than the apparent contact area. The real contact area A_r is equal to the normal load F_n divided by the yield stress of the material σ_m:

$$A_r = \frac{F_n}{\sigma_m}$$

The force required to cause tangential motion is the product of the real contact area and the shear strength of the material τ_m:

$$F_t = A_r \tau_m$$

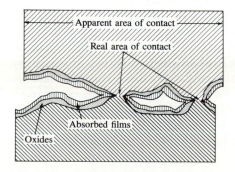

FIGURE 13-4
Schematic of two surfaces in relative motion under boundary lubrication conditions.[3]

Thus the coefficient of friction f is

$$f = \frac{F_t}{F_n} = \frac{\tau_m}{\sigma_m} \tag{13.4}$$

For dissimilar materials, the properties of the weaker material dominate the friction behavior. Since, as shown in Fig. 13-4, the surfaces are covered by oxide films and adsorbed lubricant films, the shear strength of the material in Eq. (13.4) is effectively the shear strength of the surface film.[3] Under boundary lubrication conditions, the coefficient of friction is essentially independent of speed. Boundary lubrication occurs between engine parts during starting and stopping (bearings, pistons, and rings), and during normal running at the piston ring/cylinder interface at top and bottom center crank positions, between heavily loaded parts, and between slow moving parts such as valve stems and rocker arms, and crankshaft timing gears and chains.[4]

Hydrodynamic lubrication conditions occur when the shape and relative motion of the sliding surfaces form a liquid film in which there is sufficient pressure to keep the surfaces separated. Resistance to motion results from the shear forces within the liquid film, and not from the interaction between surface irregularities, as was the case under boundary lubrication. The shear stress τ in a liquid film between two surfaces in relative motion is given by

$$\tau = \mu \left(\frac{dv}{dy} \right)$$

where μ is the fluid viscosity and (dv/dy) the velocity gradient across the film. Hence, the friction coefficient (shear stress/normal load stress) in this regime will be proportional to viscosity × speed ÷ loading: i.e., a straight line on the Stribeck diagram. Full hydrodynamic lubrication or viscous friction is independent of the material or roughness of the parts, and the only property of the lubricant involved is its viscosity. Hydrodynamic lubrication is present between two converging surfaces, moving at relatively high speed in relation to each other and withstanding a limited load, each time an oil film can be formed. This type of lubrication is encountered in engine bearings, between piston skirt and cylinder liner, and between piston rings and liners for high sliding velocities.

Hydrodynamic lubrication breaks down when the thickness of the fluid film becomes about the same as the height of the surface asperities. To the viscous friction is then added metal-to-metal solid friction at the peaks of the asperities. Both hydrodynamic and boundary conditions coexist. The surface texture controls this transition from hydrodynamic to mixed lubrication: rougher surfaces make the transition at lower loads.[3] Abrupt load or speed variations or mechanical vibration may cause this transition to occur. This phenomenon occurs in connecting rod and crankshaft bearings where periodic metal-to-metal contact results from sudden breaks in the oil film. The contact area between rings and cylinders is a zone where, due to sudden variations in speed, load, and temperature, lubrication is of the mixed type. Intermittent metal-to-metal contacts occur as the result of breaks in the oil film.

13.3.2 Turbulent Dissipation

Part of the total friction work is spent in pumping fluids through flow restrictions. The cylinder gases, cooling water, and oil are pumped through the engine; the fan pumps air over the engine block. This work is eventually dissipated in turbulent mixing processes. The pressure difference required to pump these fluids around their flow paths is proportional to ρv^2, where v is a representative fluid velocity. The proportionality constant essentially depends only on flow-path geometry. Hence the friction forces associated with fluid pumping will be proportional to N^2 (or $\bar{S}_p^2$ if the piston motion forces the flow).

13.3.3 Total Friction

The work per cycle for each component i of the total friction is given by integrating the friction force $F_{f,i}$ times its displacement dx around the cycle:

$$W_{f,i} = \int F_{f,i}(\theta) \, dx$$

The friction force components are either independent of speed (boundary friction), proportional to speed (hydrodynamic friction) or to speed squared (turbulent dissipation), or some combination of these. It follows that the total friction work per cycle (and thus the friction mean effective pressure) for a given engine geometry engine will vary with speed according to

$$W_{\text{tf}} \text{ (or tfmep)} = C_1 + C_2 N + C_3 N^2 \tag{13.5}$$

Some of the components of hydrodynamic lubrication friction and turbulent dissipation will be dependent on mean piston speed rather than crankshaft rotational speed N. Examples are piston skirt and ring friction, and the pressure losses associated with gas flow through the inlet and exhaust valves. For conventional engine geometries, crankshaft rotational speed is usually used to scale the total friction data rather than mean piston speed,[5, 6] though more detailed models must include both these variables.

13.4 MEASUREMENT METHODS

A true measurement of friction in a firing engine can only be obtained by subtracting the brake power from the indicated power determined from accurate measurements of cylinder pressure throughout the cycle. However, this method is not easy to use on multicylinder engines, both because of cylinder-to-cylinder differences in indicated power and due to the difficulties in obtaining sufficiently accurate pressure data. As a result, friction is often measured in a motored engine. Friction in a firing and a motored engine are different for the reasons outlined below. First, the common measurement methods will be described.

1. *Measurement of fmep from imep.* The gross indicated mean effective pressure is obtained from $\int p \, dV$ over the compression and expansion strokes for a four-

stroke engine, and over the whole cycle for a two-stroke engine. This requires accurate and in-phase pressure and volume data. Accurate pressure versus crank angle data must be obtained from each cylinder with a pressure transducer and crank angle indicator. Volume versus crank angle values can be calculated. Great care must be exercised if accurate imep data are to be obtained.[7] Both $imep_g$ and pmep are obtained from the p-V data. By subtracting the brake mean effective pressure, the combined rubbing friction plus auxiliary requirements, rfmep + amep, are obtained.

2. *Direct motoring tests.* Direct motoring of the engine, under conditions as close as possible to firing, is another method used for estimating friction losses. Engine temperatures should be maintained as close to normal operating temperatures as possible. This can be done either by heating the water and oil flows or by conducting a "grab" motoring test where the engine is switched rapidly from firing to motored operation. The power required to motor the engine includes the pumping power. In tests on SI engines at part-load, the throttle setting is left unchanged. "Motoring" tests on a progressively disassembled engine can be used to identify the contribution that each major component of the engine makes to the total friction losses.

3. *Willans line.* An approximate equivalent of the direct motoring test for diesel engines is the Willans line method. A plot of fuel consumption versus brake output obtained from engine tests at a fixed speed is extrapolated back to zero fuel consumption. An example is shown in Fig. 13-5. Generally, the plot has a slight curve, making accurate extrapolation difficult. Agreement with a motored test result is shown.

4. *Morse test.* In the Morse test, individual cylinders in a multicylinder engine are cut out from firing, and the reduction in brake torque is determined while maintaining the same engine speed. The remaining cylinders drive the cylinder cut out. Care must be taken to determine that the action of cutting out one cylinder does not significantly disturb the fuel or mixture flow to the others.

Only the first of these four methods has the potential for measuring the true friction of an operating engine. The last three methods measure the power requirements to motor the engine. The motoring losses are different from the firing losses for the following reasons:

1. Only the compression pressure and not the firing pressure acts on the piston, piston rings, and bearings. The lower gas loadings during motoring lower the rubbing friction.

2. Piston and cylinder bore temperatures are lower in motored operation. This results in greater viscosity of the lubricant and therefore increased viscous friction. In addition, piston-cylinder clearances are greater during motoring operation which tends to make friction lower. However, in firing operation, the lubrication of the top ring near the top of the stroke is inadequate to maintain normal hydrodynamic lubrication with the higher gas pressures

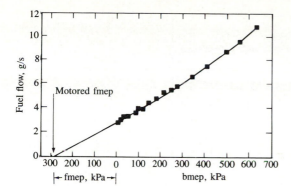

FIGURE 13-5
Willans line method for determining friction mean effective pressure.[5]

behind the ring. The resulting boundary friction in this region makes friction in the firing engine higher. Overall, the net effect of lower piston and cylinder temperatures during motoring is unclear.

3. In motored operation, the exhaust blowdown phase is missing and the gases discharged later in the exhaust stroke have a higher density than under firing conditions. These effects can result in different pumping work.

4. When motoring, net work is done during the compression and expansion strokes because of heat loss from the gas to the walls, and because of gas loss through blowby. This work is not part of the true total friction work in a firing engine and should not be deducted from the indicated work of the firing engine to obtain the brake work; heat losses and blowby are additional energy transfers to the indicated work, friction work, and brake work.

Figure 13-6 shows pumping mep, rubbing friction plus auxiliary mep, and total friction mep for an SI engine over the entire range of throttle positions for firing and motoring tests. Firing test data come from imep and bmep measurements. The engine was a special four-cylinder, in-line, overhead-valve, 3.26-dm^3 displacement tractor SI engine of rugged design and 12 : 1 compression ratio. The pmep values are closely comparable; the rubbing friction mep values diverge

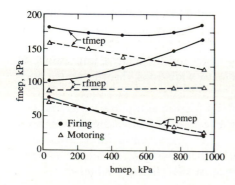

FIGURE 13-6
Total friction mean effective pressure (tfmep), rubbing friction mep (rfmep), and pumping mep (pmep) as a function of load for four-cylinder 3.26-dm^3 spark-ignition engine with bore = 95.3 mm, stroke = 114 mm, and r_c = 12, operated at 1600 rev/min. Motoring and firing conditions.[8]

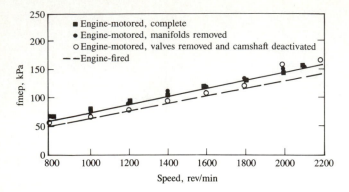

FIGURE 13-7
Rubbing friction and auxiliary mep for six-cylinder diesel engine under motored and fired conditions. Effect of removing manifolds, valves, and camshaft drive under motored conditions also shown.[9]

significantly as load increases.[8] However, the firing friction is not necessarily higher than the motoring test values. Figure 13-7 shows rubbing plus auxiliary mep for a six-cylinder diesel. The firing data are slightly lower than the motored data for this case.[9]

13.5   ENGINE FRICTION DATA

13.5.1   SI Engines

Figure 13-8 shows total motored friction mep for several four-stroke cycle four-cylinder SI engines between 845 and 2000 cm^3 displacement, at wide-open throttle, as a function of engine speed.[6] The data are well correlated by an equation of the form of (13.5):

$$\text{tfmep(bar)} = 0.97 + 0.15\left(\frac{N}{1000}\right) + 0.05\left(\frac{N}{1000}\right)^2 \tag{13.6}$$

where N is in revolutions per minute. Mean piston speed did not provide as good a correlation as rotational speed for this friction data. The importance of avoiding high engine speeds in the interests of good mechanical efficiency are evident. Under normal automobile engine operating conditions, a reduction in total friction mean effective pressure by about 10 kPa results in about a 2 percent improvement in fuel consumption.[10]

Figure 13-9 shows how mechanical efficiency and the relative importance of pumping work vary over the load range idle to wide-open throttle under mid-speed operating conditions.

The effect of compression ratio on rubbing friction and pumping losses, as a function of load at 1600 rev/min, is shown in Fig. 13-10.[8] At the same bmep, both friction and pumping mep are higher at a higher compression ratio. Friction is higher because peak cylinder pressures are higher. Pumping is higher at fixed bmep because the engine is throttled more because the efficiency is higher.

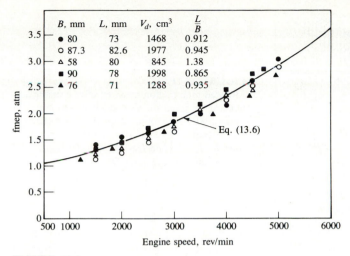

B, mm	L, mm	V_d, cm³	$\frac{L}{B}$
● 80	73	1468	0.912
○ 87.3	82.6	1977	0.945
△ 58	80	845	1.38
■ 90	78	1998	0.865
▲ 76	71	1288	0.935

Eq. (13.6)

FIGURE 13-8
Friction mean effective pressure under motored conditions at wide-open throttle for several four-cylinder spark-ignition engines.[6]

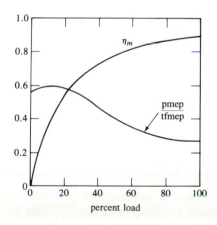

FIGURE 13-9
Mechanical efficiency η_m and ratio of pumping mep to total friction mep as a function of load for a typical spark-ignition engine at fixed speed.[3]

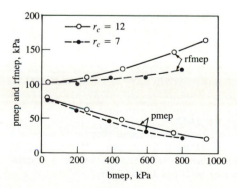

FIGURE 13-10
Pumping mep (pmep) and rubbing friction mep (rfmep) as a function of load for $r_c = 12$ and 7, four-cylinder SI engine with $B = 95.3$ mm and $L = 114$ mm. 1600 rev/min.[8]

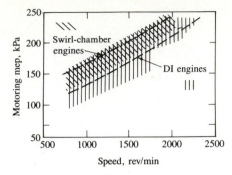

FIGURE 13-11
Motored total friction mean effective pressure as a function of speed for several DI diesels (bores in range 100 to 137 mm) and IDI swirl-chamber diesels (bores in range 100 to 121 mm). Correlations for $r_c = 15$ and $L = 142$ mm (DI engine) and $r_c = 16$ and $L = 142$ mm (IDI engine).[5]

13.5.2 Diesel Engines

Figure 13-11 shows total friction as determined from motoring tests for both direct-injection and swirl-chamber indirect-injection four- and six-cylinder CI engines in the 10 to 14 cm bore range. The higher compression ratio IDI engines lie in the upper half of the scatter band. Correlations for a typical engine of each type are shown, of the form:

$$\text{Motoring mep (kPa)} = C_1 + 48\left(\frac{N}{1000}\right) + 0.4\bar{S}_p^2 \tag{13.7}$$

where N is in revolutions per minute and $\bar{S}_p$ in meters per second. For the direct-injection engine $C_1 = 75$ kPa; for the large swirl chamber IDI engine $C_1 = 110$ kPa. Mean piston speed was found to give a better correlation for the last term in Eq. (13.5) which is mainly pumping mep. Figure 13-12 shows similar results for small swirl-chamber IDI engines. The same correlation, Eq. (13.7) with $C_1 = 144$ kPa, is a good fit to the data.

Friction mep increases as engine size decreases. Also, the *motoring* friction loss for the swirl-chamber engines is higher than for direct-injection engines, primarily because of heat transfer to the prechamber throat and not due to extra pumping losses which are small. Comparative motoring tests show the increase in motored fmep to be about 27 kPa and essentially independent of speed. This is typical of a heat-loss effect, whereas a pumping loss would increase as the square

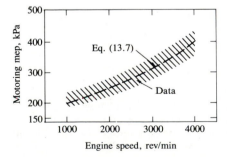

FIGURE 13-12
Motored total friction mean effective pressure as a function of speed for smaller IDI swirl-chamber diesel engines (bores in range 73 to 93 mm). Correlation for $r_c = 21$ and $L = 95.3$ mm.[5]

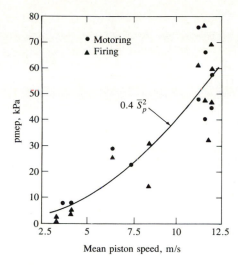

FIGURE 13-13
Pumping mean effective pressure as a function of mean piston speed for several naturally aspirated diesel engines.[5]

of the speed.[5] This extra heat loss is not part of the difference between indicated and brake output in a firing engine, as noted previously.

Pumping mean effective pressure data for a series of naturally aspirated diesels under both firing and motored conditions is shown in Fig. 13-13. The solid line is the term $0.4\bar{S}_p^2$, with $\bar{S}_p$ in meters per second; this is the last term in the overall motored engine friction correlation [Eq. (13.7)].

13.6 ENGINE FRICTION COMPONENTS

In this section, a more detailed analysis of the major components of engine friction is presented and, where possible, equations for predicting or scaling the different components are developed.

13.6.1 Motored Engine Breakdown Tests

Motored engine tests, where the engine is disassembled or broken down in stages, can be used to determine the friction associated with each major engine assembly. While this test procedure does not duplicate the combustion forces of actual engine operation, such tests are useful for assessing the relative importance of individual friction components. Figure 13-14 shows results of breakdown tests on a spark-ignition engine and DI diesel engines. These tests show the large contribution from the piston assembly (piston, rings, rod, including compression loading effects), with the valve train, crankshaft bearings, and water and oil pumps all making significant contributions to the total. An approximate breakdown of rubbing and accessory friction is: piston assembly 50 percent; valve train 25 percent; crankshaft bearings 10 percent; accessories 15 percent. Their relative importance varies over the speed range, however. In the sections that follow, total engine friction will be discussed under the following headings:

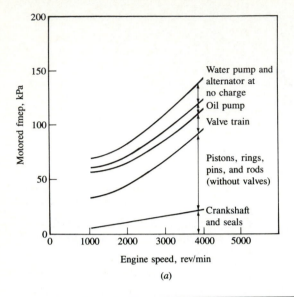

(a)

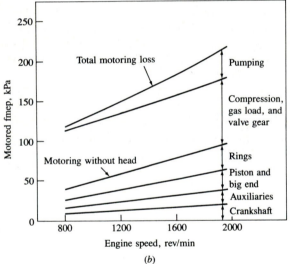

(b)

FIGURE 13-14
Motored friction mean effective pressure versus engine speed for engine breakdown tests. (a) Four-cylinder spark-ignition engine.[10] (b) Average results for several four- and six-cylinder DI diesel engines.[5]

pumping friction, piston assembly friction, valve train friction, crankshaft bearing friction, and (in Sec. 13.7) accessory power requirements.

13.6.2 Pumping Friction

Engine pumping mep data for SI and CI engines, as a function of speed and load, were given in Sec. 13.5. A more detailed breakdown of pumping work is developed here. Figure 13-15 shows the pumping loop for a firing four-stroke cycle spark-ignition engine. The pumping work per cycle (see Fig. 2-4) is the $\int p \, dV$

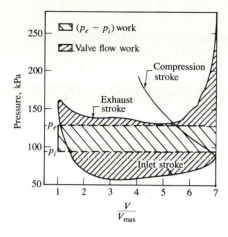

FIGURE 13-15

Pumping loop diagram for spark-ignition engine under firing conditions, showing throttling work $V_d(p_e - p_i)$ and valve flow work.[11]

over the inlet and exhaust strokes. In Fig. 13-15, the firing pumping loop is compared with the inlet and exhaust manifold pressures, p_i and p_e. The work $V_d(p_i - p_e)$ measures the effect of restrictions outside the cylinder, in the inlet and exhaust systems: air filter, carburetor, throttle valve, intake manifold (on the inlet side); exhaust manifold and tail pipe, catalytic converter, and muffler (on the exhaust side). The other area, shown as *valve flow work*, corresponds mainly to pressure losses in the inlet and exhaust valves, and to a lessor extent in the inlet and exhaust ports. As load is reduced in an SI engine, the throttle restriction is increased, the $V_d(p_e - p_i)$ term—called *throttling work*—will increase, and the valve flow work will decrease. The increase in throttling work is much more rapid than the decrease in valve flow work. Both throttling work and valve flow work increase as speed increases at constant load.

The manifold pressures in naturally aspirated engines can be related to imep through a set of equations developed by Bishop:[11, 12]

$$\text{imep}_c = 12.9 p_a \left(\frac{p_{i,a}}{p_a} - 0.1 \right) \tag{13.8}$$

where $p_{i,a}$ is the absolute inlet manifold pressure and p_a is the atmospheric pressure. (All pressures are in kilopascals.)

For SI engines,

$$p_{i,g} = p_a - \frac{\text{imep}_c}{12.9} - 10 \tag{13.9}$$

$$p_{e,g} = p'_{e,g} \left[\left(\frac{\text{imep}_c}{3904} \right) \left(\frac{N}{1000} \right) \right]^2 \tag{13.10}$$

For diesel engines (naturally aspirated),

$$p_{i,g} = 0 \quad \text{and} \quad \text{imep}_c[\text{in Eq. (13.8)}] = 972 \text{ kPa} \tag{13.11}$$

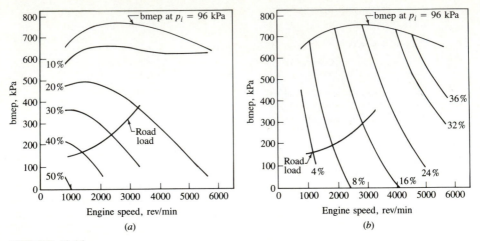

FIGURE 13-16
Relative importance of (a) throttling friction mep and (b) valve pumping friction mep, for spark-ignition engine, as percent of total friction mep on engine load versus speed map.[12]

Here $p_{i,g}$ and $p_{e,g}$ are the intake and exhaust manifold *gauge* pressures (both are positive numbers), p_a is the atmospheric pressure, and $p'_{e,g}$ is the exhaust gauge pressure (all in kilopascals) at 4000 rev/min and full load.

The *throttling mep* for firing engine operation is then given by

$$\text{mep(throttling)} = p_{i,g} + p_{e,g} \tag{13.12}$$

The *valve-pumping mep* was correlated by

$$\text{mep (valve pump)} = 8.96\left(\frac{\text{imep}_c}{1124}\right)^{0.5}\left(\frac{N}{1000}\right)^{1.7}\left(\frac{2.98}{F}\right)^{1.28} \tag{13.13}$$

where

$$F = \frac{n_{iv}\, n_c\, D_{iv}^2}{V_d} \qquad \text{m}^{-1}$$

and n_{iv} is the number of inlet valves per cylinder, n_c the number of cylinders, D_{iv} the inlet valve head diameter, and V_d the displaced volume. For diesel engines, in Eq. (13.13), $\text{imep}_c = 1124$ kPa.

Figure 13-16 shows the relative importance of the throttling and valve pumping losses as a percentage of the total friction mep over the speed and load range of a typical SI engine. The curves are obtained with the equations given above for a six-cylinder, 9 : 1 compression ratio, 3.3-liter (202 in^3) displacement engine. The trends of increasing importance of valve pumping with increasing speed and increasing importance of throttling losses with decreasing load are evident.

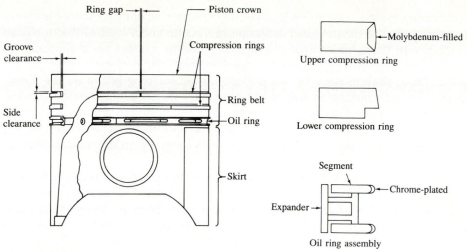

FIGURE 13-17
Construction and nomenclature of typical piston and ring assembly.[10]

13.6.3 Piston Assembly Friction

The construction and nomenclature of a typical piston and ring assembly is shown in Fig. 13-17. The piston skirt is a load-bearing surface which keeps the piston properly aligned within the cylinder bore. The piston lands and skirt carry the side load which is present when the connecting rod is at an angle to the cylinder axis. The rings control the lubrication between these surfaces and the liner. Two types of rings—compression and oil rings—perform the following tasks: (1) seal the clearance between the piston and cylinder to retain gas pressure and minimize blowby; (2) meter adequate lubricant to the cylinder surface to sustain high thrust and gas force loads at high surface speed and at the same time control oil consumption to acceptable limits; and (3) control piston temperatures by assisting in heat transfer to the cylinder walls and coolant. Automobile engines normally use three rings, though two-ring designs exist. Larger diesel engines may use four rings.

Many designs of compression ring are employed,[13] the differences between them being in the cross-sectional shapes (and hence relative flexibility) and in their use of wear-resistant surface treatments. Top compression rings are usually made of cast iron. The axial profiles are chosen to facilitate hydrodynamic lubrication. Common shapes are a rectangular cross section with inner and outer edges chamfered to prevent sticking in the groove, or with a barrel-shaped working surface which can accommodate the rotation of the piston which occurs with short piston skirts. Wear-resistant coatings (either a hard chromium-plated overlay or a molybdenum-filled inlay) are usually applied to the outer ring surface. The second compression ring serves principally to reduce the pressure drop across the top ring. Since the operating environment is less arduous, the second ring can be made more flexible to give better oil control. The objective is to compensate for the torsional deflection of the ring under load so that top-edge contact with the cylinder liner is avoided. Top-edge contact tends to pump oil toward the combustion chamber, detracting from the performance of the oil control ring. Bottom-edge contact provides an oil scraping action on the down-stroke. The oil control ring meters and distributes the oil directed onto the cylinder liner by the crankshaft system, returning excess oil to the crankcase sump. It must exert sufficient pressure against the cylinder, possess suitably shaped wiping edges (usually two thin steel rings), and provide adequate oil drainage. Slotted or composite rings are normally used.[14]

The tension in all the piston rings holds them out against the cylinder wall and hence contributes to friction. The gas pressure behind the compression rings increases this radial force. The gas pressures behind the second ring are substantially lower than behind the first ring. The gas pressures behind the rings are a function of speed and load. An approximate rule for estimating ring friction is that each compression ring contributes about 7 kPa (1 lb/in^2) mep.[5] Oil rings, due to their substantially higher ring tension, operate under boundary lubrication; they contribute about twice the friction of each compression ring.[15]

The piston assembly is the dominant source of engine rubbing friction. The components that contribute to friction are: compression rings, oil control rings, piston skirt, and piston pin. The forces acting on the piston assembly include: static ring tension (which depends on ring design and materials); the gas pressure forces (which depend on engine load); the inertia forces (which are related to component mass and engine speed). The major design factors which influence piston assembly friction are the following: ring width, ring face profile, ring tension, ring gap (which governs inter-ring gas pressure), liner temperature, ring-land width and clearances, skirt geometry, skirt-bore clearance.[3]

Piston assembly friction is dominated by the ring friction. The forces acting on a typical compression ring, lubricated by a thin oil film, are shown in Fig. 13-18. The analysis of this hydrodynamic contact is complex because the forces acting on the ring vary with time and slight changes in ring face geometry can have large effects on the computed results. Cylinder pressure p_c normally acts on the top and back of the ring. The inter-ring gas pressure p_{ir} (which depends on cylinder pressure and the geometry of the lands, ring grooves, and ring, especially the ring gap), acts on the oil film and bottom part of the ring. The character of the gas flow into and out of the inter-ring regions and its effect on ring motion were discussed in Sec. 8.6. Late in the expansion stroke, pressure reversals can occur which may cause the ring to move to the upper surface of the groove or to flutter in between. Ring tension acts to force the ring against the liner. The pressure in the lubricating oil film is generated as shown by the surface A-B in Fig. 13-18 as the ring moves downward. It is believed that the film cavitates between B and C so the pressure decreases to a low value and then increases to p_c. When the direction of motion is reversed, C-B becomes the pressure-generating surface.

Models for the ring and oil film behavior have been developed. For the practical case where the oil film thickness h is much less than the ring width, the

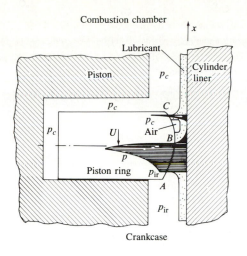

FIGURE 13-18
Schematic of pressure distribution in the lubricating oil film and around a compression ring during expansion stroke. Pressure profile in the oil film indicated by horizontal shading.[3]

Navier-Stokes equation for the liquid film motion reduces to a Reynolds equation of the form:

$$\frac{\partial}{\partial x}\left(h^3 \frac{\partial p}{\partial x}\right) = 6U\mu \frac{\partial h}{\partial x} + 12\mu \frac{\partial h}{\partial t} \tag{13.14}$$

where h is the local film thickness, μ the liquid viscosity, and U the relative velocity between the two surfaces. This equation, along with the appropriate force balances on the ring, can then be solved for the coupled film and ring behavior (e.g., see Ref. 15).

Measured oil film thicknesses in an operating direct-injection diesel engine are shown in Fig. 13-19. A capacitance technique with electrodes embedded in the top compression ring was used to make the measurements.[16] At top-center during combustion, the thickness is a minimum (≈ 1 μm); it then increases as gas loading on the rings decreases and piston velocity increases during the expansion stroke to a value an order of magnitude higher. Higher engine load results in higher gas loading on the rings. It also results in higher lubricant temperatures and lower viscosity, which reduce the film thickness during intake, compression, and exhaust. This large change in film thickness over one cycle is the reason the ring friction regime changes from boundary lubrication to thick-film hydrodynamic lubrication. When the oil film thickness drops below about 1 μm, asperity contact will begin.†

An analysis of the side thrust between the piston and cylinder wall helps explain piston design trends. A force balance on the crank/connecting rod mechanism of Fig. 2-1 leads to the following. An axial force balance relates the piston

---

† The critical film thickness depends on both the cylinder liner and ring surface finish.

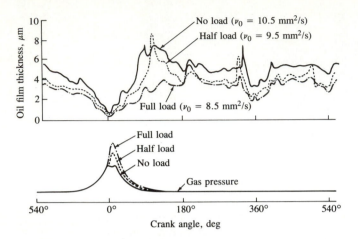

FIGURE 13-19
Measured oil film thickness between top ring and cylinder liner of a DI diesel engine, operated at 1300 rev/min. Bore = 139.7 mm, stroke = 152.4 mm. v_0 is estimated oil viscosity. Reprinted by permission of the Society of Tribologists and Lubrication Engineers (STLE), formerly the American Society of Lubrication Engineers (ASLE).[16]

mass m and acceleration to the net axial force:

$$m \frac{dS_p}{dt} = -F_r \cos \phi + \frac{\pi B^2}{4} p \mp F_f \tag{13.15}$$

where ϕ is the angle between the cylinder axis and connecting rod, and p is the cylinder *gauge* pressure. A transverse force balance gives

$$F_t = F_r \sin \phi = \left(-m \frac{dS_p}{dt} + \frac{\pi B^2}{4} p \mp F_f \right) \tan \phi \tag{13.16}$$

Here F_r is the force in the connecting rod (positive when in compression) and F_f is the friction force on the piston assembly ($-$ when piston is moving toward the crank; $+$ when piston moves away from the crank). dS_p/dt is the piston acceleration obtained by differentiating the equation for piston velocity [Eq. (2.11)]:

$$\frac{dS_p}{dt} = \frac{d^2s}{dt^2} = \pi^2 N \bar{S}_p \left[\cos \theta + \frac{R^2 \cos 2\theta + \sin^4 \theta}{(R^2 - \sin^2 \theta)^{3/2}} \right] \tag{13.17}$$

The side thrust F_t given by Eq. (13.16) is transmitted to the liner via the rings and piston skirt. It changes direction as the piston passes through top- and bottom-center positions. Since the friction force changes sign at these locations and the gas pressure during expansion is greater than during compression, the side thrust during expansion is greater.

The piston skirt carries part of this side thrust so it contributes to piston assembly friction. The large contact area between the skirt and liner, relative to

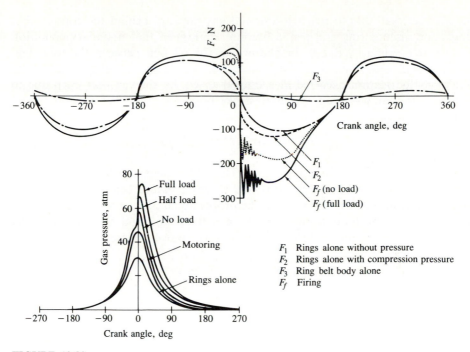

FIGURE 13-20
Measured frictional force on cylinder liner of 137 mm bore and 135 mm stroke single-cylinder DI diesel engine. 1200 rev/min, coolant temperature 80°C, cylinder liner inside temperature 97°C.[18]

the ring contact area, results in lighter loading (force/area) and promotes hydrodynamic lubrication. Piston skirt areas have been reduced substantially in recent years to reduce piston mass (which reduces side thrust) and contact area. An additional reduction in side thrust, leading to reduced skirt friction, has been achieved with the use of an offset wrist-pin. By offsetting the pin axis by 1 to 2 mm without changing its vertical location, the crank angle at which the piston traverses the bore and "slaps" the other side of the cylinder is advanced so it occurs before combustion has increased the cylinder gas pressure significantly.[17]

Direct measurements of the friction force associated with the piston assembly have been made. The most common technique involves the use of a special engine where the axial force on the cylinder liner is measured directly with a load transducer (e.g., Ref 18). Figure 13-20 shows the friction forces measured in such an engine (a DI diesel engine) through the engine's operating cycle. Friction forces are highest just before and after top-center at the end of the compression stroke. The high values at the start of the expansion stroke under firing conditions are caused by the piston slap impulse and the high side-thrust force as well as the combustion gas pressure loading on the rings.

Bishop[12] has developed correlations for piston and ring friction in the following categories: boundary condition friction (primarily between the rings and the cylinder wall due to ring tension, and gas pressure behind the compression rings) and viscous ring and piston friction. He argued that boundary condition friction was primarily due to breakdown of the oil film between the rings and cylinder wall over part of the piston travel. Assuming that the transition to boundary lubrication occurred at a critical speed, he showed that fmep due to boundary friction was proportional to stroke/bore², i.e.:

$$(\text{fmep})_{\text{boundary}} \propto \text{loading} \times \frac{L}{B^2} \qquad (13.18)$$

The ring loading has two components. The component due to ring tension is essentially constant. The component due to gas pressure behind the rings will depend on load. Bishop assumed it to be proportional to inlet manifold pressure. The viscous piston friction—friction between the piston and rings and cylinder wall under hydrodynamic lubrication conditions—was correlated by

$$(\text{fmep})_{\text{hydrodyn}} \propto \frac{\bar{S}_p A_{p,\,\text{eff}}}{LB^2} \qquad (13.19)$$

where $A_{p,\,\text{eff}}$ is the effective area of the piston skirt in contact with the cylinder liner.

The relative importance of the boundary lubrication piston and ring friction, and viscous piston and ring friction over the load and speed range, is as follows. The viscous friction component increases in importance with increasing speed. The boundary lubrication friction component increases with increasing load as the cylinder gas pressures increase.

13.6.4   Crankshaft Bearing Friction

Crankshaft friction contributions come from journal bearings (connecting rod, main and accessory or balance shaft bearings) and their associated seals. A schematic of a journal bearing operating under hydrodynamic lubrication is shown in Fig. 13-21. Large loads can be carried by journal bearings with low energy losses under normal operating conditions, due to the complete separation of the two surfaces in relative motion by the lubricant film. Loads on crankshaft journal bearings vary in magnitude and direction because they result primarily from the inertial loads of the piston/connecting rod mechanism and the cylinder gas loads [see Eq. (13.15)]. Typical loads and the resulting journal eccentricity diagram for a connecting rod bearing are shown in Fig. 13-22. From the journal eccentricity diagram the minimum oil film thickness is determined. This quantity, the minimum separation distance between the journal and bearing surfaces, is a critical bearing design parameter. If the film thickness is too low, asperities will break through the oil film and substantially increase the friction and wear. Journal bearings are usually designed to provide minimum film thicknesses of about 2 μm.

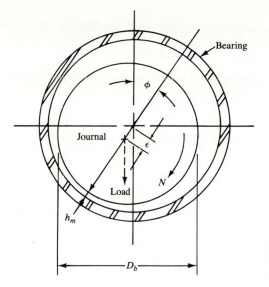

FIGURE 13-21
Schematic of hydrodynamically lubricated journal bearing.[3]

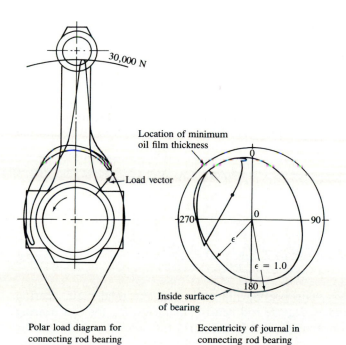

Polar load diagram for
connecting rod bearing

Eccentricity of journal in
connecting rod bearing

FIGURE 13-22
Typical engine journal bearing load and eccentricity diagrams.[3]

The friction force F_f in the bearing is given approximately by the product of the bearing area, the oil viscosity, and the mean velocity gradient in the oil:

$$F_f \approx (\pi D_b L_b)\mu \left(\frac{\pi D_b N}{\bar{h}} \right) = \frac{\pi^2 \mu D_b^2 L_b N}{\bar{h}} \tag{13.20}$$

where D_b and L_b are the bearing diameter and length, $\bar{h}$ is the mean radial clearance, and N is the shaft rotational speed. A more sophisticated analysis of the friction in a hydrodynamically lubricated bearing yields the relation[19]

$$F_f = \frac{\pi^2 \mu D_b^2 L_b N}{(1 - \varepsilon^2)^{1/2}\bar{h}} + \frac{\bar{h}\varepsilon W}{D_b}\sin\phi \tag{13.21}$$

where ε is the eccentricity ratio $(\bar{h} - h_m)/\bar{h}$ and h_m is the minimum clearance. The first term closely matches the approximation given in Eq. (13.20). The factor $1/(1 - \varepsilon^2)^{1/2}$ and the second term correct for the offset of the journal center relative to the bearing center: W is the bearing load and ϕ the attitude angle. To first order, with hydrodynamic lubrication the friction power does not depend significantly on the bearing load. If σ is the loading per unit projected area of the bearing $[W/(L_b D_b)]$, then the coefficient of friction f is given by

$$f = \frac{F_f}{W} \approx \frac{\pi^2 \mu D_b^2 L_b N}{\sigma L_b D_b \bar{h}} = \frac{\pi^2 D_b}{\bar{h}}\frac{\mu N}{\sigma} \tag{13.22}$$

For a given bearing, or series of geometrically similar bearings, the friction coefficient is proportional to $\mu N/\sigma$. However, at low values of $\mu N/\sigma$ the hydrodynamic pressure in an actual bearing will be insufficient to support the shaft load and the oil film becomes incomplete. The friction coefficient increases rapidly as mixed lubrication then occurs.

Bishop[12] summed the friction power loss in all crankshaft and con rod journal bearings and divided by the displaced volume per unit time to obtain the following correlation for bearing friction mep (in kilopascals):

$$\text{fmep (bearings)} = 41.4 \left(\frac{B}{L} \right)\left(\frac{N}{1000} \right) K \tag{13.23}$$

where

$$K = \frac{D_{mb}^2 L_{mb} + D_{rb}^2 L_{rb}/m + D_{as}^2 L_{as}}{B^3} \tag{13.24}$$

In Eq. (13.24), D_{mb} is the main bearing diameter, L_{mb} the total main bearing length $\div$ number of cylinders, D_{rb} the rod bearing diameter, L_{rb} the rod bearing length, m the number of pistons per rod bearing, D_{as} the accessory shaft bearing diameter, L_{as} the total length of all accessory shaft bearings $\div$ number of cylinders, and all dimensions are in millimeters. The similarity between engines is such that $K \approx 0.14$ for spark-ignition engines and $K \approx 0.29$ for diesel engines.

Type I:
OHC, direct-acting

Type II:
OHC, end pivot rocker

Type III:
OHC, center pivot rocker

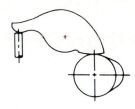

Type IV:
OHC, center pivot rocker with lifter

Type V:
push rod

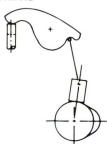

FIGURE 13-23
Different valve train configurations.[21]

The front and rear main bearing seals[20] also contribute to crankshaft assembly friction. At 1500 rev/min they are responsible for about 20 percent of the friction attributable to the crankshaft.[10]

13.6.5 Valve Train Friction

The valve train carries high loads over the entire speed range of the engine. Loads acting on the valve train at lower speeds are due primarily to the spring forces, while at higher speeds the inertia forces of the component masses dominate. Valve train designs can be classified by type of configuration, as indicated in Fig. 13-23. Large valves and high rated speeds generally increase spring and inertia loads and friction. Friction differences between these systems are difficult to quantify. For example, measurements of valve train friction mean effective pressure for several of these valve train types showed significant variations (± 30 percent): see Fig. 13-24a.[10,21] However, when the data were adjusted to a common spring load, Fig. 13-24b, the low-speed friction mep values converged and the high-speed fmep differences were reduced.[10]

The total valve train friction can be broken down by critical contact regions: camshaft journal bearings, rocker arm/fulcrum and cam/tappet interface.

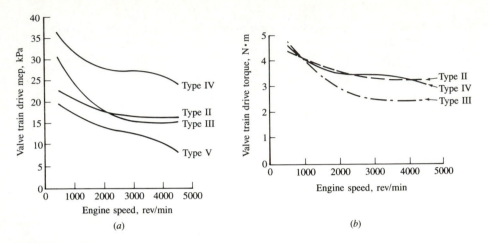

FIGURE 13-24
(*a*) Total valve train friction mean effective pressure as a function of speed for four engines with different valve configurations (see Fig. 13-23). (*b*) Valve train friction torque for three of these engines after adjusting to common valve spring load.[10]

The shape of the valve train mep versus speed curve indicates that the predominant regime of lubrication in the valve train at lower engine speeds is boundary lubrication. The cam/lifter interface usually contributes the largest friction loss due to the very high loads and small contact areas.[22]

Effective methods of reducing valve train friction are: (1) spring load and valve mass reduction; (2) use of tappet roller cam followers; (3) use of rocker arm fulcrum needle bearings. One such low-friction valve train design is shown in Fig. 13-25.[22] The roller cam-followers provide the largest benefit especially at lower speeds: reductions of order 50 percent in valve train friction can be achieved.

Bishop[12] developed a correlation for valve train friction from design data on valve spring loads and valve weights, and experimental data from dynamometer tests of push rod engines. He shows that

$$\text{fmep (valve train)} = \frac{C[1 - 0.133(N/1000)]n_{\text{iv}} D_{\text{iv}}^{1.75}}{B^2 L} \qquad (13.25)$$

where n_{iv} is the number of inlet valves per cylinder, D_{iv} is the inlet valve head diameter, and B and L are bore and stroke. This relation does not include camshaft bearing friction, which is included in Eq. (13.23). The functional form of Eq. (13.25) is an acceptable fit to more modern engine data. Bishop's value for C (1.2×10^4 with fmep in kilopascals, N in revolutions per minute, and dimensions in millimeters) gives valve train fmep values (which exclude camshaft bearing losses) about two-thirds the total valve train friction of current production engines. This is consistent with the data in Fig. 13-24.

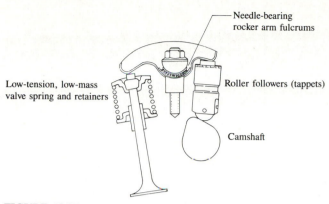

Needle-bearing
rocker arm fulcrums

Low-tension, low-mass
valve spring and retainers

Roller followers (tappets)

Camshaft

FIGURE 13-25
Low friction valve train.[22]

13.7 ACCESSORY POWER REQUIREMENTS

The coolant water pump and oil pump are built-in accessories, essential to engine operation, and are normally considered part of the basic engine.[2] A fully equipped engine usually includes additional accessories—a fan and generator; in automobile use it often includes a power-steering pump, an air conditioner, and an air pump for emission control. The power delivered by the fully equipped engine (the net power) is lower than the power delivered by the basic engine due to the power requirements of these additional accessories.

The friction mean effective pressures associated with driving the water pump and alternator, and oil pump are shown in Fig. 13-14a. Together they comprise about 20 percent of the total (motored) engine friction. The water pump is typically less than about 7 kPa at 1500 rev/min;[10] the oil pump 4 to 10 kPa at this speed;[10] the alternator requires 7 to 10 kPa.[23] These numbers vary significantly with component design details. The generator power depends on the electrical load to be met and the generator blower design. A requirement of about two-thirds of the peak is indicated for average generator power.[23]

The power requirements for a fan, generator, and power-steering pump typical of a 5.7-liter engine are shown in Fig. 13-26. The fan requirements are the largest and with a direct drive increase with the cube of the speed. Alternative couplings such as a viscous drive reduce the fan speed at high engine speed and thereby reduce its power significantly. The power-steering pump is only required to provide high pressures intermittently. Here only the fluid pumping losses are charged against the engine.

Air-conditioning is standard on a majority of U.S. cars; additional power is required for the air-conditioning compressor. Also, since the compressed refrigerant is condensed in a second radiator, a larger-than-standard fan is required to pull additional air through the combined radiator systems. An air pump which pumps air into the engine exhaust ports may be part of an SI engine emission

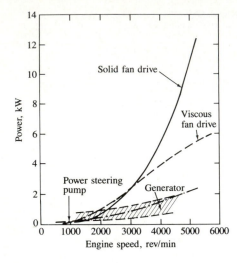

FIGURE 13-26
Power requirements for engine fan, generator, and power-steering pump typical of 5.7-liter eight-cylinder engine.[23, 24]

control system (see Sec. 11.6). Its power requirements (~ 1 kW at normal engine speeds) must then be added to the accessory friction requirements.

13.8 LUBRICATION

The lubricant and the lubricating system perform the following functions:[25]

1. Reduce the frictional resistance of the engine to a minimum to ensure maximum mechanical efficiency.
2. Protect the engine against wear.
3. Contribute to cooling the piston and regions of the engine where friction work is dissipated.
4. Remove all injurious impurities from lubricated regions.
5. Hold gas and oil leakage (especially in the ring region) at an acceptable minimum level.

13.8.1 Lubrication System

The principle moving parts of an engine are positively lubricated by introducing a supply of oil from a pressurized system. An example of a lubrication system (for an air-cooled diesel engine) is shown in Fig. 13-27. The oil pump draws oil from the engine sump and delivers it through a control valve to the oil cooler. The oil then passes through the filter to the main oil gallery. From the main oil gallery it is branched to the main, the big end, and the camshaft bearings. Oil is also ducted to the injection pump. Through a passage in the camshaft bearing the oil flows to the tappet bridges. As the oil passages of tappets and tappet bridges line up during tappet motion, rocker arms and valve stems are pulse-lubricated

TABLE 13.1

Functions and qualities required of engine oils

Main functions required	Where and when	Qualities required
Reduce frictional resistance	During cold-starting	Low enough viscosity to provide good pumping and avoid undue cranking resistance
	Between con-rod/ crankshaft bearings, and journals	Minimum viscosity without risk of metal-to-metal contact under the varying conditions of temperature, speed, and load
	Between pistons, rings, and cylinders	Sufficiently high viscosity at high temperatures; good lubrication property outside the hydrodynamic condition, especially at top-center
		Antiseizure properties, especially during the run-in period
Protect against corrosion and wear	During shut-down or when running at low temperature	Must protect metallic surfaces against corrosive action of fuel decomposition products (water, SO_2, HBr, HCl, etc.)
		Must resist degradation (resist oxidation, have good thermal stability)
	In normal running	Must counteract action of fuel and lubricant decomposition products at high temperatures, especially on non-ferrous metals
		By intervention in the friction mechanism must reduce the consequences of unavoidable metal-to-metal contact
		Must resist deposit formations which would affect lubrication (detergency or dispersive action)
		Must contribute to the elimination of dust and other contaminants (dispersive action)
Assist sealing	In the ring zone, especially at TC	Must have sufficient viscosity at high temperatures and low volatility
		Must limit ring and liner wear
		Must not contribute to formation of deposits in ring grooves and must prevent such formation
Contribute to cooling	Chiefly of pistons, rings, and con-rod bearings	Must have good thermal stability and oxidation resistance
		Must have low volatility
		Viscosity must not be too high
Facilitate the elimination of undesirable products	During oil drains to eliminate atmospheric dust, soot from diesel engines, Pb salts, wear debris, organic products from burned fuel and lubricants, and other contaminants which promote deposits or accelerate wear	Must be able to maintain in fine suspension all solid material (dispersivity) whatever the temperature and physical and chemical conditions (water)
		Must be able to solubilize certain organic compounds, particularly heavy oxidation products

Source: From Schilling.[25]

1 Sump
2 Suction pipe
3 Lube oil pump
4 Oil pressure control valve
5 Pressure pipe
6 Bypass pipe or alternative
7 Cooling coil or, alternatively:
8 Block-type oil cooler
9 Oil filter
10 Safety valve
11 Main oil gallery
12 Main bearing
13 Big end bearing
14 Camshaft bearing
15 Tappet (with timing groove
    to pulse-lubricate rocker arm)
16 Push rod (hollow, used as
    rocker arm oil feed pipe)
17 Rocker arm bearing
18 Metering plug (to control
    valve lubrication)
19 Push rod duct (used as
    cylinder-head-to-crankcase
    oil return pipe)
20 Splash hole to lubricate
    timing gears

21 Piston cooling nozzle
22 Oil pressure gauge adaptor
23 Oil pressure gauge

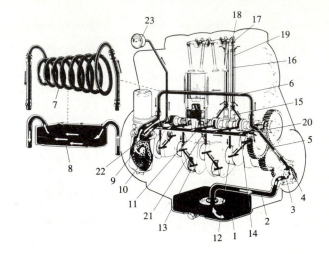

FIGURE 13-27
Lubrication system layout for air-cooled DI diesel engine. (*Courtesy Klöckner-Humboldt-Deutz AG.*)

through the tappets and pushrods. For cooling pistons and lubricating cylinders, oil is thrown against the underside of the piston through nozzles connected to the main bearings. Spring-loaded ball valves incorporated in the nozzles interrupt the jet cooling at low engine speeds to insure that the oil pressure remains above a safe level. The gears of the main timing train are splash-lubricated. The oil is returned from the injection pump and rocker chamber cover to the sump.

13.8.2 Lubricant Requirements

Table 13.1 lists the qualities required of engine oils to perform the main lubrication system functions. These qualities can be summarized under the following headings.[25]

OXIDATION STABILITY. The temperature of the oil and engine parts it contacts, the presence of oxygen, the nature of the metal surfaces and debris, and the products of the fuel combustion, all influence the oxidation of the hydrocarbon components in lubricating oil. High temperatures are the primary factor, and the top piston ring groove and the crankcase are the critical regions. The temperature of the top ring groove can easily reach 250°C. The lubricating oil when subject to these conditions must not, through oxidation, contribute to deposit formation, even after long periods of running. These deposits would eventually

lead to ring sticking which results in excessive blowby. At high temperatures, deposits are related to the oxidized fraction of the oil.

The oil temperature in the crankcase is 120 to 130°C, or higher. Oil maintained at this temperature should neither form any acid products capable of attacking the bearing alloys nor form insoluble products which form deposits. Good-quality mineral oils cannot withstand these temperatures, so antioxidant and anticorrosive additives are used to control these problems. While antioxidants help to reduce deposit formation, detergent/dispersant additives are required to maintain any insoluble materials formed through oxidation in suspension.

DETERGENCY/DISPERSION. Except for deposits formed in the combustion chamber, deposits in the oil are controlled by its detergency. The amount of deposits formed depends on the fuel used, the quality of combustion, the temperature of the lubricating oil and coolant, and on the effectiveness of gas sealing at the piston rings. The detergency property is given to straight mineral oils by additives; the function of the detergent additive is to reduce the amount of deposits formed and make their removal easier.

At low temperatures, deposits are mainly byproducts of fuel combustion, and the detergency function is to keep them in suspension or solution in the oil. At high temperatures, deposits come from the oxidized fractions of the oil. The detergency function here is both to keep these products in suspension and to inhibit the reactions that lead to the formation of varnishes and lacquers. In addition, in diesel engines, the detergency helps in neutralizing the acidic reaction products from the sulfur compounds in the fuel.

WEAR REDUCTION. Wear is due to the individual and combined effects of corrosion, adhesion (i.e., metal-to-metal contact), and abrasion.

Corrosive attack by acidic products of combustion is one of the chief causes of cylinder and ring wear. The effect is worst at low cylinder wall temperatures. In diesel engines, the sulfur in the fuel increases the corrosive wear. Corrosive wear is effectively prevented by the use of detergent oils which neutralize the corrosive acids as they form, and by designing the cooling system to give appropriate metal temperatures.

Adhesive wear affects certain parts of the engine. In the upper cylinder, metal-to-metal contact between piston, rings, and cylinder walls takes place each time the engine is started (most significant during cold-starts) because there is insufficient oil in the top portion of the engine. Oils with antiwear additives and low viscosity at low temperatures provide a partial remedy. Adhesive wear also occurs on components such as cams, tappets, drive gears, rocker arm ends, and valve stems.

Abrasion results from the presence of atmospheric dust, and metallic debris from corrosive and adhesive wear, in the lubricating oil. Efficient air filtration is therefore most important (see Ref. 26 for a discussion of air filters). Elimination of

abrasive particle impurities from the oil system by filtration and periodic oil change is essential.

VISCOSITY. For low resistance to cranking and ease of starting, and rapid distribution of the oil while the engine is cold, a low oil viscosity at low ambient temperatures is required. When the engine (and oil) is fully warmed up, viscosity in the proper range is important for adequate sealing of the piston, acceptable oil consumption, and low friction losses. The viscosity of the oil at both low and normal engine temperatures (a spread of some 200 K) is, therefore, important. The viscosity of lubricating oils decreases with increasing temperature. The *pour point, viscosity,* and *viscosity index* are used to characterize the behavior of a lubricating oil for these aspects of engine operation.

The *pour point* is determined by cooling a sample of oil in a test jar until, when the jar is rotated from the vertical to the horizontal, no perceptible movement of the oil will occur within 5 s; 5°F above this temperature is the pour point.

The viscosity of lubricating oils is determined by measuring the time required for a specified volume of oil to flow through a capillary tube or orifice, contained in a constant temperature water bath. The *kinematic viscosity*, v (v = μ/ρ), is determined by this method. Use of a Saybolt tube with an orifice of specified diameter is the standard U.S. measurement practice. The viscosity is then given by the time t (in seconds) required to flow 60 cm^3 of oil, and is expressed as Saybolt universal seconds, SUS. Approximate conversion to centistokes (1 centistoke = 10^{-6} m^2/s) can be obtained via

$$v = at - \frac{b}{t}$$

where for 115 > t > 34 s, a = 0.224 and b = 185; for 215 > t > 115 s, a = 0.223 and b = 155; and for t > 215 s, a = 0.2158 and b = 0.[27]

The viscosity of lubricating oils decreases with increasing temperature. Since engine oils must operate over a range of temperatures, a measure of the rate of decrease is important. The *viscosity index*, an empirical number indicating the effect of temperature changes on viscosity, is used for this purpose;[28] a low viscosity index indicates a relatively large change of viscosity with temperature. To increase the viscosity index, lubricating oils incorporate additives called "viscosity-index improvers." These are high molecular weight compounds (molecular weight $\approx 10^3$ to 10^4) whose primary function is to reduce the viscosity variation with temperature.

The lubricating oil classification used most extensively is the SAE classification.[29] It depends solely on the viscosity of the oil. The seven different classification numbers 5W, 10W, 20W, 20, 30, 40, and 50 correspond to viscosity ranges; increasing numbers correspond to increasing viscosity, as shown in Fig. 13-28. SAE numbers followed by W (abbreviation for winter) refer to oils for use in cold climates, and viscosity is determined at −18°C (0°F). SAE numbers without W are applied to engine oils commonly used under warmer conditions;

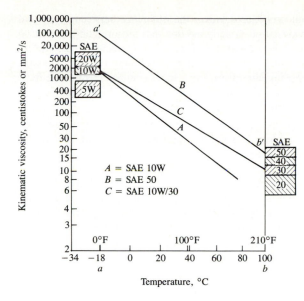

FIGURE 13-28
Viscosity versus temperature curves illustrating SAE lubricating oil classification.[25]

they are based on viscosity at 99°C (210°F). Multigrade oils (for example, 10W-40) satisfy service requirements at low as well as high temperatures in terms of the SAE classification. The first number indicates the viscosity at −18°C; the second number at 99°C. Examples are shown in Fig. 13-28. Multigrade oils have a higher viscosity index than single-grade oils, which make them more attractive for engine use.

PROBLEMS

13.1. (a) Show how friction mean effective pressure for a four-stroke cycle engine can be obtained from the brake power P_b, engine speed N, displaced volume V_d, and $\int p\,dV$ over the compression and expansion strokes ($= W_{c,ig}$).

(b) How is pumping mean effective pressure related to $\int p\,dV$ over the compression and expansion strokes and $\int p\,dV$ over the full four-stroke cycle?

(c) Find the brake power, total friction power, total friction imep, and pumping imep for a four-stroke cycle SI engine operating at 1800 rev/min with a measured brake torque of 32 N·m, a gross imep of 933 kPa, and a net imep of 922 kPa. $V_d = 0.496$ dm^3.

13.2. Three categories of friction are described in Sec. 13.3: boundary friction, hydrodynamic (or fully lubricated) friction, and turbulent dissipation. By means of Eq. (13.6), estimate the relative proportion of total friction work per cycle in each category for a four-cylinder automobile spark-ignition engine operating at 3000 rev/min.

13.3. For four-stroke cycle naturally aspirated multicylinder spark-ignition and diesel automobile engines at full load and one-third full load, at mid speed (2000 rev/min), give approximate estimates of the percentages of total friction mep in these three categories: pumping mep, rubbing friction mep, and accessory friction mep. State explicitly how you develop these estimates and what you include as accessories.

13.4. All of the friction measurement procedures except the difference between brake and gross indicated power or mep measured directly assume that motored engine friction and firing engine friction are closely comparable. This is not an accurate assumption for the pumping component. Summarize the differences between the gas exchange processes under motoring and firing conditions for a spark-ignition engine at a fixed part-load throttle setting that will result in the pumping work being significantly different under these two conditions.

13.5. On separate accurately proportioned sketches of the piston, cylinder, connecting rod, and crank mechanism (similar to Fig. 2-1), during the intake stroke (at 120° ATC), compression stroke (at 60° BTC), expansion stroke (at 60° ATC), and exhaust stroke (at 120° BTC), draw an arrow for each of the forces acting on the piston (pressure forces, force from connecting rod, friction force, inertia force). Mark clearly the *positive* direction of each force. Express each force in terms of cylinder pressure p_c, crankcase pressure p_{cc}, friction force F_f, piston area A_p, effective piston (and part of connecting rod) mass m_p, piston acceleration a, connecting rod force F_{cr}.

13.6. (a) For the DI diesel engine for the friction force data in Fig. 13-20, estimate the maximum pressure force on the piston (under full-load conditions) and the approximate magnitude of the inertia force [mass of piston plus part of the connecting rod (7 kg) × $\bar{S}_p$ × $(N/4)$]. Compare these forces with the piston friction force at time of peak pressure.

(b) Figure 13-6 shows the variation in friction force acting on the piston of a DI diesel engine under no-load and full-load firing conditions. Carefully sketch the shape (indicating direction and rough magnitude) of the cylinder pressure force on the piston, the piston velocity, and the piston acceleration, as functions of crank angle on the same graph as these friction forces. Use these graphs to explain the variation of piston friction force throughout the four strokes of the cycle.

13.7. (a) Show by dimensional analysis of the variables that govern the friction in a journal bearing (friction force F_f, oil viscosity μ, bearing diameter D_b, length L_b, mean clearance $\bar{h}$, shaft rotational speed N) that

$$\frac{F_f}{\mu D_b^2 N} = f\left(\frac{L_b}{D_b}, \frac{\bar{h}}{D_b}\right)$$

What additional physical assumptions are then required to obtain an equation of the form of (13.20)?

(b) Under what conditions can Eq. (13.23), an empirically developed relation for engine bearing fmep, be obtained from Eq. (13.20)?

13.8. Explain whether each of the following components of engine friction would be expected to depend on (1) crankshaft rotational speed N, (2) mean piston speed $\bar{S}_p$, (3) or both of these variables. Crankshaft journal bearings, connecting rod bearings, valve train, piston rings, piston skirt, water pump, fan, valve flow loss (resistance to flow through the inlet and exhaust valves).

REFERENCES

1. Ball, W. F., Jackson, N. S., Pilley, A. D., and Porter, B. C.: "The Friction of a 1.6 Litre Automotive Engine-Gasoline and Diesel," SAE paper 860418, 1986.
2. SAE Test Code J816b.

3. Rosenberg, R. C.: "General Friction Considerations for Engine Design," SAE paper 821576, 1982.

4. Schilling, A.: *Automobile Engine Lubrication*, Scientific Publication, 1972.

5. Millington, B. W., and Hartles, E. R.: "Frictional Losses in Diesel Engines," paper 680590, *SAE Trans.*, vol. 77, 1968.

6. Barnes-Moss, H. W.: "A Designer's Viewpoint," in *Passenger Car Engines, Conference Proceedings*, pp. 133–147, Institution of Mechanical Engineers, London, 1975.

7. Lancaster, D. R., Krieger, R. B., and Lienesch, J. H.: "Measurement and Analysis of Engine Pressure Data," paper 750026, *SAE Trans.*, vol. 84, 1975.

8. Gish, R. E., McCullough, J. D., Retzloff, J. B., and Mueller, H. T.: "Determination of True Engine Friction," *SAE Trans.*, vol. 66, pp. 649–661, 1958.

9. Brown, W. L.: "The Caterpillar IMEP Meter and Engine Friction," paper 730150, *SAE Trans.*, vol. 82, 1973.

10. Kovach, J. T., Tsakiris, E. A., and Wong, L. T.: "Engine Friction Reduction for Improved Fuel Economy," SAE paper 820085, 1982.

11. Cleveland, A. E., and Bishop, I. N.: "Several Possible Paths to Improved Part-Load Economy of Spark-Ignition Engines," SAE paper 150A, 1960.

12. Bishop, I. N.: "Effect of Design Variables on Friction and Economy," *SAE Trans.*, vol. 73, pp. 334–358, 1965.

13. *Piston Rings*, Mobil Technical Bulletin.

14. Nunney, M. J.: *The Automotive Engine*, Newnes-Butterworths, London, 1974.

15. Furuhama, S., Takiguchi, M., and Tomizawa, K.: "Effect of Piston and Piston Ring Designs on the Piston Friction Forces in Diesel Engines," SAE paper 810977, *SAE Trans.*, vol. 90, 1981.

16. Furuhama, S., Ashi, C., and Hiruma, M.: "Measurement of Piston Ring Oil Film Thickness in an Operating Engine," ASLE preprint 82-LC-6C-1, 1982.

17. McGeehan, J. A.: "A Literature Review of the Effects of Piston and Ring Friction and Lubricating Oil Viscosity on Fuel Economy," SAE paper 780673, *SAE Trans.*, vol. 87, 1978.

18. Furuhama, S., and Takiguchi, M.: "Measurement of Piston Frictional Force in Actual Operating Diesel Engine," SAE paper 790855, *SAE Trans.*, vol. 88, 1979.

19. Cameron, A.: *The Principles of Lubrication*, Wiley, New York, 1966.

20. McGeehan, J. A.: "A Survey of the Mechanical Design Factors Affecting Engine Oil Consumption," SAE paper 790864, *SAE Trans.*, vol. 88, 1979.

21. Armstrong, W. B., and Buuck, B. A.: "Valve Gear Energy Consumption: Effect of Design and Operational Parameters," SAE paper 810787, 1981.

22. Staron, J. T., and Willermet, P. A.: "An Analysis of Valve Train Friction in Terms of Lubrication Principles," SAE paper 830165, *SAE Trans.*, vol. 92, 1983.

23. Burke, C. E., Nagler, L. H., Campbell, E. C., Lundstrom, L. C., Zierer, W. E., Welch, H. L., Kosier, T. D., and McConnell, W. A.: "Where Does All the Power Go," *SAE Trans.*, vol. 65, pp. 713–737, 1957.

24. Dean, J. W., and Casebeer, H. M.: "Chrysler 340 Cu In. V-8 Engine Produces 275 HP at 5000 RPM," SAE paper 680019, 1968.

25. Schilling, A.: *Motor Oils and Engine Lubrication*, Scientific Publications, 1968.

26. Annand, W. J., and Roe, G. E.: *Gas Flow in the Internal Combustion Engine*, Haessner Publishing, 1974.

27. ASTM Standards, Part 17, Petroleum Products.

28. ASTM D2270-64.

29. SAE J300a.

MODELING REAL ENGINE FLOW AND COMBUSTION PROCESSES

14.1 PURPOSE AND CLASSIFICATION OF MODELS

In engineering, modeling a process has come to mean developing and using the appropriate combination of assumptions and equations that permit critical features of the process to be analyzed. The modeling of engine processes continues to develop as our basic understanding of the physics and chemistry of the phenomena of interest steadily expands and as the capability of computers to solve complex equations continues to increase. Modeling activities can make major contributions to engine engineering at different levels of generality or detail, corresponding to different stages of model development, by:

1. Developing a more complete understanding of the process under study from the discipline of formulating the model;
2. Identifying key controlling variables to provide guidelines for more rational and therefore less costly experimental development efforts;
3. Predicting engine behavior over a wide range of design and operating variables to screen concepts prior to major hardware programs, to determine

trends and tradeoffs, and, if the model is sufficiently accurate, to optimize design and control;

4. Providing a rational basis for design innovation.

Each of these contributions is valuable. Whether a model is ready to pass from one stage to the next depends on the accuracy with which it represents the actual process, the extent to which it has been tested and validated, and the time and effort required to use the model for extensive sets of calculations and to interpret the results.

This chapter reviews the types of models and their primary components that are being developed and used to describe engine operating and emissions characteristics. These models describe the thermodynamic, fluid-flow, heat-transfer, combustion, and pollutant-formation phenomena that govern these performance aspects of engines. Many of the building blocks for these models have been described in the previous chapters. The purpose here is to show how fluid dynamics, heat-transfer, thermodynamics, and kinetics fundamentals can be combined at various levels of sophistication and complexity to predict, with varying degrees of completeness, internal combustion engine combustion and emissions processes, and hence engine operating characteristics.

For the processes that govern engine performance and emissions, two basic types of models have been developed. These can be categorized as *thermodynamic* or *fluid dynamic* in nature, depending on whether the equations which give the model its predominant structure are based on energy conservation or on a full analysis of the fluid motion. Other labels given to thermodynamic energy-conservation-based models are: zero-dimensional (since in the absence of any flow modeling, geometric features of the fluid motion cannot be predicted), phenomenological (since additional detail beyond the energy conservation equations is added for each phenomenon in turn), and quasi-dimensional (where specific geometric features, e.g., the spark-ignition engine flame or the diesel fuel spray shapes, are added to the basic thermodynamic approach). Fluid-dynamic-based models are often called multidimensional models due to their inherent ability to provide detailed geometric information on the flow field based on solution of the governing flow equations.

Some general observations about models of engine processes provide a context for the details that follow. The processes themselves are extremely complex. While much is known about these processes, they are not adequately understood at a fundamental level. At present, it is not possible to construct models that predict engine operation from the basic governing equations alone. Thus the objectives of any model development effort should be clearly defined, and the structure and detailed content of the model should be appropriate to these objectives. It is impractical to construct models that attempt to describe *all* important aspects of engine operation: more limited objectives are appropriate.

Due to this complexity of engine processes and our inadequate understanding at a fundamental level, most engine models are incomplete. Empirical relations and *ad hoc* approximations are often needed to bridge gaps in our understanding of critical phenomena. Hence, since models will continue to develop greater completeness and generality, the emphasis in this chapter is on the basic relationships used in engine process models rather than the current status of these models.

Finally, an important issue in any overall engine model is balance in complexity and detail amongst the process submodels. A model is no more accurate than its weakest link. Thus critical phenomena should be described at comparable levels of sophistication.

14.2 GOVERNING EQUATIONS FOR OPEN THERMODYNAMIC SYSTEM

It is often required to model a region of the engine as an open thermodynamic system. Examples are the cylinder volume and the intake and exhaust manifolds (or portions of these volumes). Such a model is appropriate when the gas inside the open system boundary can be assumed uniform in composition and state at each point in time, and when that state and composition vary with time due to heat transfer, work transfer and mass flow across the boundary, and boundary displacement. Such an open system is illustrated in Fig. 14-1. The important equations are mass and energy conservation. These equations for an open system, with time or crank angle as the independent variable, are the building blocks for thermodynamic-based models.

14.2.1 Conservation of Mass

The rate of change of the total mass m of an open system is equal to the sum of the mass flows into and out of the system:

$$\dot{m} = \sum_j \dot{m}_j \tag{14.1}$$

Mass flows into the system are taken as positive; mass flows out are taken as negative. For conservation of the fuel chemical elements, it is convenient to use the fuel fraction f, which is defined as m_f/m, where m_f denotes the mass of fuel (or

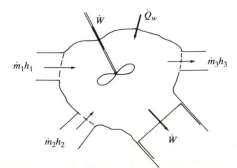

FIGURE 14-1
Open thermodynamic system.

fuel elements in the combustion products) in the open system:

$$\dot{m}_f = \frac{d}{dt}(mf) = \sum_j \dot{m}_{f,j} = \sum_j \dot{m}_j f_j \tag{14.2}$$

Differentiation of Eq. (14.2) leads to an equation for the rate of change of fuel fraction:

$$\dot{f} = \sum_j \left(\frac{\dot{m}_j}{m}\right)(f_j - f) \tag{14.3}$$

The fuel/air equivalence ratio ϕ is related to f via $\phi = f/[(F/A)_s(1-f)]$. Hence the rate of change of equivalence ratio of the material in the open system is

$$\dot{\phi} = \frac{1}{(F/A)_s}\frac{\dot{f}}{(1-f)^2} \tag{14.4}$$

14.2.2 Conservation of Energy

The first law of thermodynamics for the open system in Fig. 14-1 can be written:

$$\dot{E} = \dot{Q}_w - \dot{W} + \sum_j \dot{m}_j h_j \tag{14.5}$$

$\dot{Q}_w$ is the total heat-transfer rate into the system, across the system boundary, and equals the sum of the heat-transfer rates across each part of the boundary, $\sum_i \dot{Q}_{w,i}$. $\dot{W}$ is the work-transfer rate out of the system across the boundary; where the piston is displaced, the work-transfer rate equals $p\dot{V}$. Because all energies and enthalpies are expressed relative to the same datum (see Sec. 4.5.3), it is not necessary to include the heat released by combustion in Eq. (14-5); this is already accounted for in the energy and enthalpy terms.

The goal is to define the rate of change of state of the open system in terms of $\dot{T}$ and $\dot{p}$. Two approaches are commonly used, depending on whether the thermodynamic property routines provide values for internal energy u or enthalpy h. Thus $\dot{E}$ in Eq. (14.5) can be expressed as

$$\dot{E} = \frac{d}{dt}(mu) \quad \text{or} \quad \dot{E} = \frac{d}{dt}(mh) - \frac{d}{dt}(pV) \tag{14.6a, b}$$

It is assumed that the system can be characterized by T, p, and ϕ; thus

$$u = u(T, p, \phi) \qquad h = h(T, p, \phi) \qquad \rho = \rho(T, p, \phi) \tag{14.7}$$

and the rate of change of u, h, and ρ can be written in the form

$$\dot{\alpha} = \left(\frac{\partial\alpha}{\partial T}\right)\dot{T} + \left(\frac{\partial\alpha}{\partial p}\right)\dot{p} + \left(\frac{\partial\alpha}{\partial\phi}\right)\dot{\phi} \tag{14.8}$$

where α is u, h, or ρ. Using the ideal gas law in its two forms, $p = \rho RT$ and $pV = mRT$, and Eq. (14.8) for $\dot{\rho}$, an equation for $\dot{p}$ can be derived:

$$\dot{p} = \frac{\rho}{\partial \rho / \partial p} \left(-\frac{\dot{V}}{V} - \frac{1}{\rho} \frac{\partial \rho}{\partial T} \dot{T} - \frac{1}{\rho} \frac{\partial \rho}{\partial \phi} \dot{\phi} + \frac{\dot{m}}{m} \right) \tag{14.9}$$

Returning now to the energy conservation equation, expressing $\dot{E}$ in terms of $\dot{u}$ or $\dot{h}$, and $\dot{u}$ or $\dot{h}$ in terms of partial derivatives with respect to T, p, and ϕ, and substituting for $\dot{p}$ with Eq. (14.9), one can obtain equations for $\dot{T}$:

$$\dot{T} = \left[B - \frac{p}{D} \frac{\partial u}{\partial p} \left(\frac{\dot{m}}{m} - \frac{\dot{V}}{V} + \frac{\partial R}{\partial \phi} \frac{\dot{\phi}}{R} \right) - \frac{\partial u}{\partial \phi} \dot{\phi} \right] \bigg/ \left(\frac{\partial u}{\partial T} + \frac{C}{D} \frac{p}{T} \frac{\partial u}{\partial p} \right) \tag{14.10}$$

where

$$B = -RT \frac{\dot{V}}{V} + \frac{1}{m} \left(\dot{Q}_w + \sum_j \dot{m}_j h_j - \dot{m} u \right)$$

$$C = 1 + \frac{T}{R} \frac{\partial R}{\partial T}$$

$$D = 1 - \frac{p}{R} \frac{\partial R}{\partial p}$$

(see Ref. 1, for example). From Ref. 2,

$$\dot{T} = \frac{B'}{A'} \left[\frac{\dot{m}}{m} \left(1 - \frac{h}{B'} \right) - \frac{\dot{V}}{V} - \frac{C'}{B'} \dot{\phi} + \frac{1}{B'm} \left(\sum_j \dot{m}_j h_j - \dot{Q}_w \right) \right] \tag{14.11}$$

where

$$A' = \frac{\partial h}{\partial T} + \frac{\partial \rho / \partial T}{\partial \rho / \partial p} \left(\frac{1}{\rho} - \frac{\partial h}{\partial p} \right)$$

$$B' = \frac{1 - \rho (\partial h / \partial p)}{\partial \rho / \partial p}$$

$$C' = \frac{\partial h}{\partial \phi} + \frac{\partial \rho / \partial \phi}{\partial \rho / \partial p} \left(\frac{1}{\rho} - \frac{\partial h}{\partial p} \right)$$

Equations (14.1), (14.3), (14.4), (14.9), and (14.10) or (14.11) can now be solved to obtain the state of the open system as a function of time. $\dot{V}$ is obtained from Eq. (2.6), and the thermodynamic properties and their derivatives from the models described in Chap. 4.

Often, for specific applications, the above equations can be simplified substantially. For the intake and exhaust systems (or sections of these systems such as the manifold or plenum, etc.), $\dot{V}$ is zero and effects of dissociation (the terms $\partial u / \partial p$, $\partial h / \partial p$, and $\partial R / \partial p$) can usually be neglected. For the cylinder during compression, dissociation can usually be neglected, also. Application of these equations during combustion must be related to the combustion model used. For the single-zone model often used in diesel engine simulations (see Sec. 10.4) the whole of the combustion chamber is treated as one system. For the two-zone model used for spark-ignition engine simulations, the unburned mixture zone and the burned mixture zone are each treated as separate open systems, with volumes V_u and V_b, respectively, where $V_u + V_b = V$. If a thermal boundary-layer region is included (see Sec. 12.6.5) an additional open system must be defined.

14.3 INTAKE AND EXHAUST FLOW MODELS

14.3.1 Background

The behavior of the intake and exhaust systems are important because these systems govern the air flow into the engine's cylinders. Inducting the maximum air flow at full load at any given speed and retaining that mass within the engine's cylinders is a primary design goal. The higher the air flow, the larger the amount of fuel that can be burned and the greater the power produced. The important parameters are volumetric efficiency (for four-stroke cycle engines) or scavenging and trapping efficiencies (for two-stroke cycle engines), along with equal air flows to each engine cylinder (see Secs. 6.2, 6.6, and 7.6.2).

   The objectives of any manifold model have an important bearing on its complexity and structure. If the goal is to provide the input or boundary conditions to a detailed model of in-cylinder processes, then sophisticated intake and exhaust system models are not necessarily required. If the manifold flows are the primary focus, then models that adequately describe the unsteady gas-flow phenomena which occur are normally required. Then simple models for the in-cylinder phenomena usually suffice to connect the intake and exhaust processes. The valves and ports, which together provide the major restriction to the intake and exhaust flow, largely decouple the manifolds from the cylinders.

   Three types of models for calculating details of intake and exhaust flows have been developed and used:

1. Quasi-steady models for flows through the restrictions which the valve and port (and other components) provide
2. Filling and emptying models, which account for the finite volume of critical manifold components
3. Gas dynamic models which describe the spatial variations in flow and pressure throughout the manifolds

   Each of these types of models can be useful for analyzing engine behavior. The appropriate choice depends on objectives, and the time and effort available. Each will now be reviewed.

14.3.2 Quasi-Steady Flow Models

Here the manifolds are considered as a series of interconnected components, which each constitute a significant flow restriction: e.g., air cleaner, throttle, port, and valve for the intake system. The flow restriction each of these components represents is defined by their geometry and discharge coefficient, usually determined empirically under steady-state conditions. The gas flow rate through each component is computed using steady one-dimensional flow equations [see App. C, Eqs. (C.8) and (C.9)]: the actual flow is assumed to be quasi steady. These components are connected by the gas flow passing through them and the pressure ratios across them; mass accumulation between components is neglected.

Quasi-steady models are often used to calculate the flow into and out of the cylinder through the inlet and exhaust valves (see Secs. 6.3 and 6.5 and Fig. 6-20). If the pressure variation with time upstream of the valve is known or is small, as usually occurs with large plenums and short manifold pipe lengths, such methods are accurate enough to be useful. This approach has been used extensively with engine cycle simulations which predict engine performance characteristics from a thermodynamics-based analysis, to calculate the mass flow rates into and out of the cylinder (see Sec. 14.4). Such methods are not able to predict the variation of volumetric efficiency with engine speed, however, because many of the phenomena which govern this variation are omitted from this modelling approach (see Sec. 6.2 and Fig. 6-9).

14.3.3   Filling and Emptying Methods

In "filling and emptying" models, the manifolds (or sections of manifolds) are represented by finite volumes where the mass of gas can increase or decrease with time. Such models can range from treating the whole intake or exhaust system as a single volume to dividing these systems into many sections, with flow restrictions such as the air cleaner, throttle valve, or inlet valve at the beginning, in between volumes, or at the end. Each volume is then treated as a control volume (an open system of fixed volume) which contains gas at a uniform state. The mass and energy conservation equations developed in Sec. 14.2 [Eqs. (14.1), (14.3), (14.9), and (14.10) or (14.11)], coupled with information on the mass flow rates into and out of each volume [e.g., determined by the equations for flow through a restriction, Eqs. (C.8) and (C.9)] are used to define the gas state in each control volume. For intake and exhaust flows these equations can be simplified since the volumes are fixed ($\dot{V} = 0$), gas composition can be assumed frozen ($\partial u / \partial p$, $\partial h / \partial p$, and $\partial R / \partial p$ are then zero), unless backflow occurs or recycled exhaust is used for emission control changes in fuel fraction are not significant, and for intake systems it may be acceptable to omit heat transfer to the walls ($\dot{Q}_w$). Such methods characterize the contents of the manifold (or a region thereof) with a single gas temperature, pressure, and composition. These vary periodically with time as each cylinder in turn draws on the intake system and discharges to the exhaust system. Also, under transient conditions when engine load and/or speed change with time, manifold conditions will vary until the new engine steady-state conditions are established. Watson and Janota[3] discuss the application of filling and emptying models to manifolds in more detail. Such models can characterize these time-varying phenomena, spatially averaged over each mani-

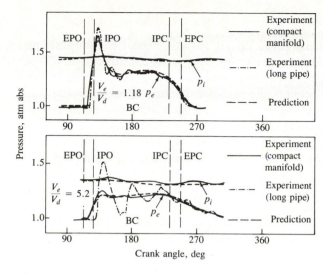

FIGURE 14-2
Comparison of intake and exhaust manifold pressures, p_i and p_e, predicted by filling and emptying model, with experimental data. Single-cylinder two-stroke loop-scavenged direct-injection diesel engine. Different ratios of exhaust system volume V_e to displaced volume V_d, and exhaust manifold shapes.[4]

EPO Exhaust port opens
IPO Inlet port opens
IPC Inlet port closes
EPC Exhaust port closes

fold region corresponding to each volume analyzed; however, they cannot describe the spatial variation of pressure (and other gas properties) due to unsteady gas dynamics in the manifolds.

A simple application of a filling and emptying model to the intake manifold of a spark-ignition engine was described in Sec. 7.6.2. The manifold was analyzed as a single control volume with the throttle plate controlling mass flow into the manifold and the engine cylinders controlling mass flow out. An equation for the rate of change of manifold pressure [Eq. (7.22)] was derived and used to explain how the air flow past the throttle varied as the throttle open angle was increased, as would occur at the start of a vehicle acceleration at part-throttle conditions (see Fig. 7-24).

A second example will illustrate the conditions under which filling and emptying models give sufficiently accurate predictions to be useful.[4] It concerns a single-cylinder two-stroke cycle loop-scavenged direct-injection diesel engine. The engine was modeled as three open systems (the intake system, the cylinder, the exhaust system) connected by flow restrictions. Various exhaust manifold volumes and shapes were examined, using nozzles at the manifold exit to simulate the exhaust-driven turbine. The in-cylinder models were calibrated to match the measured engine performance. Figure 14-2 shows the predicted and measured pressure variation at the exhaust system exit for two exhaust system volumes (V_e). With the compact manifold the measured and predicted pressures were in good agreement. With the larger exhaust system shown in the figure ($V_e/V_d = 5.2$) and the compact manifold, good agreement is again obtained. Only with the larger volume and long pipe exhaust system is there evidence in the measured pressure variation of substantial unsteady gas dynamic effects. For small manifolds, and manifolds that are compact in shape, filling and emptying models can be a useful predictive tool.

14.3.4 Gas Dynamic Models

Many induction and exhaust system design variables determine overall performance. These variables include the length and cross-sectional area of both primary and secondary runners, the volume and location of the plenums or junctions which join the various runners, the entrance or exit angles of the runners at a junction, the number of engine cylinders and their dimensions, intake and exhaust port and valve design, and valve lift and timing (see Secs. 6.2, 6.3, 6.7, and 7.6). Most of this geometric detail is beyond the level which can be incorporated into the models discussed above. Coupled with the pulsating nature of the flow into and out of each cylinder, these details create significant gas dynamic effects on intake and exhaust flows which require a more complete modeling approach.

Gas dynamic models have been in use for a number of years to study engine gas exchange processes. These models use the mass, momentum, and energy conservation equations for the unsteady compressible flow in the intake and exhaust. Normally, the one-dimensional unsteady flow equations are used.† These models often use a thermodynamic analysis of the in-cylinder processes to link the intake and exhaust flows. In the past, the method of characteristics was used to solve the gas dynamic equations. Finite difference techniques are used in more recent intake and exhaust flow models. The basic equations and assumptions of these models will now be reviewed.[5, 6]

UNSTEADY FLOW EQUATIONS. Consider the flow through the control volume within a straight duct shown in Fig. 14-3. It is assumed that the area change over the length dx of the control volume is small so the flow is essentially one-

---

† Two- and three-dimensional effects can be important and can be modeled with multidimensional flow models described in Sec. 14.5.

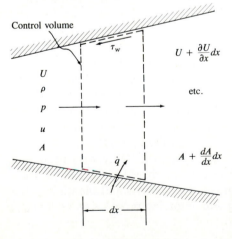

FIGURE 14-3
Control volume for unsteady one-dimensional flow analysis.

dimensional. Mass conservation requires that the rate of change of mass within the control volume equals the net flow into the control volume: i.e.,

$$\frac{\partial}{\partial t} (\rho A \ dx) = \rho A U - \left[\rho A U + \frac{\partial}{\partial x} (\rho A U) \ dx \right] \tag{14.12}$$

Retaining only first-order quantities, this equation simplifies to

$$\frac{\partial \rho}{\partial t} + \frac{\partial}{\partial x} (\rho U) + \frac{\rho U}{A} \frac{dA}{dx} = 0 \tag{14.13}$$

The momentum conservation equation states that the net pressure forces plus the wall shear force acting on the control volume surface equal the rate of change of momentum within the control volume plus the net flow of momentum out of the control volume. The net forces and momentum changes are given by:

Pressure forces:

$$pA - \left(p + \frac{\partial p}{\partial x} \ dx \right) \left(A + \frac{dA}{dx} \ dx \right) + p \frac{dA}{dx} \ dx = - A \frac{\partial p}{\partial x} \ dx$$

Shear forces:

$$- \tau_w \pi D \ dx = - \xi \frac{\rho U^2}{2} \pi D \ dx$$

where D is the equivalent diameter $(4A/\pi)^{1/2}$ and ξ is the friction coefficient given by $\tau_w / (\frac{1}{2}\rho U^2)$.

The rate of change of momentum within the control volume is

$$\frac{\partial}{\partial t} (U \rho A \ dx)$$

and the net efflux of momentum across the control volume surface is

$$\left(\rho + \frac{\partial \rho}{\partial x} \ dx \right) \left(U + \frac{\partial U}{\partial x} \ dx \right)^2 \left(A + \frac{dA}{dx} \ dx \right) - \rho U^2 A = \frac{\partial}{\partial x} (\rho U^2 A) \ dx$$

Combining these terms into the momentum equation yields

$$- A \frac{\partial p}{\partial x} \ dx - \xi \frac{\rho U^2}{2} \pi D \ dx = \frac{\partial}{\partial t} (\rho U A \ dx) + \frac{\partial}{\partial x} (\rho U^2 A) \ dx \tag{14.14}$$

This can be rearranged and combined with the mass conservation equation (14.13) to give

$$\frac{\partial U}{\partial t} + U \frac{\partial U}{\partial x} + \frac{1}{\rho} \frac{\partial p}{\partial x} + 2\xi \frac{U^2}{D} = 0 \tag{14.15}$$

ENERGY CONSERVATION. The first law of thermodynamics for a control volume states that the energy within the control volume changes due to heat and shear work transfers across the control volume surface and due to a net efflux of stagnation enthalpy resulting from flow across the control volume surface. The stagnation enthalpy h_0 is

$$h_0 = h + \frac{U^2}{2} = u + \frac{p}{\rho} + \frac{U^2}{2}$$

where u is the specific internal energy of the fluid (often approximated by $c_v T$). The shear work transfer across the control volume surface is zero.

The heat-transfer rate $\dot{Q}_w$ is given by

$$\delta\dot{Q}_w = \dot{q}\rho A \ dx$$

where $\dot{q}$ is the heat transfer per unit mass of fluid per unit time into the control volume.

The rate of change of energy within the control volume is

$$\frac{\partial E}{\partial t} = \frac{\partial}{\partial t}\left[(\rho A \ dx)\left(u + \frac{U^2}{2}\right)\right]$$

The net efflux of stagnation enthalpy is

$$\frac{\partial}{\partial x}\left[(\rho U A)\left(u + \frac{p}{\rho} + \frac{U^2}{2}\right)\right]dx$$

Hence, the equation for energy conservation becomes

$$\frac{\partial}{\partial t}\left[(\rho A \ dx)\left(u + \frac{p}{\rho} + \frac{U^2}{2}\right)\right] + \frac{\partial}{\partial x}\left[(\rho U A)\left(u + \frac{p}{\rho} + \frac{U^2}{2}\right)\right]dx - \dot{q}\rho A \ dx = 0$$

$$(14.16)$$

Additional simplifications are possible. Expanding Eq. (14.16) and using the mass and momentum conservation equations yields

$$\frac{\partial u}{\partial t} + U\frac{\partial u}{\partial x} = \dot{q} + 2\xi\frac{U^3}{D} - \frac{p}{\rho A}\frac{\partial(UA)}{\partial x} \qquad (14.17)$$

If u can be represented by $c_v T$ and $R/c_v = \gamma - 1$ is constant, Eq. (14.17) can be rearranged and simplified to give

$$\frac{\partial p}{\partial t} + U\frac{\partial p}{\partial x} - a^2\left(\frac{\partial \rho}{\partial t} + U\frac{\partial \rho}{\partial x}\right) - (\gamma - 1)\rho\left(\dot{q} + 2\xi\frac{U^3}{D}\right) = 0 \qquad (14.18)$$

where the sound speed a for an ideal gas is given by

$$a^2 = \left(\frac{\partial p}{\partial \rho}\right)_s = \gamma\frac{p}{\rho} \qquad (14.19)$$

If friction and heat-transfer effects are small enough to be neglected, Eqs. (14.15) and (14.18) can be considerably simplified. In the absence of these effects the flow is *isentropic*; it has uniform entropy which is constant with time and is often called *homentropic* flow.[6] If the duct area can be neglected then the continuity equation, (14.13), can be simplified also.

These one-dimensional unsteady flow equations have been used for a number of years to study the flow in the intake and exhaust systems of spark-ignition and diesel engines, both naturally aspirated and turbocharged. Two types of methods have been used to solve these equations: (1) the method of characteristics and (2) finite difference procedures. The characteristic methods have a numerical accuracy that is first order in space and time, and require a large number of computational points if resolution of short-wavelength variations is important. Finite difference techniques can be made higher order and prove to be more efficient:[7, 8] this approach is now preferred. Methods for treating the boundary conditions will also be described.

METHOD OF CHARACTERISTICS. The method of characteristics is a well-established mathematical technique for solving hyperbolic partial differential equations. With this technique, the partial differential equations are transformed into ordinary differential equations that apply along so-called characteristic lines. Pressure waves are the physical phenomenon of practical interest in the unsteady intake flow, and these propagate relative to the flowing gas at the local sound speed. In this particular application, the one-dimensional unsteady flow equations, (14.13) and (14.15), are rearranged so that they contain only the local fluid velocity U and local sound speed a.

Since the absolute velocity of small amplitude sound waves is $U + a$ in the direction of flow and $U - a$ opposite to the flow direction, the lines of slope $U \pm a$ are the *position* characteristics of the propagating pressure waves which define the position x of the pressure wave at time t. *Compatability conditions* accompanying the position characteristics relate U to a. The compatability relationships are expressed in terms of variables (called Riemann invariants) which are constant along the position characteristics for constant-area homentropic flow, though they vary if these restrictions do not apply. Thus, the solution of the mass and momentum conservation equations for this one-dimensional unsteady flow is reduced to the solution of a set of ordinary differential equations.

The equations are usually solved numerically using a rectangular grid in the x and t directions. The intake or exhaust system is divided into individual pipe sections which are connected at junctions. A mesh is assigned to each section of pipe between junctions. From the initial values of the variables at each mesh point at time $t = 0$, the values of the Riemann variables at each mesh point at subsequent time steps are then determined. Gas pressure, density, and temperature can then be calculated from the energy conservation equation and the ideal gas law. Additional details of the method are given by Benson *et al.*[5, 6]

FINITE DIFFERENCE METHODS. Finite difference methods for solving the one-dimensional unsteady flow equations in intake and exhaust manifolds are proving more efficient and flexible than the method of characteristics. The conservation equations, (14.13), (14.14), and (14.16), can be rearranged and written in

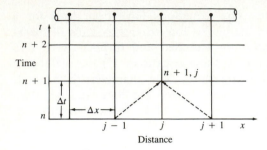

FIGURE 14-4
Mesh in time-distance plane for application of one-step Lax-Wendroff method to intake or exhaust pipe.

matrix form as

$$\frac{\partial}{\partial t}\begin{pmatrix}\rho \\ \rho U \\ \rho u\end{pmatrix}+\frac{\partial}{\partial x}\begin{pmatrix}\rho U \\ \rho U^2+p \\ pU+\rho Uu\end{pmatrix}=\begin{pmatrix}-\rho U\dfrac{dA}{dx} \\[2mm] -\rho\dfrac{U^2}{A}\dfrac{dA}{dx}-\rho\dfrac{2\xi U|U|}{D} \\[2mm] -\dfrac{4h_c(T-T_w)}{D\rho}-\dfrac{1}{A}\dfrac{dA}{dx}\left(\dfrac{1}{2}\rho U^3+\dfrac{\gamma}{\gamma-1}Up\right)\end{pmatrix}$$

(14.20)

The fluid viscous shear is small relative to friction at the wall in the momentum equation, and heat conduction and viscous dissipation prove negligible relative to convective heat transfer at the wall in the energy conservation equation. These equations have the vector form:

$$\frac{\partial F}{\partial t}+\frac{\partial G}{\partial x}=H$$

(14.21)

where G and H are functions of F only. Several finite difference methods have been used to solve Eq. (14.21) (see Refs. 7, 8, and 9). The one-step Lax-Wendroff method will be illustrated.[8] Equation (14.21) can be developed into a Taylor series with respect to time, and the time and space derivatives approximated by central differences around the mesh point, shown in Fig. 14-4, as

$$F_j^{n+1}=F_j^n-\frac{1}{2}\frac{\Delta t}{\Delta x}(G_{j+1}^n-G_{j-1}^n)+\Delta tH_j^n$$

$$+\frac{1}{4}\left(\frac{\Delta t}{\Delta x}\right)^2[(G_{j+1}'^n+G_j'^n)(G_{j+1}^n-G_j^n)-(G_j'^n+G_{j-1}'^n)(G_j^n-G_{j-1}^n)] \quad (14.22)$$

where $G'=\partial G/\partial F$. This equation is first-order accurate, unless H is small. For stability in the integration process, the time step and mesh size must satisfy the requirement that

$$C=(|U|+a)\frac{\Delta t}{\Delta x}<1$$

(14.23)

where C is the Courant number.

TABLE 14.1
Boundary conditions for unsteady one-dimensional finite element analysis[9]

Pipe ends			
Out-flow	Mass	$\rho_1 U_1 A_1 = \rho_2 U_2 A_2$	
	Energy	$c_p T_1 + \dfrac{U_1^2}{2} = c_p T_2 + \dfrac{U_2^2}{2}$	
	Isobaric	$p_2 = p_3$	
In-flow	Mass	$\rho_1 U_1 A_1 = \rho_2 U_2 A_2$	
	Energy	$c_p T_3 = c_p T_2 + \dfrac{U_2^2}{2} = c_p T_1 + \dfrac{U_1^2}{2}$	
	Isentropic	$p_2/\rho_2^\gamma = p_3/\rho_3^\gamma$	
Pipe junctions			
	Mass	$V \dfrac{\partial \rho}{\partial t} = \sum_i \rho_i U_i A_i$	
	Energy	$\rho V \dfrac{\partial u}{\partial t} = \sum_i (\rho_i U_i A_i)\left(c_p T_i + \dfrac{U_i^2}{2}\right)$	
	Pressure	$p_1 - \Delta p_1 = p_2 + \Delta p_2 = p_3 + \Delta p_3 = \cdots$ $\Delta p_i/p_i = C_i (U_i/a_i)^2$	

These finite difference solution methods usually require the introduction of some form of dissipation or damping to prevent instabilities and large non-physical oscillations from occurring with nonlinear problems with large gradients (e.g., a shock wave in the exhaust system). Amplification of the physical viscosity and the addition of artificial viscosity, damping, and smoothing terms to Eq. (14.22) are frequently used techniques.[8, 9]

The boundary conditions at pipe ends and junctions are obtained from the appropriate conservation equations and pressure relations, as illustrated in Table 14.1. Out-flows and in-flows obviously conserve mass and energy. For the flow out through a restriction, there is no pressure recovery downstream: for flow in through a restriction, the flow upstream of the restriction is isentropic. For pipe junctions, the conservation equations are applied to the control volume contained within the dashed line in the sketch in the table. The pressure boundary conditions are most easily estimated by modifying the simple constant-pressure assumption with pressure losses at each pipe exit or entry, calculated from experimentally determined loss coefficients (see Fig. 6-5).[9]

Calculations of intake and exhaust flows using these techniques predict the variations in intake and exhaust manifold pressure with crank angle (as shown, for example, in Fig. 6-7), in single and multicylinder engines, with acceptable accuracy.[7, 9] Measured volumetric efficiency variations with engine speed, manifold design, and valve dimensions and timing are adequately predicted also. Figure 14-5a shows the instantaneous exhaust and intake mass flow rates for cylinder number 1 of a four-cylinder spark-ignition engine at wide-open throttle at 1500 rev/min. Note how gas dynamic effects distort the exhaust flow. Note also the "reverse" flows into the cylinder past the exhaust valve and out of the

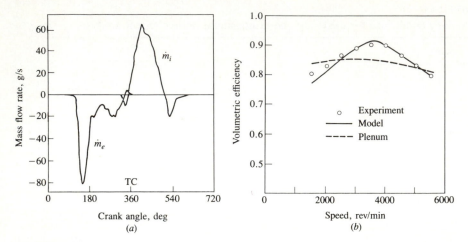

FIGURE 14-5

(a) Predicted mass flow rate through the exhaust valve $\dot{m}_e$ and through the intake valve $\dot{m}_i$ in cylinder 1, four-cylinder four-stroke-cycle spark-ignition engine at wide-open throttle and 1500 rev/min. Flows into cylinder are positive; flows out are negative. (b) Predicted and measured volumetric efficiency at wide-open throttle for four-cylinder spark-ignition engine. Solid line: one-dimensional unsteady flow model. Dashed line: quasi-steady flow calculation based on infinite plenums for manifolds.[7]

cylinder past the intake valve at the end of the exhaust process, and the larger reverse flow at the end of the intake process at this low engine speed. Figure 14-5b shows the volumetric efficiency for this engine based on these predicted mass flow rates, as a function of speed. Experimental values and values predicted with quasi-steady flow equations and infinite plenums for manifolds are also shown. These results clearly demonstrate the important role that intake and exhaust system gas dynamics play in determining both the engine speed at which peak breathing efficiency occurs and the air charging characteristics over the full engine speed range.[7]

14.4   THERMODYNAMIC-BASED IN-CYLINDER MODELS

14.4.1   Background and Overall Model Structure

If the mass transfer into and out of the cylinder during intake and exhaust, the heat transfer between the in-cylinder gases and the cylinder head, piston, and cylinder liner, and the rate of charge burning (or energy release from the fuel) are all known, the energy and mass conservation equations permit the cylinder pressure and the work transfer to the piston to be calculated. Engine models of this type have been developed and used extensively to predict engine operating characteristics (indicated power, mean effective pressure, specific fuel consumption, etc.) and to define the gas state for emission calculations. These models effectively follow the changing thermodynamic and chemical state of the working fluid through the engine's intake, compression, combustion, expansion, and exhaust processes; they are often called engine *cycle simulations*.

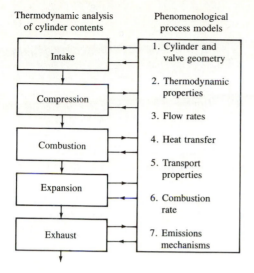

Thermodynamic analysis
of cylinder contents

Phenomenological
process models

Intake

Compression

Combustion

Expansion

Exhaust

1. Cylinder and
   valve geometry

2. Thermodynamic
   properties

3. Flow rates

4. Heat transfer

5. Transport
   properties

6. Combustion
   rate

7. Emissions
   mechanisms

FIGURE 14-6
Logic structure of thermodynamic-based simulations of internal combustion engine operating cycle.

The starting point for these cycle simulations is the first law of thermodynamics for an open system, developed in Sec. 14.2. This is applied to the cylinder volume for the intake, compression, combustion, expansion, and exhaust processes that in sequence make up the engine's operating cycle. The structure of this type of engine simulation is indicated in Fig. 14-6. Then, during each process, submodels are used to describe geometric features of the cylinder and valves or ports, the thermodynamic properties of the unburned and burned gases, the mass and energy transfers across the system boundary, and the combustion process.

During intake and compression, the cylinder volume is modeled as a single open system. Application of the conservation equations in the form of Eqs. (14.1), (14.3), and (14.10) or (14.11) for the intake and then the compression process gives[2]

Intake:

$$\dot{m} = \dot{m}_i - \dot{m}_e \tag{14.24}$$

$$\dot{f} = \frac{\dot{m}_i}{m}(f_i - f) - \frac{\dot{m}_e}{m}(f_e - f) \tag{14.25}$$

$$\dot{T} = \frac{B}{A}\left[\frac{\dot{m}}{m}\left(1 - \frac{h}{B}\right) - \frac{\dot{V}}{V} - \frac{C}{B}\dot{\phi} + \frac{1}{Bm}(\dot{m}_i h_i - \dot{m}_e h_e - \dot{Q}_w)\right] \tag{14.26}$$

where m is the mass of gas in the cylinder, $\dot{m}_i$ and $\dot{m}_e$ are the mass flow rates through the inlet valve and the exhaust valve, and f is the fuel fraction m_f/m. The subscripts i and e denote properties of the flow through the intake and exhaust valves, respectively. The thermodynamic properties for these flows are the values upstream of the valves and therefore depend on whether the flow is into or out of the cylinder.

Compression:

$$\dot{m} = 0 \qquad \dot{f} = 0 \tag{14.27a, b}$$

$$\dot{T} = \frac{B}{A}\left(-\frac{\dot{V}}{V} - \frac{Q_w}{Bm}\right) \tag{14.28}$$

The pressure is then determined from Eq. (14.9).

During intake and compression, the working fluid composition is frozen. The composition and thermodynamic properties can be determined using the models described in Secs. 4.2 and 4.7. Mass flows across open valves are usually calculated using one-dimensional compressible flow equations for flow through a restriction (see App. C and Secs. 6.3.2 and 14.3.2) or filling and emptying models (Sec. 14.3.3). The more accurate unsteady gas dynamic intake (and exhaust) flow models described in Sec. 14.3.4 are sometimes used to calculate the mass flow into the engine cylinder in complete engine cycle simulations when the variation in engine flow rate with speed is especially important:[10] the disadvantage is much increased computing time. Heat transfer during intake and compression is calculated using one of the Nusselt-Reynolds number relations for turbulent convective heat transfer described in Sec. 12.4.5. The transport properties, viscosity, and thermal conductivity used in these correlations can be obtained from relations such as Eqs. (4.52) to (4.55).

During combustion which starts with the spark discharge in spark-ignition engines and with spontaneous ignition of the developing fuel-air jets in compression-ignition engines, the actual processes to be modeled become much more complex. Many approaches to predicting the burning or chemical energy release rate have been used successfully to meet different simulation objectives. The simplest approach has been to use a one-zone model where a single thermodynamic system represents the entire combustion chamber contents and the energy release rate is defined by empirically based functions specified as part of the simulation input. At the other extreme, quasi-geometric models of turbulent premixed flames are used with a two-zone analysis of the combustion chamber contents—an unburned and a burned gas region—in more sophisticated simulations of spark-ignition engines. In compression-ignition engines, multiple-zone models of the developing fuel-air jets have been used to provide more detailed predictions of the combustion process and nonuniform cylinder composition and state. These combustion models will be reviewed in the following sections (14.4.2 and 14.4.3) and the appropriate conservation equations for the *combustion process* will be developed there. In diesels, radiation heat transfer becomes important during the combustion process (see Sec. 12.5).

The *expansion process* is either treated as a continuation of the combustion process or, once combustion is over, can use the form of the mass, fuel, and energy conservation equations which hold during compression [Eqs. (14.27) and (14.28)]. The exhaust process conservation equations for a one-zone open-system model of the cylinder contents are[2]

Exhaust:

$$\dot{m} = -\dot{m}_e \qquad \dot{f} = -\frac{\dot{m}_e}{m}(f_e - f) \qquad (14.29a, b)$$

$$\dot{T} = \frac{B}{A}\left[-\frac{\dot{m}_e}{m}\left(1 - \frac{h}{B}\right) - \frac{\dot{V}}{V} - \frac{C}{B}\dot{\phi} + \frac{1}{Bm}(-\dot{m}_e h_e - \dot{Q}_w)\right] \qquad (14.30)$$

where h_e, the enthalpy of the flow through the exhaust valve, is the cylinder average enthalpy for flow *out* of the cylinder and the exhaust system gas enthalpy if reverse flow occurs.

The engine operating cycle should end with the working fluid at the same state that it started out. For the first calculations of the sequence of processes in Fig. 14-6, property values defining the initial state of the fluid in the cylinder were assumed. If the values of these properties at the end of the first cycle differ from the assumed values, the cycle calculation is repeated with the appropriate new initial values until the discrepancy is sufficiently small. Convergence with these cycle simulations occurs within a few iterations.

The working fluid state is now defined throughout the operating cycle. The work transfer to the piston per cycle

$$W_c = \oint p \, dV \qquad (14.31)$$

can now be obtained. From W_c, the masses of fuel and air inducted, m_f and m_a, and engine speed N, all the engine *indicated* performance parameters can be calculated: power, torque, mean effective pressure, specific fuel consumption, fuel-conversion efficiency; as well as volumetric efficiency, residual gas fraction, total heat transfer, etc. With a friction model, the indicated quantities can be converted to brake quantities.

The more sophisticated of these thermodynamic-based engine cycle simulations define the working fluid state throughout the cycle in sufficient detail for useful predictions of engine emissions to be made. The discussion in Chap. 11 of emission-formation mechanisms indicates that our understanding of how some of these pollutants form (e.g., NO_x, CO) is reasonably complete, and can be modeled accurately. The formation processes of the other pollutants (unburned hydrocarbons and particulates) are not adequately understood, though modeling activities are continuing to contribute to that understanding. The key features of models for predicting engine emissions were discussed in Chap. 11.

Cycle simulations and combustion models which have been developed for spark-ignition engines, where the fuel, air, residual gas mixture is essentially uniformly mixed, are discussed in Sec. 14.4.2. Compression-ignition engine simulations and combustion models are then discussed in Sec. 14.4.3. The special features required for prechamber engine models are reviewed in Sec. 14.4.4. Finally, thermodynamic-based models for more complex engine systems—multicylinder, turbocharged, and turbocompounded engines—are discussed in Sec. 14.4.5.

14.4.2 Spark-Ignition Engine Models

These models have usually followed the conceptual structure indicated in Fig. 14-6. Our focus here is on the combustion submodels that have been developed and used successfully. Features of the spark-ignition engine combustion process that permit major simplifying assumptions for thermodynamic modeling are: (1) the fuel, air, residual gas charge is essentially uniformly premixed; (2) the volume occupied by the reaction zone where the fuel-air oxidation process actually occurs is normally small compared with the clearance volume—the flame is a thin reaction sheet even though it becomes highly wrinkled and convoluted by the turbulent flow as it develops (see Sec. 9.3); thus (3) for thermodynamic analysis, the contents of the combustion chamber during combustion can be analyzed as two zones—an unburned and a burned zone.

Useful combustion chamber design information can be generated with simple geometric models of the flame. In the absence of strong swirl, the surface which defines the leading edge of the flame can be approximated by a portion of the surface of a sphere. Thus the mean burned gas front can also be approximated by a sphere. Then, for a given combustion chamber shape and assumed flame center location (e.g., the spark plug), the spherical burning area A_b [see Eq. (9.40)], the burned gas volume V_b [see Eq. (9.39)], and the combustion chamber surface "wetted" by the burned gases can be calculated for a given flame radius r_b and piston position (defined by crank angle) from purely geometric considerations.† The practical importance of such "model" calculations is that (1) the mass burning rate for a given burning speed S_b (which depends on local turbulence and mixture composition) is proportional to the spherical burning area A_b as given by Eq. (9.44); (2) the heat transfer occurs largely between the burned gases and the walls and is proportional to the chamber surface area wetted by the burned gases $A_{b,w}$ [see Eq. (12.21)]. Using the fact that the density ratio across the flame ρ_u/ρ_b is approximately constant and equal to 4, the unburned and burned gas volumes V_u and V_b can be related to the unburned and burned mass fractions $(1 - x_b)$ and x_b, respectively.

Examples of the results of such flame geometry calculations are shown in Figs. 14-7 and 14-8.[11] Figure 14-7a shows spherical flame areas A_b as a function of flame radius r_b for two different chambers and two plug locations and the TC piston position. The much larger flame area and shorter flame travel length of the central plug location are obvious. Such area data can be plotted as a function of burned gas volume V_b, as shown in Fig. 14-7b, so that comparisons of $A_b(r_b)$ for different chambers at the same mass fraction burned can be made. The advantage of a more compact chamber with higher central clearance height is apparent. Figure 14-8 shows that burned-gas-wetted wall area on the cylinder head, cylin-

---

† Note that the center of this sphere may be convected away from the spark plug location, especially if some swirl is present. However, only strong swirling and squish flows produce major distortions to the flame surface shape.

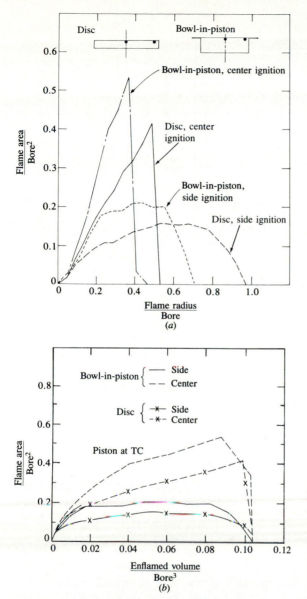

FIGURE 14-7

Calculated spark-ignition engine spherical flame surface area: (a) as a function of flame radius for different combustion chamber shapes and spark plug locations and (b) as a function of enflamed volume. Piston in top center position.[11]

der wall, and piston as a function of flame radius and crank angle for an open chamber with central ignition. The cylinder head and piston are the dominant areas early in the expansion stroke when the burned gas temperatures and heat fluxes are highest.

Mass fraction burned versus crank angle profiles determined from a first law analysis of cylinder pressure data, as shown in Figs. 9-2, 9-5, and 9-8, have an essentially universal dimensionless shape, as indicated in Fig. 9-13. Much useful

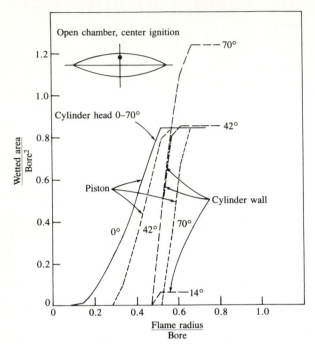

FIGURE 14-8
Calculated burned-gas-wetted wall area as a function of radius based on spherical flame model of an open-chamber SI engine with center plug location, for piston locations of 0°, 42°, and 70°.[11]

analysis has been done with engine simulations where this universal combustion profile has been used as a calculation input. The S-shaped mass fraction burned profile is often represented by the Wiebe function:

$$x_b = 1 - \exp\left[-a\left(\frac{\theta - \theta_0}{\Delta\theta}\right)^{m+1} \right]$$

(14.32)

where θ is the crank angle, θ_0 is the start of combustion, $\Delta\theta$ is the total combustion duration ($x_b = 0$ to $x_b \approx 1$), and a and m are adjustable parameters which fix the shape of the curve. Actual mass fraction burned curves have been fitted with $a = 5$ and $m = 2$.[12]

The conservation equations for an open system [Eqs. (14.1) and (14.10) or (14.11)] are now applied to the unburned gas zone ahead of the flame and to the burned zone behind the flame, in turn (see Fig. 9-4). For premixed engines, $\dot{f}$ and $\dot{\phi}$ are zero. During combustion, $\dot{m}$ and $\dot{m}_j$ in Eq. (14.10) or Eq. (14.11) are the mass flow rate across the flame sheet. This is $-\dot{m}_b$ for the unburned zone system and $+\dot{m}_b$ for the burned zone system; $\dot{m}_b$ is given by $m\dot{x}_b$, with $\dot{x}_b$ obtained by differentiating Eq. (14.32).

To calculate the effect of heat transfer on the burned gas state more accurately, the burned gas zone in Fig. 9-4 can be modeled in two parts: an adiabatic core and a boundary-layer region. The intent here is to account for the fact that heat loss to the walls primarily cools the burned gas adjacent to the wall, and only indirectly affects the core gas through the change in pressure that results

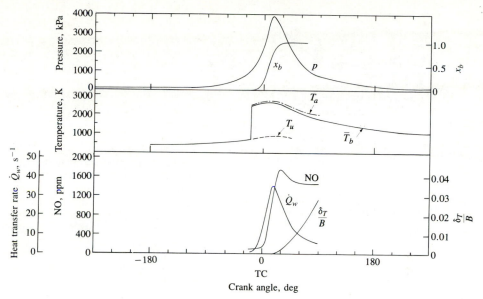

FIGURE 14-9

Cylinder pressure p, mass fraction burned x_b, unburned and burned gas temperatures (T_u = unburned, T_a = adiabatic burned core, $\bar{T}_b$ = mean burned gas temperatures), heat-transfer rate $\dot{Q}_w$ (normalized by fuel flow rate × heating value), thermal boundary-layer thickness δ_T, and mean nitric oxide concentration in the burned gases, through a four-stroke engine operating cycle, predicted by thermodynamic-based cycle simulation. 5.7-dm³ displacement eight-cylinder engine operating at wide-open throttle, 2500 rev/min, with equivalence ratio = 1.1. Gross indicated mean effective pressure is 918 kPa and specific fuel consumption is 254 g/kW·h.[13]

from heat loss. The open-system conservation equations, (14-1) and (14.10) or (14.11), are now applied to the core and boundary-layer region separately. The boundary-layer region covers that portion of the combustion chamber wall wetted by the burned gases, as shown in Fig. 9-4, and is of thickness δ_T, which increases with time. The temperature of the boundary-layer zone (assumed uniform) is usually taken to be the mean of the wall temperature and burned gas core temperature. Equation (14.10) or Eq. (14.11) is used to relate the enthalpy flux due to the mass flow across the inner edge of the boundary layer (which has an enthalpy equal to the core gas enthalpy), the heat transfer to the wall, the changing energy within the boundary-layer system due to its increasing mass and changing state, and the work transfer due to its changing volume.

An example of predictions of cylinder pressure, unburned and burned gas temperatures, heat-transfer rate, and boundary-layer thickness, based on an assumed 50° total burn duration for a 5.7-dm³ eight-cylinder engine at wide-open throttle and 2500 rev/min is shown in Fig. 14-9.[13] Appropriately based predictions of overall engine performance parameters made with this type of thermodynamic model agree well with engine data. Figure 14-10 shows predictions of indicated specific fuel consumption and exhaust gas temperature as a function of

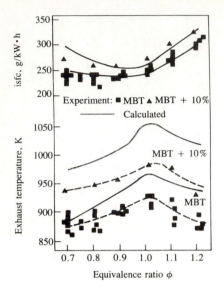

FIGURE 14-10
Predicted and measured indicated specific fuel consumption and exhaust temperature as a function of the fuel/air equivalence ratio for a spark-ignition engine operated at 1250 rev/min and imep of 379 kPa. MBT: maximum brake torque timing. MBT + 10%: combustion timing retarded to give 10 percent fuel consumption penalty.[14]

the fuel/air equivalence ratio at fixed load and speed. The isfc predictions and data agree well (except for very lean mixtures with retarded timing where cycle-by-cycle combustion variations are sufficiently large so predictions based on the average cycle lose accuracy); the predicted curves for exhaust temperature show the same trends as the experimental data. However, they are higher due to underestimation of the heat losses during the exhaust process.[14]

The output from such thermodynamic-based cycle simulations has replaced the fuel-air cycle as a predictor of effects of major variables on engine performance and efficiency. An instructive example of the value of such predictions is shown in Fig. 14-11, where fuel consumption at constant equivalence ratio, load, and speed has been computed as a function of total burn duration and heat loss to the chamber walls: increasing burn duration and heat loss both worsen fuel consumption.[15] Such data can be used to assess the efficiency improvements that should result from reduced heat transfer (e.g., reduced chamber surface area) and increased burn rate. Obviously the dependence of burn rate on engine design and operating parameters has not been modeled; the burn rate profile was a calculation input. Such models are most useful either (1) when the burn rate profile is *not critical* to the problem under study or (2) when predictions for a *range* of assumed burn rate profiles provide the required information.

So far we have discussed engine cycle simulations where details of the combustion process have been specified as input. The same thermodynamic-based simulation structure can be used in conjunction with a combustion model which *predicts* the rate of fuel burning. Various combustion models have been proposed and used for this purpose. Some of these are empirically based; some are based on the highly wrinkled, thin reaction-sheet flame model described in Sec. 9.3. All

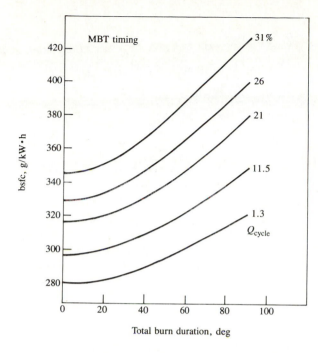

bsfc, g/kW·h

MBT timing

31%
26
21
11.5
1.3

Q_{cycle}

Total burn duration, deg

FIGURE 14-11
Predicted brake specific fuel consumption as a function of heat transfer per cycle to the combustion chamber walls (as a percent of the fuel's heating value) and total burn duration [$\Delta\theta$ in Eq. (14.32)]. 1250 rev/min, 262 kPa bmep, fuel/air equivalence ratio = 0.91, maximum brake torque spark timing.[15]

these models assume that the overall flame shape approximates a portion of a sphere centered at or near the spark plug. Empirical flame models have difficulty appropriately describing the three phases of the combustion process—flame development, rapid burning, and termination—with sufficient generality to be widely useful. One such model, based on the experimental data shown in Fig. 9-30, has been used successfully to evaluate different combustion chambers.[16, 17] The burning speed S_b [defined by Eq. (9.44)] is related empirically to the laminar flame speed S_L (see Sec. 9.3.3), the local rms velocity fluctuation u'_F [see Eq. (8.22)] under motored engine conditions, the firing and motored cylinder pressure at the same crank angle, and spark advance. While a good fit to the data in Fig. 9-30 for engine flames during their turbulent rapid-burning phase was obtained, during the flame development period a correction factor was required to fit the data.

Spark-ignition engine combustion models with a more fundamental framework have been developed and used. Based on coupled analysis of flame front location and cylinder pressure data, Keck and coworkers[18-20] have derived the following burning law:

$$\frac{dm_b}{dt} = \rho_u A_f S_L + \frac{\mu}{\tau_b} \tag{14.33}$$

$$\frac{d\mu}{dt} = \rho_u A_f u_T(1 - e^{-t/\tau_b}) - \frac{\mu}{\tau_b} \tag{14.34}$$

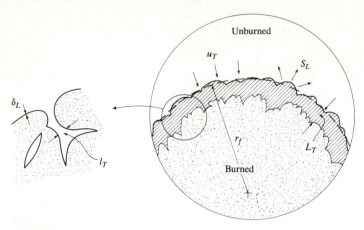

FIGURE 14-12
Schematic of turbulent premixed spark-ignition engine flame, illustrating the physical basis for burning law of Eqs. (14.33) to (14.35). The approximately spherical front of the "thick" turbulent flame (dashed line) diffuses outward at the laminar flame speed S_L. Fresh mixture also crosses this front at a characteristic velocity u_T due to turbulent convection. Schematic on left shows detailed flame structure: δ_L is a reaction-sheet thickness, l_T is characteristic scale of wrinkles in the sheet.

where

$$\mu = m_e - m_b = \rho_u(V_f - V_b) = \rho_u l_T(A_L - A_f) \tag{14.35}$$

is a parametric mass (interpreted as the mass entrained within the flame region that has yet to burn), u_T a characteristic speed, and $\tau_b = l_T/S_L$ is a characteristic burning time. l_T, A_L, V_f, A_f, V_b are defined in Sec. 9.3.4.

Figure 14-12 illustrates the physical basis for this model. The first term in Eq. (14.33) represents the laminar (diffusive) propagation forward of the approximately spherical front of the "thick" turbulent flame; the second term represents the burning of mixture already entrained within this flame front. In Eq. (14.34), which describes the rate of change of *unburned* mixture mass μ within the flame zone, the first term represents the turbulent convection of unburned mixture across the spherical front of the flame and the second term represents the mass rate of burning of entrained but not yet burned mixture which is contained within the "wrinkles" and "islands," which the distorting and stretching of the thin reaction sheet by the turbulent flow produces. This has been called an "entrainment" or "eddy-burning" model for the above reasons. The exponential term in brackets in Eq. (14.34) allows for the fact that the flame sheet initially is spherical and laminarlike: it requires a time of about τ_b to develop into a turbulent flame.

The behavior of Eqs. (14.33) and (14.34) in four important limits is:

1. For a quiescent mixture, $u_T \to 0$ or $l_T \to \infty$,

$$S_b \to S_L \tag{14.36a}$$

2. Initially, as $t \rightarrow 0$,

$$S_b \rightarrow S_L \qquad (14.36b)$$

3. Quasi-steady state, $d\mu/dt \approx 0$,

$$S_b \approx u_T + S_L \qquad (14.36c)$$

4. Final burning stage after the flame front reaches the wall, $t \geq t_w$ (when $A_f \rightarrow 0$),

$$\frac{\dot{m}_b}{\dot{m}_b(t_w)} = e^{-(t-t_w)/\tau_b} \qquad (14.36d)$$

To apply Eqs. (14.33) and (14.34), the quantities u_T and τ_b (or $l_T = \tau_b S_L$) must be evaluated. Two approaches have been taken: (1) use of empirical correlations for these variables, derived from engine flame data (such as that described in Sec. 9.3.4); (2) use of more fundamental models to predict these quantities.

Keck has derived the following correlations for u_T and l_T, based on the application of Eqs. (14.33) and (14.34) to several sets of engine combustion data:

$$u_T = 0.08\bar{u}_i \left(\frac{\rho_u}{\rho_i}\right)^{1/2} \qquad (14.37)$$

$$l_T = 0.8 L_{iv} \left(\frac{\rho_i}{\rho_u}\right)^{3/4} \qquad (14.38)$$

u_T was found to be proportional to $\sqrt{\rho_u}$ (at time of spark) and to correlate well with mean inlet gas speed $\bar{u}_i = \eta_v(A_p/A_{iv})\bar{S}_p$, where η_v is volumetric efficiency, A_p is piston area, A_{iv} is the maximum open area of the inlet valve, $\bar{S}_p$ is mean piston speed. l_T appears to scale with valve lift, L_{iv}; it decreases with increasing density at a rate proportional to $\rho_u^{-3/4}$. While u_T and l_T are not constant during the combustion process, their variation is modest.[18]

A quantitative comparison of predicted and measured flame radius as a function of time is shown in Fig. 14-13 for hydrogen and propane fuel-air mixtures which exhibit widely different behavior:[18] the figure indicates both the behavior and validity of the model. Predicted burned gas expansion speeds u_b [see Eq. (9.43)] are shown in Fig. 14-13a as a function of burned gas radius; the parameters u_T and l_T were chosen to fit the propane data. Figure 14-13b shows that the measured flame front radii, r_f, are in good agreement with the predicted flame and burned gas radii, r_f and r_b, for these two fuels. The initial expansion speed of hydrogen is about 10 times that of propane. Since $r_f \approx r_b$ for early times, $S_b \approx S_L$ and this ratio is expected. As r_b become large, $(r_f - r_b) \rightarrow u_T \tau_b$, which is several times smaller for hydrogen mixtures than for propane mixtures.

An adaption of this approach developed by Tabaczynski and coworkers[21, 22] is based on the following model of turbulent flame propagation. The vorticity in the turbulent flow field is concentrated in vortex sheets which are

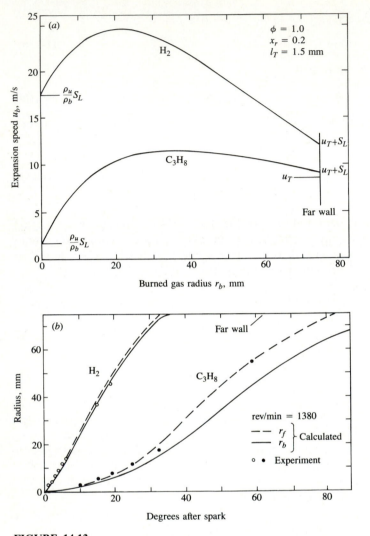

FIGURE 14-13
(a) Calculated burned gas expansion speed u_b for stoichiometric hydrogen-air and propane-air mixtures as a function of burned gas radius r_b. (b) Comparison of experimentally measured (points) and calculated (dashed curve) flame radii r_f for these mixtures as a function of crank angle. Also shown (solid curve) is the burned gas radius r_b.[18]

of a size comparable to the Kolmogorov scale l_K [see Eq. (8-11)]. These vortex sheets are assumed to have a characteristic spacing which is of the order of the Taylor microscale l_M, which is a function of the integral length scale l_I and the turbulent Reynolds number as indicated by Eq. (8.15). From these turbulence assumptions it is argued that ignition sites propagate along the vortex sheets with a velocity $u' + S_L$, where u' is the local turbulence intensity. The propagation of the reaction front between the vortex sheets is assumed to be a laminar process.

Thus, in Eqs. (14.33) and (14.34), u_T and τ_b are given by

$$u_T \approx u' \qquad \text{and} \qquad \tau_b = \frac{l_T}{S_L} \approx \frac{l_M}{S_L} \tag{14.39}$$

where l_M, the microscale, is determined from the integral scale and the turbulent Reynolds number via Eq. (8.15), assuming that the turbulence is homogeneous and isentropic. The task therefore becomes one of evaluating u' and l_I.

One approach used is to relate the turbulence intensity at the start of the combustion process to the mean intake flow velocity through the valve: e.g.,[23]

$$u'_0 = \frac{C \bar{S}_p B^2}{L_{iv} D_{iv}} \tag{14.40}$$

where $\bar{S}_p$ is the mean piston speed, B the bore, and L_{iv} and D_{iv} the lift and diameter of the inlet valve. It is assumed that the integral length scale at the start of combustion, $l_{I,0}$, is proportional to a characteristic flow dimension, usually the clearance height h. Then, during combustion, the unburned portion of the charge is assumed to undergo isentropic compression sufficiently rapidly that the angular momentum of the "eddies" is conserved and the length scale follows the eddy size, i.e., a simple rapid distortion process occurs:

$$u' = u'_0 \left(\frac{\rho}{\rho_0} \right)^{1/3} \qquad l_I = l_{I,0} \left(\frac{\rho_0}{\rho} \right)^{1/3} \tag{14.41}$$

This model predicts an increase in turbulence intensity and decrease in length scale with compression, which is only partly confirmed by experiment.

A more sophisticated approach is to describe the dynamic behavior of the turbulence with one or more rate equations for the key turbulence parameters: k the turbulent kinetic energy and ε the dissipation rate of k. Turbulence is generated, diffused, and dissipated by the flow field, so the rate of change of turbulent kinetic energy k can be written:

$$\frac{dk}{dt} = P_k + D_k - \rho \varepsilon \tag{14.42}$$

where the term P_k represents the volumetric production of turbulence and the diffusion term D_k can be modeled as a gradient diffusion with an effective turbulent viscosity which dominates the laminar diffusion process. In this application, Eq. (14.42) is integrated over the combustion chamber (or a region of the chamber) to provide spatially averaged turbulence predictions. Then the diffusion terms become boundary fluxes: e.g., the transport of kinetic energy across the combustion chamber boundary due to flow through the inlet or exhaust valve. The dissipation rate ε is related to the integral length scale via

$$\varepsilon = \frac{C_D k^{3/2}}{l_I} \qquad C_D = 0.09 \tag{14.43}$$

l_I can be taken as proportional to the clearance height ($l_I \approx 0.22\, h$), or an additional rate equation for a second turbulence parameter, the dissipation rate ε, can

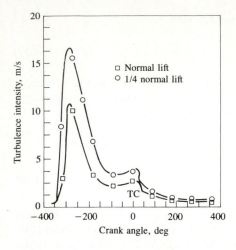

FIGURE 14-14
Predicted turbulence intensity u' as a function of crank angle and valve lift in engine operating at 1500 rev/min, 414 kPa imep, with a compression ratio of 10.[27]

be used. In the more complete of these $k - \varepsilon$ turbulence models,[24] the ε equation is similar to the k equation with production, diffusion, and dissipation terms. These $k - \varepsilon$ turbulence models are discussed more fully in Sec. 14.5.2.

The application of this turbulence model to the spark-ignition engine combustion chamber becomes complicated and the reader is referred to references for the details.[25–28] Considerable success with predicting trends in mass burning rate has been achieved with this type of model. Design variables examined include: swirl, squish, valve lift, bore/stroke ratio. The advantage of such models is that they are straightforward computationally so that extensive parametric sets of calculations are feasible. The major disadvantage is the *ad hoc* nature of the turbulence and flame models which involve plausible but arbitrary assumptions. Sample predictions are shown in Figs. 14-14 and 14-15.[27, 28] Figure 14-14 shows the variation in turbulence intensity u' in an engine with a disc-shaped combustion chamber, throughout the operating cycle. A normal valve-lift profile and

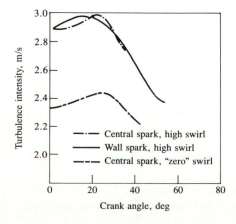

FIGURE 14-15
Predicted turbulence intensity during combustion for high and "zero" swirl levels for central and cylinder wall spark plug locations. Same engine and operating conditions as in Fig. 14-14.[28]

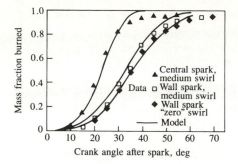

reduced maximum valve-lift profile (one-quarter normal) are shown. The high levels of turbulence generated during the first half of the intake process decay substantially before the latter stages of the compression stroke produce some amplification. Reduced valve lift produces higher levels of turbulence intensity at combustion, as is well known.[29] Figure 14-15 shows the predicted turbulence behavior during combustion for a disc-shaped chamber for different swirl levels and plug locations. Swirl is shown to increase the turbulence intensity. Comparison of predicted and measured mass fraction burned profiles versus crank angle for different swirl levels and plug locations are shown in Fig. 14-16. The large flame area effects (shown here in the two limiting plug locations: side wall and center) and significant though lesser effect of swirl are correctly modeled. Such models are useful for relating changes in spark-ignition engine design and operating variables to changes in engine performance, via predictions of changes in flame development and propagation.

The above type of combustion model has been used to obtain explicit relations for the flame development and rapid burning angles as functions of engine design and operating variables.[30] The equation for the mass burning rate, (14.33), was effectively integrated over the relevant portion of the total combustion process; the turbulent characteristic velocity was assumed proportional to $\bar{S}_p$, the mean piston speed. The flame development angle was found to vary as

$$\Delta\theta_d = C(\bar{S}_p v)^{1/3}\left(\frac{h}{S_L}\right)^{2/3} \tag{14.44}$$

where v is the kinematic viscosity ($v = \mu/\rho$) and h is the clearance height at ignition. C is a constant which depends on engine geometry and is determined by matching Eq. (14.44) with engine data. The rapid burn angle (here taken as the crank angle between $x_b = 0.01$ and 1.0) is given by

$$\Delta\theta_b = C'\left(\frac{B}{h^*}\right)\left(\frac{\rho_i}{\rho_u^*}\right)^{10/9}(\bar{S}_p v^*)^{1/3}\left(\frac{h_i}{S_L^*}\right)^{2/3} \tag{14.45}$$

where C' is a constant which depends on engine geometry, B is the bore, the subscript i denotes the value at ignition, and the superscript $*$ denotes the value at cylinder conditions where $x_b = 0.5$. These expressions show reasonable agreement with observed trends in $\Delta\theta_d$ and $\Delta\theta_b$.

14.4.3   Direct-Injection Engine Models

In direct-injection compression-ignition and stratified-charge engines, the liquid fuel is injected into the cylinder as one or several jets just prior to ignition. In large direct-injection compression-ignition engines, the air flow is essentially quiescent. However, in medium and smaller size DI engines, the air flow is usually swirling about the cylinder axis at up to 10 times the crankshaft rotational speed; this air-flow pattern increases the rate of entrainment of air into the fuel jet to increase the fuel-air mixing rate. Thus modeling of the ignition and combustion processes for direct-injection types of engines is much more complex than for premixed-charge spark-ignition engines. The unsteady liquid-fuel jet phenomena—atomization, liquid jet and droplet motion, fuel vaporization, air entrainment, fuel-air mixing, and the ignition chemistry—all play a role in the heat-release process (see Chap. 10). It is not yet possible to model all these phenomena from a fundamental basis, even with the most sophisticated fluid-dynamic-based codes now available (see Sec. 14.5), since many of these processes are not yet adequately understood. However, models at various levels of detail and empiricism have been developed and have proven useful in direct-injection diesel and stratified-charge engine analysis. This section reviews the important features of single-zone heat-release models and phenomenological jet-based combustion models. Their relative simplicity and modest computer time requirements make them especially useful for diesel cycle simulation and more complex engine system studies.

Single-zone models assume that the cylinder contents can be adequately described by property values representing the average state, and use one or more algebraic formulas to define the heat-release rate. The functional forms of these formulas are chosen to match experimentally observed heat-release profiles (see Sec. 10.4.2). Coefficients in these formulas, which may vary with engine design details and operating conditions, are determined empirically by fitting with data. The phenomenological description of diesel combustion developed by Lyn (see Sec. 10.3) comprises three primary phases: the ignition delay period, the premixed fuel-burning phase, and the mixing-controlled fuel-burning phase. Ignition delay correlations are reviewed in Sec. 10.6.6. Here models for the second and third phases, when the major heat release occurs, are summarized (see Ref. 31 for a more extensive review). The attraction of the one-zone heat-release approach is its simplicity: however, since it cannot fully describe the complex phenomena which comprise the compression-ignition engine combustion process, substantial empirical input must be used. Several one-zone heat-release models have been proposed and used (e.g., Refs. 32 to 34). These use simple equations to describe the rate of release of the fuel's energy, sometimes modeled on the presumed controlling physical or chemical process and always calibrated by comparison with data.

One extensively used model of this type developed by Watson *et al.*[35] is especially appropriate for use in total diesel system simulations where the combustion process details are not the primary focus. It is based on Lyn's description of compression-ignition combustion—a rapid premixed burning phase followed by a slower mixing controlled burning phase. The fraction of the injected fuel that burns in each of these phases is empirically linked to the duration of the ignition delay. One algebraic function is used to describe the premixed heat-release phase and a second function to describe the mixing-controlled heat-release phase. These two functions are weighted with a phase proportionality factor, β, which is largely a function of the ignition delay. Thus:

$$\frac{m_{f,b}(t')}{m_{f,0}} = \beta f_1 + (1 - \beta) f_2 \tag{14.46}$$

where $m_{f,b}$ is the mass of fuel burned, $m_{f,0}$ is the total fuel mass injected per cycle per cylinder, and t' is time from ignition non-dimensionalized by total time allowed for combustion $[=(t - t_{ign})/\Delta t_{comb}]$.† The premixed-burning function is

$$f_1 = 1 - (1 - t'^{K_1})^{K_2} \tag{14.47}$$

and the mixing-controlled function is

$$f_2 = 1 - \exp{(-K_3 t'^{K_4})} \tag{14.48}$$

where K_1, K_2, K_3, and K_4 are empirical coefficients. The proportionality factor β is given by

$$\beta = 1 - \frac{a\phi^b}{\tau_{id}^c} \tag{14.49}$$

where ϕ is the overall fuel/air equivalence ratio and a, b, and c are empirical constants.

Correlation with data from a typical turbocharged truck engine gave the following values for K_1 to K_4:

$$K_1 = 2 + 1.25 \times 10^{-8}(\tau_{id} N)^{2.4}$$

$$K_2 = 5000$$

$$K_3 = \frac{14.2}{\phi^{0.644}} \tag{14.50}$$

$$K_4 = 0.79 K_3^{0.25}$$

where τ_{id}, the ignition delay, is in milliseconds and N, engine speed, is in revolutions per minute. It also gave these ranges for a, b, c:[35]

$$0.8 < a < 0.95; \qquad 0.25 < b < 0.45; \qquad 0.25 < c < 0.5$$

---

† The combustion duration at Δt_{comb} is an arbitrary period within which combustion must be completed. A value of 125° was used above.

Such single-zone heat-release models are useful because of their simplicity. They obviously cannot relate engine design and operating variables explicitly to the details of the combustion process. Experience indicates that those models with only one function are not usually able to fit experimentally determined heat-release profiles with sufficient accuracy. All single-zone heat-release models should be checked against experimentally derived heat-release profiles, and recalibrated if necessary, before being used for predictions.

Many thermodynamic-based direct-injection engine simulations incorporate an explicit model for each fuel spray which attempts to describe how the spray develops with time. The spray starts out as a liquid fuel jet which then vaporizes, entrains air and (later) burned gases. Mixture preparation can be limited by the availability of either fuel vapor or air, the former limited by droplet evaporation and the latter by air entrainment. While there is evidence in the literature to support both of these phenomena as rate-limiting, more recent studies[36] show that most (70 to 95 percent) of the injected fuel is in the vapor phase at the start of combustion, whereas only 10 to 35 percent of the vaporized fuel is mixed to within the combustion limits (equivalence ratio between 0.3 and 3). This suggests that the combustion process in typical heavy-duty direct-injection compression-ignition engines is mixing controlled rather than vaporization controlled.

While spray geometry is an essential aspect of the fuel-air mixing process, it may not be necessary to model the precise details of the actual configuration. For the purpose of heat-release and emission analysis, it suffices in many phenomenological models to calculate the evolution of the fuel mass, composition, volume, and temperature of critical regions of the spray based on a generic spray geometry. Alternate approaches attempt to provide a detailed structure for the fuel spray to improve the modeling of air entrainment, effects of swirl/spray interaction, and heat transfer. The more commonly used approaches are illustrated in Fig. 14-17.

The schematic in Fig. 14-17a illustrates the simplest approach: it is assumed that the growth and motion of the spray or jet within the chamber can be analyzed as a quasi-steady one-dimensional turbulent gaseous jet.[38-40] The intent here is to describe the position of the jet within the combustion chamber and the overall jet size as a function of time. Entrainment of air into the jet is assumed to take place at each point along the jet surface at a rate proportional to the velocity difference between the jet and surrounding air at that point. Two empirical entrainment coefficients[41] are used for the proportionality constants for the relative motion in the jet axial and transverse directions. Conservation equations for fuel mass and total mass, and momentum (in two or three orthogonal directions) are used to determine the jet trajectory and size. The jet slows down due to air entrainment. Deflection of the jet results from the entrainment of air with a momentum component normal to the jet axis, and from drag forces due to the normal component of air flow past the jet. This approach does not define the velocity and concentration profiles across the jet: it only calculates the mean values at any jet axial position. Experimentally determined radial profiles for

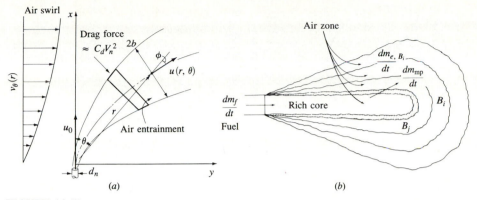

FIGURE 14-17
(a) Schematic of one-dimensional quasi-steady fuel spray model used to define spray centerline trajectory and width as radially outward-moving spray interacts with swirling air flow. (b) Schematic of multizone model for fuel spray which, based on empirically calculated spray motion and assumed concentration distributions in the spray, successfully evolves discrete combustion zones (each containing a fixed fraction of the fuel) as fuel is injected, vaporized, and mixed with air. (dm_f/dt) = rate of fuel injection into rich core; (dm_{mp}/dt) = rate of preparation of mixture for burning; $(dm_{e,\,B_i}/dt)$ = rate of entrainment of air into zone B_i.[37]

axisymmetric turbulent jets[42] are often assumed to apply. Although the fuel spray is initially pure liquid, the liquid fuel drops soon become a small fraction of the jet volume due to vaporization and air entrainment. Downstream of the initial liquid breakup region, the velocity of the small drops relative to the vaporized fuel and air is small, so the spray acts as a gas jet. Adding a combustion model to this quasi-steady gaseous jet model for fuel-air mixing is an additional major step.

A comparison between this type of gas jet model and an experimental engine spray is shown in Fig. 14-18. A single fuel jet was injected into a disc-shaped chamber in the location shown, and schlieren photography used to observe the spray trajectory. Good agreement was obtained for the spray center-line: note the significant effect of swirl. Reasonable agreement was also obtained between predicted and measured spray boundaries.

Figure 14-17b shows a multizone model for each fuel spray which has been used extensively for engine performance and emissions studies in quiescent DI diesels.[37, 43] The spray is modeled as a gas jet, with penetration, trajectory and spreading rate determined from empirical equations based on axisymmetric turbulent jet data. These equations describe the approximate spray geometry. The fuel-air distribution within the spray is determined by using a normal distribution across the spray cross section and a hyperbolic profile along the axis of the spray. Progressively evolving, discrete combustion zones, each containing a fixed fraction of the total fuel mass, are then superimposed on the geometrically defined fuel-air distribution. Outer zones are diluted with air and inner zones are added as fuel vaporizes and mixes, as injection and combustion proceeds. The model implicitly assumes that combustion does not affect the mixing rates. With careful

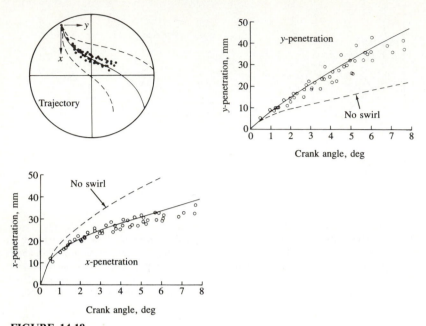

FIGURE 14-18
Spray trajectory and width calculated using one-dimensional quasi-steady spray model of type illustrated in Fig. 14-17a, compared with experimental data taken in special visualization direct-injection stratified-charge engine.[39]

adjustment of calibrating constants, this model describes engine performance variations with reasonable accuracy as major design and operating variables change.

More detailed geometric models of the fuel-air mixing and combustion processes in engine sprays have been developed (e.g., Ref. 44). The intent is to follow the spray development in a swirling air flow and the spray interaction with the combustion chamber wall. Figure 14-19 illustrates the approach. The liquid fuel which enters the chamber through the injector nozzle is divided into many small equal mass "elements." The spray motion is defined by an experimentally based correlation. Air entrainment is calculated from momentum conservation and the spray velocity decrease predicted by this correlation. The processes which occur within each element are also illustrated in Fig. 14-19. The fuel drops evaporate and fuel vapor mixes with entrained air. When ignition occurs combustible mixture prepared before ignition burns rapidly: it is assumed to burn at the stoichiometric composition. The continuation of the burning process then depends on the composition of the element: it may be limited by either the rate of production of fuel vapor by evaporation or the availability of air by the rate of entrainment (paths A and B in Fig. 14-19).

The growth of the spray is determined from the air entrainment into each element and the combustion-produced expansion of each element, as indicated in Fig. 14-20. When impingement on the wall occurs, the spray is assumed to spread

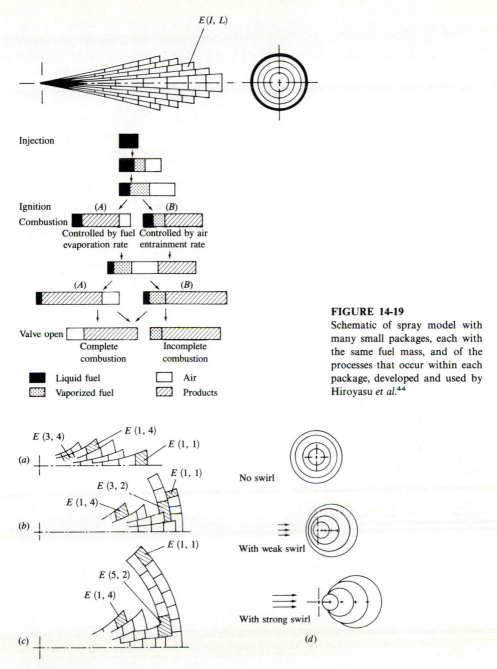

FIGURE 14-19
Schematic of spray model with many small packages, each with the same fuel mass, and of the processes that occur within each package, developed and used by Hiroyasu et al.[44]

FIGURE 14-20
Method used with model of Fig. 14-19 to compute spray and flame configuration: (a) prior to impingement of spray on wall—shaded elements indicate combustion; (b) and (c) show spray behavior following impingement on the cylindrical bowl wall of the DI diesel combustion chamber; (d) shows effect of swirl on spray and flame configuration.[44]

along the wall with a constant thickness as shown in Fig. 14-20. When the periphery of the spray reaches that of a neighboring spray the sideways growth of the spray is then prevented and the thickness of the elements along the wall increases. Swirl effects are calculated from tangential momentum considerations. Each annular cone ring element is shifted sideways by the swirl as indicated in Fig. 14-20.

The heat-release rate in the combustion chamber is obtained by summing up the heat release in each element. Nitric oxide and soot formation calculations are based on the time histories of temperature, vaporized fuel, air and combustion products in each element. The overall structure of this particular complete diesel engine performance and emissions model is indicated in Fig. 14-21: it is typical of the type of compression-ignition engine simulation used to study engine performance and emissions. Figure 14-22 shows an example of the output from the above model. The injection rate diagram, the assumed Sauter mean drop size of the spray, and the air swirl determine the spray development which leads to the heat-release rate predictions. This determines the cylinder pressure profile. Predicted engine performance results show reasonable but not precise agreement with experimental data. That is not surprising given the complexity of the phenomena being modeled. A review of these types of jet models is given by Hiroyasu.[46]

14.4.4   Prechamber Engine Models

Small high-speed compression-ignition engines use an auxiliary combustion chamber, or prechamber, to achieve adequate fuel-air mixing rates. The prechamber is connected to the main combustion chamber above the piston via a nozzle, passageway, or one or more orifices (see Secs. 1.8, 8.5, and 10.2.2). Auxiliary chambers are sometimes used in spark-ignition engines, also. The plasma and flame-jet ignition systems described in Sec. 9.5.3 enclose the spark plug in a cavity or small prechamber which connects to the main chamber via one or more orifices. The function of the prechamber is to increase the initial growth rate of the flame. Combustion in the main chamber is initiated by one or more flame jets emanating from the prechamber created by the ignition process and subsequent energy release within the prechamber. If the mixture within the prechamber is richer than in the main chamber (due to fuel injection or a separate prechamber intake valve—see Sec. 1.9) these are called stratified-charge engines.

The additional phenomena which these prechambers introduce beyond those already present in conventional chamber engines are: (1) gas flows through the nozzle or orifice between the main chamber and prechamber due to piston motion; (2) gas flows between these chambers due to the combustion-generated pressure rise; (3) heat is transferred to the nozzle or passageway walls due to these flows. The first of these phenomena results in nonuniform composition and temperature distributions between the main and prechamber due to gas displacement primarily during compression, and determines the nature of the flow field within the prechamber toward the end of compression just prior to combustion.

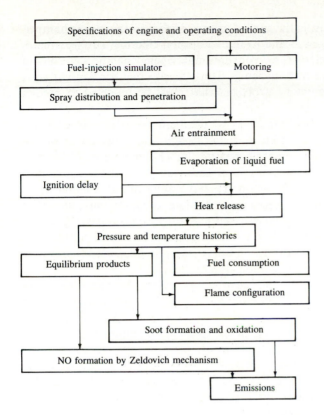

FIGURE 14-21
Structure of thermodynamic-based DI diesel simulation for predicting engine performance and emissions. Simulation incorporates spray model of type illustrated in Figs. 14-19 and 14-20.[45]

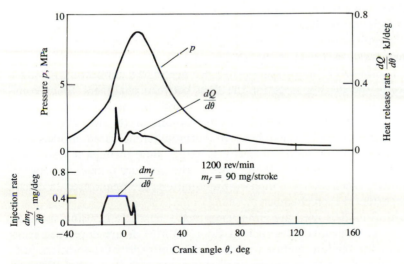

FIGURE 14-22
Fuel-injection rate, heat-release rate profile, and cylinder pressure predicted with thermodynamic-based DI diesel simulation with spray and combustion model of type shown in Fig. 14-21.[45]

The second phenomena controls the rate of energy release in the main chamber. The heat losses in the passageway and to the additional chamber surface area of the prechamber designs relative to conventional open chambers result in decreased engine performance and efficiency. Thus the prechamber concept adds additional complexity to the engine processes that must be modeled to predict engine behavior.

The following variables are important to prechamber engine performance and emissions characteristics, in addition to the design and operating variables which govern single-chamber engine behavior: prechamber geometry—size, shape, flow area and shape of connecting passageway(s); prechamber location in relation to main chamber geometry; geometry and timing of any auxiliary prechamber valve; fuel metering strategy in prechamber compression-ignition or stratified-charge engine. Thermodynamic-based models have been developed and used to examine the overall impact of these variables (see Ref. 47). Computational fluid dynamic models (see Sec. 14.5 and Fig. 8-26) have also been used to examine specific prechamber engine flow and combustion processes.

Useful predictions of fuel, air, and residual gas distributions and the corresponding temperature within the prechamber and main chamber can be obtained with simple gas displacement models. Only during combustion is the pressure difference across the nozzle or orifice sufficiently large in magnitude for its modeling to be essential; the assumption of uniform pressure during compression, the critical process for determining conditions just prior to combustion, introduces little error into calculations of the flows between the chambers. Section 8.5 develops the appropriate equations for these piston-motion driven gas displacements. Use of the conservation equations for an open system, for total mass, fuel mass, residual gas, and energy given in Sec. 14.2, for the main chamber and the prechamber, then give the mean composition and temperature variation in each chamber as a function of time due to this flow. Figure 8-25 illustrates the mean composition variation in the prechamber that results during the compression stroke of a three-valve stratified-charge engine.

During combustion, the pressure difference across the connecting passageway or orifice is the driving force for the flow between chambers. Since combustion starts in the prechamber, the initial flow is into the main chamber; later, as the heat release in the main chamber becomes dominant, the flow may reverse direction and be into the prechamber. In thermodynamic-based models, the equations for one-dimensional quasi-steady ideal gas flow through a restriction given in App. C are used to relate these flows to the pressure difference between the two chambers. Open-system conservation equations are again used to calculate mean properties in each chamber.

Combustion models used are either empirically based [e.g., using specified heat-release or mass burning rates such as Eq. (14.32)[48]] or are developed from direct-injection compression-ignition engine models with spray evaporation, fuel-air mixing, and ignition delay processes explicitly included.[49] Because of the complexity of these processes in the prechamber engine geometry, substantial simplifying assumptions and empiricism must be used.

Heat transfer to the passageway and chamber walls is affected by the flows between the chambers: high velocities within the passageway result in high heat-transfer rates to the passageway walls, and the vigorous flows set by the passage-way exit flow entering the prechamber or the main chamber increase heat-transfer rates to the walls of these chambers. The standard engine heat-transfer correlations which relate the heat-transfer coefficient to mean flow field variables via Nusselt-Reynolds number relationships (see Sec. 12.4) are normally used to describe these heat-transfer processes. The length scales are chosen to match the prechamber or main chamber or passageway dimensions. The charac-teristic velocities in these relationships are equated with velocities which are rep-resentative of the flow in each of these regions at the relevant time in the engine operating cycle.[50, 51]

The utility of the more sophisticated of these prechamber engine per-formance and emissions models is illustrated by the sample results shown in Fig. 14-23. This simulation of the indirect-injection compression-ignition engine's flow and combustion processes describes, through the use of stochastic mixing models, the development of the fuel/air ratio distribution and fuel-energy release distribu-tion, and hence the development of the gas pressure and gas temperature dis-tribution, within the prechamber and main chambers of the engine. With the (nonuniform) gas composition and state defined, the models for NO formation described in Sec. 11.2.1 was used to predict NO_x emissions. The approaches used to describe the evolution of the prechamber, main chamber, and passageway con-tents are summarized in Fig. 14-23a.

The cylinder contents were divided up into a large number of elements. Pairs of elements are selected at random to undergo "turbulent mixing" inter-actions at a frequency related to the turbulence in each region. Rate processes—evaporation, ignition, NO formation, etc.—proceed within each element between these mixing interactions. Figure 14-23b shows sample results. At about TC, after some of the injected fuel has evaporated and the ignition delay is over, com-bustion starts in the prechamber and the prechamber pressure p_p rises above the main chamber pressure p_m. This forces air, fuel, and burned gases to flow from the prechamber into the main chamber; fuel and rich products can now mix with air and burn in the main chamber. NO starts to form in each mass element, once it burns, at a rate dependent on each element's composition and state. Most of the NO forms within the prechamber and then flows into the main chamber as the expansion process proceeds. The attractive feature of this type of emission calculation is that the kinetically controlled NO formation calculations are based directly on local gas composition and temperature in a manner that approximately simulates the mean and turbulent nonuniformities in these vari-ables. Predictions of engine operation and emissions showed good agreement with data.[52]

Fluid-dynamic-based models have been used to study fluid flow, com-bustion, and pressure wave phenomena in prechamber engines. Section 14.5 reviews this type of engine model. Additional details of these applications can be found in Refs. 53 and 54.

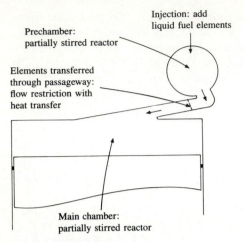

Injection: add
liquid fuel elements

Prechamber:
partially stirred reactor

Elements transferred
through passageway:
flow restriction with
heat transfer

Main chamber:
partially stirred reactor

Partially stirred reactors contain many equal
mass elements. These elements may be air
(plus residual), liquid fuel, unburned mixture
(fuel vapor, air, burned gas), and burned mixture

(a)

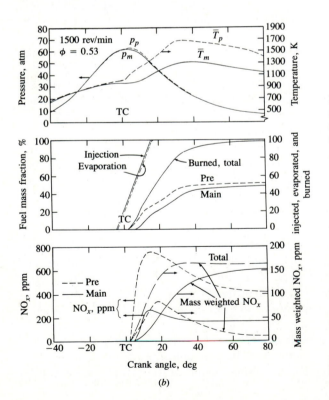

(b)

FIGURE 14-23
(a) Schematic of IDI diesel
engine illustrating how stochastic
mixing models are applied to
prechamber, main chamber, and
passageway to simulate turbulent
mixing processes and pressure-
driven flows. (b) Example of
simulation predictions through
the engine's operating cycle.
Shown are prechamber and main
chamber pressures, prechamber
and main chamber average gas
temperatures; fuel mass injected,
evaporated, and burned in prechamber and main chamber,
average NO concentration in
each chamber (and total) in ppm
by volume and mass weighted
ppm (mass in chamber × NO
concentration in chamber/total
mass in cylinder).[52]

14.4.5 Multicylinder and Complex Engine System Models

The models discussed in the previous parts of Sec. 14.4 focus on the processes occurring within *each* cylinder of an internal combustion engine. Most engines are multicylinder engines and the individual cylinders interact via the intake and exhaust manifolds. Also, many engine systems are more complex: internal combustion engines can be supercharged, turbocharged, or turbocompounded, and the manifolds then connect to the atmosphere via compressors or turbines (see Fig. 6-37 and Sec. 6.8). Thermodynamic-based simulations of the relevant engine processes, constructed from the types of model components already described, prove extremely useful for examining the behavior of these more complex engine systems. By describing the mass and energy flows between individual components and cylinders of such systems throughout the engine's operating cycle, the total system preformance can be predicted. Such models have been used to examine steady-state engine operation at constant load and speed (where time-varying conditions in the manifolds due to individual cylinder filling and emptying events affect multicylinder engine behavior), and how the total system responds to changes in load and speed during engine transients.

The block diagram of a turbocharged and turbocompounded diesel engine system in Fig. 14-24 illustrates the interactions between the system components. By describing the mass and energy flows between components and the heat and work transfers within each component, total system behavior can be studied. In such engine simulations, the reciprocator cylinders, the intake manifold, and the various sections of the exhaust system are treated as connected open systems. The flows into and out of these volumes are usually analyzed using the quasi-steady emptying and filling approach described in Sec. 14.3.3, using the open-system conservation equations of Sec. 14.2. The reciprocator cycle is treated as a sequence of processes within each cylinder: intake, compression, combustion (including expansion), and exhaust. These are modeled using the approaches described previously in Secs. 14.4.1 to 14.4.4. Heat transfer has, of course, an important effect on the in-cylinder processes. It also is important in the exhaust system since the performance of the turbocharger turbine and of any compounded turbine depends on the gas state at the turbine inlet. The performance of the turbomachinery components is normally defined by maps that interrelate efficiency, pressure ratio, mass flow rate, and shaft speed for each component (see Secs. 6.8.2 to 6.8.4). Special provisions are usually required in the logic of the turbomachinery map interpolation routines to avoid problems with the compressor surge and turbine choking operating limits of these devices.

When the reciprocator is coupled with turbomachinery its manifolds no longer connect directly with the atmosphere: matching procedures are required to ensure that the pressure levels and mass flow rates of the compressor and turbines match with those of the engine. The following matching process is typical of those used for turbocharged engines (one compressor and one turbine only). At a given time, the values of the variables describing the state of the

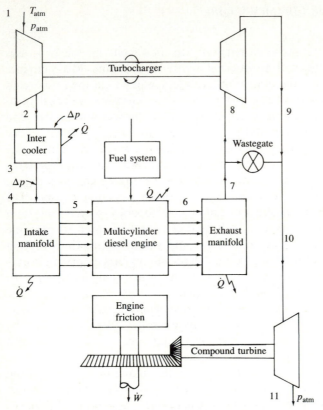

FIGURE 14-24
Block diagram of turbocharged turbocompounded diesel engine system.

various system components are known (from integration of the system governing equations over the previous time step). These include the intake and exhaust manifold pressures and the turbocharger rotor speed. The compressor inlet pressure is atmospheric pressure less the intake air-filter pressure drop. The turbine exit pressure is atmospheric plus the muffler pressure drop. By relating the compressor discharge pressure to the intake manifold pressure and the turbine inlet pressure to the exhaust manifold pressure (through suitable pressure drops) the pressure ratio across each machine is determined. Hence, the compressor and turbine maps can be entered using the calculated pressure ratios, and the rotor speed (same for both turbomachines) as inputs. The output from the map interpolation routines determines the mass flow rate and efficiency of each component for the next time step. From these the power required to drive the compressor $(-\dot{W}_C)$ and to drive the turbine $(\dot{W}_T)$ are determined from Eqs. (6.42) and (6.48), respectively. Any excess power (or power deficiency) will result in a change of rotor speed according to the turbocharger dynamics equation

$$\dot{W}_C + \dot{W}_T = I_{TC}\,\omega\,\frac{d\omega}{dt} + B\omega^2 \qquad (14.51)$$

where I_{TC} is the rotational inertia of the turbocharger, ω is angular velocity, and B is the rotational damping. The values of the other state variables for the next time step are determined from the solution of the mass and energy conservation equations for each open system, with the compressor and turbine mass flows taken from the output of the turbomachinery map interpolation routines.

This approach can be used to establish the steady-state engine operating characteristics from an assumed initial set of state variables. (Of course, due to the pulsating nature of the flows into and out of the cylinders, these state variables will vary in a periodic fashion throughout the engine cycle at a fixed engine load and speed.) This approach can also be used to follow transient engine behavior as load or speed is varied from such a steady-state condition.[35] The additional inputs required are the fuel pump delivery characteristics as a function of fuel pump rack position and speed, with the latter evaluated from an appropriate model for dynamic behavior of the governor.[55] From the brake torque of the engine (determined by subtracting friction torque from the indicated torque), the torque required by the load T_L, the inertia of the engine and load I_E and I_L, the dynamic response of the engine and load to changing fuel rate or engine speed can be obtained from

$$T_B - T_L = (I_E + I_L)\,\frac{d\omega}{dt} \qquad (14.52)$$

An example of the output from this type of engine model is shown in Fig. 14-25. The response of a turbocharged DI diesel engine to an increase in load from 0 to 95 percent of full load is shown. The predictions come from a model of the type shown in Fig. 14-24, and engine details correspond to the experimental configuration.[55] The simulation follows the data through the engine transient with reasonable accuracy. Note that with the assumed constant governor setting, during this transient the equivalence ratio of the trapped mixture rises to close to stoichiometric because the increase in air flow lags the increase in fuel flow. This would result in excessive smoke emissions. Such models prove extremely useful for exploring the effect of changes in engine system design on transient response.[56]

For two-staged turbocharged or turbocompounded systems the engine-turbocharger matching process is more complicated. The division of the pressure ratio between the exhaust manifold and atmosphere between the two turbines in Fig. 14-24 is not known a priori. Nor, with two compressors, is the intake pressure ratio distribution known. Iterative procedures based on an assumed mass flow rate are used to determine the pressure level between the two turbines such that mass flow and pressure continuity through the exhaust system is satisfied (e.g., Ref. 2).

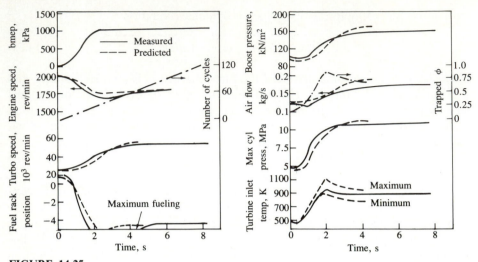

FIGURE 14-25

Predicted (- - -) and measured (——) response of a turbocharged direct-injection diesel engine to an increase in load.[55]

14.4.6 Second Law Analysis of Engine Processes

The first-law-based methods for evaluating power plant performances do not explicitly identify those processes within the engine system that cause unrecoverable degradation of the thermodynamic state of the working fluid. However, second-law-based analysis methods do provide the capability to identify and quantify this unrecoverable state degradation. Thus, cause and effect relationships which relate these losses to individual engine processes can be determined. The first law analysis approaches summarized in this section (14.4) are based on the fact that energy is conserved in every device and process. Thus, they take account of the conversion of energy from one form to another: e.g., chemical, thermal, mechanical. Although energy is conserved, second law analysis indicates that various forms of energy have differing levels of ability to do useful mechanical work. This ability to perform useful mechanical work is defined as *availability*.

The availability of a system at a given state is defined as the amount of useful work that could be obtained from the combination of the system and its surrounding atmosphere, as the system goes through reversible processes to equilibrate with the atmosphere. It is a property of the system and the environment with which the system interacts, and its value depends on both the state of the system and the properties of the atmosphere. Availability is not a conserved property; availability is destroyed by irreversibilities in any process the system undergoes. When availability destruction occurs, the potential for the system to do useful mechanical work is permanently decreased. Thus to make a proper evaluation of the processes occurring within an engine system both energy and availability must be considered concurrently.

The basis for an availability analysis of realistic models of internal com-

TABLE 14.2
Available energy equations for various processes

Mechanism	Equation
Work transfer	$dA_W = dW$
Heat transfer	$dA_Q = dQ(1 - T_0/T)$
Gas transfer	$dA_g = dm_g[(h - h_0) - T_0(s - s_0)]$
Liquid fuel transfer†	$dA_f = dm_f(1.0338Q_{\text{LHV}})$
Control volume storage	$dA_{\text{CV}} = d\{m_{\text{CV}}[(u - u_0) - T_0(s - s_0) + p_0(v - v_0)]\}$

† The availability of the fuel is 1.0338 times its lower heating value; see Sec. 5.7.

bustion engine processes has already been developed in Sec. 5.7. The change in availability of any system undergoing any process where work, heat, and mass transfers across the system boundary occur (see Fig. 5.13) can be written:

$$\Delta A = A_{\text{in}} - A_{\text{out}} - A_{\text{destroyed}} \qquad (14.53)$$

where A_{in} and A_{out} represent the availability transfers into and out of the system across the boundary. Since availability is not a conserved quantity, this equation can only be used to solve for the availability destruction term, $A_{\text{destroyed}}$. Table 14.2 summarizes the equations for the availability change of the system and the availability transfers associated with work, heat and mass transfer across the system boundary, developed in Sec. 5.7.

This availability balance is applied to the internal combustion engine operating cycle as follows. A first-law-based cycle analysis of the type described above in this section (14.4) is used to define the variation in working fluid thermodynamic state, and the work, heat, and mass transfers that occur in each of the processes that make up the total engine cycle. Integration of the availability balance over the duration of each process then defines the magnitude of the availability destruction that occurs during that process.

To illustrate this procedure, consider the operating cycle of a 10-liter six-cylinder turbocharged and aftercooled direct-injection four-stroke cycle compression-ignition engine, operating at its rated power and speed of 224 kW and 2100 rev/min. The variations in temperature, energy, and entropy are determined with a first-law-based analysis. Figure 14-26 shows the T-s diagram for the working fluid as it goes through the sequence of processes from air inlet from the atmosphere (state 1) to exhaust gas exit to the atmosphere (state 10).[57] The incoming air is compressed (with some irreversibility) in the turbocompressor to state 2 and cooled with an aftercooler to state 3. The air at state 3 is drawn into the cylinder and mixed (irreversibly) with residual gases until, at the end of the intake, the cylinder gases are represented by state 4. That mixture is subsequently compressed (with modest heat loss) to state 5. Fuel addition commences close to state 5; subsequent burning increases the combustion chamber pressure and temperature along the line 5-6. At 6 the heat release, heat transfer, and volume change rate are such that the maximum cylinder pressure is reached (a few degrees after TC). From 6 to 7 combustion continues to completion, the burned

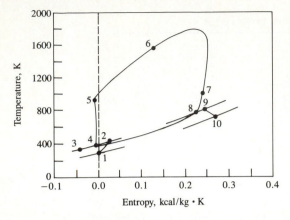

FIGURE 14-26
T-s diagram for the working fluid as it goes through the sequence of processes from air inlet to exhaust in turbo-charged aftercooled DI compression-ignition engine. The 10-liter six-cylinder engine is operated at its rated power (224 kW) and speed (2100 rev/min). The text relates processes to numbered end states.[57]

gases continue to expand, doing work on the piston and losing heat to the walls. At state 7 the exhaust valve opens initiating a rapid pressure equilibration with the exhaust manifold to a pressure corresponding to point 8. Gases are expelled from the cylinder to the exhaust manifold. After the intake valve opens, cylinder residual gases are mixed with incoming air at state 3 to yield gases at state 4 to complete the in-cylinder cycle. The exhaust gases that have been expelled from the cylinder experience additional thermodynamic losses and can be represented by state 9. These gases then pass through the turbocharger turbine to state 10 to provide the work to drive the compressor.

A first law and second law analysis of a naturally aspirated diesel engine are compared in Table 14.3. Also shown is a second law analysis of a turbocharged version of the naturally aspirated diesel. These results illustrate the value of defining the losses in availability that occur in each process.

Consider the first and second law analysis results for the naturally aspirated engine. While 25.1 percent of the fuel energy leaves the combustion chamber in the form of heat transfer, the availability transfer corresponds to 21.4 percent of the fuel's availability. It is this latter number that indicates the maximum amount of the heat transfer that can be converted to work. The table shows that 34.6 percent of the fuel energy is carried out of the engine in the exhaust gases. However, the second law analysis shows that the exhaust contains only 20.4 percent of the available energy of the fuel. The ratio of these quantities shows that only about 60 percent of the exhaust energy can be converted to work using ideal thermodynamic devices.† The exhaust gas leaves the system in a high-temperature, ambient pressure state and therefore has high entropy (relative to the p_0, T_0 reference state). This, via the gas-transfer equation in Table 14.2, reduces the available energy of the exhaust gas stream.

---

† Of course, real thermodynamic devices will produce less work than ideal devices.

TABLE 14.3

Comparison of first and second law analysis for six-cylinder 14-liter naturally aspirated and turbocharged diesel engine at 2100 rev/min[58]

	Naturally aspirated	Turbocharged	
	First law, % fuel energy	Second law, % fuel availability	
Indicated work†	40.3	39.1	43.9
Combustion loss	—	15.9	19.2
Cylinder heat transfer	25.1	21.4	17.6
Internal valve throttling	—	0.7	0.7
Exhaust valve throttling	—	2.5	2.3
Loss in compressor	—	—	1.4
Loss in turbine	—	—	0.8
Exhaust to ambient	34.6	20.4	14.1
Total	100.0	100.0	100.0
Brake power, kW	185	185	220

† Note that the indicated work for the second law balance is a lower percentage than for the first law. This occurs because the availability of the fuel is 1.0317 times the fuel's heating value.

The quantity referred to as combustion loss in Table 14.3 is determined from an availability balance for the combustion chamber over the duration of the combustion period. The "availability destroyed" term in Eq. (14.53) then represents the deviation of the actual combustion process from a completely reversible process. The second law analysis shows that the availability loss associated with the combustion irreversibilities is 15.9 percent of the fuel's availability. This loss depends on the overall equivalence ratio at which the engine is operating, as indicated in Fig. 5-17. Combustion of leaner air/fuel ratios would give a higher fractional availability loss due to mixing of the fuel combustion products with increased amounts of excess air and the lower bulk temperature.

Overall, the most important point emerging from this comparison is that the work-producing potential of the heat loss to the combustion chamber walls and the exhaust mass flow out of the engine is not as large as the magnitude of the energy transferred: some of these energy transfers, even with ideal thermodynamic work-producing devices, must ultimately be rejected to the environment as heat.

A comparison of the second and third columns in Table 14.3, both obtained with a combined first and second law analysis, illustrates how turbocharging improves the performance of a naturally aspirated engine. The brake fuel conversion efficiency of the turbocharged engine is considerably improved—from 33.9 to 39.2 percent. The table indicates that through turbocharging, the availability transfers associated with the heat loss and exhaust gas flow are reduced from 41.8

to 31.7 percent (a difference of 10.1 percentage points), while the combustion and added turbomachinery availability losses increase from 15.9 to 21.4 percent (a difference of 5.5 percentage points). By turbocharging, advantage has been taken of the following changes. While the leaner air/fuel ratio operation of the turbo-charged engine increases the combustion availability losses due to the use of a greater portion of the chemical energy of the fuel to mix with and heat excess air, the lower burned gas temperature this produces results in reduced heat losses and lower cylinder exhaust temperature. In addition, the turbocharger transfers available energy from the cylinder exhaust to the inlet air. The reduced heat loss and lower final exhaust availability level give a substantial performance improvement.[58]

To interpret the second law analysis results, one must remember that the desired output is brake work and increases in this quantity (for a given fuel flow) represent improved performance. All other availability terms represent losses or undesirable transfers from the system; decreasing these terms constitutes an improvement. These undesirable available energy transfer and destruction terms fall into five categories: (1) heat transfer, (2) combustion, (3) fluid flow, (4) exhaust to ambient, (5) mechanical friction. The available energy flows identified as heat transfer represent the summation of all availability transfers that occur due to heat transfers. The most significant of these are the in-cylinder and aftercooler heat rejection. The combustion loss represents the amount of available energy destroyed due to irreversibilities occurring in releasing the chemical potential of the fuel as thermal energy and mixing the combustion products with any excess air. The fluid flow losses include the available energy destroyed within the working fluid in the compressor, aftercooler, intake valve, exhaust valve, exhaust manifold, and turbine due to fluid shear and throttling. The availability destroyed due to fluid shear and mechanical rubbing, exterior to the working fluid, are contained in the mechanical friction category. The effect of variations in engine load and speed on these five categories of losses or transfers will now be described.

Figure 14-27 shows the availability transfers or losses in each of these categories for a turbocharged six-cylinder 10-liter displacement direct-injection diesel engine, expressed as a percentage of the fuel availability, as a function of engine load. The percentage of fuel availability associated with the heat transfers varies

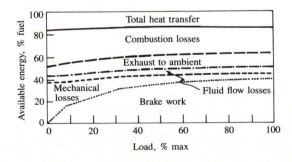

FIGURE 14-27
Distribution of available energy into major categories for the engine of Fig. 14-26 as a function of engine load.[57]

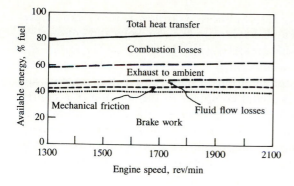

FIGURE 14-28
Distribution of available energy into major categories for the engine of Fig. 14-26 as a function of engine speed.[57]

little over the load range. The combustion loss increases from 21.8 to 32.5 percent as load is decreased due to an increasingly lean operation of the engine. Fluid friction losses, as a percentage, increase slightly as load increases due to larger mass flow rates. Since friction is approximately constant in absolute magnitude, its relative importance increases drastically as the brake output goes to zero. Exhaust flow available energy decreases from 12.2 to 8 percent as load is decreased from 100 to 0 percent.[57]

The effect of varying engine speed (at full load) is shown in Fig. 14-28. The availability associated with heat transfers changes over the speed range shown from 15.6 to 21 percent: more time during each cycle is available for heat transfer at lower speeds. Fluid flow and friction losses decrease with decreasing speed. Other availability losses remain essentially constant as a percentage of the fuel's availability.[57]

14.5 FLUID-MECHANIC-BASED MULTIDIMENSIONAL MODELS

14.5.1 Basic Approach and Governing Equations

The prediction of the details of the flow field within engines, and the heat-transfer and combustion processes that depend on those flow fields, by numerical solution of the governing conservation equations has become a realizable goal. Such methods have been under development for more than a decade, during which time they have steadily improved their ability to analyze the flow field in realistic engine geometries. While the overall dynamic characteristics of intake and exhaust flows can usefully be studied with one-dimensional unsteady fluid dynamic computer calculations (see Sec. 14.3.4), flows within the cylinder and in intake and exhaust ports are usually inherently unsteady and three dimensional. Recent increases in computing power, coupled with encouraging results with two-dimensional calculations, indicate that useful three-dimensional calculations are now feasible. However, they still do not have the capability to predict accurately all the features of real engine processes of interest. Gas-flow patterns can be predicted best; predictions of fuel spray behavior are less complete, and combustion calculations present considerable difficulties.

These computational, fluid dynamic, engine process analysis codes solve the partial differential equations for conservation of mass, momentum, energy, and species concentrations. To apply a digital computer to the solution of a continuum problem (such as the flow field inside the cylinder), the continuum must be represented by a finite number of discrete elements. The most common method of discretization is to divide the region of interest into a number of small zones or cells. These cells form a grid or mesh which serves as a framework for constructing finite volume approximations to the governing partial differential equations. The time variable is similarly discretized into a sequence of small time intervals called time steps, and the transient solution is "marched out" in time: the solution at time t_{n+1} is calculated from the known solution at time t_n. Three-dimensional formulations of the finite difference equations are required for most practical engine calculations; two-dimensional (or axisymmetric) formulations can be useful, however, under simpler flow situations, and have been more extensively used to date due to their simpler models and computer codes and requirement for less computer time and storage capacity.

The principal components of these multidimensional engine flow models are the following:[59]

1. The mathematical models or equations used to describe the flow processes. Especially important is the turbulence model, which describes the small-scale features of the flow which are not accessible to direct calculation.
2. The discretization procedures used to transform the differential equations of the mathematical model into algebraic relations between discrete values of velocity, pressure, temperature, etc., located on a computing mesh which (ideally) conforms to the geometry of the combustion chamber with its moving valves and piston.
3. The solution algorithm whose function is to solve the algebraic equations.
4. The computer codes which translate the numerical algorithm into computer language and also provide easy interfaces for the input and output of information.

The basic equations for all existing in-cylinder flow calculation methods are the differential equations expressing the conservation laws of mass, momentum (the Navier-Stokes equations—a set of three), energy, and species concentrations. These equations, in the above order, may be written:

$$\frac{D}{Dt}\begin{pmatrix} 1 \\ u_i \\ e \\ Y_\alpha \end{pmatrix} = \begin{pmatrix} 0 \\ 0 \\ Q \\ S_\alpha \end{pmatrix} - \frac{\partial}{\partial x_j}\begin{pmatrix} 0 \\ \tau_{ij} \\ q_j \\ J_{\alpha i} \end{pmatrix} \tag{14.54}$$

The first term on the right gives the source terms, the second term the diffusive transport. The D/Dt operator provides the convective transport terms and is

$$\frac{Df}{Dt} \equiv \frac{\partial(\rho f)}{\partial t} + \frac{\partial}{\partial x_j}(\rho u_j f) \tag{14.55}$$

Here, ρ is the density, u_i the ith velocity component, e the internal energy per unit mass, and Y_α the concentration of species α per unit mass.

In the IC engine context, the thermal energy source term Q involves a viscous term and source terms arising from chemical reaction of the fuel. Both Q and the species source term, S_α, will depend upon the chemical rate equations, which must be known to close the problem. Note that diffusion of the various species contributes to the diffusive flow of internal energy, q_j, in addition to conductive heat diffusion.

The fact that turbulent flows exhibit important spatial and temporal variations over a range of scales (dictated at the upper end by chamber dimensions and at the lower end by viscous dissipative processes, see Sec. 8.2.1) makes direct numerical solution of these governing equations impractical for flows of engine complexity. Recourse must therefore be made to some form of averaging or filtering which removes the need for direct calculation of the small-scale motions. Two approaches have been developed for dealing with this turbulence modeling problem: full-field modeling (FFM), sometimes called statistical flux modeling; and large-eddy simulation (LES) or subgrid-scale simulation. In FFM, one works with the partial differential equations describing suitably averaged quantities, using the same equations everywhere in the flow. For periodic engine flows, time averaging must be replaced by ensemble or phase averaging (see Sec. 8.2.1). The variables include the velocity field, thermodynamic state variables, and various mean turbulence parameters such as the turbulent kinetic energy, the turbulent stress tensor, etc. In FFM, models are needed for various averages of the turbulence quantities. These models must include the contributions of all scales of turbulent motion.[59, 60]

Large-eddy simulation (LES) is an approach in which one actually calculates the large-scale three-dimensional time-dependent turbulence structure in a single realization of the flow. Thus, only the small-scale turbulence need be modeled. Since the small-scale turbulence structure is more isotropic than the large-scale structure and responds rapidly to changes in the large-scale flow field, modeling of the statistical fluxes associated with the small-scale motions is a simpler task than that faced in FFM where the large-scale turbulence must be included.

An important difference between FFM and LES is their definition of "turbulence." In FFM the turbulence is the deviation of the flow at any instant from the average over many cycles of the flow at the same point in space and oscillation phase [i.e., the fluctuation velocity defined by Eq. (8.16) or (8.18)]. Thus, FFM "turbulence" contains some contribution from cycle-by-cycle flow variations. LES defines turbulence in terms of variations about a local average; hence in LES turbulence is related to events in the current cycle.[60]

14.5.2 Turbulence Models

In *full-field modeling* (FFM), equations for the averaged variables are formed from Eqs. (14.54). With periodic engine flows, phase or ensemble averaging must be used (see Secs. 8.2.1 and 8.2.2). Since the flow during the engine cycle is compressed and expanded, mass-weighted averaging (called Favre averaging) can be used to make the averaged compressible-flow equations look almost exactly like the averaged equations for incompressible flows. The combined ensemble-Favre averaging approach works as follows.[60]

We denote the phase-averaging process by { }, i.e.:

$$\{\rho(\mathbf{x}, t)\} = \lim_{N \to \infty} \frac{1}{N} \sum_{n=1}^{N} \rho(\mathbf{x}, t + n\tau) \tag{14.56}$$

where τ is the cycle period. We also write $\{\rho\} = \tilde{\rho}$, and decompose ρ into $\rho = \tilde{\rho} + \rho'$. The mass-weighted phase-averaged quantities (indicated by an overbar) are defined by

$$\tilde{\rho}(\mathbf{x}, t)\bar{f}(\mathbf{x}, t) = \lim_{N \to \infty} \sum_{n=1}^{N} \rho(\mathbf{x}, t + n\tau)f(\mathbf{x}, t + n\tau) \tag{14.57}$$

where all flow variables (except density and pressure) have been decomposed as $f = \bar{f} + f'$. Note that $\{\rho'\}$ is zero, $\{\bar{f}\} = \bar{f}$, the mass-weighted phase average of f' is zero, but $\{f'\}$ is not zero. With these definitions:

$$\{\rho f\} = \tilde{\rho}\bar{f}$$
$$\{\rho f'\} = 0$$
$$\{\rho f g\} = \tilde{\rho}(\bar{f}\bar{g} + \overline{f'g'})$$
$$\{\rho f g h\} = \tilde{\rho}(\bar{f}\bar{g}\bar{h} + \bar{f}\overline{g'h'} + \bar{g}\overline{f'h'} + \bar{h}\overline{f'g'} + \overline{f'g'h'}) \tag{14.58}$$

Phase-averaging Eq. (14.54), one obtains[60]

$$\frac{\bar{D}}{Dt}\begin{pmatrix} \bar{u}_i \\ \bar{e} \\ \bar{Y}_\alpha \end{pmatrix} = \begin{pmatrix} 0 \\ \{Q\} \\ \{S_\alpha\} \end{pmatrix} - \frac{\partial}{\partial x_j}\begin{pmatrix} \{\tau_{ij}\} \\ \{q_j\} \\ \{J_{\alpha j}\} \end{pmatrix} - \frac{\partial}{\partial x_j}\begin{pmatrix} \tilde{\rho}\overline{u'_i u'_j} \\ \tilde{\rho}\overline{e'u'_j} \\ \tilde{\rho}\overline{Y'_\alpha u'_j} \end{pmatrix} \tag{14.59}$$

where

$$\frac{\bar{D}\bar{f}}{Dt} = \frac{\partial}{\partial t}(\tilde{\rho}\bar{f}) + \frac{\partial}{\partial x_j}(\tilde{\rho}\bar{u}_j \bar{f}) \tag{14.60}$$

The terms on the left-hand side in Eq. (14.59) involve only the solution variables $\tilde{\rho}$, $\bar{u}_i$, $\bar{e}$, and $\bar{Y}_\alpha$, and hence require no modeling. However, all of the terms on the right, particularly the last terms that represent turbulent transport, involve turbulence fluctuation quantities and must be modeled in terms of the solution variables. The source terms $\{Q\}$ and $\{S_\alpha\}$ present special difficulties to the engine modeler. Due to the exponential dependence of the heat release Q on temperature, $\{Q\}$ will be strongly influenced by temperature fluctuations. These issues are discussed more fully in Secs. 14.5.5 and 14.5.6.

The momentum equations contain terms, $-\tilde{\rho}\overline{u_i'u_j'}$, which represent turbulent stresses (and are often called the Reynolds stresses). These terms must be modeled with additional equations before the set of equations, (14.59), is "closed" and can be solved. The most widely used *turbulence model* or equation set is the *k-ε model*.[60-63] This assumes a newtonian relationship between the turbulent stresses and mean strain rates, and computes the (fictitious) turbulent viscosity appearing in this relationship from the local turbulent kinetic energy k $(=\overline{u_i u_i}/2)$ and its dissipation rate ε. An equation governing k can be developed by multiplying the u_i equation in Eq. (14.54) by u_i, subtracting from this the equation formed by multiplying the $\bar{u}_i$ equation in Eq. (14.59) by $\bar{u}_i$, and phase-averaging the result. The equation so obtained is

$$\frac{\bar{D}\bar{k}}{Dt} = \tilde{\rho}(P - \varepsilon) - \frac{\partial}{\partial x_j} J_k \qquad (14.61a)$$

where P is the rate of turbulence production per unit mass

$$P = -\overline{u_i'u_j'}\frac{\partial \bar{u}_i}{\partial x_j} \qquad (14.61b)$$

and J_k represents diffusive transport.

In the most commonly used two-equation k-ε model, all the unknown turbulence quantities are modeled in terms of the turbulent velocity scale $k^{1/2}$ and the turbulence length scale $k^{3/2}/\varepsilon$ obtained from the definition of the energy dissipation rate, via

$$\varepsilon \propto \frac{k^{3/2}}{l} \qquad (14.62)$$

The rationale is that the rate of energy dissipation is controlled by the rate at which the large eddies feed energy to the smaller dissipative scales which in turn adjust to handle this energy.[60]

A turbulent viscosity μ_T is defined:

$$\mu_T = \frac{C_0 \tilde{\rho} k^2}{\varepsilon} \qquad (14.63)$$

where C_0 is a model constant. The turbulent stress terms appearing in Eqs. (14.59) and (14.61) are then modeled in a quasi-newtonian manner:

$$\tilde{\rho}\overline{u_i'u_j'} = \tfrac{2}{3}\tilde{\rho}k\delta_{ij} + \tfrac{2}{3}\mu_T \nabla \cdot \bar{\mathbf{u}}\delta_{ij} - 2\mu_T \bar{S}_{ij} \qquad (14.64)$$

where $\bar{S}_{ij}$ is the strain rate of the $\bar{u}_i$ field:

$$\bar{S}_{ij} = \frac{1}{2}\left(\frac{\partial \bar{u}_i}{\partial x_j} + \frac{\partial \bar{u}_j}{\partial x_i}\right) \qquad (14.65)$$

The viscous-stress terms in the momentum equations are evaluated using a new-tonian constitutive relation. The turbulent-diffusion terms in the various transport equations are modeled using the turbulent diffusivity. The diffusing flux of a quantity ϕ is given by

$$J_{\phi i} = -\frac{\mu_T}{\sigma_\phi}\frac{\partial \phi}{\partial x_i} \tag{14.66}$$

where σ_ϕ is a turbulent Prandtl number for ϕ.

The model is completed with a transport equation for ε. An exact equation can be developed by suitable manipulation of the Navier-Stokes equations. All ε equation models are of the form[60]

$$\frac{\bar{D}\varepsilon}{Dt} = W - \frac{\partial H_i}{\partial x_i} \tag{14.67}$$

where W is the source term and H_i is the diffusive flux of $\tilde{\rho}\varepsilon$ which is modeled similarly to the other diffusion terms. The appropriate form of W is the subject of much debate. For an incompressible flow, W can be modeled adequately by

$$W = \left(-C_2 + C_1 \frac{P}{\varepsilon}\right)\frac{\tilde{\rho}\varepsilon^2}{k} \tag{14.68}$$

C_1 and C_2 are constants: the C_2 term produces the proper behavior of homogeneous isotropic turbulence and the C_1 term modifies this behavior for homogeneous shear. However, for a flow with compression and expansion, an additional term in Eq. (14.68) is needed to account for changes in ε produced by dilation. Several forms for this additional term have been proposed[60, 62] (for example, $C_3 \tilde{\rho}\varepsilon\nabla \cdot \bar{u}$) and compared.[63] The goal is to construct a W that predicts the appropriate physical behavior under the relevant engine conditions. While different choices for modeling these terms do affect the results (especially the behavior of the turbulence length scale during the cycle[62]), the predictions of mean flow and turbulence intensity do not differ very significantly.[63]

One other FFM that has been applied to engines is the *Reynolds stress model* (RSM) which, in its most general form, comprises seven simultaneous partial differential equations for the six stress components and the dissipation rate ε. This obviously imposes a much greater computing burden compared with the two-equation k-ε model. The limited results available[64] indicate that RSM predictions of the flow field are closer to corresponding measured data than k-ε model predictions.[65]

The large-eddy simulation (LES) approach to turbulence modeling[66] has also been applied to engines. Since here one calculates the large-scale three-dimensional time-dependent flow structure directly, only the turbulence smaller in scale than the grid size need be modeled. Hence these are often referred to as subgrid-scale models. A new dependent variable q, which represents the kinetic energy per unit mass of the turbulent length scales that are too small to resolve in the mesh, is introduced. This variable satisfies a transport equation which contains terms for production and decay of q and for its convection and diffusion. In the KIVA engine code,[67] this equation has the form:

$$\frac{\partial}{\partial t} (\rho q) + \nabla \cdot (\rho q \mathbf{u}) = -\tfrac{2}{3}\rho q \nabla \cdot \mathbf{u} + \boldsymbol{\sigma} : \nabla \mathbf{u} + \nabla \cdot (\mu \nabla q) - C\rho L^{-1} q^{3/2} + \dot{W}_s$$

$$(14.69)$$

where $\boldsymbol{\sigma}$ is the turbulent stress tensor, μ the turbulent viscosity, C a constant of order unity, and L a characteristic length on the order of twice the mesh spacing. $\dot{W}_s$ is a source term representing the production of turbulence by the motion of fuel droplets in situations where fuel sprays are important.

The physical meaning of the terms in Eq. (14.69) are as follows. The term $\nabla \cdot (\rho q \mathbf{u})$ is the convection of the turbulence by the resolved (large-scale) velocity field. The term $-\tfrac{2}{3}\rho q \nabla \cdot \mathbf{u}$ is a compressibility term that is the turbulent analog of $p \, dV$ work. The term $\boldsymbol{\sigma} : \nabla \mathbf{u}$ represents the production of turbulence by shear in the resolved velocity field; $\nabla \cdot (\mu \nabla q)$ is the self-diffusion of the turbulence with diffusivity μ/ρ. The term $-C\rho L^{-1} q^{3/2}$ represents the decay of turbulent energy into thermal energy. This term appears with opposite sign as a source term in the thermal internal energy equation in place of $\boldsymbol{\sigma} : \nabla \mathbf{u}$, which can be thought of as the rate at which kinetic energy of the resolved motions is dissipated by the turbulence. Before it is dissipated, the kinetic energy of the resolved velocity field is first converted into subgrid scale turbulent energy q, which is then converted into heat by the decay term $C\rho L^{-1} q^{3/2}$.[67]

Under most circumstances, the velocity and temperature boundary layers in an engine cylinder will be too thin to be resolved explicitly with a computing mesh that is practical on present-day computers. However, these layers are important because they determine the wall shear and heat flux which are essential boundary conditions for the numerical simulation, and are of practical importance (see Secs. 8.3 and 12.6.5). Special submodels for these boundary layers, referred to as wall functions, are used to connect the wall shear stresses, heat fluxes, wall temperatures, etc., to conditions at the outer edge of the boundary layer. This removes the need to place grid points within the layer. Since the boundary layers are usually turbulent, the logarithmic "turbulent law of the wall" is commonly used. Key assumptions made are: that the finite difference mesh point nearest the wall lies in the law-of-the-wall region and that the law-of-the-wall relation for steady flow past a plane wall is valid under engine cylinder conditions. While these may not be valid assumptions, it is not yet feasible to resolve the flow details within the boundary layer.[68]

14.5.3   Numerical Methodology

The three important numerical features of multidimensional methods are: the computational grid arrangement, which defines the number and positions of the locations at which the flow parameters are to be calculated; the discretization practices used to transform the differential equations of the mathematical model into algebraic equations; and the solution algorithms employed to obtain the flow parameters from the discrete equations.[59, 65]

COMPUTING MESH. The requirements of the computing mesh are:

1. It adequately fits the topography of the combustion chamber and/or inlet port, including the moving components.
2. It allows control of local resolution to obtain the maximum accuracy with a given number of grid points.
3. It has the property that each interior grid point is connected to the same number of neighboring points.

The first requirement obviously follows from the need to simulate the effects of changes in engine geometry. The second requirement stems from the fact that computing time increases at least linearly with the number of mesh points. Thus it is desirable that the mesh allow concentration of grid points in regions where steep gradients exist such as jets and boundary layers. The third requirement comes from the need for the mesh to be topologically rectangular in some transformed space so that highly efficient equation solvers for such mesh systems can be utilized.

Early engine models used a grid defined by the coordinate surfaces of a cylindrical-polar frame. Such an approach is adequate provided the combustion chamber walls also coincide with coordinate surfaces. This only occurs for a restricted number of practical chambers (e.g., disc and centered cylindrical bowl-in-piston shapes); even for these, the inlet and exhaust valve circumferences would in general cut across the grid (see Fig. 14-29a). While procedures have been devised for modifying the difference equations for such grids to allow for noncoincident boundaries, the preferred approach is to employ some form of flexible "body-fitting" coordinate frame/grid whose surfaces can be shaped to the chamber geometry, as illustrated in Fig. 14-29b, which shows a diesel engine combustion chamber fitted by a mesh which is orthogonal-curvilinear in the bowl. This enables the bowl shape to be accurately represented and the boundary layers on its surfaces to be resolved in greater detail. The region between the piston crown and cylinder head surfaces is fitted with a bipolar system which expands and contracts axially to accommodate the piston motion. The orthogonality constraint of this mesh limits its usefulness: the generation of orthogonal meshes for general three-dimensional geometries is cumbersome and the resulting mesh often far from optimal. These problems are largely surmounted by "arbitrary" nonorthogonal lagrangian-eulerian meshes like that used in KIVA,[67] illustrated in Fig. 14-29c. This has the additional advantage that the mesh points in the swept volume are not constrained to move axially: their motion can be arbitrarily prescribed.[59]

DISCRETIZATION PRACTICES. These multidimensional engine flow models are time-marching programs that solve finite difference approximations to the gov-

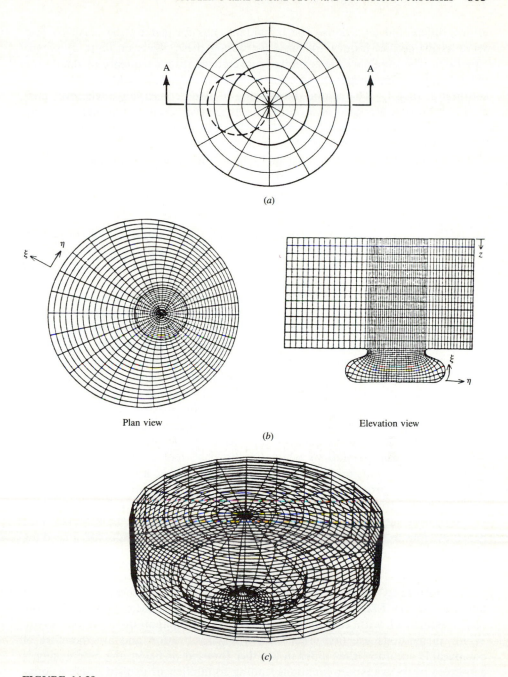

FIGURE 14-29
Different types of computing mesh arrangements for engine combustion chambers. (a) Cylindrical polar mesh: dashed line shows valve head circumference.[65] (b) Body-fitted orthogonal curvilinear mesh fitted to DI diesel combustion chamber bowl.[69] (c) Arbitrary nonorthogonal lagrangian-eulerian (ALE) mesh for offset diesel combustion bowl.[67]

erning differential equations. The individual cells formed by the mesh or grid serve as the spatial framework for constructing these algebraic finite difference equations. The time variable is similarly discretized into a sequence of small time intervals called time steps: the solution at time t_{n+1} is calculated from the known solution at time t_n. The spatial differencing is made conservative wherever possible. The procedure used is to difference the basic equations in integral form, with the volume of a typical cell used as the control volume and the divergence terms transformed into surface integrals using the divergence theorem.[67]

The discretized equations for any dependent variable ϕ are of the general form:

$$A_p \phi_p^{i+1} = \sum_n A_n \phi_n^{i+1} + S_{\phi, p} V_p + A_p^i \phi_p^i \qquad (14.70)$$

where the A's are coefficients expressing the combined influences of convection and diffusion, $S_{\phi, p} V_p$ is the source integral over the cell volume V_p, the subscript p denotes a typical node point in the mesh, the summation $\sum$ is over its (six) nearest neighbors, and the superscripts $i + 1$ and i denote "new" and "old" values, at times $t + \delta t$ and t, respectively, where δt is the size of the time step.[69]

Until recently all methods involved similar spatial approximations to calculate convective and diffusive transport, using a blend of first-order upwind differencing for the former and second-order central differencing for the latter. Unfortunately, all discretization practices introduce inaccuracies of some kind, and the standard first-order upwind scheme produces spatial diffusion errors which act in the same way as real diffusion to "smooth" the solutions. The magnitude of the numerical diffusion reduces as the mesh density is increased, but even with as many as 50 mesh points in each coordinate direction, the effect is not eliminated. A recent development has been the introduction of "higher order" spatial approximations which, in the past, had a tendency to produce spurious extrema. This problem has been overcome by the use of "flux blending" techniques. First-order upwind and higher-order approximations are blended in appropriate proportions to eliminate the overshoots of the latter. Even with these schemes, however, true mesh-independent solutions could not be achieved with densities of up to 50 nodes in each coordinate direction; so there is still a need for further improvement.[65]

SOLUTION ALGORITHMS. Numerical calculations of compressible flows are inefficient at low Mach numbers because of the wide disparity between the time scales associated with convection and with the propagation of sound waves. While all methods use first-order temporal discretization and are therefore of comparable accuracy, they differ in whether forward or backward differencing is employed in the transport equations leading to implicit or explicit discrete equations, respectively. In explicit schemes, this inefficiency occurs because the time steps needed to satisfy the sound-speed stability condition are much smaller than those needed to satisfy the convective stability condition alone. In implicit schemes, the inefficiency manifests itself in the additional computational labor needed to solve the implicit (simultaneous) system of equations at each time step. This solution is usually performed by iterative techniques.

The computing time requirements of these two approaches scale with the number of equations n and the number of mesh points m, as follows. For explicit methods, computing time scales as nm, but the time step is limited by the stability condition as summarized above. For implicit methods, computing time scales as $n^3 m$ and Δt is only limited by accuracy considerations.

One procedure used, a semi-implicit method, is the *acoustic subcycling* method. All terms in the governing equations that are not associated with sound waves are explicitly advanced with a larger time step Δt similar to that used with implicit methods. The terms associated with acoustic waves (the compression terms in the continuity and energy equations and the pressure gradient in the momentum equation) are explicitly advanced using a smaller time step δt that satisfies the sound-speed stability criterion [Eq. (14.23)], and of which the main time step is an integral multiple. While this method works well in many IC engine applications where the Mach number is not unduly low, it is unsuitable for very low Mach number flows since the number of subcycles ($\Delta t/\delta t$) tends to infinity as the Mach number tends to zero. For values of $\Delta t/\delta t$ greater than 50 an implicit scheme becomes more efficient. *Pressure gradient scaling* can be used to extend the method to lower Mach numbers. The Mach number is artificially increased to a larger value (but still small in an absolute sense) by multiplying the pressure gradient in the momentum equation by a time-dependent scaling factor $1/\alpha(t)^2$, where $\alpha(t) > 1$. This reduces the effective sound speed by the factor α. This does not significantly affect the accuracy of the solution because the pressure gradient in low Mach number flows is effectively determined by the flow field and not vice versa. Coupling pressure gradient scaling with acoustic subcycling reduces the number of subcycles by α.[67]

The implicit equations that result from forward differencing consist of simultaneous sets for all variables and thus require more elaborate methods of solution. However, they contain no intrinsic stability constraints. Fully iterative solution algorithms for solution of these equation sets are being replaced with more efficient simultaneous linear equation solvers.[65]

14.5.4 Flow Field Predictions

To illustrate the potential for multidimensional modeling of IC engine flows, examples of the output from such calculations will now be reviewed. A large amount of information on many fluid flow and state variables is generated with each calculation, and the processing, organization, and presentation of this information are tasks of comparable scope to its generation! Flow field results are usually presented in terms of the gas velocity vectors at each grid point of the mesh in appropriately selected planes. Arrows are usually used to indicate the direction and magnitude (by length) of each vector. Examples of such plots—of the flow pattern in the cylinder during the intake process—are shown in Fig. 14-30.[70] The flow field is shown 60° ATC during the intake stroke. A helical intake

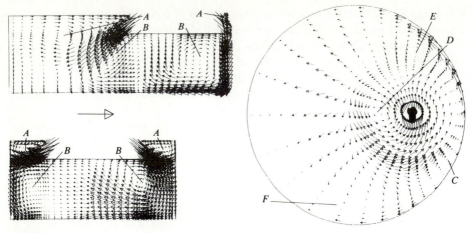

FIGURE 14-30
Computed velocity field within the cylinder at 60° ATC during the intake stroke. Top left: plane through cylinder and inlet valve axes. Bottom left: orthogonal plane through valve axis. Right: circumferential-radial plane halfway between piston and cylinder head. Reference vector arrow corresponds to velocity of 132 m/s. Letters denote centers of toroidal flow structures.[70]

port is used to general swirl, and the flow through the valve curtain area (see Sec. 6.3.2)—the inlet boundary condition for the calculation—was determined by measurement. The calculation used a curvilinear, axially expanding and contracting grid with about 16,000 mesh points of the type shown in Fig. 14-29b. It employed a fully iterative solution algorithm with standard upwind differencing and the k-ε turbulence model. Shown in Fig. 14-30 are the plane through the valve and cylinder axis (top left), the perpendicular plane through the valve axis (bottom left), and a circumferential radial plane halfway between the cylinder head and the piston (right).

The major features of the conical jet flow through the inlet valve into the cylinder are apparent (see Sec. 8.1). However, the off-cylinder-axis valve and the swirl generated by the helical port produce substantial additional complexity. The letters on the figures show regions of local recirculation. Regions A and B correspond to the rotating flow structures observed in simpler geometries (see Fig. 8-3): however, regions CF indicate that the swirling motion is far from solid-body rotation.[70]

Figures 14-31 and 14-32 show comparisons of three-dimensional predictions of in-cylinder flow fields with data. The computational and experimental geometries have been matched, as have the inlet flow velocities through the valve open area and engine speed. Figure 14-31 shows predicted and measured mean flow velocities and turbulence intensities within the cylinder, with a conventional inlet port and valve configuration, at 68° ATC during the intake stroke.[71] The experimental values come from LDA measurements (see Sec. 8.2.2). The general features of the mean flow are reproduced by the model with reasonable accuracy, though some details such as the flow along the cylinder toward the head in the

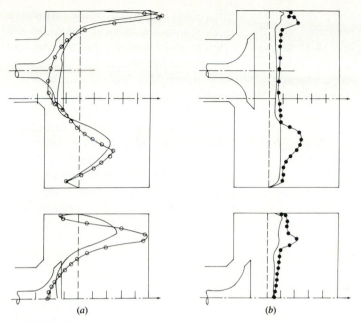

FIGURE 14-31
Comparison of (a) measured and predicted axial velocity profiles and (b) measured and predicted turbulence intensity profiles at 68° ATC during the intake stroke. Data: line with points. Predictions: line without points. Each interval on the scale on cylinder axes corresponds to 2 times the mean piston speed.[71]

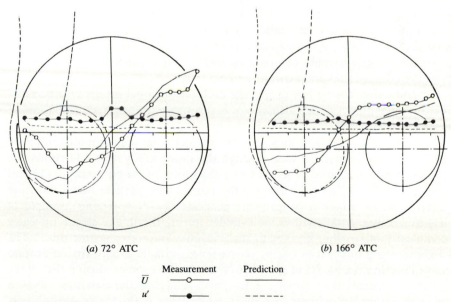

(a) 72° ATC   (b) 166° ATC

	Measurement	Prediction
$\overline{U}$	—o—	————
u'	—●—	- - - - -

FIGURE 14-32
Comparison between measured and predicted swirl velocities and turbulence intensities at 72 and 166° ATC during the intake stroke. Engine equipped with helical port.[59]

symmetry plane are not predicted. The approximate magnitude of the turbulence intensity levels are predicted, but the values within the conical intake jet are underestimated. Figure 14-32 shows in-cylinder swirl velocity predictions and measurements in an engine with a disc-shaped chamber and helical intake port, during the intake and compression strokes. Again the major features of the experimental profiles are predicted adequately, though differences in detail are significant.[59]

Comparative multidimensional modeling studies of different turbulence models,[63] differencing schemes,[59, 65] and number of grid points[59] indicate the following. Differences in the form and coefficients of the dilation term in the k-ε turbulence model have only modest effects on flow field predictions. Higher-order turbulence models might provide improved accuracy.[65] Both mesh refinement, more finely spaced grid points, and use of higher-order differencing schemes have been shown to improve significantly the accuracy of the predictions, often of course with substantial increases in computing requirements.[59]

Examples of predictions of other types of engine flow processes are the following. Squish flows into bowl-in-piston combustion chambers have been extensively analyzed. Figure 14-33a shows the flow field into and within an off-axis bowl in piston at 20° BTC of the compression stroke. The strong radially inward squish flow at the bowl lip is apparent. However, the bowl-axis offset produces a stronger flow where the squish region is greatest in extent and results in a net flow across the bowl center plane and a complex flow pattern within the bowl. Turbulence intensity results are often displayed on contour plots. Figure 14-33b shows the turbulence intensity distribution within the bowl at TC after compression. The correspondence between high-velocity regions generated by the squish flow (Fig. 14-33a) and higher turbulence intensities is apparent. A substantial variation in intensity throughout the bowl is predicted. Assimilation of detailed three-dimensional velocity data from individual two-dimensional planar vector maps is cumbersome: computer-generated three-dimensional perspective views of the velocity field are proving valuable.[72]

An alternative method of displaying multidimensional model results, especially from three-dimensional calculations, is through particle traces. Infinitesimal particles are placed at key locations in the flow field at a given crank angle (e.g., at the start of the process of interest) and their trajectories are computed from the velocity field as a function of time through the process. Figure 14-34 shows the traces of four particles, initially located near the center of the entrance to a helical inlet port at 30° ATC, as they traverse the port.[73] The particle traces illustrate the mechanism by which a helical port generates swirl. A second example of particle traces (Fig. 14-35) within the cylinder during the intake stroke indicates the complexity of swirling flows with realistic port and valve geometries.[74] The figure shows the paths traced out by six particles, initially evenly spaced around the valve curtain area at TC at the start of the intake process, during the intake stroke with a tangentially directed inlet port. While all the particles follow a helical path within the cylinder, the steepness of these paths varies substantially depending on the initial location of each particle.

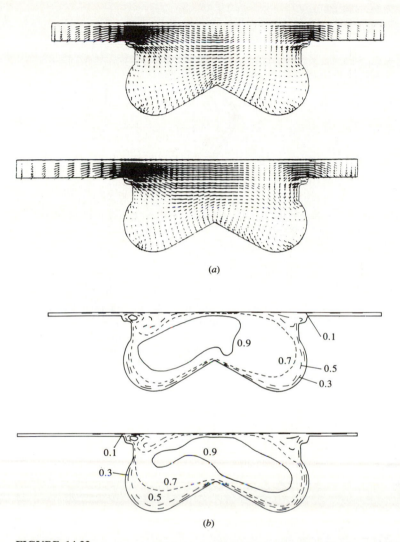

FIGURE 14-33
(a) Predicted velocity flow field within the offset bowl of DI diesel chamber in two orthogonal planes through the bowl center, at 20° BTC toward end of compression. Reference vector = 45 m/s. (b) Predicted relative turbulence intensity $u'/\bar{S}_p$ within the bowl in the same two planes at TC at the end of compression. Numbered contours show fraction of maximum value.[72]

Multidimensional models also provide local composition information. Studies have been done of two-stroke cycle scavenging flows (e.g., Ref. 75) and of the mixing between fresh mixture and residual gases in four-stroke cycle engines (e.g., Ref. 76). Figure 14-36 shows how the mixing between fresh fuel and air, and

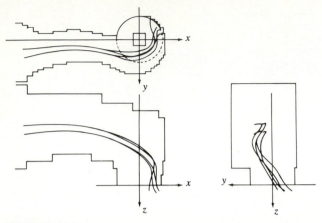

FIGURE 14-34
Computed trajectories of gas particles moving through a helical inlet port during the intake process. Particles initially located near center of port at 30° ATC.[73]

residual gases, proceeds during the intake and compression strokes of a spark-ignition engine four-stroke cycle. Concentrations (defined as fresh mixture mass/total mixture mass) at different locations within the cylinder are plotted against crank angle ($z = 2$ is near the head, $z = 7$ near the piston; $y = 2$ is near the cylinder axis, $y = 7$ near the cylinder liner). A relatively long time is required for the fresh and residual gases to mix and at 30° BTC there is still several percent

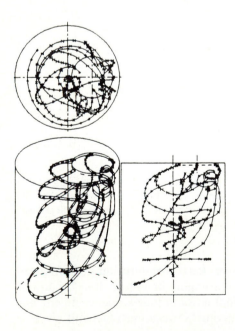

FIGURE 14-35
Computed trajectories traced out during the intake stroke by six gas particles initially evenly spaced around the valve curtain area at TC at the start of the intake process, with a tangentially directed inlet port. Cylinder shown with piston at BC, at the end of the intake stroke.[74]

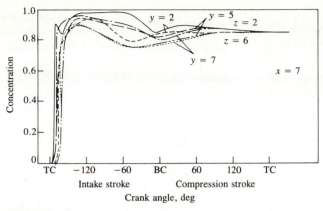

FIGURE 14-36
Computed concentration distribution of fresh fuel-air mixture and residual gas within the cylinder during the intake and compression stroke of a spark-ignition engine. Concentration expressed as fresh mixture mass/total mixture mass. $z = 2$ is near the cylinder head, $z = 7$ near the piston; $y = 2$ near the cylinder axis, $y = 7$ near the cylinder liner; $x = 7$ along the radius passing beneath the inlet valve. 2000 rev/min and wide-open throttle.[76]

nonuniformity. At part load with its higher residual fraction, one would expect these differences to be larger.[76]

14.5.5 Fuel Spray Modeling

The physical behavior of liquid fuel sprays when injected into the engine cylinder, as occurs in compression-ignition (or stratified-charge) engines, has already been described in Sec. 10.5. Here the current status of models for such spray behavior which are used with multidimensional models of gas motion within the cylinder are reviewed. Fuel-injected internal combustion engines present a particularly difficult problem for numerical simulation. The fuel spray produces an inhomogeneous fuel-air mixture: the spray interacts with and strongly affects the flow patterns and temperature distribution within the cylinder. The fuel is injected as liquid, it atomizes into a large number of small droplets with a wide spectrum of sizes, the droplets disperse and vaporize as the spray moves through the surrounding air, droplet coalescence and separation can occur, gaseous mixing of fuel vapor and air then takes place, followed, finally, by combustion. Models which explicitly treat the two-phase structure of this spray describe the spray behavior in terms of differential conservation equations for mass, momentum, and energy.

Two such classes of model exist, usually called the continuum droplet model (CDM) and the discrete droplet model (DDM). Both approaches average over flow processes occurring on a scale comparable to the droplet size, and thus require independent modeling of the interactions occurring at the gas droplet interface: typically this is done with correlations for droplet drag and heat and mass transfer. The CDM attempts to represent the motion of all droplets via an eulerian partial differential spray probability equation containing, in its most general case, eight independent variables: time, three spatial coordinates, droplet radius and the three components of the droplet velocity vector. This approach imposes enormous computational requirements. The DDM uses a statistical approach; a representative sample of individual droplets, each droplet being a member of a group of identical non-interacting droplets termed a "parcel," is tracked in a lagrangian fashion from its origin at the injector by solving ordinary differential equations of motion which have time as the independent variable.

This latter type of model is used in engine spray analysis.[77, 78] Droplet parcels are introduced continuously throughout the fuel-injection process, with specified initial conditions of position, size, velocity, number of droplets in the parcel prescribed at the "zone of atomization" according to an assumed or known size distribution, initial spray angle, fuel-injection rate, and fuel temperature at the nozzle exit. The values of these parameters are chosen to represent statistically all such values within the spray. They are then tracked in a lagrangian fashion through the computational mesh used for solving the gas-phase partial differential conservation equations.

The equations describing the behavior of individual droplets are[79]

$$\frac{d}{dt}\mathbf{x}_k = \mathbf{u}_k \tag{14.71}$$

$$\frac{d}{dt}(m_k\mathbf{u}_k) = -\frac{m_k}{\rho_k}\nabla p + F_{D,k}(\mathbf{u} - \mathbf{u}_k) \tag{14.72}$$

$$m_k\frac{dh_k}{dt} = \dot{q}_k + (h_v - h_k)\frac{dm_k}{dt} \tag{14.73}$$

where $\mathbf{x}_k$ is the position vector for droplet k and $\mathbf{u}_k$ its velocity, m_k is the droplet mass and ρ_k the droplet density, $\mathbf{u}$ is the gas velocity, h_k the droplet specific enthalpy, $\dot{q}_k$ the heat-transfer rate from the gas to the droplet, and h_v the specific enthalpy of fuel vapor. $F_{D,k}$ is the droplet drag function:

$$F_{D,k} = \tfrac{1}{2}\pi r_k^2 \rho C_D |\mathbf{u} - \mathbf{u}_k| \tag{14.74}$$

where r_k is the droplet radius, μ and ρ the gas viscosity and density, and C_D is the drag coefficient. $F_{D,k}$ is the sum of the Stokes drag and the form drag, and in the laminar limit where $C_D = 24/\text{Re}$ with $\text{Re} = 2r_k\rho|\mathbf{u} - \mathbf{u}_k|/\mu$ it goes to $6\pi r_k\mu$.

An equation for the evaporation rate completes this set: it is usually assumed that the droplet is in thermal equilibrium at its wet-bulb temperature, T_s. Then a balance between heat transfer to the droplet and the latent heat of vaporization carried away by the fuel vapor exists:

$$(h_v - h_k)\frac{dm_k}{dt} = -\dot{q}_k \tag{14.75}$$

While a large portion of the droplet lifetime is spent in this equilibrium, terms can be added to Eq. (14.75) so that it describes the heat-up phase where the droplet temperature increases from its initial value to T_s.[79] The heat and mass exchange rates are calculated from experimentally based correlations for droplet Nusselt and Sherwood numbers as functions of Reynolds, Prandtl, and Schmidt numbers.[77, 80]

Account must now be taken of the two-way nature of the coupling between the gas and the liquid. The gas velocity, density, temperature, and fuel vapor concentration required for solving the droplet equations are taken from the values prevailing in the grid cell in which the droplet parcel resides. At the same time, a field of "sources" is assembled for the interphase mass, momentum, and energy transfer, and these are subsequently fed back into the gas-phase solution preserving conservation between phases.[81] The gas-phase mass, momentum, and energy conservation equations require additional terms to account for the displacement effect of the particles, the density change associated with mixing with the fuel vapor, the drag of the droplets, the initial momentum difference and enthalpy difference between evaporated fuel at the drop surface and the surrounding gas, and heat transfer to the droplet.[79]

The above treatment is limited to "thin sprays" where the droplets are sufficiently far apart for interparticle interactions to be unimportant. This assumption is not valid in the immediate vicinity of the injector or in narrow cone sprays. In such "thick sprays" interparticle interactions—collisions which can result in coalescence or in reseparation of droplets—are important.

The most complete models of atomized fuel sprays represent the spray by a Monte Carlo–based discrete-particle technique.[67, 80] The spray is described by a droplet distribution function—a droplet number density in a phase space of droplet position, velocity, radius, and temperature. The development of this distribution function is determined by the so-called spray equation.[80] The distribution function is statistically sampled and the resulting discrete particles are followed as they locally interact and exchange mass, momentum, and energy with the gas, using the above lagrangian droplet equations. Each discrete droplet represents a group or parcel of droplets. Droplet collisions are described by appropriate terms in the spray equation.

Figure 14-37 shows the type of results such spray models can generate. The calculation involves a direct-injection stratified-charge engine with an offset bowl-in-piston combustion chamber and a tilted injector. Injection of a single hollow-cone fuel spray commences at 52° BTC. Figure 14-37*a* and *b* shows the fuel spray at 39° BTC at the end of injection and later, at 28° BTC, just before combustion. Of the 2000 computational particles injected (each representing some number of identical physical droplets), 1218 remain at 39° BTC and 773 at 28° BTC. Evaporation and coalescence have caused these decreases. Figure 14-37*c* shows the gas velocity vectors at the end of injection: the flow field set up by momentum exchange with the injected fuel spray can be seen, and the highest velocities exist in the spray region. Figure 14-37*d* shows the equivalence ratio contours at 28° BTC just prior to ignition. The highly nonhomogeneous fuel vapor distribution within the bowl is evident.[67,82]

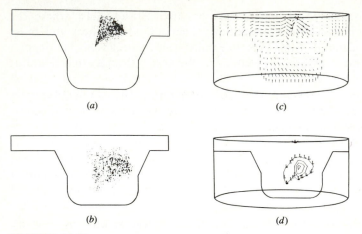

FIGURE 14-37
Predictions of single hollow-cone fuel spray behavior in direct-injection stratified-charge engine. Injection commences at 52° BTC with 2000 computational particles. (a) Location of 1218 remaining spray particles at 39° BTC at the end of injection. (b) Location of 773 spray particles at 28° BTC, just before combustion. (c) Gas velocity vectors at the end of injection at 39° BTC. (d) Fuel/air equivalence ratio contours just prior to ignition at 28° BTC. The L contour is $\phi = 0.5$, the contour interval $\Delta\phi$ is 0.5.[82]

14.5.6   Combustion Modeling

In numerical calculations of reacting flows, computer time and storage constraints place severe restrictions on the complexity of the reaction mechanisms that can be incorporated. While it is feasible to include detailed chemical mechanisms for combustion of hydrocarbon-air mixtures in one-dimensional calculations, it becomes increasingly impractical to attempt such complexity in two- and three-dimensional studies. Accordingly, engine calculations have been forced to use greatly simplified reaction schemes. In addition, detailed reaction schemes are only available for the simpler hydrocarbon fuels (e.g., methane, propane, butane): for higher hydrocarbon compounds and practical fuels which are blends of a large number of hydrocarbons, the detailed mechanisms have yet to be defined. Accordingly, multidimensional engine calculations have used highly simplified chemical kinetic schemes, with one or at most a few reactions, to represent the combustion process. While such schemes can be calibrated with experimental data to give acceptable results over a limited range of engine conditions, they lack an adequate fundamental basis.

The most common practice has been to assume the combustion process, fuel + oxidizer → products, is governed by a single rate equation of an Arrhenius form:

$$R_f = A\rho^2 x_f^a x_{ox}^b \exp\left(-\frac{E_A}{\tilde{R}T}\right) \tag{14.76}$$

where R_f is the rate of disappearance of unburned fuel, x_f and x_{ox} are unburned fuel and oxidizer mass fractions, $\tilde{R}$ is the universal gas constant, and a, b, A, and

E_A are constants (usually a and b are taken to be unity, or to be 0.5). Values for the preexperimental factor A and activation energy E_A are obtained by matching to experimentally determined rates of burning.

While this approach "works" in the sense that, when calibrated, its predictions can show reasonable agreement with data, it has three major problems. The first is the presumption that the complex hydrocarbon fuel oxidation process can be adequately represented by a single (or limited number of) overall reaction(s). The fact that it is usually necessary to adjust the constants in Eq. (14.76) as engine design and operating parameters change is one indication of this problem. Second, Eq. (14.76) uses local *average* values of concentrations and temperatures to calculate the local reaction rate, whereas the *instantaneous* local values will actually determine the reaction rate. These two rates will only be equal if the reaction time scale is much longer than that of the turbulent fluctuations, which is not the case in engine combustion. Third, the implied strong dependence of burning rate on chemical kinetics is at variance with the known experimental evidence on engine combustion (see Secs. 9.3.2 and 10.3). The effects of turbulence on the burning rate, apart from the augmentation of the thermal and mass diffusivities, are not represented by equations of the form of (14.76).[83, 84]

An alternative, equally straightforward, approach assumes that turbulent mixing is the rate-controlling process: the kinetics are sufficiently fast for chemistry modeling to be unimportant. Thus reactions proceed instantly to completion once mixing occurs at a molecular level in the smaller-scale eddies of the turbulent flow; the rate-controlling process is then the communication between and decay of the large-scale eddies. Thus the reaction rate is inversely proportional to the turbulent mixing time τ_T $(= l_I/u')$ which is equated to k/ε. Whether fuel or oxygen concentration is limiting, and the need for sufficient hot products to ensure flame spreading are also incorporated. For extremely lean (or rich) mixtures, the reaction may become kinetically controlled. A choice between Eq. (14.76) and the above mixing-controlled model can be made depending on whether the ratio of a chemical reaction time to the turbulent mixing time is greater or less than unity.[83, 84]

An example of a two-dimensional calculation of flame propagation in a premixed-charge spark-ignition engine illustrates the type of results which have

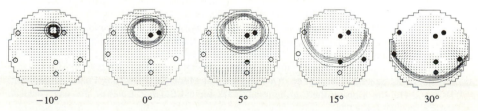

$-10°$ $0°$ $5°$ $15°$ $30°$

FIGURE 14-38
Isotherms and velocity vectors during the combustion process in premixed spark-ignition engine predicted by two-dimensional computational fluid dynamic code. Points show ionization probe locations in the cylinder head in corresponding experiment: open symbols are before flame arrival; filled symbols are after flame arrival. Crank angle values are relative to TC $= 0°$.[85]

been generated to date. Figure 14-38[85] shows computed constant-temperature lines and velocity vectors, looking down on the piston, as the flame develops from the spark. The points show ionization probe locations: open symbols denote prior to and closed symbols after flame arrival. A combustion model of the form of Eq. (14.76) was used and results in a thick "turbulent" flame with an approximately cylindrical front surface. Flame front propagation speeds are adequately predicted; the flame is not modeled in sufficient detail to describe its actual structure (see Sec. 9.3.2). Practical use is now being made of these combustion codes for both spark-ignition engine (e.g., Refs. 83, 86, and 87) and compression-ignition engine studies.[84, 88]

References

1. Streit, E. E., and Borman, G. L.: "Mathematical Simulation of a Large Turbocharged Two-Stroke Diesel Engine," SAE paper 710176, *SAE Trans.*, vol. 80, 1971.
2. Assanis, D. N., and Heywood, J. B.: "Development and Use of Computer Simulation of the Turbocompounded Diesel System for Engine Performance and Component Heat Transfer Studies," SAE paper 860329, 1986.
3. Watson, N., and Janota, M. S.: *Turbocharging the Internal Combustion Engine*, John Wiley, New York, 1982.
4. Janota, M. S., Hallam, A. J., Brock, E. K., and Dexter, S. G.: "The Prediction of Diesel Engine Performance and Combustion Chamber Component Temperatures Using Digital Computers," *Proc. Instn Mech. Engrs*, vol. 182, pt. 3L, pp. 58–70, 1967–1968.
5. Benson, R. S., Garg, R. D., and Woollatt, D.: "A Numerical Solution of Unsteady Flow Problems," *Int. J. Mech. Sci.*, vol. 6, pp. 117–144, 1964.
6. Benson, R. S.: In J. H. Horlock and D. E. Winterbone (eds.), *The Thermodynamics and Gas Dynamics of Internal Combustion Engines*, vol. 1, Clarendon Press, Oxford, 1982.
7. Chapman, M., Novak, J. M., and Stein, R. A.: "Numerical Modeling of Inlet and Exhaust Flows in Multi-Cylinder Internal Combustion Engines," in T. Uzkan (ed.), *Flows in Internal Combustion Engines*, ASME, 1982.
8. Bulaty, T., and Niessner, H.: "Calculation of 1-D Unsteady Flows in Pipe Systems of I.C. Engines," ASME paper ASME-WA7, 1984.
9. Takizawa, M., Uno, T., Oue, T., and Yura, T.: "A Study of Gas Exchange Process Simulation of an Automotive Multi-Cylinder Internal Combustion Engine," SAE paper 820410, *SAE Trans.*, vol. 91, 1982.
10. Baruah, P. C., Benson, R. S., and Balouch, S. K.: "Performance and Emission Predictions of a Multi-Cylinder Spark Ignition Engine with Exhaust Gas Recirculation," SAE paper 780663, 1978.
11. Poulos, S. G., and Heywood, J. B.: "The Effect of Chamber Geometry on Spark-Ignition Engine Combustion," SAE paper 830334, *SAE Trans.*, vol. 92, 1983.
12. Heywood, J. B., Higgins, J. M., Watts, P. A., and Tabaczynski, R. J.: "Development and Use of a Cycle Simulation to Predict SI Engine Efficiency and NO_x Emissions," SAE paper 790291, 1979.
13. Watts, P. A., and Heywood, J. B.: "Simulation Studies of the Effects of Turbocharging and Reduced Heat Transfer on Spark-Ignition Engine Operation, SAE paper 800289, 1980.
14. Lavoie, G. A., and Blumberg, P. N.: "A Fundamental Model for Predicting Fuel Consumption, NO_x and HC Emissions of the Conventional Spark-Ignited Engine," *Combust. Sci. and Technol.*, vol. 21, pp. 225–258, 1980.
15. Novak, J. M., and Blumberg, P. N.: "Parametric Simulation of Significant Design and Operating Alternatives Affecting the Fuel Economy and Emissions of Spark-Ignited Engines," SAE paper 780943, *SAE Trans.*, vol. 87, 1978.
16. Groff, E. G., and Matekunas, F. A.: "The Nature of Turbulent Flame Propagation in a Homogeneous Spark-Ignited Engine," SAE paper 800133, *SAE Trans.*, vol. 89, 1980.

17. Mattavi, J. N.: "The Attributes of Fast Burning Rates in Engines," SAE paper 800920, *SAE Trans.*, vol. 89, in SP-467, *The Piston Engine—Meeting the Challenge of the 80s*, 1980.
18. Keck, J. C.: "Turbulent Flame Structure and Speed in Spark-Ignition Engines," *Proceedings of Nineteenth Symposium (International) on Combustion*, pp. 1451–1466, The Combustion Institute, 1982.
19. Beretta, G. P., Rashidi, M., and Keck, J. C.: "Turbulent Flame Propagation and Combustion in Spark Ignition Engines," *Combust. Flame*, vol. 52, pp. 217–245, 1983.
20. Keck, J. C., Heywood, J. B., and Noske, G.: "Early Flame Development and Burning Rates in Spark-Ignition Engines," SAE paper 870164, 1987.
21. Tabaczynski, R. J., Ferguson, C. R., and Radhakrishnan, K.: "A Turbulent Entrainment Model for Spark-Ignition Engine Combustion," SAE paper 770647, *SAE Trans.*, vol. 86, 1977.
22. Tabaczynski, R. J., Trinker, F. H., and Shannon, B. A. S.: "Further Refinement and Validation of a Turbulent Flame Propagation Model for Spark-Ignition Engines," *Combust. Flame*, vol. 39, pp. 111–121, 1980.
23. Borgnakke, C.: "Flame Propagation and Heat-Transfer Effects in Spark-Ignition Engines," in J. C. Hilliard and G. S. Springer (eds.), *Fuel Economy in Road Vehicles Powered by Spark Ignition Engines*, chap. 5, pp. 183–224, Plenum Press, 1984.
24. Launder, B. E., and Spalding, D. B.: *Lectures in Mathematical Models of Turbulence*, Academic Press, 1972.
25. Borgnakke, C., Arpaci, V. S., and Tabaczynski, R. J.: "A Model for the Instantaneous Heat Transfer and Turbulence in a Spark Ignition Engine," SAE paper 800287, 1980.
26. Borgnakke, C., Davis, G. C., and Tabaczynski, R. J.: "Predictions of In-Cylinder Swirl Velocity and Turbulence Intensity for an Open Chamber Cup in Piston Engine," SAE paper 810224, *SAE Trans.*, vol. 90, 1981.
27. Davis, G. C., Tabaczynski, R. J., and Belaire, R. C.: "The Effect of Intake Valve Lift on Turbulence Intensity and Burnrate in S.I. Engines," SAE paper 840030, *SAE Trans.*, vol. 93, 1984.
28. Davis, G. C., Mikulec, A., Kent, J. C., and Tabaczynski, R. J.: "Modeling the Effect of Swirl on Turbulence Intensity and Burn Rate in S.I. Engines and Comparison with Experiment," SAE paper 860325, 1986.
29. Stivender, D. L.: "Sonic Throttling Intake Valves Allow Spark-Ignition Engine to Operate with Extremely Lean Mixtures," SAE paper 680399, *SAE Trans.*, vol. 77, 1968.
30. Hires, S. D., Tabaczynski, R. J., and Novak, J. M.: "The Prediction of Ignition Delay and Combustion Intervals for a Homogeneous Charge, Spark Ignition Engine," SAE paper 780232, *SAE Trans.*, vol. 87, 1978.
31. Primus, R. J., and Wong, V. W.: "Performance and Combustion Modeling of Heterogeneous Charge Engines," SAE paper 850343, 1985.
32. Shipinski, J., Uyehara, O. A., and Myers, P. S.: "Experimental Correlation between Rate of Injection and Rate of Heat Release in Diesel Engine," ASME paper 68-DGP-11, 1968.
33. Whitehouse, N. D., and Way, R. J. B.: "Simple Method for the Calculation of Heat Release Rates in Diesel Engines Based on the Fuel Injection Rate," SAE paper 710134, 1971.
34. Woschni, G., and Anisits, F.: "Experimental Investigation and Mathematical Presentation of Rate of Heat Release in Diesel Engines Dependent Upon Engine Operating Conditions," SAE paper 740086, 1974.
35. Watson, N., Pilley, A. D., and Marzouk, M.: "A Combustion Correlation for Diesel Engine Simulation," SAE paper 800029, 1980.
36. Kuo, T., Yu, R. C., and Shahed, S. M.: "A Numerical Study of the Transient Evaporating Spray Mixing Process in the Diesel Environment," SAE paper 831735, *SAE Trans.*, vol. 92, 1983.
37. Chiu, W. S., Shahed, S. M., and Lyn, W. T.: "A Transient Spray Mixing Model for Diesel Combustion," SAE paper 760128, *SAE Trans.*, vol. 85, 1976.
38. Rife, J. M., and Heywood, J. B.: "Photographic and Performance Studies of Diesel Combustion with a Rapid Compression Machine," SAE paper 740948, *SAE Trans.*, vol. 83, 1974.
39. Sinnamon, J. F., Lancaster, D. R., and Steiner, J. C.: "An Experimental and Analytical Study of Engine Fuel Spray Trajectories," SAE paper 800135, *SAE Trans.*, vol. 89, 1980.
40. Kobayashi, H., Yagita, M., Kaminimoto, T., and Matsuoka, S.: "Prediction of Transient Diesel

Sprays in Swirling Flows Via a Modified 2-D Model," SAE paper 860332, 1986.

41. Ricou, F. P., and Spalding, D. B.: "Measurements of Entrainment by Axisymmetric Turbulent Jets," *J. Fluid Mech.*, vol. 9, pp. 21–32, 1961.

42. Abramovich, G. M.: *The Theory of Turbulent Jets*, MIT Press, Cambridge, Mass., 1963.

43. Shahed, S. M., Flynn, P. F., and Lyn, W. T.: "A Model for the Formation of Emissions in a Direct-Injection Diesel Engine," in J. N. Mattavi and C. A. Amann (eds.), *Combustion Modeling in Reciprocating Engines*, pp. 345–368, Plenum Press, 1980.

44. Hiroyasu, H., Kadota, T., and Arai, M.: "Development and Use of a Spray Combustion Modeling to Predict Diesel Engine Efficiency and Pollutant Emission," paper 214–12, *Bull. JSME*, vol. 26, no. 214, pp. 569–575, 1983.

45. Hiroyasu, H., Kadota, T., and Arai, M.: "Development and Use of a Spray Combustion Modeling to Predict Diesel Engine Efficiency and Pollutant Emissions (Part 2. Computational Procedure and Parametric Study), "paper 214-13, *Bull. JSME*, vol. 26, no. 214, pp. 576–583, 1983.

46. Hiroyasu, H.: "Diesel Engine Combustion and Its Modeling," in *Proceedings of International Symposium on Diagnostics and Modeling of Combustion in Reciprocating Engines*, COMODIA 85, pp. 53–75, Tokyo, September 4–6, 1985.

47. Blumberg, P. N., Lavoie, G. A., and Tabaczynski, R. J.: "Phenomenological Models for Reciprocating Internal Combustion Engines," *Prog. Energy Combust. Sci.*, vol. 5, pp. 123–167, 1979.

48. Hires, S. D., Ekchian, A., Heywood, J. B., Tabaczynski, R. J., and Wall, J. C.: "Performance and NO_x Emissions Modeling of Jet Ignition Prechamber Stratified Charge Engine," SAE paper 760161, *SAE Trans.*, vol. 85, 1976.

49. Hiroyasu, H., Yoshimatsu, A., and Arai, M.: "Mathematical Model for Predicting the Rate of Heat Release and Exhaust Emissions in IDI Diesel Engines," paper C102/82, *Proceedings of Conference on Diesel Engines for Passenger Cars and Light Duty Vehicles*, Institution of Mechanical Engineers, London, 1982.

50. Watson, N., and Kamel, M.: "Thermodynamic Efficiency Evaluation of an Indirect Injection Diesel Engine," SAE paper 790039, *SAE Trans.*, vol. 88, 1979.

51. Mansouri, S. H., Heywood, J. B., and Radhakrishnan, K.: "Divided-Chamber Diesel Engine, Part I: A Cycle-Simulation which Predicts Performance and Emissions," SAE paper 820273, *SAE Trans.*, vol. 91, 1982.

52. Mansouri, S. H., Heywood, J. B., and Ekchian, J. A., "Studies of NO_x and Soot Emissions from an IDI Diesel using an Engine Cycle Simulation," paper C120/82, in *Diesel Engines for Passenger Cars and Light Duty Vehicles*, Institution of Mechanical Engineers Conference Publication 1982-8, pp. 215–227, 1982.

53. Syed, S. A., and Bracco, F. V.: "Further Comparisons of Computed and Measured Divided-Chamber Engine Combustion," SAE paper 790247, 1979.

54. Meintjes, K., and Alkidas, A. C.: "An Experimental and Computational Investigation of the Flow in Diesel Prechambers," SAE paper 820275, *SAE Trans.*, vol. 91, 1982.

55. Watson, N., and Marzouk, M.: "A Non-Linear Digital Simulation of Turbocharged Diesel Engines under Transient Conditions," SAE paper 770123, *SAE Trans.*, vol. 86, 1977.

56. Marzouk, M., and Watson, N.: "Load Acceptance of Turbocharged Diesel Engines," paper C54/78, *Proceedings of Conference on Turbocharging and Turbochargers*, Institution of Mechanical Engineers, London, 1978.

57. Primus, R. J., and Flynn, P. F.: "Diagnosing the Real Performance Impact of Diesel Engine Design Parameter Variation (A Primer in the Use of Second Law Analysis)," in *Proceedings of International Symposium on Diagnostics and Modeling of Combustion in Reciprocating Engines*, COMODIA 85, pp. 529–538, Tokyo, September 4–6, 1985.

58. Primus, R. J., Hoag, K. L., Flynn, P. F., and Brands, M. C.: "An Appraisal of Advanced Engine Concepts Using Second Law Analysis Techniques," SAE paper 841287, *SAE Trans.*, vol. 93, 1984.

59. Gosman, A. D.: "Computer Modeling of Flow and Heat Transfer in Engines, Progress and Prospects," in *Proceedings of International Symposium on Diagnostics and Modeling of Combustion in Reciprocating Engines*, COMODIA 85, pp. 15–26, Tokyo, September 4–6, 1985.

60. Reynolds, W. C.: "Modeling of Fluid Motions in Engines—An Introductory Overview," in J. N.

Mattavi and C. A. Amann (eds.), *Combustion Modeling in Reciprocating Engines*, pp. 41–68, Plenum Press, 1980.

61. El Tahry, S. H.: "k-ε Equation for Compressible Reciprocating Engine Flows," *J. Energy*, vol. 7, no. 4, pp. 345–353, 1983.

62. Morel, T., and Mansour, N. N.: "Modeling of Turbulence in Internal Combustion Engines," SAE paper 820040, 1982.

63. Ahmadi-Befrui, B., Gosman, A. D., and Watkins, A. P.: "Prediction of In-Cylinder Flow and Turbulence with Three Versions of k-ε Turbulence Model and Comparison with Data," in T. Uzkan (ed.), *Flows in Internal Combustion Engines—II*, FED—vol. 20, p. 27, ASME, New York, 1984.

64. El Tahry, S. H.: "Application of a Reynolds Stress Model to Engine Flow Calculations," in T. Uzkan (ed.), *Flows in Internal Combustion Engines—II*, FED—vol. 20, pp. 39–46, ASME, New York, 1984.

65. Gosman, A. D.: "Multidimensional Modeling of Cold Flows and Turbulence in Reciprocating Engines," SAE paper 850344, 1985.

66. Ferzieger, J. H.: "Large Eddy Simulations of Turbulent Flows," AIAA paper 76-347, 1976.

67. Amsden, A. A., Butler, T. D., O'Rourke, P. J., and Ramshaw, J. D.: "KIVA—A Comprehensive Model for 2-D and 3-D Engine Simulations," SAE paper 850554, 1985.

68. Butler, T. D., Cloutman, L. D., Dukowicz, J. K., and Ramshaw, J. D.: "Multidimensional Numerical Simulation of Reactive Flow in Internal Combustion Engines," in *Prog. Energy Combust. Sci.*, vol. 7, pp. 293–315, 1981.

69. Gosman, A. D., Tsui, Y. Y., and Watkins, A. P.: "Calculation of Three Dimensional Air Motion in Model Engines," SAE paper 840229, *SAE Trans.*, vol. 93, 1984.

70. Brandstatter, W., Johns, R. J. R., and Wigley, G.: "The Effect of Inlet Port Geometry on In-Cylinder Flow Structure," SAE paper 850499, 1985.

71. Gosman, A. D., Tsui, Y. Y., and Watkins, A. P.: "Calculation of Unsteady Three-Dimensional Flow in a Model Motored Reciprocating Engine and Comparison with Experiment," presented at Fifth International Turbulent Shear Flow Meeting, Cornell University, August 1985.

72. Schapertons, H., and Thiele, F.: "Three-Dimensional Computations for Flowfields in DI Piston Bowls," SAE paper 860463, 1986.

73. Isshiki, Y., Shimamoto, Y., and Wakisaka, T.: "Numerical Prediction of Effect of Intake Port Configurations on the Induction Swirl Intensity by Three-Dimensional Gas Flow Analysis," in *Proceedings of International Symposium on Diagnostics and Modeling of Combustion in Reciprocating Engines*, COMODIA 85, pp. 295–304, Tokyo, September 4–6, 1985.

74. Wakisaka, T., Shimamoto, Y., and Isshiki, Y.: "Three-Dimensional Numerical Analysis of In-Cylinder Flows in Reciprocating Engines," SAE paper 860464, 1986.

75. Diwakar, R.: "Multidimensional Modeling of the Gas Exchange Processes in a Uniflow-Scavenged Two-Stroke Diesel Engine," in T. Uzkan, W. G. Tiederman, and J. M. Novak (eds.), *International Symposium on Flows in Internal Combustion Engines—III*, FED—vol. 28, pp. 125–134, ASME, New York, 1985.

76. Yamada, T., Inoue, T., Yoshimatsu, A., Hiramatsu, T., and Konishi, M.: "In-Cylinder Gas Motion of Multivalve Engine—Three Dimensional Numerical Simulation," SAE paper 860465, 1986.

77. Gosman, A. D., and Johns, R. J. R.: "Computer Analysis of Fuel-Air Mixing in Direct-Injection Engines," SAE paper 800091, *SAE Trans.*, vol. 89, 1980.

78. Watkins, A. P., Gosman, A. D., and Tabrizi, B. S.: "Calculation of Three Dimensional Spray Motion in Engines," SAE paper 860468, 1986.

79. Butler, T. D., Cloutman, L. D., Dukowicz, J. K., and Ramshaw, J. D.: "Toward a Comprehensive Model for Combustion in a Direct-Injection Stratified-Charge Engine," in J. N. Mattavi and C. A. Amann (eds.), *Combustion Modelling in Reciprocating Engines*, pp. 231–260, Plenum Press, 1980.

80. Bracco, F. V.: "Modeling of Engine Sprays," SAE paper 850394, 1985.

81. Cartellieri, W., and Johns, R. J. R.: "Multidimensional Modeling of Engine Processes: Progress and Prospects," paper presented at the Fifteenth CIMAC-Congress, Paris, June 1, 1983.

82. Amsden, A. A., Ramshaw, J. D., O'Rourke, P. J., and Dukowicz, J. K.: "KIVA: A Computer Program for Two- and Three-Dimensional Fluid Flows with Chemical Reactions and Fuel Sprays," report LA-10245-MS, Los Alamos National Laboratory, Los Alamos, New Mexico, February 1985.
83. Ahmadi-Befrui, B., Gosman, A. D., Lockwood, F. C., and Watkins, A. P.: "Multidimensional Calculation of Combustion in an Idealized Homogeneous Charge Engine: A Progress Report," SAE paper 810151, *SAE Trans.*, vol. 90, 1981.
84. Gosman, A. D., and Harvey, P. S.: "Computer Analysis of Fuel-Air Mixing and Combustion in an Axisymmetric D.I. Diesel," SAE paper 820036, *SAE Trans.*, vol. 91, 1982.
85. Basso, A., and Rinolfi, R.: "Two-Dimensional Computations of Engine Combustion: Comparisons of Measurements and Predictions," SAE paper 820519, 1982.
86. Basso, A.: "Optimization of Combustion Chamber Design for Spark Ignition Engines," SAE paper 840231, 1984.
87. Schapertons, H., and Lee, W.: "Multidimensional Modeling of Knocking Combustion in SI Engines," SAE paper 850502, 1985.
88. Cheng, W. K., and Theobald, M. A.: "A Numerical Study of Diesel Ignition," paper 87-FE-2, presented at the ASME Energy-Sources Technology Conference, Dallas, February 1987.

CHAPTER
15

ENGINE
OPERATING
CHARACTERISTICS

This chapter reviews the operating characteristics of the common types of spark-ignition and compression-ignition engines. The performance, efficiency, and emissions of these engines, and the effect of changes in major design and operating variables, are related to the more fundamental material on engine combustion, thermodynamics, fluid flow, heat transfer, and friction developed in earlier chapters. The intent is to provide data on, and an explanation of, actual engine operating characteristics.

15.1 ENGINE PERFORMANCE PARAMETERS

The practical engine performance parameters of interest are power, torque, and specific fuel consumption. Power and torque depend on an engine's displaced volume. In Chap. 2 a set of normalized or dimensionless performance and emissions parameters were defined to eliminate the effects of engine size. Power, torque, and fuel consumption were expressed in terms of these parameters (Sec. 2.14) and the significance of these parameters over an engine's load and speed range was discussed (Sec. 2.15). Using these normalized parameters, the effect of engine size can be made explicit. The power P can be expressed as:

$$P = \text{mep } A_p \bar{S}_p/4 \quad \text{(four-stroke cycle)}$$

$$P = \text{mep } A_p \bar{S}_p/2 \quad \text{(two-stroke cycle)}$$

(15.1)

The torque T is given by

$$T = \text{mep } V_d/(4\pi) \qquad \text{(four-stroke cycle)}$$

$$T = \text{mep } V_d/(2\pi) \qquad \text{(two-stroke cycle)} \tag{15.2}$$

Thus for well-designed engines, where the maximum values of mean effective pressure and piston speed are either flow limited (in naturally aspirated engines) or stress limited (in turbocharged engines), power is proportional to piston area and torque to displaced volume. Mean effective pressure can be expressed as

$$\text{mep} = \eta_f \, \eta_v \, Q_{\text{HV}} \, \rho_{a,i}\left(\frac{F}{A}\right) \tag{15.3}$$

for four-stroke cycle engines [Eq. (2.41)], and as

$$\text{mep} = \eta_f \, \eta_{\text{tr}} \, \Lambda Q_{\text{HV}} \, \rho_{a,i}\left(\frac{F}{A}\right) \tag{15.4}$$

for two-stroke cycle engines [Eqs. (2.19), (2.38), and (6.25)]. The importance of high fuel conversion efficiency, breathing capacity, and inlet air density is clear. Specific fuel consumption is related to fuel conversion efficiency by Eq. (2.24):

$$\text{sfc} = \frac{1}{\eta_f \, Q_{\text{HV}}} \tag{15.5}$$

These parameters have both brake and indicated values (see Secs. 2.3, 2.4, and 2.5). The difference between these two quantities is the engine's friction (and pumping) requirements and their ratio is the mechanical efficiency η_m.

   The relative importance of these parameters varies over an engine's operating speed and load range. The maximum or normal rated brake power (see Sec. 2.1) and the quantities such as bmep derived from it (see Sec. 2.7) define an engine's full potential. The maximum brake torque (and bmep derived from it), over the full speed range, indicates the ability of the designer to obtain a high air flow through the engine over the full speed range and use that air effectively. Then over the whole operating range, and most especially those parts of that range where the engine will operate for long periods of time, engine fuel consumption and efficiency, and engine emissions are important. Since the operating and emissions characteristics of spark-ignition and compression-ignition engines are substantially different, each engine type is dealt with separately.

15.2 INDICATED AND BRAKE POWER AND MEP

The wide-open-throttle operating characteristics of a production spark-ignition automotive engine are shown in Fig. 15-1. The power shown is the gross power for the basic engine; this includes only the built-in engine accessories.[2] The maximum net power for the fully equipped engine with the complete intake and exhaust system and full cooling system is about 14 percent lower. The indicated

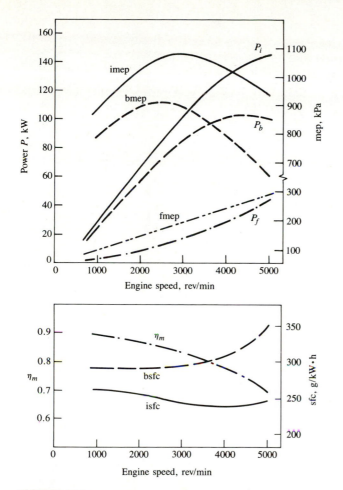

FIGURE 15-1

Gross indicated, brake, and friction power (P_i, P_b, P_f), indicated, brake, and friction mean effective pressure, indicated and brake specific fuel consumption, and mechanical efficiency for 3.8-dm³ six-cylinder automotive spark-ignition engine at wide-open throttle. Bore = 96.8 mm, stroke = 86 mm, $r_c = 8.6$.[1]

power was obtained by adding the friction power to the brake power; it is the average rate of work transfer from the gases in the engine cylinders to the pistons during the compression and expansion strokes of the engine cycle (see Sec. 2.4). The indicated mean effective pressure shows a maximum in the engine's mid-speed range, just below 3000 rev/min. The shape of the indicated power curve follows from the imep curve. Since the full-load indicated specific fuel consumption (and hence indicated fuel conversion efficiency) varies little over the full speed range, this variation of full-load imep and power with speed is primarily due to the variation in volumetric efficiency, η_v [see Eq. (15.3)]. Since friction mean effective pressure increases almost linearly with increasing speed, friction

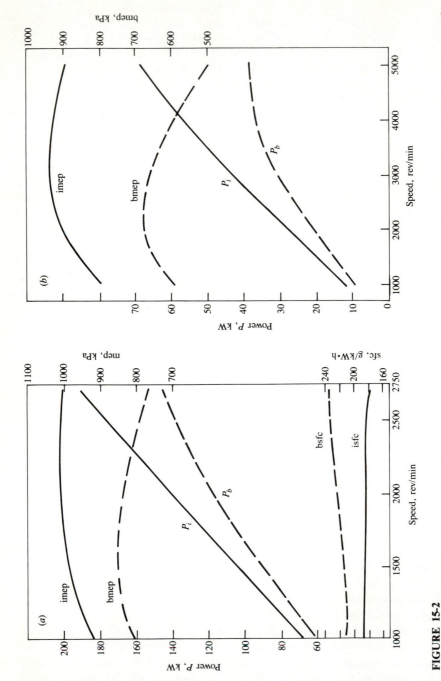

FIGURE 15-2

Gross indicated and brake power (P_i, P_b), mean effective pressure (imep, bmep), and specific fuel consumption (isfc, bsfc) for: (a) 8.4-dm^3 six-cylinder naturally aspirated direct-injection diesel engine: bore = 115 mm, stroke = 135 mm, r_c = 16;^{3} (b) 1.8-dm^3 four-cylinder naturally aspirated indirect-injection swirl-chamber diesel engine: bore = 84 mm, stroke = 82 mm, r_c = 22.^{4}

power will increase more rapidly. Hence mechanical efficiency decreases with increasing speed from a maximum of about 0.9 at low speed to 0.7 at 5000 rev/min. Thus bmep peaks at a lower speed than imep. The brake power shows a maximum at about 4300 rev/min; increases in speed above this value result in a decrease in P_b. The indicated fuel conversion efficiency increases by about 10 percent from 0.31 to 0.34 over the speed range 1000 to 4000 rev/min. This is primarily due to the decreasing importance of heat transfer per cycle with increasing speed.

At part load at fixed throttle position, these parameters behave similarly; however, at higher speeds torque and mean effective pressure decrease more rapidly with increasing speed than at full load. The throttle chokes the flow at lower and lower speeds as the throttle open area is reduced, increasingly limiting the air flow (see Fig. 7-22). The pumping component of total friction also increases as the engine is throttled, decreasing mechanical efficiency (see Figs. 13-9 and 13-10).

Figure 15-2 shows full-load indicated and brake power and mean effective pressure for naturally aspirated DI and IDI compression-ignition engines. Except at high engine speeds, brake torque and mep vary only modestly with engine speed since the intake system of the diesel can have larger flow areas than the intake of SI engines with their intake-system fuel transport requirements. The part-load torque and bmep characteristics (at fixed amount of fuel injected per cycle) have a similar shape to the full-load characteristics in Fig. 15-2. The decrease in torque and bmep with increasing engine speed is due primarily to the increase in friction mep with speed (see Figs. 13-7, 13-11, and 13-12). Decreasing engine heat transfer per cycle and decreasing air-flow rate, as speed increases, have modest additional impacts.

15.3 OPERATING VARIABLES THAT AFFECT SI ENGINE PERFORMANCE, EFFICIENCY, AND EMISSIONS

The major operating variables that affect spark-ignition engine performance, efficiency, and emissions at any given load and speed are: spark timing, fuel/air or air/fuel ratio relative to the stoichiometric ratio, and fraction of the exhaust gases that are recycled for NO_x emission control. Load is, of course, varied by varying the inlet manifold pressure. The effect of these variables will now be reviewed.

15.3.1 Spark Timing

Figure 9-3 and the accompanying text explain how variations in spark timing relative to top-center affected the pressure development in the SI engine cylinder. If combustion starts too early in the cycle, the work transfer from the piston to the gases in the cylinder at the end of the compression stroke is too large; if combustion starts too late, the peak cylinder pressure is reduced and the expansion stroke work transfer from the gas to the piston decreases. There exists a particular spark timing which gives maximum engine torque at fixed speed, and mixture composition and flow rate. It is referred to as MBT—maximum brake torque—timing. This timing also gives maximum brake power and minimum brake specific fuel consumption. Figure 15-3a shows the effect of spark advance variations on wide-open-throttle brake torque at selected speeds between 1200 and 4200 rev/min for a production eight-cylinder engine. At each speed, as spark is advanced from an initially retarded setting, torque rises to a maximum and then decreases. MBT timing depends on speed; as speed increases the spark must be advanced to maintain optimum timing because the duration of the combustion process in crank angle degrees increases. Optimum spark timing also depends on load. As load and intake manifold pressure are decreased, the spark timing must be further advanced to maintain optimum engine performance.

The maximum in each brake torque curve in Fig. 15-3a is quite flat. Thus accurate determination of MBT timing is difficult, but is important because NO and HC emissions vary significantly with spark timing. In practice, to permit a more precise definition of spark timing, the spark is often retarded to give a 1 or 2 percent reduction in torque from the maximum value.

In Fig. 15-3a the mixture composition and flow rate were held constant at each engine speed. If the mixture flow rate is adjusted to maintain constant brake

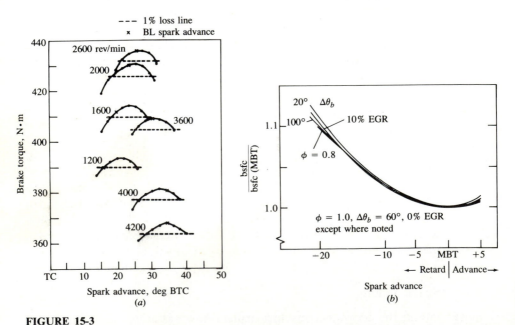

FIGURE 15-3
(a) Variation in brake torque with spark advance, eight-cylinder automotive spark-ignition engine at wide-open throttle, at engine speeds from 1200 to 4200 rev/min. 1 percent torque loss from MBT and spark advance for borderline knock are shown.[5] (b) Predicted variation in brake specific fuel consumption (normalized by MBT value) with spark retard at several different part-load engine conditions.[6, 7]

torque, the effect of spark timing variations on fuel consumption at constant engine load can be evaluated. Figure 15-3b shows results obtained with a computer simulation of the engine operating cycle.[6, 7] The curves for several different part-load operating conditions and burn durations (from fast to slow) have been normalized and fall essentially on top of each other. Five degrees of retard in spark timing have only a modest effect on fuel consumption; for 10 to 20° retard, the impact is much more significant.

Spark timing affects peak cylinder pressure and therefore peak unburned and burned gas temperatures (see Sec. 9.2.1). Retarding spark timing from the optimum reduces these variables. Retarded timing is sometimes used therefore for NO_x emission control (see Fig. 11-13 and accompanying text) and to avoid knock (see Sec. 9.6.1). The exhaust temperature is also affected by spark timing. Retarding timing from MBT increases exhaust temperature; both engine efficiency and heat loss to the combustion chamber walls (see Fig. 12-27) are decreased. Retarded timing is sometimes used to reduce hydrocarbon emissions by increasing the fraction oxidized during expansion and exhaust due to the higher burned gas temperatures that result (see Sec. 11.4.3). Retarded timing may be used at engine idle to bring the ignition point closer to TC where conditions for avoiding misfire are more favorable.

15.3.2 Mixture Composition

The unburned mixture in the engine cylinder consists of fuel (normally vaporized), air, and burned gases. The burned gas fraction is the residual gas plus any recycled exhaust used for NO control. Mixture composition during combustion is most critical, since this determines the development of the combustion process which governs the engine's operating characteristics. While substantial efforts are made to produce a uniform mixture within the cylinder, some nonuniformities remain (see Sec. 9.4.2). In a given cylinder, cycle-by-cycle variations in average charge composition exist. Also, within each cylinder in a given engine cycle, the fuel, air, EGR, and residual gas are not completely mixed, and composition nonuniformities across the charge may be significant.† These together produce variations in composition at the spark plug location (the critical region since the early stages of flame development influence the rest of the combustion process) which can be of order ± 10 percent peak-to-peak (see Fig. 9-34). In addition, in multicylinder engines, the average air, fuel, and EGR flow rates to each cylinder are not identical. Typical cylinder-to-cylinder variations have standard deviations of ±5 percent of the mean for air flow rate and fuel flow rate (giving a

---

† This aspect of mixture nonuniformity is least well defined. Mixing of the fresh mixture (fuel, air, and EGR) with residual gas is likely to be incomplete (see Fig. 14-36), especially at light load when the residual gas fraction is highest. With intake-port fuel-injection systems, there is evidence of incomplete fuel-air mixing due to the fact that the air flow and fuel flow processes are not in phase.[9] When the engine is cold, fuel distribution within the cylinder is known to be nonuniform.

± 7 percent variation in the air/fuel ratio) for steady-state engine operation. EGR cylinder-to-cylinder flow rates may have higher variability. Under unsteady engine operating conditions all these variations can be higher.

It is necessary to consider the effect of mixture composition changes on engine operating and emissions characteristics in two regimes: (1) wide-open throttle (WOT) or full load and (2) part throttle or load. At WOT, the engine air flow is the maximum that the engine will induct.† Fuel flow can be varied, but air flow is set by engine design variables and speed. At part throttle, air flow, fuel flow, and EGR flow can be varied. Evaluation of mixture composition changes at part load should be done at fixed (brake) load and speed, i.e., under conditions where the engine provides the desired torque level at the specified speed. To maintain torque (or load or bmep) constant as mixture composition is varied normally requires changes in throttle setting (and if EGR is varied, changes in EGR flow-control valve setting). This distinction between part-load comparisons at specified torque or bmep, rather than at constant throttle settings (which gives essentially constant air flow), is important because the pumping work component of engine friction will vary at constant engine load as mixture composition changes. At constant throttle setting and speed, the pumping work remains essentially unchanged.

AIR/FUEL OR EQUIVALENCE RATIO CHANGES. Mixture composition effects are usually discussed in terms of the air/fuel ratio (or fuel/air ratio) because in engine tests, the air and fuel flow rates to the engine can be measured directly and because the fuel metering system is designed to provide the appropriate fuel flow for the actual air flow at each speed and load. However, the relative proportions of fuel and air can be stated more generally in terms of the fuel/air equivalence ratio ϕ [the actual fuel/air ratio normalized by dividing by the stoichiometric fuel/air ratio, see Eq. (3.8)] or the relative air/fuel ratio λ [see Eq. (3.9)]. The combustion characteristics of fuel-air mixtures and the properties of combustion products, which govern engine performance, efficiency, and emissions, correlate best for a wide range of fuels relative to the stoichiometric mixture proportions. Where appropriate, therefore, the equivalence ratio will be used as the defining parameter. Equation (7.1) converts the air/fuel ratio with gasoline to the equivalence ratio.

The theoretical basis for understanding the effect of changes in the equivalence ratio is the fuel-air cycle results in Figs. 5-9 and 5-10, where the indicated fuel conversion efficiency and mean effective pressure are shown as a function of the fuel/air equivalence ratio, ϕ. The mean effective pressure peaks slightly rich of stoichiometric, between $\phi = 1$ and 1.1. Due to dissociation at the high temperatures following combustion, molecular oxygen is present in the burned gases under stoichiometric conditions, so some additional fuel can be added and par-

---

† EGR is normally zero at WOT, since maximum torque is usually desired.

tially burned. This increases the temperature and the number of moles of the burned gases in the cylinder. These effects increase the pressure to give increased power and mep. Fuel conversion efficiency decreases approximately as $1/\phi$, as the mixture is richened above stoichiometric ($\phi > 1$) due to the decreasing combustion efficiency associated with the richening mixture.

For mixtures lean of stoichiometric, the theoretical fuel conversion efficiency increases linearly as ϕ decreases below 1.0. Combustion of mixtures leaner than stoichiometric produces products at lower temperature, and with less dissociation of the triatomic molecules CO_2 and H_2O. Thus the *fraction* of the chemical energy of the fuel which is released as sensible energy near TC is greater; hence a greater *fraction* of the fuel's energy is transferred as work to the piston during expansion, and the fraction of the fuel's available energy rejected to the exhaust system decreases (see Sec. 5.7). There is a discontinuity in the fuel conversion efficiency and imep curves at the stoichiometric point; the burned gas composition is substantially different on the rich and the lean sides of $\phi = 1$.

Figure 15-4 shows gross indicated specific fuel consumption data for a six-cylinder spark-ignition engine at wide-open throttle and 1200 rev/min,[9] and values of gross indicated mean effective pressure and fuel conversion efficiency derived from the isfc data. In these engine tests, the fuel-air mixture was prepared in two different ways: (1) with the normal carburetor and (2) with a heated vaporizing tank to ensure intake-mixture uniformity. Shapes of the practical efficiency curves and the theoretical curves in Fig. 5-9 differ. Cylinder-to-cylinder air/fuel ratio maldistribution prevents the carbureted engine operating leaner than $\phi \approx 0.85$ ($A/F \approx 17$) without misfire under these conditions. While use of a fuel vaporizing and mixing tank essentially removes this maldistribution and extends the lean misfire limit, $\eta_{f,i}$ does not continue to increase as ϕ decreases. The reasons for this are that cycle-to-cycle pressure fluctuations and the total dura-

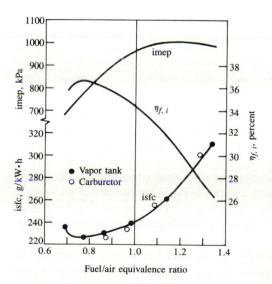

FIGURE 15-4

Effect of the fuel/air equivalence ratio variations on indicated mean effective pressure, specific fuel consumption, and fuel conversion efficiency of six-cylinder spark-ignition engine at wide-open throttle and 1200 rev/min. Data for standard carbureted engine, and engine equipped with vapor tank which extends the lean operating limit, are shown.[9]

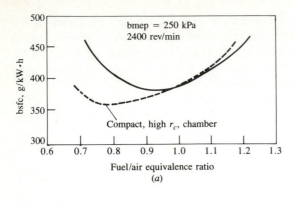

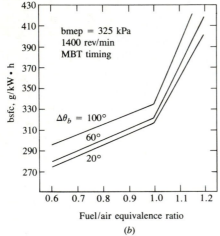

FIGURE 15-5
Effect of combustion chamber design and burn rate on spark-ignition engine brake specific fuel consumption. (a) 1.6-dm³ four-cylinder engine with conventional combustion chamber and 1.5-dm³ four-cylinder engine with compact fast-burning high-compression-ratio chamber beneath the exhaust valve with $r_c = 13$, both at bmep of 250 kPa and 2400 rev/min.[10] (b) Predictions from thermodynamic-based computer simulation of engine cycle for 5.7-dm³ eight-cylinder engine at bmep of 325 kPa and 1400 rev/min with MBT spark timing.[6]

tion of the burning process increase as the mixture becomes leaner: both these factors degrade engine efficiency. Since the spark advance is set for the average cycle, increasing cycle-to-cycle dispersion produces increasing imep (and hence $\eta_{f,i}$) losses in "nonaverage" cycles due to nonoptimum timing. The lengthening burn duration directly decreases efficiency, even in the absence of cyclic variations.

Engine fuel consumption and efficiency well lean of stoichiometric depend strongly on the engine combustion chamber design. Figure 15-5 shows two sets of engine bsfc data, for a conventional combustion chamber and a compact high-compression-ratio chamber, at constant load and speed (250 kPa bmep and 2400 rev/min) as a function of equivalence ratio. Also shown are bsfc results obtained from a thermodynamic-based computer cycle simulation of the spark-ignition engine operating cycle (at 325 kPa bmep and 1400 rev/min).[6] Though the load and speed are different, the behavior of the data and predictions for rich mixtures, $\phi > 1$, are comparable. On the lean side of stoichiometric, however, fuel consumption depends on the combustion characteristics of the chamber. The faster-burning compact high-compression-ratio chamber shows decreasing bsfc until the lengthening burn duration and larger cycle-by-cycle variations cause bsfc to increase. For the slower-burning conventional chamber, this deterioration in combustion starts to occur almost immediately on the lean side of stoichiometric, and fuel consumption worsens for $\phi \leq 0.9$.

Thus the equivalence ratio for optimum fuel consumption at a given load depends on the details of chamber design (including compression ratio) and mixture preparation quality. It also varies for a given chamber over the part-throttle load and speed range. For lighter loads and lower speeds it is closer to stoichiometric since the residual gas fraction is higher and combustion quality is poorer with greater dilution and at lower speeds.

At part load, as the air/fuel ratio is varied at constant brake load, the pumping work varies, and this also contributes to the brake specific fuel consumption and efficiency variation with equivalence ratio. Figure 15-6 shows the gross and net indicated fuel conversion efficiencies and brake efficiency as a function of equivalence ratio at a part-throttle constant load and speed point (325 kPa bmep and 1400 rev/min), calculated using a thermodynamic-based computer simulation of the engine's operating cycle. The difference between the net and gross indicated curves illustrates the magnitude of the effect of the pumping work changes. Part-throttle comparisons of different operating conditions should be done at constant brake load (torque or bmep) and speed: the task the engine is required to perform is then the same. At constant bmep and speed, the mechanical rubbing friction is essentially fixed; thus *net* imep is constant (and *gross* imep will vary if the pumping mep varies).

Note that all the engine data show a smooth transition between the rich and lean characteristics at the stoichiometric point, whereas the calculated sfc and

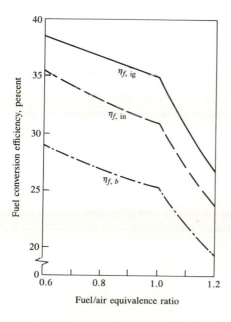

FIGURE 15-6

Gross and net indicated, and brake, fuel conversion efficiencies predicted by thermodynamic-based cycle simulation at constant part-load bmep (325 kPa) and speed (1400 rev/min) for a fixed burn duration (0–100 percent, 60° CA).[6]

efficiency characteristics show a discontinuity in slope. The difference is due to cylinder-to-cylinder and cycle-by-cycle mixture composition variations[7] and to cycle-by-cycle cylinder pressure variations which exist (though to a lesser extent) even in the absence of these mixture variations. Averaging over these variations smooths out the theoretical discontinuity in slope at $\phi = 1.0$.

The equivalence ratio requirements of a spark-ignition engine over the full load and speed range can now be explained from the point of view of performance and efficiency. However, since emissions depend on ϕ also, emission control requirements may dictate a different engine calibration, as will be discussed later. The mixture requirements in the induction system are usually discussed in relation to *steady* and *transient* engine operation. Steady operation includes operation at a given speed and load over several engine cycles with a warmed-up engine. Transient operation includes engine starting, engine warm-up to steady-state temperatures, and changing rapidly from one engine load and speed to another. The mixture requirements of the engine as defined by the composition of the combustible mixture at the time of ignition, while they vary somewhat with speed and load, are essentially the same for all these operating modes.† However, the methods used to prepare the mixture prior to entry to the cylinder must be modified in the transient modes when liquid fuels are used, to allow for variations in the liquid fuel flow and fuel evaporation rate in the intake manifold as the air flow varies and as the manifold and inlet port pressure and temperature change. The transient fuel metering requirements for adequate mixture preparation are discussed in Chap. 7.

At all load points at a given speed, the ideal equivalence ratio is that which gives minimum brake specific fuel consumption at the required load. However, once wide-open-throttle air flow has been reached, increases in power can only be obtained by increasing the fuel flow rate. The equivalence ratio requirements for optimum-efficiency steady-state engine operation can be summarized on a plot of equivalence ratio versus percent of maximum air flow at any given speed. A typical plot was shown in Fig. 7-1. For part-throttle operation, unless dictated otherwise by emission control requirements, the equivalence ratio is set close to the equivalence ratio for minimum fuel consumption consistent with avoiding partial burning or misfire in one or more cylinders. At very light load the best bsfc mixture is richer to compensate for slower flame speeds at lower mixture density and increased residual fraction. As wide-open throttle is approached, the mixture is richened to obtain maximum power.

The exhaust gas temperature varies with the equivalence ratio. The exhaust gas temperature also varies continuously as the gas leaves the engine cylinder and flows through the exhaust port and the manifold and pipe (see Sec. 6.5), so an appropriate definition of an average exhaust gas temperature should be used

---

† Except during start-up and cold engine operation, when a substantial part of the fuel within the cylinder can be in the liquid phase.

in quantifying this variation. However, time-averaged thermocouple measurements from specific locations in the exhaust system can provide useful information on trends. Figure 14-10 shows examples of predictions of the
enthalpy-averaged exhaust gas temperature at the exhaust port exit as a function
of equivalence ratio compared with time-averaged measurements. The enthalpy-
averaged temperature is defined by Eq. (6.19). These are typically 50 to 100 K
higher than time-averaged measurements. The exhaust temperature peaks at the
stoichiometric point and decreases as the mixture is richened and leaned on
either side.

The fuel/air equivalence ratio is an important parameter controlling spark-
ignition engine emissions. The critical factors affecting emissions, that are governed by the equivalence ratio, are the oxygen concentration and the temperature of
the burned gases. Excess oxygen is available in the burned gases lean of stoichiometric. The maximum burned gas temperatures occur slightly rich of stoichiometric at the start of the expansion stroke, and at the stoichiometric composition
at the end of expansion and during the exhaust process. Figure 11-2 illustrates
the general trends in emissions with equivalence ratio which have already been
discussed.

Figure 15-7 shows the effect of variations in fuel/air equivalence ratio on
NO_x and HC emissions and fuel consumption when a special fuel vapor generator was used to produce a uniform fuel-air mixture. As explained in Sec. 11.2.3,
the formation rate of NO depends on the gas temperature and oxygen concentration. While maximum burned gas temperatures occur at $\phi \approx 1.1$, at this equiva-

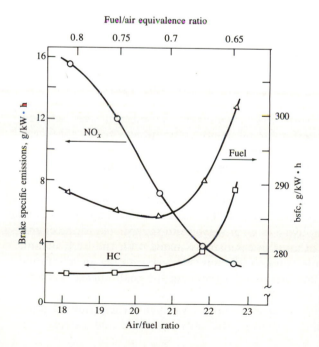

FIGURE 15-7
Variation of brake specific HC and
NO_x emissions and fuel consumption with (A/F) and fuel/air equivalence ratio. 5.7-dm^3 eight-cylinder
spark-ignition engine at 385 kPa
bmep and 1400 rev/min with uniform vaporized fuel-air mixture.[11]

lence ratio oxygen concentrations are low. As the mixture is leaned out, increasing oxygen concentration initially offsets the falling gas temperatures and NO emissions peak at $\phi \approx 0.9$. Then, decreasing temperatures dominate and NO emissions decrease to low levels.

Figure 15-7 also shows the effect of variations in equivalence ratio for lean mixtures on unburned hydrocarbon emissions. For rich mixtures, Fig. 11-2 shows that emissions are high. This is primarily due to the lack of oxygen for afterburning of any unburned hydrocarbons that escape the primary combustion process, within the cylinder and the exhaust system. HC emissions decrease as the stoichiometric point is approached: increasing oxygen concentration and increasing expansion and exhaust stroke temperatures result in increasing HC burnup. For moderately lean mixtures, HC emission levels vary little with equivalence ratio. Decreasing fuel concentration and increasing oxygen concentration essentially offset the effect of decreasing bulk gas temperatures. As the lean operating limit of the engine is approached, combustion quality deteriorates significantly and HC emissions start to rise again due to the occurrence of occasional partial-burning cycles. For still leaner mixtures, HC emissions rise more rapidly due to the increasing frequency of partial-burning cycles, and even the occurrence of completely misfiring cycles (see Sec. 9.4.3). The equivalence ratio at which partial-burning and misfiring cycles just start to appear depends on details of the engine combustion and fuel preparation systems, as well as the load and speed point.

The effect of equivalence ratio variations on CO emissions has already been explained in Sec. 11.3 (see Fig. 11-20). For rich mixtures, CO levels are high because complete oxidation of the fuel carbon to CO_2 is not possible due to insufficient oxygen. For lean mixtures, CO levels are approximately constant at a low level of about 0.5 percent or less.

Figure 15-7 indicates that if an engine can be designed and operated so that its stable operating limit under the appropriate part-load conditions is sufficiently lean, excellent fuel consumption and substantial control of engine NO, HC, and CO emissions can be achieved. Such an approach requires good control of mixture preparation and a fast-burning combustion chamber design (see Sec. 15.4.1). However, this lean-engine approach is not compatible with the three-way catalyst system (see Sec. 11.6.2) which, with close-to-stoichiometric mixtures, achieves substantial additional reductions in NO, HC, and CO emissions.

EXHAUST GAS RECYCLE. Exhaust gas recycle (EGR) is the principal technique used for control of SI engine NO_x emissions (see Sec. 11.2.3). A fraction of the exhaust gases are recycled through a control valve from the exhaust to the engine intake system. The recycled exhaust gas is usually mixed with the fresh fuel-air mixture just below the throttle valve. EGR acts, at part load, as an additional diluent in the unburned gas mixture, thereby reducing the peak burned gas temperatures and NO formation rates. Note that it is the total burned gas fraction in the unburned mixture in the cylinder that acts as a diluent. These burned gases are comprised of both residual gas from the previous cycle and exhaust gas recycled to the intake. As described in Sec. 6.4, the residual gas fraction is influenced by load and valve timing (especially the extent of valve overlap) and, to a lesser degree, by the air/fuel ratio and compression ratio. The total burned gas mass fraction is given by Eq. (4.3). Since the burned gases dilute the unburned mixture, the absolute temperature reached after combustion varies inversely with the burned gas mass fraction. Hence increasing the burned gas fraction reduces the rate of formation of NO emissions.

Figure 11-10 shows the effect on NO emissions of increasing the burned gas fraction by recycling exhaust gases to the intake system. Substantial reductions in NO concentrations are achieved with 10 to 25 percent EGR. However, EGR also reduces the combustion rate which makes stable combustion more difficult to achieve (see Sec. 9.4.3 and Fig. 9-36). The amount of EGR a particular combustion chamber design will tolerate depends on its combustion characteristics, the speed and load, and the equivalence ratio. EGR percentages in the 15 to 30 range are about the maximum amount of EGR a spark-ignition engine will tolerate under normal part-throttle conditions. Faster-burning engines will tolerate more EGR than slower-burning engines. Because of the decrease in burn rate and increase in cycle-by-cycle combustion variations, hydrocarbon emissions increase with increasing EGR, as shown in Fig. 11-29. At first the increase in HC is modest and is due primarily to decreased HC burnup due to lower expansion and exhaust stroke temperatures. The HC increase becomes more rapid as slow combustion, partial burning, and even misfire, in turn, occur with increasing frequency. EGR has no significant effect on engine CO emissions.

The effect of exhaust gas recycle on engine performance and efficiency, for mixtures with $\phi \leq 1.0$, is similar to the addition of excess air. Both EGR and excess air dilute the unburned mixture. In practice since EGR is only used at part-throttle conditions, $\phi \leq 1.0$ is the region of interest. Because three-way catalysts are now used where NO_x emission constraints are severe, greatest attention has focused on dilution with EGR at $\phi \approx 1.0$. Figure 15-8 shows the effect of increasing EGR on bsfc and enthalpy-mean exhaust temperature [defined by Eq. (6.19)] at constant bmep, predicted using a thermodynamic-based computer simulation of the engine's operating cycle. Predictions made for different burn durations are shown, at MBT timing for a stoichiometric mixture. At constant burn duration, bsfc and exhaust temperature decrease with increasing EGR. Only for very long combustion processes is the burn rate especially significant. This improvement in fuel consumption with increasing EGR is due to three factors: (1) reduced pumping work as EGR is increased at constant brake load (fuel and air flows remain almost constant; hence intake pressure increases); (2) reduced heat loss to the walls because the burned gas temperature is decreased significantly; and (3) a reduction in the degree of dissociation in the high-temperature burned gases which allows more of the fuel's chemical energy to be converted to sensible energy near TC. The first two of these are comparable in magnitude and each is about twice as important as the third.[12]

Figure 15-9 shows experimental bsfc versus EGR data for two combustion chambers: a combustion chamber with a moderate burning rate and a faster-

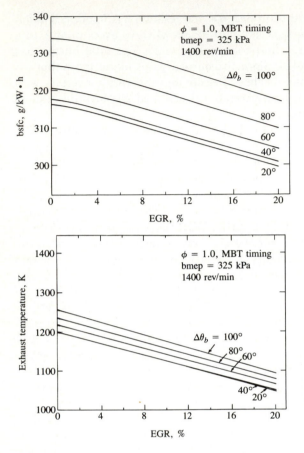

FIGURE 15-8
Effect of recycled exhaust on brake specific fuel consumption and exhaust temperature at constant bmep and speed, stoichiometric mixture, and various burn durations (0–100 percent). Predictions from thermodynamic-based cycle simulation.[6]

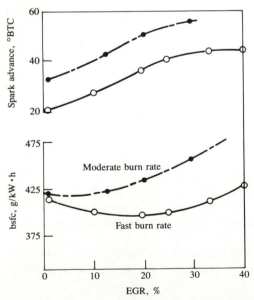

FIGURE 15-9
Brake specific fuel consumption and MBT spark advance as a function of percent recycled exhaust, for four-cylinder spark-ignition engine with a moderate burn rate combustion chamber and a fast burn rate combustion chamber. 1400 rev/min, 324 kPa bmep, equivalence ratio 1.0.[12]

burning chamber with open geometry and with induction-generated swirl. Though addition of EGR lengthens both the flame development and propagation processes (as indicated by the increasing MBT spark advance requirement with increasing EGR), the faster-burning chamber follows the anticipated pattern of significant bsfc reductions until, at about 20 percent EGR, the combustion quality deteriorates. For the slower-burning combustion chamber, the tolerance to dilution with EGR is much less.

15.3.3   Load and Speed

One common way to present the operating characteristics of an internal combustion engine over its full load and speed range is to plot brake specific fuel consumption contours on a graph of brake mean effective pressure versus engine speed. Operation of the engine coupled to a dynamometer on a test stand, over its load and speed range, generates the torque and fuel flow-rate data from which such a *performance map* is derived. Equation (2.20) relates bmep to torque, and bsfc values are obtained from Eq. (2.22) at each operating point. Figure 15-10 shows an example of such a performance map for a four-cylinder spark-ignition engine. The upper envelope of the map is the wide-open-throttle performance curve. Points below this curve define the part-load operating characteristics. While details differ from one engine to another, the overall shapes of these maps for spark-ignition engines are remarkably similar. When mean piston speed $\bar{S}_p$ is used instead of crankshaft speed for the abscissa, the quantitative similarity of such maps over a wide range of engine sizes is more apparent.

Maximum bmep occurs in the mid-speed range; the minimum bsfc island is located at a slightly lower speed and at part load. These map characteristics can be understood in terms of variations in volumetric efficiency η_v, gross indicated fuel conversion efficiency $\eta_{f,ig}$ and mechanical efficiency η_m as A/F, EGR (if used), and the importance of heat losses and friction change, via Eqs. (15.3) and (15.5).

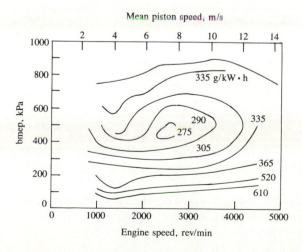

FIGURE 15-10

Performance map for 2-dm³ four-cylinder fast-burn spark-ignition engine showing contours of constant bsfc in grams per kilowatt-hour.[13]

The maximum bmep curve reflects the variation with speed of η_v, the decrease of η_m as $\bar{S}_p$ increases, and the increase of $\eta_{f,ig}$ as $\bar{S}_p$ increases due to decreasing importance of heat transfer per cycle. The bsfc contours have the following explanation. Starting at the minimum bsfc point, increasing speed at constant load increases bsfc due primarily to the increasing friction mep at higher speeds (which decreases η_m). While $\eta_{f,ig}$ increases as speed increases, friction increases dominate. Decreasing speed at constant load increases bsfc due primarily to the increasing importance of heat transfer per cycle (which decreases $\eta_{f,ig}$). Friction decreases, increasing η_m, but this is secondary. Any mixture enrichment required to maintain a sufficiently repeatable combustion process at low engine speeds (see Fig. 7-1) contributes too. Increasing load at constant speed from the minimum bsfc point increases bsfc due to the mixture enrichment required to increase torque as the engine becomes increasingly air-flow limited. Decreasing load at constant speed increases bsfc due to the increased magnitude of friction (due to increased pumping work), the increased relative importance of friction, and increasing importance of heat transfer (which decreases $\eta_{f,ig}$).

The effects of speed and load variations on NO and HC emissions are shown in Fig. 15-11.[14] NO concentrations increase moderately with increasing speed at constant load. At lower loads, the proportional increase in NO is greater

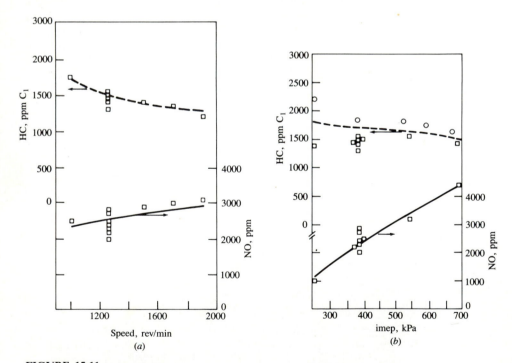

FIGURE 15-11
Variation in spark-ignition engine HC and NO_x emissions with (a) engine speed at 379 kPa imep and (b) load (or imep) at 1250 rev/min. Equivalence ratio = 0.9, MBT spark timing, $r_c = 7$.[14]

than at higher loads.[6] The residual gas fraction decreases as speed increases, this effect being greater at lower inlet manifold pressures (lighter loads) (see Fig. 6-19). Also, the relative importance of heat transfer per cycle is less as speed increases (see Fig. 12-25), which would also be expected to increase NO concentration. With increasing load (at constant speed), NO concentrations also increase. Again, as inlet manifold pressure and load increase, the residual gas fraction decreases (Fig. 6-19); also, the relative importance of heat transfer per cycle decreases with increasing load (Fig. 12-25).

The hydrocarbon concentration trends with speed and load changes are the opposite of the NO concentration trends. As indicated in Table 11.7, speed and load are likely to affect several of the HC formation mechanisms, the in-cylinder mixing of unburned hydrocarbons which escape combustion with the bulk gases, and the fraction of the in-cylinder HC which escape into the exhaust. However, not enough is yet known about the details of these processes to make these dependencies explicit. If oxygen is available, oxidation of unburned hydrocarbons both within the cylinder and in the exhaust system will be significantly enhanced by increases in speed since the expansion stroke and exhaust process gas temperatures increase substantially, due to the reduced significance of heat transfer per cycle with increasing speed. This more than offsets the reduced residence time in the cylinder and in the exhaust. Measurements of the percent HC reacted in the exhaust port as a function of engine speed show the same proportional reduction in the exhaust emissions data in Fig. 15-11.[15] The rationale for the variation with load is less clear. As load increases at constant speed, expansion and exhaust stroke temperatures increase, and the in-cylinder oxidation rate, if oxygen is available, will increase. However, as the exhaust gas flow rate increases, the residence time in critical sections of the exhaust system decreases and a reduction in exhaust port HC oxidation occurs.[16] The net trend is for HC concentration to decrease modestly as load is increased.

15.3.4 Compression Ratio

The ideal cycle analysis of Chap. 5 showed that indicated fuel conversion efficiency increased continuously with compression ratio according to Eq. (5.31). With $\gamma = 1.3$, this relation also matches closely the fuel-air cycle predictions with $\phi \approx 1.0$. However, in an actual engine other processes which influence engine performance and efficiency vary with changes in compression ratio: namely, combustion rate and stability, heat transfer, and friction. Over the load and speed range, the relative impact that these processes have on power and efficiency varies also. Hence, the applicability of Eq. (5.31) is open to question. Also, while the geometric compression ratio (ratio of maximum to minimum cylinder volume) is well defined, the actual compression and expansion processes in engines depend on valve timing details and the importance of flow through the valves while they are opening or closing (which depends on engine speed). Of course, our ability to increase the compression ratio is limited by the octane quality of available fuels and knock (see Sec. 9.6.1).

Only a few studies have examined the effect of compression ratio on spark-ignition engine performance and efficiency over a wide range of compression ratios. Figure 15-12 shows results obtained at wide-open throttle at 2000 rev/min with a series of eight-cylinder 5.3-dm³ displacement engines, from the most extensive of these studies.[17] Gross-indicated and brake fuel conversion efficiencies and mean effective pressures are shown. Indicated mep was obtained by adding motoring friction mep to brake mep. The mep data were obtained with (A/F) and spark timing adjusted to give maximum torque; for the efficiency data, (A/F) and spark timing were adjusted to give maximum efficiency. The mechanical efficiency remained essentially constant at 0.89 over the full compression ratio range. The volumetric efficiency was also constant at 0.825. Both $\eta_{f,\text{ig}}$ and mep show a maximum at a compression ratio of about 17; for higher compression ratios efficiency and mep decrease slightly. This trend was explained as being due to increasing surface/volume ratio and slower combustion, and is also due to the increasing importance of crevice volumes: at the higher compression ratios studied the combustion chamber height became very small.

To assess more broadly the effect of compression ratio variations on fuel conversion efficiency, several data sets have been normalized and compared in Fig. 15-13 which shows the ratio of fuel conversion efficiency at the given compression ratio divided by the efficiency at $r_c = 8$, for wide-open-throttle engine operation. The agreement for $r_c \leq 14$ is good. Over the compression ratio range that is accessible to SI engines with available fuels ($r_c \leq 12$), fuel conversion efficiency increases by about 3 percent per unit of compression ratio increase. Note, of course, that engine power increases by about the same amount.

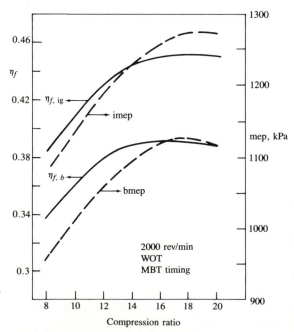

FIGURE 15-12
Effect of compression ratio on indicated mean effective pressure and fuel conversion efficiency. 5.3-dm³ eight-cylinder spark-ignition engine at 2000 rev/min and wide-open throttle. Equivalence ratio and spark timing adjusted for maximum torque for mep data; adjusted for minimum fuel consumption for efficiency data.[17]

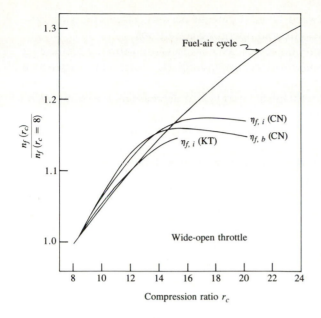

FIGURE 15-13
Relative fuel conversion efficiency improvement with increasing compression ratio, spark-ignition engines at wide-open throttle: CN,[17] KT.[18]

A similar comparison of the effect of compression ratio increases on efficiency at part load is shown in Fig. 15-14.[19] The figure shows brake fuel conversion efficiency data from engines of different cylinder volume. Both the compression ratio for maximum efficiency and the maximum efficiency depend on cylinder size. The wide-open-throttle and road-load data (top two curves[17]) confirm that the increase in efficiency with an increase in the compression ratio at part load apparently depends on the details of engine operation to a significant degree also. For the important compression ratio range of 9 to 11, the relative

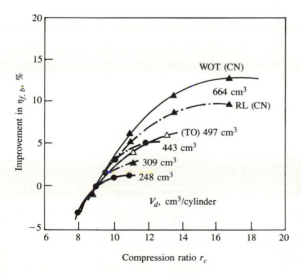

FIGURE 15-14
Relative brake fuel conversion efficiency improvement with increasing compression ratio of spark-ignition engines of different displaced volume per cylinder at part throttle (except top curve at WOT).[19] RL road load. CN,[17] TO.[10]

efficiency improvement is between 1 and 3 percent per unit of compression ratio increase, depending on cylinder size and operating conditions.

The exhaust temperature decreases as compression ratio and efficiency increase until the compression ratio corresponding to maximum efficiency is reached. It has also been shown that heat losses to the combustion chamber walls, as a fraction of the fuel's chemical energy, also decrease as the compression ratio and efficiency both increase.[17]

The effect of compression ratio changes on NO emissions is small. Some studies show a modest increase in specific NO emissions as the compression ratio increases at constant load and speed; other studies show a slight decrease. Increasing the compression ratio increases exhaust hydrocarbon emissions. Several trends could contribute: increased importance of crevice volumes at high r_c; lower gas temperatures during the latter part of the expansion stroke, thus producing less HC oxidation in the cylinder; decreasing residual gas fraction, thus increasing the fraction of in-cylinder HC exhausted; lower exhaust temperatures, hence less oxidation in the exhaust system.

15.4 SI ENGINE COMBUSTION CHAMBER DESIGN

15.4.1 Design Objectives and Options

There has always been extensive debate over the optimum SI engine combustion chamber design. There are a large number of options for cylinder head and piston crown shape, spark plug location, size and number of valves, and intake port design.[20] Debate revolves around issues such as chamber compactness, surface/volume ratio, flame travel length, and use of swirl and squish types of mixture motion. Figure 15-15 shows examples of several common types of combustion chamber shapes. Over the past few years a consensus has developed which favors faster-burning combustion-chamber designs. A chamber design where the fuel burning process takes place faster, i.e., occupies a shorter crank angle interval at a given engine speed, produces a more robust and repeatable combustion pattern that provides emission control and efficiency gains simultaneously. A faster-burning chamber with its shorter burn time permits operation with substantially higher amounts of EGR, or with very lean mixtures, within the normal constraints of engine smoothness and response. Thus greater emissions control within the engine can be achieved, and at part load at this higher level of dilution a faster-burning chamber shows an improvement in fuel consumption due to the reduced pumping work, reduced heat transfer (due to lower burned gas temperatures), and reduced amount of dissociation in the burned gases.[22]

The major combustion chamber design objectives which relate to engine performance and emissions are: (1) a fast combustion process, with low cycle-by-cycle variability, over the full engine operating range; (2) a high volumetric efficiency at wide-open throttle; (3) minimum heat loss to the combustion chamber walls; (4) a low fuel octane requirement.

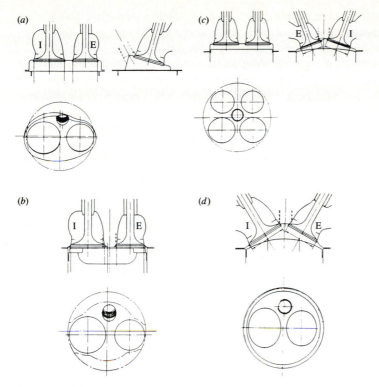

FIGURE 15-15
Examples of common spark-ignition engine combustion chamber shapes: (a) bathtub and wedge: (b) bowl-in-piston; (c) four-valve pent roof; (d) hemispherical.[21]

Many methods for producing a "fast burn" have been proposed. These include ways of making the combustion chamber shape more compact, moving the spark plug to a more central location within the chamber, using two plugs, and increasing in-cylinder gas motion by creating swirl during the induction process or during the latter stages of compression.

A faster combustion process relative to more moderate burn rate engines does result in a direct engine efficiency gain, other factors being equal. The magnitude of this direct gain is relatively modest. Experimental studies of the effect of an increase in burn rate from moderate to fast at constant engine load, speed, and mixture composition show that this effect is a few percent at most.[23] Computer simulations of the engine operating cycle confirm these experimental observations: while a decrease in total burn duration from 100 to 60° (slow to moderate burn) does result in a 4 percent decrease in bsfc, a decrease in burn duration from 60 to 20° gives only a further 1.5 percent bsfc decrease.[6]

Of greater importance is the fact that the faster burn process is more robust and results in the engine being able to operate satisfactorily with much more EGR, or much leaner, without a large deterioration in combustion quality. Faster burning chamber designs exhibit much less cycle-by-cycle variability. This ability to operate with greater dilution at part load while maintaining a short burn duration and low cycle-by-cycle variability, permits much greater control of NO_x within the engine with 20 or more percent EGR without any substantial increase in HC emissions (see Fig. 11-29), or permits very lean operation. In both cases the efficiency gain relative to moderate burn rate engines, *which must operate with less dilution,* is sizeable.[24]

High volumetric efficiency is required to obtain the highest possible power density. The shape of the cylinder head affects the size of valves that can be incorporated into the design. Effective valve open area, which depends on valve diameter and lift, directly affects volumetric efficiency. Swirl is used in many modern chamber designs to speed up the burning process and achieve greater combustion stability. Induction-generated swirl appears to be a particularly stable in-cylinder flow. Swirl results in higher turbulence inside the chamber during combustion, thus increasing the rate of flame development and propagation. Generating swirl during the intake process decreases volumetric efficiency.

Heat transfer to the combustion chamber walls has a significant impact on engine efficiency. It is affected by cylinder head and piston crown surface area, by the magnitude of in-cylinder gas velocities during combustion and expansion, by the gas temperatures and the wall temperatures. The heat-transfer implications of a combustion chamber should be included in the design process.

Knock effectively limits the maximum compression ratio that can be used in any combustion chamber; it therefore has a direct impact on efficiency. Knock is affected by all the factors discussed above. It is the hardest of all the constraints to incorporate into the design process because of its obvious complexity.

Knowledge of the fundamentals of spark-ignition engine combustion, in-cylinder gas motion, and heat transfer has developed to the point where a rational procedure for evaluating these factors for optimum combustion chamber development and design can be defined. The next two sections develop such a procedure.

15.4.2   Factors That Control Combustion

Our understanding of the structure of the spark-ignition engine flame as it develops and propagates across the combustion chamber (see Secs. 9.3 and 9.4) allows us to relate the physical and chemical factors that control this process to the relevant engine design and operating parameters. The following factors affect the flame development and propagation processes:

1. *Geometry.* Combustion chamber shape and spark plug location.
2. *Flow field characteristics.* Mean velocity, turbulence intensity, and characteristic turbulence length scale in the unburned mixture during combustion.
3. *Unburned mixture composition and state.* Fuel, equivalence ratio, burned gas fraction, mixture pressure and temperature.

Geometry primarily affects combustion through the flame front surface area. It has a lesser effect on combustion development through its influence on in-cylinder motion. Geometric calculations (see Sec. 14.4.2), based solely on the assumption that the front surface of the flame can be modeled as a portion of a sphere centered at the spark plug, provide data on flame front area and the volume behind the flame front surface (the enflamed volume), contained within the combustion chamber at the appropriate flame radii and piston positions.

Flame area varies significantly from one chamber shape to another for a given enflamed volume. In the example shown in Fig. 14-7, the bowl-in-piston chamber gives flame surface areas 30 to 45 percent larger than those for the disc chamber under equivalent conditions around top-center. Hemispherical and open or clamshell chambers showed gains of about 30 percent relative to the equivalent disc configuration. For a given chamber shape, flame area depends even more significantly on plug location. Figure 14-7 shows that shifting the plug from a side to a center location for the bowl-in-piston chamber increased the peak flame area by 150 percent. For hemispherical and open chambers, the increases for a similar shift in plug location were 75 and 90 percent, respectively.[25]

Maps of flame area as a function of radius at different crank angle locations indicate the following pattern. For chamber geometries with side ignition, as flame radius increases, the flame area first rises slowly, then remains approximately constant, and then decreases slowly to zero. In contrast, chambers with central ignition show, as flame radius increases, a rise in flame area to a peak during the major part of the flame travel followed by a rapid decrease as the flame encounters the chamber walls. Moving the plug location toward the center of the chamber produces a larger increase in flame front area than does making the chamber shape more compact (though this has a positive impact too).

The effect of chamber geometry on burn rate has been examined using thermodynamic-based engine cycle simulations with various types of combustion model (e.g., the type developed by Keck and coworkers, see Sec. 14.4.2). Figure 15-16 shows results from one such study.[25] The combustion characteristics of ten different chamber geometries were compared at fixed part-load engine operating conditions. The flame development and propagation phases were separated into 0 to 10 and 10 to 90 percent mass fraction burned times. These were then normalized by the equivalent burn times of the slowest burning chamber—the disc with side ignition. Chamber geometry has the greatest impact on the 10 to 90 percent burn time; its effect on 0 to 10 percent time is significant but substantially smaller. Total burn times can be reduced by between 20 to 30 percent by optimizing spark plug location—comparing worst to best location for each chamber shape. Comparing worst and best chamber shapes, total burn time with fixed plug location can be reduced by about 10 percent.

Increased turbulence in the unburned mixture at the time of combustion increases the burning rate. Turbulence is usually increased by generating swirl during the induction process (see Sec. 8.3.2 and below). Cycle simulation studies[25] indicate that both the duration of the early stage of the burning process and of the main stage decrease when the turbulent velocity at the start of com-

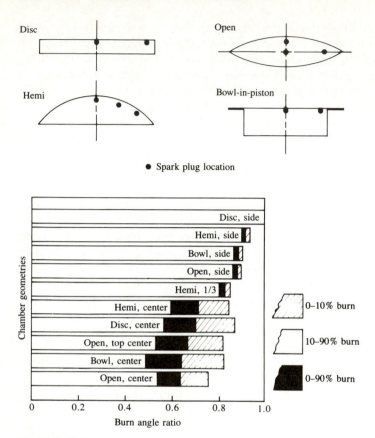

FIGURE 15-16
Comparison of burn angles (0–10 percent burned, 10–90 percent burned, 0–90 percent burned; see Fig. 9-13) for ten different spark-ignition engine combustion chamber geometries and spark plug locations. Burn angles are normalized by angles for slowest burning chamber: disc with side plug.[26]

bustion is increased. The faster combustion process comes primarily from the higher turbulence intensity; however, the decreased characteristic turbulence scale that accompanies the increased turbulence is also significant since it results in a shorter characteristic burning time [see Eq. (14.39) and the accompanying text]. It is important to note that the fuel conversion efficiency of higher-turbulence chambers at the same operating conditions can be lower than for normal chambers, despite the faster burn rates, due to the higher heat transfer that accompanies the higher in-cylinder velocities. For example, predictions based on the combustion model defined by Eqs. (14.33) to (14.35), where the characteristic mixture speed u_T was increased by a factor of two, showed that the 0 to 10 percent and 10 to 90 percent burn durations decreased by about one-third. However, the indicated fuel conversion efficiency decreased by about 6 percent due to the predicted 15 percent increase in heat transfer.[25]

Mixture composition and state affect the burn rate through the dependence of laminar flame speed on temperature, pressure, fuel/air equivalence ratio, and burned gas fraction (residual gas and EGR): see Sec. 9.3.3 and Eqs. (14.33) to (14.35). Table 15.1 compares the burn durations for a stoichiometric mixture, a lean mixture with $\phi = 0.8$, and a stoichiometric mixture with 20 percent EGR. The values of the laminar flame speed at the time of spark are also given (conditions at spark as well as composition are different in each case). The longer burn durations of the more dilute mixtures are clear. Note that EGR as a diluent has a much more deleterious effect on combustion than does air at these approximately equal levels of dilution.

All the above-described factors—flame geometry, fluid motion, and mixture composition—can vary cycle-by-cycle, and therefore contribute to combustion variability (see Sec. 9.4). Cyclic differences in gas motion in the vicinity of the spark plug result in differences in motion of the flame kernel during its early stages of development. Differences in turbulence result in differences in the rates at which the initially smooth surface of the flame kernel becomes wrinkled and convoluted by the flow. Different initial flame center motions change the geometrical interaction of the flame front with the combustion chamber walls later in the flame propagation process. Differences in the amount of fuel, air, and EGR which enter each cylinder cycle-by-cycle, the nonuniformity in composition of the entering charge, and any incomplete mixing of that entering charge with the residual gases in the cylinder also contribute to combustion variability. These composition nonuniformities lead to differences in the early stages of flame development. The variations in the amounts of fuel, air, and EGR that enter each cylinder cycle-by-cycle and in the uniformity of that mixture are factors within the direct control of the engine designer.

A fast combustion process reduces cyclic combustion variability for the following reasons. With a faster burn, optimum spark timing is closer to top-center: mixture temperature and pressure at the time of spark are higher, so the laminar flame speed at the start of combustion is greater. This, combined with the higher turbulence of most fast-burn concepts, results in faster flame kernel development.

TABLE 15.1
Effect of excess air and recycled exhaust on burn duration

ϕ	EGR, %	θ_s, degree	Burn durations, degree		S_L at θ_s, cm/s
			0–10%	10–90%	
1.0	0	340	22	17	75
0.8	0	336	26	21	52
1.0	20	324	31	28	23

400 cm³ per cylinder displaced volume, 80 mm bore, 8.5 compression ratio, disc chamber, center plug location. 1500 rev/min, stoichiometric operation, θ_s = spark timing (MBT), inlet pressure 0.5 atm, inlet temperature 350 K, S_L = laminar flame speed.[26]

More rapid initial flame growth results in less variation in flame center motion during the critical flame-development phase. The resulting geometric variations in the flame front/chamber wall interaction are therefore reduced; this decreases the variations in burn rate that result from these geometric variations. Also, the faster burning process ends earlier in the expansion stroke. Thus the problem of occasional slow burning cycles, partial burning cycles, and eventually misfire, which occurs with dilute mixtures under normal burning conditions due to quenching of the combustion process as gas temperatures fall during expansion, is largely avoided (see Sec. 9.4.3).

15.4.3   Factors That Control Performance

VOLUMETRIC EFFICIENCY. Combustion chamber shape affects volumetric efficiency through its constraints on maximum valve size and through the degree of swirl (if any) that the chamber and port designs produce to achieve the desired combustion characteristics. To obtain maximum performance and to reduce pumping losses, the size of the valve heads should be as large as practical; the valve sizes that can be accommodated depend on cylinder head layout. Table 6.1 lists the typical maximum valve sizes that can be accommodated into several common chamber shapes (see Fig. 15-15). The approximate mean piston speed at maximum power is a measure of the maximum air flow that each engine design can pump. Note that of the two-valve configurations, the designs with inclined valve stems permit substantially greater maximum air flow. The four-valve pent-roof design, which also has inclined valve stems, is the best of those listed since it accommodates the largest valve and port areas (there are other four-valve head designs which are comparable).

Swirl can be generated during the intake process through suitable port, valve, and head design. It requires either that the flow through the intake valve be directed tangentially into the cylinder so that gas flows through one side of the valve opening preferentially (e.g., through the use of masks to restrict flow at the mask location or through the use of a tangentially directed port or a flow deflector in the port just upstream of the valve), or requires the use of a helical intake port that imparts an angular velocity to the flow before it enters the cylinder. In either case the inlet flow enters the cylinder with higher velocity than it would have in the absence of swirl; hence the pressure drop across the valve is increased, and maximum air flow through the cylinder is reduced. Well-designed helical swirl-generating ports (see Sec. 8.3.2) appear to be the best way to create swirl. However, geometric and production constraints often prevent the incorporation of such ports into the cylinder head design, and other swirl-generating methods must be used. The engine maximum-power penalty associated with generating significant swirl is of order 5 to 10 per cent.

Since swirl is only required at part-throttle operation when enhancement of the burn rate is most critical and is not usually required at full throttle when the flow restriction penalty is most significant, induction systems with a separate passage for the part-throttle air flow, where only this separate passage generates swirl, are an attractive option. However, the gains in volumetric efficiency are offset by a higher cost due to the additional complexity in port and manifold of the double passage and the individual throttle valves required in each port for flow control.

Swirl can be *intensified* during compression with bowl-in-piston combustion chambers by decreasing the moment of inertia of the in-cylinder charge as the piston moves toward top-center, and thereby increasing its angular velocity (see Sec. 8.3.3). An advantage here is that the swirl level generated during induction is less than would be required without the compression-produced radially inward motion of the charge. This approach can be used with combustion chamber designs that are axisymmetric and compact. Swirl can also be *generated* by squish motion toward the end of compression with a suitable design of chamber. The advantage of this approach is that there is no induction-stroke swirl-generating volumetric efficiency penalty. However, the cylinder head geometries proposed to date for either intensifying or generating swirl have vertical valve stems, and hence have smaller valve sizes which in themselves restrict air flow. Also, the cylinder head geometry required to generate swirl during compression has a larger surface area than more open chamber designs and, therefore, has significantly higher heat losses.

The impact of conventional radially inward squish motion (see Sec. 8.4) on in-cylinder turbulence, and hence combustion, is unclear. Chambers with significant squish are also more compact; for this reason alone they would be faster burning.

HEAT TRANSFER. The convective engine heat transfer to the combustion chamber walls is described by equations of the form of (12.2): e.g., Eq. (12.21). The heat-transfer coefficient is usually correlated by expressions of the form of Eq. (12.3), which relate the Nusselt, Reynolds, and Prandtl numbers (see Sec. 12.4). Thus combustion chamber surface area, and especially the surface area in contact with the burned gases, is important. Gas velocity is also important; it influences the heat-transfer rate through the Reynolds number. Various characteristic velocities have been used in the Reynolds number to scale heat transfer: mean piston speed, mean in-cylinder gas velocity, turbulence intensity, either individually or in combination. Both of these variables, area and velocity, are affected by combustion chamber design.

Studies of engine performance using thermodynamic-based simulations of the engine's operating cycle (see Sec. 14.4) provide data that indicate the importance of changes in heat transfer. At part-throttle operating conditions, such simulation calculations show that a 10 percent change in combustion chamber heat losses results in a change of between 2 and 5 percent in brake specific fuel consumption; an average fuel consumption change of about one-third the magnitude of the heat-transfer change (and of opposite sign) is an appropriate rule of thumb.[25, 27] At wide-open throttle, the effect on mean effective pressure is comparable: a 10 percent change in heat transfer results in about a 3 percent change in bmep.

This impact of heat transfer on engine efficiency and performance underlines the importance of combustion chamber details that affect heat transfer. For the chamber shapes shown in Fig. 15-16, the total heat losses as a fraction of the fuel's energy, at fixed engine speed and intake conditions, were also calculated. Both chamber shape and spark plug details affect heat losses since together these govern the surface area of the hot burned gases in contact with the walls. The open and hemispherical chambers had least heat transfer. Geometries such as the bowl-in-piston, which obviously have a higher surface area, had about 10 percent higher heat transfer. The effect of shifting the plug from a side to center location depended on chamber shape. Open and bowl-in-piston chambers showed little change; the hemispherical chamber showed a 4 percent reduction. Given a general chamber shape choice, the details of the actual design are important also; it is easy to add substantial surface area with piston cutouts, plug bosses, and cylinder head masking or squish regions which will deteriorate chamber performance to a measurable degree.

Higher in-cylinder velocities affect heat-transfer rates through the Reynolds number term in the heat-transfer coefficient correlation. Swirl- and squish-generated flows increase in-cylinder gas velocities and will, therefore, increase heat-transfer rates.

15.4.4   Chamber Octane Requirement

Knock limits an engine's compression ratio, and hence its performance and efficiency. The more fundamental aspects of knock were reviewed in Sec. 9.6. Knock occurs when the end-gas autoignites prior to its being burned up by the normal flame-propagation process. The tendency to knock depends on engine design and operating variables which influence end-gas temperature, pressure, and time spent at high values of these two properties before flame arrival.

The presence or absence of knock in an engine depends primarily on the antiknock quality of the fuel, which is defined by the fuel's *octane number* (see Sec. 9.6.3). It determines whether or not a fuel will knock in a given engine under given operating conditions: the higher the octane number, the higher the resistance to knock. The *octane number requirement* of an engine is defined as the minimum fuel octane number that will resist knock throughout its speed and load range. The following factors affect an engine's octane requirement: (1) composition of the fuel; (2) chamber geometry and size; (3) charge motion; (4) spark-advance curve; (5) inlet air, intake manifold, and water jacket temperatures; (6) carburetor or fuel-injector air-fuel ratio calibration; (7) the ambient conditions—pressure, temperature, and relative humidity—during the requirement determination.

The following illustrates the interaction between fuel factors and engine operating variables. Figure 15-17 shows the relation between spark advance, torque, and speed in an engine operating at wide-open throttle. The dashed lines, determined with a fuel of sufficiently high octane rating to avoid knock, show MBT timing as a function of speed, along with the spark-advance limits for con-

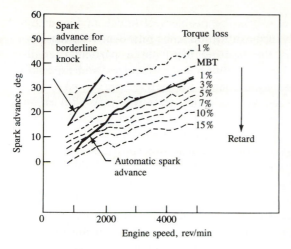

FIGURE 15-17
Relation between spark advance, speed, and torque loss, for spark-ignition engine at wide-open throttle, showing knock limit for specific gasoline and typical spark-advance schedule that avoids knock problems.[28]

stant specified percentage torque reductions. The upper solid line traces the spark advance for borderline knock with a particular commercial gasoline. To avoid knock with this fuel, the spark advance must be set to lose one percent of engine torque at 800 rev/min, with the torque loss diminishing to zero at 1200 rev/min. Above that speed this particular fuel allows operation at MBT timing without knocking. The lower solid curve represents a typical spark-advance schedule at WOT. It lies below the borderline knock advance (and results in a significant torque loss) for the following reasons. One is that different commercial gasolines with the same research octane number can respond differently to variations in engine operating conditions. Calibrating the engine (i.e., specifying the schedules for spark advance, A/F, and EGR) must be done with sufficient margin of conservatism to avoid objectionable knock with the normal range of commercial gasolines over the full operating conditions of the engine. A second reason is engine-to-engine production variability despite the close dimensional tolerances of modern production engineering. For example, the effective compression pressure in each cylinder of a multicylinder engine is not identical, due to geometric and ring-pack behavior differences. The cylinder with the highest compression pressure is most knock-prone. Allowing for corresponding effects of cylinder-to-cylinder variations in A/F, EGR rates, and spark timing, it is obvious that for a given operating condition in a multicylinder engine, one cylinder is more likely to knock than the others. It is that cylinder which limits the spark advance.† A third reason for the discrepancy between actual spark-advance calibration and the knock limit for a given engine and fuel is the octane requirement increase associated with the buildup of deposits on the combustion chamber walls over extended mileage (see Sec. 9.6.3).

---

† There is no assurance that the same cylinder will be the principal offender in all engines of the same model, nor in a given engine at all operating conditions.

In the example shown above, it was the problem of knock at low engine speed which required the spark advance calibration to be retarded. Whether low-, medium-, or high-speed knock is the limiting factor in a particular engine depends on the sensitivity of the fuel, on engine design features, and especially upon the engine's spark-advance requirements for MBT. The knock-limited spark advance determined from road octane rating tests will vary with engine speed and fuel sensitivity, as shown in Fig. 15-18. Low sensitivity fuels will tolerate more severe engine operating conditions and vice versa. Figure 15-18b, c, and d shows a typical engine spark-advance characteristic superposed on the knock-limited spark-advance plot. Depending on the fuel sensitivity and shape of the spark-advance curve, the knock region may occur at low, medium, or high speed (or not at all).

It will be apparent from the above discussion that defining the effect of combustion chamber geometry on knock can only be done in an approximate fashion. The importance of fuel composition details, differences in engine design, the variability between engines of the same type, and the effect of deposits all make the quantification of trends as chamber design is varied extremely difficult. One of the most important chamber variables is the compression ratio. Figure 15-19 shows the relationship between the octane requirement and compression ratio for a number of combustion chambers. The octane requirement was defined as the research octane number of the fuel required to operate the engine at WOT with the weakest mixture for maximum power with borderline

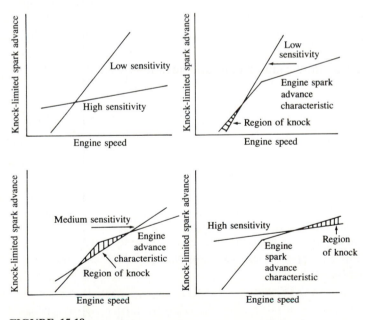

FIGURE 15-18
Diagrams showing knock-limited spark-advance curves for fuels of different sensitivity and how these can give low-, medium-, and high-speed knock in the same engine.[29]

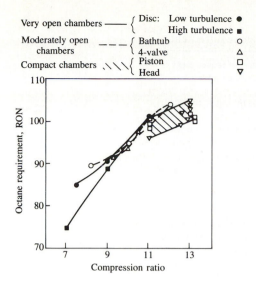

Very open chambers ———— { Disc: Low turbulence ●
                                    High turbulence ■
Moderately open   — — — { Bathtub   ○
chambers   4-valve   △
Compact chambers ＼＼＼ { Piston   □
                           Head   ▽

Octane requirement, RON (vertical axis): 70, 80, 90, 100, 110

Compression ratio (horizontal axis): 7, 9, 11, 13

FIGURE 15-19

Octane requirement (gasoline research octane number), at wide-open throttle and MBT timing, to avoid knock as a function of compression ratio for various combustion chamber designs.[10]

(or light) knock coinciding with MBT timing at the given speed. As is well known, the octane requirement increases with increasing compression ratio; there are, however, differences in the octane requirement between different types of chamber at the same compression ratio. The chambers studied were disc-shaped chambers, bathtub and four-valve (open chambers with squish) and compact high compression ratio chambers (bowl or cup-type chambers in the piston crown or in the cylinder head around one of the valves). In the 9 to 11 compression ratio range there are only modest differences between the chambers studied. At higher compression ratios, 11 to 13, the compact chambers show a lower octane requirement which gives them a 1 to 2 compression ratio advantage over the more open chambers. This advantage for the compact (and high-turbulence) chambers comes largely from the increased heat-transfer rates in these chambers. Whether the higher turbulence is generated during intake or at the end of the compression stroke, it increases the heat transfer from the end-gas, reducing its temperature and therefore its propensity to knock. However, this higher heat transfer also reduces engine power and efficiency, so the benefits of the compression ratio advantage are reduced. There is some increase in the knock-limited compression ratio with a given fuel as burn time is decreased by using one, two, three, and then four spark plugs simultaneously, with a given chamber geometry, but the effect is much smaller than the differences suggested by Fig. 15-19.[23]

Spark plug location within the chamber is an important factor affecting octane requirement. More centrally located plug positions shorten the flame travel path to the cylinder walls and decrease the time between spark discharge and flame arrival at the end-gas location. This decreases the octane requirement. The position of the spark plug in relation to the exhaust valve is also important: it is advantageous to burn the unburned mixture which has been heated by contact with the hot exhaust valve early in the combustion process.

TABLE 15.2

Engine conditions affecting octane number requirement

Octane number requirement tends to go *up* when:	Octane number requirement tends to go *down* when:
1. Ignition timing is advanced.	1. Ignition timing is retarded.
2. Air density rises due to supercharging or a larger throttle opening or higher barometric pressure.	2. Engine is operated at higher altitudes or smaller throttle opening or lower barometric pressure.
3. Humidity or moisture content of the air decreases.	3. Humidity of the air increases.
4. Inlet air temperature is increased.	4. Inlet air temperature is decreased.
5. Coolant temperature is raised.	5. Fuel/air ratio is richer or leaner than that producing maximum knock.
6. Antifreeze (glycol) engine coolant is used.	6. Exhaust gas recycle system operates at part throttle.
7. Engine load is increased.	7. Engine load is reduced.

Operating variables that affect the temperature or pressure time histories of the end-gas during combustion or the basic autoignition characteristics of the unburned fuel, air, residual mixture will also affect the engine's octane requirement. The most important additional variables which increase or decrease octane requirement in a consistent manner are listed in Table 15.2. Relative spark

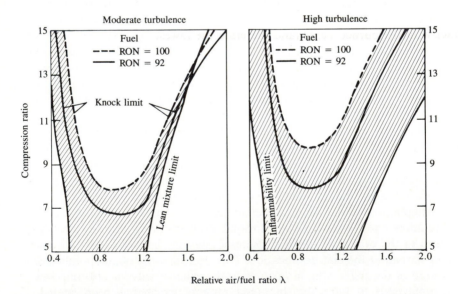

FIGURE 15-20

Knock limits and lean engine operating limits as function of compression ratio and relative air/fuel ratio λ ($\lambda = 1/\phi$) for moderate and high-turbulence engine combustion chambers.[32]

advance has a major impact on knock; since it is also easy to adjust, it is the engine variable most commonly used to control knock. Studies show that typically 0.5 to 1.0 RON reduction is achieved per degree of retard.[30] Atmospheric conditions—pressure, temperature, and humidity—all affect the octane number requirement.[31]

The fuel/air equivalence ratio affects the octane requirement of an engine. The highest requirement is for slightly rich mixtures; increasing richness and leanness about this point decrease the octane requirement substantially. Figure 15-20 shows the knock-limited compression ratio as a function of the relative air/fuel ratio ($\lambda = 1/\phi$; $\lambda > 1$ for lean mixtures) for conventional and high-turbulence chambers, for two fuels with different octane ratings. Substantially higher compression ratios can be used with lean mixtures, especially with the high-turbulence chamber which extends the lean limit. The coolant temperature affects the octane requirement. Higher coolant temperature increases the inlet mixture temperature, and reduces heat losses from the end-gas to a modest degree.

15.4.5 Chamber Optimization Strategy

The discussion in the previous sections suggests that the following sequence of steps in a combustion chamber development process is most logical. First should come the selection of the best chamber geometry. Geometric optimization can result in substantial benefits and carries no significant penalties. Chamber geometry involves cylinder head and piston crown shape, and plug location. Open chambers such as the hemispherical or pent-roof cylinder head, and clam-shell, with near central plug location, give close to the maximum flame front surface area (and hence a faster burn), have the lowest chamber surface area in contact with the burned gases (and therefore the lowest heat transfer), and have inclined valves which give high volumetric efficiency. Spark plug location close to the center of the chamber is especially important in obtaining a fast burn rate. More compact chamber shapes than the open chambers listed above, such as the bowl-in-piston or chamber-in-head designs, do produce a somewhat faster burn, but with lower volumetric efficiency and higher heat losses.

Following this first step, two problem areas may remain: the chamber may have a higher octane requirement than is desired and the burn rate may not be fast enough to absorb the high dilution required at part load to meet the emissions and fuel consumption goals.

Positioning the spark plug as close to the center as possible will have reduced the octane requirement for that particular chamber shape. Depending on chamber design details, some squish area could be introduced. However, the perceived octane advantage of chamber designs which contain substantial squish is offset, at least in part, by their higher heat losses. A unit compression ratio increase results in a 3 percent or less increase in efficiency at part load. However, if the measures required to increase the compression ratio from, say, 9 to 10 resulted in a 10 percent increase in heat transfer, engine efficiency would not improve.

The next step should be to reduce the cyclic variability in the combustion process to the maximum extent feasible, by improving the uniformity of the intake fuel, air, and EGR mixture. Delivery of equal amounts of each of these constituents to each cylinder, provision for good mixing between constituents in the intake manifold and port, and accurate control of mixture composition during engine transients are all especially important. Also important is achieving closely similar flow patterns within each engine cylinder during intake so as to obtain equal burn rates in all cylinders. Attention to these intake process and mixture preparation details will always improve engine operation and carries no performance penalties.

However, the burn rate may still not be fast enough, especially during the critical early stages of flame development, and cyclic variability may still be too high to meet the engine's performance goals. Then higher turbulence levels during combustion must be achieved. This is usually best done by creating swirl during the induction process. The appropriate method for introducing swirl will depend on any geometric manufacturing constraints and cost issues. With no geometric constraints, use of helical swirl-generating ports or a divided intake-port system with valves to control the flow at light load are likely to have the lowest power penalties. It is especially important that only the minimum additional turbulence required to achieve the performance objectives be added at this stage. Higher than necessary gas velocities within the cylinder result in excessive heat losses and low volumetric efficiency.

In summary, to meet the objectives of a fast, repeatable, and robust combustion process with high volumetric efficiency, low heat transfer, and acceptable octane requirement, combustion chamber development should proceed through the following steps.

1. Optimize the chamber geometry within the design constraints for the maximum flame front area, minimum burned gas/chamber wall contact area, and largest valve size.

2. Obtain additional reductions in the cyclic combustion variability by improving mixture distribution and uniformity and by creating flow patterns into each cylinder that are essentially identical.

3. Achieve any additional improvement in burn rate and cyclic variability required to meet objectives by increasing turbulence to the minimum extent. This is usually best done by creating swirl during the induction process.

15.5   VARIABLES THAT AFFECT CI ENGINE PERFORMANCE, EFFICIENCY, AND EMISSIONS

15.5.1   Load and Speed

The performance of a naturally aspirated DI heavy-duty truck diesel engine and a small IDI engine at full load over the engine speed range have already been

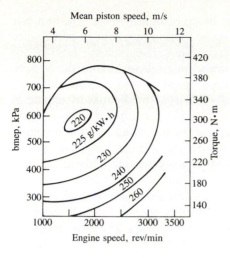

FIGURE 15-21
Performance map for 6.54-dm³ eight-cylinder air-cooled naturally aspirated medium-swirl DI diesel engine. Contours of constant bsfc in grams per kilowatt-hour shown. Bore = 102 mm, stroke = 100 mm, r_c = 18. Multihole fuel nozzle.[33]

discussed in Sec. 15.2. Here we examine the part-load behavior of various types of naturally aspirated diesel engines.

As with SI engines (see Sec. 15.3.3), performance maps where bsfc contours are plotted on a graph of bmep versus engine speed are commonly used to describe the effects of load and speed variations. Figure 15-21 shows the performance map for an air-cooled four-stroke cycle medium-swirl naturally aspirated DI diesel (similar to the engine in Fig. 1-23). Maximum rated power for this 6.54-dm³ displacement engine at 3200 rev/min is 119 kW, maximum bmep at 2000 rev/min is 784 kPa, and minimum bsfc (at 1600 rev/min and 580 kPa bmep) is 220 g/kW·h, which corresponds to a brake fuel conversion efficiency of 38.5 percent. The gross indicated fuel conversion efficiency would be about 48 percent.

Figure 15-22 shows the performance map for a small high-swirl DI diesel which uses the M.A.N. combustion system with a single fuel jet sprayed tangentially into the swirling air flow. Due to the higher speed and higher swirl than the

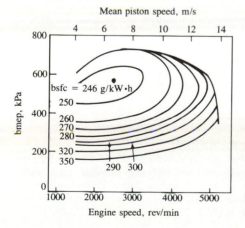

FIGURE 15-22
Performance map for 1.47-dm³ four-cylinder naturally aspirated DI diesel engine with high-swirl single-hole-nozzle M.A.N. combustion system. Contours of constant bsfc in grams per kilowatt-hour shown. Bore = 76.5 mm, stroke = 80 mm, r_c = 18.5.[34]

larger DI engine in Fig. 15-21, the maximum bmep is slightly lower. The best bsfc is about 10 percent higher largely due to higher friction mep, but in part due to higher heat losses resulting from the less favorable surface/volume ratio of the smaller bore engine and higher swirl, and lower heat-release rate of the M.A.N. system. Note that the maximum mean piston speed for this engine, 13.3 m/s at 5000 rev/min, is comparable to that of the larger medium-swirl engine in Fig. 15-21 (10.7 m/s).

Figure 15-23 gives the performance characteristics of an automotive naturally aspirated swirl-chamber IDI diesel engine. Maximum bmep values are usually higher than those of equivalent size DI engines because without the need to generate swirl during the intake process, the intake port and valve are less restrictive and volumetric efficiency is higher, and because the IDI engine can be run at lower A/F without smoking. The best bsfc values are usually some 15 percent higher than values typical of equivalent DI engines. The best brake fuel conversion efficiency of the engine of Fig. 15-23 is 32.5 percent.

Comparisons between naturally aspirated DI and IDI diesel engines of closely comparable design and size indicate that the DI engine is always more efficient, though the benefit varies with load. At full load, differences of up to 20 percent in bsfc have been noted, especially in engines with larger displacement per cylinder. At part load, the gain is less—of order 10 percent. Comparisons should be made at equal emission levels, a task that is difficult to accomplish in practice. Emission control with the DI engine is more difficult, so this constraint reduces the benefit somewhat. Figure 15-24 shows a breakdown of the indicated efficiency differences between the two systems. At full load ($A/F = 18$ to 20) the IDI suffers a penalty of about 15 to 17 percent due in large part to the retarded timing of the IDI combustion process and its long, late-burning, heat-release profile. At light load, about 300 kPa bmep ($A/F = 50$), these combustion effects are small and the indicated efficiency penalty of the IDI (around 5 to 7 percent) is due to the higher heat losses associated with the larger surface area and high-velocity flow through the connecting nozzle of the divided-chamber geometry and due to the pumping pressure loss between the main and auxiliary chambers.[36]

Note that all these diesel engine performance maps are similar in general

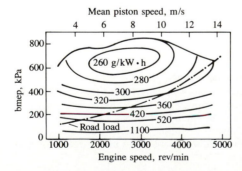

FIGURE 15-23
Performance map for 1.987-dm³ five-cylinder naturally aspirated IDI swirl-chamber diesel engine. Contours of constant bsfc in grams per kilowatt-hour shown. Bore = 76.5 mm, stroke = 86.4 mm, $r_c = 23$.[35]

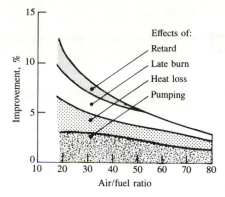

Effects of:
Retard
Late burn
Heat loss
Pumping

FIGURE 15-24
Factors which improve the indicated efficiency of naturally aspirated small DI diesel combustion systems relative to IDI swirl-chamber combustion system, as a function of A/F or load.[36]

shape and when plotted against $\bar{S}_p$ are quantitatively comparable. The increase in bsfc from the minimum value with increasing speed at constant load is due to the increase in friction mep, partly offset by the effect of decreasing importance of heat losses per cycle on efficiency. The increase in bsfc with decreasing load at constant speed is dominated by the decreasing mechanical efficiency as bmep is reduced. The indicated fuel conversion efficiency increase as the fuel/air equivalence ratio is decreased partly offsets this. The trends in bsfc when increasing load at constant speed and increasing speed at constant load from the minimum are more modest. They are the net results of (1) the increase in mechanical efficiency and decrease in indicated fuel conversion efficiency as the load increases and (2) decreasing indicated efficiency due to increasing importance of heat losses and increasing mechanical efficiency as the speed decreases. The enrichment of the mixture at high load and low speed of spark-ignition engines is, of course, absent.

Figure 15-25 shows the effect of load on NO_x and HC emissions for naturally aspirated DI and IDI diesel engines. For the DI engine NO_x concentrations rise steadily as the fuel/air ratio increases with increasing bmep at constant injection timing. The increasing quantity of fuel injected per cycle results in an increasing amount of close-to-stoichiometric combustion products near the peak pressure and temperature (see Sec. 11.2.4). The IDI engine shows a similiar trend except that, at high load, NO_x concentrations level off. These characteristics do not change substantially with engine speed. The IDI engine shows significantly lower HC emissions than the DI engine. The high HC at idle and light load are thought to result from fuel mixing to too lean an equivalence ratio. If diesel engines are overfueled at high load, HC emissions then rise rapidly. These HC mechanisms are described in Sec. 11.4.4. Injection timing affects NO_x and HC emissions significantly, as discussed in Sec. 15.5.2 below.

Figure 15-26 shows smoke and particulate mass emissions from a naturally aspirated IDI engine. Rapidly increasing black smoke at very high load limits the maximum bmep that a diesel engine can produce. On a specific emission basis [Eq. (2.36)], the particulates typically show a U-shaped behavior due to the predominance of hydrocarbons in their composition at light load and of carbon at high load.[38]

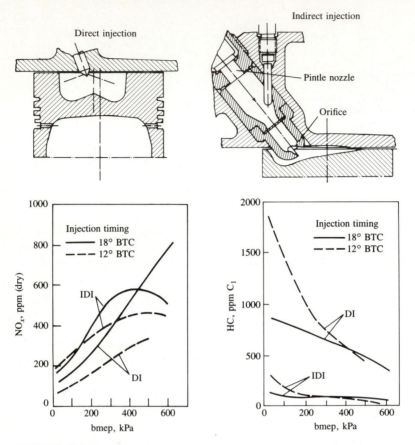

FIGURE 15-25
Effect of load on naturally aspirated diesel engine NO_x and HC emissions at rated speed, with two injection timings. Direct-injection and indirect-injection (prechamber) combustion systems. Six-cylinder, 5.9-dm^3 displaced volume, engine. DI: $r_c = 17$, rated speed = 2800 rev/min; IDI: $r_c = 16.7$, rated speed = 3000 rev/min.[37]

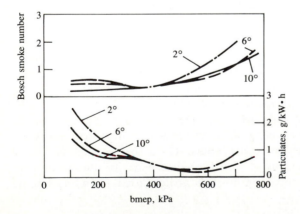

FIGURE 15-26
Smoke (Bosch smoke number) and particulate mass emissions (in grams per kilowatt-hour) as a function of load and injection timing for six-cylinder 3.7-dm^3 IDI swirl-chamber diesel engine at 1600 rev/min (no EGR).[38]

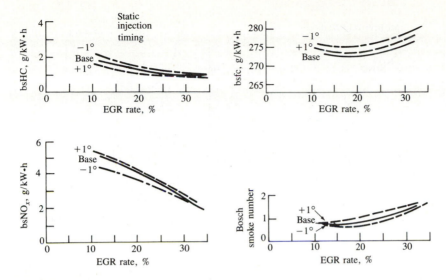

FIGURE 15-27
Brake specific HC, NO_x and fuel consumption, and smoke emissions, as a function of percent recycled exhaust for 2.4-dm^3 four-cylinder high-swirl DI diesel engine at 1250 rev/min and 255 kPa bmep.[39]

Recycled exhaust gas, at part load, can be used to reduce diesel engine NO_x emissions. Note that since diesel engines operate with the air flow unthrottled, at part load the CO_2 and H_2O concentrations in exhaust gas are low; they are essentially proportional to the fuel/air ratio. Because of this, high EGR levels are required for significant reductions in NO_x emissions. Figure 11-18 shows how NO_x concentrations decrease as a DI diesel engine inlet air flow is diluted at a constant fueling rate. The dilution is expressed in terms of oxygen concentration in the mixture after dilution. Figure 15-27 shows how the EGR affects specific NO_x and HC, fuel consumption, and smoke for a small high-swirl DI diesel engine at typical automobile engine part-load conditions. Effective reduction of $bsNO_x$ is achieved and modest reductions in bsHC, with only a slight increase in bsfc. However, smoke increased as the EGR rate increased.[39]

15.5.2 Fuel-Injection Parameters

Fuel-injection timing essentially controls the crank angle at which combustion starts. While the state of the air into which the fuel is injected changes as injection timing is varied, and thus ignition delay will vary, these effects are predictable (see Sec. 10.6.4). The fuel-injection rate, fuel nozzle design (including number of holes), and fuel-injection pressure all affect the characteristics of the diesel fuel spray and its mixing with air in the combustion chamber.

Figure 15-28 shows the effect on performance and emissions of varying injection timing, in (a) a medium-swirl DI diesel engine and (b) an IDI engine. At fixed speed and constant fuel delivery per cycle, the DI engine shows an optimum

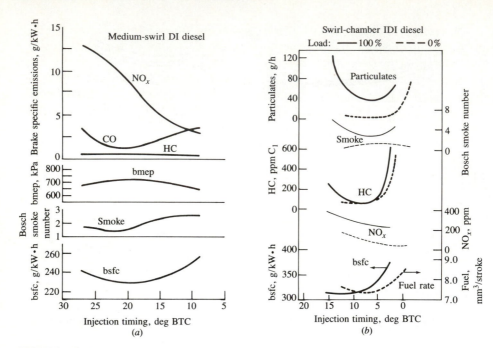

FIGURE 15-28

Effect of start-of-injection timing on diesel engine performance and emissions. (a) Medium-swirl DI diesel engine with deep combustion bowl and four-hole injection nozzle, 2600 rev/min, fuel delivery 75 mm³/cycle, fuel/air equivalence ratio 0.69.[37] (b) Swirl-chamber IDI engine, 2500 rev/min, 0 and 100 percent load.[40]

bsfc and bmep at a specific start of injection for a given injection duration.† The IDI engine experiments are at fixed bmep; here, bsfc at full load and fueling rate at idle show a minimum at specific injection timings. Injection timing which is more advanced than this optimum results in combustion starting too early before TC; injection retarded from this optimum results in combustion starting too late.

Injection timing variations have a strong effect on NO_x emissions for DI engines: the effect is significant but less for IDI engines. Retarded injection is commonly used to help control NO_x emissions. It gives substantial reductions, initially with only modest bsfc penalty. For the DI engine, at high load, specific HC emissions are low and vary only modestly with injection timing. At lighter loads, HC emissions are higher and increase as injection becomes significantly retarded from optimum. This trend is especially pronounced at idle. For IDI diesel engines HC emissions show the same trends but are much lower in magnitude than DI engine HC emissions.[41] Figure 15-25 supports this discussion.[37]

---

† This optimum injection timing gives maximum brake torque, though the designation MBT timing is less commonly used with diesels than with SI engines.

Retarding timing generally increases smoke, though trends vary significantly between different types and designs of diesel engine. Mass particulate emissions increase as injection is retarded.

The injection rate depends on the fuel-injector nozzle area and injection pressure. Higher injection rates result in higher fuel-air mixing rates, and hence higher heat-release rates (see Sec. 10.7.3). For a given amount of fuel injected per cylinder per cycle, as the injection rate is increased the optimum injection timing moves closer to TC. The effects of injection rate and timing on bsfc in a naturally aspirated DI diesel engine are shown in Fig. 15-29. The higher heat-release rates and shorter overall combustion process that result from the increased injection rate decrease the minimum bsfc at optimum injection timing: however, a limit to these benefits is eventually reached.

Increasing the injection rate increases NO_x emissions and decreases smoke or particulate emissions. The controlling physical process is the rate of fuel-air mixing in the combustion chamber so, at constant fuel injected per cylinder per cycle, both increased injection pressure at fixed nozzle orifice area (which reduces injection duration) and reduced nozzle area at fixed injection duration produce these trends.[42]

The engine designer's goal is obviously to achieve the best bsfc possible

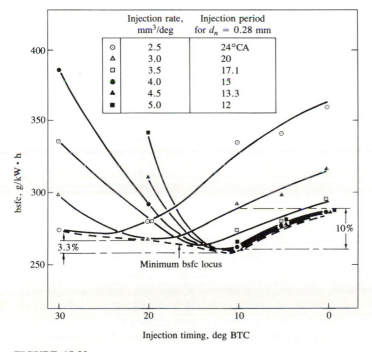

FIGURE 15-29
Effect of injection timing and injection rate on bsfc for 0.97-dm³ single-cylinder naturally aspirated DI diesel engine with swirl. 2000 rev/min, 60 mm³ per stroke fueling rate.[42]

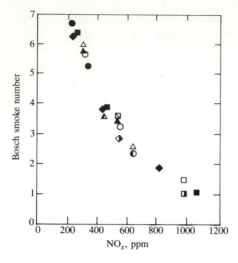

FIGURE 15-30
Tradeoff between NO_x and smoke emissions for quiescient single-cylinder DI diesel engine with bore = 140 mm, stroke = 152 mm, $r_c = 14.3$, eight-hole injector nozzle. Various speeds, fueling rates, injection timings, injection pressures, % EGR; constant $A/F = 25$.[43]

with emission levels low enough to satisfy the constraints imposed by emission standards. The variations of bsfc, NO_x, and particulate emissions described above involve tradeoffs that make achieving this goal especially difficult. One well-established tradeoff is between bsfc and $bsNO_x$. Injection retard from optimum injection timing decreases $bsNO_x$ at the expense of an increase in bsfc. A second important tradeoff is that between NO_x and particulate emissions, illustrated for a DI diesel engine in Fig. 15-30. Smoke is plotted versus NO_x for a range of speeds, loads (fuel per cycle), injection timings, injection pressures, and EGR rates. The air/fuel ratio was maintained constant at 25 ($\phi = 0.58$). The figure indicates that for a well-optimized DI diesel engine, the smoke nitric oxide tradeoff is relatively independent of engine speed, injection rate, injection timing, and amount of EGR. A given reduction in one of these pollutants through changing any one of these variables results in a given increase in the other pollutant. This tradeoff exists for essentially all types of diesel engine, though the magnitude depends on engine details.

15.5.3 Air Swirl and Bowl-in-Piston Design

Increasing amounts of air swirl within the cylinder (see Sec. 8.3) are used in direct-injection diesel engines, as engine size decreases and maximum engine speed increases, to achieve adequately fast fuel-air mixing rates (see Sec. 10.2.1). In these medium-to-small size engines, use of a bowl-in-piston combustion chamber (Fig. 10-1b and c) results in substantial swirl amplification at the end of the compression process (Sec. 8.3.3). Here, the impacts of varying air swirl on the performance and emissions characteristics of this type of DI engine are reviewed.

Since air swirl is used to increase the fuel-air mixing rate, one would expect the overall duration of the combustion process to shorten as swirl increases and emissions that depend on the local fuel/air equivalence ratio to be dependent on swirl level. Figure 15-31 shows the effects of swirl and injection-timing variations on bsfc and emissions of a DI engine of 1.36 dm³ per cylinder displacement with a toroidal bowl-in-piston chamber (see Fig. 10-3b). The swirl ratio [Eq. (8.28)] was varied using shrouded inlet valves with shrouds of different subtended angle (60 to 120°). The injection timing which gives minimum bsfc shifts toward TC as the swirl ratio increases due to the decreasing total combustion duration. The minimum bsfc was achieved with a swirl ratio of 6 to 7: while higher swirl levels continue to increase fuel-air mixing rates, heat transfer increases also and eventually offsets the mixing rate gain. Particulate and CO emissions decrease as swirl increases due to more rapid fuel-air mixing. NO_x emissions increase with increasing swirl. At constant injection timing, however, about half the increase is due to the effect of injection advance relative to the optimum timing and half to the shorter combustion process.[44] Similar trends have been observed as swirl is varied with the M.A.N. single-hole-nozzle diesel combustion system of Fig. 10-1c.

In production engines, the various types of port design shown in Fig. 8-13 can be used to generate swirl during the induction process. Of these, the helical ports are most effective at producing relatively uniform high swirl with the minimum loss in volumetric efficiency.

The geometry of the bowl-in-piston combustion chamber governs the extent to which induction-generated swirl is amplified during compression. The flow field in the bowl during fuel injection is also dependent on the interaction between this swirling flow and the squish motion which occurs as the top of the piston crown approaches the cylinder head (see Sec. 8.4). Various types of bowl-in-piston design for multihole fuel nozzle DI engines are shown in Fig. 15-32 (for the M.A.N. single-hole-nozzle system a spherical bowl is used; see Fig. 10-1c). More conventional designs (e.g., Fig. 15-32a) have the bowl sides essentially parallel to the cylinder liner. Note that it is often necessary to offset the bowl axis from the cylinder axis and the injector nozzle hole locations from the bowl axis,

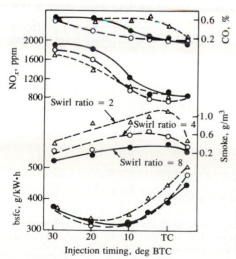

FIGURE 15-31
Effect of air swirl on bsfc and emissions of single-cylinder DI diesel engine with toroidal bowl-in-piston chamber. 1.36-dm³ displacement, $r_c = 16$, bowl diameter/bore = 0.5, 2000 rev/min, full load. Swirl ratio measured in bowl-in-piston at injection.[44]

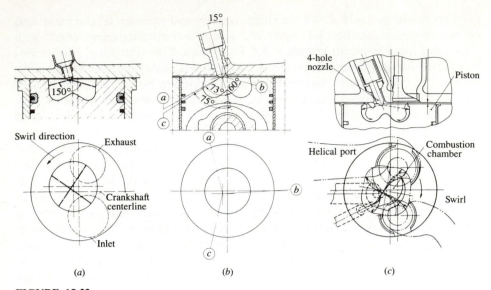

FIGURE 15-32
Various bowl-in-piston chamber designs for DI diesel engines with swirl: (a) conventional straight-sided bowl,[37] (b) reentrant bowl,[45] (c) square reentrant bowl.[46]

due to the geometric constraints imposed by the valves. An alternative design with a reentrant bowl (Fig. 15-32b) is sometimes used to promote more rapid fuel-air mixing within the bowl. The squish-swirl interaction with highly reentrant bowl designs differs markedly from the interaction in nonreentrant bowls. Figure 15-33 shows the two different flow patterns set up in a diametral plane. With a conventional bowl, the swirling air entering the bowl flows down to the base of the bowl, then inward and upward in a toroidal motion. In reentrant bowls the swirling air entering the bowl spreads downward and outward into the undercut region, and then divides into a stream rising up the bowl sides and a stream flowing along the bowl base. Reentrant chambers generally produce higher swirl at the end of compression, and maintain a high swirl level further

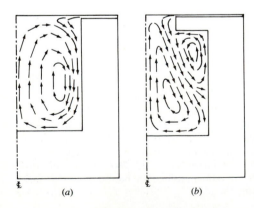

FIGURE 15-33
Flow pattern set up in diametral plane by squish-swirl interaction in (a) conventional and (b) reentrant bowl-in-piston combustion chambers. ₵ cylinder axis.[47]

into the expansion stroke.[47] Reentrant chambers usually achieve lower HC and smoke emissions and slightly lower bsfc, especially at retarded injection timings.

Square cavity chambers (see Fig. 15-32c) are also used with swirl to achieve low emissions in smaller-size DI diesel engines. The interaction between the swirl and the chamber corners produces additional turbulence which, with fuel injected into the corners as shown, achieves a more uniform mixture within the bowl.

The air flow field within bowl-in-piston combustion chambers when fuel injection occurs is highly complex. Certain generalizations hold: e.g., reducing the bowl diameter at a constant compression ratio increases the swirl levels in the bowl at TC [see Eq. (8.35) and the accompanying text] which decreases smoke and increases NO_x and HC emissions.[37] However, the squish-swirl interaction is difficult to unravel, especially with the off-center bowls often required due to the constraints on injector location caused by the valves. Figure 14-33 gives an example of such a flow. It shows velocity vectors and turbulence intensities in two orthogonal bowl-diametral planes within an off-center reentrant bowl as TC is approached in a small high-swirl DI engine. The off-center bowl location coupled with the swirl-squish interaction cause substantial asymmetry in the flow within the bowl.

15.6 SUPERCHARGED AND TURBOCHARGED ENGINE PERFORMANCE

The equations for power, torque, and mep in Sec. 2.14 show that these engine performance parameters are proportional to the mass of air inducted per cycle. This depends primarily on inlet air density. Thus the performance of an engine of given displacement can be increased by compressing the inlet air prior to entry to the cylinder. Methods for achieving higher inlet air density in the gas exchange processes—mechanical supercharging, turbocharging, and pressure-wave supercharging—are discussed in Sec. 6.8. The arrangements of the various practical supercharging and turbocharging configurations are shown in Fig. 6-37. Figures 1-11, 6-40, 6-43, 6-49, 6-53, and 6-58 show examples of the different devices used to achieve higher inlet air densities. In this section the effects of boosting air density on engine performance are examined. Spark-ignition and compression-ignition engines are dealt with separately. Power boosting via supercharging and/or turbocharging is common in diesel engines: few spark-ignition engines are turbocharged. Knock prevents the full potential of boosting from being realized in the latter type of engine. A more extensive discussion of turbocharged engine operation is provided by Watson and Janota.[48]

15.6.1 Four-Stroke Cycle SI Engines

The bmep of most production spark-ignition engines at wide-open throttle is knock-limited over part of the engine speed range (see Sec. 15.4.4). The compression ratio is usually set at a sufficiently high value so that some spark retard from

MBT timing is needed to avoid knock for the expected range of available fuel octane rating and sensitivity (see Fig. 15-17). The propensity of the end-gas to knock is increased by increases in end-gas temperature and pressure (see Sec. 9.6.2). Hence attempts to boost the output of a given size spark-ignition engine by an inlet air compression device that increases air pressure and temperature will aggravate the knock problem, since end-gas pressure and temperature will increase. However, the potential advantages of power boosting are significant. The higher output for a given displaced volume will decrease engine specific weight and volume (Sec. 2.11). Also, if the power requirements in a specific application (such as an automobile) can be met with either a naturally aspirated SI engine of a certain size or with a *smaller* size engine which is turbocharged to the same maximum power, the smaller turbocharged engine should offer better fuel economy at part load. At a given part-load torque requirement, the mechanical efficiency of the smaller turbocharged engine is higher, and if the gross indicated efficiencies of the engines are the same, the smaller engine will show a brake efficiency benefit. In practice, it proves difficult to realize much of this potential efficiency gain for the reasons described below.

While a naturally aspirated spark-ignition engine may have sufficient margin of safety relative to knock to allow modest inlet-air boost, any substantial air compression prior to cylinder entry will require changes in engine design and/or operating variables to offset the negative impact on knock. The variables which are adjusted to control knock in turbocharged SI engines are: compression ratio, spark retard from optimum, charge air temperature, and fuel/air equivalence ratio.† Figure 15-34 shows how the knock limits depend on charge pressure, temperature, fuel/air equivalence ratio and compression ratio for given octane rating fuels. The difference in boost achievable with the premium and the regular quality gasoline is significant, as expected (Sec. 9.6.3). Charge-air temperature has a strong influence on allowable boost levels: lowering the compressed air temperature prior to entry to the cylinder with a charge-air cooler allows a substantially higher compression ratio to be used at a given boost level, with a corresponding impact on engine efficiency.‡ The boost pressure benefits of the richer mixtures in Fig. 15-34a ($\phi = 1.1$ compared with 0.9) are largely due to the cooling effect of the additional fuel on the air charge. For example, Fig. 15-34b shows that, with a rich mixture and charge cooling to 60°C, a charge pressure of 1.5 atm can be utilized at optimum spark timing with a compression ratio of 8. Without charge cooling, the same charge pressure can only be used with a compression ratio of 6.[49]

In turbocharged SI engines, the knock limit is usually reached at spark timings retarded from the MBT optimum. Figure 15-35 shows the brake mean

---

† Valve timing changes are often made too. These are done primarily to improve low-speed torque where turbocharging has a limited impact.

‡ The turbocharged engine in Fig. 1-10 has an intercooler to reduce the inlet charge temperature.

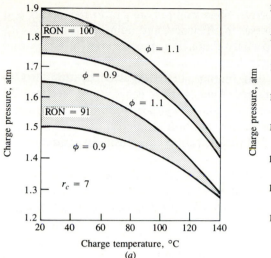

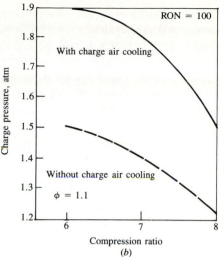

(a)

(b)

FIGURE 15-34
Dependence of SI engine knock limits on: (a) charge pressure, temperature, and equivalence ratio ϕ, with $r_c = 7$, 2500 rev/min, MBT timing, 91 and 100 research octane number fuel; (b) charge pressure and compression ratio, without and with (to 60°C) charge air cooling, 2500 rev/min, MBT timing, $\phi = 1.1$, 100 RON fuel.[49]

effective pressure achievable at a fixed compression ratio as a function of charge pressure and ignition timing with and without charge-air cooling. Additional retard allows higher boost pressures to be utilized; however, at a constant safety margin from the knock limit, the resulting gains in bmep decrease as retard is increased. To avoid an unnecessary fuel consumption penalty, retarded timing should only be used when the turbocharger does develop a high boost pressure.

The above discussion illustrates why turbocharged spark-ignition engines normally have lower compression ratios than naturally aspirated engines, use substantial mixture enrichment (up to $\phi \approx 1.3$) at high boost to cool the charge, often use an intercooler to reduce the charge-air temperature, and operate with

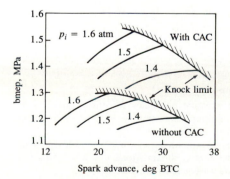

Spark advance, deg BTC

FIGURE 15-35
Brake mean effective pressure and knock limits for turbocharged SI engines as a function of spark advance and inlet pressure p_i (in atmospheres). 2500 rev/min, $r_c = 7$, $\phi = 1.1$, 99 RON fuel, without and with ($\Delta T = 45°C$) charge-air cooling.[49]

retarded timing at high boost pressures. Since compression ratio reductions and retarded ignition timings result in losses in efficiency, and unintended knock with high boost pressures would be especially damaging, precise control of ignition timing is critical. Most turbocharged SI engines now use a knock sensor and ignition-timing control system so that timing can be adjusted continuously to avoid knock without unnecessary retard. The sensor is usually an accelerometer which senses above-normal vibration levels on the cylinder head at the characteristic knock frequency. With a knock sensor, ignition timing can be automatically adjusted in response to changes in fuel octane rating and sensitivity, and ambient conditions.

Turbocharged SI engines where fuel is mixed with the air upstream or downstream of the compressor, using carburetors or fuel-injection systems, have been developed and used. Most modern turbocharged engines use port fuel injection. This provides easier electronic control of fuel flow, avoids filling most of the pressurized manifold volume with fuel-air mixture, and improves the dynamic response of the system by reducing fuel transport delays.

We now consider the performance of actual turbocharged spark-ignition engines. Examples of compressor outlet or boost pressure schedules as a function of speed at wide-open throttle for three turbocharged engines are shown in Fig. 15-36. The essential features of the curves are the same. Below about 1000 engine rev/min the turbocharger achieves negligible boost. Boost pressure then rises with increasing speed to 1.4 to 1.8 atm (absolute pressure) at about 2000 rev/min. Boost pressure then remains essentially constant with increasing engine speed. The rising portion of the curve is largely governed by the relative size of the turbine selected for a given engine. This is usually expressed in terms of the A/R ratio of the turbine—the ratio of the turbine's inlet casing or volute area A to the radius of the centroid of that area. Lower A/R values (smaller-capacity turbines) give a more rapid boost pressure rise with increasing speed; however, they give higher boost pressures at high engine speed, which is undesirable.[48, 50]

Avoidance of knock is the reason why boost must be limited at medium to

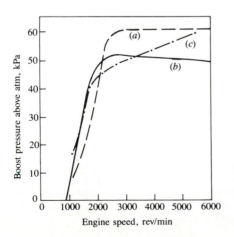

FIGURE 15-36
Boost pressure schedules for three turbocharged spark-ignition engines: (a) 3.8-dm³ V-6 engine, 86.4 mm stroke, $r_c = 8$;[50] (b) 2.2-dm³ four-cylinder engine, 92 mm stroke, $r_c = 8.1$;[51] (c) 2.32-dm³ four-cylinder engine, 80 mm stroke, $r_c = 8.7$.[52] All schedules are wastegate controlled.

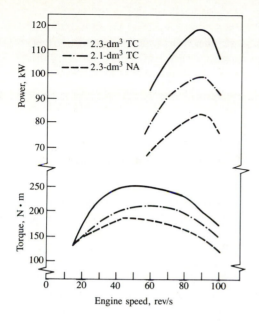

FIGURE 15-37
Power and torque as a function of engine speed for two turbocharged and one naturally aspirated four-cylinder spark-ignition engine. See Table 15.3.[52]

high engine speed; the details of this problem have already been discussed above. Even with the use of very rich mixtures and spark retard at WOT, lower compression ratios for turbocharged engines, and intercooling, knock avoidance requires that boost pressures (which would continue to rise with increasing engine speed in the absence of any control) be maintained approximately constant. This is normally achieved by reducing the exhaust flow through the turbine as speed increases by bypassing a substantial fraction of the exhaust around the turbine through the *wastegate* or flow control valve (see Sec. 6.8.4). A wastegate is a spring-loaded valve acting in response to the inlet manifold pressure on a controlling diaphragm. Although other methods of controlling boost can be used,[48] the wastegate is the most common. About 30 to 40 percent of the exhaust bypasses the turbine at maximum engine speed and load.

Figure 15-37 compares the performance of two turbocharged spark-ignition engines (four-cylinder, 2.1- and 2.3-dm³ displacement) with that of the base 2.3-dm³ engine in its naturally aspirated form. Table 15.3 gives details of these

TABLE 15.3
Turbocharged spark-ignition engine performance[52]

Type	2.1-dm³ TC	2.3-dm³ NA	2.3-dm³ TC/AC
Displacement, dm³	2.127	2.316	2.316
Bore × stroke, mm	92 × 80	96 × 80	96 × 80
Compression ratio	7.5	9.5	8.7
Maximum power, kW at rev/min	98 at 5400	83 at 5400	117 at 5300
Maximum torque, N·m at rev/min	210 at 3800	184 at 2800	250 at 2900
Maximum bmep, kPa	1241	998	1356

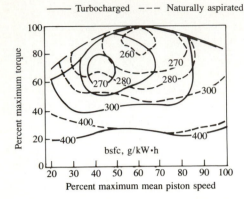

──── Turbocharged   ─ ─ ─ Naturally aspirated

FIGURE 15-38
Comparison of bsfc contours (in grams per kilowatt-hour) on performance maps of turbocharged and naturally aspirated versions of the same spark-ignition engine, scaled to the same maximum torque and mean piston speed.[53]

three engines. The 2.1-dm^3 turbocharged but *not* intercooled engine (which also does not have a knock sensor to control spark advance) requires a lower compression ratio and achieves less of a bmep gain than the 2.3-dm^3 turbocharged intercooled engine with its knock-sensor spark-advance control, which together permit use of a higher compression ratio. Turbocharging the naturally aspirated 2.3-dm^3 engine, with the modifications indicated, results in a 36 percent increase in maximum engine torque and a flatter torque-versus-speed profile.

The brake specific fuel consumption contours of an engine produced in both naturally aspirated and turbocharged versions are shown in Fig. 15-38. The data have been scaled to represent engines of different displaced volume but the same maximum engine torque. The smaller-displacement low-compression-ratio turbocharged engine ($r_c = 6.9$) shows a reduction in bsfc at low speed and part load due to improved mechanical efficiency. At high speed and load the larger-displacement naturally aspirated engine has an advantage in bsfc due to its higher compression ratio (8.2), less enrichment, and more optimum timing.[53] In a vehicle context, the low-speed part-load advantage of the smaller size but equal power turbocharged engine should result in an average fuel economy benefit relative to the larger naturally aspirated engine. This benefit has been estimated as a function of load. At full load the average efficiencies should be comparable; at half load, the turbocharged engine should show a benefit of about 10 percent, the benefit increasing as load is decreased.[49]

15.6.2 Four-Stroke Cycle CI Engines

The factors that limit turbocharged diesel engine performance are completely different to those that limit turbocharged spark-ignition engines. The output of naturally aspirated diesel engines is limited by the maximum tolerable smoke emission levels, which occur at overall equivalence ratio values of about 0.7 to 0.8. Turbocharged diesel engine output is usually constrained by stress levels in critical mechanical components. These maximum stress levels limit the maximum cylinder pressure which can be tolerated under continuous operation, though the thermal loading of critical components can become limiting too. As boost pressure is raised, unless engine design and operating conditions are changed, maximum pressures and thermal loadings will increase almost in proportion. In practice, the compression ratio is often reduced and the maximum fuel/air equivalence ratio must be reduced in turbocharged engines (relative to naturally aspirated engines) to maintain peak pressures and thermal loadings at acceptable levels. The fuel flow rate increases at a much lower rate than the air flow rate as boost pressure is increased. Limitations on turbocharged engine performance are discussed more fully by Watson and Janota.[48]

Small automotive indirect-injection (IDI) turbocharged engines are limited by structural and thermal considerations to about 130 atm maximum swirl- or pre-chamber pressure, 14 m/s maximum mean piston speed, and 860°C maximum exhaust temperature.[54] Smoke and NO_x emission standards are additional constraints. Figure 15-39 shows the full-load engine and turbocharger performance characteristics of a six-cylinder 2.38-dm³ displacement Comet V swirl-chamber automobile diesel engine. The maximum boost pressure is controlled by a poppet-valve-type wastegate to 0.75 bar above atmospheric. The fuel consumption map for this engine is shown in Fig. 15-40. Superimposed on the turbocharged engine map is the map for the base naturally aspirated swirl-chamber IDI engine of the same geometry and compression ratio ($r_c = 23$). The turbocharged engine has a maximum torque 46 percent higher and a maximum power 33 percent higher than the naturally aspirated engine. The best bsfc values are closely comparable.

The different methods of supercharging internal combustion engines were reviewed in Sec. 6.8. Turbocharging, mechanical supercharging with a Roots blower, and pressure wave supercharging with the Comprex are alternative methods of boosting the performance of a small automotive swirl-chamber IDI

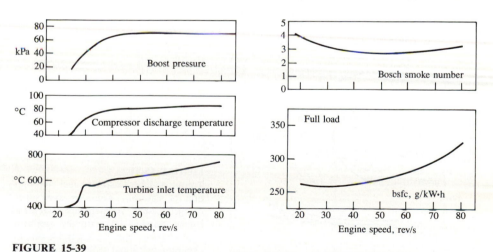

FIGURE 15-39
Engine and turbocharger characteristics of six-cylinder 2.38-dm³ swirl-chamber IDI automotive diesel engine at full load.[54]

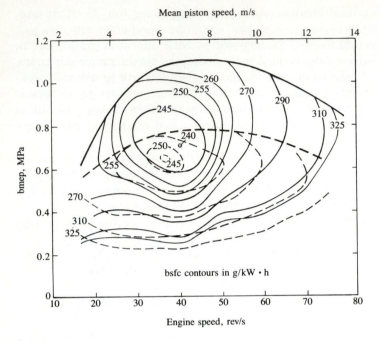

FIGURE 15-40
Fuel consumption map (bsfc in grams per kilowatt-hour) for turbocharged (——) and naturally aspirated (- - -) versions of 2.38-dm³ six-cylinder swirl-chamber IDI diesel engine.[54]

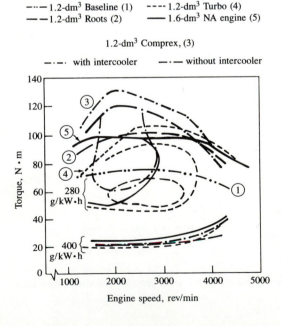

FIGURE 15-41
Torque and brake specific fuel consumption of naturally aspirated and supercharged 1.2-dm³ swirl-chamber IDI diesel engine. Baseline (1): naturally aspirated. Supercharged with (2) Roots blower; (3) Comprex (with and without intercooler); (4) turbocharger. Larger displacement 1.6-dm³ naturally aspirated engine (5).[55]

diesel engine. Figure 15-41 compares the torque and bsfc values obtained with each of these supercharging methods on a performance map for a 1.2-dm³ engine. Values for a 1.6-dm³ naturally aspirated IDI diesel engine are also shown. All three approaches achieve close to the desired maximum power of the 1.6-dm³ NA engine (40 kW at 4800 rev/min): e.g., 1.2-dm³ turbo, 41.2 kW at 4500 rev/min; 1.2-dm³ Comprex with intercooler, 42.3 kW at 3500 rev/min; 1.2-dm³ Roots, 37.6 kW at 4000 rev/min. The Comprex system produces the highest torque at low engine speeds, even under unsteady engine operating conditions. The density of the charge air determines the amount of charge, and hence the torque. Charge-air pressure and temperature for the three supercharging systems are shown in Fig. 15-42. The Comprex (here without an intercooler) must have the highest charge pressure because it has the highest charge temperature. Intercooling would be particularly effective in this case.[55]

Small high-speed high-swirl turbocharged direct-injection diesel engines (e.g., suitable for automobile or light-truck applications) have similar performance maps to those of equivalent IDI engines (Figs. 15-39 and 15-40). Maximum bmep values are closely comparable: usually slightly higher boost is required to offset the lower volumetric efficiency of the high-swirl-generating port and valve of the DI engine. Best bsfc values for the DI engine are usually about 15 percent lower than of comparable IDI engines (see Ref. 56).

The operating characteristics of larger medium-swirl turbocharged DI diesel engines are illustrated by the data shown in Fig. 15-43. The engine is a 12-dm³ displacement six-cylinder heavy-duty truck engine. The combustion chamber is similar to that shown in Fig. 15-32c, with a square combustion cavity and relatively low levels of swirl. The swirl is generated by a helical port in one of the two intake ports and a tangential port in the other in the four-valve cylinder head. Both the engine's operating map and the turbocharger compressor map with the boost pressure curve superposed are shown for two different compressor impellors. The adoption of the backward-vaned rake-type impellor compared to a more conventional design significantly increases low- and medium-speed per-

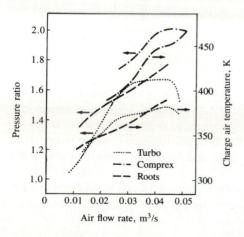

FIGURE 15-42
Charge pressure and temperature with the IDI diesel engine and different supercharging methods of Fig. 15-41.[55]

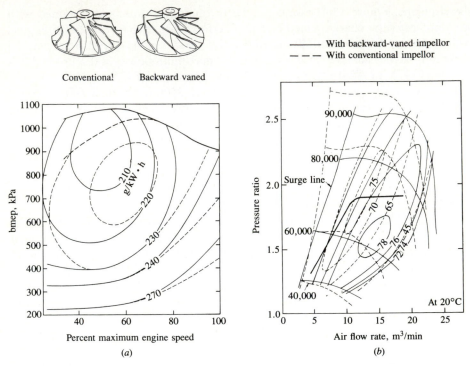

FIGURE 15-43
Performance characteristics of turbocharged 12-dm³ six-cylinder medium-swirl heavy-duty truck DI diesel engine, with two different compressor impellors: (a) fuel consumption maps; (b) compressor maps with full-load boost operating line for engine with backward-vaned impellor superposed. Bore = 135 mm, stroke = 140 mm, r_c = 16.[57]

formance by improving the compressor efficiency over the engine's boost pressure curve (Fig. 15-43b). A wastegate is then used to control the boost level at high engine speeds. The improvement in low-speed engine torque is apparent in Fig. 15-43a. The dependence of the maximum torque curve on both engine and turbocharger design details is clear. With boost pressure ratios limited to below 2, in the absence of air-charge cooling, maximum bmep values of 1.1 MPa are typical of this size and type of diesel engine.

With structurally more rugged component designs, aftercooled turbocharged medium-speed diesel engines with swirl in this cylinder size range can utilize higher boost and generate much higher bmep. Wastegate control of boost is no longer required. Figure 15-44 shows the performance characteristics of a V-8 cylinder engine with its compressor map and full-load boost characteristic. This turbocharged intercooled engine achieves a maximum bmep of about 1.5 MPa and bsfc below 200 g/kW·h between the maximum torque speed and rated power. Boost pressure at full load increases continuously over the engine speed range.[58]

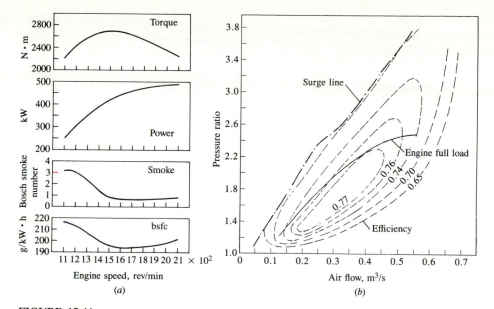

FIGURE 15-44
Performance characteristics of medium-speed turbocharged aftercooled DI diesel engine. (*a*) Torque, power, smoke number, and bsfc for V twelve-cylinder version. (*b*) Compressor characteristics and engine full-load line for V-8 cylinder version. Bore = 128 mm, stroke = 140 mm, r_c = 15.[58]

Examples of values of combustion-related parameters for this type of engine over the load range at its maximum rated speed are shown in Fig. 15-45 for a 14.6-dm^3 six-cylinder turbocharged aftercooled DI diesel engine with a boost pressure ratio of 2 at rated power. The ignition delay decreases to about 10° (0.9 ms at 1800 rev/min) as load is increased. The bmep at 100 percent rated load at this speed is 1.2 MPa. Exhaust temperature increases substantially with increasing load: maximum cylinder pressure increases to about 10 MPa at the rated load. In this particular study it was found that these operating parameters were relatively insensitive to fuel variations. The cross-hatched bands show data for an additional nine fuels of varying sulfur content, aromatic content, 10 and 90 percent distillation temperatures.[59]

Higher outputs can be obtained with two-stage turbocharged aftercooled diesel engines, the arrangement shown in Fig. 6-37*d*. The performance character-istics of such a high bmep (1.74 MPa) six-cylinder engine of 14-dm^3 displacement are shown in Fig. 15-46. The high air flow requires an overall pressure ratio of 3 at sea level ambient conditions (rising to 4 at 3658 m altitude). This was obtained at lower cost with two turbochargers in series than with a multistage single tur-bocharger. At rated conditions, the maximum cylinder pressure is 12.7 MPa and the maximum mean piston speed is 10.6 m/s.

Additional gains in efficiency with these heavy-duty automotive diesel engines can be achieved with turbocompounding: some of the available energy in the exhaust gases is captured in a turbine which is geared directly to the engine

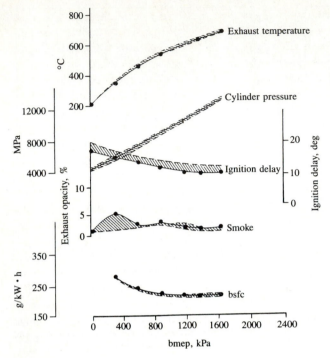

FIGURE 15-45
Operating parameters of 14.6-dm³ six-cylinder turbocharged aftercooled DI diesel engine as a function of load at maximum rated speed of 1800 rev/min. Maximum rated power = 261 kW at bmep = 1192 kPa. Points: standard diesel fuel. Shaded band: nine fuels of varying sulfur content, aromatic content, 10 and 90 per cent distillation temperatures.[59]

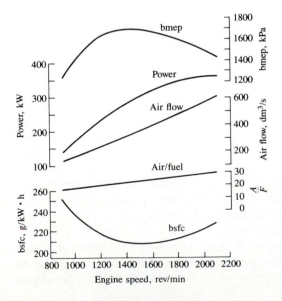

FIGURE 15-46
Operating characteristics of 14-dm³ six-cylinder two-stage turbocharged aftercooled quiescent-chamber DI diesel engine. Maximum bmep = 1.74 MPa. Boost pressure ratio at rated power = 3. Bore = 140 mm, stroke = 152 mm.[60]

drive shaft. The above discussion indicates that typical turbocharged DI diesel engines achieve bsfc levels of 210 to 220 g/kW · h (brake fuel conversion efficiencies of 0.4 to 0.38). With the increased cylinder pressure capability, higher fuel-injection pressures, and lower-temperature aftercooling of the above higher bmep engines, bsfc values of 200 g/kW · h (0.42 brake efficiency) or lower can be achieved. With turbocompounding, bsfc values can be reduced another 5 to 6 percent to about 180 g/kW · h, or a brake efficiency of 0.47, at rated power.[61]

The largest four-stroke cycle DI diesel engines are used for marine propulsion. An example is the Sulzer 400 mm bore 480 mm stroke engine which produces 640 kW per cylinder at 580 rev/min ($\bar{S}_p = 9.3$ m/s). Very high bmep levels (2.19 MPa) are achieved at maximum continuous rated power through progress in turbocharger design and engine improvements which allow higher maximum cylinder pressures. These, combined with optimization of gas exchange and combustion processes, achieve bsfc values of 185 to 190 g/kW · h (45 to 46 percent brake efficiency).[62]

Many diesel system concepts are being examined which promise even higher output and/or efficiency. Variable-geometry turbocharger-turbine nozzles improve utilization of exhaust gas available energy at low engine speeds. The hyperbar turbocharging system—essentially a combination of a diesel engine with a free-running gas turbine (a combustion chamber is placed between the engine and the turbocharger turbine)—has the potential of much higher bmep. Diesel systems with thermally insulated combustion chambers which reduce heat losses and increase the available exhaust energy have the potential for improving efficiency and for increasing power through additional exhaust energy recovery in devices such as compounded turbines and exhaust-heated Rankine cycle systems.[48]

15.6.3 Two-Stroke Cycle SI Engines

The two-stroke cycle spark-ignition engine in its standard form employs sealed crankcase induction and compression of the fresh charge prior to charge transfer, with compression and spark ignition in the engine cylinder after charge transfer. The fresh mixture must be compressed to above exhaust system pressures, prior to entry to the cylinder, to achieve effective scavenging of the burned gases. Two-stroke cycle scavenging processes were discussed in Sec. 6.6. The two-stroke spark-ignition engine is an especially simple and light engine concept and finds its greatest use as a portable power source or on motorcycles where these advantages are important. Its inherent weakness is that the fresh fuel-air mixture which short-circuits the cylinder directly to the exhaust system during the scavenging process constitutes a significant fuel consumption penalty, and results in excessive unburned hydrocarbon emissions.

This section briefly discusses the performance characteristics of small crankcase compression two-stroke cycle SI engines. The performance characteristics (power and torque) of these engines depend on the extent to which the displaced volume is filled with fresh mixture, i.e., the charging efficiency [Eq. (6.24)]. The fuel consumption will depend on both the trapping efficiency [Eq. (6.21)] and the charging efficiency. Figure 15-47a shows how the trapping efficiency η_{tr} varies with increasing delivery ratio Λ at several engine speeds for a two-cylinder 347-cm³ displacement motorcycle crankcase compression engine. The delivery ratio increases from about 0.1 at idle conditions to 0.7 to 0.8 at wide-open throt-

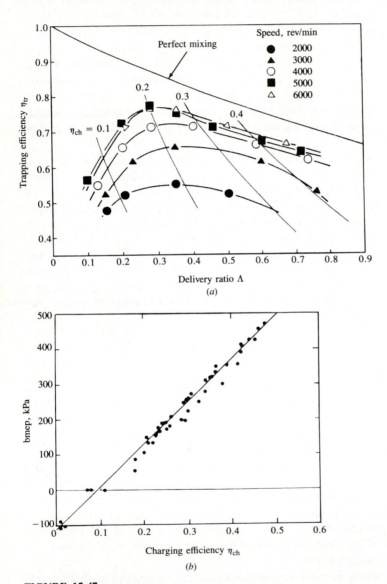

FIGURE 15-47

(a) Trapping and charging efficiencies as a function of the delivery ratio. (b) Dependence of brake mean effective pressure on fresh-charge mass defined by charging efficiency. Two-cylinder 347-cm³ displacement two-stroke cycle spark-ignition engine.[63]

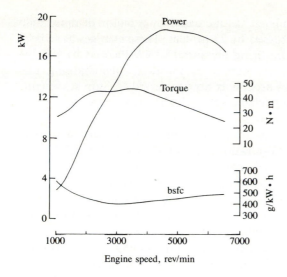

FIGURE 15-48
Performance characteristics of a three-cylinder 450-cm^3 two-stroke cycle spark-ignition engine. Maximum bmep = 640 kPa. Bore = 58 mm, stroke = 56 mm.[64]

tle. Lines of constant charging efficiency η_{ch} [which equals $\Lambda\eta_{tr}$; see Eq. (6.25)] are shown. Figure 15-47b shows bmep plotted against these charging efficiency values and the linear dependence on fresh charge mass retained is clear.

Performance curves for a three-cylinder 450-cm^3 two-stroke cycle minicar engine are shown in Fig. 15-48. Maximum bmep is 640 kPa at about 4000 rev/min. Smaller motorcycle engines can achieve slightly higher maximum bmep at higher speeds (7000 rev/min). Fuel consumption at the maximum bmep point is about 400 g/kW·h. Average fuel consumption is usually one-and-a-half to two times that of an equivalent four-stroke cycle engine.

CO emissions from two-stroke cycle engines vary primarily with the fuel/air equivalence ratio in a manner similar to that of four-stroke cycle engines (see Fig. 11-20). NO$_x$ emissions are significantly lower than from four-stroke engines due to the high residual gas fraction resulting from the low charging efficiency. Unburned hydrocarbon emissions from carbureted two-stroke engines are about five times as high as those of equivalent four-stroke engines due to fresh mixture short-circuiting the cylinder during scavenging. Exhaust mass hydrocarbon emissions vary approximately as $\Lambda(1 - \eta_{tr})\phi$, where ϕ is the fuel/air equivalence ratio.[63]

15.6.4 Two-Stroke Cycle CI Engines

Large marine diesel engines (0.4 to 1 m bore) utilize the two-stroke cycle. These low-speed engines with relatively few cylinders are well suited to marine propulsion since they are able to match the power/speed requirements of ships with simple direct-drive arrangements. These engines are turbocharged to achieve high brake mean effective pressures and specific output. The largest of these engines can achieve brake fuel conversion efficiencies of up to 54 percent. An example of a large marine two-stroke engine is shown in Fig. 1-24. Over the past 25 years the output per cylinder of such engines has increased by a factor of more than two, and fuel consumption has decreased by 25 percent. These changes have been achieved by increasing the maximum firing pressure to 13 MPa, and by refining critical engine processes such as fuel injection, combustion, supercharging, and scavenging. The uniflow-scavenging process is now preferred to loop scavenging since it achieves higher scavenging efficiency at high stroke/bore ratios and allows increases in the expansion stroke.[62]

The performance characteristics of a 580 mm bore Sulzer two-stroke marine diesel engine with a stroke/bore ratio of 2.9 are shown in Fig. 15-49. The solid lines show the standard turbocharged engine characteristics. The rated speed for the engine is 125 rev/min, corresponding to a maximum mean piston speed of 7.2 m/s. The rated bmep is 1.66 MPa. The minimum bsfc is 175 g/kW · h which equals a brake fuel conversion efficiency of 48 percent. For larger lower-speed engines, the efficiency is higher. The dashed lines show how the performance of this engine can be improved by turbocompounding. A proportion of the engine's exhaust flow, at loads higher than 50 percent, is diverted from the turbocharger inlet to a separate turbine coupled to the engine power takeoff gear via an epicyclic speed-reduction gear and hydraulic coupling. The additional power recovered in this manner from the engine exhaust flow improves bsfc by 5 g/kW · h. At part load, when the full exhaust flow passes through the turbocharger, an efficiency gain is also obtained, due to the higher scavenging pressure (and therefore increased cylinder pressure) obtained with the full exhaust flow.

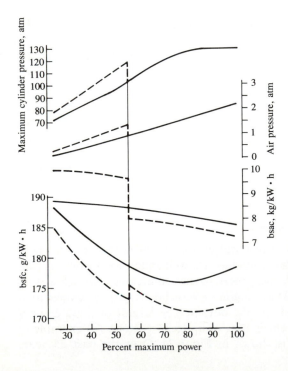

FIGURE 15-49
Performance characteristics of large marine two-stroke cycle uniflow-scavenged DI diesel engine. Bore = 580 mm, stroke/bore = 2.9, maximum rated speed = 125 rev/min (mean piston speed = 7.2 m/s), bmep (at rated power) = 1.66 MPa. Solid line: standard turbocharged configuration. Dashed lines: parallel turbocompounded configuration at greater than 50 percent load. bsac: brake specific air consumption.[62]

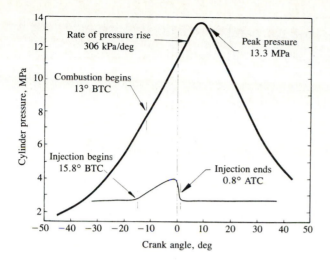

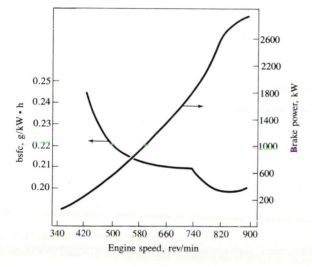

FIGURE 15-50
Injection, combustion, and performance characteristics of intermediate-size turbocharged two-stroke cycle uniflow-scavenged DI diesel engine. Bore = 230.2 mm, stroke = 279.4 mm and $r_c = 16$. Shallow dish-in-piston combustion chamber with swirl. At maximum rated power at 900 rev/min, bmep = 0.92–1.12 MPa depending on application.[65]

Both two-stroke and four-stroke cycle diesel engines of intermediate size (200 to 400 mm bore) are used in rail, industrial, marine, and oil drilling applications. The performance characteristics of a turbocharged two-stroke cycle uniflow-scavenged DI diesel engine (similar to the engine in Fig. 1-5), with 230.2 mm bore, 279.4 mm stroke, and a compression ratio of 16, are shown in Fig. 15-50. Combustion in the shallow dish-in-piston chamber with swirl occurs smoothly yielding a relatively low rate of pressure rise. The pressure curve shown with peak pressure of 13.3 MPa is for full-load operation. The bmep at rated power at 900 rev/min is 0.92 to 1.12 MPa depending on application. The maximum mean piston speed is 8.4 m/s. The bsfc of 200 g/kW · h corresponds to $\eta_{f,b} = 0.42$.

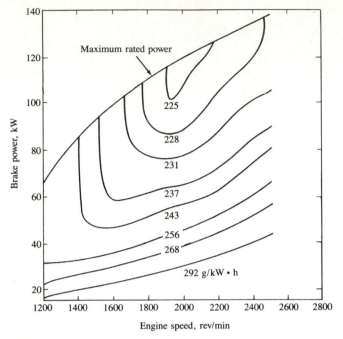

FIGURE 15-51
Brake power and specific fuel consumption (grams per kilowatt-hour) map of four-cylinder 3.48-dm³ uniflow-scavenged two-stroke cycle DI diesel engine. Engine turbocharged at mid and high loads; Roots blown at low loads. Maximum boost pressure ratio = 2.6. Bore = 98.4 mm, stroke = 114.3 mm, r_c = 18.[66]

Smaller turbocharged two-stroke cycle DI diesel engines also compete with four-stroke cycle engines in the marine, industrial, and construction markets. The fuel consumption map of such a four-cylinder 3.48-dm³ displacement uniflow-scavenged two-stroke cycle diesel engine is shown in Fig. 15-51. The engine uses a Roots blower to provide the required scavenging air pressure for starting and light-load operation. At moderate and high loads the turbocharger supplies sufficient boost and the blower is not needed; the blower is unloaded (air flow is bypassed around the blower) under these conditions. The engine generates 138 kW at its rated speed of 2500 rev/min (mean piston speed of 9.5 m/s) and a maximum bmep of 951 kPa at 1500 rev/min. The best bsfc is 225 g/kW·h and the maximum boost pressure ratio is 2.6.

15.7 ENGINE PERFORMANCE SUMMARY

The major performance characteristics of the spark-ignition and compression-ignition engines described in previous sections of this chapter are summarized here to highlight the overall trends. Table 15.4 lists the major design features of these engines, the bmep at maximum engine torque, bmep and the value of the mean piston speed $\bar{S}_p$ at maximum rated power, and the minimum value of bsfc

TABLE 15.4
Performance of representative engines in different categories

Engine type†	Bore, mm	Stroke, mm	Stroke/bore	r_c	Volume per cylinder, dm³	Number of cylinders	Maximum torque		Rated maximum power				Maximum efficiency		Reference
							bmep, kPa	Speed, rev/min	bmep, kPa	Speed, rev/min	Boost pressure ratio	$\bar{S}_p$, m/s	bsfc, g/kW·h	$\eta_{f,b}$	
SI/4S/NA	96.8	86	0.88	8.6	0.632	6	910	2500	750	4300	—	12.3			1
SI/4S/NA	84.5	88	1.04	8.5	0.494	4	966	2800	767	5200	—	15.3			67
SI/4S/NA	86*	86*	1*	8.5*	0.5	4	910	3500	758	5000	—	14.3	274	0.30	13
SI/4S/NA	96	80	0.83	9.5	0.579	4	998	2800	796	5400	—	14.4			52
SI/4S/TC	92	80	0.87	7.5	0.532	4	1241	3800	1024	5400	1.6*	14.4			52
SI/4S/TCAC	96	80	0.83	8.7	0.579	4	1356	2900	1144	5300	1.6	14.1			52
SI/2S/C	58	56	0.97		0.144	3	654	3500	575	4500		8.4	~400*	~0.2	64
SI/2S/C	64	54	0.84		0.174	2	686	7000	590	8200		14.8	~340*	~0.24	63
IDI/4S/NA	76.5	86.4	1.13	23	0.397	5	850	3100	670	4800	—	13.8	280	0.30	35
IDI/4S/NA	84	82	0.98	22	0.454	4	675	2000	502	5000	—	13.7			4
IDI/4S/NA	102	100	0.98	19	0.817	4	848	2200	743	3500	—	11.7	251	0.34	46
IDI/4S/TC	76.5	86.4	1.13	23	0.397	6	1080	2400	840	4800	1.7	13.8	240	0.35	54
DI/4S/NA	76.5	80	1.05	18.5	0.368	4	735	2800	600	5000	—	13.3	246	0.34	34
DI/4S/NAA	102	100	0.98	18	0.817	8	784	2000	682	3200	—	10.7	220	0.39	33
DI/4S/NA	102	100	0.98	17	0.817	4	886	2200	782	3500	—	11.7	221	0.38	46
DI/4S/NA	115	135	1.17	16	1.40	6	851	1400	777	2700	—	12.2	204	0.42	3
DI/4S/NA	135	140	1.04		2.00	6	862	1400	763	2500	—	11.7			57
DI/4S/TC	115	135	1.17		1.40	6	1098	1500	941	2500	—	11.2	203	0.42	3
DI/4S/TCAC	115	135	1.17		1.40	6	1344	1600	1240	2300	—	10.4			3
DI/4S/TCAC	128	140	1.09	15	1.8	6–16	1560	1500	1280	2100	2.5	9.8	195	0.43	58
DI/4S/TC	135	140	1.04	16	2.00	6	1087	1300	911	2300	1.9	10.7	210	0.40	57
DI/4S/2TCAC	140	152	1.09		2.33	6	1740	1400	1445	2100	3	10.6	207	0.41	60
DI/4S/TCAC	400	480	1.20		60.3	6–18			2190	580		9.3	185	0.46	62
DI/2S/TC	98.4	114.3	1.16	18	0.870	3, 4, 6	1065	1500	952	2500	2.6	9.5	226	0.37	66
DI/2S/TC	230	279.4	1.21	16	11.6	8–20			920–1122	900	2.8	8.4	200	0.42	65
DI/2S/TCAC	380–840	1100–2900	2.9–3.4		125–1607	4–12			1660	196–90	3.5	7.2	180–160	0.47–0.53	62

† Engine type: SI = spark-ignition; IDI = indirect-injection compression-ignition; DI = direct-injection compression-ignition; 4S = four-stroke; 2S = two-stroke; NA = naturally aspirated; NAA = NA and air-cooled; C = crankcase compression of scavenging mixture; TC = turbocharged; TCAC = turbocharged and aftercooled; 2TC = two-stage turbocharged.
* Denotes estimated value.

and the corresponding brake fuel conversion efficiency. It should be stressed that there are many different engine configurations and uses, and that for each of these there are variations in design and operating characteristics. However, these representative values of performance parameters illustrate the following trends:

1. Within a given category of engines (e.g., naturally aspirated four-stroke SI engines) the values of maximum bmep, and bmep and $\bar{S}_p$ at maximum rated power, are closely comparable. Within an engine category where the range in size is substantial, there is an increase in maximum bmep and a decrease in minimum bsfc as size increases due to the decreasing relative importance of friction and heat loss per cycle. There is also a decrease in $\bar{S}_p$ at maximum power as engine size increases. Note the higher bmep of naturally aspirated SI engines compared to equivalent NA diesels due to the fuel-rich operation of the former at wide-open throttle.

2. Two-stroke cycle spark-ignition engines have significantly lower bmep and higher bsfc than four-stroke cycle SI engines.

3. The effect of increasing inlet air density by increasing inlet air pressure increases maximum bmep values substantially. Turbocharging with after-cooling gives increased bmep gains relative to turbocharging without after-cooling at the same pressure level. The maximum bmep of turbocharged SI engines is knock-limited. The maximum bmep of turbocharged compression-ignition engines is stress-limited. The larger CI engines are designed to accept higher maximum cylinder pressures, and hence higher boost.

4. The best efficiency values of modern automobile SI engines and IDI diesel engines are comparable. However, the diesel has a significant advantage at lower loads due to its low pumping work and leaner air/fuel ratio. Small DI diesels have comparable (or slightly lower) maximum bmep to equivalent IDI diesels. The best bsfc values for DI diesels are 10 to 15 percent better, however.

5. In the DI diesel category (which is used over the largest size range—less than 100 mm bore to almost 1 m), maximum bmep and best brake fuel conversion efficiency steadily improve with increasing engine size due to reduced impact of friction and heat loss per cycle, higher allowable maximum cylinder pressure so higher boost can be used, and (additionally in the larger engines) through turbocompounding.

PROBLEMS

15.1. The schematics show three different four-stroke cycle spark-ignition engine combustion chambers. A and B are two-valve engines, C is a four-valve engine (two inlet valves which open simultaneously, two exhaust valves). Dimensions in millimeters are indicated. A and C have normal inlet ports and do not generate any swirl, B has a helical inlet port and generates substantial swirl. Spark plug locations are indicated. All three engines operate at the same speed (3000 rev/min), with the same inlet mixture composition, temperature, and pressure, and have the same displaced volume.

(a) Rank the chambers 1, 2, 3 in the order of their volumetric efficiency (1 = highest η_v).

(b) Rank the chambers in order (1, 2, 3) of their flame frontal area (1 = highest) when the mass fraction burned is about 0.2 and the piston is at TC.

(c) Given this relative flame front area ranking, discuss whether the ranking by mass burning rate dm_b/dt will be different from the flame area ranking.

(d) Briefly discuss the knock implications of these three chamber designs. Which is likely to have the worst knock problem?

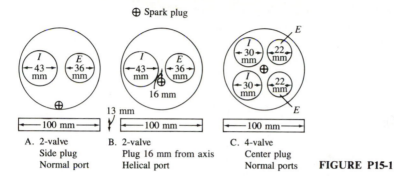

A. 2-valve
   Side plug
   Normal port

B. 2-valve
   Plug 16 mm from axis
   Helical port

C. 4-valve
   Center plug
   Normal ports

FIGURE P15-1

15.2. Figures 15-23 and 15-10 show the variation in brake specific fuel consumption (bsfc) for a swirl-chamber IDI automobile diesel (D) and a conventional automobile spark-ignition (SI) engine as a function of load and speed, respectively. From these graphs determine, and then plot, brake fuel conversion efficiency: (1) as a function of speed at *full load* and (2) as a function of load at a mid-speed of 2500 rev/min. Both engines are naturally aspirated. Assume the engine details are:

	Compression ratio	Equivalence ratio range	Displacement, dm³
Diesel	22	0.3–0.8	2.3
SI engine	9	1.0–1.2	1.6

(a) List the major engine design and operating variables that determine brake fuel conversion efficiency.

(b) Explain briefly the reasons for the shapes of the curves you have plotted and the relative relationship of the D and SI curves.

(c) At 2500 rev/min, estimate which engine will give the higher maximum brake power.

15.3. The diesel system shown in the figure consists of a multicylinder reciprocating diesel engine, a turbocharger (with a compressor C and turbine T_T mechanically connected to each other), an intercooler (I), and a power turbine (T_P) which is geared to the engine drive shaft. The gas and fuel flow paths and the gas states at the numbered points are shown. You can assume that the specific heat at constant pressure c_p of the gas throughout the entire system is 1.2 kJ/kg·K and $\gamma = c_p/c_v = 1.333$. The engine operates at 1900 rev/min. The fuel has a lower heating value of 42 MJ/kg of fuel.

(a) What is the power (in kilowatts) which the turbocharger turbine (T_T) must produce? What is the gas temperature at exit to the turbocharger turbine?

(b) What is the power turbine power output?

(c) The heat losses in the engine are 15 percent of the fuel's chemical energy ($\dot{m}_f Q_{LHV}$). Find the engine power output, the total system power output, and the total system brake fuel conversion efficiency (friction effects in the engine and power turbine are internal to these devices and do not need to be explicitly evaluated).

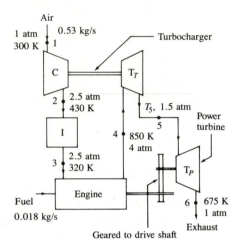

FIGURE P15-3

15.4. The attached graph shows how the brake power and specific fuel consumption of a four-stroke cycle single-cylinder spark-ignition engine vary with the fuel/air equivalence ratio at wide-open throttle. It also shows how the following efficiencies vary with equivalence ratio:

The volumetric efficiency: η_v

The mechanical efficiency: η_m [Eq. (2.17)]

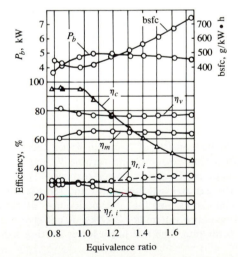

FIGURE P15-4

The combustion efficiency: η_c [Eq. (3.27)]
The indicated fuel conversion efficiency: $\eta_{f,i}$ [Eq. (2.23)]
The indicated thermal conversion efficiency: $\eta_{t,i}$ [Eq. (3.31)]

(a) Derive a relation between the variables $\eta_{f,i}$, η_c, and $\eta_{t,i}$.
(b) Derive an equation which relates the brake power P_b to η_v, η_m, η_c, $\eta_{t,i}$, and any other engine and fuel parameters required.
(c) Explain briefly why the variations of η_v, η_m, η_c, $\eta_{f,i}$, $\eta_{t,i}$ with equivalence ratio in the figure have the form shown (e.g., why the parameter is approximately constant, or has a maximum/minimum, or decreases/increases with increasing richness or leanness, etc.).

15.5. The diagram shows the layout of a low heat loss turbocharged turbocompounded diesel engine. The engine and exhaust system is insulated with ceramics to reduce heat losses to a minimum. Air flows steadily at 0.4 kg/s and atmospheric conditions into the compressor C, and exits at 445 K and 3 atm. The air is cooled to 350 K in the intercooler I. The specific heat of air, c_p, is 1 kJ/kg·K. In the reciprocating diesel engine, the fuel flow rate is 0.016 kg/s, the fuel heating value is 42.5 MJ/kg, and the heat lost through the ceramic walls is 60 kW.

The exhaust gases leave the reciprocating engine at 1000 K and 3 atm, and enter the first turbine T_A, which is mechanically linked to the compressor. The pressure between the two turbines is 1.5 atm. The second turbine T_B is mechanically coupled to the engine drive shaft and exhausts to the atmosphere at 800 K. The specific heat of exhaust gases, c_p, is 1.1 kJ/kg·K.

(a) Analyze the reciprocating diesel engine E and determine the indicated power obtained from this component of the total system. If the engine mechanical efficiency is 0.9 what is the brake power obtained from component E?
(b) Determine the power obtained from the power turbine T_B.
(c) Determine the total brake power obtained from the complete engine system and the fuel conversion efficiency of the system. You can neglect mechanical losses in the coupling between the power turbine and the engine drive shaft.

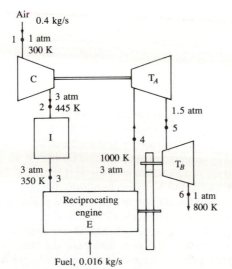

FIGURE P15-5

15.6. New automobile spark-ignition engines employ "fast-burn technology" to achieve an improvement in fuel consumption and reductions in hydrocarbon (HC) and oxides of nitrogen (NO_x) emissions. This question asks you to explain the experimental data which shows that faster-burning combustion chambers do provide these benefits relative to more moderate burn-rate chambers.

(a) Figure 9-36b shows the effect of increasing the percent of the exhaust gas recycled to the intake (for NO_x control) in a moderate burn-rate engine at constant speed and load, stoichiometric air/fuel ratio, with timing adjusted for maximum brake torque at each condition. COV_{imep} is the standard deviation in imep divided by the average imep, in percent. The different types of combustion are: *misfire, partial burn, slow burn, normal burn,* defined in Sec. 9.4.3. Frequency is percent of cycles in each of these categories. Use your knowledge of the spark-ignition engine flame-propagation process and HC emission mechanism to explain these trends in COV_{imep}, HC, and frequency as EGR is increased.

(b) The fast-burn combustion chamber uses two spark plugs and generates swirl inside the chamber by placing a vane in the inlet port to direct the air to enter the chamber tangentially. The swirl angular velocity in the cylinder at the end of intake is six times the crankshaft angular velocity. There is no swirl in the moderate burn-rate chamber which has a single spark plug and a relatively quiescent in-cylinder flow. The table shows spark timing, average time of peak pressure, average flame-development angle (0 to 10 percent mass burned) and rapid burning period (10 to 90 percent mass burned) for these two engines. Figures 11-29 and 15-9 show how the operating and emission characteristics of the fast burn and moderate burn-rate engines change as percent EGR is increased. Explain the reasons for the differences in these trends in COV_{imep}, bsfc (brake specific fuel consumption), and HC, and similarity in NO_x. The operating conditions are held constant at the same values as before.

	Fast burn	Moderate burn	
Spark timing	18°	40°	BTC
Crank angle for average p_{max}	15°	16°	ATC
0–10% burned	24°	35°	
10–90% burned	20°	50°	

15.7. Two alternative fuels, methanol and hydrogen, are being studied as potential future spark-ignition engine fuels which might replace gasoline (modeled by isooctane C_8H_{18}). The table gives some of the relevant properties of these fuels.

(a) For each fuel calculate the energy content per unit volume (in joules per cubic meter) of a stoichiometric mixture of fuel vapor and air at 1 atm and 350 K. The universal gas constant is 8314 J/kmol·K. What implications can you draw from these numbers regarding the maximum power output of an engine of fixed geometry operating with these fuels with stoichiometric mixtures?

(b) The octane rating of each fuel, and hence the knock-limited compression ratio of an engine optimized for each fuel, is different. Estimate the ratio of the maximum indicated mean effective pressure for methanol- and hydrogen-fueled engines to that of the gasoline-fueled engine, allowing for energy density effects at intake (at 1 atm and 350 K), at the knock-limited compression ratio for each fuel, for stoichiometric mixtures. You can assume that the fuel-air cycle results for isooctane apply also for methanol and hydrogen cycles to a good approximation, when the energy density is the same.

(c) The lean operating limit for the three fuels is different as indicated. Estimate the ratio of indicated fuel conversion efficiency for methanol and hydrogen at their lean limit and knock-limited compression ratio, relative to gasoline at its lean limit and knock-limited compression ratio, at the same inlet pressure (0.5 atm). Under these conditions, rank the fuel-engine combinations in order of decreasing power output.

	Gasoline (isooctane) C_8H_{18}	Methanol CH_3OH	Hydrogen H_2
Stoichiometric F/A	0.066	0.155	0.0292
Lower heating value, MJ/kg	44.4	20.0	120.1
Molecular weight of fuel	114	32	2
Molecular weight of stoichiometric mixture	30.3	29.4	21
Research octane number	95	106	~90
Knock-limited compression ratio	9	12	8
Equivalence ratio at lean misfire limit	0.9	0.8	0.6

15.8. Small-size direct-injection (DI) diesel engines are being developed as potential replacements for indirect-injection (IDI) or prechamber engines in automobile applications. Figures 10-1b and 10-2 show the essential features of these two types of diesel. The DI engine employs high air swirl, which is set up with a helical swirl-generating inlet port (Fig. 8-13). The injector is centrally located over the bowl-in-piston combustion chamber and the injector nozzle has four holes, one in each quadrant. The IDI engine (a Ricardo Comet swirl chamber), in contrast, has no swirl in the main chamber, but generates high velocities and a rotating flow in the prechamber during compression.

Figures 15-21 and 15-23 show performance maps for typical versions of these two types of engines. Bmep, brake mean effective pressure, is plotted against engine speed. Brake specific fuel consumption contours are shown with the numbers in grams per kilowatt-hour.

The heat-release-rate profiles for these two types of engine at a typical mid-load mid-speed point are shown versus crank angle in the sketch. $\dot{Q}$ has units of joules per second.

(a) Explain the reasons for the differences in shape and relative timing in the cycle of the heat-release-rate profiles.

(b) Suggest reasons for the differences (magnitude and shape) in the maximum bmep versus mean-piston-speed line for the DI and IDI engines.

(c) Evaluate the brake fuel conversion efficiency of each engine at its maximum efficiency point, and at 2000 rev/min and road load (road load is the power requirement to maintain a vehicle at constant speed; it is 2 bar bmep at 2000 rev/min). Explain the origin of the observed differences in efficiency at these two operating conditions.

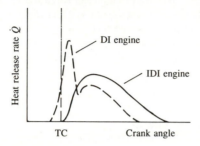

FIGURE P15-8

15.9. A four-stroke cycle naturally aspirated direct-injection diesel is being developed to provide 200 kW of power at the engine's maximum rated speed. Using information available in Chaps. 2, 5, and 15, on typical values of critical engine operating parameters at maximum power and speed for good engine designs, estimate the following:

(a) The compression ratio, the number of cylinders, the cylinder bore and stroke, and the maximum rated speed of an appropriate engine design that would provide this maximum power.

(b) The brake specific fuel consumption of this engine design at the maximum power operating point.

(c) The approximate increase in brake power that would result if the engine was turbocharged.

15.10. Natural gas (which is close to 100 percent methane, CH_4) is being considered as a spark-ignition engine fuel. The properties of methane and gasoline (assume the same properties as isooctane) and the engine details for each fuel are summarized below (ϕ is the fuel/air equivalence ratio).

	Natural gas	Gasoline
Composition	CH_4	C_8H_{18}
Heating value, MJ/kg	50.0	44.3
Research octane number	120	94
Compression ratio	14	8
Displaced volume, dm³	2	2
Lean misfire limit	$\phi = 0.5$	$\phi = 0.8$
Part-load equivalence ratio	$\phi = 0.6$	$\phi = 0.9$
Full-load equivalence ratio	$\phi = 1.1$	$\phi = 1.2$

As indicated in the table, the displaced volume of the engine is unchanged when the conversion for natural gas is made; however, the clearance height is reduced to increase the compression ratio.

(a) Estimate the ratio of the volumetric efficiency of the engine operating on natural gas to the volumetric efficiency with gasoline, at wide-open throttle and 2000 rev/min. Both fuels are in the gaseous state in the intake manifold.

(b) Estimate the ratio of the maximum indicated power of the engine operating with natural gas to the maximum power of the gasoline engine.

(c) Estimate the ratio of the gross indicated fuel conversion efficiency of the natural gas engine to that of the gasoline engine, at the part-load conditions given.

(d) Explain whether the NO, CO, and hydrocarbon specific emissions (grams of pollutant per hour, per unit indicated power) at part-load conditions of the natural gas engine will be higher, about the same, or lower than the NO, CO, and HC emissions from the gasoline engine. Explain briefly why.

You can assume that the fuel-air cycle results derived for isooctane-air mixtures are also appropriate for methane-air mixtures.

15.11. Spark-ignition and prechamber diesel engines are both used as engines for passenger cars. They must meet the same exhaust emission requirements. Of great importance are their emission characteristics when optimized for maximum power at wide-open throttle (WOT) and when optimized at cruise conditions for maximum efficiency.

(a) Give typical values for the equivalence ratio for a passenger car spark-ignition engine and a prechamber diesel optimized for maximum power at WOT and 2000 rev/min, and optimized for maximum efficiency at part load (bmep = 300 kPa) and 1500 rev/min. Briefly explain the values you have chosen.

(b) Construct a table indicating whether at these two operating conditions the specific emissions of CO, HC, NO_x, and particulates are low (L), medium (M), or high (H) relative to the other load point and to the other engine. Explain your reasoning for each table entry.

15.12. For a naturally aspirated four-stroke cycle diesel engine:

(a) Show from the definition of mean effective pressure that

$$\text{bmep} \propto \eta_m \eta_{f,i} \eta_v (F/A)$$

where   bmep = brake mean effective pressure
         η_m = mechanical efficiency
         $\eta_{f,i}$ = indicated fuel conversion efficiency
         η_v = volumetric efficiency
         F/A = fuel/air ratio

(b) Sketch carefully proportioned qualitative graphs of η_m, $\eta_{f,i}$, η_v, and $(F/A)/(F/A)_{\text{stoich}}$ versus speed N at full load, and explain the reasons for the shapes of the curves. Then explain why the maximum bmep versus speed curve has the shape shown in Fig. P15-12.

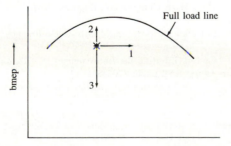

FIGURE P15-12

(c) The minimum brake specific fuel consumption point is indicated by the asterisk (*) in Fig. P15-12 (see Figs. 15-21 and 15-22). Explain why brake specific fuel consumption *increases* with (1) increasing speed, (2) increasing bmep, (3) decreasing bmep.

REFERENCES

1. Armstrong, D. L., and Stirrat, G. F.: "Ford's 1982 3.8L V6 Engine," SAE paper 820112, 1982.
2. "Engine Rating Code—Spark-Ignition," SAE Standard J245, in *SAE Handbook*.
3. Okino, M., Okada, K., and Abe, M.: "Isuzu New 8.4L Diesel Engine," SAE paper 850258, 1985.
4. Higashisono, M., Takeuchi, K., and Hara, H.: "The New Isuzu 1.8 Liter 4-Cylinder Diesel Engine for the United States Market," SAE paper 820116, *SAE Trans.*, vol. 91, 1982.
5. *General Motors Automotive Engine Test Code For Four Cycle Spark Ignition Engines*, 6th ed., 1975.
6. Heywood, J. B., Higgins, J. M., Watts, P. A., and Tabaczynski, R. J.: "Development and Use of a Cycle Simulation to Predict SI Engine Efficiency and NO_x Emissions," SAE paper 790291, 1979.
7. Heywood, J. B., and Watts, P. A.: "Parametric Studies of Fuel Consumption and NO Emissions of Dilute Spark-Ignition Engine Operation Using a Cycle Simulation," paper C98/79, in *Proceedings of Conference on Fuel Economy and Emissions of Lean Burn Engines*, Institution of Mechanical Engineers, London, 1979.
8. Quader, A. A.: "The Axially-Stratified-Charge Engine," SAE paper 820131, *SAE Trans.*, vol. 91, 1982.
9. Robison, J. A., and Brehob, W. M.: "The Influence of Improved Mixture Quality on Engine Exhaust Emissions and Performance," *J. Air Pollution Control Ass.*, vol. 17, no. 7, pp. 446–453, July 1967.
10. Thring, R. H., and Overington, M. T.: "Gasoline Engine Combustion—The High Ratio Compact Chamber," SAE paper 820166, *SAE Trans.*, vol. 91, 1982.
11. Hamburg, D. R., and Hyland, J. E.: "A Vaporized Gasoline Metering System for Internal Combustion Engines," SAE paper 760288, 1976.
12. Nakajima, Y., Sugihara, K., and Takagi, Y.: "Lean Mixture or EGR—Which is Better for Fuel Economy and NO_x Reduction?," paper C94/79, in *Proceedings of Conference on Fuel Economy and Emissions of Lean Burn Engines*, Institution of Mechanical Engineers, London, 1979.
13. Wade, W., and Jones, C.: "Current and Future Light Duty Diesel Engines and Their Fuels," SAE paper 840105, *SAE Trans.*, vol. 93, 1984.
14. Lavoie, G. A., and Blumberg, P. N.: "A Fundamental Model for Predicting Fuel Consumption, NO_x and HC Emissions of a Conventional Spark-Ignited Engine," *Combust. Sci. Technol.*, vol. 21, pp. 225–258, 1980.
15. Caton, J. A., Heywood, J. B., and Mendillo, J. V.: "Hydrocarbon Oxidation in a Spark-Ignition Engine Exhaust Port," *Combust. Sci. Technol.*, vol. 37, nos. 3 and 4, pp. 153–169, 1984.
16. Caton, J. A., and Heywood, J. B.: "Models for Heat Transfer, Mixing and Hydrocarbon Oxidation in an Exhaust Port of a Spark-Ignited Engine," SAE paper 800290, 1980.
17. Caris, D. F., and Nelson, E. E., "A New Look at High Compression Engines," *SAE Trans.*, vol. 67, pp. 112–124, 1959.
18. Kerley, R. V., and Thurston, K. W.: "The Indicated Performance of Otto-Cycle Engines," *SAE Trans.*, vol. 70, pp. 5–30, 1962.
19. Muranaka, S., Takagi, Y., and Ishida, T.: "Factors Limiting the Improvement in Thermal Efficiency of S.I. Engine at Higher Compression Ratio," SAE paper 870548, 1987.
20. Gruden, D. O.: "Combustion Chamber Layout for Modern Otto Engines," SAE paper 811231, 1981.
21. Barnes-Moss, H. W.: "A Designers Viewpoint," paper C343/73, in *Proceedings of Conference on Passenger Car Engines*, pp. 133–147, Institution of Mechanical Engineers, Conference publication 19, London, 1973.

22. Kuroda, H., Nakajima, Y., Sugihara, K., Takagi, Y., and Maranaka, S.: "Fast Burn with Heavy EGR Improves Fuel Economy and Reduces NO_x Emission," *JSAE Rev.*, no. 5, pp. 63–69, 1980.

23. Thring, R. H.: "The Effects of Varying Combustion Rate in Spark Ignited Engines," SAE paper 790387, 1979.

24. Harada, M., Kadota, T., and Sugiyama, Y.: "Nissan NAPS-Z Engine Realizes Better Fuel Economy and Low NO_x Emission," SAE paper 810010, 1981.

25. Poulos, S. G., and Heywood, J. B.: "The Effect of Chamber Geometry on Spark-Ignition Engine Combustion," SAE paper 830334, *SAE Trans.*, vol. 92, 1983.

26. Heywood, J. B.: "Combustion Chamber Design for Optimum Spark-Ignition Engine Performance," *Int. J. Vehicle Des.*, vol. 5, no. 3, pp. 336–357, 1984.

27. Novak, J. M., and Blumberg, P. N.: "Parametric Simulation of Significant Design and Operating Alternatives Affecting the Fuel Economy and Emissions of Spark-Ignited Engines," SAE paper 780943, *SAE Trans.*, vol. 87, 1978.

28. Amann, C. A.: "Control of the Homogeneous-Charge Passenger-Car Engine: Defining the Problem," SAE paper 801440, 1980.

29. Bell, A. G.: "The Relationship between Octane Quality and Octane Requirement," SAE paper 750935, 1975.

30. Leppard, W. R.: "Individual-Cylinder Knock Occurrence and Intensity in Multicylinder Engines," SAE paper 820074, 1982.

31. Ingamells, J. C., Stone, R. K., Gerber, N. H., and Unzelman, G. H.: "Effects of Atmospheric Variables on Passenger Car Octane Number Requirements," SAE paper 660544, *SAE Trans.*, vol. 75, 1966.

32. Gruden, D.: "Performance, Exhaust Emissions and Fuel Consumption of an IC Engine Operating with Lean Mixtures," paper C111/79, in *Proceedings of Conference on Fuel Economy and Emissions of Lean Burn Engines*, Institution of Mechanical Engineers, London, 1979.

33. Slezak, P. J., and Vossmeyer, W.: "New Deutz High Performance Diesel Engine," SAE paper 810905, 1981.

34. Neitz, A., and D'Alfonso, N.: "The M.A.N. Combustion System with Controlled Direct Injection for Passenger Car Diesel Engines," SAE paper 810479, 1981.

35. Sator, K., Buttgereit, W., and Sturzebecher, U.: "New 5- and 6-Cylinder VW Diesel Engines for Passenger Cars and Light Duty Trucks," SAE paper 790206, 1979.

36. Monaghan, M. L.: "The High Speed Direct Injection Diesel for Passenger Cars," SAE paper 810477, 1981.

37. Pischinger, R., and Cartellieri, W.: "Combustion System Parameters and Their Effect upon Diesel Engine Exhaust Emissions," SAE paper 720756, *SAE Trans.*, vol. 81, 1972.

38. Ball, W. F., and Hil, R. W.: "Control of a Light Duty Indirect Injection Diesel Engine for Best Trade-Off between Economy and Emissions," paper C122/82, in *Proceedings of Conference on Diesel Engines for Passenger Cars and Light Duty Vehicles*, Publication 1982-8, Institution of Mechanical Engineers, London, 1982.

39. Wade, W. R., Idzikowski, T., Kukkonen, C. A., and Reams, L. A.: "Direct Injection Diesel Capabilities for Passenger Cars," SAE paper 850552, 1985.

40. Greeves, G., and Wang, C. H. T.: "Origins of Diesel Particulate Mass Emission," SAE paper 810260, *SAE Trans.*, vol. 90, 1981.

41. Greeves, G., Khan, I. M., and Wang, C. H. T.: "Origins of Hydrocarbon Emissions from Diesel Engines," SAE paper 770259, *SAE Trans.*, vol. 86, 1977.

42. Greeves, G.: "Response of Diesel Combustion Systems to Increase of Fuel Injection Rate," SAE paper 790037, *SAE Trans.*, vol. 88, 1979.

43. Yu, R. C., and Shahed, S. M.: "Effects of Injection Timing and Exhaust Gas Recirculation on Emissions from a D.I. Diesel Engine," SAE paper 811234, *SAE Trans.*, vol. 90, 1981.

44. Khan, I. M., Greeves, G., and Wang, C. H. T.: "Factors Affecting Smoke and Gaseous Emissions from Direct Injection Engines and a Method of Calculation," SAE paper 730169, 1973.

45. Bassoli, C., Cornetti, G. M., and Cuniberti, F.: "IVECO Diesel Engine Family for Medium Duty Vehicles," SAE paper 820031, 1982.

46. Kawamura, H., Kihara, R., and Kinbara, M.: "Isuzu's New 3.27L Small Direct Injection Diesel," SAE paper 820032, 1982.

47. Arcoumanis, C., Bicen, A. F., and Whitelaw, J. H.: "Squish and Swirl-Squish Interaction in Motored Model Engines," *ASME Trans., J. Fluids Engng*, vol. 105, pp. 105–112, 1983.

48. Watson, N., and Janota, M. S.: *Turbocharging the Internal Combustion Engine*, Wiley-Interscience Publications, John Wiley, New York, 1982.

49. Hiereth, H., and Withalm, G.: "Some Special Features of the Turbocharged Gasoline Engine," SAE paper 790207, 1979.

50. Wallace, T. F.: "Buick's Turbocharged V-6 Powertrain for 1978," SAE paper 780413, *SAE Trans.*, vol. 87, 1978.

51. Allen, F. E., and Rinschler, G. L.: "Turbocharging the Chrysler 2.2 Liter Engine," SAE paper 840252, *SAE Trans.*, vol. 93, 1984.

52. Andersson, J., and Bengtsson, A.: "The Turbocharged and Intercooled 2.3 Liter Engine for the Volvo 760," SAE paper 840253, *SAE Trans.*, vol. 93, 1984.

53. Watson, N.: "Turbochargers for the 1980s—Current Trends and Future Prospects," SAE paper 790063, *SAE Trans*, vol. 88, 1979.

54. Grandinson, A., and Hedin, I.: "A Turbocharged Engine for a Growing Market," paper C119/82, in *Diesel Engines for Passenger Cars and Light Duty Vehicles*, Institution of Mechanical Engineers, Conference publication 1982–8, London, 1982.

55. Walzer, P., and Rottenkolber, P.: "Supercharging of Passenger Car Diesels," paper C117/82, in *Diesel Engines for Passenger Cars and Light Duty Vehicles*, Institution of Mechanical Engineers, Conference publication 1982–8, London, 1982.

56. Carstens, U. G., Isik, T., Biaggini, G., and Cornetti, G.: "Sofim Small High-Speed Diesel Engines—D.I. Versus I.D.I.," SAE paper 810481, 1981.

57. Okada, K., and Takatsuki, T.: "Isuzu's New 12.0L Turbocharged Diesel with Wastegate Boost Control for Fuel Economy," SAE paper 820029, 1982.

58. Schittler, M.: "MWM TBD 234 Compact High-Output Engines for Installation in Heavy Equipment and Military Vehicles," SAE paper 850257, 1985.

59. Barry, E. G., McCabe, L. J., Gerke, D. H., and Perez, J. M.: "Heavy-Duty Diesel Engine/Fuels Combustion Performance and Emissions—A Cooperative Research Program," SAE paper 852078, 1985.

60. Robinson, R. H., and Schnapp, J. P.: "Cummins NTC-475 Series Turbocharged Engine," SAE paper 820982, 1982.

61. Wilson, D. E.: "The Design of a Low Specific Fuel Consumption Turbocompound Engine," SAE paper 860072, 1986.

62. Lustgarten, G. A.: "The Latest Sulzer Marine Diesel Engine Technology," SAE paper 851219, 1985.

63. Tsuchiya, K., and Hirano, S.: "Characteristics of 2-Stroke Motorcycle Exhaust HC Emission and Effects of Air-Fuel Ratio and Ignition Timing," SAE paper 750908, 1975.

64. Uchiyama, H., Chiku, T., and Sayo, S.: "Emission Control of Two-Stroke Automobile Engine," SAE paper 770766, *SAE Trans.*, vol. 86, 1977.

65. Kotlin, J. J., Dunteman, N. R., Chen, J., and Heilenbach, J. W.: "The GM/EMD Model 710 G Series Turbocharged Two-Stroke Cycle Engine," ASME paper 85-DGP-24, 1985.

66. Fellberg, M., Huber, J. W., and Duerr, J. W.: "The Development of Detroit Diesel Allison's New Generation Series 53 Engines," SAE paper 850259, 1985.

67. Hisatomi, T., and Iida, H.: "Nissan Motor Company's New 2.0 Liter Four-Cylinder Gasoline Engine," SAE paper 820113, *SAE Trans.*, vol. 91, 1982.

UNIT
CONVERSION
FACTORS

This table provides conversion factors for common units of measure for physical quantities to the International System (SI) units. The conversion factors are presented in two ways: columns 2 and 3 give the conversion to the base or derived SI unit with the conversion factor as a number between one and ten with six or fewer decimal places, followed by the power of ten that the number must be multiplied by to obtain the correct value; columns 4 and 5 provide conversion to a recommended multiple or submultiple of the SI unit with the conversion factor given as a four-digit number between 0.1 and 1000.

1 To convert from	2 To	3 Multiply by	4 To	5 Multiply by
Area				
foot^2	m^2	$9.290\,304 \times 10^{-2}$	cm^2	929.0
inch^2	m^2	$6.451\,600 \times 10^{-4}$	cm^2	6.452
Energy, heat, and work				
Btu (International Table)	J	$1.055\,056 \times 10^{3}$	kJ	1.055
calorie (thermochemical)	J	$4.184\,000 \times 10^{0}$	J	4.184
erg	J	$1.000\,000 \times 10^{-7}$	μJ	0.1000
foot pound-force (ft · lbf)	J	$1.355\,818 \times 10^{0}$	J	1.356
horsepower-hour (hp · h)	J	$2.684\,520 \times 10^{6}$	MJ	2.685
kilowatt-hour (kW · h)	J	$3.600\,000 \times 10^{6}$	MJ	3.600
metre kilogram-force (m · kgf)	J	$9.806\,650 \times 10^{0}$	J	9.807

1 To convert from	2 To	3 Multiply by	4 To	5 Multiply by
Energy (specific, specific heat)				
Btu (IT)/lb	J/kg	$2.326\,000 \times 10^3$	kJ/kg	2.326
Btu (IT)/lb · °F	J/kg · K	$4.186\,800 \times 10^3$	kJ/kg · K	4.187
calorie (thermo.)/g	J/kg	$4.184\,000 \times 10^3$	kJ/kg	4.184
calorie (thermo.)/g · °C	J/kg · K	$4.184\,000 \times 10^3$	kJ/kg · K	4.184
Force				
dyne	N	$1.000\,000 \times 10^{-5}$	μN	10.00
kilogram-force	N	$9.806\,650 \times 10^0$	N	9.807
pound-force	N	$4.448\,222 \times 10^0$	N	4.448
Force per unit length (includes surface tension)				
dyne/centimeter	N/m	$1.000\,000 \times 10^{-3}$	mN/m	1.000
pound-force/inch	N/m	$1.751\,268 \times 10^2$	N/m	175.1
pound-force/foot	N/m	$1.459\,390 \times 10^1$	N/m	14.59
Fuel consumption (economy)				
pound/horsepower-hour	kg/J	$1.689\,660 \times 10^{-7}$	g/kW · h	608.3
gram/kilowatt-hour	kg/J	$2.777\,778 \times 10^{-10}$	μg/J	0.2778
mile/gallon (U.S.)	m/m³	$4.251\,437 \times 10^5$	km/dm³	0.4251
mile/gallon (Imp.)	m/m³	$3.540\,060 \times 10^5$	km/dm³	0.3540
Heat flux (includes thermal conductivity)				
Btu (IT) · in/h · ft² · °F	W/m · K	$1.442\,279 \times 10^{-1}$	W/m · K	0.1442
Btu (IT)/ft²	J/m²	$1.135\,653 \times 10^4$	kJ/m²	11.36
Btu (IT)/h · ft² · °F	W/m² · K	$5.678\,263 \times 10^0$	W/m² · K	5.678
calorie (thermo.)/cm²	J/m²	$4.184\,000 \times 10^4$	kJ/m²	41.84
Length				
foot	m	$3.048\,000 \times 10^{-1}$	m	0.3048
inch	m	$2.540\,000 \times 10^{-2}$	mm	25.40
micron	m	$1.000\,000 \times 10^{-6}$	μm	1.000
mile	m	$1.609\,344 \times 10^3$	km	1.609
Mass				
ounce	kg	$2.834\,952 \times 10^{-2}$	g	28.35
pound	kg	$4.535\,924 \times 10^{-1}$	kg	0.4536
ton (long or Imp., 2240 lb)	kg	$1.016\,047 \times 10^3$	Mg	1.016
ton (short, 2000 lb)	kg	$9.071\,847 \times 10^2$	Mg	0.9072
tonne (metric)	kg	$1.000\,000 \times 10^3$	Mg	1.000
Mass per unit time (flow)				
pound/second	kg/s	$4.535\,924 \times 10^{-1}$	kg/s	0.4536
pound/minute	kg/s	$7.559\,873 \times 10^{-3}$	g/s	7.560
pound/hour	kg/s	$1.259\,979 \times 10^{-4}$	g/s	0.1260
Mass per unit volume				
gram/gallon (U.S.)	kg/m³	$2.641\,724 \times 10^{-1}$	g/dm³	0.2642
pound/foot³	kg/m³	$1.601\,846 \times 10^1$	kg/m³	16.02
pound/inch³	kg/m³	$2.767\,990 \times 10^4$	kg/dm³	27.68
pound/gallon (Imp.)	kg/m³	$9.977\,644 \times 10^1$	kg/dm³	0.0998
pound/gallon (U.S.)	kg/m³	$1.198\,264 \times 10^2$	kg/dm³	0.1198
Power, heat flow				
Btu (IT)/hour	W	$2.930\,711 \times 10^{-1}$	W	0.2931
horsepower (550 ft · lbf/s)	W	$7.456\,999 \times 10^2$	kW	0.7457
horsepower (metric, CV, PS)	W	$7.354\,99\ \times 10^2$	kW	0.7355

1 To convert from	2 To	3 Multiply by	4 To	5 Multiply by
Pressure, stress (force per unit area)				
atmosphere (normal, 760 torr)	Pa	$1.013\,250 \times 10^5$	kPa	101.3
inch of mercury (60°F)	Pa	$3.376\,85\ \times 10^3$	kPa	3.377
kilogram-force/centimeter^2	Pa	$9.806\,650 \times 10^4$	kPa	98.07
mm of mercury, 0°C (torr)	Pa	$1.333\,224 \times 10^2$	Pa	133.3
pound-force/foot^2	Pa	$4.788\,026 \times 10^1$	Pa	47.88
pound-force/inch^2 (psi)	Pa	$6.894\,757 \times 10^3$	kPa	6.895
Temperature interval				
degree Celsius	K	$1.000\,000 \times 10^0$		
degree Fahrenheit	K	$5.555\,556 \times 10^{-1}$	K	0.5556
Temperature				
temperature (°C)	K	°C + 273.15		
temperature (°F)	K	(°F + 459.67)/1.80	°C	(°F − 32)/1.80
Torque				
kilogram-force meter	N·m	$9.806\,650 \times 10^0$	N·m	9.807
pound-force foot	N·m	$1.355\,818 \times 10^0$	N·m	1.356
Velocity				
foot/second	m/s	$3.048\,000 \times 10^{-1}$	m/s	0.3048
kilometer/hour	m/s	$2.777\,778 \times 10^{-1}$	m/s	0.2778
mile/hour	m/s	$4.470\,400 \times 10^{-1}$	km/h	1.609
Viscosity				
centipoise	Pa·s	$1.000\,000 \times 10^{-3}$	mPa·s	1.000
centistoke	m^2/s	$1.000\,000 \times 10^{-6}$	mm^2/s	1.000
poise	Pa·s	$1.000\,000 \times 10^{-1}$	Pa·s	0.1000
stoke	m^2/s	$1.000\,000 \times 10^{-4}$	mm^2/s	100.0
Volume				
barrel (42 U.S. gallon)	m^3	$1.589\,873 \times 10^{-1}$	m^3	0.1590
foot^3	m^3	$2.831\,685 \times 10^{-2}$	dm^3	28.32
gallon (Imp.)	m^3	$4.546\,092 \times 10^{-3}$	dm^3	4.546
gallon (U.S.)	m^3	$3.785\,412 \times 10^{-3}$	dm^3	3.785
inch^3	m^3	$1.638\,706 \times 10^{-5}$	cm^3	16.39
liter	m^3	$1.000\,000 \times 10^{-3}$	dm^3	1.000
Volume per unit time				
foot^3/minute (cfm)	m^3/s	$4.719\,474 \times 10^{-4}$	dm^3/s	0.4719
foot^3/second	m^3/s	$2.831\,685 \times 10^{-2}$	dm^3/s	28.32
gallon (U.S.)/minute (gpm)	m^3/s	$6.309\,020 \times 10^{-5}$	cm^3/s	63.09

Notes:
1. Derived units such as that for torque (newton-metre, N·m) are written with a period between each component unit for clarity. In practice, the period is often omitted.
2. Derived from Mobil Technical Bulletin *SI Units, The Modern Metric System*. Copyright Mobil Oil Corporation, 1974. Sections reproduced courtesy Mobil Oil Corporation.

APPENDIX
B

IDEAL
GAS
RELATIONSHIPS

B.1 IDEAL GAS LAW

The gas species which make up the working fluids in internal combustion engines (e.g., oxygen, nitrogen, carbon dioxide, etc.) can usually be treated as ideal gases. This Appendix reviews the relationships between the thermodynamic properties of ideal gases.

The pressure p, specific volume v, and absolute temperature T of an ideal gas are related by the ideal gas law

$$pv = RT \tag{B.1}$$

For each gas species, R is a constant (the gas constant). It is different for each gas and is given by

$$R = \frac{\tilde{R}}{M} \tag{B.2}$$

where $\tilde{R}$ is the universal gas constant (for all ideal gases) and M is the molecular weight of the gas. Since v is given by V/m, where V is the volume of a mass of gas m, Eq. (B.1) can be rewritten as

$$pV = mRT = \frac{m\tilde{R}T}{M} \tag{B.3}$$

B.2   THE MOLE

It is convenient to introduce a mass unit based on the molecular structure of matter, the mole:

> The mole is the amount of substance which contains as many molecules as there are carbon atoms in 12 grams of carbon-12.†

Thus, the number of moles n of gas is given by

$$n = \frac{m}{M} \tag{B.4}$$

and Eq. (B.3) becomes

$$pV = n\tilde{R}T \tag{B.5}$$

Values for the universal gas constant in different units are given in Table B.1. In the SI system, the value is 8314.3 J/kmol · K.

TABLE B.1
Values of universal gas constant $\tilde{R}$

8314.3 J/kmol · K
8.3143 J/mol · K
1.9859 Btu/lb-mole · °R
1543.3 ft · lbf/lb-mole · °R

B.3   THERMODYNAMIC PROPERTIES

It follows from Eq. (B.1) that the internal energy u‡ of an ideal gas is a function of temperature only:

$$u = u(T) \tag{B.6}$$

Since the enthalpy h is given by $u + pv$, it follows also that

$$h = h(T) \tag{B.7}$$

---

† This is the SI system definition of the mole; it was formerly called the gram-mole. The kilogram-mole (kmol) is also used; it is 1000 times as large as the mole.

‡ The symbol u will be used for internal energy per unit mass, $\tilde{u}$ for internal energy per mole, and U for internal energy of a previously defined system of mass m. Similar notation will be used for enthalpy, entropy, and specific heats, per unit mass and per mole.

The specific heats at constant volume and constant pressure of an ideal gas, c_v and c_p, respectively, are defined by

$$c_v = \left(\frac{\partial u}{\partial T}\right)_v = \frac{du}{dT} \tag{B.8}$$

$$c_p = \left(\frac{\partial h}{\partial T}\right)_p = \frac{dh}{dT} \tag{B.9}$$

From Eq. (B.1) it follows that

$$c_p - c_v = R \tag{B.10}$$

The ratio of specific heats, γ, is a useful quantity:

$$\gamma = \frac{c_p}{c_v} \tag{B.11}$$

An *additional* restrictive assumption is often made that the specific heats are constants. This is not a necessary part of the ideal gas relationships.

In general, the internal energy and enthalpy of an ideal gas at a temperature T relative to its internal energy and enthalpy at some reference temperature T_0 are given by

$$u = u_0 + \int_{T_0}^{T} c_v(T)\, dT \tag{B.12}$$

and

$$h = h_0 + \int_{T_0}^{T} c_p(T)\, dT \tag{B.13}$$

The entropy at T, v, and p, relative to the entropy at some reference state T_0, v_0, p_0, can be obtained from the relationships

$$ds = \frac{c_v}{T}\, dT + R\frac{dv}{v} = \frac{c_p}{T}\, dT - R\frac{dp}{p} \tag{B.14}$$

which integrate to give

$$s = s_0 + \int_{T_0}^{T} \frac{c_v}{T}\, dT + R \ln \frac{v}{v_0} \tag{B.15a}$$

and

$$s = s_0 + \int_{T_0}^{T} \frac{c_p}{T}\, dT - R \ln \frac{p}{p_0} \tag{B.15b}$$

The properties u, h, and s can be evaluated on a per unit mass or per mole basis. On a mass basis, c_v, c_p, and R would have the units J/kg·K (Btu/lbm·°R); on a mole basis u, h, and s are replaced by $\tilde{u}$, $\tilde{h}$, and $\tilde{s}$. R is then the universal gas constant $\tilde{R}$, c_v and c_p are replaced by $\tilde{c}_v$ and $\tilde{c}_p$, and $\tilde{c}_v$, $\tilde{c}_p$, and $\tilde{R}$ would have the units J/kmol·K (Btu/lb-mol·°R).

B.4 MIXTURES OF IDEAL GASES

The working fluids in engines are mixtures of gases. The composition of a mixture of ideal gases can be expressed in terms of the following properties of each component:

Partial pressure p_i. The pressure each component would exert if it alone occupied the volume of the mixture at the temperature of the mixture.

Parts by volume V_i/V. The fraction of the total mixture volume each component would occupy if separated from the mixture, at the mixture temperature and pressure.

Mass fraction x_i. The mass of each component m_i, divided by the total mass of mixture m.

Mole fraction $\tilde{x}_i$. The number of moles of each component n_i, divided by the total number of moles of mixture n.

From Eq. (B.5) it follows that

$$\frac{p_i}{p} = \frac{V_i}{V} = x_i \frac{M}{M_i} = \tilde{x}_i \tag{B.16}$$

The thermodynamic properties of mixtures of ideal gases can be computed from the following relationships:

Molecular weight

$$M = \frac{1}{n} \sum_i n_i M_i = \sum_i \tilde{x}_i M_i \tag{B.17}$$

Internal energy, enthalpy, and entropy
On a mass basis:

$$u = \sum_i x_i u_i \qquad h = \sum_i x_i h_i \qquad s = \sum_i x_i s_i \tag{B.18a, b, c}$$

On a mole basis:

$$\tilde{u} = \sum_i \tilde{x}_i \tilde{u}_i \qquad \tilde{h} = \sum_i \tilde{x}_i \tilde{h}_i \qquad \tilde{s} = \sum_i \tilde{x}_i \tilde{s}_i \tag{B.19a, b, c}$$

EQUATIONS FOR FLUID FLOW THROUGH A RESTRICTION

In many parts of the engine cycle, fluid flows through a restriction or reduction in flow area. Real flows of this nature are usually related to an equivalent ideal flow. The equivalent ideal flow is the steady adiabatic reversible (frictionless) flow of an ideal fluid through a duct of identical geometry and dimensions. For a real fluid flow, the departures from the ideal assumptions listed above are taken into account by introducing a flow coefficient or discharge coefficient C_D, where

$$C_D = \frac{\text{actual mass flow}}{\text{ideal mass flow}}$$

Alternatively, the flow or discharge coefficient can be defined in terms of an effective cross-sectional area of the duct and a reference area. The reference area A_R is usually taken as the minimum cross-sectional area. The effective area of the flow restriction A_E is then the cross-sectional area of the throat of a frictionless nozzle which would pass the measured mass flow between a large upstream reservoir at the upstream stagnation pressure and a large downstream reservoir at the downstream measured static pressure. Thus

$$C_D = \frac{A_E}{A_R}$$

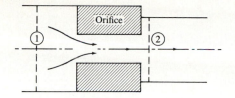

FIGURE C-1
Schematic of liquid flow through orifice.

C.1 LIQUID FLOW

Consider the flow of a liquid through an orifice as shown in Fig. C-1. For the ideal flow, Bernoulli's equation can be written

$$p_1 + \rho \frac{V_1^2}{2} = p_2 + \rho \frac{V_2^2}{2}$$

For an incompressible flow, continuity gives $V_1 A_1 = V_2 A_2$ and the ideal mass flow rate through an orifice is given by

$$\dot{m}_{\text{ideal}} = A_2 \left[\frac{2\rho(p_1 - p_2)}{1 - (A_2/A_1)^2} \right]^{1/2} \qquad \text{(C.1)}$$

The real mass flow rate is obtained by introducing the discharge coefficient:

$$\dot{m}_{\text{real}} = C_D A_2 \left[\frac{2\rho(p_1 - p_2)}{1 - (A_2/A_1)^2} \right]^{1/2} \qquad \text{(C.2)}$$

The discharge coefficient is a function of orifice dimensions, shape and surface roughness, mass flow rate, and fluid properties (density, surface tension, and viscosity). The use of the orifice Reynolds number

$$\text{Re}_o = \frac{\rho V_2 D_2}{\mu} = \frac{V_2 D_2}{\nu}$$

as a correlating parameter for the discharge coefficient accounts for the effects of $\dot{m}$, ρ, ν, and D_2 to a good approximation.[1]

C.2 GAS FLOW

Consider the flow of an ideal gas with constant specific heats through the duct shown in Fig. C-2. For the *ideal flow*, the stagnation temperature and pressure, T_0 and p_0, are related to the conditions at other locations in the duct by the steady flow energy equation

$$T_0 = T + \frac{V^2}{2c_p}$$

and the isentropic relation

$$\left(\frac{T}{T_0} \right) = \left(\frac{p}{p_0} \right)^{(\gamma - 1)/\gamma}$$

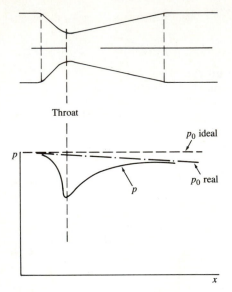

FIGURE C-2
Pressure distribution for gas flow through a nozzle.

By introducing the Mach number $M = V/a$, where a is the sound speed $(= \sqrt{\gamma R T})$, the following equations are obtained:

$$\frac{T_0}{T} = 1 + \frac{\gamma - 1}{2} M^2 \tag{C.3}$$

$$\frac{p_0}{p} = \left(1 + \frac{\gamma - 1}{2} M^2\right)^{\gamma/(\gamma - 1)} \tag{C.4}$$

The mass flow rate $\dot{m}$ is

$$\dot{m} = \rho A V$$

With the ideal gas law and the above relations for p and T, this can be rearranged as

$$\frac{\dot{m}_{\text{ideal}} \sqrt{\gamma R T_0}}{A p_0} = \gamma M \left(1 + \frac{\gamma - 1}{2} M^2\right)^{-(\gamma + 1)/2(\gamma - 1)} \tag{C.5}$$

or

$$\frac{\dot{m}_{\text{ideal}} \sqrt{\gamma R T_0}}{A p_0} = \gamma \left(\frac{p}{p_0}\right)^{1/\gamma} \left\{\frac{2}{\gamma - 1}\left[1 - \left(\frac{p}{p_0}\right)^{(\gamma - 1)/\gamma}\right]\right\}^{1/2} \tag{C.6}$$

For given values of p_0 and T_0, the maximum mass flow occurs when the velocity at the minimum area or throat equals the velocity of sound. This condition is called choked or critical flow. When the flow is choked the pressure at the throat, p_T, is related to the stagnation pressure p_0 as follows:

$$\frac{p_T}{p_0} = \left(\frac{2}{\gamma + 1}\right)^{\gamma/(\gamma - 1)}$$

This ratio is called the critical pressure ratio. For (p_T/p_0) less than or equal to the critical pressure ratio,

$$\frac{\dot{m}_{ideal}\sqrt{\gamma R T_0}}{A_T p_0} = \gamma \left(\frac{2}{\gamma + 1}\right)^{(\gamma + 1)/2(\gamma - 1)} \tag{C.7}$$

The critical pressure ratio is 0.528 for $\gamma = 1.4$ and 0.546 for $\gamma = 1.3$.

For a real gas flow, the discharge coefficient is introduced. Then, for subcritical flow, the real mass flow rate is given in terms of conditions at the minimum area or throat by

$$\dot{m}_{real} = \frac{C_D A_T p_0}{\sqrt{RT_0}} \left(\frac{p_T}{p_0}\right)^{1/\gamma} \left\{\frac{2\gamma}{\gamma - 1}\left[1 - \left(\frac{p_T}{p_0}\right)^{(\gamma - 1)/\gamma}\right]\right\}^{1/2} \tag{C.8}$$

For a choked flow,

$$\dot{m}_{real} = \frac{C_D A_T p_0}{\sqrt{RT_0}} \gamma^{1/2}\left(\frac{2}{\gamma + 1}\right)^{(\gamma + 1)/2(\gamma - 1)} \tag{C.9}$$

Equation (C.8) can be rearranged in the form of Eq. (C.2) (with $A_2 \ll A_1$) as

$$\dot{m}_{real} = C_D A_R [2\rho_0(p_0 - p_T)]^{1/2}\Phi \tag{C.10}$$

where Φ is given by

$$\Phi = \left\{\frac{[\gamma/(\gamma - 1)][(p_T/p_0)^{2/\gamma} - (p_T/p_0)^{(\gamma + 1)/\gamma}]}{1 - p_T/p_0}\right\}^{1/2} \tag{C.11}$$

Figure C-3 shows the variation of Φ and $(\dot{m}/\dot{m}^*)_{ideal}$ with $(p_0 - p_T)/p_0$. $\dot{m}^*$ is the mass flow rate through the restriction under choked flow conditions (when the Mach number at the throat is unity). For flow rates less than about 60 percent of the choked flow, the effects of compressibility on the mass flow rate are less than 5 percent.

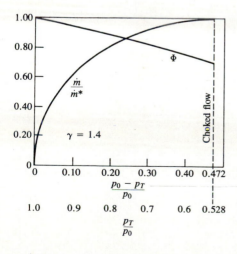

FIGURE C-3
Relative mass flow rate $\dot{m}/\dot{m}^*$ and compressible flow function Φ [Eq. (C.11)] as function of nozzle or restriction pressure ratio for ideal gas with $\gamma = 1.4$. (*From Taylor.*[2])

Flow coefficients are determined experimentally and are a function of the shape of the passage, the Reynolds number and Mach number of the flow, and the gas properties. For a Mach number at the throat less than about 0.7 and for passages of similar shape, the flow coefficient is essentially a function of Reynolds number only.

Orifice plates are frequently used to measure gas flow rates. Standard methods for determining flows through orifice plates can be found in Ref. 3.

REFERENCES

1. Lichtarowicz, A., Duggins, R. K., and Markland, E.: "Discharge Coefficients for Incompressible Non-Cavitating Flow through Long Orifices," *J. Mech. Eng. Sci.*, vol. 7, no. 2, pp. 210–219, 1965.
2. Taylor, C. F.: *The Internal Combustion Engine in Theory and Practice*, vol. I, p. 506, MIT Press, 1966.
3. *Marks' Standard Handbook for Mechanical Engineers*, 8th ed., McGraw-Hill, 1978.

APPENDIX
D

DATA
ON
WORKING
FLUIDS

TABLE D.1

Thermodynamic properties of air at low density†

T, K	h, kJ/kg	u, kJ/kg	Ψ kJ/(kg·K)	Φ	p_r	v_r	c_p kJ/(kg·K)	c_v	γ
250	409.9	338.1	4.4505	7.6603	38.81	1849.0	1.003	0.715	1.401
275	435.0	356.0	4.5187	7.7559	54·14	1458.0	1.003	0.716	1.401
300	460.1	374.0	4.5811	7.8432	73.39	1173.0	1.004	0.717	1.400
325	485.2	391.9	4.6385	7.9236	97.13	960.6	1.006	0.718	1.400
350	510.4	409.9	4.6919	7.9982	125.9	797.8	1.007	0.720	1.399
375	535.6	427.9	4.7416	8.0678	160.5	670.8	1.010	0.723	1.397
400	560.8	446.0	4.7884	8.1330	201.4	570.0	1.013	0.725	1.396
425	586.2	464.2	4.8324	8.1945	249.6	488.9	1.016	0.729	1.394
450	611.6	482.5	4.8742	8.2527	305.6	422.7	1.020	0.733	1.392
475	637.2	500.8	4.9139	8.3079	370.4	368.1	1.024	0.737	1.390
500	662.8	519.3	4.9518	8.3606	445.0	322.6	1.028	0.741	1.387
525	688.6	537.9	4.9881	8.4109	530.2	284.3	1.033	0.746	1.385
550	714.5	556.6	5.0229	8.4590	627.1	251.8	1.039	0.752	1.382
575	740.5	575.5	5.0565	8.5053	736.8	224.0	1.044	0.757	1.379
600	766.7	594.5	5.0888	8.5499	860.6	200.1	1.050	0.763	1.376
625	793.0	613.6	5.1201	8.5929	999.5	179.5	1.056	0.768	1.374
650	819.5	632.9	5.1503	8.6344	1155.0	161.5	1.061	0.774	1.371
675	846.1	652.3	5.1796	8.6745	1329.0	145.9	1.067	0.780	1.368
700	872.9	671.9	5.2081	8.7135	1521.0	132.1	1.073	0.786	1.365
725	899.8	691.7	5.2358	8.7512	1735.0	119.9	1.079	0.792	1.362
750	926.8	711.5	5.2628	8.7879	1972.0	109.2	1.085	0.798	1.360
775	954.0	731.6	5.2891	8.8236	2233.0	99.63	1.091	0.804	1.357
800	981.4	751.7	5.3147	8.8584	2520.0	91.12	1.097	0.810	1.354
825	1008.9	772.1	5.3397	8.8922	2836.0	83.52	1.103	0.816	1.352
850	1036.5	792.5	5.3641	8.9252	3181.0	76.71	1.108	0.821	1.350
875	1064.3	813.1	5.3880	8.9574	3559.0	70.58	1.114	0.827	1.347
900	1092.2	833.8	5.4114	8.9889	3971.0	65.07	1.119	0.832	1.345
925	1120.2	854.7	5.4342	9.0196	4419.0	60.08	1.124	0.837	1.343
950	1148.4	875.7	5.4566	9.0496	4907.0	55.58	1.129	0.842	1.341
975	1176.7	896.8	5.4786	9.0790	5436.0	51.49	1.134	0.847	1.339
1000	1205.1	918.1	5.5001	9.1078	6009.0	47.77	1.139	0.852	1.337
1025	1233.7	939.4	5.5212	9.1360	6629.0	44.39	1.144	0.856	1.335
1050	1262.3	960.9	5.5419	9.1636	7299.0	41.30	1.148	0.861	1.333
1075	1291.1	982.5	5.5622	9.1907	8020.0	38.48	1.152	0.865	1.332
1100	1319.9	1004.1	5.5821	9.2172	8797.0	35.90	1.157	0.870	1.330
1125	1348.9	1025.9	5.6017	9.2432	9632.0	33.53	1.161	0.874	1.329
1150	1378.0	1047.8	5.6209	9.2688	10529.0	31.35	1.165	0.878	1.327
1175	1407.1	1069.8	5.6399	9.2939	11490.0	29.36	1.168	0.881	1.326
1200	1436.4	1091.9	5.6585	9.3185	12520.0	27.51	1.172	0.885	1.324

† Abstracted with permission from *Thermodynamic Properties in SI* (Graphs, Tables, and Computational Equations for Forty Substances), by W. C. Reynolds, Published by the Department of Mechanical Engineering, Stanford University, Stanford, CA 94305, 1979.

TABLE D.2

Standard enthalpy of formation and molecular weight of species

Species	Formula	Molecular weight g/mole	State†	$\Delta \bar{h}_f^{\circ}$ MJ/kmol	$\Delta \bar{h}_f^{\circ}$ kcal/mol
Oxygen	O_2	32.00	gas	0	0
Nitrogen	N_2	28.01	gas	0	0
Carbon	C	12.011	solid	0	0
Carbon monoxide	CO	28.01	gas	−110.5	−26.42
Carbon dioxide	CO_2	44.01	gas	−393.5	−94.05
Hydrogen	H_2	2.016	gas	0	0
Water	H_2O	18.02	gas	−241.8	−57.80
Water	H_2O	18.02	liquid	−285.8	−68.32
Methane	CH_4	16.04	gas	−74.9	−17.89
Propane	C_3H_8	44.10	gas	−103.8	−24.82
Isooctane	C_8H_{18}	114.23	gas	−224.1	−53.57
Isooctane	C_8H_{18}	114.23	liquid	−259.28	−61.97
Cetane	$C_{16}H_{34}$	226.44	liquid	−454.5	−108.6
Methyl alcohol	CH_3OH	32.04	gas	−201.2	−48.08
Methyl alcohol	CH_3OH	32.04	liquid	−238.6	−57.02
Ethyl alcohol	C_2H_5OH	46.07	gas	−234.6	−56.08
Ethyl alcohol	C_2H_5OH	46.07	liquid	−277.0	−66.20

† At 298.15 K (25°C) and 1 atm.

TABLE D.3

Enthalpy of C, CO, CO$_2$, H$_2$, H$_2$O, N$_2$, O$_2$

	$\tilde{h}°(T) - \tilde{h}°(298.15)$, kcal/mol						
T(K)	C	CO	CO$_2$	H$_2$	H$_2$O	N$_2$	O$_2$
298	0.000	0.000	0.000	0.000	0.000	0.000	0.000
300	0.004	0.013	0.016	0.013	0.015	0.013	0.013
400	0.250	0.711	0.958	0.707	0.825	0.710	0.724
500	0.569	1.417	1.987	1.406	1.654	1.413	1.455
600	0.947	2.137	3.087	2.106	2.509	2.125	2.210
700	1.372	2.873	4.245	2.808	3.390	2.853	2.988
800	1.831	3.627	5.453	3.514	4.300	3.596	3.786
900	2.318	4.397	6.702	4.226	5.240	4.355	4.600
1000	2.824	5.183	7.984	4.944	6.209	5.129	5.427
1100	3.347	5.983	9.296	5.670	7.210	5.917	6.266
1200	3.883	6.794	10.632	6.404	8.240	6.718	7.114
1300	4.432	7.616	11.988	7.148	9.298	7.529	7.971
1400	4.988	8.446	13.362	7.902	10.384	8.350	8.835
1500	5.552	9.285	14.750	8.668	11.495	9.179	9.706
1600	6.122	10.130	16.152	9.446	12.630	10.015	10.583
1700	6.696	10.980	17.565	10.233	13.787	10.858	11.465
1800	7.275	11.836	18.987	11.030	14.964	11.707	12.354
1900	7.857	12.697	20.418	11.836	16.160	12.560	13.249
2000	8.442	13.561	21.857	12.651	17.373	13.418	14.149
2100	9.029	14.430	23.303	13.475	18.602	14.280	15.054
2200	9.620	15.301	24.755	14.307	19.846	15.146	15.966
2300	10.212	16.175	26.212	15.146	21.103	16.015	16.882
2400	10.807	17.052	27.674	15.993	22.372	16.886	17.804
2500	11.403	17.931	29.141	16.848	23.653	17.761	18.732
2600	12.002	18.813	30.613	17.708	24.945	18.638	19.664
2700	12.602	19.696	32.088	18.575	26.246	19.517	20.602
2800	13.203	20.582	33.567	19.448	27.556	20.398	21.545
2900	13.807	21.469	35.049	20.326	28.875	21.280	22.493
3000	14.412	22.357	36.535	21.210	30.201	22.165	23.446

Source: *JANAF Thermochemical Tables*, National Bureau of Standards Publication NSRDS-NBS37, 1971.

TABLE D.4
Data on fuel properties

Fuel	Formula (phase)	Molecular weight	Specific gravity;† (density)† kg/dm³	Heat of vaporization, kJ/kg‡	Specific heat — Liquid, kJ/kg·K	Specific heat — Vapor c_p, kJ/kg·K	Higher heating value, MJ/kg	Lower heating value, MJ/kg	LHV of stoich. mixture, MJ/kg	$(A/F)_s$	$(F/A)_s$	Fuel octane rating — RON	Fuel octane rating — MON
Practical fuels§													
Gasoline	$C_nH_{1.87n}$(l)	~110	0.72–0.78	350	2.4	~1.7	47.3	44.0	2.83	14.6	0.0685	91–99	82–89
Light diesel	$C_nH_{1.8n}$(l)	~170	0.78–0.84	270	2.2	~1.7	46.1	43.2	2.79	14.5	0.0690	—	—
Heavy diesel	$C_nH_{1.7n}$(l)	~200	0.82–0.88	230	1.9	~1.7	45.5	42.8	2.85	14.4	0.0697	—	—
Natural gas	$C_nH_{3.8n}N_{0.1n}$(g)	~18	(~0.79†)	—	—	~2	50	45	2.9	14.5	0.069	—	—
Pure hydrocarbons													
Methane	CH_4(g)	16.04	(0.72†)	509	0.63	2.2	55.5	50.0	2.72	17.23	0.0580	120	120
Propane	C_3H_8(g)	44.10	0.51 (2.0†)	426	2.5	1.6	50.4	46.4	2.75	15.67	0.0638	112	97
Isooctane	C_8H_{18}(l)	114.23	0.692	308	2.1	1.63	47.8	44.3	2.75	15.13	0.0661	100	100
Cetane	$C_{16}H_{34}$(l)	226.44	0.773	358	—	1.6	47.3	44.0	2.78	14.82	0.0675	—	—
Benzene	C_6H_6(l)	78.11	0.879	433	1.72	1.1	41.9	40.2	2.82	13.27	0.0753	—	115
Toluene	C_7H_8(l)	92.14	0.867	412	1.68	1.1	42.5	40.6	2.79	13.50	0.0741	120	109
Alcohols													
Methanol	CH_4O(l)	32.04	0.792	1103	2.6	1.72	22.7	20.0	2.68	6.47	0.155	106	92
Ethanol	C_2H_6O(l)	46.07	0.785	840	2.5	1.93	29.7	26.9	2.69	9.00	0.111	107	89
Other fuels													
Carbon	C(s)	12.01	~2§	—	—	—	33.8	33.8	2.70	11.51	0.0869	—	—
Carbon monoxide	CO(g)	28.01	(1.25†)	—	—	1.05	10.1	10.1	2.91	2.467	0.405	—	—
Hydrogen	H_2(g)	2.015	(0.090†)	—	—	1.44	142.0	120.0	3.40	34.3	0.0292	—	—

(l) liquid phase; (g) gaseous phase; (s) solid phase.

† Density in kg/m³ at 0°C and 1 atm.

‡ At 1 atm and 25°C for liquid fuels: at 1 atm and boiling temperature for gaseous fuels.

§ Typical values.

RON, research octane number; MON, motor octane number.

Sources:

E. M. Goodger, *Hydrocarbon Fuels: Production, Properties and Performance of Liquids and Gases*, Macmillan, London, 1975.

E. F. Obert, *Internal Combustion Engines and Air Pollution*, Intext Educational Publishers, 1973 edition.

C. F. Taylor, *The Internal Combustion Engine in Theory and Practice*, vol. I, MIT Press, 1966.

J. W. Rose and J. R. Cooper (eds.), *Technical Data on Fuel*, 7th ed., British National Committee, World Energy Conference, London, 1977.

INDEX